Modeling and Simulation in Science, Engineering and Technology

More information about this series at http://www.springer.com/series/4960

Antonio Romano · Addolorata Marasco

Classical Mechanics with *Mathematica*®

Second Edition

 Birkhäuser

may generate large effects and that chaos lurks in the corner of many mechanical systems.

An elementary introduction to Maxwell's (James Clerk Maxwell, 1831–1879) kinetic theory of perfect gases and to Gibbs's (Josiah Willard Gibbs, 1839–1903) formulation of the statistical mechanics of equilibrium can be found in Chapter 21. In this new edition, the ergodic approach to statistical mechanics of equilibrium has been further extended. In Chapter 22, the balance equations and the Lagrange equations are extended to impulsive dynamics. Chapter 23 contains the basic elements of the dynamics of a perfect fluid. This chapter appears in the book simply to show how we must modify the axioms of mechanics of rigid bodies to take into account the deformability of real bodies when they are acted upon by forces. In this edition we added Chapters 24, 25, and 27. In Chapter 24, there is an introduction to celestial mechanics; further, the Newtonian gravitational field of extended mass distribution is analyzed, as well as a sketch about the form of self-gravitating bodies. In Chapter 25, the reader can find a simplified analysis of one-dimensional systems with applications to beams and strings. In particular, the wave propagation along a string is studied. An introduction to special relativity and its four-dimensional formulation can be found in Chapter 26. Finally, Chapter 27 contains an introduction to variational calculus with applications.

The book includes approximately 200 exercises. Further, it provides the names of many notebooks, written using *Mathematica*, that are relevant to several of the book's chapters. These notebooks have the twofold aim of showing the possibilities of this software and of helping the reader to manage some difficult problems of dynamics.

The notebooks referring to the first part are **Geometry**, **Weierstrass**, **Phase2D**, and **LinStab**. In particular, the notebook **Phase2D** contains two programs: **Phase2D** and **PolarPhase**. The other notebooks relate to the second part.

Naples, Italy   Antonio Romano
                                                                                                 Addolorata Marasco

Contents

Part I
Introduction to Linear Algebra and Differential Geometry

and satisfying the following properties:

$$x + (y + z) = (x + y) + z, \tag{1.2}$$
$$x + y = y + x, \tag{1.3}$$

$$\exists 0 \in E : x + 0 = x, \tag{1.4}$$
$$\forall x \in E, \exists (-x) \in E : x + (-x) = 0, \tag{1.5}$$

$\forall x, y, z \in E$.

Besides the addition, we suppose that another composition law is defined by

$$(a, x) \in E \times \Re \rightarrow ax \in E, \tag{1.6}$$

which associates to any pair (a, x), with $a \in \Re$ and $x \in E$, a new element $ax \in E$ in such a way that the following properties are satisfied:

$$(ab)x = a(bx), \tag{1.7}$$
$$a(x + y) = ax + ay, \tag{1.8}$$
$$(a + b)x = ax + bx, \tag{1.9}$$
$$1x = x, \tag{1.10}$$

$\forall a, b \in \Re$ and $\forall x, y \in E$.

Definition 1.1. The set E, equipped with the above operations, is said to be a **real vector space** or a **vector space on** $\Re$. The elements of E and $\Re$ are, respectively, called **vectors** and **scalars**. The vector $x + y$ is the **sum** of the vectors x and y; the vector ax is the **product** of the scalar a by the vector x. Finally, the vector 0 is said to be the **zero vector** and $-x$ is the **opposite vector** of x.

Henceforth, the sum $x + (-y)$ will be denoted by $x - y$.

Proposition 1.1. *The zero vector is unique.*

Proof. If there are two zero vectors 0 and $0'$, then the result is

$$x + 0 = x, \quad \forall x \in E,$$

so that, in particular, it is $0' + 0 = 0'$. On the other hand, from the relation

$$x + 0' = x, \quad \forall x \in E,$$

it follows that $0 + 0' = 0$. From (1.3) we conclude that $0' = 0$. $\square$

Proposition 1.2. *The opposite vector* $-\mathbf{x}$ *of any* $\mathbf{x} \in E$ *is uniquely determined.*

Proof. If $\mathbf{x}'$ and $\mathbf{x}''$ are two opposite vectors of $\mathbf{x}$, then it is

$$\mathbf{x}' + \mathbf{x} = \mathbf{x}'' + \mathbf{x} = \mathbf{0}.$$

On the other hand, in view of (1.2), we have that

$$(\mathbf{x}' + \mathbf{x}) + \mathbf{x}'' = \mathbf{x}' + (\mathbf{x} + \mathbf{x}''),$$

i.e., $\mathbf{x}' = \mathbf{x}''$. □

In a similar way, it is possible to prove the following propositions.

Proposition 1.3. $\forall \mathbf{x} \in E$ *and* $a \in \mathfrak{R}$, *it is*

$$0\mathbf{x} = \mathbf{0}, \quad a\mathbf{0} = \mathbf{0}, \quad (-a)\mathbf{x} = -(a\mathbf{x}).$$

Proposition 1.4. $\forall \mathbf{x}_1, \dots, \mathbf{x}_n \in E$ *and* $a \in \mathfrak{R}$, *we have*

$$a(\mathbf{x}_1 + \cdots + \mathbf{x}_n) = a\mathbf{x}_1 + \cdots + a\mathbf{x}_n.$$

1.2 Dimension of a Vector Space

Definition 1.2. Let E be a vector space and let $\mathbf{x}_1, \dots, \mathbf{x}_n$ be vectors of E. The vectors $\mathbf{x}_1, \dots, \mathbf{x}_n$ are said to be *linearly dependent* if there exist nonzero real numbers $a^1, \dots, a^n$ such that

$$a^1\mathbf{x}_1 + \cdots + a^n\mathbf{x}_n = \mathbf{0}. \tag{1.11}$$

In contrast, if (1.11) implies $a^1 = \cdots = a^n = 0$, then we say that $\mathbf{x}_1, \dots, \mathbf{x}_n$ are *linearly independent* or, equivalently, that they form a *free system of order* n.

It is evident that the vector $\mathbf{0}$ does not belong to a free system.

Definition 1.3. The vector space E is said to have a *finite dimension* $n > 0$ on $\mathfrak{R}$ if

1. There exists at least a free system of vectors of order n;
2. Every system of vectors of order $n + 1$ is linearly dependent.

If E contains free systems of vectors of any order, then E is said to have *infinite dimension*.

Throughout the book we only consider vector spaces with finite dimension.

Definition 1.4. Let E_n be a vector space with finite dimension n. Any free vector system $\{\mathbf{e}_1, \dots, \mathbf{e}_n\}$ of order n is a *basis* of E_n.

Remark 1.1. Henceforth, we denote by $(\mathbf{e}_i)$ the vector set $\{\mathbf{e}_1, \ldots, \mathbf{e}_n\}$; further, we adopt the Einstein convention according to which a summation is understood over a pair of upper and lower indices denoted by the same symbol. For instance, the relation $a^1\mathbf{x}_1 + \cdots + a^n\mathbf{x}_n$ will be written in the compact form $a^i\mathbf{x}_i$. The index i, which is summed over, is called a *dummy* index. In a computation, such an index can be denoted by any Latin letter.

Theorem 1.1. *A vector system* $(\mathbf{e}_i)$ *is a basis of the vector space* E_n *if and only if for any* $\mathbf{x} \in E_n$ *it yields*

$$\mathbf{x} = x^i \mathbf{e}_i, \tag{1.12}$$

where the coefficients x^i *are uniquely determined.*

Proof. If $(\mathbf{e}_i)$ is a basis, then the system $(\mathbf{x}, (\mathbf{e}_i))$ is linearly dependent so that $n+1$ scalars (a, a^i) exist, some of which are different from zero, such that

$$a\mathbf{x} + a^i \mathbf{e}_i = \mathbf{0}. \tag{1.13}$$

But $a \neq 0$ since, otherwise, we would have $a^i \mathbf{e}_i = \mathbf{0}$, with some scalars a^i different from zero, against the hypothesis that $(\mathbf{e}_i)$ is a basis. Dividing (1.13) by a, we obtain (1.12), where $x^i = -a_i/a$. To prove that the coefficients x^i in (1.12) are uniquely determined, we suppose that, besides the decomposition (1.12), we also have

$$\mathbf{x} = \overline{x}^i \mathbf{e}_i.$$

Subtracting this decomposition from (1.12), we obtain the relation $\mathbf{0} = (x^i - \overline{x}^i)\mathbf{e}_i$, which implies $x^i = \overline{x}^i$, since $(\mathbf{e}_i)$ is a basis.

If (1.12) is supposed to hold $\forall \mathbf{x} \in E_n$, then, for $\mathbf{x}=\mathbf{0}$, we obtain

$$\mathbf{0} = x^i \mathbf{e}_i.$$

Since the quantities x^i are uniquely determined, this relation implies that $x^1 = \cdots = x^n = 0$ and the vector system $(\mathbf{e}_i)$ is free. To prove that $(\mathbf{e}_i)$ is a basis, we must verify that any vector system $(\mathbf{u}_1, \ldots, \mathbf{u}_n, \mathbf{x})$ of order $n+1$ is linearly dependent. That is true if $(\mathbf{u}_i)$ is linearly dependent. By contrast, if $(\mathbf{u}_i)$ is free, then from (1.12) we derive the relation

$$\mathbf{u}_1 = a^i \mathbf{e}_i,$$

in which at least one of the scalars a^i does not vanish since $\mathbf{u}_1 \neq \mathbf{0}$. Supposing that $a_1 \neq 0$, the preceding relation leads us to

$$\mathbf{e}_1 = \frac{1}{a^1}\mathbf{u}_1 - \frac{1}{a^1}\sum_{i=2}^{n} a^i \mathbf{e}_i.$$

In view of this result and (1.12), we can state that any vector $\mathbf{x} \in E_n$ can be represented as a linear combination of the vectors $(\mathbf{u}_1, \mathbf{e}_2, \ldots, \mathbf{e}_n)$. Consequently, we can say that

$$\mathbf{u}_2 = b^1\mathbf{u}_1 + b^2\mathbf{e}_2 + \cdots + b^n\mathbf{e}_n,$$

where at least one of the scalars $b^2, \ldots, b^n$ does not vanish. Indeed, if all these coefficients were equal to zero, then we would have $\mathbf{u}_2 = b^1\mathbf{u}_1$, and the vector system $(\mathbf{u}_1, \ldots, \mathbf{u}_n)$ would be linearly dependent. If $b^2 \neq 0$, we can write that

$$\mathbf{e}_2 = c^1\mathbf{u}_1 + c^2\mathbf{u}_2 + c^3\mathbf{e}_3 + \cdots + \mathbf{e}_n.$$

In view of (1.12), this relation implies that any vector $\mathbf{x} \in E_n$ is a linear combination of $(\mathbf{u}_1, \mathbf{u}_2, \mathbf{e}_3, \ldots, \mathbf{e}_n)$. Iterating the procedure, we conclude that (1.12) and the hypothesis that $(\mathbf{u}_i)$ is free imply the relation $\mathbf{x} = x^i\mathbf{u}_i$, i.e., $\mathbf{x} - x^i\mathbf{u}_i = \mathbf{0}$. Then, the vector system $(\mathbf{u}_1, \ldots, \mathbf{u}_n, \mathbf{x})$ is linearly dependent and the theorem is proved. $\qquad \square$

1.3 Basis Changes

Definition 1.5. The unique coefficients expressing any vector $\mathbf{x}$ as a linear combination of the basis $(\mathbf{e}_i)$ are called the ***contravariant components*** of $\mathbf{x}$ in the basis $(\mathbf{e}_i)$ or relative to the basis $(\mathbf{e}_i)$.

Let $(\mathbf{e}_i)$ and $(\mathbf{e}_i')$ be two bases of the n-dimensional vector space E_n. Then, $\mathbf{x}$ admits the two representations

$$\mathbf{x} = x^i\mathbf{e}_i = x'^i\mathbf{e}_i'. \tag{1.14}$$

Further, since $(\mathbf{e}_i)$ and $(\mathbf{e}_i')$ are bases of E_n, any vector of the basis $(\mathbf{e}_i')$ can be expressed as a linear combination of the vectors belonging to the basis $(\mathbf{e}_i)$, i.e.,

$$\mathbf{e}_j' = A_j^i\mathbf{e}_i, \quad \mathbf{e}_i = (A^{-1})_i^j\mathbf{e}_j', \tag{1.15}$$

where $((A^{-1})_i^j)$ is the inverse matrix of (A_j^i). Introducing (1.15) into (1.14), we obtain the condition

$$\mathbf{x} = x'^j\mathbf{e}_j' = (A^{-1})_i^j x^i\mathbf{e}_j',$$

which, given the uniqueness of the representation of a vector with respect to a basis, implies the equations

$$x'^j = (A^{-1})_i^j x^i \tag{1.16}$$

relating the contravariant components of the same vector evaluated in two different bases. It is evident that the inverse formulae of (1.16) are

$$x^i = A^i_j x'^j. \tag{1.17}$$

Notice that the contravariant components of the vector $\mathbf{x}$ are transformed with the inverse matrix of the basis change (1.15).

1.4   Vector Subspaces

Definition 1.6. A subset $V \subset E_n$ is a ***vector subspace*** of E_n if

$$\forall \mathbf{x}, \mathbf{y} \in V, \forall a, b \in \Re \Rightarrow a\mathbf{x} + b\mathbf{y} \in V. \tag{1.18}$$

It is evident that V is a vector space.

Let $(\mathbf{e}_1, ..., \mathbf{e}_m)$, $m \leq n$, be a free vector system of E_n. It is plain to verify that the totality V of the vectors

$$\mathbf{x} = x^1 \mathbf{e}_1 + \cdots + x^m \mathbf{e}_m, \tag{1.19}$$

obtained on varying the coefficients x^i in $\Re$ represents a vector subspace of E_n. Such a subspace is said to be the ***subspace generated*** or ***spanned*** by the system $(\mathbf{e}_1, ..., \mathbf{e}_m)$. In particular, any vector space E_n is generated by any basis of it.

Definition 1.7. Let U_p and V_q be two vector subspaces of E_n, having dimensions p and q, $p + q \leq n$, respectively, and such that $U_p \cap V_q = \{\mathbf{0}\}$. Then, we denote by

$$W = U_p \oplus V_q \tag{1.20}$$

the ***direct sum*** of the subspaces U_p and V_q, that is, the vector set formed by all the vectors $\mathbf{u} + \mathbf{v}$, where $\mathbf{u} \in U_p$ and $\mathbf{v} \in V_q$.

Theorem 1.2. *The direct sum $W = U_p \oplus V_q$ of the vector subspaces U_p and V_q is a vector subspace of dimension $m = p + q$.*

Proof. Let $(\mathbf{u}_1, ..., \mathbf{u}_p)$ and $(\mathbf{v}_1, ..., \mathbf{v}_q)$ be bases of U_p and V_q, respectively. By Definition 1.7, we can state that none of these vectors belongs to both bases. Moreover, any $\mathbf{w} \in W$ can be written in the form $\mathbf{w} = \mathbf{u} + \mathbf{v}$, where $\mathbf{u}$ and $\mathbf{v}$ are unique vectors belonging to U_p and V_q, respectively. Therefore, we have that

$$\mathbf{w} = \sum_{i=1}^{p} u^i \mathbf{u}_i + \sum_{i=1}^{q} v^i \mathbf{v}_i,$$

where the contravariant components u^i and v^i are uniquely determined. In conclusion, the vectors $(\mathbf{u}_1, ..., \mathbf{u}_p, \mathbf{v}_1, ..., \mathbf{v}_q)$ form a basis of W. $\qquad\square$

1.5   Algebras

Definition 1.8. Let E be a vector space. Let us assign an internal mapping $(\mathbf{x}, \mathbf{y}) \in E \times E \to \mathbf{xy} \in E$, called a **product**, that for any choice of the scalars a and b and any choice of the vectors $\mathbf{x}$, $\mathbf{y}$, and $\mathbf{z}$ verifies the following properties:

$$
\begin{aligned}
a(\mathbf{xy}) &= (a\mathbf{x})\mathbf{y} = \mathbf{x}(a\mathbf{y}), \\
(a\mathbf{x} + b\mathbf{y})\mathbf{z} &= a(\mathbf{xz}) + b(\mathbf{yz}), \\
\mathbf{x}(a\mathbf{y} + b\mathbf{z}) &= a(\mathbf{xy}) + b(\mathbf{xz}).
\end{aligned}
\tag{1.21}
$$

The vector space E equipped with a product is an example of an **algebra** on E. This algebra is **commutative** if

$$
\mathbf{xy} = \mathbf{yx}.
\tag{1.22}
$$

A vector $\mathbf{e}$ is the **unit vector** of the product if

$$
\mathbf{xe} = \mathbf{ex} = \mathbf{x}.
\tag{1.23}
$$

It is simple to prove that (1.23) implies that $\mathbf{e}$ is unique.

Definition 1.9. We say that the preceding algebra is a **Lie algebra** if the product

- Is skew-symmetric, i.e.,

$$
\mathbf{xy} = -\mathbf{yx};
\tag{1.24}
$$

- Verifies Jacobi's identity

$$
\mathbf{x}(\mathbf{yz}) + \mathbf{z}(\mathbf{xy}) + \mathbf{y}(\mathbf{zx}) = \mathbf{0}.
\tag{1.25}
$$

1.6   Examples

In this section we present some interesting examples of vector spaces and algebras.

- *The vector space of oriented segments.* Let $\mathcal{E}_3$ be a three-dimensional Euclidean space. We denote by $\overrightarrow{AB}$ an oriented segment starting from the point $A \in \mathcal{E}_3$ and ending at the point $B \in \mathcal{E}_3$. In the set of all oriented segments of $\mathcal{E}_3$ we introduce the following equivalence relation $\mathcal{R}$: two oriented segments are equivalent if they are equipollent. Then we consider the set $E = \mathcal{E}_3/\mathcal{R}$ and denote by $\mathbf{x} = [\overrightarrow{AB}]$ the equivalence class of an arbitrary oriented segment $\overrightarrow{AB}$. Since all these equivalence classes are in one-to-one correspondence with the oriented segments starting from a fixed point $O \in \mathcal{E}_3$, we can introduce the notation $\mathbf{x} = [\overrightarrow{OA}]$. If $\mathbf{y} = [\overrightarrow{OB}]$, then we introduce the following operations in E:

$$a\mathbf{x} = [a\overrightarrow{OA}], \tag{1.26}$$

$$\mathbf{x} + \mathbf{y} = [\overrightarrow{OA} + \overrightarrow{OB}], \tag{1.27}$$

where in (1.27) the oriented segment $\overrightarrow{OA} + \overrightarrow{OB}$ is the diagonal of a parallelogram with sides $\overrightarrow{OA}$ and $\overrightarrow{OB}$. It is plain to verify that the foregoing operations equip E with the structure of a three-dimensional vector space since three unit vectors, which are orthogonal to each other, form a basis of E.

It is also easy to verify that if we introduce into E the cross product $\mathbf{x} \times \mathbf{y}$ of two vectors, the vector space E becomes a Lie algebra.

- *The vector space of polynomials of degree* $\leq n$. Denote by E the set of polynomials $P(x)$ of degree r, where $0 \leq r \leq n$. Any polynomial can be written as $P(x) = p_0 + p_1 x + \cdots + p_n x^n$, where some of its coefficients p_i may vanish. If, for any $a \in \Re$, we define

$$aP(x) = a(p_0 + p_1 x + \cdots + p_n x^n) = ap_0 + ap_1 x + \cdots + ap_n x^n,$$
$$P(x) + Q(x) = (p_0 + p_1 x + \cdots + p_n x^n) + (q_0 + q_1 x + \cdots + q_n x^n)$$
$$= (p_0 + q_0) + (p_1 + q_1)x + \cdots + (p_n + q_n)x^n,$$

then E becomes a vector space on $\Re$. Moreover, the $n+1$ polynomials

$$P_0 = 1, \ldots, P_n = x^n \tag{1.28}$$

are linearly independent since, from the linear combination

$$\lambda_0 P_0(x) + \cdots + \lambda_n P_n(x) = 0$$

and the fundamental theorem of algebra, it follows that all the real numbers $\lambda_0, \ldots, \lambda_n$ vanish. On the other hand, any polynomial of degree n can be expressed as a linear combination of the $n+1$ polynomials (1.28). We can conclude that E is an $(n+1)$-dimensional vector space.
- *The vector space of continuous functions on the interval* $[a, b]$. Denote by $C^0[a, b]$ the set of continuous functions on the closed and bounded interval $[a, b]$. With the ordinary operations of multiplication of a function by a number and the addition of two functions, $C^0[a, b]$ becomes a vector space. Since the polynomials of any degree are continuous functions, the vector space of the polynomial of degree n, where n is an arbitrary integer, is a vector subspace of $C^0[a, b]$. Consequently, E has infinite dimension since it contains free vector systems of any order.
- *The vector space of* $n \times n$ *matrices.* Let E be the set of the $n \times n$ matrices $\mathbb{A} = (a_{ij})$. The operations

$$a\mathbb{A} = (aa_{ij}), \quad \mathbb{A} + \mathbb{B} = (a_{ij} + b_{ij}) \tag{1.29}$$

equip E with the structure of a vector space. It is a very simple exercise to verify that the n^2 matrices

$$\begin{pmatrix} 1 & 0 & \cdots & 0 \\ \cdots & \cdots & \cdots & \cdots \\ 0 & 0 & 0 & 0 \end{pmatrix}, \dots, \begin{pmatrix} 0 & 0 & \cdots & 0 \\ \cdots & \cdots & \cdots & \cdots \\ 0 & 0 & 0 & 1 \end{pmatrix}$$

form a basis of E that is an n^2-dimensional vector space. If we add to the operations (1.29), consisting in the multiplication of rows by columns the two matrices $n \times n$

$$\mathbb{AB} = (a_{ik}b_{kj}), \tag{1.30}$$

then E becomes a noncommutative algebra since

$$\mathbb{AB} \neq \mathbb{BA}. \tag{1.31}$$

Finally, with the operation

$$[\mathbb{A}, \mathbb{B}] = \mathbb{AB} - \mathbb{BA}, \tag{1.32}$$

E is a Lie algebra. In fact, operation (1.32) is skew-symmetric, and it is a simple exercise to prove Jacobi's identity (1.25) (see Exercise 1).

1.7 Linear Maps

Definition 1.10. Let E_n and G_m be two vector spaces on $\Re$ with dimensions n and m, respectively. We say that the map

$$F : E_n \to G_m \tag{1.33}$$

is a *linear map* or a *morphism* if

$$F(a\mathbf{x} + b\mathbf{y}) = aF(\mathbf{x}) + bF(\mathbf{y}), \tag{1.34}$$

$\forall a, b \in \Re$, and $\forall \mathbf{x}, \mathbf{y} \in E_n$.

Henceforth, we denote by $\mathrm{Lin}(E_n, G_m)$ the set of all linear maps of E_n into G_m.

Theorem 1.3. *A linear map $F \in \mathrm{Lin}(E_n, G_m)$ is determined by the vectors $F(\mathbf{e}_i) \in G_m$ corresponding to the vectors of a basis $(\mathbf{e}_i)$ of E_n.*

Proof. In fact, $\forall \mathbf{x} \in E_n$ we have that

$$\mathbf{y} = F(\mathbf{x}) = F(x^i \mathbf{e}_i) = x^i F(\mathbf{e}_i). \tag{1.35}$$

$\square$

Since the vectors $F(\mathbf{x})$ belong to G_m, they can be represented as a linear combination of the vectors of a basis $(\mathbf{g}_i)$ of G_m, i.e.,

$$F(\mathbf{e}_i) = F_i^h \mathbf{g}_h, \tag{1.36}$$

where, for any fixed i, the scalars F_i^h denote the components of the vector $F(\mathbf{e}_i)$ relative to the basis $(\mathbf{g}_h)$. Considering the upper index of F_i^h as a row index and the lower index as a column index, the numbers F_i^h become the elements of an $m \times n$ matrix $\mathbb{F}$ with m rows and n columns. When the bases $(\mathbf{e}_i)$ in E_n and $(\mathbf{g}_h)$ in G_m are chosen, such a matrix uniquely determines the linear map F since (1.35) and (1.36) imply that

$$\mathbf{y} = F(\mathbf{x}) = F_i^h x^i \mathbf{g}_h. \tag{1.37}$$

For this reason, $\mathbb{F} = (F_i^h)$ is called the **matrix of the linear map** F relative to the bases $(\mathbf{e}_i)$ and $(\mathbf{g}_h)$. Adopting the matrix notation, (1.37) becomes

$$\mathbb{Y} = \mathbb{F}\mathbb{X}, \tag{1.38}$$

where $\mathbb{Y}$ is the $m \times 1$ matrix formed with the components of $\mathbf{y}$ and $\mathbb{X}$ is the $n \times 1$ matrix formed with the components of $\mathbf{x}$.

Definition 1.11. $\forall F \in \mathrm{Lin}(E_n, G_m)$ the subset

$$\mathrm{Im}(F) = \{\mathbf{y} \in G_m | \exists \mathbf{x} \in E_n : \mathbf{y} = F(\mathbf{x})\} \tag{1.39}$$

is called the **image** of F, whereas the subset

$$\ker(F) = \{\mathbf{x} \in E_n | F(\mathbf{x}) = \mathbf{0}\} \tag{1.40}$$

is the **kernel** of F.

A linear map (1.33) is an **epimorphism** if $\mathrm{Im}(F) = G$; further, F is called a **monomorphism** if no two different elements of E are sent by F into the same element of G. In particular, an epimorphism $F : E_n \to E_n$ is called **endomorphism** and the set of all endomorphisms is denoted by $\mathrm{Lin}(E_n)$. The linear map (1.33) is an **isomorphism** if it is a monomorphism and $F(E_n) = G_m$. In this case, there exists the inverse linear map $F^{-1} : G_m \to E_n$. Finally, an isomorphism $F : E_n \to E_n$ is said to be an **automorphism**.

Theorem 1.4. $\mathrm{Im}(F)$ and $\ker(F)$ are vector subspaces of G_m and E_n, respectively. Moreover, F is a monomorphism if and only if $\ker(F) = \{\mathbf{0}\}$.

Proof. If $\mathbf{y}_1, \mathbf{y}_2 \in \mathrm{Im}(F)$, then there are $\mathbf{x}_1, \mathbf{x}_2 \in E_n$ such that $\mathbf{y}_1 = F(\mathbf{x}_1)$ and $\mathbf{y}_2 = F(\mathbf{x}_2)$. Then, $\forall a, b \in \Re$ the linearity of F implies that

$$a\mathbf{y}_1 + b\mathbf{y}_2 = aF(\mathbf{x}_1) + bF(\mathbf{x}_2) = F(a\mathbf{x}_1 + b\mathbf{x}_2)$$

and $a\mathbf{y}_1 + b\mathbf{y}_2 \in \mathrm{Im}(F)$. Similarly, if $\mathbf{x}_1$, $\mathbf{x}_2 \in \ker(F)$, then it follows that $F(\mathbf{x}_1) = F(\mathbf{x}_2) = \mathbf{0}$, so that

$$\mathbf{0} = a F(\mathbf{x}_1) + b F(\mathbf{x}_2) = F(a\mathbf{x}_1 + b\mathbf{x}_2),$$

and we conclude that $a\mathbf{x}_1 + b\mathbf{x}_2 \in \ker(F)$.

The second part of the theorem remains to be proved. Since $F(\mathbf{0}) = \mathbf{0}$, if the inverse map F^{-1} of F exists, then the subspace $F^{-1}(\mathbf{0})$ reduces to the vector $\mathbf{0}$, that is, $\ker(F) = \mathbf{0}$. In contrast, supposing that this last condition is satisfied, if it is possible to find two vectors $\mathbf{x}'$ and $\mathbf{x}'' \in E_n$ such that $F(\mathbf{x}') = F(\mathbf{x}'')$, then it is also true that $F(\mathbf{x}' - \mathbf{x}'') = \mathbf{0}$. But this condition implies that

$$\mathbf{x}' - \mathbf{x}'' = \mathbf{0} \Rightarrow \mathbf{x}' - \mathbf{x}'' \in \ker(F) \Rightarrow \mathbf{x}' = \mathbf{x}'',$$

and the theorem is proved.   □

In particular, from the preceding theorem it follows that

$$F \text{ is an isomorphism} \Leftrightarrow \mathrm{Im}(F) = G_m, \ker(F) = \mathbf{0}. \qquad (1.41)$$

Theorem 1.5. *Let $(\mathbf{e}_i)$ be a basis of the vector space E_n and let $(\mathbf{g}_h)$ be a basis of the vector space G_m. For any $F \in \mathrm{Lin}(E_n, G_m)$, the following statements are equivalent:*

(a) F is an isomorphism;
(b) $m = n$ and the vectors $F(\mathbf{e}_i)$ form a basis of G_m;
(c) $m = n$ and the matrix $\mathbb{F}$ of F relative to the pair of bases $(\mathbf{e}_i)$ and $(\mathbf{g}_h)$ is not singular.

Proof. (a)$\Rightarrow$(b). The condition $\lambda^i F(\mathbf{e}_i) = F(\lambda^i \mathbf{e}_i) = \mathbf{0}$ and (1.41) imply that $\lambda^i \mathbf{e}_i = \mathbf{0}$. Since $(\mathbf{e}_i)$ is a basis of E_n, we have that $\lambda^i = 0$, $i = 1, \ldots, n$, and the vectors $(F(\mathbf{e}_i))$ are independent. Moreover, if F is an isomorphism, $\forall \mathbf{y} \in G_m$ there exists one and only one $\mathbf{x} \in G_m$ such that $\mathbf{y} = F(\mathbf{x}) = x^i F(\mathbf{e}_i)$. Consequently, the vectors $F(\mathbf{e}_i)$ form a basis of G_m and $m = n$.

(b)$\Rightarrow$(c). In view of (1.37), the relation $\lambda^i F(\mathbf{e}_i) = \mathbf{0}$ becomes $\lambda^i F_i^h \mathbf{g}_h = \mathbf{0}$. But *(b)* implies that the linear system $\lambda^i F_i^h = 0$ has only the solution $\lambda^i = 0$, $i = 1, \ldots, n$. Consequently, $m = n$ and the matrix $\mathbb{F}$ is not singular.

(c)$\Rightarrow$(a). Equation (1.38) is the matrix form of the mapping F with respect to the bases $(\mathbf{e}_i)$ and $(\mathbf{g}_h)$. For any choice of the vector $\mathbf{y} \in G_m$, (1.38) is a linear system on n equations in the n unknown x^i. Owing to *(c)*, this system admits one and only one solution so that F is an isomorphism.   □

We conclude this section with the following remark. Consider a basis change

$$\mathbf{g}_h' = G_h^k \mathbf{g}_k, \quad \mathbf{e}_i' = E_i^j \mathbf{e}_j \qquad (1.42)$$

Fig. 1.1 Two bases in a plane

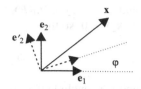

in the two vector spaces E_n and G_m. Then, the matrix $\mathbb{F}$ relative to the bases $(\mathbf{e}_h)$ and $(\mathbf{g}_h)$ and the matrix $\mathbb{F}'$ relative to the bases $(\mathbf{e}'_h)$ and $(\mathbf{g}'_h)$ are related by the following equation:

$$\mathbb{F}' = \mathbb{G}^{-1}\mathbb{F}\mathbb{E} \tag{1.43}$$

(see Exercise 4).

1.8   Exercises

1. In the vector space E_{n^2} of the $n \times n$ matrices, prove that Jacobi's identity is verified.

   We start by noting that

   $$\begin{aligned}[\mathbb{A}, [\mathbb{B}, \mathbb{C}]] &= \mathbb{A}[\mathbb{B}, \mathbb{C}] - [\mathbb{B}, \mathbb{C}]\mathbb{A} \\ &= \mathbb{A}(\mathbb{B}\mathbb{C} - \mathbb{C}\mathbb{B}) - (\mathbb{B}\mathbb{C} - \mathbb{C}\mathbb{B})\mathbb{A} \\ &= \mathbb{A}\mathbb{B}\mathbb{C} - \mathbb{A}\mathbb{C}\mathbb{B} - \mathbb{B}\mathbb{C}\mathbb{A} + \mathbb{C}\mathbb{B}\mathbb{A}.\end{aligned}$$

   Jacobi's identity is proved applying the foregoing identity to the three terms of the expression $[\mathbb{A}, [\mathbb{B}, \mathbb{C}]] + [\mathbb{C}, [\mathbb{A}, \mathbb{B}]] + [\mathbb{B}, [\mathbb{C}, \mathbb{A}]]$.

2. Let

   $$\begin{aligned}\mathbf{e}'_1 &= \cos\varphi\,\mathbf{e}_1 + \sin\varphi\,\mathbf{e}_2, \\ \mathbf{e}'_2 &= -\sin\varphi\,\mathbf{e}_1 + \cos\varphi\,\mathbf{e}_2\end{aligned} \tag{1.44}$$

   be a basis change in the vector space E_2 (Fig. 1.1). Evaluate the components (x'^i) in the basis $(\mathbf{e}'_i)$ of the vector that has the components $(1, 2)$ in the basis $(\mathbf{e}_i)$. *Hint*: It is sufficient to verify that the inverse matrix of the basis change is

   $$(A^{-1})^i_j = \begin{pmatrix} \cos\varphi & -\sin\varphi \\ \sin\varphi & \cos\varphi \end{pmatrix}$$

   and then to apply (1.16).

3. Let $(\mathbf{e}_i)$, $i=1,2$, be a basis of the two-dimensional vector space E_2 and let $(\mathbf{g}_h)$, $h=1,2,3$, be a basis of the three-dimensional vector space G_3. Verify that the ker and the Im of the linear mapping $F : E_2 \rightarrow G_3$, whose matrix relative to the bases $(\mathbf{e}_i)$ and $(\mathbf{g}_h)$ is

$$\begin{pmatrix} 1 & 2 \\ 0 & 1 \\ \alpha & \alpha \end{pmatrix}, \quad \alpha \in \Re,$$

are given by the following subspaces:

$$\ker(F) = \{\mathbf{0}\}, \quad \mathrm{Im}(F) = \{\mathbf{y} \in G_3 : \mathbf{y} = (y^1, y^2, \alpha(y^1 - y^2))\},$$

where (y^h) are the contravariant components of any vector $\mathbf{y} \in G_3$.

4. Prove formula (1.43).

Let $(\mathbf{e}_i)$, $(\mathbf{g}_h)$ be a pair of bases in the vector space E_n and denote by $(\mathbf{e}'_i)$, $(\mathbf{g}'_h)$ a pair of bases in the vector spaces G_m. We have that, for any $\mathbf{x} \in E_n$, the vector $\mathbf{y} = F(\mathbf{x}) \in G_m$ can be written [see (1.36), (1.37)] in the following forms:

$$\begin{aligned} \mathbf{y} &= y^h \mathbf{g}_h = F_i^h x^i \mathbf{g}_h, \\ \mathbf{y} &= y'^h \mathbf{g}'_h = F_i'^h x^i \mathbf{g}'_h. \end{aligned}$$

From the foregoing representations of $\mathbf{y}$ we have that

$$F_i^h x^i \mathbf{g}_h = F_i'^h x'^i \mathbf{g}'_h.$$

Taking into account the basis changes (1.42), this condition becomes

$$F_i^h E_l^i x'^l (G^{-1})_h^k \mathbf{g}'_k = F_l'^k x'^l \mathbf{g}'_k.$$

From the arbitrariness of $\mathbf{x}$ there follows (1.43).

5. Determine the image and the kernel of the following linear mapping:

$$\begin{pmatrix} y^1 \\ y^2 \\ y^3 \end{pmatrix} = \begin{pmatrix} 1 & 0 & -1 \\ 2 & 1 & 0 \\ 1 & a & -2 \end{pmatrix} \begin{pmatrix} x^1 \\ x^2 \\ x^3 \end{pmatrix}.$$

6. Determine the vectors (a,b,c) of the vector space V generated by the vectors $u = (2, 1, 0)$, $v = (1, -1, 2)$, and $w = (0, 3, -4)$.

  Hint: We must find the real numbers x, y, and z such that

$$(a, b, c) = x(2, 1, 0) + y(1, -1, 2) + z(0, 3, -4) = (2x + y, x - y + 3z, 2y - 4z),$$

i.e., we must solve the linear system

$$2x + y = a,$$
$$3y - 6z = b,$$
$$0 = 2a - 4b - 3c.$$

Analyze this system.

7. Verify if the vectors (a, b, c) generate a subspace W when they satisfy one of the following conditions:

   - $a = 2b$;
   - $ab = 0$;
   - $a = b = c$;
   - $a = b^2$.

8. Prove that the polynomials $(1 - x)^3$, $(1 - x)^2$, and $(1, -x)$ generate the vector space of polynomials of the third degree.

9. Let U be the subspace of $\mathfrak{R}^3$ generated by the vectors $(a, b, 0)$, with a and b arbitrary real numbers. Further, let W be the subspace of $\mathfrak{R}^3$ generated by the two vectors $(1, 2, 3)$ and $(1, -1, 1)$. Determine vectors generating the intersection $U \bigcap W$.

10. Determine the image and the kernel of the following linear map:

$$(x, y, z) \in \mathfrak{R}^3 n \to (x + 2y - z, y + z, x + y - 2z) \in \mathfrak{R}^3.$$

Chapter 2
Tensor Algebra

Abstract This chapter contains an introduction to tensor algebra. After defining covectors and dual bases, the space of covariant 2-tensor is introduced. Then, the results derived for this space are extended to the general space of the (r, s)-tensors. Finally, the operations of contraction and contracted multiplication are introduced.

2.1 Linear Forms and Dual Vector Space

Definition 2.1. Let E be a vector space on $\Re$. The map

$$\omega : E \to \Re \qquad (2.1)$$

is said to be a **_linear form_**, a **1-form**, or a **_covector_** on E if

$$\omega(a\mathbf{x} + b\mathbf{y}) = a\omega(\mathbf{x}) + b\omega(\mathbf{y}), \qquad (2.2)$$

$\forall a, b \in \Re$ and $\forall \mathbf{x}, \mathbf{y} \in E$.[1]

The set E^* of all linear forms on E becomes a vector space on $\Re$ when we define the sum of two linear forms $\omega, \sigma \in E^*$ and the product of the scalar $a \in \Re$ with the linear form ω in the following way:

$$(\omega + \sigma)(\mathbf{x}) = \omega(\mathbf{x}) + \sigma(\mathbf{x}), \quad (a\omega)(\mathbf{x}) = a\,\omega(\mathbf{x}), \ \forall \mathbf{x} \in E. \qquad (2.3)$$

---

[1] For the contents of Chaps. 2 –9, see [7, 8, 10, 11, 13, 14].

© Springer International Publishing AG, part of Springer Nature 2018
A. Romano and A. Marasco, *Classical Mechanics with Mathematica®*,
Modeling and Simulation in Science, Engineering and Technology,
https://doi.org/10.1007/978-3-319-77595-1_2

Theorem 2.1. *Let E_n be a vector space with finite dimension n. Then, E^* has the same dimension n. Moreover, if $(\mathbf{e}_i)$ is a basis of E_n, then the n covectors θ^i such that[2]*

$$\theta^i(\mathbf{e}_j) = \delta^i_j \tag{2.4}$$

define a basis of E^*.

Proof. First, we remark that, owing to (2.4), the linear forms θ^i are defined over the whole space E_n since, $\forall \mathbf{x} = x^i \mathbf{e}_i \in E_n$, we have that

$$\theta^i(\mathbf{x}) = \theta^i(x^j \mathbf{e}_j) = x^j \theta^i(\mathbf{e}_j) = x^i. \tag{2.5}$$

To show that E_n^* is n-dimensional, it is sufficient to verify that any element of E_n^* can be written as a unique linear combination of the covectors (θ^i). Owing to the linearity of any $\omega \in E_n^*$, we have that

$$\omega(\mathbf{x}) = x^i \omega(\mathbf{e}_i) = x^i \omega_i, \tag{2.6}$$

where we have introduced the notation

$$\omega_i = \omega(\mathbf{e}_i). \tag{2.7}$$

On the other hand, (2.5) allows us to write (2.6) in the following form:

$$\omega(\mathbf{x}) = \omega_i \theta^i(\mathbf{x}), \tag{2.8}$$

from which, owing to the arbitrariness of $\mathbf{x} \in E_n$, it follows that

$$\omega = \omega_i \theta^i. \tag{2.9}$$

To prove the theorem, it remains to verify that the quantities ω_i in (2.9) are uniquely determined. If another representation $\omega = \omega'_i \theta^i$ existed, then we would have $\mathbf{0} = (\omega_i - \omega'_i)\theta^i$, i.e.,

$$(\omega_i - \omega'_i)\theta^i(\mathbf{x}) = 0, \quad \forall \mathbf{x} \in E_n.$$

Finally, from (2.4) it follows that $\omega'_i = \omega_i$.   □

------

[2] Here δ^i_j is the Kronecker symbol

$$\delta^i_j = \begin{cases} 0, i \neq j, \\ 1, i = j. \end{cases}$$

Remark 2.1. Note that (2.6) gives the *value* of the linear map ω when it is applied to the vector $\mathbf{x}$, whereas (2.9) supplies the linear map ω as a linear combination of the n linear maps θ^i.

By (2.4), the dual basis θ^i of E_n^* is associated with the basis $(\mathbf{e}_i)$ of E_n. Consequently, a basis change $(\theta^i) \rightarrow (\theta'^i)$ in E_n^* corresponds to a basis change $(\mathbf{e}_i) \rightarrow (\mathbf{e}_i')$ in E_n expressed by (1.15). To determine this basis change, since it is

$$\theta'^i(\mathbf{x}) = x'^i, \quad \theta^i(\mathbf{x}) = x^i, \tag{2.10}$$

and (1.16) holds, we have that

$$\theta'^i(\mathbf{x}) = (A^{-1})_j^i x^j = (A^{-1})_j^i \theta^j(\mathbf{x}).$$

In view of the arbitrariness of the vector $\mathbf{x} \in E_n$, from the preceding relation we obtain the desired transformation formulae of the dual bases:

$$\theta'^i = (A^{-1})_j^i \theta^j, \quad \theta^i = A_j^i \theta'^j. \tag{2.11}$$

The transformation formulae of the components ω_i of the linear form ω corresponding to the basis change (2.11) are obtained recalling that, since ω is a vector of E_n^*, its components are transformed with the inverse matrix of the dual-basis change. Therefore, it is

$$\omega_i' = A_i^j \omega_j, \quad \omega_i = (A^{-1})_i^j \omega_j'. \tag{2.12}$$

Remark 2.2. Bearing in mind the foregoing results, we can state that the components of a covector relative to a dual basis are transformed according to the *covariance law* (i.e., as the bases of the vector space E_n). By contrast, the dual bases are transformed according to the *contravariance law* (i.e., as the components of a vector $\mathbf{x} \in E_n$).

Remark 2.3. Since the vector spaces E_n and E_n^* have the same dimension, it is possible to build an isomorphism between them. In fact, by choosing a basis $(\mathbf{e}_i)$ of E_n and a basis (θ^i) of E_n^*, an isomorphism between E_n and E_n^* is obtained by associating the covector $\omega = \sum_i x^i \theta^i$ with the vector $\mathbf{x} = x^i \mathbf{e}_i \in E_n$. However, owing to the different transformation character of the components of a vector and a covector, the preceding isomorphism *depends on the choice of the bases* $(\mathbf{e}_i)$ and (θ^i). In Section 5.1 we will show that, when E_n is a Euclidean vector space, it is possible to define an isomorphism between E_n and E_n^* that does not depend on the aforementioned choice, i.e., it is *intrinsic*.

2.2 Biduality

We have already proved that E_n^* is itself a vector space. Consequently, it is possible to consider its dual vector space E_n^{**} containing all the linear maps $G : E_n^* \rightarrow \Re$. Moreover, to any basis $(\theta^i) \in E_n^*$ we can associate the dual basis $(\mathbf{f}_i) \in E_n^{**}$ defined by the conditions [see (2.4)]

$$\mathbf{f}_i(\theta^j) = \delta_i^j, \tag{2.13}$$

so that any $\mathbf{F} \in E_n^{**}$ admits a unique representation in this basis:

$$\mathbf{F} = F^i \mathbf{f}_i. \tag{2.14}$$

At this point we can consider the idea of generating "ad libitum" vector spaces by the duality definition. But this cannot happen since E_n and E_n^{**} are isomorphic, i.e., they can be identified. In fact, let us consider the linear map

$$\mathbf{x} \in E_n \rightarrow \mathbf{F_x} \in E_n^{**}$$

such that

$$\mathbf{F_x}(\omega) = \omega(\mathbf{x}), \quad \forall \omega \in E_n^*. \tag{2.15}$$

To verify that (2.15) is an isomorphism, denote by $(\mathbf{e}_i)$ a basis of E_n, (θ^i) its dual basis in E_n^*, and $\mathbf{f}_i$ the dual basis of (θ^i) in E_n^{**}. In these bases, (2.15) assumes the form

$$F^i \omega_i = x^i \omega_i, \tag{2.16}$$

which, owing to the arbitrariness of ω, implies that

$$F^i = x^i, \quad i = 1, \dots, n. \tag{2.17}$$

The correspondence (2.17) refers to the bases $(\mathbf{e}_i)$, (θ^i), and $(\mathbf{f}_i)$. To prove that the isomorphism (2.17) does not depend on the basis, it is sufficient to note that the basis change (1.15) in E_n determines the basis change (2.11) in E_n^*. But $(\mathbf{f}_i)$ is the dual basis of (θ^i), so that it is transformed according to the formulae

$$\mathbf{f}_i' = A_i^j \mathbf{f}_j.$$

Consequently, the components x^i of $\mathbf{x} \in E_n$ and the components F^i of $\mathbf{F} \in E_n^{**}$ are transformed in the same way under a basis change.

We conclude by remarking that the foregoing considerations allow us to look at a vector as a linear map on E_n^*. In other words, we can write (2.15) in the following way:

$$\mathbf{x}(\omega) = \omega(\mathbf{x}). \tag{2.18}$$

2.3   Covariant 2-Tensors

Definition 2.2. A bilinear map

$$\mathbf{T} : E_n \times E_n \to \mathfrak{R}$$

is called a *covariant* 2-*tensor* or a $(0, 2)$-*tensor*.

With the following standard definitions of addition of two covariant 2-tensors and multiplication of a real number a by a covariant 2-tensor:

$$(\mathbf{T}_1 + \mathbf{T}_2)(\mathbf{x}, \mathbf{y}) = \mathbf{T}_1(\mathbf{x}, \mathbf{y}) + \mathbf{T}_2(\mathbf{x}, \mathbf{y}),$$
$$(a\mathbf{T})(\mathbf{x}, \mathbf{y}) = a\mathbf{T}(\mathbf{x}, \mathbf{y}),$$

the set $T_2(E_n)$ of all covariant 2-tensors on E_n becomes a vector space.

Definition 2.3. The *tensor product* $\omega \otimes \sigma$ of $\omega, \sigma \in E_n^*$ is a covariant 2-tensor such that

$$\omega \otimes \sigma(\mathbf{x}, \mathbf{y}) = \omega(\mathbf{x})\sigma(\mathbf{y}), \quad \forall \mathbf{x}, \mathbf{y} \in E_n. \tag{2.19}$$

Theorem 2.2. *Let* $(\mathbf{e}_i)$ *be a basis of* E_n *and let* (θ^i) *be the dual basis in* E_n^*. *Then* $(\theta^i \otimes \theta^j)$ *is a basis of* $T_2(E_n)$, *which is a* n^2-*dimensional vector space.*

Proof. Since $\mathbf{T} \in T_2(E_n)$ is bilinear, we have that

$$\mathbf{T}(\mathbf{x}, \mathbf{y}) = \mathbf{T}(x^i \mathbf{e}_i, y^j \mathbf{e}_j) = x^i y^j \mathbf{T}(\mathbf{e}_i, \mathbf{e}_j).$$

Introducing the *components* of the covariant 2-tensor $\mathbf{T}$ in the basis $(\theta^i \otimes \theta^j)$

$$T_{ij} = \mathbf{T}(\mathbf{e}_i, \mathbf{e}_j), \tag{2.20}$$

the foregoing relation becomes

$$\mathbf{T}(\mathbf{x}, \mathbf{y}) = T_{ij} x^i y^j. \tag{2.21}$$

On the other hand, in view of (2.19) and (2.5), it also holds that

$$\theta^i \otimes \theta^j (\mathbf{x}, \mathbf{y}) = x^i y^j, \tag{2.22}$$

and (2.21) assumes the form

$$\mathbf{T}(\mathbf{x}, \mathbf{y}) = T_{ij} \theta^i \otimes \theta^j (\mathbf{x}, \mathbf{y}).$$

Since this identity holds for any $\mathbf{x}, \mathbf{y} \in E_n$, we conclude that the set $(\theta^i \otimes \theta^j)$ generates the whole vector space $T_2(E_n)$:

$$\mathbf{T} = T_{ij} \theta^i \otimes \theta^j. \tag{2.23}$$

Then, the covariant 2-tensors $(\theta^i \otimes \theta^j)$ form a basis of $T_2(E_n)$ if they are linearly independent. To prove this statement, it is sufficient to note that from the arbitrary linear combination

$$a_{ij} \theta^i \otimes \theta^j = 0$$

we obtain

$$a_{ij} \theta^i \otimes \theta^j (\mathbf{e}_h, \mathbf{e}_k) = a_{ij} \delta_h^i \delta_k^j = 0,$$

so that $a_{hk} = 0$ for any choice of the indices h and k. $\square$

The basis change (1.15) in E_n determines the basis change (2.11) in E_n^* and a basis change

$$\theta'^i \otimes \theta'^j = (A^{-1})_h^i (A^{-1})_k^j \theta^h \otimes \theta^k, \quad \theta^i \otimes \theta^j = A_h^i A_k^j \theta'^h \otimes \theta'^k \tag{2.24}$$

in $T_2(E_n)$. On the other hand, it also holds that

$$\mathbf{T} = T_{ij} \theta^i \otimes \theta^j = T'_{ij} \theta'^i \otimes \theta'^j, \tag{2.25}$$

and taking into account (2.24) we obtain the following transformation formulae of the components of a covariant 2-tensor $\mathbf{T} \in T_2(E_n)$ under a basis change (2.24):

$$T'_{ij} = A_i^h A_j^k T_{hk}. \tag{2.26}$$

Remark 2.4. If $\mathbb{T} = (T_{ij})$ is a matrix whose elements are the components of $\mathbf{T}$, and $\mathbb{A} = (A_j^i)$ is a matrix of the basis change (1.15), then the matrix form of (2.26) is

$$\mathbb{T}' = \mathbb{A}^T \mathbb{T} \mathbb{A}. \tag{2.27}$$

Definition 2.4. A *contravariant* 2-*tensor* or a (2, 0)-*tensor* is a bilinear map

$$\mathbf{T} = E_n^* \times E_n^* \rightarrow \mathfrak{R}. \tag{2.28}$$

It is evident that the set $T^2(E_n)$ of all contravariant 2-tensors becomes a vector space by the introduction of the standard operations of addition of two contravariant 2-tensors and the product of a 2-tensor by a real number.

Definition 2.5. The *tensor product* of two vectors $\mathbf{x}, \mathbf{y} \in E_n$ is the contravariant 2-tensor

$$\mathbf{x} \otimes \mathbf{y}(\omega, \sigma) = \omega(\mathbf{x})\sigma(\mathbf{y}), \quad \forall \omega, \sigma \in E_n^*. \tag{2.29}$$

Theorem 2.3. *Let* $(\mathbf{e}_i)$ *be a basis of the vector space* E_n. *Then* $(\mathbf{e}_i \otimes \mathbf{e}_j)$ *is a basis of* $T^2(E_n)$, *which is an* n^2-*dimensional vector space.*

Proof. $\forall \omega, \sigma \in E_n^*$,

$$\mathbf{T}(\omega, \sigma) = \mathbf{T}(\omega_i \theta^i, \sigma_j \theta^j) = \omega_i \sigma_j \mathbf{T}(\theta^i, \theta^j).$$

By introducing the *components* of the contravariant 2-tensor $\mathbf{T}$ relative to the basis $(\mathbf{e}_i \otimes \mathbf{e}_j)$

$$T^{ij} = \mathbf{T}(\theta^i, \theta^j), \tag{2.30}$$

we can write

$$\mathbf{T}(\omega, \sigma) = T^{ij} \omega_i \sigma_j. \tag{2.31}$$

Since

$$\mathbf{e}_i \otimes \mathbf{e}_j(\omega, \sigma) = \omega_i \sigma_j, \tag{2.32}$$

and ω, σ in (2.31) are arbitrary, we obtain the result

$$\mathbf{T} = T^{ij} \mathbf{e}_i \otimes \mathbf{e}_j, \tag{2.33}$$

which shows that $(\mathbf{e}_i \otimes \mathbf{e}_j)$ generates the whole vector space $T^2(E_n)$. Further, it is a basis of $T^2(E_n)$ since, in view of (2.32), any linear combination

$$a^{ij} \mathbf{e}_i \otimes \mathbf{e}_j = 0$$

implies that $a^{ij} = 0$ for all the indices $i, j = 1, \ldots, n$. $\qquad \square$

The basis change (1.15) in E_n determines the basis change

$$\mathbf{e}'_i \otimes \mathbf{e}'_j = A^h_i A^k_j \mathbf{e}_h \otimes \mathbf{e}_k \tag{2.34}$$

in $T^2(E_n)$. The contravariant 2-tensor $\mathbf{T}$ can be represented in both bases by

$$\mathbf{T} = T'^{ij} \mathbf{e}'_i \otimes \mathbf{e}'_j = T^{ij} \mathbf{e}_h \otimes \mathbf{e}_k, \tag{2.35}$$

so that, taking into account (2.34), we derive the following transformation formulae for the components of $\mathbf{T}$:

$$T'^{ij} = (A^{-1})^i_h (A^{-1})^j_k T^{hk}. \tag{2.36}$$

Exercise 2.1. Verify that the matrix form of (2.36) is [see (2.27)]

$$\mathbb{T}' = (\mathbb{A}^{-1})\mathbb{T}(\mathbb{A}^{-1})^T. \tag{2.37}$$

Definition 2.6. A *mixed* 2-*tensor* or a $(1, 1)$-*tensor* is a bilinear map

$$\mathbf{T} : E^*_n \times E_n \rightarrow \mathfrak{R}. \tag{2.38}$$

Once again, the set $T^1_1(E_n)$ of all mixed 2-tensors becomes a vector space by the introduction of the standard operations of addition of two mixed 2-tensors and the product of a mixed 2-tensor by a real number.

Definition 2.7. The *tensor product* of a vector $\mathbf{x}$ and a covector $\sigma \in E^*_n$ is the mixed 2-tensor

$$\mathbf{x} \otimes \omega(\sigma, \mathbf{y}) = \mathbf{x}(\sigma)\omega(\mathbf{y}) = \sigma(\mathbf{x})\omega(\mathbf{y}), \quad \forall \sigma, \mathbf{y} \in E_n. \tag{2.39}$$

Theorem 2.4. *Let* $(\mathbf{e}_i)$ *be a basis of the vector space* E_n *and let* (θ^i) *be the dual basis in* E^*_n. *Then* $(\mathbf{e}_i \otimes \theta^j)$ *is a basis of* $T^1_1(E_n)$, *which is an* n^2 *-dimensional vector space.*

In view of this theorem we can write

$$\mathbf{T} = T^i_j \mathbf{e}_i \otimes \theta^j, \tag{2.40}$$

where the components of the mixed 2-tensor are

$$T^i_j = \mathbf{T}(\theta^i, \mathbf{e}_j). \tag{2.41}$$

Moreover, in the basis change $(\mathbf{e}_i \otimes \theta^j) \rightarrow (\mathbf{e}'_i \otimes \theta'^j)$ given by

$$\mathbf{e}_i \otimes \theta^j = A^h_i (A^{-1})^j_k \mathbf{e}_h \otimes \theta^k, \tag{2.42}$$

the components of $\mathbf{T}$ are transformed according to the formulae

$$T_j^{\prime i} = (A^{-1})_h^i A_j^k T_k^h. \tag{2.43}$$

Exercise 2.2. Verify that the matrix form of (2.43) is [see (2.27)]

$$\mathbb{T}' = (\mathbb{A}^{-1})\mathbb{T}\mathbb{A}. \tag{2.44}$$

2.4 (r,s)-Tensors

Definition 2.8. An (r,s)-tensor is a multilinear map

$$\mathbf{T} : E_n^{*r} \times E_n^s \to \mathfrak{R}. \tag{2.45}$$

It is quite clear how to transform the set $T_s^r(E_n)$ of all (r,s)-tensors (2.45) in a real vector space.

Definition 2.9. Let $\mathbf{T} \in T_s^r(E_n)$ be an (r,s)-tensor and let $\mathbf{L} \in T_q^p(E_n)$ be a (p,q)-tensor. Then the **tensor product** of these two tensors is the $(r+p, s+q)$-tensor $\mathbf{T} \otimes \mathbf{L} \in T_{s+q}^{r+p}(E_n)$ given by

$$\mathbf{T} \otimes \mathbf{L}(\sigma_1, \ldots, \sigma_{r+p}, \mathbf{x}_1, \ldots, \mathbf{x}_{s+q})$$
$$= \mathbf{T}(\sigma_1, \ldots, \sigma_r, \mathbf{x}_1, \ldots, \mathbf{x}_s) \mathbf{L}(\sigma_{r+1}, \ldots, \sigma_{r+p}, \mathbf{x}_{r+1}, \ldots, \mathbf{x}_{s+q}). \tag{2.46}$$

By imposing that the associative property holds, the tensor product can be extended to any number of factors. Then we introduce the following definition of the tensor product of vectors and covectors:

$$\mathbf{x}_1 \otimes \cdots \otimes \mathbf{x}_r \otimes \omega_1 \otimes \cdots \otimes \omega_s(\sigma_1, \ldots, \sigma_r, \mathbf{u}_1, \ldots, \mathbf{u}_s) =$$
$$\mathbf{x}_1(\sigma_1) \cdots \mathbf{x}_r(\sigma_r)\omega_1(\mathbf{u}_1) \cdots \omega_s(\mathbf{u}_s). \tag{2.47}$$

With the procedure revealed in the previous section it is possible to prove the following theorem.

Theorem 2.5. *The dimension of the vector space $T_s^r(E_n)$ of the $(r+s)$-tensors on E_n is n^{r+s}, and*

$$\mathbf{e}_{i_1} \otimes \cdots \otimes \mathbf{e}_{i_r} \otimes \theta^{j_1} \otimes \cdots \otimes \theta^{j_s} \tag{2.48}$$

is a basis of it.

In (2.48), $(\mathbf{e}_i)$ is a basis of E_n and (θ^j) the dual basis in E_n^*. Instead of Eqs. (2.40)–(2.43), we now obtain

$$\mathbf{T} = T^{i_1 \cdots i_r}_{j_1 \cdots j_s} \mathbf{e}_{i_1} \otimes \cdots \otimes \mathbf{e}_{i_r} \otimes \boldsymbol{\theta}^{j_1} \otimes \cdots \otimes \boldsymbol{\theta}^{j_s}, \tag{2.49}$$

$$T^{j_1, \ldots, j_s}_{i^1, \ldots, i_n} = \mathbf{T}(\boldsymbol{\theta}^{j_1}, \ldots, \boldsymbol{\theta}^{j_s}, \mathbf{e}_{i_1}, \ldots, \mathbf{e}_{i_r}), \tag{2.50}$$

$$\mathbf{e}'_{i_1} \otimes \cdots \otimes \mathbf{e}'_{i_r} \otimes \boldsymbol{\theta}'^{j_1} \otimes \cdots \otimes \boldsymbol{\theta}'^{j_s}$$
$$= A^{h_1}_{i_1} \cdots A^{h_r}_{i_r} (A^{-1})^{j_1}_{k_1} \cdots (A^{-1})^{j_s}_{k_s} \mathbf{e}_{h_1} \otimes \cdots \otimes \mathbf{e}_{h_r} \otimes \boldsymbol{\theta}^{k_1} \otimes \cdots \otimes \boldsymbol{\theta}^{k_s}, \tag{2.51}$$

$$T'^{i_1 \cdots i_r}_{j_1 \cdots j_s} = (A^{-1})^{i_1}_{h_1} \cdots (A^{-1})^{i_r}_{h_r} A^{k_1}_{j_1} \cdots A^{k_s}_{j_s} T^{h_1 \cdots h_r}_{k_1 \cdots k_s}. \tag{2.52}$$

2.5 Tensor Algebra

In the previous sections we defined the addition of tensors belonging to the *same* tensor space. On the other hand, the tensor product of two tensors that might belong to different tensor spaces defines a new tensor that belongs to *another* tensor space. In conclusion, the tensor product is not an internal operation. However, it is possible to introduce a suitable set that, equipped with the aforementioned operations, becomes an algebra.

Let us consider the *infinite* direct sum (Sect. 1.4)

$$T E_n = \bigoplus_{r,s \in \mathbb{N}} T^r_s E_n \tag{2.53}$$

whose elements are *finite* sequences $\{a, \mathbf{x}, \omega, \mathbf{T}, \mathbf{L}, \mathbf{K}, \ldots\}$, where $a \in \Re$, $\mathbf{x} \in E_n$, $\omega \in E^*_n$, $\mathbf{T} \in T^2_0(E_n)$, $\mathbf{K} \in T^0_2(E_n)$, $\mathbf{L} \in T^1_1(E_n)$, etc. With the introduction of this set, the multiplication by a scalar, the addition, and the tensor product become internal operations and the set $T E_n$, equipped with them, is called a ***tensor algebra***.

2.6 Contraction and Contracted Multiplication

In the tensor algebra $T E_n$ we can introduce two other internal operations: the contraction and the contracted product.

Theorem 2.6. *Denote by $(\mathbf{e}_i)$ a basis of the vector space E_n and by $(\boldsymbol{\theta}^i)$ the dual basis in E^*_n. Then, for any pair of integers $1 \le h,k \le n$, the linear map*

$$C_{h,k} : \mathbf{T} = T^{i_1 \cdots h \cdots i_r}_{j_1 \cdots k \cdots j_s} \mathbf{e}_{i_1} \otimes \cdots \otimes \mathbf{e}_{i_r} \otimes \boldsymbol{\theta}^{j_1} \otimes \cdots \otimes \boldsymbol{\theta}^{j_s} \in T^r_s(E_n)$$
$$\rightarrow C_{h,k}(\mathbf{T}) = T^{i_1 \cdots h \cdots i_r}_{j_1 \cdots h \cdots j_s} \mathbf{e}_{i_1} \otimes \cdots \mathbf{e}_{i_{h-1}} \otimes \mathbf{e}_{i_{h+1}} \otimes \cdots \otimes \mathbf{e}_{i_r} \otimes \tag{2.54}$$
$$\boldsymbol{\theta}^{j_1} \otimes \cdots \otimes \boldsymbol{\theta}^{j_{h-1}} \otimes \boldsymbol{\theta}^{j_{h+1}} \otimes \cdots \otimes \boldsymbol{\theta}^{j_s} \in T^{r-1}_{s-1}(E_n),$$

which is called a **contraction**, *to any tensor* $\mathbf{T} \in T_s^r(E_n)$ *associates a tensor* $C_{h,k}(\mathbf{T}) \in T_{s-1}^{r-1}(E_n)$, *which is obtained by equating the contravariant index h and the covariant index k and summing over these indices.*

Proof. To simplify the notations, we prove the theorem for a $(2, 1)$-tensor $\mathbf{T} = T_h^{ij}\mathbf{e}_i \otimes \mathbf{e}_j \otimes \boldsymbol{\theta}^h$. For this tensor $C_{1,1}(\mathbf{T}) = T_h^{hj}\mathbf{e}_j$, and it will be sufficient to prove that, under a basis change, the quantities T_h^{hj} are transformed as vector components. Since in a basis change we have that

$$T_h'^{ij} = (A^{-1})_l^i (A^{-1})_m^j A_h^p T_p^{lm},$$

it also holds that

$$T_h'^{hj} = (A^{-1})_l^h (A^{-1})_m^j A_h^p T_p^{lm} = (A^{-1})_m^j T_h^{hm},$$

and the theorem is proved. $\qquad\square$

The preceding theorem makes it possible to give the following definition.

Definition 2.10. A **contracted multiplication** is a map

$$(\mathbf{T}, \mathbf{L}) \in T_s^r(E_n) \times T_q^p(E_n) \to C_{h,k}(\mathbf{T} \otimes \mathbf{L}) \in T_{s+q-1}^{r+p-1}(E_n). \tag{2.55}$$

For instance, if $\mathbf{T} = T_h^{ij}\mathbf{e}_i \otimes \mathbf{e}_j \otimes \boldsymbol{\theta}^h$ and $\mathbf{L} = L_m^l \mathbf{e}_l \otimes \boldsymbol{\theta}^m$, then $C_{i,m}(\mathbf{T} \otimes \mathbf{L}) = T^{hj}L_h^l \mathbf{e}_j \otimes \mathbf{e}_l$.

Theorem 2.7. *The n^{r+s} quantities*

$$T_{j_1 \cdots j_s}^{i_1 \cdots i_r} \tag{2.56}$$

are the components of an $(r+s)$-tensor $\mathbf{T}$ *if and only if any contracted multiplication of these quantities by the components of a (p, q)-tensor* $\mathbf{L}$, $p \le s, q \le r$ *generates an $(r-q, s-p)$-tensor.*

Proof. The definition of contracted multiplication implies that the condition is necessary. For the sake of simplicity, we prove that the condition is sufficient considering the quantities T_h^{ij} and a $(0, 1)$-tensor $\mathbf{L}$. In other words, we suppose that the quantities $T_h^{ij}L_i$ are transformed as the components of a $(1, 1)$-tensor. Then we have that

$$T_h'^{ij} L_i' = (A^{-1})_l^j A_h^m T_m^{kl} L_k.$$

Since $L_k = (A^{-1})_k^r L_r'$, the theorem is proved. $\qquad\square$

Remark 2.5. Let $\mathbf{T} : \mathbf{x} \in E_n \to \mathbf{y} \in E_n$ be a linear map and denote by $(\mathbf{e}_i)$ a basis of E_n and by (T_j^i) the matrix of $\mathbf{T}$ relative to the basis $(\mathbf{e}_i)$. In terms of components, the map $\mathbf{T}$ becomes

$$y^i = T^i_j x^j, \tag{2.57}$$

where y^i are the components of the vector $\mathbf{y} = \mathbf{T}(\mathbf{x})$ in the basis $(\mathbf{e}_i)$. Owing to the previous theorem, the quantities (T^i_j) are the components of a $(1, 1)$-tensor relative to the basis $(\mathbf{e}_i \otimes \theta^j)$. It is easy to verify that any $(1, 1)$-tensor $\mathbf{T}$ determines a linear map $\mathbf{T} : E_n \to E_n$.

2.7 Exercises

1. Determine all the linear maps that can be associated with an (r, s)-tensor, where $r + s \le 3$.
2. Let $(\mathbf{e}_i)$ be a basis of a vector space E_n, and let (θ^i) be the dual basis. Verify that a tensor that has the components (δ^i_j) in the basis $(\mathbf{e}_i \otimes \theta^j)$ has the same components in any other basis.
3. Let $(\mathbf{e}_i)$ be a basis of a two-dimensional vector space E_2, and denote by (θ^i) the dual basis. Determine the value of the covariant 2-tensor $\theta^1 \otimes \theta^2 - \theta^2 \otimes \theta^1$ when it is applied to the pairs of vectors $(\mathbf{e}_1, \mathbf{e}_2)$, $(\mathbf{e}_2, \mathbf{e}_1)$, and $(\mathbf{x}, \mathbf{y})$ and recognize geometric meaning of each result.
4. The components of a $(1, 1)$-tensor $\mathbf{T}$ of the vector space E_3 relative to a basis $(\mathbf{e}_i \otimes \theta^j)$ are given by the matrix

$$\begin{pmatrix} 1 & 2 & 1 \\ 2 & 1 & 2 \\ 1 & 2 & 1 \end{pmatrix}.$$

   Determine the vector corresponding to $\mathbf{x} = (1, 0, 1)$ by the linear endomorphism determined by $\mathbf{T}$.
5. In the basis $(\mathbf{e}_i)$ of the vector space E_3, two vectors $\mathbf{x}$ and $\mathbf{y}$ have the components $(1, 0, 1)$ and $(2, 1, 0)$, respectively. Determine the components of $\mathbf{x} \otimes \mathbf{y}$ relative to the basis $(\mathbf{e}'_i \otimes \mathbf{e}'_j)$, where

$$\begin{aligned} \mathbf{e}'_1 &= \mathbf{e}_1 + \mathbf{e}_3, \\ \mathbf{e}'_2 &= 2\mathbf{e}_1 - \mathbf{e}_2, \\ \mathbf{e}'_3 &= \mathbf{e}_1 + \mathbf{e}_2 + \mathbf{e}_3. \end{aligned}$$

6. Given the $(0, 2)$-tensor $T_{ij}\theta^i \otimes \theta^j$ of $T_2(E_2)$, where

$$(T_{ij}) = \begin{pmatrix} 1 & 2 \\ 1 & 0 \end{pmatrix},$$

   determine if there exists a new basis in which its components become

$$(T'_{ij}) = \begin{pmatrix} 1 & 1 \\ 0 & 1 \end{pmatrix}.$$

7. For which $(1,1)$-tensor (T^i_j) of $T^1_1(E_3)$ does the linear map F defined by the matrix (T^i_j) satisfy the condition $F(\mathbf{x}) = a\mathbf{x}$, $\forall \mathbf{x} \in E_3$ and $\forall a \in \Re$?

8. Given the $(1,1)$-tensor $T^i_j \mathbf{e}_i \otimes \theta^j$ of $T^1_1(E_2)$, verify that $T^1_1 + T^2_2$ and $\det(T^i_j)$ are invariant with respect to a change of basis.

   Hint: Use formula $\det\left(T^j_i\right) = \epsilon_{hk} T^h_1 T^k_2$, where ϵ_{hk} is the Levi–Civita symbol (see Section 5.3).

9. Prove that if the components of a $(0,2)$-tensor $\mathbf{T}$ belonging to $T_2(E_n)$ satisfy either of the conditions

$$T_{ij} = T_{ji}, \quad T_{ij} = -T_{ji}$$

   in a given basis, then they satisfy the same conditions in any other basis.

10. Given $(0,2)$-tensors that in the basis $(\mathbf{e}_i)$, $i = 1,2,3$, have the components

$$\mathbf{T}_1 = \begin{pmatrix} 1 & 0 & -1 \\ 0 & -1 & 2 \\ -1 & 2 & 1 \end{pmatrix},$$

$$\mathbf{T}_2 = \begin{pmatrix} 0 & 1 & -1 \\ -1 & 0 & 2 \\ 1 & -2 & 0 \end{pmatrix},$$

    determine the covector $\omega_\mathbf{u}$ depending on $\mathbf{u}$ such that

$$\begin{aligned} \omega_\mathbf{u}(\mathbf{v}) &= \mathbf{T}_1(\mathbf{u}, \mathbf{v}), \\ \omega_\mathbf{u}(\mathbf{v}) &= \mathbf{T}_2(\mathbf{u}, \mathbf{v}), \end{aligned} \quad \forall \mathbf{v}.$$

    Further, find the vectors $\mathbf{u}$ such that

$$\mathbf{T}_1(\mathbf{u}, \mathbf{v}) = 0, \quad \mathbf{T}_2(\mathbf{u}, \mathbf{v}) = 0, \quad \forall \mathbf{v}.$$

Chapter 3
Skew-Symmetric Tensors and Exterior Algebra

Abstract This chapter is devoted to $(0, r)-$ skew-symmetric tensors. After determining their strict components, it is proved that they form a vector space $\Lambda_r(E_n)$. Then, the dimension of this space is determined together with the bases of this space and the transformation properties of strict components. Finally, the exterior algebra is formulated and oriented vector spaces are introduced.

3.1 Skew-Symmetric (0, 2)-Tensors

Definition 3.1. A tensor $\mathbf{T} \in T_2(E_n)$ is *skew-symmetric* or *alternating* if

$$\mathbf{T}(\mathbf{x}, \mathbf{y}) = -\mathbf{T}(\mathbf{y}, \mathbf{x}), \tag{3.1}$$

$\forall \mathbf{x}, \mathbf{y} \in E_n$. In particular, (3.1) implies $\mathbf{T}(\mathbf{x}, \mathbf{x}) = 0$.

Denote by $\mathbf{x} = x^i \mathbf{e}_i$ and $\mathbf{y} = y^i \mathbf{e}_i$, respectively, the representations of $\mathbf{x}$ and $\mathbf{y}$ relative to a basis $(\mathbf{e}_i)$ of E_n. If (θ^i) is the dual basis of $(\mathbf{e}_i)$ and $\mathbf{T} = T_{ij}\theta^i \otimes \theta^j$, then (3.1) becomes

$$T_{ij}x^i y^j = -T_{ji}x^i y^j.$$

This equality is identically satisfied $\forall \mathbf{x}, \mathbf{y} \in E_n$ if and only if the components T_{ij} of $\mathbf{T}$ verify the conditions

$$T_{ij} = -T_{ji}, \quad i \neq j, \quad T_{ii} = 0. \tag{3.2}$$

Exercise 3.1. Prove that if conditions (3.2) hold in a basis, then they hold in any basis.

© Springer International Publishing AG, part of Springer Nature 2018

A. Romano and A. Marasco, *Classical Mechanics with Mathematica®*,
Modeling and Simulation in Science, Engineering and Technology,
https://doi.org/10.1007/978-3-319-77595-1_3

If we denote by $\mathbb{T}$ the matrix of the components of $\mathbf{T}$, then conditions (3.2) assume the following matrix form:

$$\mathbb{T} = -\mathbb{T}^T. \tag{3.3}$$

Definition 3.2. The *exterior product* of the covectors $\omega, \sigma \in E_n^*$ is the $(0,2)$-tensor such that

$$\omega \wedge \sigma = \omega \otimes \sigma - \sigma \otimes \omega. \tag{3.4}$$

Theorem 3.1. *The exterior product (3.4) has the following properties:*

- $\omega \wedge \sigma$ *is a skew-symmetric tensor;*
- $\omega \wedge \sigma = -\sigma \wedge \omega$;
- *If* $\sigma = a\omega, a \in \Re$, *then* $\omega \wedge \sigma = 0$.

Proof. From (3.4), $\forall \mathbf{x}, \mathbf{y} \in E_n$, we have that

$$\begin{aligned}
\omega \wedge \sigma(\mathbf{x}, \mathbf{y}) &= \omega \otimes \sigma(\mathbf{x}, \mathbf{y}) - \sigma \otimes \omega(\mathbf{x}, \mathbf{y}) \\
&= \omega(\mathbf{x})\sigma(\mathbf{y}) - \sigma(\mathbf{x})\omega(\mathbf{y}) \\
&= -\omega \wedge \sigma(\mathbf{y}, \mathbf{x}),
\end{aligned} \tag{3.5}$$

and the skew symmetry of $\omega \wedge \sigma$ is proved. Starting from (3.5) it is simple to verify the other two properties. $\qquad \square$

The set $\Lambda_2(E_n)$ of the skew-symmetric $(0,2)$-tensors is a subspace of the vector space $T_2(E_n)$. The dimension and the bases of $\Lambda_2(E_n)$ are determined by the following theorem.

Theorem 3.2. *Let* $(\mathbf{e}_i)$ *be a basis of* E_n *and denote by* (θ^i) *its dual basis in* E_n^*. *The dimension of* $\Lambda_2(E_n)$ *is* $\begin{pmatrix} n \\ 2 \end{pmatrix}$, *and the vectors*

$$(\theta^i \wedge \theta^j), \quad i < j,$$

form a basis of $\Lambda_2(E_n)$.

Proof. For any $\mathbf{T} \in \Lambda_2(E_n)$ we have that

$$\begin{aligned}
\mathbf{T} = T_{ij}\theta^i \otimes \theta^j &= \sum_{i<j} T_{ij}\theta^i \otimes \theta^j + \sum_{i>j} T_{ij}\theta^i \otimes \theta^j \\
&= \sum_{i<j} T_{ij}\theta^i \otimes \theta^j + \sum_{j>i} T_{ji}\theta^j \otimes \theta^i \\
&= \sum_{i<j} T_{ij}\theta^i \otimes \theta^j - \sum_{i<j} T_{ij}\theta^j \otimes \theta^i.
\end{aligned}$$

Therefore, we can write

$$\mathbf{T} = T_{(ij)}\theta^i \wedge \theta^j, \tag{3.6}$$

where

$$T_{(ij)} = T_{ij}, \quad i < j. \tag{3.7}$$

Relation (3.6) shows that the set of the $(0, 2)$-tensors $(\theta^i \wedge \theta^j) \in \Lambda_2(E_n)$ generates the whole subspace $\Lambda_2(E_n)$. Consequently, to verify that they form a basis of $\Lambda_2(E_n)$, it is sufficient to verify their linear independence. Now, from any linear combination

$$a_{(ij)}\theta^i \wedge \theta^j = \mathbf{0}, \ i < j,$$

we have the condition

$$a_{(ij)}\theta^i \wedge \theta^j(\mathbf{e}_h, \mathbf{e}_k) = 0,$$

which, in view of (3.5) and (2.4), implies $a_{(ij)} = 0$, and the proof is complete. □

Definition 3.3. The quantities $T_{(ij)}$, $i < j$, are called *strict components* of **T** relative to the basis $(\theta^i \wedge \theta^j)$.

To determine the transformation formulae of the strict components, we recall that they are components of a $(0, 2)$-tensor. Consequently, in view of (2.26), we can write

$$\begin{aligned}
T'_{(ij)} &= A^h_{(i} A^k_{j)} T_{hk} \\
&= \sum_{h<k} A^h_{(i} A^k_{j)} T_{hk} + \sum_{h>k} A^h_{(i} A^k_{j)} T_{hk} \\
&= \sum_{h<k} (A^h_{(i} A^k_{j)} - A^k_{(i} A^h_{j)}) T_{(hk)},
\end{aligned}$$

and the transformation formulae of the strict components of $\mathbf{T} \in \Lambda_2(E_n)$ are

$$T'_{(ij)} = \begin{vmatrix} A^h_i & A^h_j \\ A^k_i & A^k_j \end{vmatrix} T_{(hk)}. \tag{3.8}$$

Starting from the inverse formula of (2.26) we also have

$$T_{(ij)} = \begin{vmatrix} (A^{-1})^h_i & (A^{-1})^h_j \\ (A^{-1})^k_i & (A^{-1})^k_j \end{vmatrix} T'_{(hk)}. \tag{3.9}$$

The transformation formulae of the bases $\theta^i \wedge \theta^j$ of $\Lambda_2(E_n)$ can be obtained by noting that

$$\mathbf{T} = T'_{(ij)}\theta'^i \wedge \theta'^j = T_{(hk)}\theta^h \wedge \theta^k. \tag{3.10}$$

Introducing (3.8) and (3.9) into (3.10) we obtain

$$\theta'^i \wedge \theta'^j = \begin{vmatrix} (A^{-1})^i_h & (A^{-1})^i_k \\ (A^{-1})^j_h & (A^{-1})^j_k \end{vmatrix} \theta^h \wedge \theta^k, \quad i < j, h < k, \tag{3.11}$$

$$\theta^i \wedge \theta^j = \begin{vmatrix} A^i_h & A^i_k \\ A^j_h & A^j_k \end{vmatrix} \theta'^h \wedge \theta'^k, \quad i < j, h < k. \tag{3.12}$$

Example 3.1. Let E_2 be a two-dimensional vector space and denote by $(\mathbf{e}_1, \mathbf{e}_2)$ and (θ^1, θ^2) a basis of E_2 and the dual basis of E_2^*, respectively. The dimension of the vector space $T_2(E_2)$ is 4, whereas the subspace $\Lambda_2(E_2)$ of the skew-symmetric $(0,2)$-tensors is one-dimensional. Consequently, any $(0,2)$-tensor of $\Lambda_2(E_2)$ can be written as follows:

$$\mathbf{T} = T_{12}\theta^1 \wedge \theta^2.$$

The skew-symmetric basis $\theta^1 \wedge \theta^2$ has a remarkable geometric meaning. In fact, $\forall \mathbf{x}, \mathbf{y} \in E_2$,

$$\theta^1 \wedge \theta^2(\mathbf{x}, \mathbf{y}) = \theta^1 \otimes \theta^2(\mathbf{x}, \mathbf{y}) - \theta^2 \otimes \theta^1(\mathbf{x}, \mathbf{y})$$
$$= (x^1 y^2 - x^2 y^1) = \begin{vmatrix} x^1 & x^2 \\ y^1 & y^2 \end{vmatrix}, \tag{3.13}$$

and we can state that $\theta^1 \wedge \theta^2(\mathbf{x}, \mathbf{y})$ measures the area of the parallelogram determined by the vectors $\mathbf{x}$ and $\mathbf{y}$. In particular, from (3.13) we obtain

$$\theta^1 \wedge \theta^2(\mathbf{e}_1, \mathbf{e}_2) = 1,$$

and the parallelogram determined by the vectors $\mathbf{e}_1$ and $\mathbf{e}_2$ has a unit area. We note that the area of the parallelogram formed by the vectors $(\mathbf{e}'_1, \mathbf{e}'_2)$ of another basis of E_2, where

$$\mathbf{e}'_1 = A^1_1\mathbf{e}_1 + A^2_1\mathbf{e}_2,$$
$$\mathbf{e}'_2 = A^1_2\mathbf{e}_1 + A^2_2\mathbf{e}_2,$$

has the value

$$\theta^1 \wedge \theta^2(\mathbf{e}'_1, \mathbf{e}'_2) = \begin{vmatrix} A^1_1 & A^2_1 \\ A^1_2 & A^2_2 \end{vmatrix} = \det A.$$

In conclusion, by choosing a basis $(\mathbf{e}_1, \mathbf{e}_2)$ of E_2 and the corresponding skew-symmetric $(0,2)$-tensor $\theta^1 \wedge \theta^2$, we introduce a criterion to evaluate the areas of parallelograms *without resorting to a metric*. This criterion does not depend on the basis, provided that the basis changes satisfy the condition $\det A = 1$.

Exercise 3.2. Verify that in a vector space E_3, related to the basis $(\mathbf{e}_i)$, a skew-symmetric $(0,2)$-tensor $\mathbf{T}$ has the following form:

$$\mathbf{T} = T_{12}\theta^1 \wedge \theta^2 + T_{13}\theta^1 \wedge \theta^3 + T_{23}\theta^2 \wedge \theta^3,$$

where (θ^i) is the dual basis in E_3^* of $(\mathbf{e}_i)$. Prove that the skew-symmetric $(0,2)$-tensor

$$\mathbf{T} = \theta^1 \wedge \theta^2 + \theta^1 \wedge \theta^3 + \theta^2 \wedge \theta^3$$

associates to any pair of vectors $\mathbf{x}$, $\mathbf{y}$ of E_3 the sum of the areas of the projections of the parallelogram formed by the vectors $\mathbf{x}$, $\mathbf{y}$ onto the subspaces generated by $(\mathbf{e}_1, \mathbf{e}_2)$, $(\mathbf{e}_1, \mathbf{e}_3)$, $(\mathbf{e}_2, \mathbf{e}_3)$, respectively.

In the next section, the content of this section will be extended to $(0, r)$-tensors.

3.2 Skew-Symmetric $(0, r)$-Tensors

Definition 3.4. Let $S_r = \{\, 1, 2, \ldots, r \}$ be the set of the first r integer numbers. A *permutation* of S_r is any one-to-one map

$$\pi : S_r \to S_r.$$

We denote by $\{\pi(1), \pi(2), \ldots, \pi(r)\}$ the set formed by the same numbers of S_r placed in a different order. It is well known that the set Π_r of all the permutations of S_r contains $r!$ one-to-one maps. This set can be equipped with the structure of a group by the usual composition of maps

$$\sigma, \pi \in \Pi_r \to \sigma \circ \pi \in \Pi_r.$$

The identity of this group is a map that does not modify the position of the numbers of S_r. Finally, the opposite of π is the inverse map π^{-1}. Let i, j, $i < j$, be two numbers of S_r. We say that the permutation π contains an *inversion* with respect to S_r if

$$\pi(i) > \pi(j).$$

A permutation π is said to be *even* or *odd* according to whether the total number of inversions contained in S_π is even or odd. In the sequel, we denote by $m(\pi)$ the total inversions of π.

Definition 3.5. A tensor $\mathbf{T} \in T_r(E_n)$, $r > 2$, is *skew-symmetric* or *alternating* if

$$\mathbf{T}(\mathbf{x}_1, \ldots, \mathbf{x}_r) = (-1)^{m(\pi)} \mathbf{T}(\mathbf{x}_{\pi(1)}, \ldots, \mathbf{x}_{\pi(r)}) \qquad (3.14)$$

$\forall \mathbf{x}_1, \ldots, \mathbf{x}_r \in E_n$ and $\forall \pi \in \Pi_r$.

The preceding definition implies that the value of $\mathbf{T}$ vanishes every time $\mathbf{T}$ is evaluated on a set of vectors containing two equal vectors. In fact, it is sufficient to consider a permutation that exchanges the position of these two vectors without modifying the position of the others and then to apply (3.14). From this remark it easily follows that the value of $\mathbf{T}$ vanishes if the vectors $\{\mathbf{x}_1, \ldots, \mathbf{x}_r\}$ are linearly dependent.

If the r vectors $\{\mathbf{x}_1, \ldots, \mathbf{x}_r\}$ belong to a basis $(\mathbf{e}_i)$ of E_n, then, in view of (2.50), we express the skew symmetry of $\mathbf{T}$ in terms of its components:

$$T_{i_1 \cdots i_r} = (-1)^{m(\pi)} T_{\pi(i_1) \cdots \pi(i_r)}. \qquad (3.15)$$

In particular, this condition implies that all the components of $\mathbf{T}$ in which two indices have the same value vanish. It is quite obvious that the set of all the skew-symmetric tensors form a subspace $\Lambda_r(E_n)$ of $T_r(E_n)$.

Remark 3.1. Since r vectors, when $r > n$, cannot be linearly independent, we can state that

$$\Lambda_r(E_n) = \{\mathbf{0}\}, \quad n < r.$$

Definition 3.6. The *exterior product* of r covectors $\boldsymbol{\omega}^1, \ldots, \boldsymbol{\omega}_r$ is the $(0, r)$-tensor

$$\boldsymbol{\omega}^1 \wedge \cdots \wedge \boldsymbol{\omega}^r = \sum_{\pi \in \Pi_r} (-1)^{m(\pi)} \boldsymbol{\omega}^{\pi(1)} \otimes \cdots \otimes \boldsymbol{\omega}^{\pi(r)}. \qquad (3.16)$$

The tensor (3.16) is skew-symmetric since, in view of (2.47), we have that

$$\boldsymbol{\omega}^1 \wedge \cdots \wedge \boldsymbol{\omega}^r(\mathbf{x}_1, \ldots, \mathbf{x}_r) = \sum_{\pi \in \Pi_r} (-1)^{m(\pi)} \boldsymbol{\omega}^{\pi(1)}(\mathbf{x}_1) \cdots \otimes \boldsymbol{\omega}^{\pi(r)}(\mathbf{x}_r)$$

$$= \begin{vmatrix} \boldsymbol{\omega}^1(\mathbf{x}_1) & \cdots & \boldsymbol{\omega}^1(\mathbf{x}_r) \\ \cdots & \cdots & \cdots \\ \boldsymbol{\omega}^r(\mathbf{x}_1) & \cdots & \boldsymbol{\omega}^r(\mathbf{x}_r) \end{vmatrix}. \qquad (3.17)$$

A permutation π of the vectors $\mathbf{x}_1, \ldots, \mathbf{x}_r$ corresponds to a permutation of the columns of the determinant in (3.17). This operation changes or does not change the sign of the determinant according to whether the permutation is odd or even. But this is just the property expressed by (3.14). Again recalling the properties of a determinant, we can easily prove the following properties of the exterior product of r covectors:

- It is a $(0, r)$-skew-symmetric tensor.
- It vanishes if one vector linearly depends on the others.

Now we extend Theorem 3.2 to skew-symmetric $(0,r)$-tensors.

Theorem 3.3. *Let $(\mathbf{e}_i)$ be a basis of the vector space E_n and denote by (θ^i) the dual basis in E_n^*. Then, if $r \leq n$, then $\binom{n}{r}$ is the dimension of the vector space $\Lambda_r(E_n)$ of the skew-symmetric $(0,r)$-tensors. Further, the skew-symmetric tensors*

$$\theta^{i_1} \wedge \cdots \wedge \theta^{i_r}, \tag{3.18}$$

where $i_1 < \cdots < i_r$ is an arbitrary r-tuple of integer numbers chosen in the set of indices $\{1, \ldots, n\}$, form a basis of $\Lambda_r(E_n)$.

Proof. First, when $r \leq n$, any $\mathbf{T} \in \Lambda_r(E_n)$ can be written as follows:

$$\mathbf{T} = T_{j_1 \cdots j_r} \theta^{j_1} \otimes \cdots \otimes \theta^{j_r}.$$

The summation on the right-hand side contains only terms in which all the indices are different from each other since $\mathbf{T}$ is skew-symmetric. This circumstance makes it possible to group all the terms as follows: first, we consider the terms obtained by extracting arbitrarily r different indices $i_1 < \cdots < i_r$ from the set $\{1,...,n\}$. We recall that there are $\binom{n}{r}$ different possible choices of these indices. Then, for each choice, which is characterized by the indices $i_1 < \cdots < i_r$, we consider all the terms obtained by permuting in all the possible ways the indices $i_1 < \cdots < i_r$. In view of (3.15) and (3.16), we can write

$$\begin{aligned}
\mathbf{T} &= \sum_{i_1 < \cdots < i_r} \sum_{\pi \in \Pi_r} T_{\pi(i_1) \cdots \pi(i_r)} \theta^{\pi(i_1)} \otimes \cdots \otimes \theta^{\pi(i_r)} \\
&= \sum_{i_1 < \cdots < i_r} T_{(i_1 \cdots i_r)} \sum_{\pi \in \Pi_r} (-1)^{m(\pi)} \theta^{\pi(i_1)} \otimes \cdots \otimes \theta^{\pi(i_r)} \\
&= \sum_{i_1 < \cdots < i_r} T_{i_1 \cdots i_r} \theta^{i_1} \wedge \cdots \wedge \theta^{i_r},
\end{aligned}$$

that is,

$$\mathbf{T} = T_{(i_1 \cdots i_r)} \theta^{i_1} \wedge \cdots \wedge \theta^{i_r}. \tag{3.19}$$

This result shows that the skew-symmetric $(0,r)$-tensors $\{\theta^{i_1} \wedge \cdots \wedge \theta^{i_r}, i_1 < \cdots i_r\}$ generate the whole vector space $\Lambda_r(E_n)$. We can easily verify that they are linearly independent, so that they form a basis of $\Lambda_r(E_n)$. $\square$

The transformation formulae of the strict components $T_{(i_1 \cdots i_r)}$ of a skew-symmetric tensor [see (3.8)–(3.12)] can be found by noting that

$$
\begin{aligned}
T'_{(i_1 \cdots i_r)} &= A^{h_1}_{(i_1} \cdots A^{h_r}_{i_r)} T_{h_1 \cdots h_r} \\
&= \sum_{j_1 < \cdots < j_r} \sum_{\pi \in \Pi_r} (-1)^{m(\pi)} A^{\pi(j_1)}_{(i_1} \cdots A^{\pi(j_r)}_{i_r)} T_{\pi(j_1) \cdots \pi(j_r)} \\
&= \sum_{j_1 < \cdots < j_r} \sum_{\pi \in \Pi_r} A^{\pi(j_1)}_{(i_1} \cdots A^{\pi(j_r)}_{i_r)} T_{(j_1 \cdots j_r)},
\end{aligned}
$$

where the indices $j_1 < \cdots < j_r$ are chosen in the set $1, \ldots, n$ in all possible ways. Finally, we have that

$$
T'_{(i_1 \cdots i_r)} = \begin{vmatrix} A^{j_1}_{i_1} & \cdots & A^{j_1}_{i_r} \\ \cdots & \cdots & \cdots \\ A^{j_r}_{i_1} & \cdots & A^{j_r}_{i_r} \end{vmatrix} T_{(j_1 \cdots j_r)}. \tag{3.20}
$$

We can easily verify that the transformation formula of the bases of $\Lambda_r(E_n)$ [see (3.11)] is

$$
\theta'^{i_1} \wedge \cdots \wedge \theta'^{i_r} = \begin{vmatrix} (A^{-1})^{i_1}_{j_1} & \cdots & (A^1)^{i_1}_{j_r} \\ \cdots & \cdots & \cdots \\ (A^{-1})^{i_r}_{j_1} & \cdots & A^{i_r}_{j_r} \end{vmatrix} \theta^{j_1} \wedge \cdots \wedge \theta^{j_r}. \tag{3.21}
$$

In particular, if $\mathbf{T} \in \Lambda_n(E_n)$, then (3.20) and (3.21) give

$$
T'_{1 \cdots n} = \det \mathbb{A}\, T_{1 \cdots n}, \tag{3.22}
$$

$$
\theta'^1 \wedge \cdots \wedge \theta'^n = \frac{1}{\det \mathbb{A}} \theta^1 \wedge \cdots \wedge \theta^n, \tag{3.23}
$$

where $\mathbb{A} = \det(A^i_j)$.

3.3 Exterior Algebra

Definition 3.7. We define the *exterior product* of the skew-symmetric tensors

$$
\mathbf{T} = T_{(i_1 \cdots i_r)} \theta^{i_1} \wedge \cdots \wedge \theta^{i_r} \in \Lambda_r(E_n)
$$

and

$$
\mathbf{L} = L_{(j_1 \cdots j_s)} \theta^{j_1} \wedge \cdots \wedge \theta^{j_s} \in \Lambda_s(E_n),
$$

as the skew-symmetric tensor of $\Lambda_{r+s}(E_n)$ given by

$$\mathbf{T} \wedge \mathbf{L} = \sum_{h_1 < \cdots < h_{r+s}} \sum_{\pi \in \Pi_{r+s}} (-1)^{m(\pi)} T_{\pi(h_1) \cdots \pi(h_r)} L_{\pi(h_{r+1}) \cdots \pi(h_{r+s})} \theta^{h_1} \wedge \cdots \wedge \theta^{h_{r+s}},$$

(3.24)

where π is any permutation of the indices $h_1 < \cdots < h_{r+s}$, in which $\pi(h_1) < \cdots < \pi(h_r)$, $\pi(h_{r+1}) < \cdots < \pi(h_{r+s})$, and $m(\pi)$ is the number of inversions of $\pi(h_1), \ldots, \pi(h_{r+s})$.

Example 3.2. The external product of the two skew-symmetric tensors

$$\mathbf{T} = T_{12}\theta^1 \wedge \theta^2 + T_{13}\theta^1 \wedge \theta^3 + T_{23}\theta^2 \wedge \theta^3 \in \Lambda_2(E_3),$$
$$\omega = \omega_1\theta^1 + \omega_2\theta^2 + \omega_3\theta^3 \in \Lambda_1(E_3)$$

is the skew-symmetric tensor of $\Lambda_3(E_3)$ given by

$$\mathbf{T} \wedge \omega = (T_{12}\omega_3 - T_{13}\omega_2 + T_{23}\omega_1)\theta^1 \wedge \theta^2 \wedge \theta^3.$$

Exercise 3.3. Prove that the exterior product of $\mathbf{T} \in \Lambda_2(E_5)$ and $\mathbf{L} \in \Lambda_2(E_5)$ has the component

$$T_{12}L_{45} - T_{14}L_{25} + T_{15}L_{24} + T_{24}L_{15} - T_{25}L_{14} + T_{45}L_{12}$$

along the basis vector $\theta^1 \wedge \theta^2 \wedge \theta^4 \wedge \theta^5$.

It is not difficult to verify that $\mathbf{T} \wedge \mathbf{L}$ is skew-symmetric and independent of the basis. It can also be proved that

$$\mathbf{T} \wedge \mathbf{L} = (-1)^{rs} \mathbf{L} \wedge \mathbf{T}.$$

(3.25)

We can define the *exterior algebra* as we did for the tensor algebra. First, we introduce the notations

$$\Lambda_0(E_n) = \Re, \quad \Lambda_1(E_n) = E_n^*.$$

Then we consider the set

$$\Lambda(E_n) = \bigoplus_{k \in N} \Lambda_k(E_n),$$

(3.26)

whose elements are formed by finite sequences $(a, \omega, \mathbf{T}, \ldots)$, where $a \in \Re$, $\omega \in E_n^*$, $\mathbf{T} \in \Lambda_2(E_n)$, etc., and we recall that $\Lambda_r(E_n) = \{0\}$ for $r > n$. This set, equipped with multiplication by a scalar, addition, and an exterior product, is the exterior algebra over E_n.

It is evident that what we have proved for the skew-symmetric tensors $\mathbf{T}_s^0$ can be repeated for the skew-symmetric tensors $\mathbf{T}_0^r$. In this way, we can define the exterior algebra $\Lambda^r(E_n)$.

3.4   Oriented Vector Spaces

Let $\mathbb{B}$ be the set of the bases of a vector space E_n, and let

$$\mathbf{e}'_j = A^i_j \mathbf{e}_i \tag{3.27}$$

be a basis change. Introduce in $\mathbb{B}$ the relation $\mathfrak{R}$ such that

$$(\mathbf{e}'_i) \mathfrak{R}(\mathbf{e}_i) \Leftrightarrow A > 0, \tag{3.28}$$

where $A = \det(A^i_j) \neq 0$.

Theorem 3.4. $\mathfrak{R}$ *is an equivalence relation partitioning* $\mathbb{B}$ *into two equivalence classes.*

Proof. From the evident conditions $\mathbf{e}_i = \delta^j_i \mathbf{e}_j$, $\det(\delta^j_i) = 1 > 0$, it follows that any basis is in the relation $\mathfrak{R}$ with itself and $\mathfrak{R}$ is reflexive. If (3.28) holds, then we have

$$\mathbf{e}_i = (A^{-1})^j_i \mathbf{e}'_j,$$

where $\det((A^{-1})^j_i) = 1/A > 0$. Therefore, $(\mathbf{e}_i)$ is in the relation $\mathfrak{R}$ with $(\mathbf{e}'_i)$ and $\mathfrak{R}$ is symmetric. It remains to prove that $\mathfrak{R}$ is transitive. To this end, we consider a third basis $(\mathbf{e}''_i)$ such that

$$\mathbf{e}''_h = B^j_h \mathbf{e}'_j, \tag{3.29}$$

where $\det(B^j_h) > 0$. Then we also have

$$\mathbf{e}''_h = B^j_h \mathbf{e}'_j = B^j_h A^k_j \mathbf{e}_k \equiv C^k_h \mathbf{e}_k. \tag{3.30}$$

Since $\det(C^k_h) = \det(B^j_h)\det(A^k_j) > 0$, $\mathfrak{R}$ is an equivalence relation. To verify that $\mathfrak{R}$ partitions $\mathbb{B}$ into two equivalence classes, we first note that these equivalence classes are at least two since the two bases $(\mathbf{e}_i)$ and $(\mathbf{e}'_i)$, such that

$$\mathbf{e}'_i = C^h_i \mathbf{e}_h, \ (C^h_i) = \begin{pmatrix} -1 & 0 & \cdots & 0 \\ 0 & 1 & \cdots & 0 \\ \cdots\cdots\cdots\cdots \\ 0 & 0 & \cdots & 1 \end{pmatrix}, \ \det(C^h_i) = -1,$$

are not equivalent. If $(\mathbf{e}''_i)$ is another arbitrary basis of E_n, then

$$\mathbf{e}''_i = A^j_i \mathbf{e}'_j = A^j_i C^h_j \mathbf{e}_h,$$

so that the basis $(\mathbf{e}''_i)$ is equivalent either to $(\mathbf{e}'_i)$ or to $(\mathbf{e}_i)$. $\square$

Definition 3.8. A vector space E_n is said to be *oriented* if one of the two equivalence classes of $\mathfrak{R}$ is chosen. In this case, the bases belonging to this class are called *positive*, whereas the bases belonging to the other class are said to be *negative*.

Let (e_i) be a positive basis of the vector space E_n, and denote by (θ^i) its dual basis. If $x_i = x_i^j e_j$, $i = 1, \dots, n$, are n vectors of E_n, from (3.17) and (2.5) we obtain that

$$\theta^1 \wedge \cdots \wedge \theta^n(x_1, \dots, x_n) = \det \begin{pmatrix} x_1^1 & \cdots & x_n^1 \\ \cdots & \cdots & \cdots \\ x_1^n & \cdots & x_n^n \end{pmatrix}. \tag{3.31}$$

In particular, we have that

$$\theta^1 \wedge \cdots \wedge \theta^n(e_1, \dots, e_n) = 1. \tag{3.32}$$

Further, if (e_i') is another basis of E_n, related to (e_i) by (3.27), then in view of (3.31), we have that

$$\theta^1 \wedge \cdots \wedge \theta^n(e_1', \dots, e_n') = \det \mathbb{A}. \tag{3.33}$$

In conclusion, we have proved (Example 3.1).

Theorem 3.5. *The* $(0, n)$*-covector* $\theta^1 \wedge \cdots \wedge \theta^n$ *of the vector space* $\Lambda_n(E_n)$ *associates to any n-tuple of vectors* $x_1, \dots, x_n$ *the volume of the parallelepiped having these vectors as wedges.*

3.5 Exercises

1. Considering in E_3 the 1-forms

$$\begin{aligned} \alpha &= \theta^1 - \theta^2, \\ \beta &= \theta^1 - \theta^2 + \theta^3, \\ \sigma &= \theta^3, \end{aligned}$$

the 2-form

$$\eta = \theta^1 \wedge \theta^3 + \theta^2 \wedge \theta^3,$$

and the 3-form

$$\Omega = \theta^1 \wedge \theta^2 \wedge \theta^3,$$

calculate the exterior products

$$\alpha \wedge \beta, \ \alpha \wedge \beta \wedge \sigma, \ \alpha \wedge \eta, \ \alpha \wedge \Omega.$$

2. Evaluate the components of the forms of the preceding exercise under the basis change

$$\theta'^1 = \theta^1 - 2\theta^2,$$
$$\theta'^2 = \theta^1 + \theta^3,$$
$$\theta'^3 = \theta^3.$$

3. Let $(\mathbf{e}_i)$ be a basis of the vector space E_3 and denote by (θ^i) the dual basis. Given the skew-symmetric tensors

$$\mathbf{T} = T_{12}\theta^1 \wedge \theta^2 + T_{13}\theta^1 \wedge \theta^3 + T_{23}\theta^2 \wedge \theta^3,$$
$$\mathbf{L} = T_{123}\theta^1 \wedge \theta^2 \wedge \theta^3,$$

and the basis change

$$\mathbf{e}'_1 = \mathbf{e}_1 - \mathbf{e}_3,$$
$$\mathbf{e}'_2 = \mathbf{e}_1 + 2\mathbf{e}_2,$$
$$\mathbf{e}'_3 = \mathbf{e}_2 - \mathbf{e}_3,$$

determine the components of the preceding tensors in the corresponding new basis of $\Lambda_2(E_3)$ and $\Lambda_3(E_3)$.

4. Determine the ratio between the volumes of the parallelepipeds formed by the two foregoing bases.

5. Write arbitrary skew-symmetric tensors of $\Lambda_2(E_4)$ and $\Lambda_2(E_5)$.

6. Multiply a skew-symmetric tensor of $\Lambda_2(E_4)$ by a skew-symmetric tensor of $\Lambda_3(E_4)$.

7. Given the volume form $\theta^1 \wedge \theta^2 \wedge \theta^3$ in a three-dimensional space E_3, determine the volume of a parallelepiped whose edges are the vectors $(1,0,2)$, $(-1,2,1)$, and $(1,1,0)$.

8. Evaluate the volume of the parallelepiped of the previous exercise adopting the volume form $\theta'^1 \wedge \theta'^2 \wedge \theta'^3$, where

$$\theta'^1 = \theta^1 + 2\theta^2,$$
$$\theta'^2 = \theta^2 + \theta^3,$$
$$\theta'^3 = \theta^1 - 2\theta^3.$$

Chapter 4
Euclidean and Symplectic Vector Spaces

Abstract In the preceding chapters we analyzed some properties of a vector space E_n. In this chapter we introduce into E_n two other operations: the scalar product and the antiscalar product. A vector space equipped with the first operation is called a Euclidean vector space, whereas when it is equipped with the second operation, it is said to be a symplectic vector space. These operations allow us to introduce into E_n many other geometric and algebraic concepts such as length of a vector, orthogonality between two vectors, etc. Further, eigenvalues and eigenvectors of a linear map are analyzed together with orthogonal transformations of E_n. Finally, symplectic vector spaces are introduced and some their properties studied.

4.1 Representation Theorems for Symmetric and Skew-Symmetric (0, 2)-Tensors

Definition 4.1. Let E_n be an n-dimensional vector space. A tensor $\mathbf{T} \in T_2(E_n)$ is said to be *symmetric* if

$$\mathbf{T}(\mathbf{x}, \mathbf{y}) = \mathbf{T}(\mathbf{y}, \mathbf{x}), \ \forall \mathbf{x}, \mathbf{y} \in E_n. \tag{4.1}$$

It is easy to prove that (4.1) is equivalent to the condition

$$T_{ij} = T_{ji}, \tag{4.2}$$

where $T_{ij} = \mathbf{T}(\mathbf{e}_i, \mathbf{e}_j)$ are the components of $\mathbf{T}$ in an arbitrary base $(\mathbf{e}_i)$ of E_n.

Theorem 4.1. *If* $\mathbf{T} \in T_2(E_n)$ *is symmetric, then there exists a basis* $(\overline{\mathbf{e}}_i)$ *of* E_n *in which the components of* $\mathbf{T}$ *are given by the matrix*

© Springer International Publishing AG, part of Springer Nature 2018
A. Romano and A. Marasco, *Classical Mechanics with Mathematica®*,
Modeling and Simulation in Science, Engineering and Technology,
https://doi.org/10.1007/978-3-319-77595-1_4

$$\mathbf{u} = T(\mathbf{x}, \mathbf{u}_2)\mathbf{u}_1 - T(\mathbf{x}, \mathbf{u}_1)\mathbf{u}_2,$$
$$\mathbf{v} = \mathbf{x} - \mathbf{u}.$$

Since $\mathbf{u} \in U^1$ and $\mathbf{x} = \mathbf{u} + \mathbf{v}$, it remains to prove that $\mathbf{v} \in V^1$ and the vectors $\mathbf{u}, \mathbf{v}$ are uniquely determined. But, in view of (4.8), for $i = 1, 2$ it is

$$\begin{aligned}
T(\mathbf{v}, \mathbf{u}_i) &= T(\mathbf{x} - \mathbf{u}, \mathbf{u}_i) = T(\mathbf{x}, \mathbf{u}_i) - T(\mathbf{u}, \mathbf{u}_i) \\
&= T(\mathbf{x}, \mathbf{u}_i) - T(T(\mathbf{x}, \mathbf{u}_2)\mathbf{u}_1 - T(\mathbf{x}, \mathbf{u}_1)\mathbf{u}_2, \mathbf{u}_i) \\
&= T(\mathbf{x}, \mathbf{u}_i) - T(\mathbf{x}, \mathbf{u}_2)T(\mathbf{u}_1, \mathbf{u}_i) + T(\mathbf{x}, \mathbf{u}_1)T(\mathbf{u}_2, \mathbf{u}_i) = 0,
\end{aligned}$$

so that $\mathbf{v} \in V^1$. Further, if there is another decomposition $\mathbf{x} = \mathbf{u}' + \mathbf{v}'$, $\mathbf{u}' \in U^1$, and $\mathbf{v}' \in V^1$, we have $\mathbf{u}' - \mathbf{u} = \mathbf{v}' - \mathbf{v}$. But $\mathbf{v}' - \mathbf{v} \in V^1$, and then

$$T(\mathbf{v}' - \mathbf{v}, \mathbf{u}_i) = T(\mathbf{u}' - \mathbf{u}, \mathbf{u}_i) = 0.$$

Since $\mathbf{u}' - \mathbf{u} \in U^1$, there are two real numbers a and b such that $\mathbf{u}' - \mathbf{u} = a\mathbf{u}_1 + b\mathbf{u}_2$. Consequently, from the foregoing result we obtain the condition

$$a T(\mathbf{u}_1, \mathbf{u}_i) + b T(\mathbf{u}_2, \mathbf{u}_i) = 0,$$

which implies $a = b = 0$ since $T(\mathbf{u}_1, \mathbf{u}_2) \neq 0$ and $T(\mathbf{u}_i, \mathbf{u}_i) = 0$. The restriction T^1 of T onto V^1 is still a skew-symmetric tensor. Then, if $n = 1$ or $T^1 = \mathbf{0}$, the theorem is proved; otherwise we can repeat for T^1 the foregoing reasoning. Iterating the procedure, we determine a basis $(\mathbf{u}_i)$ of E_n in which T is represented by the matrix

$$\begin{pmatrix}
0 & 1 & 0 & \cdots & \cdots & \cdots & 0 \\
-1 & 0 & 0 & \cdots & \cdots & \cdots & \\
\cdots & \cdots & \cdots & 0 & 1 & \cdots & \cdots \\
\cdots & \cdots & \cdots & -1 & 0 & \cdots & \cdots \\
\cdots & \cdots & \cdots & \cdots & \cdots & \cdots & \\
0 & 0 & \cdots & \cdots & \cdots & \cdots & 0
\end{pmatrix}.$$

In the new basis,

$$\begin{aligned}
\bar{\mathbf{e}}_i &= \mathbf{u}_{2i-1}, & i &= 1, \ldots, r, \\
\bar{\mathbf{e}}_{r+i} &= \mathbf{u}_{2i}, & i &= 1, \ldots, r, \\
\bar{\mathbf{e}}_k &= \mathbf{u}_k, & k &= 2r+1, \ldots, n,
\end{aligned}$$

the representative matrix of T becomes (4.7), and the theorem is proved. $\quad\square$

The basis $(\bar{\mathbf{e}}_i)$, in which the representative matrix of the skew-symmetric tensor T has the form (4.7), is called a **canonical basis** of E_n. Let $(\bar{\theta}^i)$ be the dual basis of $(\bar{\mathbf{e}}_i)$. Then the foregoing theorem supplies the following **canonical form** of the skew-symmetric tensor T in the basis $(\bar{\theta}^i \otimes \bar{\theta}^j)$ of $T_2(E_n)$:

$$\mathbf{T} = \sum_{i=1}^{r} (\overline{\theta}^i \otimes \overline{\theta}^{r+i} - \overline{\theta}^{r+i} \otimes \overline{\theta}^i),$$

(4.9)

that contains only the components different from zero.

4.2 Degenerate and Nondegenerate (0, 2)-Tensors

For any $\mathbf{T} \in T_2(E_n)$ we consider the vector subspace of E_n

$$E_0 = \{\mathbf{x} \in E_n, \mathbf{T}(\mathbf{x}, \mathbf{y}) = 0, \forall \mathbf{y} \in E_n\}.$$

Definition 4.2. The tensor $\mathbf{T} \in T_2(E_n)$ is said to be **nondegenerate** if $E_0 = \{\mathbf{0}\}$, **degenerate** if $\dim(E_0) = k \geq 1$.

In components, the condition $\mathbf{T}(\mathbf{x}, \mathbf{y}) = 0$, $\forall \mathbf{y} \in E_n$, is expressed by the system

$$T_{ij}x^i = 0, \quad j = 1, \ldots, n,$$

(4.10)

of n linear equations in n unknowns (x^i). It is well known that, denoting by $\mathbb{T}$ the representative matrix of $\mathbf{T}$ in the basis $(\mathbf{e}_i)$ of E_n, system (4.10) admits a zero solution when the rank p of $\mathbb{T}$ is equal to n, and infinite solutions, which belong to a vector subspace of E_n, when $p < n$. Consequently, $\dim(E_0) = n - p$.

For a symmetric tensor $\mathbf{T} \in T_2(E_n)$, resorting to its canonical representation, we can state that the rank p coincides with the number $r + s$ of the elements of the principal diagonal of matrix (4.3) that assume the values ± 1, whereas $\dim(E_0)$ is equal to the number of zeros contained into the principal diagonal. For a symmetric nondegenerate (0, 2)-tensor $\mathbf{T}$ there is no zero in the principal diagonal of (4.3). Starting from (4.7), we conclude that for a skew-symmetric (0, 2)-tensor $\mathbf{T}$, it results that the rank $(\mathbf{T}) = 2r$ and $\mathbf{T}$ is nondegenerate if and only if $2r = n$. In particular, if n is odd, there is no nondegenerate tensor.

Theorem 4.3. *The numbers r and s appearing in the canonical representation (4.6) of a symmetric tensor $\mathbf{T} \in T_2(E_n)$ do not depend on the canonical basis of E_n.*

Proof. Let $(\overline{\mathbf{e}}_i)$ be the canonical basis in which $\mathbf{T}$ is represented by the matrix (4.3). The vector subspaces E^+ and E^- of E_n generated by $(\overline{\mathbf{e}}_1, \ldots, \overline{\mathbf{e}}_r)$ and $(\overline{\mathbf{e}}_{r+1}, \ldots, \overline{\mathbf{e}}_{r+s})$, respectively, satisfy the condition

$$E_n = E^+ \oplus E^- \oplus E_0.$$

(4.11)

Moreover, for any choice of

$$\mathbf{x} = \sum_{i=1}^{r} \overline{x}^i \overline{\mathbf{e}}_i \in E^+,$$

$$\mathbf{y} = \sum_{i=r+1}^{r+s} \overline{y}^i \overline{\mathbf{e}}_i \in E^-,$$

we have that

$$\mathbf{T}(\mathbf{x}, \mathbf{x}) = \sum_{i=1}^{r} (\overline{x}^i)^2 > 0, \quad \mathbf{T}(\mathbf{y}, \mathbf{y}) = - \sum_{i=r+1}^{r+s} (\overline{y}^i)^2 < 0. \tag{4.12}$$

In another canonical basis $(\overline{\mathbf{e}}'_i)$ of E_n, the representative matrix $\mathbb{T}$ of $\mathbf{T}$ has again the form (4.3), with r' positive numbers and s' negative numbers in the principal diagonal such that $r' + s' = r + s = n - p$. Denoting by E'^+ and E'^- the vector subspaces of E_n generated, respectively, by $(\overline{\mathbf{e}}'_1, \ldots, \overline{\mathbf{e}}'_{r'})$ and $(\overline{\mathbf{e}}'_{r'+1}, \ldots, \overline{\mathbf{e}}'_{r'+s'})$, again we have the decomposition

$$E_n = E'^+ \oplus E'^- \oplus E_0,$$

and (4.12) gives

$$\mathbf{T}(\mathbf{x}, \mathbf{x}) = \sum_{i=1}^{r'} (\overline{x}'^i)^2 > 0, \quad \mathbf{T}(\mathbf{y}, \mathbf{y}) = - \sum_{i=r'+1}^{r'+s'} (\overline{y}'^i)^2 < 0. \tag{4.13}$$

We can easily prove that the intersection $E^+ \cap (E'^- \oplus E_0) = \{\mathbf{0}\}$. In fact, if $\mathbf{x} \in E^+$ and $\mathbf{x} \neq \mathbf{0}$, then $(4.12)_1$ is satisfied; further, any $\mathbf{y} \in E'^-$, $\mathbf{y} \neq \mathbf{0}$, verifies $(4.13)_2$. Consequently, the subspace $E^+ \oplus E'^- \oplus E_0$ of E_n has a dimension $r + n - r' \leq n$, so that $r \leq r'$. Applying this reasoning to $E'^+ \oplus E^- \oplus E_0$, we obtain $r = r'$. $\quad \square$

Definition 4.3. The integer number r is called the *index* of $\mathbf{T}$, whereas the difference $r - s$ is the *signature* of $\mathbf{T}$.

Definition 4.4. Let $\mathbf{T}$ belong to $T_2(E_n)$. The map $q : E_n \rightarrow \mathfrak{R}$ such that

$$q(\mathbf{x}) = \mathbf{T}(\mathbf{x}, \mathbf{x})$$

is said to be the *quadratic form* associated with $\mathbf{T}$. Since $q(\mathbf{x})$ vanishes identically when $\mathbf{T}$ is skew-symmetric, throughout the following sections we refer only to symmetric tensors.

Definition 4.5. A symmetric tensor $\mathbf{T} \in T_2(E_n)$ is said to be *positive semidefinite* if $\forall \mathbf{x} \in E_n$ the following results are obtained:

$$q(\mathbf{x}) \equiv \mathbf{T}(\mathbf{x}, \mathbf{x}) \geq 0. \tag{4.14}$$

If (4.14) assumes a zero value if and only if $\mathbf{x} = \mathbf{0}$, then $\mathbf{T}$ is *positive definite*.

By adopting a canonical basis in E_n and resorting to Theorem 4.3, we conclude that $s=0$ when $\mathbf{T}$ is positive semidefinite. Further, $\mathbf{T}$ is positive definite if and only if $r=n$. The character of positive definiteness of $\mathbf{T}$ does not depend on the basis $(\mathbf{e}_i)$ of E_n. In fact, in any basis (4.14) can be written as

$$T_{ij}x^i x^j \geq 0, \quad \forall(x^i) \in \mathfrak{R}^n. \tag{4.15}$$

It is well known that quadratic forms are positive semidefinite (positive definite) if and only if all the principal minors T^i, $i=1,...,n$, of the representative matrix $\mathbb{T} = (T_{ij})$ of $\mathbf{T}$ with respect to the basis $(\mathbf{e}_i)$ satisfy the following conditions:

$$T^i \geq 0, \quad (T^i > 0), \quad i = 1, \ldots, n. \tag{4.16}$$

Theorem 4.4. *Let* $\mathbf{T} \in T_2(E_n)$ *be a symmetric and positive semidefinite* $(0, 2)$-*tensor. Then,* $\forall \mathbf{x}, \mathbf{y} \in E_n$ ***Schwarz's inequality***

$$|\mathbf{T}(\mathbf{x}, \mathbf{y})| \leq \sqrt{\mathbf{T}(\mathbf{x}, \mathbf{x})}\sqrt{\mathbf{T}(\mathbf{y}, \mathbf{y})} \tag{4.17}$$

*and **Minkowski's inequality***

$$\sqrt{\mathbf{T}(\mathbf{x} + \mathbf{y}, \mathbf{x} + \mathbf{y})} \leq \sqrt{\mathbf{T}(\mathbf{x}, \mathbf{x})} + \sqrt{\mathbf{T}(\mathbf{y}, \mathbf{y})} \tag{4.18}$$

hold.

Proof. Note that $\forall a \in \mathfrak{R}$ and $\forall \mathbf{x}, \mathbf{y} \in E_n$ the following results are obtained:

$$\mathbf{T}(a\mathbf{x} + \mathbf{y}, a\mathbf{x} + \mathbf{y}) = a^2\mathbf{T}(\mathbf{x}, \mathbf{x}) + 2a\mathbf{T}(\mathbf{x}, \mathbf{y}) + \mathbf{T}(\mathbf{y}, \mathbf{y}) \geq 0. \tag{4.19}$$

Since the left-hand side of the preceding inequality is a second-degree polynomial in the variable a, we can state that its discriminant is not positive, that is,

$$|\mathbf{T}(\mathbf{x}, \mathbf{y})|^2 - \mathbf{T}(\mathbf{x}, \mathbf{x})\mathbf{T}(\mathbf{y}, \mathbf{y}) \leq 0, \tag{4.20}$$

and (4.17) is proved. To prove (4.18), we start from the inequality

$$\begin{aligned}\mathbf{T}(\mathbf{x} + \mathbf{y}, \mathbf{x} + \mathbf{y}) &= \mathbf{T}(\mathbf{x}, \mathbf{x}) + 2\mathbf{T}(\mathbf{x}, \mathbf{y}) + \mathbf{T}(\mathbf{y}, \mathbf{y}) \\ &\leq \mathbf{T}(\mathbf{x}, \mathbf{x}) + 2|\mathbf{T}(\mathbf{x}, \mathbf{y})| + \mathbf{T}(\mathbf{y}, \mathbf{y}),\end{aligned}$$

which in view of (4.17) implies

$$\begin{aligned}\mathbf{T}(\mathbf{x} + \mathbf{y}, \mathbf{x} + \mathbf{y}) &\leq \mathbf{T}(\mathbf{x}, \mathbf{x}) + 2\sqrt{\mathbf{T}(\mathbf{x}, \mathbf{x})\mathbf{T}(\mathbf{y}, \mathbf{y})} + \mathbf{T}(\mathbf{y}, \mathbf{y}) \\ &= [\sqrt{\mathbf{T}(\mathbf{x}, \mathbf{x})} + \sqrt{\mathbf{T}(\mathbf{x}, \mathbf{x})}]^2,\end{aligned}$$

and (4.18) is proved.   □

$$|\mathbf{x}| = \sqrt{g_{ij}x^i x^j}, \tag{4.35}$$

$$\cos\varphi = \frac{g_{ij}x^i y^j}{\sqrt{g_{ij}x^i x^j}\sqrt{g_{ij}y^i y^j}}. \tag{4.36}$$

Definition 4.8. Let $(\mathbf{e}_i)$ be a basis of the Euclidean vector space E_n. We call *covariant components* of the vector $\mathbf{x}$ relative to the basis $(\mathbf{e}_i)$ the quantities

$$x_i = \mathbf{x} \cdot \mathbf{e}_i = g_{ij}x^j. \tag{4.37}$$

When a basis $(\mathbf{e}_i)$ is given, there is a one-to-one map between vectors and their contravariant components. The same property holds for the covariant components. In fact, it is sufficient to note that $\det(g_{ij}) \neq 0$ and refer to linear relations (4.37).

All the preceding formulae assume their simplest form relative to an orthonormal basis $(\overline{\mathbf{e}}_i)$. In fact, since in such a basis

$$\overline{\mathbf{e}}_i \cdot \overline{\mathbf{e}}_j = \delta_{ij}, \tag{4.38}$$

we obtain also

$$\mathbf{x} \cdot \mathbf{y} = \sum_{i=1}^{n} \overline{x}^i \overline{y}^i, \tag{4.39}$$

$$|\mathbf{x}| = \sqrt{\sum_{i=1}^{n} (\overline{x}^i)^2}, \tag{4.40}$$

$$\overline{x}_i = \overline{x}^i. \tag{4.41}$$

After checking the advantage of the orthonormal bases, we understand the importance of *Schmidt's orthonormalization procedure*, which allows us to obtain an orthonormal basis $(\mathbf{u}_i)$ starting from any other basis $(\mathbf{e}_i)$. First, we set

$$\mathbf{u}_1 = \mathbf{e}_1. \tag{4.42}$$

Then, we search for a vector $\mathbf{u}_2$ such that

$$\mathbf{u}_2 = a_2^1 \mathbf{u}_1 + \mathbf{e}_2, \tag{4.43}$$

$$\mathbf{u}_1 \cdot \mathbf{u}_2 = 0. \tag{4.44}$$

Introducing (4.44) into (4.43), we obtain the condition

$$a_2^1 \mathbf{u}_1 \cdot \mathbf{u}_1 + \mathbf{u}_1 \cdot \mathbf{e}_2 = 0.$$

Since $\mathbf{u}_1 \neq \mathbf{0}$, the preceding condition allows us to determine a_2^1 and the vector $\mathbf{u}_2$ is not zero owing to the linear independence of $\mathbf{e}_1$ and $\mathbf{e}_2$. Then, we search for a vector $\mathbf{u}_3$ such that

$$\mathbf{u}_3 = a_3^1 \mathbf{u}_1 + a_2^2 \mathbf{u}_2 + \mathbf{u}_3$$
$$\mathbf{u}_1 \cdot \mathbf{u}_3 = 0,$$
$$\mathbf{u}_2 \cdot \mathbf{u}_3 = 0.$$

These relations imply the linear system

$$a_3^1 \mathbf{u}_1 \cdot \mathbf{u}_1 + \mathbf{u}_1 \cdot \mathbf{e}_3 = 0,$$
$$a_3^2 \mathbf{u}_2 \cdot \mathbf{u}_2 + \mathbf{u}_2 \cdot \mathbf{e}_3 = 0,$$

which determines the unknowns a_3^1 and a_3^2 since the vectors $\mathbf{u}_1$ and $\mathbf{u}_2$ do not vanish. Finally, the system $(\mathbf{u}_1, \mathbf{u}_2, \mathbf{u}_3)$ is orthogonal. After n steps, an orthogonal system $(\mathbf{u}_1, ..., \mathbf{u}_n)$ is determined. Dividing each vector of this system by its length, we obtain an orthonormal system.

Definition 4.9. Let V be a vector subspace of E_n. Then, the set

$$V_\perp = \{\mathbf{x} \in E_n, \mathbf{x} \cdot \mathbf{y} = 0, \forall \mathbf{y} \in V\}, \tag{4.45}$$

containing all the vectors that are orthogonal to any vector of V, is said to be the *orthogonal complement* of V.

Theorem 4.5. *If E_n is a Euclidean vector space and V any vector subspace of E_n, then $V_\perp$ is a vector subspace of E_n; in addition, the following results are obtained:*

$$E_n = V \oplus V_\perp. \tag{4.46}$$

Proof. If $\mathbf{x}_1, \mathbf{x}_2 \in V_\perp, a_1, a_2 \in \Re$, and $\mathbf{y} \in V$, then we have that

$$(a_1 \mathbf{x}_1 + a_2 \mathbf{x}_2) \cdot \mathbf{y} = a_1 \mathbf{x}_1 \cdot \mathbf{y} + a_2 \mathbf{x}_2 \cdot \mathbf{y} = 0,$$

and $a_1 \mathbf{x}_1 + a_2 \mathbf{x}_2 \in V_\perp$. Further, we note that if $(\mathbf{e}_1, ..., \mathbf{e}_m)$ is an orthonormal basis of V and $(\mathbf{e}_1, ..., \mathbf{e}_m, \mathbf{e}_{m+1}, ..., \mathbf{e}_n)$ is an orthonormal basis of E_n, then any vector that is orthogonal to all the vectors of this basis is an element of $V_\perp$. In fact, when $\mathbf{x} \cdot \mathbf{e}_i = 0, i = 1, ..., n$, for any $\mathbf{y} \in V$ we obtain that

$$\mathbf{x} \cdot \mathbf{y} = \mathbf{x} \cdot \sum_{i=1}^m y^i \mathbf{e}_i = \sum_{i=1}^m y^i \mathbf{x} \cdot \mathbf{e}_i = 0,$$

and then $\mathbf{x} \in V_\perp$. Now, $\forall \mathbf{x} \in E_n$ we set

$$\mathbf{x}' = (\mathbf{x} \cdot \mathbf{e}_1)\mathbf{e}_1 + \cdots (\mathbf{x} \cdot \mathbf{e}_m)\mathbf{e}_m,$$
$$\mathbf{x}'' = \mathbf{x} - \mathbf{x}'.$$

Since the vector $\mathbf{x}'' \in V_\perp$, the decomposition $\mathbf{x} = \mathbf{x}' + \mathbf{x}''$ is such that $\mathbf{x}' \in V$ and $\mathbf{x}'' \in V_\perp$. To prove that this decomposition is unique, we suppose that there is another decomposition $\mathbf{y}' + \mathbf{y}''$. Then, it must be that $(\mathbf{y}' - \mathbf{x}') + (\mathbf{y}'' - \mathbf{x}'') = \mathbf{0}$, where the vector inside the first parentheses belongs to V, whereas the vector inside the other parentheses belongs to $V_\perp$. Finally, in a Euclidean space the sum of two orthogonal vectors vanishes if and only if each of them vanishes and the theorem is proved.   □

4.5 Eigenvectors of Euclidean 2-Tensors

Definition 4.10. Let $\mathbf{T}$ be a $(1, 1)$-tensor of a Euclidean vector space E_n. We say that the number $\lambda \in \Re$ and the vector $\mathbf{x} \neq \mathbf{0}$ are, respectively, an **eigenvalue** of $\mathbf{T}$ and an **eigenvector** of $\mathbf{T}$ belonging to λ if λ and $\mathbf{x}$ satisfy the **eigenvalue equation**

$$\mathbf{T}(\mathbf{x}) = \lambda\mathbf{x}. \tag{4.47}$$

The property that $\mathbf{T}$ is linear implies that the set V_λ of all the eigenvectors belonging to the same eigenvalue λ form a vector subspace of E_n. In fact, if $a, b \in \Re$ and $\mathbf{x}$, $\mathbf{y} \in V_\lambda$, then we have that

$$\mathbf{T}(a\mathbf{x} + b\mathbf{y}) = a\mathbf{T}(\mathbf{x}) + b\mathbf{T}(\mathbf{y}) = \lambda(a\mathbf{x} + b\mathbf{y}),$$

so that $a\mathbf{x} + b\mathbf{y} \in V_\lambda$.

Definition 4.11. The dimension of the vector subspace associated with the eigenvalue λ is called a **geometric multiplicity** of the eigenvalue λ; in particular, an eigenvalue with multiplicity 1 is also said to be **simple**. The set of all the eigenvalues of $\mathbf{T}$ is called the **spectrum** of $\mathbf{T}$. Finally, the **eigenvalue problem** relative to $\mathbf{T}$ consists in determining the whole spectrum of $\mathbf{T}$.

To find the eigenvalues of $\mathbf{T}$, we start out by noting that in a basis $(\mathbf{e}_i)$ of E_n, (4.47) is written as

$$\left(T^i_j - \lambda\delta^i_j\right) x^j = 0, \ i = 1, \ldots, n. \tag{4.48}$$

This is a homogeneous linear system of n equations in n unknowns $x^1, \ldots, x^n$, which admits a solution different from zero if and only if

$$P_n(\lambda) \equiv \det(T^i_j - \lambda\delta^i_j) = 0. \tag{4.49}$$

Now we show a fundamental property of the preceding equation: although the components T^i_j of tensor $\mathbf{T}$ depend on the choice of the basis $(\mathbf{e}_i)$, the coefficients of (4.49) do not depend on it. In fact, in the basis change

$$\mathbf{e}'_i = A^j_i \mathbf{e}_j,$$

with the usual meaning of the symbols, we have that

$$\begin{aligned} P'_n(\lambda) &= \det(\mathbb{T}' - \lambda\mathbb{I}') = [\det \mathbb{A}^{-1}(\mathbb{T} - \lambda\mathbb{I})\mathbb{A}] \\ &= \det \mathbb{A}^{-1} \det \mathbb{A} P_n(\lambda), \end{aligned}$$

and then

$$P'_n(\lambda) = P_n(\lambda). \tag{4.50}$$

Since the polynomial $P_n(\lambda)$ does not depend on the basis $(\mathbf{e}_i)$, it is called the *characteristic polynomial* of $\mathbf{T}$. Denoting by I_i the coefficient of the power λ^{n-i} and noting that $I_0 = (-1)^n$, we can write $P_n(\lambda)$ as follows:

$$P_n(\lambda) = (-1)^n \lambda^n + I_1 \lambda^{n-1} + \cdots + I_n. \tag{4.51}$$

Remark 4.1. It is possible to verify that

$$I_i = (-1)^i J_i, \; i = 1, \ldots, n, \tag{4.52}$$

where J_i is the sum of all determinants of the principal minors of order i of matrix $\mathbb{T}$. In particular, $I_1 = T^1_1 + \cdots + T^n_n$ and $I_n = \det \mathbb{T}$.

In conclusion, we have proved what follows.

Theorem 4.6. *The eigenvalues of a $(1, 1)$-tensor $\mathbf{T}$ are the real roots of the characteristic polynomial*

$$P_n(\lambda) = (-1)^n \lambda^n + I_1 \lambda^{n-1} + \cdots + I_n = 0. \tag{4.53}$$

Definition 4.12. Equation (4.53) is the *characteristic equation* of the tensor $\mathbf{T}$. Further, the multiplicity of a root λ of (4.53) is called the *algebraic multiplicity* of the eigenvalue λ.

Let λ be any real roots of (4.53), i.e., an eigenvalue of the spectrum of $\mathbf{T}$. Introducing λ into (4.48), we obtain a linear homogeneous system whose solutions form a subspace V_λ of eigenvectors. The dimension of V_λ is equal to $k = n - p$, where p is the rank of the matrix $\mathbb{T} - \lambda\mathbb{I}$. In other words, we can find k independent eigenvectors $\mathbf{u}_1, \ldots, \mathbf{u}_k$, belonging to V_λ, that form a basis of V_λ. In particular, if there exists a basis of E_n formed by eigenvectors of $\mathbf{T}$ belonging to the eigenvalues $\lambda_1, \ldots, \lambda_n$ enumerated with their algebraic multiplicity, then the corresponding matrix $\mathbb{T}$ representative of $\mathbf{T}$ assumes the following diagonal form:

$$\mathbb{T} = \begin{pmatrix} \lambda_1 \cdots 0 \\ \cdots\cdots\cdots \\ 0 \cdots \lambda_n \end{pmatrix}. \tag{4.54}$$

The following theorem, whose proof we omit, is very useful in applications.

Theorem 4.7. *Let* **T** *be a symmetric tensor of a Euclidean vector space* E_n. *Then, all the eigenvalues of* **T** *are real and the dimension of the subspace* V_λ *associated with the eigenvalue* λ *is equal to the multiplicity of* λ. *Further, eigenvectors belonging to different eigenvalues are orthogonal to each other, and there exists at least a basis of eigenvectors of* **T** *relative to which the matrix* $\mathbb{T}$, *representative of* **T**, *is diagonal.*

4.6 Orthogonal Transformations

Definition 4.13. Let E_n be a Euclidean n-dimensional vector space. An endomorphism $\mathbf{Q} : E_n \rightarrow E_n$ is an **orthogonal transformation** if

$$\mathbf{Q}(\mathbf{x}) \cdot \mathbf{Q}(\mathbf{y}) = \mathbf{x} \cdot \mathbf{y}, \ \forall \mathbf{x}.\mathbf{y} \in E_n. \tag{4.55}$$

If the basis $(\mathbf{e}_i)$ is orthonormal, then the n vectors $\mathbf{Q}(\mathbf{e}_i)$ are independent and consequently form a basis of E_n. Therefore, $\mathbf{Q}$ is an isomorphism and the matrix $\mathbb{Q}$, representative of $\mathbf{Q}$ in any basis $(\mathbf{e}_i)$, is not singular. The condition (4.55) can be written in one of the following forms:

$$g_{hk} Q_i^k Q_j^k = g_{ij}, \tag{4.56}$$

$$\mathbb{Q}^T \mathbb{G} \mathbb{Q} = \mathbb{G}. \tag{4.57}$$

In particular, relative to the orthonormal basis $(\mathbf{e}_i)$, in which $\mathbb{G} = \mathbb{I}$, the foregoing relations can be written as follows:

$$\mathbb{Q}^T \mathbb{Q} = \mathbb{I} \Leftrightarrow \mathbb{Q}^T = \mathbb{Q}^{-1}. \tag{4.58}$$

A matrix satisfying one of conditions (4.58) is said to be **orthogonal**.

Since the composite of two orthogonal transformations is still orthogonal, and such are the identity transformation and the inverse transformation, the set of all orthogonal transformations is a group $O(n)$, which is called **orthogonal group**. In view of (4.58) it follows that

$$\det \mathbb{Q} = \pm 1. \tag{4.59}$$

The orthogonal transformations of a three-dimensional Euclidean space E_3 are also called **rotations**; in particular, the rotations for which $\det \mathbb{Q} = 1$ are called **proper rotations**. A group of rotations of E_3 is denoted by $O(3)$, whereas the subgroup of proper rotations is denoted by $SO(3)$. Finally, the orthogonal transformation $-\mathbf{I}$ is called the **central inversion**.

The following theorem is fundamental.

Theorem 4.8. (Euler). *A proper rotation* $\mathbf{Q} \neq \mathbf{I}$ *of the three-dimensional vector space* E_3 *always has the simple eigenvalue* $\lambda = 1$ *whose corresponding eigenspace* V_1 *is one-dimensional and invariant under* $\mathbf{Q}$. *Further, if the restriction* $\mathbf{Q}_\perp$ *of* $\mathbf{Q}$ *to the orthogonal complementary space* $V_{1\perp}$ *is different from the identity, then* $\lambda = 1$ *is the only eigenvalue of* $\mathbf{Q}$.

Proof. Relative to an orthonormal basis $(\mathbf{e}_i)$ of E_3 the eigenvalue equation of $\mathbf{Q}$ is

$$(Q^i_j - \lambda \delta^i_j)x^j = 0,$$

where $\mathbb{Q} = (Q^i_j)$ is an orthogonal matrix. On the other hand, the characteristic equation is

$$P_3(\lambda) = -\lambda^3 + I_1\lambda^2 + I_2\lambda + I_3 = 0,$$

with [see (4.52)]

$$I_1 = Q^1_1 + Q^2_2 + Q^3_3, \tag{4.60}$$

$$I_2 = -\begin{pmatrix} Q^2_2 & Q^2_3 \\ Q^3_2 & Q^3_3 \end{pmatrix} - \begin{pmatrix} Q^1_1 & Q^1_3 \\ Q^3_1 & Q^3_3 \end{pmatrix} - \begin{pmatrix} Q^1_1 & Q^1_2 \\ Q^2_1 & Q^2_2 \end{pmatrix}, \tag{4.61}$$

$$I_3 = \det \mathbb{Q} = 1. \tag{4.62}$$

In view of (4.62), the characteristic equation can also be written as

$$-\lambda^3 + I_1\lambda^2 + I_2\lambda + 1 = 0.$$

Denoting by A^i_j the cofactor of Q^i_j, we have that

$$(Q^{-1})^i_j = \frac{A^i_j}{\det \mathbb{Q}},$$

and recalling that $\mathbb{Q}$ is orthogonal and $\det \mathbb{Q} = 1$, the preceding equation gives

$$(Q^{-1})^i_j = A^i_j = (Q^T)^i_j = Q^j_i \tag{4.63}$$

from which we obtain

$$A^1_1 = Q^1_1, \quad A^2_2 = Q^2_2, \quad A^3_3 = Q^3_3. \tag{4.64}$$

This result shows that $I_2 = -I_1$ and the characteristic equation becomes

$$-\lambda^3 + I_1\lambda^2 - I_1\lambda + 1 = 0,$$

so that $\lambda = 1$ is a solution of the characteristic equation. This eigenvalue can have a multiplicity of $3 - p$, where p is the rank of the matrix $\mathbf{Q} - \mathbf{I}$. A priori p can assume the values $0, 1, 2$. The value 0 must be excluded since it implies $\mathbf{Q} = \mathbf{I}$, against the hypothesis of the theorem. If $p = 1$, then all the minors of order two of $\mathbf{Q} - \mathbf{I}$ vanish; in particular, the following minors vanish:

$$\begin{pmatrix} Q_1^1 - 1 & Q_2^1 \\ Q_1^2 & Q_2^2 - 1 \end{pmatrix} = A_3^3 - Q_1^1 - Q_2^2 + 1 = 0,$$

$$\begin{pmatrix} Q_1^1 - 1 & Q_3^1 \\ Q_1^3 & Q_3^3 - 1 \end{pmatrix} = A_2^2 - Q_1^1 - Q_3^3 + 1 = 0,$$

$$\begin{pmatrix} Q_2^2 - 1 & Q_3^2 \\ Q_2^3 & Q_3^3 - 1 \end{pmatrix} = A_1^1 - Q_2^2 - Q_3^3 + 1 = 0.$$

In view of (4.64), the preceding equations imply $Q_1^1 = Q_2^2 = Q_3^3 = 1$. Taking into account this result and the orthogonality conditions

$$(Q_1^1)^2 + (Q_1^2)^2 + (Q_1^3)^2 = 1,$$

$$(Q_2^1)^2 + (Q_2^2)^2 + (Q_2^3)^2 = 1,$$

$$(Q_3^1)^2 + (Q_3^2)^2 + (Q_3^3)^2 = 1,$$

we obtain $Q_j^i = 0$, $i \neq j$, and again we have that $\mathbb{Q} = \mathbb{I}$, against the hypothesis. Consequently, $p = 2$ and the eigenvalue $\lambda = 1$ is simple.

Let V_1 be the eigenspace belonging to the eigenvalue $\lambda = 1$, and denote by $V_\perp$ the two-dimensional vector space of all the vectors that are orthogonal to V_1. The vectors $(\mathbf{u}_1, \mathbf{u}_2, \mathbf{u}_3)$, where $(\mathbf{u}_1, \mathbf{u}_2)$ is an orthonormal basis of $V_\perp$ and $\mathbf{u}_3$ a unit vector of V_1, form a basis of E_3. Since $\mathbf{Q}$ is orthogonal, $\mathbf{Q}(\mathbf{v}) \in V_\perp, \forall \mathbf{v} \in V_\perp$. In particular, we have that

$$\mathbf{Q}(\mathbf{u}_1) = \overline{Q}_1^1 \mathbf{u}_1 + \overline{Q}_1^2 \mathbf{u}_2,$$

$$\mathbf{Q}(\mathbf{u}_2) = \overline{Q}_2^1 \mathbf{u}_1 + \overline{Q}_2^2 \mathbf{u}_2,$$

$$\mathbf{Q}(\mathbf{u}_3) = \mathbf{u}_3,$$

and the representative matrix of $\mathbf{Q}$ relative to the basis $(\mathbf{u}_1, \mathbf{u}_2, \mathbf{u}_3)$ gives

$$\overline{\mathbb{Q}} = \begin{pmatrix} \overline{Q}_1^1 & \overline{Q}_2^1 & 0 \\ \overline{Q}_1^2 & \overline{Q}_2^2 & 0 \\ 0 & 0 & 1 \end{pmatrix},$$

whereas the orthogonality conditions become

$$\left(\overline{Q}_1^1\right)^2 + \left(\overline{Q}_1^2\right)^2 = 1,$$
$$\left(\overline{Q}_2^1\right)^2 + \left(\overline{Q}_2^2\right)^2 = 1,$$
$$\overline{Q}_1^1 \overline{Q}_2^1 + \overline{Q}_1^2 \overline{Q}_2^2 = 0.$$

These relations imply the existence of an angle $\varphi \in (0, 2\pi)$ such that

$$\overline{\mathbb{Q}} = \begin{pmatrix} \cos\varphi & -\sin\varphi & 0 \\ \sin\varphi & \cos\varphi & 0 \\ 0 & 0 & 1 \end{pmatrix}.$$

$\square$

4.7 Symplectic Vector Spaces

Definition 4.14. A *symplectic vector space* is a pair $(E_{2n}, \boldsymbol{\Omega})$, where E_{2n} is a vector space with even dimension and $\boldsymbol{\Omega}$ a skew-symmetric nondegenerate $(0, 2)$-tensor.

Remark 4.2. It is fundamental to require that the dimension of the vector space be even. In fact, owing to Theorem 4.2, in a vector space with odd dimension, any skew-symmetric $(0, 2)$-tensor is always degenerate.

Definition 4.15. In the symplectic vector space $(E_{2n}, \boldsymbol{\Omega})$, the *antiscalar product* is the map $(\mathbf{x}, \mathbf{y}) \in E_{2n} \times E_{2n} \rightarrow [\mathbf{x}, \mathbf{y}] \in \Re$ such that

$$[\mathbf{x}, \mathbf{y}] = \boldsymbol{\Omega}(\mathbf{x}, \mathbf{y}). \tag{4.65}$$

An antiscalar product has the following properties:

$$[\mathbf{x}, \mathbf{y}] = -[\mathbf{y}, \mathbf{x}], \tag{4.66}$$
$$\left[\mathbf{x}, \mathbf{y} + \mathbf{z}\right] = \left[\mathbf{x}, \mathbf{y}\right] + [\mathbf{x}, \mathbf{z}], \tag{4.67}$$
$$a\left[\mathbf{x}, \mathbf{y}\right] = \left[a\mathbf{x}, \mathbf{y}\right], \tag{4.68}$$
$$\left[\mathbf{x}, \mathbf{y}\right] = 0, \forall \mathbf{y} \in E_{2n} \Rightarrow \mathbf{x} = \mathbf{0}. \tag{4.69}$$

In fact, the first property follows from the skew symmetry of $\boldsymbol{\Omega}$. The second and third properties follow from the bilinearity of $\boldsymbol{\Omega}$. Finally, the fourth property is due to the fact that $\boldsymbol{\Omega}$ is nondegenerate.

We say that the vectors $\mathbf{x}$ and $\mathbf{y}$ are antiorthogonal if

$$[\mathbf{x}, \mathbf{y}] = 0. \tag{4.70}$$

In view of (4.69), we can state that the only vector that is antiorthogonal to any other vector is **0**. Further, any vector is antiorthogonal to itself. In any basis $(\mathbf{e}_i)$ of E_{2n}, the representative matrix of $\boldsymbol{\Omega}$ is skew-symmetric and the antiscalar product can be written as

$$[\mathbf{x}, \mathbf{y}] = \Omega_{ij} x^i y^j, \tag{4.71}$$

where x^i and y^i are, respectively, the components of $\mathbf{x}$ and $\mathbf{y}$ relative to $(\mathbf{e}_i)$. In a canonical basis $(\mathbf{u}_i)$ (Sect. 4.1), which is also called a *symplectic basis*, $\boldsymbol{\Omega}$ is given by the matrix

$$\boldsymbol{\Omega} = \begin{pmatrix} \mathbb{O} & \mathbb{I} \\ -\mathbb{I} & \mathbb{O} \end{pmatrix}, \tag{4.72}$$

where $\mathbb{O}$ is a zero $n \times n$ matrix and $\mathbf{I}$ is a unit $n \times n$ matrix. Moreover, in a symplectic basis, in view of (4.71) and (4.72), we obtain

$$[\mathbf{u}_i, \mathbf{u}_j] = [\mathbf{u}_{n+i}, \mathbf{u}_{n+j}], \tag{4.73}$$

$$[\mathbf{u}_i, \mathbf{u}_{n+j}] = \delta_{ij}, \tag{4.74}$$

$$[\mathbf{x}, \mathbf{y}] = \sum_{i=1}^{n} (\overline{x}^i \overline{y}^{n+i} - \overline{x}^{n+i} \overline{y}^i). \tag{4.75}$$

An automorphism $\mathbf{S} : E_{2n} \rightarrow E_{2n}$ of a symplectic vector space E_{2n} is a *symplectic transformation* if the antiscalar product does not change, that is,

$$[\mathbf{S}(\mathbf{x}, \mathbf{y})] = [\mathbf{x}, \mathbf{y}], \ \forall \mathbf{x}, \mathbf{y} \in E_{2n}. \tag{4.76}$$

The symplectic transformations correspond to the orthogonal transformations of a Euclidean space. If we denote by $\mathbb{S} = (S_k^h)$ the matrix representative of $\mathbf{S}$ in any basis $(\mathbf{e}_i)$, then condition (4.76) assumes the form

$$\Omega_{hk} S_i^h S_j^k = \Omega_{ij}. \tag{4.77}$$

In matrix form this condition can be written as

$$\mathbb{S}^T \boldsymbol{\Omega} \mathbb{S} = \boldsymbol{\Omega}. \tag{4.78}$$

In a *symplectic* basis, when we adopt the notation

$$\mathbb{S} = \begin{pmatrix} \mathbb{A} & \mathbb{B} \\ \mathbb{C} & \mathbb{D} \end{pmatrix}, \tag{4.79}$$

condition (4.78) can explicitly be written as

$$\begin{pmatrix} A^T & C^T \\ B^T & D^T \end{pmatrix} \begin{pmatrix} O & I \\ -I & O \end{pmatrix} \begin{pmatrix} A & B \\ C & D \end{pmatrix} = \begin{pmatrix} O & I \\ -I & O \end{pmatrix}.$$

In conclusion, a transformation S is symplectic if and only if

$$A^T C - C^T A = O, \tag{4.80}$$
$$A^T D - C^T B = I, \tag{4.81}$$
$$B^T C - D^T A = -I, \tag{4.82}$$
$$B^T D - D^T B = O. \tag{4.83}$$

The set $Sp(E_{2n}, \boldsymbol{\Omega})$ of the symplectic transformations of E_{2n} is called the **symplectic transformation group** of E_{2n}. To verify that $Sp(E_{2n}, \boldsymbol{\Omega})$ is a group, we start out by noting that, $\forall S_1, S_2 \in Sp(E_{2n}, \boldsymbol{\Omega})$,

$$\big[S_1(S_2(\mathbf{x})), S_1(S_2(\mathbf{y}))\big] = \big[S_2(\mathbf{x}), S_2(\mathbf{y})\big] = \big[\mathbf{x}, \mathbf{y}\big]$$

and $S_1 S_2 \in Sp(E_{2n})$. Further, from (4.78), since $\det \boldsymbol{\Omega} = 1$, it follows that

$$(\det S)^2 = 1,$$

and there is the inverse automorphism S^{-1}. Finally,

$$\big[\mathbf{x}, \mathbf{y}\big] = \big[S(S^{-1}(\mathbf{x})), S(S^{-1}(\mathbf{y}))\big] = \big[S^{-1}(\mathbf{x}), S^{-1}(\mathbf{y})\big],$$

and $S^{-1} \in Sp(E_{2n})$.

4.8 Exercises

1. Let V be a subspace of the Euclidean vector space $\Re^4$ generated by the vectors $(1, 0, 1, 3)$ and $(0, 1, 1, 2)$. Determine the subspace $V_\perp$ of the vectors that are orthogonal to V.
2. Let E be the four-dimensional vector space of the 2×2 matrices with real coefficients. If A and B are two arbitrary matrices of E, show that

$$A \cdot B = \mathrm{tr}(B^T A)$$

is a scalar product and that

$$\left(\begin{pmatrix} 1 & 0 \\ 0 & 0 \end{pmatrix}, \begin{pmatrix} 0 & 1 \\ 0 & 0 \end{pmatrix}, \begin{pmatrix} 0 & 0 \\ 1 & 0 \end{pmatrix}, \begin{pmatrix} 0 & 0 \\ 0 & 1 \end{pmatrix} \right)$$

is an orthonormal basis of E.

3. Find an orthogonal 2×2 matrix whose first row is either $(1/\sqrt{5}, 2/\sqrt{5})$ or $(1, 2)$.
4. Let E_n be a vector space, and denote by $\mathbf{u}$ a given vector of E_n. Prove that

   i)

$$\Sigma = \{\mathbf{v} \in E_n | \mathbf{u} \cdot \mathbf{v} = 0\}$$

   is an $(n-1)$-dimensional subspace of E_n;
   ii)

$$E_n = \Sigma \oplus \Theta,$$

   where $\Theta = \{\mathbf{v} = a\mathbf{u}, a \in \mathfrak{R}\}$;
   iii) The orthogonal projection $P_\Sigma(\mathbf{v})$ of $\mathbf{v} \in E_n$ onto Σ is given by

$$P_\Sigma(\mathbf{v}) = \mathbf{v} - (\mathbf{v} \cdot \mathbf{u})\mathbf{u}.$$

5. Let E_n be the n-dimensional vector space E_n of the polynomials $P(x)$ of degree n-1 in the interval $(0, 1)$ (Sect. 1.6). Prove that

$$P(x) \cdot Q(x) = \int_0^1 P(x)Q(x)\mathrm{d}x, \quad P(x), Q(x) \in E_n$$

defines a Euclidean scalar product in E_n.

Chapter 5
Duality and Euclidean Tensors

Abstract In this chapter we show that when E_n is a Euclidean vector space, there is an isomorphism among the tensor spaces $T_s^r(E_n)$ for which $r+s$ has a given value. In other words, we show the existence of an isomorphism between E_n and E_n^*, of isomorphisms between T_0^2, T_1^1, and T_2^0, and so on. These isomorphisms lead to the definitions of Euclidean vectors, tensors, and to the covariant, contravariant, and mixed components of a tensor.

5.1 Duality

Theorem 5.1. *Let $(E_n, \mathbf{g})$ be a Euclidean vector space. Then, the map $\tau : \mathbf{x} \in E_n \rightarrow \omega_{\mathbf{x}} \in E_n^*$, such that*

$$\omega_{\mathbf{x}}(\mathbf{y}) = \mathbf{g}(\mathbf{x}, \mathbf{y}) = \mathbf{x} \cdot \mathbf{y}, \quad \forall \mathbf{y} \in E_n \tag{5.1}$$

*defines an isomorphism that is called a **duality**.*

Proof. First, (5.1) defines a covector since $\mathbf{g}$ is linear with respect to $\mathbf{y}$. Moreover, the linearity of $\mathbf{g}$ with respect to $\mathbf{x}$ implies that the mapping τ is linear. Finally, $\ker \tau = \{\mathbf{x} \in E_n, \tau(\mathbf{x}) = \omega_{\mathbf{x}} = \mathbf{0}\}$ contains the vectors $\mathbf{x}$ such that

$$\omega_{\mathbf{x}}(\mathbf{y}) = \mathbf{g}(\mathbf{x}, \mathbf{y}) = \mathbf{x} \cdot \mathbf{y} = 0, \quad \forall \mathbf{y} \in E_n.$$

But $\mathbf{g}$ is a nondegenerate 2-tensor and the preceding condition implies that $x = 0$. Therefore, $\ker \tau = \{\mathbf{0}\}$, and τ is an isomorphism. $\qquad \square$

Remark 5.1. Let $(\mathbf{e}_i)$ be a basis of E_n and let (θ^i) be its dual basis. The basis $(\theta_{\mathbf{e}_i})$ of E_n^* corresponding to $(\mathbf{e}_i)$ in the isomorphism τ is not equal to (θ^i). In fact, it is such that

© Springer International Publishing AG, part of Springer Nature 2018
A. Romano and A. Marasco, *Classical Mechanics with Mathematica®*,
Modeling and Simulation in Science, Engineering and Technology,
https://doi.org/10.1007/978-3-319-77595-1_5

$$\theta_{\mathbf{e}_i}(\mathbf{y}) = \mathbf{g}(\mathbf{e}_i, \mathbf{y}) = \mathbf{g}(\mathbf{e}_i, \mathbf{e}_j)y^j = g_{ij}y^j = g_{ij}\theta^j(\mathbf{y}),$$

i.e.,

$$\theta_{\mathbf{e}_i} = g_{ij}\theta^j. \tag{5.2}$$

In the bases $(\mathbf{e}_i)$ and (θ^i), isomorphism (5.1) has the following representation:

$$\omega_i = g_{ij}x^j, \tag{5.3}$$

whereas the inverse map is

$$x^i = g^{ij}\omega_j. \tag{5.4}$$

In other words, isomorphism (5.1) associates with any vector $\mathbf{x}$ a covector whose components in the dual basis are equal to the covariant components of $\mathbf{x}$ in the basis $(\mathbf{e}_i)$. We can also state that (5.1) allows us to identify the vectors of E_n and the covectors of E_n^*.

Definition 5.1. Any pair $(\mathbf{x}, \omega_{\mathbf{x}})$ in which $\omega_{\mathbf{x}}$ is given by (5.1) is said to be a *Euclidean vector*. Further, the real numbers x^i and ω_i satisfying (5.3) are, respectively, called *contravariant components* and *covariant components* of the Euclidean vector $\mathbf{x}$.

Henceforth a Euclidean vector will be denoted by the first element of the pair $(\mathbf{x}, \omega_{\mathbf{x}})$.

5.2 Euclidean Tensors

The isomorphism (5.1) can be extended to the tensor spaces $T_s^r(E_n)$ and $T_q^p(E_n)$, for which $r + s = p + q$. For the sake of simplicity, we prove the preceding statement for the spaces $T_2^0(E_n)$, $T_1^1(E_n)$, and $T_2^0(E_n)$.

Consider the maps

$$\mathbf{T} \in T_2^0(E_n) \rightarrow \mathbf{T}' \in T_1^1(E_n) \rightarrow \mathbf{T}'' \in T_0^2(E_n)$$

such that

$$\mathbf{T}(\mathbf{x}, \mathbf{y}) = \mathbf{T}'(\mathbf{x}, \omega_{\mathbf{y}}) = \mathbf{T}''(\omega_{\mathbf{x}}, \omega_{\mathbf{y}}), \quad \forall \mathbf{x}, \mathbf{y} \in E_n. \tag{5.5}$$

These linear maps are isomorphisms since, adopting a basis in E_n and its dual basis in E_n^*, they have the following coordinate representations:

$$T_j'^i = g^{ih}T_{hj}, \quad T''^{ij} = g^{ih}g^{jk}T_{hk}. \tag{5.6}$$

Definition 5.2. The triad $(\mathbf{T}, \mathbf{T}', \mathbf{T}'')$ is called a *Euclidean double tensor*. Further, the components T_{ij} of $\mathbf{T}$ in the basis $(\theta^i \otimes \theta^j)$, the components T^i_j of $\mathbf{T}'$ in the basis $(\mathbf{e}_i \otimes \theta^j)$, and the components T^{ij} of $\mathbf{T}''$ in the basis $(\mathbf{e}_i \otimes \mathbf{e}_j)$ are, respectively, called *covariant, mixed, and contravariant components* of the Euclidean tensor $(\mathbf{T}, \mathbf{T}', \mathbf{T}'')$.

Henceforth we will denote a Euclidean tensor by the first element of a triad. It is evident how to extend the preceding considerations to any tensor space $T^r_s(E_n)$.

In view of the foregoing results, we conclude that the tensor spaces $T^r(E_n)$ and $T_r(E_n)$ are isomorphic. We now verify that their subspaces $\Lambda^r E_n$ and $\Lambda_r E_n$ are also isomorphic, and we determine the form of this isomorphism. For the sake of simplicity, we refer to the case $r = 2$. If $\mathbf{T} \in \Lambda_2 E_n$, we have that

$$
\begin{aligned}
T_{(ij)} &= \sum_{i<j} \sum_{h<k} g_{ih} g_{jk} T^{hk} + \sum_{i<j} \sum_{h>k} g_{ih} g_{jk} T^{hk} \\
&= \sum_{i<j} \sum_{h<k} g_{ih} g_{jk} T^{hk} + \sum_{i<j} \sum_{h<k} g_{ik} g_{jh} T^{kh} \\
&= (g_{ih} g_{jk} - g_{ik} g_{jh}) T^{(hk)}.
\end{aligned}
$$

In conclusion, we have

$$
T_{(ij)} = \det \begin{pmatrix} g_{ih} & g_{ik} \\ g_{jh} & g_{jk} \end{pmatrix} T^{(hk)}. \tag{5.7}
$$

More generally, we could prove that

$$
T_{(i_1 \ldots i_r)} = \det \begin{pmatrix} g_{i_1 h_1} & \cdots & g_{i_1 h_r} \\ \cdots & \cdots & \cdots \\ g_{i_r h_1} & \cdots & g_{i_r h_r} \end{pmatrix} T^{(h_1 \ldots h_r)}. \tag{5.8}
$$

In particular, when $r = n$, (5.8) gives

$$
T_{1 \ldots n} = \det(\mathbf{g}) T^{1 \ldots n}. \tag{5.9}
$$

5.3   The Levi–Civita Tensor

Definition 5.3. Let E_n be a vector space. A $(0, k)$- *pseudotensor density of weight p* is a multilinear map $\mathbf{T} : T^0_k(E_n) \to \Re$ such that under a basis change

$$
\mathbf{e}'_j = A^i_j \mathbf{e}_i \tag{5.10}
$$

of E_n, the components of $\mathbf{T}$ are transformed according to the law

$$
T'_{j_1 \ldots j_k} = \operatorname{sgn}(A) |A|^p A^{i_1}_{j_1} \cdots A^{i_k}_{j_k} T_{i_1 \ldots i_k}, \tag{5.11}
$$

where

$$A = \det(A^i_j)$$ (5.12)

and p is a nonnegative rational number. In particular, if $p = 0$, then **T** is said to be a *pseudotensor*.

Starting from the formula

$$g'_{ij} = A^h_i A^k_j g_{hk},$$

and introducing the notations

$$g' = \det(g'_{ij}), \quad g = \det(g_{ij}),$$ (5.13)

we obtain

$$g' = A^2 g, \quad |g'| = A^2 |g|,$$

so that

$$\sqrt{|g'|} = \pm A \sqrt{|g|},$$ (5.14)

where we take the $+$ sign when $A > 0$ and the $-$ sign when $A < 0$. This formula shows that $\sqrt{|g|}$ is a pseudoscalar density of weight 1.

Introduce the skew-symmetric *Levi–Civita symbol*

$$\epsilon_{i_1 \ldots i_n} = \begin{cases} 0, \\ 1, \\ -1, \end{cases}$$ (5.15)

where the value 0 corresponds to a permutation $i_1, \ldots, i_n$ of $1, \ldots, n$ with two or more indices equal, the value 1 to an even permutation $i_1, \ldots, i_n$, and the value -1 to an odd permutation $i_1, \ldots, i_n$. Before proceeding, we recall the following formula relative to the development of a determinant A:

$$A\epsilon_{j_1 \ldots j_n} = \epsilon_{i_1 \ldots i_n} A^{i_1}_{j_1} \cdots A^{i_n}_{j_n}.$$ (5.16)

Let E_n be a Euclidean vector space. We prove that the multilinear map η, which in a basis $(\mathbf{e}_i)$ of E_n has the following components:

$$\eta_{i_1 \ldots i_n} = \sqrt{|g|} \epsilon_{i_1 \ldots i_n},$$ (5.17)

is a $(0, n)$-pseudotensor. In fact, in view of (5.14) and (5.16), under the basis change (5.10), we have that

$$\eta'_{j_1\dots j_n} = \sqrt{|g'|}\epsilon_{j_1\dots j_n} = \pm A\sqrt{|g|}\epsilon_{j_1\dots j_n}$$
$$= \pm A^{i_1}_{j_1}\cdots A^{i_n}_{j_n}\sqrt{|g|}\epsilon_{i_1\dots i_n} \tag{5.18}$$
$$= \pm A^{i_1}_{j_1}\cdots A^{i_n}_{j_n}\eta_{i_1\dots i_n},$$

and our statement is proved. In conclusion, η is a skew-symmetric tensor only under congruent basis changes.

The skew-symmetric tensor η is an element of $\Lambda_n(E_n)$. Since this space has dimension 1 (Chap. 3), we can write

$$\eta = \sqrt{|g|}\theta^1 \wedge \cdots \wedge \theta^n, \tag{5.19}$$

where (θ^i) is a basis of the dual vector space E_n^*. Taking into account what we have just proved about the n-form η and recalling the results of Sect. 3.4, we can state that η is a volume form that is invariant with respect to congruent basis changes.

We conclude this section with the following definition.

Definition 5.4. The *Hodge star operator* is a linear mapping

$$\star : \Lambda_k(E_n) \to \Lambda_{n-k}(E_n) \tag{5.20}$$

such that

$$\star T^{(j_1\dots j_k)} = T_{(i_1\dots i_{(n-k)})} = \eta_{i_1\dots i_{(n-k)}j_1\dots j_k}T^{(j_1\dots j_k)}. \tag{5.21}$$

The skew-symmetric tensor $\star T$ is called the *adjoint* or the *dual form* of **T**.

5.4 Exercises

1. Verify that in an arbitrary base $(\mathbf{e}_i)$ of the Euclidean vector space E_n, the eigenvalue equation (4.49) relative to the Euclidean tensor **T** assumes the form

$$(T_{ij} - \lambda g_{ij})u^j = 0, \tag{5.22}$$

where $g_{ij} = \mathbf{e}_i \cdot \mathbf{e}_j$.
2. Show that the components of the cross-product $\mathbf{u} \times \mathbf{v}$ of the vectors $\mathbf{u}$ and $\mathbf{v}$ of the three-dimensional Euclidean space E_3 are

$$(\mathbf{u} \times \mathbf{v})_i = \eta_{ijk}u^j v^k. \tag{5.23}$$

3. Using the Levi–Civita pseudotensor, evaluate the following vector operations in the Euclidean space E_3:

$$\mathbf{u} \times (\mathbf{v} \times \mathbf{w}), \quad (\mathbf{u} \times \mathbf{v}) \cdot (\mathbf{w} \times \mathbf{w}).$$

Chapter 6
Differentiable Manifolds

Abstract This chapter is an introduction to the wide subject of differentiable manifolds. A differentiable manifold, which generalizes the ordinary definitions of curves and surfaces, is presented here from an intrinsic point of view, that is, without referring to an ambient space in which it is embedded. Differentiable functions, tangent vectors, tensor fields, and r-forms are defined. Further, differential and codifferential of a map between manifolds are studied. In order to introduce metric evaluation on a manifold, the Riemann manifolds are introduced and the geodesics are defined. Many interesting exercises conclude the chapter.

6.1 Differentiable Manifolds

Let U be an open set of $\Re^n$. The real-valued function $f : U \to \Re$ is said to be of **class** $C^k(U)$ or a C^k **function** in U, where $k \geq 0$, if it is continuous with its partial derivatives up to the order k. In particular, a C^0 function in U is a continuous one.

A map

$$f : (x^1, \ldots, x^n) \in U \to (y^1, \ldots, y^m) \in \Re^m$$

is of **class** C^k if any ith projection $pr^i \circ f$

$$y^i = pr^i \circ f(x^1, \ldots, x^n) \equiv y^i(x^1, \ldots, x^n)$$

is a C^k function.

A **homeomorphism** $f : U \to V$, where V is an open set of $\Re^n$, is a continuous map with its inverse. Finally, the map f is a **diffeomorphism** of class C^k if both f and f^{-1} are C^k maps. A diffeomorphism is represented by an invertible system of functions $y^i(x^1, \ldots, x^n)$, $(x^1, \ldots, x^n) \in U, i = 1, \ldots, n$, of class C^k together with the inverse functions. It is well known that the condition

© Springer International Publishing AG, part of Springer Nature 2018   69
A. Romano and A. Marasco, *Classical Mechanics with Mathematica®*,
Modeling and Simulation in Science, Engineering and Technology,
https://doi.org/10.1007/978-3-319-77595-1_6

$$\det \left(\frac{\partial y^i}{\partial x^j} \right)_0 \neq 0$$

at the point $(x_0^1, \ldots, x_0^n) \in U$ is a sufficient condition for the invertibility of these functions in a neighborhood of the point $(x_0^1, \ldots, x_0^n)$.

A differentiable manifold can roughly be defined as an n-dimensional surface embedded in $\Re^m$, $n < m$. This approach to the analysis of differentiable manifolds is more intuitive but not convenient for the following reasons. First, determining the lowest-dimension m of the space $\Re^m$ in which we can embed the manifold is not an easy task. For instance, the plane curves can be embedded in $\Re^2$, whereas the skew curves can be embedded in $\Re^3$. Further, in this approach the geometric objects on the manifold are defined starting from the space $\Re^m$ in which they are embedded. It is much more interesting to build the geometry of a manifold in an intrinsic way. This approach makes it possible to answer questions like the following ones: Is it possible to recognize if a manifold is a sphere staying on it? Is it possible to recognize the geometric structure of the three-dimensional space in which we live by measures that are necessarily internal to our space?

Definition 6.1. Let n be a positive integer number and denote by X a Hausdorff[1] paracompact topological space.[2] X is said to be an n-***dimensional manifold*** if, $\forall x \in X$, there exist an open neighborhood U of x and a homeomorphism $\varphi : U \to \varphi(U) \subseteq \Re^n$. The pair (U, φ) is called a ***chart*** of ***domain*** U and ***coordinate map*** φ. Finally, the n numbers

$$(x^1, \ldots, x^n) = \varphi(x) \in \varphi(U)$$

are the ***coordinates*** in the chart (U, φ) (Fig. 6.1).

Definition 6.2. An ***atlas*** of class C^k on an n-dimensional manifold X is a collection α of charts on X satisfying the following conditions:

- The collection of the domains of the charts of α is an open covering of X;
- $\forall (U, \varphi), (V, \psi) \in \alpha$ the map

$$\psi \circ \varphi^{-1} : \varphi(U \cap V) \to \psi(U \cap V) \tag{6.1}$$

is a C^k diffeomorphism, called a ***coordinate transformation*** (See Fig. 6.2).

A chart (V, ψ) is ***compatible*** with the atlas α if, $\forall (U, \varphi) \in \alpha$, map (6.1) is of class C^k. We call the collection of all the charts that are compatible with α a ***maximal atlas*** $\overline{\alpha}$ of X.

---

[1] A topological space X is a Hausdorff space if, $\forall x, y \in X, x \neq y$, there are neighborhoods U, V of x, y, respectively, such that $U \cap V = \emptyset$.

[2] A Hausdorff space is paracompact if every open covering contains a subcovering that is locally finite.

Fig. 6.1 *n*-dimensional manifold

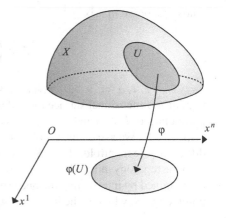

Fig. 6.2 Coordinate transformation

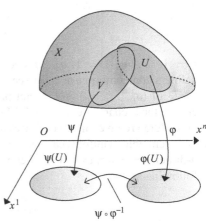

Definition 6.3. The pair $V_n = (X, \overline{\alpha})$ is called an *n-dimensional differentiable manifold of class* C^k.

A difficult theorem of Whitney proves that any C^k manifold $V_n, k \geq 1$, becomes an analytic manifold (i.e., the coordinate transformations between charts of an atlas are analytic diffeomorphisms) by discarding a suitable collection of C^k charts belonging to the original maximal atlas. It is even more difficult to show that a C^0 manifold may fail to become a C^1 manifold.

Now we show how to obtain differentiable manifolds.

- Let $U \subset \mathfrak{R}^n$ be an open set, and denote by $(u^1, \ldots, u^n)$ a point of U. Let S be the locus of the points $(x^1, \ldots, x^l) \in \mathfrak{R}^l, n < l$, given by the set of C^1 functions

$$
\begin{aligned}
x^1 &= x^1(u^1, \ldots, u^n), \\
&\cdots\cdots\cdots\cdots\cdots\cdots\cdots\cdots \\
x^l &= x^l(u^1, \ldots, u^n),
\end{aligned} \tag{6.2}
$$

whose Jacobian matrix

$$J = \left(\frac{\partial x^i}{\partial u^\alpha} \right),$$

$i = 1, \ldots, l, \alpha = 1, \ldots, n$, has rank n at any point of U. In other words, S is a regular n-dimensional surface of $\Re^l$ defined by the parametric equations (6.2). In particular, for $l = 3$ and $n = 1, 2$, we obtain regular curves and surfaces of $\Re^3$, respectively. We sketch the proof that all these regular surfaces are n-dimensional differentiable manifolds. First, S becomes a topological space when it is equipped with the topology induced by $\Re^l$. It is well known that the open sets of this topology are obtained by intersecting the open sets of $\Re^l$ with S. Further, suppose that at the point $x_0 \in S$, which is the image of $(u_0^1, \ldots, u_0^n)$, the determinant of the minor

$$\left(\frac{\partial x^i}{\partial u^\alpha} \right),$$

$i = 1, \ldots, n, \alpha = 1, \ldots, n$, does not vanish. Then, the first n equations (6.2) define a homeomorphism φ between a neighborhood U of x_0 and a neighborhood $\varphi(U)$ of $(u_0^1, \ldots, u_0^n)$. We do not prove that all the coordinate transformations among these charts are of class C^k. Some examples of manifolds obtained by this procedure are given in the exercises at the end of this chapter.

- Differentiable manifolds can also be obtained by the implicit representation of C^k n-dimensional surfaces of $\Re^l$. Let S be such a surface implicitly defined by the following system:

$$f_1(x^1, \ldots, x^l) = 0,$$
$$\ldots\ldots\ldots\ldots\ldots\ldots\ldots, \tag{6.3}$$
$$f_m(x^1, \ldots, x^l) = 0,$$

where $m < l$, the functions $f_\alpha, \alpha = 1, \ldots, m$, are of class C^k, and the Jacobian matrix

$$J = \left(\frac{\partial f_\alpha}{\partial x^i} \right), \tag{6.4}$$

$i = 1, \ldots, l$, has rank equal to m. Again, S becomes a topological space with the topology induced by $\Re^l$. Further, let $x_0 = (x_0^1, \ldots, x_0^l)$ be a point of S, and suppose that the determinant of the minor formed with the first m rows and m columns of (6.4) does not vanish at x_0. Then, the m equations (6.3) can be written as

$$x^1 = x^1(x^{m+1}, \ldots, x^l),$$
$$\ldots\ldots\ldots\ldots\ldots\ldots\ldots\ldots$$
$$x^m = x^m(x^{m+1}, \ldots, x^l),$$

in a neighborhood $V \subset \mathfrak{R}^n, n = l - m$, of $(x_0^1, \ldots, x_0^l)$. The preceding equations define a homeomorphism between V and the neighborhood

$$U = \{(x^1(x^{m+1}, \ldots, x^l), \ldots, x^m(x^{m+1}, \ldots, x^l), x^{m+1}, \ldots, x^l),$$
$$|(x^{m+1}, \ldots, x^l) \in V\} \tag{6.5}$$

on S (see exercises at the end of the chapter).

- A manifold can be obtained by the topological product of two manifolds. Let V_n be a C^k n-dimensional manifold, and let W_m be a C^k m-dimensional manifold. First, we equip $V_n \times W_m$ with the product topology. If $\alpha = \{(U_i, \varphi_i)\}_{i \in I}$ is a C^k atlas of V_n and $\beta = \{(V_j, \psi_j)\}_{j \in J}$ is a C^k atlas of W_m, then it is easy to verify that $\{U_i \times V_j, (\varphi_i, \psi_j)\}_{(i, j) \in I \times J}$ is a C^k atlas of $V_n \times W_m$, which becomes a $C^k(n + m)$-dimensional manifold (see exercises at the end of the chapter).
- A manifold can be defined by a collection $(U_i)_{i \in I}$ of an open set of $\mathfrak{R}^n$ and a set of diffeomorphisms among their parts (see exercises at the end of the chapter).

6.2 Differentiable Functions and Curves on Manifolds

Definition 6.4. Let V_n be an n-dimensional differentiable manifold, and denote by α an atlas on V_n. We say that the real-valued function $f : V_n \to \mathfrak{R}$ is a C^k **function on V_n** if the function

$$f \circ \varphi^{-1} : \varphi(U) \to \mathfrak{R} \tag{6.6}$$

is a C^k function $\forall (U, \varphi) \in \alpha$.

Definition 6.5. A C^h**curve** γ on the C^k manifold $V_n, h \leq k$, is a map $\gamma : [a, b] \subseteq \mathfrak{R} \to V_n$ such that, $\forall (U, \varphi) \in \alpha$, where α is a C^k atlas of V_n, the map

$$\varphi \circ \gamma : [a, b] \to \mathfrak{R}^n \tag{6.7}$$

is a C^h map (Fig. 6.3). The real-valued functions

$$x^i(t) = \mathrm{pr}^i \circ \varphi \circ \gamma(t), \quad t \in [a, b],$$

are the **parametric equations** of γ in the chart (U, φ). The curve γ is closed if $\gamma(a) = \gamma(b)$.

Let (x^i) be the coordinates defined by the chart (U, φ). If $x_0 = (x_0^i) \in U$ is a point of U, the ith coordinate curve at x_0 is a curve with the following parametric equations:

Fig. 6.3 Curve on a
manifold

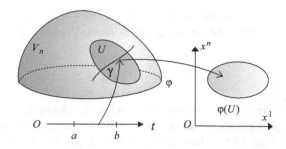

$$x^1 = x_0^1,$$
$$\dots\dots\dots,$$
$$x^i = t, \tag{6.8}$$
$$\dots\dots\dots,$$
$$x^n = x_0^n.$$

Definition 6.6. Let V_n be an n-dimensional manifold with a C^k atlas α, and let W_m be an m-dimensional manifold with a C^k atlas β. The map

$$F : V_n \to W_m$$

is of class C^k if

$$\psi \circ F \circ \varphi^{-1} : \varphi(U) \subseteq \mathfrak{R}^n \to \psi(V) \subseteq \mathfrak{R}^m, \tag{6.9}$$

is C^k, $\forall (U, \varphi) \in \alpha$ and $\forall (V, \psi) \in \beta$ (Fig. 6.4).

If (x^i), $i = 1, \dots, n$, are the coordinates relative to the chart (U, φ) and (y^α), $\alpha = 1, \dots, m$, the coordinates relative to (V, ψ), then map (6.9) is equivalent to a system of $m\, C^k$ functions of n real variables:

$$y^\alpha = y^\alpha(x^1, \dots, x^n), \quad \alpha = 1, \dots, m. \tag{6.10}$$

6.3 Tangent Vector Space

We denote by $\gamma : [a, b] \to V_n$ a C^k curve on a differentiable manifold V_n and by $\mathcal{F}(x)$ the $\mathfrak{R}$-vector space of the C^k functions in a neighborhood of a point $x = \gamma(t)$.

Definition 6.7. The tangent vector to the curve γ at x is the map

$$\mathbf{X}_x : \mathcal{F}(x) \to \mathfrak{R} \tag{6.11}$$

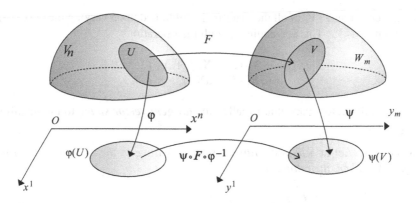

Fig. 6.4 Map between manifolds

such that

$$\mathbf{X}_x f = \left(\frac{\mathrm{d}}{\mathrm{d}t}(f \circ \gamma(t))\right)_t. \tag{6.12}$$

In other words, a tangent vector is defined as an operator that associates to any C^k function about the point x of the curve γ the directional derivative along $\gamma(t)$ at the point x. In the coordinates (x^i) relative to the chart (U, φ) on V_n, we have that

$$f \circ \gamma(t) = f \circ \varphi^{-1} \circ \varphi \circ \gamma(t) = f \circ \varphi^{-1}(x^1(t), \dots, x^n(t)),$$

where $(x^1(t), \dots, x^n(t))$ are the parametric equations of γ in the chart (U, φ). Then, (6.12) gives

$$\mathbf{X}_x f = \left(\frac{\partial}{\partial x^i}(f \circ \varphi^{-1})\right)_{\varphi(x)} \frac{\mathrm{d}x^i}{\mathrm{d}t}. \tag{6.13}$$

To better understand the preceding definition, we consider a curve $\gamma(t) = (x^1(t), x^2(t), x^3(t))$ in the Euclidean three-dimensional space $\mathcal{E}_3$. The directional derivative of a C^1 function $f(x^1, x^2, x^3)$ along $\gamma(t)$ is given by

$$\frac{\mathrm{d}}{\mathrm{d}t}(f(x^1(t), x^2(t), x^3(t))) = (\mathbf{t} \cdot \nabla)_x f \equiv \mathbf{X}_x f, \tag{6.14}$$

where $\mathbf{t} = (\mathrm{d}x^i(t)/\mathrm{d}t)$ is the tangent vector to γ at the point $x = (x^i(t))$. In other words, by (6.14), a derivation operator corresponds to any vector $\mathbf{t}$. It is evident that, if the directional derivatives along three independent functions are given at x, then (6.14) leads to a unique vector $\mathbf{t}$. We note that our definition of tangent vector as derivative operator does not require an environment containing the manifold V_n.

Consider the set $T_x V_n$ of all the maps (6.12) obtained upon varying the curve $\gamma(t)$ at the point $x \in V_n$. This set, equipped with the operations

$$(\mathbf{X}_x + \mathbf{Y}_x)f = \mathbf{X}_x f + \mathbf{Y}_x f,$$
$$(a\mathbf{X}_x)f = a\mathbf{X}_x f, \quad a \in \Re$$

becomes an $\Re$-vector space that is called the **tangent vector space** to the manifold V_n at point x.

Theorem 6.1. *Let (x^i) be a coordinate system relative to the chart (U, φ) of the manifold V_n. Then, the relations*

$$\left(\frac{\partial}{\partial x^i}\right)_{x_0} f = \left(\frac{\partial}{\partial x^i}(f \circ \varphi^{-1})\right)_{\varphi(x_0)} \tag{6.15}$$

define n independent vectors tangent to the coordinate curves.

Proof. Denoting by $\gamma^i(t)$ the ith coordinate curve crossing x_0, that is, the curve with the following parametric equations:

$$x^1 = x_0^1, \ldots, x^i = t, \ldots, x^n = x_0^n,$$

the directional derivative along $\gamma^i(t)$ is

$$\left(\frac{\mathrm{d}}{\mathrm{d}t}(f \circ \gamma^i(t))\right)_{t_0} = \left(\frac{\partial}{\partial x^j}(f \circ \varphi^{-1})\right)_{\varphi(x_0)} \left(\frac{\mathrm{d}x^j}{\mathrm{d}t}\right)_{t_0},$$
$$\left(\frac{\partial}{\partial x^j}(f \circ \varphi^{-1})\right)_{\varphi(x_0)} \delta_i^j = \left(\frac{\partial}{\partial x^i}(f \circ \varphi^{-1})\right)_{\varphi(x_0)},$$

and relations (6.15) define n vectors tangent to the coordinate curves. Their linear independence is proved applying the linear combination

$$\lambda^j \left(\frac{\partial}{\partial x^j}\right)_{x_0} = 0$$

to the coordinate function $x^i = pr^i \circ \varphi$ and recalling (6.15). In fact, we obtain

$$\lambda^j \left(\frac{\partial}{\partial x^j}\right)_{x_0} x^i = \lambda^j \delta_j^i = \lambda^i = 0.$$

$\square$

From this result and (6.13) comes the following theorem.

Theorem 6.2. *The tangent space $T_x V_n$ is an n-dimensional $\Re$-vector space, and the vectors $(\partial/\partial x^i)_x$ form a basis of $T_x V_n$, which is called a **holonomic basis** or **natural***

basis relative to the coordinates (x^i). *Therefore,* $\forall \mathbf{X}_x \in T_x V_n$ *the following result is obtained:*

$$\mathbf{X}_x = X^i \left(\frac{\partial}{\partial x^i} \right)_x, \tag{6.16}$$

where the real numbers X^i *are the components of* $\mathbf{X}_x$ *relative to the natural basis.*

It is fundamental to determine the transformation formulae of the natural bases and the components of a tangent vector for a change $(x^i) \rightarrow (x'^j)$ of the local coordinates. From (6.15) we obtain that

$$\left(\frac{\partial}{\partial x'^i} \right)_x = \frac{\partial x^j}{\partial x'^i} \left(\frac{\partial}{\partial x^j} \right)_x \equiv A_i^j \left(\frac{\partial}{\partial x^j} \right)_x, \tag{6.17}$$

and, consequently,

$$X'^i = \frac{\partial x'^i}{\partial x^j} X^j \equiv (A^{-1})_j^i X^j. \tag{6.18}$$

6.4 Cotangent Vector Space

Definition 6.8. The dual vector space of $T_x V_n$ (Sect. 2.1) is called the *cotangent vector space* $T_x^* V_n$.

If $T_x V_n$ is referred to as the natural basis $(\partial/\partial x^i)_x$ relative to the coordinates (x^i), the dual basis (θ_x^i) is characterized by the conditions

$$\theta_x^i (\mathbf{X}_x) = X_x^i, \tag{6.19}$$

where X_x^i are the components of $\mathbf{X}_x$ relative to the basis $(\partial/\partial x^i)_x$.

Definition 6.9. If $f \in \mathcal{F}(x)$, the *differential* $(\mathbf{d}f)_x$ of f at the point $x \in V_n$ is the linear map

$$(\mathbf{d}f)_x : T_x V_n \rightarrow \mathfrak{R}, \tag{6.20}$$

such that

$$(\mathbf{d}f)_x \mathbf{X}_x = \mathbf{X}_x f, \quad \mathbf{X}_x \in T_x V_n. \tag{6.21}$$

In a natural basis relative to the coordinates (x^i) of the chart (U, φ) of V_n, (6.21) gives

$$(\mathbf{d}f)_x \mathbf{X}_x = a_i X_x^i, \tag{6.22}$$

where

$$a_i = \left(\frac{\partial}{\partial x^i}\right)_x f. \tag{6.23}$$

In particular, for the differentials of the coordinate functions we obtain

$$(\mathbf{d}x^i)_x \mathbf{X}_x = \mathbf{X}_x x^i = X_x^h \left(\frac{\partial}{\partial x^h}\right)_x x^i = X_x^h \delta_h^i = X_x^i. \tag{6.24}$$

Comparing (6.24) and (6.20), we can state the following theorem.

Theorem 6.3. *The differentials* $(\mathbf{d}x^i)_x$ *of the coordinate functions form the dual basis of the cotangent space* $T_x^* V_n$. *Consequently, any covector* $\boldsymbol{\omega}_x \in T_x^* V_n$ *can be written as*

$$\boldsymbol{\omega}_x = \omega_i (\mathbf{d}x^i)_x, \tag{6.25}$$

where

$$\omega_i = \boldsymbol{\omega}_x \left(\left(\frac{\partial}{\partial x^i}\right)_x\right). \tag{6.26}$$

Owing to the results of Sect. 2.2, we can state that under the coordinate change $(x^i) \rightarrow (x'^i)$, the following transformation formulae of the dual bases and the components of a covector hold [see (6.17)]:

$$(\mathbf{d}x'^i)_x = \frac{\partial x'^i}{\partial x^j}(\mathbf{d}x^j)_x = (A^{-1})_j^i (\mathbf{d}x^j)_x, \tag{6.27}$$

$$\omega_i' = \frac{\partial x^j}{\partial x'^i}\omega_j = A_i^j \omega_j. \tag{6.28}$$

Starting from $T_x V_n$ and $T_x^* V_n$ it is possible to build the whole tensor algebra $(T_s^r)_x V_n$ as well as the exterior algebra $(\wedge_s)_x V_n$ at any point $x \in V_n$ (Chap. 2). In particular, the transformation formulae under a change of local coordinates $(x^i) \rightarrow (x'^i)$ of the components $T_{j_1 \cdots j_s}^{i_1 \cdots i_r}$ of any (r, s)-tensor belonging to $(T_s^r)_x V_n$,

$$\mathbf{T} = T_{j_1 \cdots j_s}^{i_1 \cdots i_r} \frac{\partial}{\partial x^{i_1}} \otimes \cdots \otimes \frac{\partial}{\partial x^{i_r}} \otimes \mathbf{d}x^{j_1} \otimes \cdots \otimes \mathbf{d}x^{j_s}, \tag{6.29}$$

are

$$T_{j_1 \cdots j_s}^{'i_1 \cdots i_r} = \frac{\partial x'^{i_1}}{\partial x^{h_1}} \cdots \frac{\partial x'^{i_r}}{\partial x^{h_r}} \frac{\partial x^{k_1}}{\partial x'^{j_1}} \cdots \frac{\partial x^{k_s}}{\partial x'^{j_s}} T_{k_1 \cdots k_s}^{h_1 \cdots h_r}. \tag{6.30}$$

The preceding definitions can be extended to the whole manifold. A C^k **vector field** is a map

$$\mathbf{X} : x \in V_n \to \mathbf{X}_x \in T_x V_n. \tag{6.31}$$

In local coordinates (x^i), map (6.31) assumes the form

$$\mathbf{X} = X^i(x^1 \ldots x^n)\frac{\partial}{\partial x^i}, \tag{6.32}$$

which differs from (6.16) since the components X^i are C^k functions of the coordinates.

Similarly, a C^k **tensor field** is a map

$$\mathbf{T} : x \in V_n \to \mathbf{T}_x \in (T_x)^r_s V_n. \tag{6.33}$$

In local coordinates (x^i), map (6.33) assumes the form

$$\mathbf{T} = T^{i_1\cdots i_r}_{j_1\cdots j_s}(x^1,\ldots,x^n)\frac{\partial}{\partial x^{i_1}}\otimes\cdots\frac{\partial}{\partial x^{i_r}}\otimes \mathbf{d}x^{j_1}\otimes\cdots\otimes\mathbf{d}x^{j_s}, \tag{6.34}$$

where the components $T^{i_1\cdots i_r}_{j_1\cdots j_s}(x^1,\ldots,x^n)$ are C^k functions of the coordinates. In particular, an s-**form** is a map

$$\boldsymbol{\Omega} : x \in V_n \to \boldsymbol{\Omega}_x \in (\wedge_x)_s V_n, \tag{6.35}$$

which locally has the coordinate form

$$\boldsymbol{\Omega} = \Omega_{(j_1\cdots j_s)}(x^1,\ldots,x^n)\mathbf{d}x^{j_1}\wedge\cdots\wedge\mathbf{d}x^{j_s}. \tag{6.36}$$

We conclude this section by introducing the Lie algebra of vector fields on a manifold V_n.

Let V_n be a C^∞ n-dimensional manifold and denote by $\mathcal{F}^\infty V_n$ the $\mathfrak{R}$-vector space of the C^∞ functions on V_n and by $\chi^\infty V_n$ the $\mathfrak{R}$-vector space of the C^∞ vector fields on V_n. Then, to any C^∞ vector field

$$\mathbf{X} : x \in V_n \to \mathbf{X}_x \in T_x V_n$$

we can associate the linear map

$$\mathbf{X} : f \in \mathcal{F}^\infty V_n \to \mathbf{X}f \in \mathcal{F}^\infty V_n \tag{6.37}$$

such that

$$(\mathbf{X}f)(x) = \mathbf{X}_x f. \tag{6.38}$$

It can be easily proved that map (6.37) verifies the following derivation property:

$$\mathbf{X}(fg) = g\mathbf{X}f + f\mathbf{X}g, \quad \forall f, g \in \mathcal{F}^\infty V_n. \tag{6.39}$$

Definition 6.10. Let $\mathbf{X}$, $\mathbf{Y} \in \chi^\infty V_n$ be two C^∞ vector fields. The **bracket** of $\mathbf{X}$ and $\mathbf{Y}$ is the C^∞ vector field $[\mathbf{X}, \mathbf{Y}]$ such that

$$[\mathbf{X}, \mathbf{Y}] f = (\mathbf{XY} - \mathbf{YX}) f, \quad \forall f \in \mathcal{F}^\infty V_n. \tag{6.40}$$

From (6.38) and (6.32) we obtain the following coordinate form of (6.40):

$$[\mathbf{X}, \mathbf{Y}] = \left(X^j \frac{\partial Y^i}{\partial x^j} - Y^j \frac{\partial X^i}{\partial x^j} \right) \frac{\partial}{\partial x^i}. \tag{6.41}$$

It is not difficult to prove the following theorem.

Theorem 6.4. *The bracket operation verifies the following properties:*

$$[\mathbf{X}, \mathbf{Y}] = -[\mathbf{Y}, \mathbf{X}],$$
$$a[\mathbf{X}, \mathbf{Y}] = [a\mathbf{X}, \mathbf{Y}] = [\mathbf{X}, a\mathbf{Y}],$$
$$[\mathbf{X}, \mathbf{Y} + \mathbf{Z}] = [\mathbf{X}, \mathbf{Y}] + [\mathbf{X}, \mathbf{Z}],$$
$$[\mathbf{X}, [\mathbf{Y}, \mathbf{Z}]] + [\mathbf{Y}, [\mathbf{Z}, \mathbf{X}]] + [\mathbf{Z}, [\mathbf{X}, \mathbf{Y}]] = 0, \tag{6.42}$$

$\forall a \in \Re$ *and* $\forall \mathbf{X}, \mathbf{Y}, \mathbf{Z} \in \chi^\infty V_n$.

This theorem proves that $\chi^\infty V_n$, equipped with the addition of vector fields, the product of a real number by a vector field, and the bracket operation, is a Lie algebra.

6.5 Differential and Codifferential of a Map

Definition 6.11. Let V_n and W_m be two C^k manifolds with dimensions n and m, respectively, and let $\gamma(t)$ be an arbitrary curve on V_n containing the point $x \in V_n$ (Fig. 6.5). The **differential** at the point $x \in V_n$ of the C^k map

$$F : V_n \to W_m$$

is the linear map

$$F_{*x} : \mathbf{X}_x \in T_x V_n \to \mathbf{Y}_{F(x)} \in T_{F(x)} W_m \tag{6.43}$$

such that the image of the tangent vector $\mathbf{X}_x$ at x to the curve $\gamma(t)$ is the tangent vector to the curve $F(\gamma(t))$ at the point $F(x)$. Formally,

$$\mathbf{Y}_{F(x)} g = \frac{\mathrm{d}}{\mathrm{d}t} g \circ F \circ \gamma(t), \quad \forall g \in \mathcal{F}(F(x)). \tag{6.44}$$

To find the coordinate representation of (6.44), we introduce a chart (U, φ) with coordinates (x^i), $i = 1, \dots, n$, in a neighborhood U of $x \in V_n$ and a chart (V, ψ),

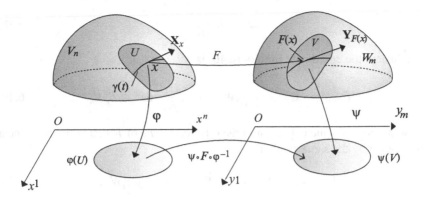

Fig. 6.5 Differential of a map F

with coordinates (y^α), $\alpha = 1, \ldots, m$, in neighborhood V of $F(x)$. We denote by $y^\alpha = y^\alpha(x^1, \ldots, x^n)$ the coordinate form of the map F and by $x^i(t)$ the parametric equations of the curve $\gamma(t)$ in the coordinates (x^i). Then, (6.44) can be written as follows:

$$Y^\alpha(y^\beta(x^i))\frac{\partial}{\partial y^\alpha}g = \frac{dx^i}{dt}\frac{\partial y^\alpha}{\partial x^i}\frac{\partial}{\partial y^\alpha}g \equiv X^i\frac{\partial y^\alpha}{\partial x^i}\frac{\partial}{\partial y^\alpha}g, \quad \forall g \in \mathcal{F}(F(x)).$$

In conclusion, the coordinate form of (6.43) is

$$F_{*x} : \mathbf{X}_x = X^i\left(\frac{\partial}{\partial x^i}\right)_x \in T_x V_n \to \frac{\partial y^\alpha}{\partial x^i}X^i\left(\frac{\partial}{\partial y^\alpha}\right)_{F(x)} \in T_{F(x)} W_m. \qquad (6.45)$$

Starting from linear map (6.43), we can give the following definition.

Definition 6.12. The linear map

$$F^*_{F(x)} : \sigma_{F(x)} \in T^*_{F(x)} W_m \to \omega_x \in T^*_x V_n, \qquad (6.46)$$

between the dual spaces $T^*_{F(x)} W_m$ and $T^*_x V_n$, defined by the condition

$$\omega_x(\mathbf{X}_x) = \sigma_{F(x)}(F_{*x}\mathbf{X}_x), \quad \forall \mathbf{X}_x \in T_x V_n, \qquad (6.47)$$

is called *codifferential* of $F : V_n \to W_m$ at $F(x)$.

Adopting the coordinates (x^i) on V_n and (y^α) on W_m, we can write (6.47) as follows:

$$\omega_i X^i = \sigma_\alpha\frac{\partial y^\alpha}{\partial x^i}X^i,$$

and, taking into account the arbitrariness of $\mathbf{X}_x$, we obtain

$$\omega_i = \sigma_\alpha \frac{\partial y^\alpha}{\partial x^i}. \tag{6.48}$$

In conclusion, the coordinate form of (6.46) is

$$F^*_{F(x)} : \sigma_\alpha \mathbf{d}y^\alpha \in T^*_{F(x)} W_m \rightarrow \sigma_\alpha \frac{\partial y^\alpha}{\partial x^i} \mathbf{d}x^i \in T^*_x V_n. \tag{6.49}$$

Now we consider the extension of the differential F_* of F that is the new linear map, denoted by the same symbol,

$$F_{*x} : \mathbf{T}_x \in (T^r_0)_x V_n \rightarrow \hat{\mathbf{T}}_{F(x)} \in (T^r_0)_{F(x)} W_m, \tag{6.50}$$

such that

$$F_{*x}(\mathbf{X}_1 \otimes \ldots \otimes \mathbf{X}_r) = F_{*x}\mathbf{X}_1 \otimes \ldots \otimes F_{*x}\mathbf{X}_r, \tag{6.51}$$

$\forall \mathbf{X}_1, \ldots, \mathbf{X}_r \in T_x V_n$. It can be easily verified that map (6.50) transforms the $(r,0)$-tensor

$$\mathbf{T} = T^{i_1 \cdots i_r} \frac{\partial}{\partial x^{i_1}} \otimes \ldots \otimes \frac{\partial}{\partial x^{i_r}}$$

into the $(r,0)$-tensor

$$\hat{\mathbf{T}} = \frac{\partial y^{\alpha_1}}{\partial x^{i_1}} \cdots \frac{\partial y^{\alpha_r}}{\partial x^{i_r}} T^{i_1 \cdots i_r} \frac{\partial}{\partial y^{\alpha_1}} \otimes \ldots \otimes \frac{\partial}{\partial y^{\alpha_r}}. \tag{6.52}$$

Similarly, we can extend (6.46) by the linear map

$$F^*_{F(x)} : \mathbf{T}_{F(x)} \in (T^0_s)_{F(x)} W_m \rightarrow \hat{\mathbf{T}}_x \in (T^0_s)_x V_n \tag{6.53}$$

such that

$$F^*_{F(x)}(\boldsymbol{\sigma}_1 \otimes \cdots \otimes \boldsymbol{\sigma}_s) = F^*_{F(x)}\boldsymbol{\sigma}_1 \otimes \cdots \otimes F^*_{F(x)}\boldsymbol{\sigma}_s \tag{6.54}$$

$\forall \boldsymbol{\sigma}_1, \ldots, \boldsymbol{\sigma}_s \in T^*_{F(x)}$. It is evident that $F^*_{F(x)}$ maps the $(0,s)$-tensor

$$\mathbf{T} = T_{\alpha_1 \cdots \alpha_s} \mathbf{d}y^{\alpha_1} \otimes \cdots \otimes \mathbf{d}y^{\alpha_s} \in (T^0_s)_{F(x)} W_m$$

into the $(0,s)$-tensor

$$\hat{\mathbf{T}} = \frac{\partial y^{\alpha_1}}{\partial x^{i_1}} \cdots \frac{\partial y^{\alpha_s}}{\partial x^{i_s}} T_{\alpha_1 \cdots \alpha_s} \mathbf{d}x^{i_1} \otimes \cdots \otimes \mathbf{d}x^{i_s} \in (T^0_s)_x V_n. \tag{6.55}$$

Finally, we can define the linear map

$$F^*_x : \boldsymbol{\Omega}_{F(x)} \in (\Lambda_s)_x W_m \rightarrow \hat{\boldsymbol{\Omega}} \in (\Lambda_s)_x V_n \tag{6.56}$$

such that

$$F^*_{F(x)}(\sigma_1 \wedge \cdots \wedge \sigma_s) = F^*_{F(x)}\sigma_1 \wedge \cdots \wedge F^*_{F(x)}\sigma_s \tag{6.57}$$

$\forall \sigma_1, \ldots, \sigma_s \in T^*_{F(x)}$.

Again it is simple to verify that (6.56) maps

$$\mathbf{\Omega} = \Omega_{(\alpha_1 \cdots \alpha_s)}\mathbf{dy}^{\alpha_1} \wedge \cdots \wedge \mathbf{dy}^{\alpha_s} \in (\Lambda_s)_{F(x)}W_m$$

into

$$\hat{\mathbf{\Omega}} = \frac{\partial(y^{\alpha_1}, \ldots, y^{\alpha_s})}{\partial(x^{i_1}, \ldots, x^{i_s})}\Omega_{(\alpha_1 \cdots \alpha_s)}\mathbf{dx}^{i_1} \wedge \cdots \wedge \mathbf{dx}^{i_s} \in (\Lambda_s)_x V_n. \tag{6.58}$$

Remark 6.1. It is important to note that, in general, none of the preceding linear maps can be extended over the entire manifolds V_n or W_m because $F : V_n \to W_m$ can be neither one-to-one nor onto. For instance, the vector field $\mathbf{X} : x \in V_n \to F_{*x}\mathbf{X}_x$ is defined on $F(V_n)$, and it could assume more values at the same point if F were not one-to-one.

If $F : V_n \to W_m$ is a diffeomorphism, then $n = m$, and we can define an isomorphism

$$F_{*x} : (T^r_s)_x V_n \to (T^r_s)_{F(x)}W_n \tag{6.59}$$

such that

$$F_{*x}(\mathbf{X}_1 \otimes \cdots \otimes \mathbf{X}_r \otimes \omega_1 \otimes \cdots \otimes \omega_s) = F_{*x}\mathbf{X}_1 \otimes \cdots \otimes F^{-1}_{*x}\omega_s. \tag{6.60}$$

Once again, it is simple to verify that the linearity of (6.59) implies that (6.60) maps the tensor

$$\mathbf{T} = T^{i_1 \cdots i_r}_{j_1 \cdots j_s}\frac{\partial}{\partial x^{i_1}} \otimes \cdots \otimes \frac{\partial}{\partial x^{i_r}} \otimes \mathbf{dx}^{j_1} \otimes \cdots \otimes \mathbf{dx}^{j_s} \in (T^r_s)_x V_n$$

into the tensor

$$F_{*x}\mathbf{T} = \frac{\partial y^{\alpha_1}}{\partial x^{i_1}} \cdots \frac{\partial y^{\alpha_r}}{\partial x^{i_r}}\frac{\partial x^{j_1}}{\partial y^{\beta_1}} \cdots \frac{\partial x^{j_s}}{\partial y^{\beta_s}}T^{i_1 \cdots i_r}_{j_1 \cdots j_s}\frac{\partial}{\partial y^{\alpha_1}} \otimes$$

$$\cdots \otimes \frac{\partial}{\partial y^{\alpha_r}} \otimes \mathbf{dy}^{\beta_1} \otimes \cdots \otimes \mathbf{dy}^{\beta_s} \in (T^r_s)_{F(x)}W_n. \tag{6.61}$$

In conclusion, if $F : V_n \to W_m$ is a diffeomorphism, then $n = m$, and the tensor and exterior algebras on V_n and W_n at the points $x \in V_n$ and $F(x) \in W_n$ are isomorphic. Further, upon varying $x \in V_n$, F_{*x} maps (r, s)-tensorial fields of V_n onto (r, s)-tensorial fields of W_n.

6.6   Tangent and Cotangent Fiber Bundles

Given the C^k manifold V_n, consider the set

$$TV_n = \{(x, \mathbf{X}_x), x \in V_n, \mathbf{X}_x \in T_x V_n\}. \tag{6.62}$$

The map

$$\pi : (x, \mathbf{X}_x) \in TV_n \rightarrow x \in V_n \tag{6.63}$$

is called a **projection map**, and the counterimage $\pi^{-1}(x) = x \times T_x V_n$ is called a **fiber** on x.

If (U, φ) is a chart on V_n, then any $x \in U$ is determined by its coordinates (x^i). Further, any vector $\mathbf{X}_x \in T_x V_n$, where $x \in U$, is determined by its components X^i relative to the natural basis $(\partial/\partial x^i)_x$. In this way, we have defined a one-to-one correspondence

$$\phi : \mathcal{U} \rightarrow \varphi(U) \times \mathfrak{R}^n,$$

where

$$\mathcal{U} = \{(x, \mathbf{X}_x), x \in U, \mathbf{X}_x \in T_x V_n\} \subseteq TV_n.$$

TV_n becomes a Hausdorff topological space when we equip it with a topology whose open sets have the form $U \times I^n$, with U an open set of V_n and I^n an open set of $\mathfrak{R}^n$. It is not difficult to verify that the map ϕ is a homeomorphism so that TV_n is a topological manifold and $(\mathcal{U} \times \mathfrak{R}^n, \phi)$ a chart of TV_n. Collecting the domains U of the charts of an atlas of V_n, we define an atlas of TV_n, which becomes a $2n$-dimensional topological manifold. We do not prove that, if V_n is a C^k manifold, then TV_n is a C^k manifold.

Definition 6.13. The $2n$-dimensional manifold TV_n is called a **tangent fiber bundle** of V_n, and the preceding coordinates (x^i, X^i) are called the **natural coordinates** of TV_n.

For instance, the tangent fiber bundle TS^1 of a circumference S^1 is the collection of all pairs $(x, \mathbf{X}_x)$, where $x \in S^1$ and $\mathbf{X}_x$ is a tangent vector to S^1 at point x. The fiber at x is the tangent straight line to S^1 at x. It is evident that TS^1 is diffeomorphic to a cylinder $S^1 \times \mathfrak{R}$.

All that has been said about TV_n can be repeated starting from the set

$$T^*V_n = \left\{(x, \omega_x), x \in V_n, \omega_x \in T_x^* V_n\right\}. \tag{6.64}$$

The corresponding C^k $2n$-dimensional manifold is called the **cotangent fiber bundle** of V_n. Natural coordinates in the open set $U \times \mathfrak{R}^n$, where U is an open set of a chart of an atlas of V_n, are given by (x^i, ω_i), with (x^i) the coordinates of a point $x \in U$ and (ω_i) the components of a covector $\omega_x \in T_x^* V_n$ in the dual basis $(\mathbf{d}x^i)$.

Fig. 6.6 Sphere in $\mathcal{E}_3$

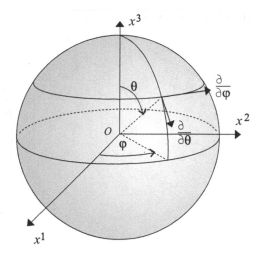

6.7 Riemannian Manifolds

Before giving the definition of a Riemannian manifold, we consider two examples. First, let S^2 be the unit sphere embedded in the three-dimensional Euclidean space $\mathcal{E}_3$ referred to Cartesian coordinates (x^i) (Fig. 6.6). The parametric equations of S^2 are

$$x^1 = \sin\theta\cos\varphi,$$
$$x^2 = \sin\theta\sin\varphi,$$
$$x^3 = \cos\theta.$$

We have already said that any regular surface in Euclidean space is a differentiable manifold. Now we wish to equip a sphere with the metrics induced by the metrics of $\mathcal{E}_3$. This means that the square distance ds^2 between two points (x^i) and $(x^i + dx^i)$ of S^2 is assumed to be equal to the Pythagorean distance in $\mathcal{E}_3$

$$ds^2 = \sum_{i=1}^{3} \left(dx^i\right)^2, \tag{6.65}$$

but expressed in terms of the coordinates $(\varphi,\ \theta)$ on S^2. To this end, we differentiate the parametric equations of the surface and obtain

$$ds^2 = \sin^2\theta\, d\varphi^2 + d\theta^2, \tag{6.66}$$

where $\varphi \in [0, 2\pi]$ and $\theta \in [0, \pi]$.

As a second example we consider the ellipsoid Σ^2 in $\mathcal{E}_3$ with parametric equations

Fig. 6.7 Ellipsoid in $\mathcal{E}_3$

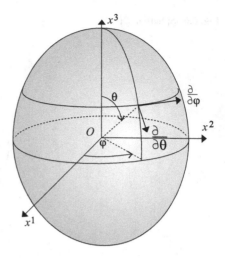

$$x^1 = \sin\theta\cos\varphi,$$
$$x^2 = \sin\theta\sin\varphi,$$
$$x^3 = a\cos\theta,$$

where the axes have lengths 2, 2, and $2a$. Again adopting the foregoing viewpoint, the square distance between two points (x^i) and $(x^i + \mathrm{d}x^i)$ of Σ^2 becomes

$$\mathrm{d}s^2 = \sin^2\theta\mathrm{d}\varphi^2 + (\cos^2\theta + a\sin^2\theta)\mathrm{d}\theta^2. \tag{6.67}$$

The sphere S^2 and the ellipsoid Σ^2 are diffeomorphic so that they are essentially the same manifold. In particular, the variables φ and θ are local coordinates for both surfaces. These two manifolds become different from each other when we introduce a metric structure by the square distance $\mathrm{d}s^2$ (Fig. 6.7).

Definition 6.14. A C^k-differentiable n-dimensional manifold V_n is a **Riemannian manifold** when on V_n a **metric tensor**, that is, a C^k $(0,2)$-symmetric, nondegenerate tensor field **g**, is given.

Owing to the properties of **g**, we can introduce into any tangent space $T_x V_n$ a scalar product of two arbitrary tangent vectors $\mathbf{X}_x$ and $\mathbf{Y}_x$

$$\mathbf{X}_x \cdot \mathbf{Y}_x = \mathbf{g}_x(\mathbf{X}_x, \mathbf{Y}_x), \tag{6.68}$$

and $T_x V_n$ becomes a pseudo-Euclidean vector space. In other words, the whole tensor algebra at any point of a Riemannian manifold becomes a pseudo-Euclidean tensor algebra.

In local coordinates (x^i) relative to a chart (U, φ) of V_n, the tensor **g** assumes the following representation:

$$\mathbf{g} = g_{ij}\mathbf{dx}^i \otimes \mathbf{dx}^j, \quad g_{ij} = g_{ji}, \tag{6.69}$$

where $\det(g_{ij}) \neq 0$, and the scalar product (6.68) becomes

$$\mathbf{X}_x \cdot \mathbf{Y}_x = g_{ij}X^iY^j. \tag{6.70}$$

In view of the results of Chap. 4 regarding the symmetric $(0,2)$-tensors, it is always possible to find, about any point $x \in V_n$, a coordinate system such that at x

$$g_{ij} = \frac{\partial}{\partial x^i} \cdot \frac{\partial}{\partial x^j} = \pm\delta_{ij}. \tag{6.71}$$

The set $\{1, \ldots, 1, -1, \ldots, -1\}$ is called the **signature** of the metric tensor and is independent of the coordinates.

Finally, we define as the **square distance** ds^2 between two points (x^i) and $(x^i + dx^i)$ the quantity

$$ds^2 = g_{ij}dx^idx^j. \tag{6.72}$$

6.8 Geodesics of a Riemannian Manifold

Definition 6.15. Let V_n be a Riemannian manifold and suppose that *the tensor* $\mathbf{g}_x$ *is positive definite at any point* $x \in V_n$. The **length** of a C^1 curve $\gamma(t) : [a,b] \subseteq \mathfrak{R} \to V_n$ is the real number

$$l(\gamma) = \int_\gamma ds. \tag{6.73}$$

Denote by $x^i(t)$, $t \in [a,b]$, the parametric equations of $\gamma(t)$ in an arbitrary system of coordinates (x^i) whose domain contains the curve $\gamma(t)$. Then the length $l(\gamma)$ of $\gamma(t)$ assumes the form

$$l(\gamma) = \int_a^b \sqrt{g_{ij}(x^h)\dot{x}^i\dot{x}^j}\,dt. \tag{6.74}$$

It is evident that $l(\gamma)$ depends neither on the choice of the coordinates nor on the parameterization of the curve.

Let $(x_a^i) = (x^i(a))$ and $(x_b^i) = (x^i(b))$ be the initial and final points of $\gamma(t)$, and consider the one-parameter family Γ of curves

$$f^i(s,t) : (-\epsilon, \epsilon) \times [a,b] \to V_n, \tag{6.75}$$

- Including $\gamma(t)$ for $s = 0$:

$$f^i(0,t) = x^i(t); \tag{6.76}$$

- In addition, they are such that any curve starts from (x_a^i) and ends at (x_b^i), i.e.,

$$f^i(s, a) = x_a^i, \quad f^i(s, b) = x_b^i, \quad \forall s \in (-\epsilon, \epsilon). \tag{6.77}$$

It is evident that the length of any curve of Γ is given by the integral

$$I(s) = \int_a^b \sqrt{g_{ij}(f^h(s, t)) \dot{f}^i \dot{f}^j} \, dt. \tag{6.78}$$

Definition 6.16. The curve $\gamma(t)$ between the points (x_a^i) and (x_b^i) is a **geodesic** of the metric **g** if the function $I(s)$ is stationary for $s=0$ *for any family* Γ of curves satisfying (6.76) and (6.77).

To determine the parametric equations of the geodesic between the points (x_a^i) and (x_b^i), we start by analyzing the stationarity condition

$$\frac{dI}{ds}(0) = 0. \tag{6.79}$$

Introducing the position

$$L(f^h, \dot{f}^h) = \sqrt{g_{ij} \dot{f}^i \dot{f}^j}, \tag{6.80}$$

and taking into account (6.78), we obtain that

$$\frac{dI}{ds}(s) = \int_a^b \left(\frac{\partial L}{\partial f^h} \frac{\partial f^h}{\partial s} + \frac{\partial L}{\partial \dot{f}^h} \frac{\partial \dot{f}^h}{\partial s} \right) dt.$$

Since

$$\frac{\partial L}{\partial \dot{f}^h} \frac{\partial \dot{f}^h}{\partial s} = \frac{\partial}{\partial t} \left(\frac{\partial L}{\partial \dot{f}^h} \frac{\partial f^h}{\partial s} \right) - \frac{\partial}{\partial t} \left(\frac{\partial L}{\partial \dot{f}^h} \right) \frac{\partial f^h}{\partial s},$$

we have that

$$\frac{dI}{ds}(s) = \int_a^b \left(\frac{\partial L}{\partial f^h} - \frac{\partial}{\partial t} \frac{\partial L}{\partial \dot{f}^h} \right) \frac{\partial f^h}{\partial s}(s, t) dt + \left[\frac{\partial L}{\partial \dot{f}^h} \frac{\partial f^h}{\partial s} \right]_a^b. \tag{6.81}$$

From this formula, when we recall (6.76) and (6.77), we deduce that the condition $dI/ds(0) = 0$ is equivalent to requiring that the equation

$$\int_a^b \left(\frac{\partial L}{\partial x^h}(x^j, \dot{x}^j) - \frac{d}{dt} \frac{\partial L}{\partial \dot{x}^h}(x^j, \dot{x}^j) \right) \frac{\partial f^h}{\partial s}(0, t) dt = 0 \tag{6.82}$$

be satisfied for any choice of the functions f^h and then of $\partial f^h/\partial s(0, t), h =$
$1, \ldots, n$. It is possible to prove that this happens if and only if the parametric
equations $x^i(t)$ of the curve $\gamma(t)$ satisfy the **Euler–Lagrange equations**

$$\frac{\partial L}{\partial x^h}(x^j, \dot{x}^j) - \frac{d}{dt}\frac{\partial L}{\partial \dot{x}^h}(x^j, \dot{x}^j) = 0, \quad h = 1, \ldots, n, \tag{6.83}$$

and the boundary conditions

$$x^h(a) = x^h_a, \quad x^h(b) = x^h_b, \quad h = 1, \ldots, n. \tag{6.84}$$

Remark 6.2. In the Cauchy problem relative to (6.83), we must assign the initial
conditions $x^h(0) = x^h_a$ and $\dot{x}^h(0) = X^h$, $h=1,\ldots,n$. It is well known that, under
general hypotheses on the function $L(x^i, \dot{x}^i)$, there is only one solution satisfying
the Euler–Lagrange equations and the initial data, that is, there is only one geodesic
starting from the point (x^i_a) and having at this point the tangent vector (X^i). In contrast,
there is no general theorem about the boundary problem (6.83), (6.84), so perhaps
there is one and only one geodesic between the points (x^h_a) and (x^h_b), no geodesic or
infinite geodesics. In this last case we say that (x^h_a) and (x^h_b) are *focal points*.

Remark 6.3. A curve $\gamma(t)$ has been defined as a map $\gamma : t \in [a, b] \to V_n$.
Consequently, a change in the parameter into the parametric equations $x^h(t)$ leads
to a different curve, although the locus of points is the same. On the other hand, the
presence of the square root under integral (6.74) implies that the value of the length
$l(\gamma)$ is not modified by a change in the parameter t; in other words, $l(\gamma)$ has the same
value for all curves determining the same set of points of V_n. This remark implies
that (6.83) *cannot be independent*. To show that only $n-1$ of these equations are
independent, it is sufficient to note that from the identity

$$L = \sum_{i=1}^{n} \dot{x}^i \frac{\partial L}{\partial \dot{x}^i} \tag{6.85}$$

it follows that

$$\frac{\partial L}{\partial x^h} = \sum_{i=1}^{n} \dot{x}^i \frac{\partial^2 L}{\partial \dot{x}^i \partial x^h}, \tag{6.86}$$

$$\frac{\partial L}{\partial \dot{x}^h} = \frac{\partial L}{\partial \dot{x}^h} + \sum_{i=1}^{n} \dot{x}^i \frac{\partial^2 L}{\partial \dot{x}^i \partial \dot{x}^h}, \tag{6.87}$$

and then

$$\sum_{i=1}^{n} \dot{x}^i \frac{\partial^2 L}{\partial \dot{x}^i \partial \dot{x}^h} = 0. \tag{6.88}$$

From the foregoing conditions there follows the identity

$$\sum_{i=1}^{n} \left(\frac{\partial L}{\partial x^i} - \frac{d}{dt} \frac{\partial L}{\partial \dot{x}^i} \right) \dot{x}^i$$

$$= \sum_{i=1}^{n} \dot{x}^i \frac{\partial L}{\partial x^i} - \sum_{i,h=1}^{n} \dot{x}^i \frac{\partial^2 L}{\partial \dot{x}^i \partial x^h} \dot{x}^h - \sum_{i,h=1}^{n} \dot{x}^i \frac{\partial^2 L}{\partial \dot{x}^i \partial \dot{x}^h} \ddot{x}^h,$$

which, in view of (6.86) and (6.87), becomes

$$\sum_{i=1}^{n} \left(\frac{\partial L}{\partial x^i} - \frac{d}{dt} \frac{\partial L}{\partial \dot{x}^i} \right) \dot{x}^i = 0. \tag{6.89}$$

This relation shows that if the parametric equations $x^h(t)$ satisfy the first $n-1$ Euler–Lagrange equations, then they satisfy the last one.

We use the arbitrariness of the parameter t to obtain a new form of the equations of a geodesic. Setting

$$L = \sqrt{\varphi}, \tag{6.90}$$

we can give (6.83) the following form:

$$\frac{1}{2\sqrt{\varphi}} \frac{\partial \varphi}{\partial x^h} - \frac{d}{dt} \left(\frac{1}{2\sqrt{\varphi}} \frac{\partial \varphi}{\partial \dot{x}^h} \right) = 0, \quad h = 1, \dots, n. \tag{6.91}$$

If we choose a parameter s for which the tangent vector $(\dot{x}^h)$ to the geodesic has unit length, i.e., a parameter s such that

$$\varphi = g_{ij} \dot{x}^i \dot{x}^j = 1, \quad \dot{x}^h = \frac{dx^h}{ds}, \tag{6.92}$$

then the Euler–Lagrange equations assume the form

$$\frac{1}{2} \frac{\partial}{\partial x^h} (g_{ij} \dot{x}^i \dot{x}^j) - \frac{d}{ds} (g_{hj} \dot{x}^j) = 0, \quad h = 1, \dots, n. \tag{6.93}$$

6.9 Exercises

1. Find an atlas for the sphere S^2 in three-dimensional Euclidean space starting from the parametric equations

$$x^1 = \sin\theta\cos\varphi,$$
$$x^2 = \sin\theta\sin\varphi,$$
$$x^3 = \cos\theta,$$

where $0 \le \varphi \le 2\pi$ is the longitude and $0 \le \theta \le \pi$ the colatitude.

2. Find an atlas for the cylinder in three-dimensional Euclidean space starting from its parametric equations

$$x^1 = \cos\varphi,$$
$$x^2 = \sin\varphi,$$
$$x^3 = z,$$

where $0 \le \varphi \le 2\pi$ is the longitude and $z \in \Re$ is the ordinate along the axis of the cylinder.

3. Find an atlas for the torus T^2 in three-dimensional Euclidean space starting from its parametric equations

$$x^1 = \sin\theta\cos\varphi,$$
$$x^2 = \sin\theta\sin\varphi,$$
$$x^3 = \cos\theta,$$

where $0 \le \varphi \le 2\pi$ and $0 \le \theta \le 2\pi$ are shown in Fig. 6.8.

Fig. 6.8 Torus T^2 in $\mathcal{E}_3$

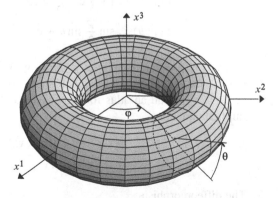

Fig. 6.9 Moebius strip

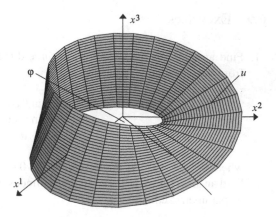

4. Find an atlas for the Moebius strip with parametric equations

$$x^1 = \left(1 + u\cos\frac{\varphi}{2}\right)\cos\varphi,$$

$$x^2 = \left(1 + u\cos\frac{\varphi}{2}\right)\sin\varphi,$$

$$x^3 = u\sin\frac{\varphi}{2},$$

where $0 \le \varphi \le 2\pi$ and $u \in [-1, 1]$ are shown in Fig. 6.9.

5. Find an atlas for a Klein bottle with parametric equations

$$x(\varphi, \theta) = \left(1 + \cos\frac{\varphi}{2}\sin\theta - \sin\frac{\varphi}{2}\sin 2\theta\right)\cos\varphi,$$

$$y(\varphi, \theta) = \left(1 + \cos\frac{\varphi}{2}\sin\theta - \sin\frac{\varphi}{2}\sin 2\theta\right)\sin\varphi,$$

$$z(\varphi, \theta) = \sin\frac{\varphi}{2}\sin\theta + \cos\frac{\varphi}{2}\sin 2\theta,$$

where $0 \le \varphi \le 2\pi$ and $0 \le \theta \le 2\pi$ (Fig. 6.10).

6. Let $\mathcal{E}_3$ be the three-dimensional space referred to Cartesian coordinates (x^1, x^2, x^3). Find an atlas for the sphere and the cylinder starting from their implicit equations

$$\left(x^1\right)^2 + \left(x^2\right)^2 + \left(x^3\right)^2 - 1 = 0,$$

$$\left(x^1\right)^2 + \left(x^2\right)^2 - 1 = 0.$$

7. The diffeomorphism

$$\varphi : (x^1, x^2) \in (1 - \Delta, 1) \times (0, 1) \rightarrow (x'^1 = x^1 + \Delta, x'^2 = x^2) \in (1 + \Delta, 1) \times (0, 1)$$

Fig. 6.10 Klein's bottle

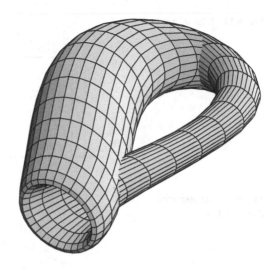

overlaps the open subset $(1 - \Delta, 1) \times (0, 1)$ of the square $U = (0, 1) \times (0, 1)$ and the open subset $(1 + \Delta, 1) \times (0, 1)$ of the square $V = (1, 2) \times (0, 1)$ in the plane Ox^1x^2. Further, the diffeomorphism

$$\psi : (x^1, x^2) \in (0, \Delta) \times (0, 1) \rightarrow (x'^1 = 2 + x^1 - \Delta, x'^2 = x^2) \in (2 - \Delta, 2) \times (0, 1)$$

overlaps the open subsets $(0, \Delta) \times (0, 1)$ and $(2 - \Delta, 2) \times (0, 1)$ of the same squares (Fig. 6.11).

Show that U, V and ϕ, ψ are an atlas for the cylinder $S^1 \times (0, 1)$. Essentially, a cylinder is obtained from a rectangle by identifying the points of two opposite sides.

8. The diffeomorphism

$$\varphi : (x^1, x^2) \in (1 - \Delta, 1) \times (0, 1) \rightarrow (x'^1 = x^1 + \Delta, x'^2 = x^2) \in (1 + \Delta, 1) \times (0, 1)$$

overlaps the open subset $(1 - \Delta, 1) \times (0, 1)$ of the square $U = (0, 1) \times (0, 1)$ and the open subset $(1 + \Delta, 1) \times (0, 1)$ of the square $V = (1, 2) \times (0, 1)$ in the plane Ox^1x^2. Further, the diffeomorphism

$$\psi : (x^1, x^2) \in (0, \Delta) \times (0, 1) \rightarrow (x'^1 = 2 + x^1 - \Delta, x'^2 = 1 - x^2) \in (2 - \Delta, 2) \times (0, 1)$$

overlaps the open subsets $(0, \Delta) \times (0, 1)$ and $(2 - \Delta, 2) \times (0, 1)$ of the same squares (Fig. 6.11). Show that U, V and ϕ, ψ are an atlas for the Moebius strip. Essentially, this surface is obtained by identifying the points of two opposite sides after a torsion of $180°$.

Fig. 6.11 Atlas of a cylinder

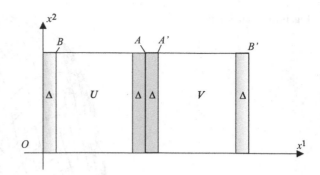

Fig. 6.12 Stereographic
projection

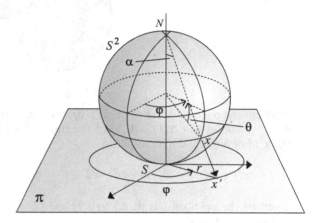

9. Determine an atlas with four coordinate domains and the relative coordinate
   transformations to obtain a torus. Show that the torus is obtained from a rectangle
   by identifying the points of the two pairs of opposite sides.

10. This exercise shows how we can obtain a geographic chart using a map between
    two manifolds. Let $S^2 - N$ be the unit sphere minus the North Pole. The dif-
    feomorphism F, which is called a *stereographic projection*, between $S^2 - N$
    and the tangent plane π at the South Pole S, is shown in Fig. 6.12. F maps the
    point $x \in S^2 - N$ into the point $x' \in \pi$ corresponding to the intersection of
    the straight line Nx with π. If we adopt spherical coordinates (φ, θ) on S^2 and
    polar coordinates (φ, r) with its center at the South Pole on π, the map F has the
    following coordinate form (Fig. 6.13):

$$\varphi = \varphi, r = 2 \tan\left(\frac{\pi}{4} - \frac{\theta}{2}\right).$$

Determine how F transforms the coordinate curves, the tangent vectors, and the
covectors on S^2. Analyze how it transforms the distances and the angles between
tangent vectors.

Fig. 6.13 Section of
stereographic projection

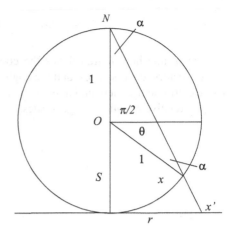

Fig. 6.14 Central projection

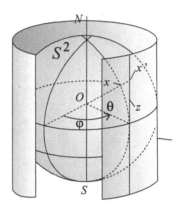

11. As another example of a geographic chart, we consider the diffeomorphism
    between $F : S^2 - \{S, N\} \rightarrow C$, where C is the cylinder in Fig. 6.14 and F
    maps $x \in S^2 - \{S, N\}$ into the point x' of C belonging to the straight line Ox.
    Adopting spherical coordinates (φ, θ) on S^2 and cylindrical coordinates (φ, z)
    on C, the map F assumes the following coordinate form:

$$\varphi = \varphi, \quad z = \tan \theta.$$

    Determine how F transforms the coordinate curves, the tangent vectors, the
    covectors, the distances, and the angles between tangent vectors.

12. We conclude the examples of geographic charts with *Mercator's projection*
    (1569). This representation is obtained considering a one-to-one map between
    $S^2 - \{S, N\}$ and a cylinder C that, adopting the same coordinates of Fig. 6.14,
    has the form

$$\varphi = \varphi, \quad z = \ln \tan \left(\frac{\pi}{4} + \frac{\theta}{2} \right).$$

Determine how F transforms the coordinate curves, the tangent vectors, the covectors, the distances, and the angles between tangent vectors.

13. Determine the Riemannian metrics on a sphere, a cylinder, and a torus, and find the relative equations of geodesics.

Chapter 7
One-Parameter Groups
of Diffeomorphisms

Abstract In this chapter the one-parameter transformation groups and Lie's derivative are defined.

7.1 Global and Local One-Parameter Groups

Definition 7.1. A *one-parameter global group of diffeomorphisms* G on a manifold V_n of class C^k, $k > 0$, is a C^k map

$$\phi : (t, x) \in \Re \times V_n \to \phi_t(x) \in V_n \tag{7.1}$$

such that

1. $\forall t \in \Re$ the map $\phi_t : x \in V_n \to \phi_t(x) \in V_n$ is a C^k diffeomorphism of V_n;
2. $\forall t, s \in \Re, \forall x \in V_n, \phi_{t+s}(x) = \phi_t \circ \phi_s(x)$.

In particular, from the second property we have $\phi_t(x) = \phi_{0+t}(x) = \phi_0 \circ \phi_t(x)$, so that

$$\phi_0(x) = x, \quad \forall x \in V_n. \tag{7.2}$$

Similarly, from $x = \phi_0(x) = \phi_t(x) \circ \phi_{-t}(x)$ we obtain that

$$\phi_{-t}(x) = (\phi_t)^{-1}(x), \quad \forall x \in V_n. \tag{7.3}$$

Definition 7.2. $\forall x_0 \in V_n$, the C^k curve $\phi_t(x_0) : \Re \to V_n$ on V_n is called the *orbit* of G determined by x_0. In view of (7.2), the orbit contains x_0.

Theorem 7.1. *Any point $x_0 \in V_n$ belongs to one and only one orbit, i.e., the orbits determine a partition of V_n.*

© Springer International Publishing AG, part of Springer Nature 2018
A. Romano and A. Marasco, *Classical Mechanics with Mathematica®*,
Modeling and Simulation in Science, Engineering and Technology,
https://doi.org/10.1007/978-3-319-77595-1_7

Proof. First, we show that if $x_2 \in \phi_t(x_1)$, then the orbit determined by x_2 coincides with $\phi_t(x_1)$ up to a change of the parameter. In other words, an orbit is determined by any of its points. In fact, if $x_2 \in \phi_t(x_1)$, then there exists a value t_2 of t such that $x_2 = \phi_{t_2}(x_1)$. In view of property (7.3), $x_1 = \phi_{-t_2}(x_2)$. Consequently, the orbit $\phi_t(x_1)$ can also be written in the form $\phi_{t-t_2}(x_2)$, which, up to the change $t \to t - t_2$ of the parameter, gives the orbit determined by x_2. Now it remains to prove that if x_2 does not belong to $\phi_t(x_1)$, then the orbits $\phi_t(x_1)$ and $\phi_t(x_2)$ do not intersect each other. In fact, if there is a point x_0 belonging to both orbits, then there exist two values $t_1, t_2 \in \Re$ such that

$$x_0 = \phi_{t_1}(x_1), \quad x_0 = \phi_{t_2}(x_2).$$

In view of (7.3), $x_2 = \phi_{-t_2}(x_0)$, and then

$$x_2 = \phi_{-t_2}(x_0) = \phi_{-t_2}(\phi_{t_1}(x_1)) = \phi_{t_1-t_2}(x_1).$$

In conclusion, x_2 belongs to the orbit determined by x_1, against the hypothesis.   □

Theorem 7.2. *Let $\mathbf{X}_x$ be the tangent vector to the orbit $\phi_t(x)$ at point x. The map $x \in V_n \to \mathbf{X}_x \in T_x(V_n)$ defines a C^{k-1} vector field $\mathbf{X}$ over V_n.*

Proof. Let (U, x^i) be a chart of V_n and denote by $\phi^i(t, x^1, \dots, x^n)$ the representation of the group in these coordinates. Then, $\forall x \in U$,

$$\mathbf{X}_x = \left(\frac{\partial \phi^i}{\partial t} \right)_{t=0} \frac{\partial}{\partial x^i},$$

and the theorem is proved.   □

Definition 7.3. The vector field $\mathbf{X}$ is called the *infinitesimal generator* of the group of diffeomorphisms.

Example 7.1. Let $\mathcal{E}_2$ be the Euclidean plane. It is easy to verify that the differential map

$$\phi : \Re \times \mathcal{E}_2 \to \mathcal{E}_2$$

such that

$$x' \equiv \phi_t(x) = x + t\mathbf{u},$$

where $\mathbf{u}$ is a constant vector in $\mathcal{E}_2$, is a one-parameter global group of diffeomorphisms of $\mathcal{E}_2$, called the *group of translations*. Further, the orbit determined by the point x_0 is $x' = x_0 + t\mathbf{u}$. Finally, the constant vector field $\mathbf{u} = \partial \phi_t(x)/\partial t$ is the infinitesimal generator of the one-parameter group.

Example 7.2. Let $\mathcal{E}_3$ be the Euclidean three-dimensional space referred to the cylindrical coordinates r, α, and z. Then, the family of diffeomorphisms

$$
\begin{aligned}
r' &= r, \\
\alpha' &= \alpha + t, \\
z' &= z
\end{aligned}
$$

is a one-parameter group of diffeomorphisms, called the *group of rotations* about the Oz-axis of the cylindrical coordinates. The orbits of the group are circumferences having the center on the Oz-axis, and the infinitesimal generator is the vector field

$$
\mathbf{X} = \frac{\partial}{\partial \alpha}.
$$

Definition 7.4. Let U be an open region of V_n. A **one-parameter local group of diffeomorphisms** is a differential map

$$
\phi : (-\epsilon, \epsilon) \times U \to V_n, \tag{7.4}
$$

where $\epsilon > 0$, such that

1. $\forall t \in (-\epsilon, \epsilon)$, $\phi_t : x \in U \to \phi_t(x) \in \phi_t(U)$ is a diffeomorphism;
2. If $t, s, t + s \in (-\epsilon, \epsilon)$, then $\phi_{t+s}(x) = \phi_t(\phi_s(x))$.

Theorem 7.3. *Let X be a differentiable vector field on V_n. For any $x \in V_n$, there exist an open region $U \subset V_n$, a real number $\epsilon > 0$, and a one-parameter local group of diffeomorphisms $\phi : (-\epsilon, \epsilon) \times U \to V_n$ whose infinitesimal generator is X.*

Proof. Consider the system of ordinary differential equations (ODEs)

$$
\frac{\mathrm{d}x^i}{\mathrm{d}t} = X^i(x^1, \ldots, x^n), \quad i = 1, \ldots, n.
$$

If the vector field is differentiable, from known theorems of analysis, $\forall x_0 \in V_n$ there exist a neighborhood U of x_0 and an interval $(-\epsilon, \epsilon)$ such that one and only one solution $x^i = \phi(t, x_0)$, $t \in (-\epsilon, \epsilon)$, of the preceding system exists satisfying the initial data $\phi^i(0, x_0) = x_0$, $\forall x_0 \in U$. For the uniqueness theorem, $x^i = \phi(t, x_0)$ is a diffeomorphism $\forall t \in (-\epsilon, \epsilon)$. We omit the proof of the property $\phi(t + s, x_0) = \phi(t, \phi(s, x_0))$. $\square$

Definition 7.5. The vector field $\mathbf{X}$ is said to be **complete** if it is an infinitesimal generator of a global one-parameter group of diffeomorphisms.

Theorem 7.4. *On a compact manifold V_n, every differential vector field is complete.*

Proof. The preceding theorem allows one to state that, $\forall x \in V_n$, a map

$$
\phi : (-\epsilon(x), \epsilon(x)) \times U_x \to V_n
$$

exists verifying conditions 1. and 2. of Definition 7.4. Since V_n is compact, the covering $(U_x)_{x \in V_n}$ admits a finite subcovering $(U_{x_i})_{i=1,...,s}$.

Setting by $\epsilon = \min \{\epsilon(x_1), \ldots, \epsilon(x_s)\}$, the map ϕ is defined on $(-\epsilon, \epsilon) \times V_n$. $\square$

7.2 Lie's Derivative

In Chap. 6, we proved that if the differentiable map $F : V_n \to V_n$ is a diffeomorphism, then its differential $(F_*)_x : T_x V_n \to T_{F(x)} V_n$ is an isomorphism that can be extended to the tensorial algebra and the exterior algebra at the point $x \in V_n$. In this section, we show that this result, which holds for any diffeomorphism ϕ_t of a one-parameter transformation group, makes it possible to introduce a meaningful derivation operator on the manifold V_n.

We denote by $\mathbf{X}$ a differentiable vector field on V_n, by $\mathcal{F}(V_n)$ the vector space of the differentiable functions $f : V_n \to \Re$, and by ϕ_t the one-parameter (local or global) transformation group generated by $\mathbf{X}$.

Definition 7.6. The *Lie derivative* of the function f with respect to the vector field $\mathbf{X}$ is the map

$$L_{\mathbf{X}} : \mathcal{F}(V_n) \to \mathcal{F}(V_n)$$

such that

$$(L_{\mathbf{X}} f)_x = \lim_{t \to 0} \frac{f(\phi_t(x)) - f(x)}{t} = \left(\frac{d}{dt} f(\phi_t(x)) \right)_{t=0}. \tag{7.5}$$

Since $\mathbf{X}$ is tangent to the orbits of ϕ_t, the Lie derivative of f at x is the directional derivative of f along the vector $\mathbf{X}_x$, that is,

$$L_{\mathbf{X}} f = \mathbf{X} f. \tag{7.6}$$

Henceforth, we denote by χ, χ^*, and χ_s^r the $F(V_n)$ modules of C^∞-vector fields, 1-forms, and (r, s)-tensor fields of V_n, respectively.

Definition 7.7. The *Lie derivative* on χ with respect to the vector field $\mathbf{X}$ is the map (Fig. 7.1)

$$L_{\mathbf{X}} : \mathbf{Y} \in \chi \to L_{\mathbf{X}} \mathbf{Y} \in \chi$$

such that, $\forall x \in V_n$,

$$(L_{\mathbf{X}} \mathbf{Y})_x = \lim_{t \to 0} \frac{1}{t} [(\phi_{-t})_* (\phi_t(x)) \mathbf{Y}_{\phi_t(x)} - \mathbf{Y}_x]. \tag{7.7}$$

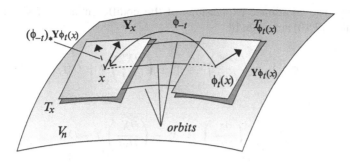

Fig. 7.1 Lie's derivative of a vector field

Definition 7.8. The *Lie derivative* on χ^* with respect to the vector field $\mathbf{X}$ is the map (Fig. 7.2)

$$L_{\mathbf{X}} : \omega \in \chi^* \to L_{\mathbf{X}}\omega \in \chi^*$$

such that, $\forall x \in V_n$,

$$(L_{\mathbf{X}}\omega)_x = \lim_{t \to 0} \frac{1}{t}[(\phi_t)^*(x)\omega_{\phi_t(x)} - \omega_x]. \qquad (7.8)$$

It is evident how Lie's derivative can be extended to the tensor fields of χ_s^r.

To find the coordinate expression of Lie's derivative, we introduce a chart (U, x^i) on V_n and denote by $y^i = \phi_t^i(x^1, \ldots, x^n)$ the coordinate expression of the one-parameter group of the diffeomorphisms $\phi_t(x)$ and by X^i the components of the infinitesimal generator of the group $\mathbf{X}$. Then, the maps $\phi_t(x)$ and $\phi_{-t}(x)$ in a neighborhood of x are given by the expressions

$$y^i = x^i + X^i(x)t + O(t), \qquad (7.9)$$
$$x^i = y^i - X^i(y)t + O(t), \qquad (7.10)$$

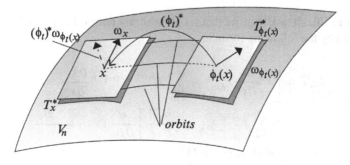

Fig. 7.2 Lie's derivative of a 1-form

respectively. The coordinate expressions of the codifferential of (7.9) and of the differential of (7.10) are given by

$$((\phi_t)^*)^i_j = \left(\frac{\partial y^i}{\partial x^j}\right)_x = \delta^i_j + \left(\frac{\partial X^i}{\partial x^j}\right)_x t + O(t), \tag{7.11}$$

$$((\phi_{-t})_*)^i_j = \left(\frac{\partial x^i}{\partial y^j}\right)_y = \delta^i_j - \left(\frac{\partial X^i}{\partial y^j}\right)_y t + O(t). \tag{7.12}$$

From the preceding relations we have that

$$((\phi_{-t})_* \mathbf{Y}_{\phi_t(x)})^i = \left[\delta^i_j - \left(\frac{\partial X^i}{\partial y^j}\right)_y t\right]\left[Y^j(x) + \left(\frac{\partial Y^j}{\partial x^h}\right)_x X^h(x)t\right] + O(t). \tag{7.13}$$

In view of (7.13), Lie's derivative (7.7) assumes the following coordinate form:

$$L_\mathbf{X}\mathbf{Y} = \left[X^j\frac{\partial Y^i}{\partial x^j} - Y^j\frac{\partial X^i}{\partial x^j}\right]\frac{\partial}{\partial x^i}. \tag{7.14}$$

By comparing (7.14) and (6.41), we derive the important result

$$L_\mathbf{X}\mathbf{Y} = [\mathbf{X}, \mathbf{Y}]. \tag{7.15}$$

Starting from (7.9), we derive the coordinate form of the codifferential $(\phi_t(x))^*$

$$((\phi_t)^*\boldsymbol{\omega}_{\phi_t(x)})_i = \left[\delta^j_i + \left(\frac{\partial X^j}{\partial x^i}\right)_x t\right]\left[\omega_j(x) + \left(\frac{\partial \omega_j}{\partial x^h}\right)_x X^h(x)t\right] + O(t), \tag{7.16}$$

so that (7.8) becomes

$$L_\mathbf{X}\omega = \left[X^j\frac{\partial \omega_i}{\partial x^j} + \omega_j\frac{\partial X^j}{\partial x^i}\right]dx^i. \tag{7.17}$$

It is not a difficult task to prove that for an arbitrary tensor $\mathbf{T} \in \chi^r_s$, Lie's derivative is expressed by the following formula:

$$L_\mathbf{X}\mathbf{T} = \left[X^h\frac{\partial}{\partial x^h}T^{i_1\cdots i_r}_{j_1\cdots j_s} - \sum_{k=1}^r T^{i_1\cdots i_{k-1}hi_{k+1}\cdots i_r}_{j_1\cdots j_s}\frac{\partial X^{i_k}}{\partial x^h}\right.$$
$$\left. + \sum_{k=1}^s T^{i_1\cdots i_r}_{j_1\cdots j_{k-1}hj_{k+1}\cdots j_s}\frac{\partial X^h}{\partial x^{j_h}}\right]\frac{\partial}{\partial x^{i_1}}\otimes\cdots\otimes\frac{\partial}{\partial x^{i_r}} \tag{7.18}$$
$$\otimes dx^{j_1}\otimes\cdots\otimes dx^{j_s}.$$

In particular, for an s-form Ω, the preceding formula gives

$$
L_X\Omega = \sum_{j_1 < \cdots < j_s} \left[X^h \frac{\partial}{\partial x^h} \Omega_{(j_1 \cdots j_s)} + \Omega_{h(j_2 \cdots j_s)} \frac{\partial X^h}{\partial x^{j_1}} \right.
$$
$$
\left. + \cdots + \Omega_{(j_1 \cdots j_{s-1})h} \frac{\partial X^h}{\partial x^{j_s}} \right] dx^{j_1} \wedge \cdots \wedge dx^{j_s}. \tag{7.19}
$$

From the definition of Lie's derivative, it follows that L_X is linear with respect to X, and the derivation property with respect to the tensorial product and the exterior product holds

$$
L_X(\mathbf{T} \otimes \mathbf{S}) = L_X\mathbf{T} \otimes \mathbf{S} + \mathbf{T} \otimes L_X\mathbf{S}, \tag{7.20}
$$

$$
L_X(\mathbf{T} \wedge \mathbf{S}) = L_X\mathbf{T} \wedge \mathbf{S} + \mathbf{T} \wedge L_X\mathbf{S}. \tag{7.21}
$$

Definition 7.9. A tensor field $\mathbf{T} \in \chi_s^r V_n$ is *invariant* under a one-parameter group $\phi_t(x)$ of diffeomorphisms if

$$
\mathbf{T}_{\phi_t(x)} = (\phi_t^* \mathbf{T})_{\phi_t(x)}, \tag{7.22}
$$

where $(\phi_t)_*$ denotes the extension of the differential $(\phi_t)_*$ of the diffeomorphisms of the group to the tensor algebra (Chap. 6).

In particular, a vector field $\mathbf{Y}$ and a 1-form ω are invariant under the group $\phi_t(x)$ if

$$
\mathbf{Y}_{\phi_t(x)} = (\phi_t)_*(x)\mathbf{Y}_x. \tag{7.23}
$$
$$
\omega_{\phi_t(x)} = (\phi_{-t})^*(x)\omega_x. \tag{7.24}
$$

We omit the proof of the following theorem.

Theorem 7.5. *The tensor field* $\mathbf{T} \in \chi_s^r V_n$ *is invariant under the one-parameter group* $\phi_t(x)$ *of diffeomorphisms if and only if*

$$
L_X\mathbf{T} = 0, \tag{7.25}
$$

where X *is the infinitesimal generator of* $\phi_t(x)$.

Definition 7.10. Let (U, x^i) be a chart of V_n, and let

$$
\omega = \omega_{(i_1 \cdots i_s)} dx^{i_1} \wedge \cdots dx^{i_s}
$$

be the coordinate representation of an s-form ω in the chart (U, x^i). The *interior product* of $\omega \in \chi_s V_n$ by the vector field X is the $(s-1)$-form $i_X \omega$, which in the chart (U, x^i) of V_n has the following representation:

$$i_{\mathbf{X}}\omega = X^h \omega_{h(i_1 \cdots i_s)} dx^{i_2} \wedge \cdots dx^{i_s}. \tag{7.26}$$

Starting from (7.26), it is easy to prove the following theorem.

Theorem 7.6. *The interior product has the following properties:*

1. *It is $\Re$-linear with respect to $\mathbf{X}$;*
2. *If $\omega \in \chi_s\, V_n$ and $\sigma \in \chi_p\, V_n$, then*

$$i_{\mathbf{X}}(\omega \wedge \sigma) = i_{\mathbf{X}}\omega \wedge \sigma + (-1)^s \omega \wedge i_{\mathbf{X}}\sigma; \tag{7.27}$$

3. *If $\omega \in \chi^* V_n$, then*

$$i_{\mathbf{X}}\omega = \omega(\mathbf{X}).$$

7.3 Exercises

1. Determine the one-parameter global group acting on $\Re^2$ whose infinitesimal generator in the coordinates x, y is

$$\mathbf{X} = x\frac{\partial}{\partial x} + y\frac{\partial}{\partial y}.$$

Hint: The parametric equations of the group's orbits are solutions of the system of ODEs

$$\frac{dx}{dt} = x, \quad \frac{dy}{dt} = y.$$

The solutions are

$$x = x_0 e^t, \quad y = y_0 e^t,$$

and it is evident that $\phi_t(x_0, y_0) = (x_0 e^t, y_0 e^t)$ is the global group generated by $\mathbf{X}$.

2. Determine the one-parameter *local* group acting on $\Re^2$ whose infinitesimal generator in the coordinates x, y is

$$\mathbf{X} = \frac{\partial}{\partial x} + e^{-y}\frac{\partial}{\partial y}.$$

Hint: The parametric equations of the group's orbits are solutions of the system of ODEs

$$\frac{dx}{dt} = 1, \quad \frac{dy}{dt} = e^{-y}.$$

Consequently, the parametric equations of the orbits are

$$x(t) = x_0 + t, \quad y(t) = \ln(t + e^{y_0}),$$

where $t > -e^{y_0}$. The family of diffeomorphisms $\phi_t(x_0, y_0) = (x_0 + t, \ln(t + e^{y_0}))$ defines a local group generated by $\mathbf{X}$.

3. Determine the one-parameter global group acting on $\Re^2$ whose infinitesimal generator in the coordinates x, y is

$$\mathbf{X} = -y\frac{\partial}{\partial x} + x\frac{\partial}{\partial y}.$$

Hint: The parametric equations of the group's orbits are solutions of the system of ODEs

$$\frac{dx}{dt} = -y, \quad \frac{dy}{dt} = x,$$

i.e., they are given by the functions

$$x(t) = x_0 \cos t - y_0 \sin t, \quad y(t) = x_0 \sin t + y_0 \cos t.$$

This family of diffeomorphisms of $\Re^2$ is the global group generated by $\mathbf{X}$.

4. Using the theory of linear differential equations, verify that the global group of $\Re^2$ generated by the vector field

$$\mathbf{X} = (x - y)\frac{\partial}{\partial x} + (x + y)\frac{\partial}{\partial y}$$

is

$$x(t) = e^t(x_0 \cos t - y_0 \sin t), \quad y(t) = e^t(y_0 \cos t + x_0 \sin t).$$

Verify the result using the built-in function DSolve of Mathematica.

5. Given the one-parameter global group $\phi_t(x)$ of $\Re^2$

$$x' = x + \alpha t, \quad y' = y + \beta t,$$

where x and y are Cartesian coordinates, determine the vector fields that are invariant under the action of the group.

Hint: The infinitesimal generator of the group $\phi_t(x)$ is the vector field $\mathbf{X} = (\alpha, \beta)$. The differential $(\phi_t)_*$ of the diffeomorphisms $\phi_t(x)$ is represented by the unit matrix. Further, in view of Theorem 7.5, we can determine the invariant fields $\mathbf{Y}$ either by (7.23) or by (7.25). The last condition gives

$$\alpha \frac{\partial Y^1}{\partial x} + \beta \frac{\partial Y^1}{\partial y} = \frac{dY^1}{dt} = 0,$$

$$\alpha \frac{\partial Y^2}{\partial x} + \beta \frac{\partial \dot{Y}^2}{\partial y} = \frac{dY^2}{dt} = 0,$$

whereas the former gives

$$Y^1(x + \alpha t, y + \beta t) = Y^1(x, y),$$
$$Y^2(x + \alpha t, y + \beta t) = Y^2(x, y).$$

Both the results show that the components of $\mathbf{Y}$ must be constant along the group's orbits.

6. On the unit sphere S referred to the spherical coordinates (φ, θ), consider the one-parameter group G of diffeomorphisms

$$\varphi' = \varphi + \alpha t, \quad \theta' = \theta + \beta t,$$

where α and β are constant. Determine the orbits of G and the vector fields that are invariant under the action of G.

Chapter 8
Exterior Derivative and Integration

Abstract Two fundamental topics of differential geometry are presented in this chapter in introductory form: exterior derivative and integration of r-forms. The exterior derivative extends to r-forms the elementary definitions of gradient of a function, curl, and divergence of a vector field as well as the meaning of exact and closed 1-forms. The integration of r-forms allows to extend the definitions of surface and volume integrals as well as the Gauss and Stokes theorems.

8.1 Exterior Derivative

We denote by $\wedge_r V_n$ the set of differential r-forms on the manifold V_n. It is evident that $\wedge_r V_n$ is both an $\Re$-vector space and an $\mathcal{F}V_n$ module. In the sequel, we use the notation $\wedge_0 V_n = \mathcal{F}V_n$.

Definition 8.1. The *exterior derivative* is an $\Re$-linear map

$$\mathbf{d} : \wedge_r V_n \to \wedge_{r+1} V_n, \quad (r = 0, 1, \ldots, n) \tag{8.1}$$

such that

1. $\forall f \in \wedge_0 V_n$ the exterior derivative of f is the differential of f;
2. $\forall \, \omega \in \wedge_r V_n, \forall \, \sigma \in \wedge_s V_n$, the map $\mathbf{d}$ is an *antiderivation*, i.e.,

$$\mathbf{d}(\omega \wedge \sigma) = \mathbf{d}\omega \wedge \sigma + (-1)^r \omega \wedge \mathbf{d}\sigma; \tag{8.2}$$

3. $\mathbf{d}^2 = 0$.

When the manifold V_n is paracompact (Chap. 6), it is possible to prove the existence and uniqueness of the map $\mathbf{d}$. Here, we limit ourselves to proving its local existence in a chart (U, x^i) of V_n.

© Springer International Publishing AG, part of Springer Nature 2018

A. Romano and A. Marasco, *Classical Mechanics with Mathematica®*,
Modeling and Simulation in Science, Engineering and Technology,
https://doi.org/10.1007/978-3-319-77595-1_8

In the domain U of the chart, any r-differential form can be written as

$$\omega = \omega_{(i_1,\cdots,i_r)}dx^{i_1} \wedge \cdots \wedge dx^{i_r}.$$

If the exterior derivative exists, then, applying properties 2 and 3, we have that

$$d\omega = d\omega_{(i_1,\cdots,i_r)} \wedge dx^{i_1} \wedge \cdots \wedge dx^{i_r}.$$

Finally, taking into account property 1, we obtain the coordinate form of the exterior derivative

$$d\omega = \frac{\partial \omega_{(i_1,\cdots,i_r)}}{\partial x^h}dx^h \wedge dx^{i_1} \wedge \cdots \wedge dx^{i_r}. \tag{8.3}$$

It is plain to verify that (8.3) defines an exterior derivative in region U.

Before proceeding, we show that the preceding definition leads to familiar concepts in a Euclidean space $\mathcal{E}_3$. In fact, if $f \in \mathcal{F}(\mathcal{E}_3)$, then (8.3) gives

$$df = \frac{\partial f}{\partial x^i}dx^i. \tag{8.4}$$

It is evident that the components of this differential form are covariant with respect to a coordinate change. Denoting by (e_i) the fields of the natural bases relative to the coordinates (x^i) and by g_{ij} the components of the scalar product, we can say that the vector field

$$\nabla f = g^{ij}\frac{\partial f}{\partial x^j}e_i \tag{8.5}$$

defines the gradient of f.

Applying (8.3) to the differential form $\omega = \omega_i dx^i \in \wedge_1 \mathcal{E}_3$, we obtain

$$d\omega = \frac{\partial \omega_i}{\partial x^j}dx^j \wedge dx^i = \sum_{j<i}\frac{\partial \omega_i}{\partial x^j}dx^j \wedge dx^i + \sum_{j>i}\frac{\partial \omega_i}{\partial x^j}dx^j \wedge dx^i.$$

The preceding relation can be written in the form

$$d\omega = \sum_{j<i}\left(\frac{\partial \omega_i}{\partial x^j} - \frac{\partial \omega_j}{\partial x^i}\right)dx^j \wedge dx^i, \tag{8.6}$$

and the quantities

$$r_{ij} = \frac{\partial \omega_i}{\partial x^j} - \frac{\partial \omega_j}{\partial x^i} \tag{8.7}$$

define the covariant components of a skew-symmetric tensor. The adjoint (Chap. 5) of this tensor is the pseudovector $\mathbf{r}$ whose contravariant components are

$$r^i = \frac{1}{2}\eta^{ijk}r_{jk} = \frac{1}{2\sqrt{|g|}}\epsilon^{ijk}r_{jk}, \tag{8.8}$$

that is,

$$\mathbf{r} = \frac{1}{\sqrt{|g|}}\left(\frac{\partial\omega_3}{\partial x^2} - \frac{\partial\omega_2}{\partial x^3}, \frac{\partial\omega_1}{\partial x^3} - \frac{\partial\omega_3}{\partial x^1}, \frac{\partial\omega_2}{\partial x^1} - \frac{\partial\omega_1}{\partial x^2}\right). \tag{8.9}$$

In other words, the components of $\mathbf{d}\omega$ are proportional to the components of $\nabla \times \mathbf{v}$, where $\mathbf{v}$ is a vector field whose *covariant* components v_i are equal to ω_i.

Finally, the exterior derivative of the 2-differential form

$$\mathbf{\Omega} = \Omega_{12}\mathbf{d}x^1 \wedge \mathbf{d}x^2 + \Omega_{13}\mathbf{d}x^1 \wedge \mathbf{d}x^3 + \Omega_{23}\mathbf{d}x^2 \wedge \mathbf{d}x^3$$

is

$$\mathbf{d}\mathbf{\Omega} = \left(\frac{\partial\Omega_{23}}{\partial x^1} + \frac{\partial\Omega_{31}}{\partial x^2} + \frac{\partial\Omega_{12}}{\partial x^3}\right)\mathbf{d}x^1 \wedge \mathbf{d}x^2 \wedge \mathbf{d}x^3. \tag{8.10}$$

Introducing the pseudovector adjoint of $\mathbf{\Omega}$

$$u^i = \frac{1}{2\sqrt{|g|}}\epsilon^{ijk}\Omega_{jk}, \tag{8.11}$$

i.e., a pseudovector vector with components

$$u^1 = \frac{1}{\sqrt{|g|}}\Omega_{23}, \quad u^2 = \frac{1}{\sqrt{|g|}}\Omega_{31}, \quad u^3 = \frac{1}{\sqrt{|g|}}\Omega_{12}, \tag{8.12}$$

and defining the **divergence** of the vector field $\mathbf{u}$ (Chap. 9)

$$\nabla \cdot \mathbf{u} = \frac{1}{\sqrt{|g|}}\frac{\partial}{\partial x^i}(\sqrt{|g|}\,u^i), \tag{8.13}$$

relation (8.10) becomes

$$\mathbf{d}\mathbf{\Omega} = \nabla \cdot \mathbf{u}\sqrt{|g|}\mathbf{d}x^1 \wedge \mathbf{d}x^2 \wedge \mathbf{d}x^3, \tag{8.14}$$

where

$$\sqrt{|g|}\mathbf{d}x^1 \wedge \mathbf{d}x^2 \wedge \mathbf{d}x^3 \tag{8.15}$$

is the volume 3-form of $\mathcal{E}_3$ (Chap. 5).

We can summarize the preceding results as follows. In the three-dimensional Euclidean space $\mathcal{E}_3$, we consider a function $f : \mathcal{E}_3 \to \mathfrak{R}$, a differential form $\omega = \omega_i dx^i$, and a 2-form $\mathbf{\Omega}$. Further, we introduce the vector field $\mathbf{v}$ whose covariant components v_i are equal to the components ω_i of ω, and the pseudovector field $\mathbf{u}$, which is the adjoint of the skew-symmetric tensor $\mathbf{\Omega}$. Then, the following results hold:

$$\mathbf{d}f = \omega \quad \Leftrightarrow \quad \nabla f = \mathbf{v}, \tag{8.16}$$

$$\mathbf{d}\omega = \mathbf{\Omega} \quad \Leftrightarrow \quad \nabla \times \mathbf{v} = \mathbf{u}, \tag{8.17}$$

$$\mathbf{d}\mathbf{\Omega} = \mathbf{0} \quad \Leftrightarrow \quad \nabla \cdot \mathbf{u} = 0. \tag{8.18}$$

8.2   Closed and Exact Differential Forms

Definition 8.2. A differential r-form $\mathbf{\Omega} \in \wedge_r V_n$ is **exact** if a differential $(r-1)$-form $\omega \in \wedge_{r-1} V_n$ exists such that

$$\mathbf{d}\omega = \mathbf{\Omega}. \tag{8.19}$$

Definition 8.3. A differential r-form $\mathbf{\Omega} \in \wedge_r V_n$ is **closed** if

$$\mathbf{d}\mathbf{\Omega} = \mathbf{0}. \tag{8.20}$$

It is evident that the exact differential r-forms belong to the vector subspace

$$E_r = \mathbf{d}(\wedge_{r-1} V_n) = \mathrm{Im}(\mathbf{d}), \tag{8.21}$$

where $\mathbf{d} : \wedge_{r-1} V_n \to \wedge_r V_n$. In contrast, the closed differential r-forms belong to the vector subspace

$$C_r = \ker(\mathbf{d}), \tag{8.22}$$

where $\mathbf{d} : \wedge_r V_n \to \wedge_{r+1} V_n$. In view of the third property of Definition 8.1, any exact differential r-form is closed, and we have that

$$E_r \subset C_r \subset \wedge_r V_n. \tag{8.23}$$

Definition 8.4. We call **cohomology relation** the following equivalence relation in the vector subspace $C_r \subset \wedge_r V_n$:

$$\forall \mathbf{\Omega}, \mathbf{\Omega}' \in C_r, \quad \mathbf{\Omega} \sim \mathbf{\Omega}' \text{ if } \mathbf{\Omega} - \mathbf{\Omega}' \in E_r. \tag{8.24}$$

This equivalence relation generates a partition of C_r into equivalence classes that are called **cohomology classes** of order r. If the differential r-form $\mathbf{\Omega}$ belongs to the cohomology class $[\boldsymbol{\Omega}]$, then to $[\boldsymbol{\Omega}]$ belong all the differential r-forms $\boldsymbol{\Omega} + \mathbf{d}\omega$, where $\omega \in \wedge_{r-1} V_n$. Two closed differential r-forms belonging to the same class are said to be *cohomologs*. In particular, $[\mathbf{0}] = E_r$.

The set C_r/E_r of these equivalence classes becomes a vector space when it is equipped with the following operations:

$$[\boldsymbol{\Omega}_1] + [\boldsymbol{\Omega}_2] = [\boldsymbol{\Omega}_1 + \boldsymbol{\Omega}_2], \quad a[\boldsymbol{\Omega}] = [a\boldsymbol{\Omega}],$$

$\forall \boldsymbol{\Omega}_1, \boldsymbol{\Omega}_2 \in C_r$ and $\forall a \in \mathfrak{R}$.

Definition 8.5. The vector space $H_r = C_r/E_r$ is called the *cohomology space* of order r. If we only refer to an addition, C_r/E_r becomes the *cohomology group* of order r. The integer number

$$b_r = \dim(H_r) \tag{8.25}$$

is the *rth-Betti number* of V_n. It is evident that $b_r = 0$ if and only if H_r reduces to the cohomology class $[\mathbf{0}]$. In such a case, any r-closed form is also exact.

Definition 8.6. A subset W of a Euclidean space $\mathcal{E}_n$ is said to be *star-shaped* if there exists a point $x_0 \in W$ such that, $\forall x \in W$, the points of the segment $x_0 + \overrightarrow{x_0 x}\, t, t \in [0, 1]$, belong to W.

An open set $U \subset V_n$ is *smoothly contractible* if there exists a point $x_0 \in U$ and a differential map $\phi : [0, 1] \times U \to U$ such that

$$\phi(1, x) = x, \quad \phi(0, x) = x_0, \quad \forall x \in U.$$

For instance, a sphere of $\mathcal{E}_3$ is star-shaped and smoothly contractible to any point of it. A region contained between two spheres with the same center and different radii is not smoothly contractible.

We omit the proof of the following theorem.

Theorem 8.1 (Poincaré's Lemma). *If A is a star-shaped subset W of a Euclidean space $\mathcal{E}_n$ or an open smoothly contractible subset U of a manifold V_n, then*

$$b_r = \dim(A) = 0,$$

that is, any closed form is exact.

8.3 Properties of Exterior Derivative

In this section, we prove some relations existing among the exterior derivative, the differential of a map, the Lie derivative, and the inner product of a differential r-form by a vector field.

In this regard the following theorem holds.

Theorem 8.2. *For $\alpha \in \Lambda_k V_n$, $f \in \mathcal{F}V_n$, and $\mathbf{X} \in \chi V_n$,*

$$i_{f\mathbf{X}}\alpha = f i_{\mathbf{X}}\alpha, \tag{8.26}$$

$$i_{\mathbf{X}}\mathbf{d}f = L_{\mathbf{X}}f, \tag{8.27}$$

$$L_{\mathbf{X}}\alpha = i_{\mathbf{X}}\mathbf{d}\alpha + \mathbf{d}(i_{\mathbf{X}}\alpha), \tag{8.28}$$

$$L_{f\mathbf{X}}\alpha = f L_{\mathbf{X}}\alpha + \mathbf{d}f \wedge i_{\mathbf{X}}\alpha, \tag{8.29}$$

$$L_{\mathbf{X}}(\mathbf{d}\alpha) = \mathbf{d}(L_{\mathbf{X}}\alpha). \tag{8.30}$$

Further, if $F : V_n \to W_m$ is a C^1 map and $\alpha \in \Lambda_k(W_m)$, $\beta \in \Lambda_s(W_m)$, then

$$F^*(\alpha \wedge \beta) = F^*(\alpha) \wedge F^*(\beta), \tag{8.31}$$

$$F^*(\mathbf{d}\alpha) = \mathbf{d}(F^*(\alpha)). \tag{8.32}$$

Proof. Equation (8.26) follows at once from (7.26). Further, $i_{\mathbf{X}}\mathbf{d}f = X^h \partial f/\partial x^h$, and, taking into account (7.6), the identity (8.27) is proved. To verify (8.28), we start supposing that α is a 1-form. Then, in view of (7.19), (7.26), (8.3), and (8.6), we have

$$L_{\mathbf{X}}\alpha = \left(X^j \frac{\partial \alpha_i}{\partial x^j} + \alpha_j \frac{\partial X^j}{\partial x^i} \right) \mathbf{d}x^i,$$

$$i_{\mathbf{X}}\mathbf{d}\alpha = X^j \left(\frac{\partial \alpha_i}{\partial x^j} - \frac{\partial \alpha_j}{\partial x^i} \right) \mathbf{d}x^i,$$

$$\mathbf{d}(i_{\mathbf{X}}\alpha) = \mathbf{d}(X^j \alpha_j) = \alpha_j \frac{\partial X^j}{\partial x^i} \mathbf{d}x^i + X^j \frac{\partial \alpha_j}{\partial x^i} \mathbf{d}x^i,$$

and (8.28) is proved for 1-forms. Since any r-form α is a linear combination of exterior products of 1-forms, and $L_{\mathbf{X}}$, $i_{\mathbf{X}}$, and $\mathbf{d}$ are linear maps, to prove the theorem, it is sufficient to verify that (8.28) is satisfied for the exterior product $\alpha_1 \wedge \alpha_2$ of two 1-forms. Now, owing to (7.21) and recalling that (8.28) has been proved for 1-forms, we have that

$$\begin{aligned}
L_{\mathbf{X}}(\alpha_1 \wedge \alpha_2) &= L_{\mathbf{X}}\alpha_1 \wedge \alpha_2 + \alpha_1 \wedge L_{\mathbf{X}}\alpha_2 \\
&= (i_{\mathbf{X}}\mathbf{d}\alpha_1 + \mathbf{d}(i_{\mathbf{X}}\alpha_1)) \wedge \alpha_2 + \alpha_1 \wedge (i_{\mathbf{X}}\mathbf{d}\alpha_2 + \mathbf{d}(i_{\mathbf{X}}\alpha_2)) \\
&= (i_{\mathbf{X}}\mathbf{d}\alpha_1) \wedge \alpha_2 + \mathbf{d}(i_{\mathbf{X}}\alpha_1) \wedge \alpha_2 \\
&\quad + \alpha_1 \wedge (i_{\mathbf{X}}\mathbf{d}\alpha_2) + \alpha_1 \wedge \mathbf{d}(i_{\mathbf{X}}\alpha_2).
\end{aligned}$$

Similarly, in view of (8.2) and (7.27), we have also

$$\begin{aligned}
i_{\mathbf{X}}\mathbf{d}(\alpha_1 \wedge \alpha_2) &= i_{\mathbf{X}}(\mathbf{d}\alpha_1 \wedge \alpha_2 - \alpha_1 \wedge \mathbf{d}\alpha_2) \\
&= i_{\mathbf{X}}\mathbf{d}\alpha_1 \wedge \alpha_2 + \mathbf{d}\alpha_1 \wedge i_{\mathbf{X}}\alpha_2 - i_{\mathbf{X}}\alpha_1 \wedge \mathbf{d}\alpha_2 + \alpha_1 \wedge i_{\mathbf{X}}\mathbf{d}\alpha_2; \\
\mathbf{d}(i_{\mathbf{X}}(\alpha_1 \wedge \alpha_2)) &= \mathbf{d}(i_{\mathbf{X}}\alpha_1 \wedge \alpha_2 - \alpha_1 \wedge i_{\mathbf{X}}\alpha_2) \\
&= \mathbf{d}(i_{\mathbf{X}}\alpha_1) \wedge \alpha_2 + i_{\mathbf{X}}\alpha_1 \wedge \mathbf{d}\alpha_2 - \mathbf{d}\alpha_1 \wedge i_{\mathbf{X}}\alpha_2 + \alpha_1 \wedge \mathbf{d}(i_{\mathbf{X}}\alpha_2),
\end{aligned}$$

and (8.28) is proved. In the same way, we can prove (8.29) and (8.30). We omit the proof of (8.31) and (8.32). □

8.4   An Introduction to the Integration of r-Forms

In Sect. 3.4, we showed that one of the two possible orientations of $\mathfrak{R}^r$ is determined by choosing a basis $(\mathbf{e}_i)$. Denote by $\mathbf{dx}^1 \wedge \cdots \wedge \mathbf{dx}^r$ the volume r-form of $\mathfrak{R}^r$.

Definition 8.7. The *integral* of the r-form

$$\omega = \omega_{(i_1 \cdots i_r)} \mathbf{dx}^{i_1} \wedge \cdots \wedge \mathbf{dx}^{i_r}$$

on the compact set $U \subset \mathfrak{R}^r$ is the number

$$\int_U \omega = \int_U \omega_{(i_1 \cdots i_r)} \mathrm{d}x^1 \cdots \mathrm{d}x^r. \tag{8.33}$$

Let U' be another compact region of $\mathfrak{R}^r$, and denote by $x' = x'(x_i')$ a diffeomorphism of U onto U'. From well-known analytical results we have that

$$\int_U J\omega_{(i_1 \cdots i_r)} \mathrm{d}x^1 \cdots \mathrm{d}x^r = \int_{U'} \omega'_{(i_1 \cdots i_r)} \mathrm{d}x'^1 \cdots \mathrm{d}x'^r, \tag{8.34}$$

where

$$J = \det \left(\frac{\partial x'^i}{\partial x^j} \right). \tag{8.35}$$

Henceforth, we identify the compact set U with the r-*cube* defined as follows:

$$U = \left\{ (x^1, \ldots, x^r) \in \mathfrak{R}^r, a^i \le x^i \le a^i + c^i \right\}. \tag{8.36}$$

We call *faces* of the r-cube U the $(r-1)$-cubes $U_{i\epsilon}, i = 1, \ldots, r-1, \epsilon = 0, 1$, defined by the conditions

$$U_{i\epsilon} = \left\{ a^1 \le x^1 \le a^1 + c^1, \ldots, x^i = a^i + \epsilon c^i, \ldots, a^r \le x^r \le a^r + c^r \right\}. \tag{8.37}$$

Starting from the $(r-1)$-cubes we can define the $(r-2)$-cubes, which are the faces of the $(r-1)$-cubes, and so on. For instance, if $r=2$, then the 2-cubes are rectangles and the 1-cubes are the sides of the rectangles. Further, if $r=3$, then a 3-cube is a parallelepiped, the 2-cubes are its faces, and the 1-cubes are its edges.

Now we consider the vectors $\mathbf{n}_{i\epsilon}, i = 1, \ldots, r, \epsilon = 0, 1$, which in the basis $(\mathbf{e}_i)$, chosen to determine an orientation of $\mathfrak{R}^r$, have the components

Fig. 8.1 An oriented 2-cube

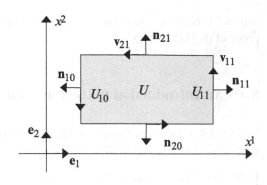

$$\mathbf{n}_{i0} = -\delta_i^j \mathbf{e}_j, \quad \mathbf{n}_{i1} = \delta_i^j \mathbf{e}_j. \tag{8.38}$$

In particular, if $r = 2$, then

$$\mathbf{n}_{10} = -\mathbf{e}_1, \quad \mathbf{n}_{11} = \mathbf{e}_1, \quad \mathbf{n}_{20} = -\mathbf{e}_2, \quad \mathbf{n}_{21} = \mathbf{e}_2.$$

It is evident that the vectors $\mathbf{n}_{i\epsilon}$ are normal to the faces of the r-cube U and outward oriented (Fig. 8.1 is relevant to the case $r = 2$). On any face $U_{i\epsilon}$ of the r-cube U, we consider a vector set $\Sigma_{i\epsilon} = (\mathbf{n}_{i\epsilon}, \mathbf{v}_{i\epsilon}^{(1)}, \ldots, \mathbf{v}_{i\epsilon}^{(r-1)})$, where the vectors $\mathbf{v}_{i\epsilon}^{(1)}, \ldots, \mathbf{v}_{i\epsilon}^{(r-1)}$ are independent vectors belonging to $U_{i\epsilon}$, chosen in such a way that the bases $\Sigma_{i\epsilon}$ are congruent with the basis $(\mathbf{e}_i)$. When U is equipped with these bases on its faces, we say that U is an oriented cube of $\mathfrak{R}^r$ and we write (U, σ), where σ is the orientation determined by $(\mathbf{e}_i)$.

Definition 8.8. We define an *oriented r-cube* on a differentiable manifold V_n, a term $C = (U, \sigma, F)$, where (U, σ) is an oriented r-cube of $\mathfrak{R}^r$ and $F : U \to V_n$ is a differentiable map.

Example 8.1. Given the 2-cube of $\mathfrak{R}^2$ defined by the inequalities

$$0 \le r \le 1, \quad 0 \le \theta \le 2\pi,$$

consider the map $F : (r, \theta) \in U \to (x^1, x^2) \in \mathcal{E}_2$, where $\mathcal{E}_2$ is two-dimensional Euclidean space such that (Fig. 8.2)

$$x^1 = r \cos \theta, \quad x^2 = r \sin \theta.$$

Example 8.2. Given the 2-cube of $\mathfrak{R}^2$ defined by the inequalities

$$0 \le \varphi \le 2\pi, \quad 0 \le \theta \le \pi/2,$$

consider the map $F : (\varphi, \theta) \in U \to (x^1, x^2, x^3) \in \mathcal{E}_3$, where $\mathcal{E}_3$ is three-dimensional Euclidean space such that (Fig. 8.3)

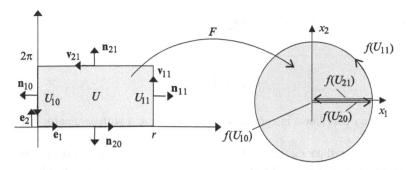

Fig. 8.2 An oriented 2-cube of $\mathcal{E}_2$

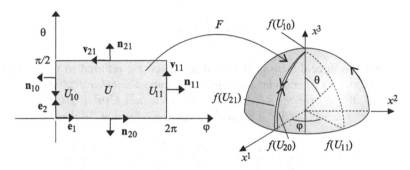

Fig. 8.3 An oriented 2-cube of $\mathcal{E}_3$

$$x^1 = r\cos\theta\sin\varphi, \ \ x^2 = r\sin\theta\sin\varphi, \ \ x^3 = \cos\theta.$$

Definition 8.9. An r-**chain** C on a manifold V_n is a formal summation $\Sigma_i a_i C_i$ of oriented r-cubes C_i, where a_i are relative integer numbers. Denoting by ∂C_i the union of the faces of U_i, we call **cochain** ∂C of the chain C the formal summation $\Sigma_i a_i \partial C_i$.

Definition 8.10. Let $\omega \in \Lambda_r V_n$ be an r-form on a differentiable manifold V_n. Denote by $C = \Sigma a_i C_i$ a chain of oriented r-cubes $C_i = (U_i, \sigma_i, F_i)$. Then, the **integral of** ω **on the chain** is given by

$$\int_C \omega = \sum_i a_i \int_{C_i} \omega = \sum_i a_i \int_{U_i} F_i^*(\omega). \tag{8.39}$$

Example 8.3. In three-dimensional Euclidean space $\mathcal{E}_3$, referred to the Cartesian coordinates (x^1, x^2, x^3), consider the oriented 1-cube $C = ([a,b], \sigma, F)$, where $[a,b] \subset \Re$, $F: t \in [a,b] \to (x^1(t), x^2(t), x^3(t)) \in \mathcal{E}_3$. If $\omega = \omega_i dx^i$ is a 1-form of $\mathcal{E}_3$, then

$$F^*\omega = \omega_i \frac{dx^i}{dt}$$

and

$$\int_C \omega = \int_a^b \omega_i \frac{dx^i}{dt} dt$$

is the ordinary curvilinear integral of ω along the curve γ with parametric equations $x^i(t)$.

If s is the curvilinear abscissa along γ and we denote by $\mathbf{t} = (dx^i/ds)$ the unit tangent to γ and by $\mathbf{v}$ a vector whose covariant components v_i are equal to ω_i, then the preceding integral assumes the familiar form

$$\int_C \omega = \int_\gamma \mathbf{v} \cdot \mathbf{t} ds. \tag{8.40}$$

Example 8.4. In three-dimensional Euclidean space $\mathcal{E}_3$, referred to the Cartesian coordinates (x^1, x^2, x^3), consider the oriented 2-cube $C = (U, \sigma, F)$, where U is a rectangle of $\mathfrak{R}^2$, $F : (u^1, u^2) \in U \to (x^1(u^1, u^2), x^2(u^1, u^2), x^3(u^1, u^2)) \in \mathcal{E}_3$. If $\omega = \omega_{12} \mathbf{dx}^1 \wedge \mathbf{dx}^2 + \omega_{13} \mathbf{dx}^1 \wedge \mathbf{dx}^3 + \omega_{23} \mathbf{dx}^2 \wedge \mathbf{dx}^3$ is a 2-form of $\mathcal{E}_3$, then [see (6.58)]

$$F^*\omega = \omega_{12} \frac{\partial(x^1, x^2)}{\partial(u^1, u^2)} + \omega_{13} \frac{\partial(x^1, x^3)}{\partial(u^1, u^2)} + \omega_{23} \frac{\partial(x^2, x^3)}{\partial(u^1, u^2)},$$

and

$$\int_C \omega = \int_U \left(\omega_{12} \frac{\partial(x^1, x^2)}{\partial(u^1, u^2)} + \omega_{13} \frac{\partial(x^1, x^3)}{\partial(u^1, u^2)} + \omega_{23} \frac{\partial(x^2, x^3)}{\partial(u^1, u^2)} \right) du^1 du^2 \tag{8.41}$$

is the integral of ω on the surface S with parametric equations $x^i(u^1, u^2)$.

In arbitrary curvilinear coordinates (x^1, x^2, x^3) of $\mathcal{E}_3$, we consider the adjoint vector

$$\mathbf{u} = \frac{1}{\sqrt{g}} (\omega_{23} \mathbf{e}_1 + \omega_{31} \mathbf{e}_2 + \omega_{12} \mathbf{e}_3),$$

where $(\mathbf{e}_1, \mathbf{e}_2, \mathbf{e}_3)$ is the natural basis relative to the coordinates (x^1, x^2, x^3). On analysis it is proved that the Jacobian minors appearing in (8.41) coincide with the components of the unit normal $\mathbf{n}$ to the surface S. Finally, the elementary area $d\sigma$ of S is equal to $\sqrt{g} du^1 du^2$, and (8.41) gives the flux of $\mathbf{u}$ across the oriented surface S

$$\int_C \omega = \int_S \mathbf{u} \cdot \mathbf{n} d\sigma. \tag{8.42}$$

We conclude this section stating without the difficult proof the following *generalized Stokes theorem*:

Theorem 8.3. *If C is an oriented r-chain of the differentiable manifold V_n and ∂C its cochain, then, $\forall \omega \in \Lambda_{r-1} V_n$, we obtain*

$$\int_C d\omega = \int_{\partial C} \omega. \tag{8.43}$$

8.5 Exercises

1. Prove the equivalence between (8.3) and the following formula:

$$d\omega = \sum_{i_1 < \cdots < i_{r+1}} \sum_{\pi \in \Pi_{r+1}} (-1)^{m(\pi)} \frac{\partial \omega_{\pi(i_2)\cdots\pi(i_{r+1})}}{\partial x^{\pi(i_1)}} dx^{i_1} \wedge \cdots dx^{i_{r+1}}, \tag{8.44}$$

   where $\Pi(r+1)$ is the set of all permutations of the indices $i_1 < i_2 < \cdots < i_{r+1}$, for which $i_2 < \cdots < i_{r+1}$, and $m(\pi)$ is the number of inversions in the permutation $\{\pi(i_1), \pi(i_2), \ldots, \pi(i_{r+1})\}$.
2. Evaluate by (8.44) the exterior derivative $d\Omega$ of the 2-form

$$\Omega = \Omega_{12} dx^1 \wedge dx^2 + \Omega_{13} dx^1 \wedge dx^3 + \Omega_{23} dx^2 \wedge dx^3$$

   on a manifold V_3.
   Hint: Since $n=3$, the only choice $i_1 < i_2 < i_3$ is $1,2,3$. On the other hand, the possible permutations $\pi(1), \pi(2), \pi(3)$ of these indices for which $\pi(2) < \pi(3)$ are $\{1, 2, 3\}, \{2, 1, 3\}, \{3, 1, 2\}$. Therefore, (8.44) gives

$$d\Omega = \left(\frac{\partial \Omega_{23}}{\partial x^1} - \frac{\partial \Omega_{13}}{\partial x^2} + \frac{\partial \Omega_{12}}{\partial x^3} \right) dx^1 \wedge dx^2 \wedge dx^3,$$

   and we again obtain (8.10).
3. On a manifold V_4, apply (8.3) and (8.44) to the 2-form

$$\Omega = \Omega_{12} dx^1 \wedge dx^2 + \Omega_{13} dx^1 \wedge dx^3 + \Omega_{14} dx^1 \wedge dx^4 + \\ \Omega_{23} dx^2 \wedge dx^3 + \Omega_{24} dx^2 \wedge dx^4 + \Omega_{34} dx^3 \wedge dx^4,$$

   and verify that they give the same result.
4. Adopting the volume form

$$\sqrt{g} d\varphi d\theta$$

[see (8.15)] evaluate the area of the triangle and the parallelogram of Figs. 9.1 and 9.2, on the unit sphere referred to the spherical coordinates φ, θ.

5. Integrate on the unit sphere the 1-form

$$x\,\mathrm{d}x + \mathrm{d}y + \mathrm{d}z$$

of the three-dimensional Euclidean space $\mathcal{E}_3$.

Chapter 9
Absolute Differential Calculus

Abstract In this chapter, we address the fundamental problem of extending differential calculus to manifolds. This extension requires the introduction of a criterion to compare vectors belonging to different tangent spaces of the manifold. This criterion, except for some general rules that it must satisfy, is quite arbitrary. The choice of such a criterion corresponds to the introduction of an affine connection on the manifold. In this chapter we define the affine connections and study their properties. On a Riemannian manifold it is possible to define only one affine connection that is compatible with the metric. This means that the connection does not modify the scalar product between two vectors when they undergo a parallel transport along an arbitrary curve, provided that the parallelism is evaluated by the affine connection.

9.1 Preliminary Considerations

In this chapter, we extend differential calculus to manifolds. To understand the problem we are faced with, consider a C^1 vector field $\mathbf{Y}(t)$ assigned along the curve $x(t)$ on the manifold V_n. We recall that on an arbitrary manifold the components $Y^i(t)$ of $\mathbf{Y}(t)$ are evaluated with respect to the local natural bases of local charts (U, x^i), $U \subset V_n$. Consequently, when we try to define the derivative of $\mathbf{Y}$ along $x(t)$, we must compare the vector $\mathbf{Y}(t + \Delta t) \in T_{x(t+\Delta t)} V_n$, referred to the local basis $\mathbf{e}_i(t + \Delta t)$, with the vector $\mathbf{Y}(t) \in T_{x(t)} V_n$, referred to the local basis $\mathbf{e}_i(t)$. Since we do not know how to relate the basis $\mathbf{e}_i(t + \Delta t)$ to the basis $\mathbf{e}_i(t)$, we are not in a position to compare the two preceding vectors; consequently, we cannot assign a meaning to the derivative of the vector field $\mathbf{Y}(t)$ along the curve $x(t)$ of V_n.

To give a reasonable solution to the preceding problem, we start with some elementary considerations in a Euclidean space $\mathcal{E}_n$ before giving an abstract definition of an affine connection on a manifold. In a Euclidean space $\mathcal{E}_n$ it is always possible to find rectilinear coordinates (y^i). We denote by $(\mathbf{u}_i)$ the constant unit vectors along

© Springer International Publishing AG, part of Springer Nature 2018   119
A. Romano and A. Marasco, *Classical Mechanics with Mathematica®*,
Modeling and Simulation in Science, Engineering and Technology,
https://doi.org/10.1007/978-3-319-77595-1_9

the axes y^i axes. Then we introduce an arbitrary local chart (U, x^i) in $\mathcal{E}_n$ and denote by $(x, \mathbf{e}_i)$ the natural bases relative to the *curvilinear* coordinates x^i. If $\mathbf{X}(x) = X^i(x)\mathbf{e}_i$ is a C^1 vector field in region U, then the family Γ of the integral curves of $\mathbf{X}$ is given by curves whose parametric equations $x^i(t)$ in the chart (U, x^i) are a solution of the first-order differential system

$$\frac{dx^i}{dt} = X^i(x^1, \ldots, x^n), \quad i = 1, \ldots, n. \tag{9.1}$$

It is evident that $\mathbf{X}(x^i(t))$ is the tangent vector to the curve $x^i(t)$ at any point. When a point $x_0(x_0^i) \in U$ is given, there is one and only one (local) integral curve of $\mathbf{X}(x)$ containing point x_0. We denote by $x^i(t, x_0)$ the general integral of (9.1).

Let $\mathbf{Y}(x) = Y^i(x)\mathbf{e}_i$ be another C^1 vector field on region U. Then, the directional derivative of $\mathbf{Y}(x)$ along the integral curves Γ of the vector field $\mathbf{X}(x)$ is

$$\begin{aligned}
\frac{d\mathbf{Y}}{dt} &= \frac{dY^i}{dt}\mathbf{e}_i + Y^i\frac{d\mathbf{e}_i}{dt} \\
&= \left(\frac{\partial Y^i}{\partial x^k}\mathbf{e}_i + Y^h\frac{\partial \mathbf{e}_h}{\partial x^k}\right)\frac{dx^k}{dt}.
\end{aligned} \tag{9.2}$$

To attribute a meaning to the derivatives $\partial \mathbf{e}_i / \partial x^k$ that express the variation of the basis vectors on varying the coordinates, we start by noting that if the functions

$$y^i = y^i(x^1, \ldots, x^n)$$

define the coordinate transformation $(x^i) \to (y^i)$, then we also have that

$$\mathbf{e}_i = \frac{\partial y^m}{\partial x^i}\mathbf{u}_m.$$

Since the vectors $\mathbf{u}_i$ are constant, the derivation in rectilinear coordinates is possible. Then, from the preceding relations we have that

$$\frac{\partial \mathbf{e}_i}{\partial x^j} = \frac{\partial^2 y^m}{\partial x^i \partial x^j}\mathbf{u}_m = \frac{\partial^2 y^m}{\partial x^i \partial x^j}\frac{\partial x^k}{\partial y^m}\mathbf{e}_k,$$

and introducing the notations

$$\Gamma_{ij}^k = \Gamma_{ji}^k \equiv \frac{\partial^2 y^m}{\partial x^i \partial x^j}\frac{\partial x^k}{\partial y^m}, \tag{9.3}$$

we obtain

$$\frac{\partial \mathbf{e}_i}{\partial x^j} = \Gamma_{ij}^k \mathbf{e}_k, \tag{9.4}$$

and we can give (9.2) the form

$$\frac{d\mathbf{Y}}{dt} = \left(\frac{\partial Y^i}{\partial x^k} + \Gamma^i_{kh} Y^h \right) X^k \mathbf{e}_i. \tag{9.5}$$

This formula is not satisfactory since the quantities Γ^i_{kh} are given in terms of the rectilinear coordinates (y^i), as is shown by (9.3), not in terms of the coordinates (x^i). But this further problem can be solved easily because we are operating in a Euclidean space. In fact, a scalar product

$$\mathbf{X} \cdot \mathbf{Y} = g_{ij} X^i Y^j$$

between two arbitrary vectors $\mathbf{X}$ and $\mathbf{Y}$ of the same tangent space to $\mathcal{E}_n$ is defined. In particular, $\mathbf{e}_i \cdot \mathbf{e}_j = g_{ij}$. Consequently, we have that

$$\frac{\partial \mathbf{e}_i}{\partial x^k} \cdot \mathbf{e}_j + \mathbf{e}_i \cdot \frac{\partial \mathbf{e}_j}{\partial x^k} = \frac{\partial g_{ij}}{\partial x^k}.$$

In view of (9.4), the preceding relation becomes

$$\Gamma^n_{ik} g_{nj} + \Gamma^n_{kj} g_{ni} = \frac{\partial g_{ij}}{\partial x^k}. \tag{9.6}$$

Cyclically permuting the indices, we obtain

$$\Gamma^n_{ji} g_{nk} + \Gamma^n_{ik} g_{nj} = \frac{\partial g_{jk}}{\partial x^i}, \tag{9.7}$$

$$\Gamma^n_{kj} g_{ni} + \Gamma^n_{ji} g_{nk} = \frac{\partial g_{ki}}{\partial x^j}. \tag{9.8}$$

Adding (9.6) and (9.7) and subtracting (9.8), we express the quantities Γ^n_{ij} in terms of the metric coefficients g_{ij} and their first derivatives in the coordinates (x^i):

$$\Gamma^n_{hj} = \frac{1}{2} g^{ni} \left(\frac{\partial g_{ij}}{\partial x^h} + \frac{\partial g_{ih}}{\partial x^j} - \frac{\partial g_{jh}}{\partial x^i} \right). \tag{9.9}$$

It is an easy exercise to verify that the map

$$\nabla_\mathbf{X} : \mathbf{Y} \to \frac{d\mathbf{Y}}{dt},$$

where $d\mathbf{Y}/dt$ is given by (9.5), has the following properties:

$$\nabla_{f\mathbf{X}+g\mathbf{Y}} = f\nabla_{\mathbf{X}} + g\nabla_{\mathbf{Y}}, \tag{9.10}$$

$$\nabla_{\mathbf{X}}(f\mathbf{Y}) = (\mathbf{X}f)\mathbf{Y} + f\nabla_{\mathbf{X}}\mathbf{Y}, \tag{9.11}$$

for any choice of C^1 real functions f and g and the C^1 vector fields $\mathbf{X}$ and $\mathbf{Y}$ on $\mathcal{E}_n$.

We conclude this section by noting that the expression (9.5) of the derivative of $\mathbf{Y}(t)$ along the curve $x^i(t)$, where the quantities are given by (9.9), was obtained under the following two conditions:

- There exists a rectilinear system of coordinates.
- In any tangent space to the manifold V_n, there is a scalar product.

In the following sections, we show how to extend the results of this section to arbitrary manifolds.

9.2   Affine Connection on Manifolds

Let V_n be a C^∞ n-dimensional manifold, and denote by $\mathcal{F}(V_n)$ the vector space of the real C^∞ functions on V_n. Then, the set $\chi(V_n)$ of the C^∞ vector fields on V_n is a $\mathcal{F}(V_n)$ module.

Definition 9.1. An *affine connection* on V_n is a map

$$\nabla : (\mathbf{X}, \mathbf{Y}) \in \chi(V_n) \times \chi(V_n) \rightarrow \nabla_{\mathbf{X}}\mathbf{Y} \in \chi(V_n) \tag{9.12}$$

such that

$$\nabla_{f\mathbf{X}+g\mathbf{Y}} = f\nabla_{\mathbf{X}} + g\nabla_{\mathbf{Y}}, \tag{9.13}$$

$$\nabla_{\mathbf{X}}(f\mathbf{Y}) = (\mathbf{X}f)\mathbf{Y} + f\nabla_{\mathbf{X}}\mathbf{Y}, \tag{9.14}$$

$\forall f, g \in \mathcal{F}(V_n)$ and $\mathbf{X}, \mathbf{Y} \in \chi(V_n)$ [see (9.10) and (9.11)].

To find the coordinate representation of the map (9.12), we introduce a local chart (U, x^i) on V_n and consider two C^∞ vector fields

$$\mathbf{X} = X^i\mathbf{e}_i, \quad \mathbf{Y} = Y^i\mathbf{e}_i, \tag{9.15}$$

where $(\mathbf{e}_i)$ are the vector fields defining a natural basis of the tangent space at any point of the open set U. From the properties (9.13) and (9.14) there follows

$$\nabla_{\mathbf{X}}\mathbf{Y} = \nabla_{X^k\mathbf{e}_k}(Y^i\mathbf{e}_i) = \left(\frac{\partial Y^i}{\partial x^k}\mathbf{e}_i + Y^i\nabla_{\mathbf{e}_k}(\mathbf{e}_i)\right)X^k. \tag{9.16}$$

Introducing the **connection coefficients** Γ^h_{ki} by the relations

$$\nabla_{\mathbf{e}_k}(\mathbf{e}_i) = \Gamma^h_{ki}\mathbf{e}_h, \tag{9.17}$$

(9.16) becomes

$$\nabla_{\mathbf{X}}\mathbf{Y} = \left(\frac{\partial Y^i}{\partial x^k} + \Gamma^i_{kh}Y^h\right)X^k\mathbf{e}_i \equiv \nabla_k Y^i X^k\mathbf{e}_i. \tag{9.18}$$

Although (9.18) and (9.5) are formally identical, in (9.18) the n^3 C^∞ functions Γ^i_{kh} are arbitrary. In particular, they could not verify the symmetry conditions $\Gamma^i_{kh} = \Gamma^i_{hk}$. In any case, if we give an affine connection on V_n, then its coefficients Γ^i_{kh} are determined in every coordinate domain. Conversely, a connection is defined in every coordinate domain U by giving the coefficients Γ^i_{kh} in U. If we introduce a new chart (V, x'^i) such that $U \cap V \neq \varnothing$, then the connection coefficients in the new chart cannot be arbitrarily assigned. In fact, in the coordinate transformation $(x^i) \to (x'^i)$, the natural bases are transformed according to the rule

$$\mathbf{e}'_j = \frac{\partial x^i}{\partial x'^j}\mathbf{e}_i \equiv A^i_j\mathbf{e}_i. \tag{9.19}$$

Consequently, introducing the notation $\partial f/\partial x^i = f_{,i}$, we have that

$$\begin{aligned}
\nabla_{\mathbf{X}}\mathbf{Y} &= (Y,^i_k + \Gamma^i_{kh}Y^h)X^k\mathbf{e}_i = (Y,'^l_m + \Gamma'^l_{mn}Y'^n)X'^m\mathbf{e}'_l \\
&= \left[A^k_m((A^{-1})^l_i Y^i)_{,k} + \Gamma'^l_{mn}(A^{-1})^n_h Y^h\right](A^{-1})^m_p X^p A^r_l\mathbf{e}_r \\
&= \left[A^k_m(A^{-1})^m_p A^r_l(A^{-1})^l_i Y^i_{,k} + A^k_m(A^{-1})^m_p A^r_l(A^{-1})^l_{i,k}Y^i \right. \\
&\quad \left. + (A^{-1})^n_h A^r_l(A^{-1})^m_p \Gamma'^l_{mn}Y^h\right]X^p\mathbf{e}_r.
\end{aligned}$$

But $A^k_m(A^{-1})^m_p = \delta^k_p$, and then the preceding formula gives

$$\nabla_{\mathbf{X}}\mathbf{Y} = \left[Y,^i_k + A^i_l(A^{-1})^l_h + (A^{-1})^m_k A^i_l(A^{-1})^n_h \Gamma'^l_{mn})Y^h\right]X^k\mathbf{e}_i.$$

Finally, we obtain the following transformation formulae for the connection coefficients:

$$\Gamma^i_{kh} = A^i_l(A^{-1})^m_k(A^{-1})^n_h \Gamma'^l_{mn} + A^i_l(A^{-1})^l_{h,k}, \tag{9.20}$$

whose inverse formulae are

$$\Gamma'^i_{kh} = (A^{-1})^i_l A^m_k A^n_h \Gamma^l_{mn} + (A^{-1})^i_l A^l_{h,k}. \tag{9.21}$$

These formulae show that the connection coefficients are not transformed like the components of a $(1,2)$-tensor unless the coordinate transformation is linear since, in this case, $A^l_{h,k} = 0$.

We do not prove that an affine connection can be assigned on any paracompact manifold.

9.3   Parallel Transport and Autoparallel Curves

Let (U, x^i) be a chart on a manifold V_n, equipped with an affine connection, and let $(x^i(t))$ be the parametric equations of a curve γ contained in region U. Finally, we denote by

$$\mathbf{X} = \frac{\mathrm{d}x^i}{\mathrm{d}t} \mathbf{e}_i \tag{9.22}$$

the tangent vector to γ evaluated in the natural bases $(\mathbf{e}_i)$ along γ.

Definition 9.2. The vector $\mathbf{Y}(t)$ is said to be *parallel transported along* γ if

$$\frac{\mathrm{d}\mathbf{Y}}{\mathrm{d}t} \equiv \nabla_{\mathbf{X}}\mathbf{Y} = \mathbf{0}. \tag{9.23}$$

In view of (9.18), the coordinate form of (9.23) is

$$\frac{\mathrm{d}Y^i}{\mathrm{d}t} + \Gamma^i_{kh} Y^h \frac{\mathrm{d}x^k}{\mathrm{d}t} = 0, \quad i = 1, \dots, n. \tag{9.24}$$

This is a first-order differential system in the unknowns $Y^i(t)$. If the functions $\Gamma^i_{kh} Y^h \frac{\mathrm{d}x^k}{\mathrm{d}t}$ are differentiable along γ and the initial vector $\mathbf{Y}(x(t_0))$ is given by the initial data $Y^i(x(t_0)) = Y^i_0$, then one and only one solution $Y^i(t, Y^i_0)$ exists of system (9.24). It is fundamental to remark that the vector $\mathbf{Y}(x^i(t))$, obtained by a parallel transport of $\mathbf{Y}(x(t_0))$ along γ, *depends on* γ.

Definition 9.3. A vector field $\mathbf{Y}$ on a manifold V_n equipped with an affine connection is *uniform* if

$$\nabla_{\mathbf{X}}\mathbf{Y} = \mathbf{0} \tag{9.25}$$

for any vector field $\mathbf{X}$.

In view of (9.18), condition (9.25) can be expressed in the following coordinate form:

$$\nabla_k Y^i \equiv Y^i{}_{,k} + \Gamma^i_{kh} Y^h = 0, \quad i, k = 1, \dots, n, \tag{9.26}$$

in any domain U of a local chart (U, x^i). When the connection is given, the connection coefficients Γ^i_{kh} are known functions of the coordinates in region U. Therefore, finding a uniform field $\mathbf{Y}$ in U requires the integration of system (9.26) with respect to the unknowns $Y^i(x^j)$. In other words, we must solve a system of n^2 equations in n unknowns. It is well known that this is possible if and only if suitable integrability conditions are satisfied. We analyze these conditions in a subsequent section. In conclusion, on an arbitrary manifold V_n with an affine connection uniform vector fields could not exist.

Definition 9.4. A curve $\gamma(s)$ on the manifold V_n, equipped with an affine connection, is said to be *autoparallel* if its tangent vector $\mathbf{X}(s)$ satisfies the condition

$$\nabla_\mathbf{X} \mathbf{X} = \mathbf{0}, \tag{9.27}$$

that is, if $\mathbf{X}$ is parallel along $\gamma(s)$.

In any chart (U, x^i), we have that $X^i(s) = \mathrm{d}x^i/\mathrm{d}s$, and (9.27) can be written as follows:

$$\frac{\mathrm{d}^2 x^i}{\mathrm{d}s^2} + \Gamma^i_{kh} \frac{\mathrm{d}x^k}{\mathrm{d}s} \frac{\mathrm{d}x^h}{\mathrm{d}s} = 0, \quad i = 1, \ldots, n. \tag{9.28}$$

This is a second-order differential system in the unknowns $x^i(s)$. Consequently, if a point $x_0 \in V_n$ is fixed together with a vector $(X^i(0)) \in T_{x_0} V_n$, then one and only one autoparallel curve exists that contains x_0 and is tangent to $(X^i(0))$ at x_0.

Remark 9.1. We recall that a curve γ of V_n is defined as a C^1 map $\gamma : s \in [a, b] \rightarrow \gamma(s) \in V_n$. This means that by changing the parameter, we obtain another curve, even if the locus of its points does not change. As a consequence, we have that, if $\gamma(s)$ is an autoparallel curve, $\gamma(s(t)) \equiv \mu(t)$ could not be an autoparallel curve. To identify the parameter changes that do not modify the autoparallelism of a curve, we start by noting that

$$\frac{\mathrm{d}x^k}{\mathrm{d}s} = \frac{\mathrm{d}x^k}{\mathrm{d}t} \frac{\mathrm{d}t}{\mathrm{d}s},$$
$$\frac{\mathrm{d}^2 x^k}{\mathrm{d}s^2} = \frac{\mathrm{d}}{\mathrm{d}s} \frac{\mathrm{d}x^k}{\mathrm{d}s} = \frac{\mathrm{d}^2 x^k}{\mathrm{d}t^2} \left(\frac{\mathrm{d}t}{\mathrm{d}s} \right)^2 + \frac{\mathrm{d}x^k}{\mathrm{d}t} \frac{\mathrm{d}^2 t}{\mathrm{d}s^2}.$$

Consequently, under the parameter change $s = s(t)$, system (9.28) becomes

$$\frac{\mathrm{d}^2 x^i}{\mathrm{d}t^2} + \Gamma^i_{kh} \frac{\mathrm{d}x^k}{\mathrm{d}t} \frac{\mathrm{d}x^h}{\mathrm{d}t} = - \left(\frac{\mathrm{d}s}{\mathrm{d}t} \right)^2 \frac{\mathrm{d}^2 t}{\mathrm{d}s^2} \frac{\mathrm{d}x^i}{\mathrm{d}t}. \tag{9.29}$$

This result allows us to state that *the autoparallel character of the curve $\gamma(s)$ is not modified by the parameter change $s = s(t)$ if and only if*

$$t = as + b, \tag{9.30}$$

where a and b are arbitrary constants. Parameters that do not modify the autoparallelism of a curve are called *canonical parameters*.

9.4   Covariant Differential of Tensor Fields

We denote by $\chi_s^r(V_n)$ the set of (r, s)-tensor fields of class C^∞ on the manifold V_n equipped with an affine connection. In particular, we denote by $\chi(V_n) = \chi_0^1(V_n)$ the set of C^∞ vector fields on V_n. We remark that $\chi_s^r(V_n)$ is also a $\mathcal{F}(V_n)$ module. Finally, we denote by $\hat{\mathcal{F}}$ the set of all the maps $\chi_s^r(V_n) \to \chi_s^r(V_n)$.

Theorem 9.1. *On the manifold V_n there is only one $\Re$-linear map*

$$\hat{\nabla} : \mathbf{X} \in \chi(V_n) \to \hat{\nabla}_{\mathbf{X}} \in \hat{\mathcal{F}} \tag{9.31}$$

with the following properties:

1. $\hat{\nabla}_{\mathbf{X}} f = \mathbf{X} f, \forall f \in \mathcal{F}(V_n)$;
2. $\hat{\nabla}_{\mathbf{X}} \mathbf{Y} = \nabla_{\mathbf{X}} \mathbf{Y}, \forall \mathbf{Y} \in \chi(V_n)$;
3. $\hat{\nabla}_{\mathbf{X}}(C\mathbf{T}) = C\hat{\nabla}_{\mathbf{X}}\mathbf{T}$, *where C is the contraction operator (Sect. 2.6) and $\mathbf{T} \in \chi_s^r(V_n)$;*
4. *It verifies the derivation property*

$$\hat{\nabla}_{\mathbf{X}}(\mathbf{T} \otimes \mathbf{S}) = (\hat{\nabla}_{\mathbf{X}}\mathbf{T}) \otimes \mathbf{S} + \mathbf{T} \otimes (\hat{\nabla}_{\mathbf{X}}\mathbf{S}). \tag{9.32}$$

Proof. We prove the *local* existence of the preceding map in the domain of a chart (U, x^i). We denote by $(\mathbf{e}_i)$ the fields of the natural bases in U and by $(\mathbf{d}x^i)$ the differential forms of the dual bases. If $\omega \in \chi_1^0(V_n)$ and $\mathbf{Y} \in \chi(V_n)$, then $C(\omega \otimes \mathbf{Y})$ is a function. From property 1 we obtain

$$\hat{\nabla}_{\mathbf{X}}(C(\omega \otimes \mathbf{Y}) = \mathbf{X}(C(\omega \otimes \mathbf{Y})).$$

On the other hand, in view of properties 2–4, we also have that

$$\hat{\nabla}_{\mathbf{X}}(C(\omega \otimes \mathbf{Y})) = C(\hat{\nabla}_{\mathbf{X}}\omega \otimes \mathbf{Y} + \omega \otimes \nabla_{\mathbf{X}}\mathbf{Y}).$$

Expressing the preceding results in components, we can write that

$$X^k(\omega_i Y^i),_k = (\hat{\nabla}_{\mathbf{X}}\omega)_i Y^i + \omega_i(Y^i,_k + \Gamma_{kh}^i Y^h)X^k. \tag{9.33}$$

Expanding the left-hand side of (9.33), simplifying the obtained relation and recalling that it must hold for any vector field $\mathbf{Y}$, we obtain the coordinate expression of the covariant derivative of a 1-form:

$$(\hat{\nabla}_{\mathbf{X}}\boldsymbol{\omega})_i = (\omega_{,i} - \Gamma^h_{ik}\omega_h)X^k. \tag{9.34}$$

Starting from the coordinate expression of an arbitrary (r, s)-tensor $\mathbf{T}$

$$\mathbf{T} = T^{i_1 \cdots i_r}_{j_1 \cdots i_s}\mathbf{e}_{i_1} \otimes \cdots \otimes \mathbf{e}_{i_r} \otimes \mathbf{dx}^{j_1} \otimes \cdots \otimes \mathbf{dx}^{j_s}$$

and applying properties 1–4, it is easy to verify that the coordinate expression of the covariant derivative of $\mathbf{T}$ is

$$(\hat{\nabla}_{\mathbf{X}}\mathbf{T})^{i_1 \cdots i_r}_{j_1 \cdots i_s} = X^k \left[T^{i_1 \cdots i_r}_{j_1 \cdots i_s},_k + \Gamma^{i_1}_{kh} T^{h \cdots i_r}_{j_1 \cdots i_s} + \cdots + \Gamma^{i_r}_{kh} T^{i_1 \cdots h}_{j_1 \cdots i_s} \right.$$
$$\left. - \Gamma^h_{kj_1} T^{i_1 \cdots i_r}_{h \cdots i_s} - \cdots - \Gamma^h_{kj_s} T^{i_1 \cdots i_r}_{j_1 \cdots h} \right].$$

$\square$

Henceforth, we use the symbol $\nabla_{\mathbf{X}}$ instead of $\hat{\nabla}_{\mathbf{X}}$.

9.5 Torsion and Curvature Tensors

Let $f(x^i)$ be a C^1 function in the chart (U, x^i). Then, owing to Schwarz's theorem,

$$f_{,ij} = f_{,ji}.$$

However, the n^2 quantities $f_{,ij}$ do not represent the components of a $(0, 2)$-tensor. In contrast, the quantities $\nabla_i f_{,j}$ can be regarded as the components of a $(0, 2)$-tensor, but we cannot change the derivation order since, in view of (9.34), we have that

$$\nabla_i f_{,j} - \nabla_j f_{,i} = (\Gamma^h_{ji} - \Gamma^h_{ij})f_{,h}. \tag{9.35}$$

The left-hand side gives the components of a $(0, 2)$-tensor and $f_{,j}$ are the components of a covector; consequently, owing to the criteria given in Sect. 2.6, we can state that the quantities

$$S^h_{ji} = \Gamma^h_{ji} - \Gamma^h_{ij} \tag{9.36}$$

are the components of a $(1, 2)$-tensor, which is called the **torsion tensor** of the affine connection of V_n. This tensor field vanishes if and only if the connection coefficients are symmetric with respect to the lower indices.

Consider a vector field $\mathbf{X}$, which in the chart (U, x^i) on the manifold V_n has the coordinate representation $\mathbf{X} = X^i \mathbf{e}_i$, where $(\mathbf{e}_i)$ is the natural basis of the coordinates x^i. Then, with simple calculations, it is possible to verify that

$$\nabla_j \nabla_i X^k - \nabla_i \nabla_j X^k = R^k_{lji}X^l - S^l_{ji}\nabla_l X^k, \tag{9.37}$$

where

$$R^k_{lji} = \Gamma^k_{il,j} - \Gamma^k_{jl,i} + \Gamma^k_{jh}\Gamma^h_{il} - \Gamma^k_{ih}\Gamma^h_{jl} \tag{9.38}$$

is the **Riemann curvature tensor**.

Now, we prove the following fundamental theorem.

Theorem 9.2. *Let V_n be a manifold equipped with an affine connection. Then the following statements are equivalent:*

(a) *There exists an atlas β on V_n in each chart of which the connection coefficients vanish;*
(b) *The torsion vanishes and, $\forall x, y \in V_n$, the parallel transport along any curve connecting x and y does not depend on the curve;*
(c) *The torsion vanishes and there are n vector fields $\mathbf{u}_1, \ldots, \mathbf{u}_n$ that are independent and uniform;*
(d) *The curvature and the torsion vanish.*

Proof. $(a) \Rightarrow (b)$.   If (U, x^i), $(\overline{U}, \overline{x}^i) \in \beta$, then in the intersection $U \cap \overline{U}$ of the domains, we obtain the results $\Gamma^i_{kh} = \overline{\Gamma}^i_{kh} = 0$. Consequently, in view of (9.20), the coefficients A^i_j of transformation matrix (9.19) verify the conditions

$$A^i_l(A^{-1})^l_{h,k} = 0,$$

that is,

$$(A^{-1})^m_i A^i_l (A^{-1})^l_{h,k} = (A^{-1})^l_{h,k} = 0,$$

and we conclude that the elements A^l_h of the transformation matrix are constant and the coordinate transformation $x^i \to \overline{x}^i$ is linear. Further, the parallel transport (9.24) along any curve contained in a domain of a chart of the atlas β reduces to $dX^i/dt = 0$. Let x, y be two arbitrary points of V_n. If both points belong to the same domain U of a chart of β, then the parallel transport of a vector $\mathbf{Y}_x = Y^i_x \mathbf{e}_i(x)$ along a curve $\gamma(t) \subset U$, connecting x and y, is obtained integrating the equation $dY^i/dt = 0$ along $\gamma(t)$, and it is independent of $\gamma(t)$. Suppose that x belongs to the chart (U, x^i) of β and y belongs to another chart $(\overline{U}, \overline{x}^i)$, with $U \cap \overline{U} \neq \varnothing$. Let $\gamma_1(t)$ and $\gamma_2(t)$ be two curves going from x to y. The parallel transport of $\mathbf{Y}_x$ along $\gamma_1(t)$ up to a point $p_1 = \gamma(t_1) \in U \cap \overline{U}$ gives the vector $Y^i_x \mathbf{e}_i(p_1)$, whereas the parallel transport of the same vector along $\gamma_2(t)$ up to a point $p_2 = \gamma(t_2) \in U \cap \overline{U}$ gives the vector $Y^i_x \mathbf{e}_i(p_2)$ since this transport for both the curves is expressed by the equation $dY^i/dt = 0$. Now we regard p_1 and p_2 as points belonging to $\overline{U}$. In the coordinates $(\overline{x}^i)$, the vectors $Y^i_x \mathbf{e}_i(p_1)$ and $Y^i_x \mathbf{e}_i(p_2)$ have, respectively, the representations $A^i_j Y^j_x \overline{\mathbf{e}}_i(p_1)$ and $A^i_j Y^j_x \overline{\mathbf{e}}_i(p_2)$ since the quantities A^i_j are constant all over $U \cap \overline{U}$. To prove condition *(b)* we must verify that the parallel transport of $A^i_j Y^j_x \overline{\mathbf{e}}_i(p_1)$ along $\gamma_1(t)$ from p_1 to y and of the vector $A^i_j Y^j_x \overline{\mathbf{e}}_i(p_2)$ along $\gamma_2(t)$

from p_2 to y leads us to the same vector $\mathbf{Y}_y$. But this is evident since we must integrate the equations $\mathrm{d}\overline{Y}^i/\mathrm{d}t = 0$ of the parallel transport along $\gamma_1(t)$ and $\gamma_2(t)$ with the same initial conditions.

$(b) \Rightarrow (c)$. Let $\mathbf{Y}_1, \ldots, \mathbf{Y}_n$ be n independent vectors at the arbitrary point $x \in V_n$. Since condition (b) holds, by a parallel transport of these vectors to any other point of V_n, we obtain n vector fields $\mathbf{u}_1, \ldots, \mathbf{u}_n$ whose values at a point do not depend on the curve along which the vectors are transported; in other words, they verify the condition

$$\nabla_\mathbf{A} \mathbf{u}_i = \mathbf{0}$$

for any vector field $\mathbf{A}$ tangent to the curve along which $\mathbf{u}_i$ is transported. From the arbitrariness of the curve, i.e., of $\mathbf{A}$, it follows that the fields $\mathbf{u}_i$ are uniform. It remains to prove that they are independent at any point. In this regard, it is sufficient to note that the parallel transport, when it is independent on the curves, defines an isomorphism $\phi : T_x(V_n) \to T_y(V_n)$ between the tangent spaces corresponding to points x and y owing to the linearity of the parallel transport equations and the existence and uniqueness theorems.

$(c) \Rightarrow (d)$. Since the torsion tensor vanishes, condition (9.37) becomes

$$\nabla_j \nabla_i Y^k - \nabla_i \nabla_j Y^k = R^k_{lji} Y^l.$$

Now we consider the vector field $\mathbf{Y} = a^i \mathbf{u}_i$, where the quantities $a^1, \ldots, a^n$ are arbitrary constants and $\mathbf{u}_1, \ldots, \mathbf{u}_n$ are uniform vector fields existing owing to property (c). For any field $\mathbf{Y}$ the preceding relation gives

$$R^k_{lji} a^l = 0,$$

and the arbitrariness of the quantities a^l implies $R^k_{lji} = 0$.

$(d) \Rightarrow (a)$. We must prove that, for any $x \in V_n$, there exists a chart (U, x'^i) such that in these coordinates the connection coefficients vanish. Let (V, x^i) be another chart containing x. Then, in view of (9.21), we must find a coordinate transformation $(x^i) \to (x'^i)$ such that

$$\Gamma'^p_{jh} = (A^{-1})^p_q A^m_j A^n_h \Gamma^q_{mn} + (A^{-1})^p_q A^q_{h,j} = 0,$$

where $(A^i_j) = (\partial x^i / \partial x'^j)$ and the connection coefficients Γ^q_{mn} are symmetric with respect to m and n since the torsion tensor vanishes. Multiplying the preceding equation by A^l_p, we obtain the system

$$A^l_{h,j} = -\Gamma^l_{mn} A^m_j A^n_h, \tag{9.39}$$

$$\frac{\partial x^l}{\partial x'^j} = A^l_j. \tag{9.40}$$

To find the $n^2 + n$ unknowns $A_j^l(x)$ and $x^{\prime i}(x)$ of the preceding $n^2 + n^3$ equations, the integrability conditions

$$\frac{\partial A_h^l}{\partial x^{\prime j} \partial x^{\prime i}} = \frac{\partial A_h^l}{\partial x^{\prime i} \partial x^{\prime j}},$$ (9.41)

$$\frac{\partial^2 x^l}{\partial x^{\prime j} x^{\prime i}} = \frac{\partial^2 x^l}{\partial x^{\prime i} x^{\prime j}}$$ (9.42)

must be satisfied. We write these conditions in the more compact form

$$A_{h,[ji]}^l = 0, \quad x^l{}_{,[ji]} = A_{[j,i]}^l = 0,$$ (9.43)

where we have used the notation $f_{[ij]} = f_{ij} - f_{ji}$. The integrability conditions $(9.43)_2$ are satisfied since, from (9.39) and the hypothesis that the torsion tensor vanishes, we obtain

$$\begin{aligned}
A_{[i,j]}^l &= -\Gamma_{mn}^l A_j^m A_i^n + \Gamma_{mn}^l A_i^m A_j^n \\
&= -\Gamma_{mn}^l A_j^m A_i^n + \Gamma_{nm}^l A_i^n A_j^m = A_i^n A_j^m \Gamma_{[n,m]}^l = 0.
\end{aligned}$$

Again for (9.39), the integrability conditions of $(9.43)_1$ can be written as follows:

$$\begin{aligned}
A_{h,[ji]}^l &= -\frac{\partial}{\partial x^{\prime i}} \left(\Gamma_{mn}^l A_j^m A_h^n \right) + \frac{\partial}{\partial x^{\prime i}} \left(\Gamma_{mn}^l A_i^m A_h^n \right) \\
&= -\frac{\partial \Gamma_{mn}^l}{\partial x^{\prime i}} A_j^m A_h^n - \Gamma_{mn}^l A_{j,i}^m A_h^n - \Gamma_{mn}^l A_j^m A_{h,i}^n \\
&\quad + \frac{\partial \Gamma_{mn}^l}{\partial x^{\prime j}} A_i^m A_h^n + \Gamma_{mn}^l A_{i,j}^m A_h^n + \Gamma_{mn}^l A_i^m A_{h,j}^n.
\end{aligned}$$

When we note that $\partial/\partial x^{\prime j} = A_j^k \partial/\partial x^k$, take into account (9.39), and recall the already proven result $A_{[i,j]}^l = 0$, the preceding equations become

$$A_{h,[ji]}^l = -A_i^m A_h^n A_j^k \left(\Gamma_{mn,k}^l - \Gamma_{kn,m}^l + \Gamma_{kp}^l \Gamma_{mn}^p - \Gamma_{mp}^l \Gamma_{kn}^p \right)$$

or, in equivalent form,

$$A_{h,[ji]}^l = -A_i^m A_h^n A_j^k R_{nmk}^l = 0,$$ (9.44)

and the theorem is proved.

$\square$

9.6 Riemannian Connection

Theorem 9.3. *A Riemannian manifold V_n admits one and only one affine connection ∇ such that*

- *The torsion tensor vanishes and*
- *The scalar product is invariant under a parallel transport.*

Proof. Let (U, x^i) be a chart on V_n, and denote by Γ^i_{kh} the connection coefficients whose existence we must prove. If $\gamma(t)$ is any curve connecting two arbitrary points $x, y \in U$ and $\mathbf{X}, \mathbf{Y}$ two vector fields that are parallel along $\gamma(t)$, then we have

$$\frac{dX^i}{dt} + \Gamma^i_{kh} X^k \frac{dx^h}{dt} = 0, \tag{9.45}$$

$$\frac{dY^i}{dt} + \Gamma^i_{kh} Y^k \frac{dx^h}{dt} = 0, \tag{9.46}$$

where $x^h(t)$ are the parametric equations of $\gamma(t)$ in the chart (U, x^i). The torsion tensor vanishes if and only if the connection coefficients are symmetric with respect to the lower indices. Further, the scalar product is invariant under the parallel transport if and only if the condition

$$\frac{d}{dt}(g_{ij} X^i Y^j) = 0 \tag{9.47}$$

is satisfied for any pair of vector fields $\mathbf{X}, \mathbf{Y}$ that verify solutions of Eqs. (9.45) and (9.46). Expanding (9.47) and taking into account (9.45) and (9.46), we obtain

$$\left(\frac{dg_{ij}}{dt} - g_{lj} \Gamma^l_{ih} \frac{dx^h}{dt} - g_{il} \Gamma^l_{jh} \frac{dx^h}{dt} \right) X^i Y^j = 0.$$

Owing to the arbitrariness of $\mathbf{X}$ and $\mathbf{Y}$ and the curve $x^h(t)$, the following conditions must be verified:

$$g_{ij,h} = g_{lj} \Gamma^l_{ih} + g_{il} \Gamma^l_{jh}. \tag{9.48}$$

This formula is identical to (9.4). Therefore, proceeding as in Section 9.1 we obtain again the formula (9.9)

$$\Gamma^n_{hj} = \frac{1}{2} g^{ni} \left(\frac{\partial g_{ij}}{\partial x^h} + \frac{\partial g_{ih}}{\partial x^j} - \frac{\partial g_{jh}}{\partial x^i} \right). \tag{9.49}$$

It is a simple exercise to verify that the connection defined by the preceding coefficients is without torsion and leaves invariant the scalar product under a parallel transport.   □

Definition 9.5. The *Riemannian connection* is the connection whose coefficients are given by (9.49).

Theorem 9.4 (Bianchi's theorem). *The metric tensor* **g** *is uniform, that is, its covariant derivative vanishes:*

$$\nabla_h g_{ij} = 0. \tag{9.50}$$

Proof. It is sufficient to refer to (9.48) and recall the expression of the covariant derivative of a $(0,2)$-tensor.   □

Theorem 9.5. *In a neighborhood of any point $x_0$of a Riemannian manifold, it is possible to define a local chart $(\overline{U}, \overline{x}^i)$ whose domain contains x_0 such that*

$$\overline{g}_{ij}(x_0) = \epsilon_i \delta_{ij}, \quad \overline{g}_{ij,h} = 0, \tag{9.51}$$

where $\epsilon = \pm 1$ according to the signature of V_n (Sect. 6.7).

Proof. Let (U, x^i) be a chart in a neighborhood of x_0, and denote by (x_0^i) the coordinates of x_0 in this chart. If $\Gamma^i_{kh}(x_0^i)$ are the Riemannian connection coefficients evaluated at x_0, then the functions

$$\overline{x}^i = x^i - x_0^i - \frac{1}{2}\Gamma^i_{kh}(x_0^i)(x^k - x_0^k)(x^h - x_0^h) \tag{9.52}$$

define a coordinate transformation in a neighborhood of x_0 since $(\partial \overline{x}^i / \partial x^j (x_0)) = (\delta_{ij})$. Further, it is

$$\frac{\partial \overline{x}^i}{\partial x^h \partial x^k}(x_0) = -\Gamma^i_{kh}(x_0). \tag{9.53}$$

From the transformation formulae (9.21), when we take into account that $A^i_j(x_0) = \delta^i_j$ and recall (9.53), we obtain

$$\overline{\Gamma}^i_{kh}(x_0) = \Gamma^i_{kh}(x_0) - \Gamma^i_{kh}(x_0) = 0.$$

Since the connection is Riemannian, (9.49) holds, and then we have $g_{ij,h}(x_0) = 0$. Finally, by a linear transformation $(\overline{x}^i) \to (\overline{x}^{i})$, which does not modify the preceding results, we can reduce the metric coefficients at x_0 to the form (9.51).   □

The Riemannian manifold V_n is said to be locally Euclidean if in the neighborhood of any point there is a system of coordinates (U, x^i) in which $g_{ij} = \epsilon_i \delta_{ij}, \forall x \in U$.

9.7 Differential Operators on a Riemannian Manifold

Let f be a C^1 real function on the Riemannian manifold V_n. We call **gradient** of f the vector field ∇f whose *covariant* components in a chart (U, x^i) are

$$(\nabla f)_k = f_{,k}. \tag{9.54}$$

Consequently, in the natural basis $(\mathbf{e}_i)$, relative to the chart (U, x^i), we can write

$$\nabla f = g^{ij} f_{,j} \mathbf{e}_i. \tag{9.55}$$

Let $\mathbf{X}$ be a vector field on V_n. The skew-symmetric 2-tensor

$$(\nabla \times \mathbf{X})_{ij} = \nabla_j X_i - \nabla_i X_j = X_{i,j} - X_{j,i} \tag{9.56}$$

is called a **curl tensor**. From (9.55) and (9.56) it follows that

$$\nabla \times \nabla f = \mathbf{0}. \tag{9.57}$$

Finally, we call the function

$$\nabla \cdot \mathbf{X} = \nabla_i X^i \tag{9.58}$$

the **divergence** of the vector field $\mathbf{X}$.

To determine a useful form of (9.58), we start by noting that

$$\nabla \cdot \mathbf{X} = X^i_{,i} + \Gamma^i_{ih} X^h. \tag{9.59}$$

On the other hand, in view of (9.48), we have that

$$g^{ij} g_{ij,h} - \Gamma^i_{ih} - \Gamma^j_{jh} = 0,$$

so that we can write

$$\Gamma^i_{ih} = \frac{1}{2} g^{ij} g_{ij,h}. \tag{9.60}$$

Further, if we set $g = \det(g_{ij})$, it is well known that $g_{,h} = g g^{ij} g_{ij,h}$. Consequently, (9.60) becomes

$$\Gamma^i_{ih} = \frac{1}{2g} g_{,h} = \frac{1}{\sqrt{|g|}} (\sqrt{|g|})_{,h} \tag{9.61}$$

and (9.59) assumes the final form

$$\nabla \cdot \mathbf{X} = \frac{1}{\sqrt{|g|}}(\sqrt{|g|}X^i)_{,i}. \tag{9.62}$$

The **Laplace operator** of the C^2 real function f is

$$\Delta f = \nabla \cdot \nabla f. \tag{9.63}$$

In local coordinates (x^i), in view of (9.63) and (9.55), we can write the Laplace operator in the form

$$\Delta f = \nabla_i(g^{ij} f_{,j}) = g^{ij}(f_{,ij} - \Gamma^k_{ij} f_{,k}) = \frac{1}{\sqrt{|g|}}(\sqrt{|g|}g^{ij} f_{,j})_{,i}. \tag{9.64}$$

Any function f on V_n verifying the Laplace equation

$$\Delta f = 0 \tag{9.65}$$

is called a **harmonic function**. A system of coordinates (U, x^i) is harmonic in the metric g_{ij} if the coordinate functions $x^i, i = 1, \ldots, n$, are harmonic. Taking into account (9.64), we can state that the coordinates x^i are harmonic if they satisfy the equations

$$F^i \equiv -\Delta x^i = g^{hk}\Gamma^i_{hk} = 0. \tag{9.66}$$

9.8   Exercises

The notebook **Geometry** can help the reader to solve the following exercises. This notebook, *given the metric of a manifold*, allows to determine all the geometric quantities that are related to the metric.

1. Let C be a cylinder on which we introduce the cylindrical coordinates φ, z. Determine the geodesics of the metrics on C and compare the parallel transport of the vector $(1,0)$ along the curve $z = \varphi, 0 \le \varphi \le \pi/4$, and along the curve union of $0 \le \varphi \le \pi/4, z = 0$, and $\varphi = \pi/4, 0 \le z \le \pi/4$.
2. Let S be a unit sphere on which we introduce the spherical coordinates φ, θ. The equator γ of S has parametric equations $(\varphi, 0), 0 < \varphi \le 2\pi$. Show that the equator is a geodesic, and evaluate the parallel transport along γ of the tangent vector $(0, 1)$.
3. On the unit sphere S consider the triangle ABC (Fig. 9.1) in which AC and BC are arcs of meridians, AB an arc of the equator, and B a pole. Let α be the angle at the vertex C. Show that the three sides of the triangle are arcs of geodesics and evaluate the parallel transport of the vector $(1,0)$ along the sides of ABC from A to A.

Fig. 9.1 Geodesic triangle

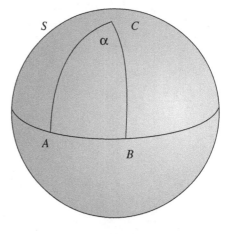

Fig. 9.2 Parallelogram on a sphere

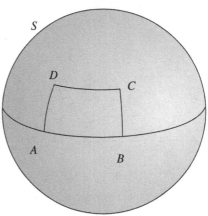

4. Using the notations of the preceding exercise, evaluate the parallel transport of the vector $(1,0)$ from A to A along the curve $ABCA$ in Fig. 9.2, where AD and BC are arcs of meridians and AB and CD are arcs of parallels.

Chapter 10
An Overview of Dynamical Systems

Abstract This chapter is devoted to an overview of dynamical systems that play a fundamental role in building mathematical models of reality. After a brief introduction of modeling, we present some theorems of existence and uniqueness as well as the definitions of first integral and phase portrait. Then, we define Liapunov's stability for autonomous systems together with some theorems of stability and instability of equilibrium. Poincare's perturbation method is described with some applications. Finally, Weierstass's qualitative analysis is presented.

10.1 Modeling and Dynamical Systems

In previous chapters, some fundamental concepts of algebra and differential geometry were presented. This chapter is devoted to an overview of dynamical systems that play a fundamental role in building mathematical models of reality.[1]

Several interesting behaviors of physical, biological, economical, and chemical systems can be described by ordinary differential equations (ODEs). Applied scientists are interested in mathematical problems of models stated by ODEs. The explicit form of their solutions can be found when the ODEs are linear, but often nonlinearity represents an inner unavoidable feature of the model, and in this case we cannot exhibit the explicit solutions. Mathematical methods must be developed to tackle these difficulties. We usually resort to qualitative analysis, which supplies important aspects of solutions such as their asymptotic behavior, stability properties, the existence of limit cycles, and the possibility of bifurcation on varying a parameter contained in the equations. When the applications require a quantitative description of solutions, we can use known procedures that supply the approximate time evolution of the dependent variable obtained by numerical integration, power series, or expansion in one or more parameters.

---

[1] The topics contained in this chapter can also be found in [5, 6, 9, 25, 28, 29, 35, 40, 55].

© Springer International Publishing AG, part of Springer Nature 2018   137
A. Romano and A. Marasco, *Classical Mechanics with Mathematica®*,
Modeling and Simulation in Science, Engineering and Technology,
https://doi.org/10.1007/978-3-319-77595-1_10

In attempting to describe reality, people resort to **models** representing simplified but useful mathematical descriptions of phenomena we are interested in. It is not a simple task to identify the main characteristics of a phenomenon, describe it in terms of mathematical variables, recognize the mathematical relations among these, and finally verify if the expectations of the model agree with observation. Solving these difficult problems is the main goal of the mathematical modeling of nature.

In this difficult process, the transcription of an aspect of reality is often represented by a mathematical object called a (scalar or vector) **ODE**. To understand why modeling leads us to this object, and what kinds of problems one finds at the end of this process, we start with a very simple example containing the fundamental ingredients of the problems we face.

Suppose that one wishes to describe the evolution of a population consisting of people, animals, bacteria, radioactive atoms, etc. The importance of this problem is clear: its resolution could allow decisive forecasts on the destiny of the examined population. For example, the growth velocity of a population of infectious bacteria in an organism could suggest the appropriate dosage of antibiotic. Similarly, if the growth law of a population is known, the right amount of food for its survival can be evaluated.

To formulate a growth model for a population, the factors that positively or negatively influence the aforementioned process must first be identified and a mathematical relation supplying the number $N(t)$ of living individuals at instant t as a function of those factors must be formulated. Proceeding from the simplest situation, all external influences on growth are neglected and the food source is assumed unlimited. In this situation, it is quite reasonable to suppose that the variation $\Delta N(t)$ of the number $N(t)$ in the time interval $(t, t + \Delta t)$ is proportional both to $N(t)$ and Δt

$$\Delta N(t) = \alpha N(t) \Delta t,$$

where α is a coefficient depending on the kind of population. Of course, this coefficient is positive if the population increases and negative in the opposite case. In the limit $\Delta t \to 0$, the number $N(t)$ satisfies the equation

$$\frac{dN(t)}{dt} = \alpha N(t),$$

in which $N(t)$ is the unknown appearing in the equation with its first derivative. Such a mathematical object is just an ODE of the first order in the unknown $N(t)$. We underline that the hypotheses leading us to the previous equation are very spontaneous, but this is not the case when forecasting the form of its solution[2]

$$N(t) = Ce^{\alpha t},$$

---

[2]This solution is obtainable at once by the method of variable separation. However, the reader can easily verify that it is really a solution for any C.

where C is an arbitrary positive constant. Now the main reason why differential equations frequently appear in modeling natural phenomena can be recognized: *It is much easier to formulate a reasonable relation between the unknown of a given problem and its variations than to imagine the form of the function itself.* The same kind of population admits infinite evolutions depending on the *initial datum* or *condition* $N(0)$ that assigns a value to C.

The previous model, where $\alpha > 0$, implies that the number of individuals goes to infinity with time, and this is absurd. To improve the model, the habitat in which the population lives is not supposed to support a population level greater than M; this means that when $N(t)$ reaches this value, the growth rate becomes zero. The limit M is called the *carrying capacity*. To take into account this constraint, the coefficient α appearing in the preceding equation is assumed depending on the quantity $(M - N(t))/M$. The simplest dependency on this variable is the direct proportionality

$$\alpha = \beta \left(1 - \frac{N(t)}{M} \right),$$

where $\beta > 0$ is constant, and we arrive at the relation

$$\dot{N}(t) = \beta \left(1 - \frac{N(t)}{M} \right) N(t),$$

which is called the *logistic equation*. This *nonlinear* equation gives a more accurate description of the evolution of a population since it introduces an upper bound to its growth. Further, it represents the starting point of many models describing the competition among species living in the same habitat or the modified prey–predator model of Lotka–Volterra (see, for instance, [28, 29]). When the variables are separated, the family of solutions

$$N(t) = \frac{M}{1 + Ce^{-\beta t}}$$

is easily obtained, where C is an arbitrary constant determined by assigning the initial value $N(0)$. It is a very lucky circumstance that the solution of the logistic equation can be exhibited. However, more frequently, one must resort to other procedures if information about the unknown solutions is needed.[3]

Before we continue, some consequences of the previous considerations must be underlined: (1) it is much easier to model by a differential equation than directly finding the finite relation between the involved variables; (2) when solutions exist, they are infinite in number, but one of them is assigned by giving the value of the unknown at a certain instant; and (3) it might be impossible to find the closed form of these solutions.

---

[3]For the application of differential equations to economy, see, for instance, [55].

In all the examined cases, we were led to an ordinary first-order differential equation, i.e., to an equation containing an unknown function and its first derivative. In many other cases, we could be compelled to model our phenomenon by a system of two or more higher order ODEs with two or more unknowns. This means that we are faced with a system of equations containing more unknowns and their higher order derivatives. In the chapters devoted to dynamics, we shall see that describing the evolution of mechanical systems almost always leads to a system of differential equations. In this regard, here we only recall that Galileo and Newton proved that the wide variety of possible motions of material bodies confirm the three fundamental principles of dynamics (Chap. 13). It is an everyday experience that a body may fall in many different ways under the influence of its weight. Similarly, the planets describe different orbits around the Sun and traverse them at different velocities. Galileo and Newton discovered that a very simple relation of proportionality relates the acceleration $\mathbf{a}$ of a material point P to the acting force $\mathbf{F}$

$$m\mathbf{a} = \mathbf{F},$$

where m is the mass of the body. If $(x(t), y(t), z(t))$ are the coordinates in a frame of reference $Oxyz$ of the moving point P as a function of time, then the components of the acceleration vector are $(\ddot{x}(t), \ddot{y}(t), \ddot{z}(t))$. On the other hand, the force depends on the position of P (that is, on $(x(t), y(t), z(t))$) as well as on its velocity $(\dot{x}(t), \dot{y}(t), \dot{z}(t))$. Consequently, the fundamental equation of dynamics is essentially a system of three ordinary second-order differential equations in the unknowns $(x(t), y(t), z(t))$. By the auxiliary variables $(u, v, w) = (\dot{x}, \dot{y}, \dot{z})$, this reduces to a system of six first-order differential equations.

10.2   General Definitions and Cauchy's Problem

Let $\Re^n$ be the vector space of the ordered n-tuples of real numbers $\mathbf{x} = (x_1, \dots, x_n)$ and $\|\mathbf{x}\|$ the Euclidean norm in $\Re^n$. Moreover, let $\mathbf{f} : \Omega \to \Re^n$ be a map defined on the open subset Ω of $\Re^n$. We call an *autonomous first-order (vector) differential equation in the normal form* the following equation:

$$\dot{\mathbf{x}} = \mathbf{f}(\mathbf{x}). \tag{10.1}$$

Essentially, (10.1) is an abbreviation of the system of differential equations

$$\dot{x}_i = f_i(x_1, \dots, x_n), \ i = 1, 2, \dots, n.$$

The vector space $\Re^n$, to which $\mathbf{x}$ belongs, is said to be the **state space**.

A smooth function $\mathbf{x} : I \to \Omega \subset \Re^n$, where I is a nonempty interval of $\Re$, is a *solution* of (10.1) if

$$\mathbf{x}(t) \in \Omega, \quad \dot{\mathbf{x}}(t) = \mathbf{f}(\mathbf{x}(t)), \quad \forall t \in I.$$

The set $\mathbf{x}(t)$, $\forall t \in I$ is called an **orbit**, and it represents a curve in the state space.

The **initial-value problem** or **Cauchy problem** for (10.1) consists in searching for the solution $\mathbf{x}(t, t_0, \mathbf{x}_0)$ of (10.1) that satisfies the following **initial condition** or **initial datum**:

$$\mathbf{x}_0 = \mathbf{x}(t_0), \quad \mathbf{x}_0 \in \Omega. \tag{10.2}$$

It is possible to prove that the solution of an autonomous equation depends on $t - t_0$. For this reason, henceforth we denote by $\mathbf{x}(t, \mathbf{x}_0)$ the solution of the Cauchy problem (10.1) and (10.2). In a geometric language we search for a curve $\mathbf{x}(t, \mathbf{x}_0)$ containing $\mathbf{x}_0$ for $t = t_0$.

A solution $\mathbf{x}(t) : I \to \mathfrak{R}^n$ of (10.1) is a **maximal solution** if there is no other solution $\mathbf{y}(t) : J \to \mathfrak{R}^n$, where $I \subset J$ such that $\mathbf{x}(t) = \mathbf{y}(t)$, $\forall t \in I$. It is possible to prove that the interval in which a maximal solution is defined is always open.

The function $\mathbf{f}(\mathbf{x})$ is a **Lipschitz function** in Ω if it satisfies a **Lipschitz condition** in Ω with respect to $\mathbf{x}$, i.e., if a positive constant K exists such that

$$\| \mathbf{f}(\mathbf{x}) - \mathbf{f}(\mathbf{y}) \| \leq K \| \mathbf{x} - \mathbf{y} \| \tag{10.3}$$

$\forall \mathbf{x}, \mathbf{y} \in \Omega$. Moreover, we say that $\mathbf{f}(\mathbf{x})$ satisfies a *local Lipschitz condition* in Ω with respect to $\mathbf{x}$ if each point of Ω has a neighborhood A such that the restriction of $\mathbf{f}$ to $A \cap \Omega$ is a Lipschitz function with respect to $\mathbf{x}$. The fundamental theorem of ODEs is as follows, which ensures the existence and uniqueness of the solution to the Cauchy problem locally in time.[4]

Theorem 10.1 (Piccard–Lindelöf). *Let* $\mathbf{f} : \Omega \to \mathfrak{R}^n$ *be a continuous and locally Lipschitz function with respect to* $\mathbf{x}$. *For any* $(t_0, \mathbf{x}_0) \in \mathfrak{R} \times \Omega$, *one and only one maximal solution of (10.1) exists* $\mathbf{x}(t) : I(t_0, \mathbf{x}_0) \to \mathfrak{R}^n$ *such that* $t_0 \in I(t_0, \mathbf{x}_0)$ *and* $\mathbf{x}_0 = \mathbf{x}(t_0)$. *Moreover, if the notation*

$$A = \{(t, t_0, \mathbf{x}_0) : (t_0, \mathbf{x}_0) \in \mathfrak{R} \times \Omega, \quad t \in I(t_0, \mathbf{x}_0)\}$$

is introduced, the maximal solution $\mathbf{x}(t, \mathbf{x}_0) : A \to \mathfrak{R}^n$ *of (10.1) is continuous.*

The following examples explain Theorem 10.1.

Example 10.1. The right-hand side of the equation $\dot{x} = \sqrt{x}$ does not satisfy a Lipschitz condition in any interval $[0, a]$, $a > 0$. It is easy to verify that the Cauchy problem obtained by associating to the equation the initial datum $x(0) = 0$ admits the infinite solutions

---

[4]For a proof, see, for instance, [40].

$$x = \begin{cases} 0, & -a \le t \le a, \\ \frac{1}{4}(t-a)^2, & a \le t, \\ \frac{1}{4}(t+a)^2, & t \le -a, \end{cases} \qquad \forall a \in \Re.$$

Example 10.2. The right-hand side of the equation

$$\dot{x} = \frac{1}{x}$$

does not satisfy a Lipschitz condition in the interval $(0,1]$. However, the solution

$$x = \sqrt{x_0{}^2 + 2t},$$

corresponding to the initial condition $x_0 > 0$, is unique.

Another very important concept is that of the *general integral* of (10.1). This is a function $\mathbf{x}(t, \mathbf{c}), \mathbf{c} \in \Re^n$ (1) that is a solution of (10.1) for any choice of $\mathbf{c}$ and (2) for any initial datum $(t_0, \mathbf{x}_0)$ only a choice $\mathbf{c}_0$ of $\mathbf{c}$ exists for which $\mathbf{x}_0 = \mathbf{x}(t_0, \mathbf{c}_0)$. We conclude this section by recalling that a *first integral* of (10.1) is a smooth real function $g(\mathbf{x}), (\mathbf{x}) \in \Omega$ that is constant along any solution of (10.1). Formally,

$$g(\mathbf{x}(t)) = \text{const}, \forall t \in I, \tag{10.4}$$

for any $\mathbf{x}(t) : I \to \Re^n$ such that $\dot{\mathbf{x}} = \mathbf{f}(\mathbf{x}(t))$.

In geometrical language, we can say that a function $g(\mathbf{x})$ in the state space is a first integral when an orbit starting from a point of the surface $g(\mathbf{x}) = \text{const}$ of $\Re^n$ lies completely on it. We can introduce the vector field $\mathbf{X}(\mathbf{x}) = \mathbf{f}(\mathbf{x})$. From (10.1), the orbits at any point $\mathbf{x}$ are tangent to the field $\mathbf{X}(\mathbf{x})$ or, equivalently, the orbits are the integral curves of the field $\mathbf{X}(\mathbf{x})$. This geometric interpretation will be used throughout this chapter.

Finally, we call the *phase portrait* the family of the orbits of (10.1). The notebook **Phase2D** supplies the phase portrait of (10.1) when $n = 2$ and Cartesian coordinates are adopted in the plane. The program **PolarPlot**, that is contained in the notebook **Phase2D**, draws the $2D$-phase portrait when polar coordinates are used.

10.3   Preliminary Considerations About Stability

Let a real system S be modeled by an equation like (10.1). A particular evolution $\mathbf{x}(t, \mathbf{x}_0)$ of S is completely determined by assigning the Cauchy datum $(t_0, \mathbf{x}_0)$. However, this datum is obtained by an experimental procedure and is therefore affected by an error. If a "small" difference in the initial data leads to a new solution that is "close" to the previous one, then $\mathbf{x}(t, \mathbf{x}_0)$ is said to be stable in the sense of Lyapunov (see, e.g., [9, 25, 28, 35]). To make precise the notion of Lyapunov stability, one must attribute a meaning to terms like small and close. By considering that $\mathbf{x}$ is a point of $\mathfrak{R}^n$, the Euclidean norm for evaluating the difference between two solutions or two initial data can be used. In particular, the stability property can always be expressed for an equilibrium solution. In fact, by introducing the new unknown $\mathbf{x} - \mathbf{x}(t, \mathbf{x}_0)$, the analysis of stability of any solution can always be reduced to an analysis of the equilibrium stability of the origin of a system, which, in general, becomes nonautonomous. An equilibrium position that is not stable is called unstable. In such a case, near the equilibrium there are initial data whose corresponding solutions go definitively away from the equilibrium. Moreover, the equilibrium position is said to be asymptotically stable if it is stable, and the solutions associated to the initial data in a neighborhood of the equilibrium tend to the equilibrium position when the independent variable goes to infinity. In this section, we refer to the stability of the origin of an autonomous system.

Although the concrete meaning and the importance of the stability theory are plain, at first sight one might think that, to check the equilibrium stability, knowledge of all solutions whose initial data are close to equilibrium is required. If this were true, it would be almost impossible to recognize this property because of the difficulty of obtaining the closed form of solutions. In this chapter, the Lyapunov direct method is described; it overcomes this difficulty by introducing a suitable function (called a *Lyapunov function*) that verifies the solutions. To understand this idea, we consider the system

$$\begin{cases} \dot{x} = y, \\ \dot{y} = -x, \end{cases}$$

whose solutions are curves $(x(t), y(t))$ of the plane x, y. It is very easy to verify that the function

$$V = \frac{1}{2}(x^2 + y^2)$$

is a first integral of the preceding system because along all the solutions it is $\dot{V} = 0$. The level curves $V(x, y) = \text{const}$ are circles, so that the solutions starting near the origin remain near it because they must belong to the circle to which the initial datum belongs. This, in turn, means that the origin is stable; this property was deduced without solving the system itself.

We conclude by remarking that the stability concept is much richer because it includes many other aspects. For example, what happens to the solutions of (10.1) if the function on the right-hand side is slightly changed? This is a very important problem because the function $\mathbf{f}$ is a mathematical transcription of the system we are describing. Further, $\mathbf{f}$ includes parameters that are the results of measures and therefore are again affected by errors. The analysis of the behavior of the solution on varying the form of $\mathbf{f}$ or of the parameters it includes is carried out by the *total stability* and *bifurcation theory* (see, for instance, [6, 28, 29, 35]). However, this subject will not be considered in this chapter.

10.4   Definitions of Stability

We suppose that the function $\mathbf{f} : D \subset \mathfrak{R}^n \to \mathfrak{R}^n$ of the autonomous system (10.1) satisfies all the conditions that ensure the existence and uniqueness of maximal solutions, namely, the vector function $\mathbf{f}$ satisfies a local Lipschitz condition (Theorem 10.1). In the sequel, $\mathbf{x} = \mathbf{x}(t, \mathbf{x}_0)$, $t \in I$, denotes the maximal solution of (10.1) corresponding to the initial datum $\mathbf{x}_0$.

Any constant function $\mathbf{x}(t) = \mathbf{x}^*$ that is a solution of (10.1) is called an ***equilibrium solution***. Equivalently, $\mathbf{x}^*$ is a root of the equation $\mathbf{f}(\mathbf{x}) = 0$. By a suitable axis translation, it is always possible to reduce any equilibrium solution at the origin. For this reason, in the sequel it is assumed that $\mathbf{f}(0) = 0$. Moreover, χ will denote the distance between the origin $\mathbf{x} = 0$ and the boundary ∂D of D.

The origin is a ***stable*** equilibrium solution if a positive real number χ exists such that $\forall \epsilon \in]0, \chi[$, $\exists \delta(\epsilon) \in]0, \epsilon[$ for which

$$\| \mathbf{x}_0 \| < \delta(\epsilon) \implies \| \mathbf{x}(t, \mathbf{x}_0) \| < \epsilon, \qquad \forall t > 0. \tag{10.5}$$

It is evident that $\delta(\epsilon) < \epsilon$; otherwise, condition (10.5) should not be verified at the initial instant. It is important to note the following. First, the stability property is equivalent to the continuity of solutions of (10.1) with respect to the initial datum $\mathbf{x}_0$ *uniformly in the unbounded time interval* $[0, \infty]$. If (10.1) is interpreted as a mathematical model of a real system S, its solutions represent the possible evolutions of the variable $\mathbf{x}$ describing the state of S. To have S at the equilibrium state $\mathbf{x}(t) = 0$, S must initially be put at 0. But this operation is physically impossible because it is necessarily realized by procedures that introduce measure errors, however accurate they may be. This means that the system is put at an initial state $\mathbf{x}_0$ next to 0, and the corresponding solution could lead the state of the system definitively far from the equilibrium. If this happened for some initial data, the equilibrium state itself would not be physically observable. In contrast, if 0 is stable, this situation is not confirmed. In fact, if the notation $U_\delta = \{\mathbf{x} \in \mathfrak{R}^n : \| \mathbf{x} \| < \delta\}$ is used, the stability notion can be formulated in the following terms. If any region $U_\epsilon \subset D$ is fixed around the origin, it is possible to find a whole neighborhood $U_\delta \subset U_\epsilon$ of initial conditions whose corresponding solutions are fully contained in U_ϵ.

The equilibrium solution $\mathbf{x} = 0$ is as follows:

1. ***Attractive*** if

$$\exists \Omega_0 \subset \mathfrak{R}^n : \forall \mathbf{x}_0 \in \Omega_0 \implies \begin{array}{l} (a) \quad \mathbf{x}(t, \mathbf{x}_0) \text{ exists } \forall t \geq 0 \\[2mm] (b) \quad \forall \epsilon > 0, \exists T(\epsilon) : \| \mathbf{x}(t, \mathbf{x}_0) \| < \epsilon, \forall t \geq T(\epsilon); \end{array} \tag{10.6}$$

2. ***Asymptotically stable*** if it is stable and

$$\lim_{t\to\infty} \mathbf{x}(t, \mathbf{x}_0) = \mathbf{0}; \tag{10.7}$$

3. **Unstable** if it is not stable, i.e., if

$$\exists \epsilon \in]0, \chi[: \forall \delta \in]0, \epsilon], \exists \mathbf{x}_0 \in U_\delta, t^* > 0 : \parallel \mathbf{x}(t^*, \mathbf{x}_0) \parallel \geq \epsilon. \tag{10.8}$$

The set Ω_0 of initial data $\mathbf{x}_0$ for which (10.6) is satisfied is called the **domain of attraction**. In particular, if $\Omega_0 = D$, and the origin is stable, $\mathbf{x} = \mathbf{0}$ is **globally asymptotically stable**. Figures 10.1–10.3 supply a rough representation in $\mathfrak{R}^2$ of a stable origin, attractive origin, asymptotically stable origin, and unstable origin, respectively. In Fig. 10.1, the origin is stable since there are two neighborhoods U_δ and U_ϵ of the origin, where $U_\delta \subset U_\epsilon$, of the origin such that the orbits starting from any point in U_δ are always contained in U_ϵ. In Fig. 10.2, the origin is attractive since the solutions corresponding to the initial data belonging to Ω_0 are contained in a neighborhood of the origin after a suitable value of time. In Fig. 10.3, the origin is asymptotically stable because the solutions corresponding to the initial data

Fig. 10.1 Stable origin

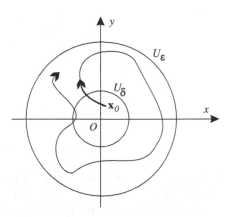

Fig. 10.2 Attractive origin

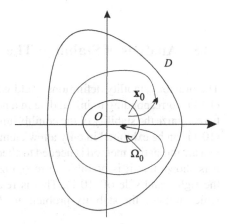

Fig. 10.3 Asymptotically
stable origin

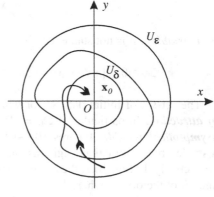

Fig. 10.4 Unstable origin

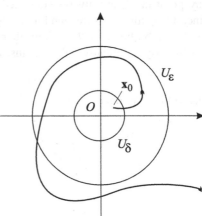

belonging to U_δ are contained in U_ϵ and, moreover, tend to the origin when the
variable t goes to infinity. Finally, in Fig. 10.4, the origin is unstable because there
are initial data whose corresponding solutions go definitively away from the origin.

10.5 Analysis of Stability: The Direct Method

The previous stability definitions could generate the wrong idea that all solutions of
(10.1), corresponding to initial data in a neighborhood of **0**, must be known in order
to recognize the stability of the equilibrium position. It is evident that the solutions of
(10.1) can be exhibited in only a few elementary cases (linear or special systems) so
that an alternative method is needed to check the stability of the origin. In this section,
it is shown that it is possible to recognize the stability of the origin by analyzing
the right-hand side of (10.1). This is realized in the ***direct method of Lyapunov***,
which reduces the stability problem to that of the existence of suitable functions

(*Lyapunov functions*) having, with their derivatives, determined properties along
the solutions of (10.5). This method is called direct because, under the hypothesis
that these functions are of class C^1, the corresponding derivatives on the solutions of
(10.1) can be directly expressed by the right-hand side of (10.1) without knowing the
solutions themselves. The determination of a Lyapunov function, in turn, is not easy.
Therefore, the original idea was developed in different directions to make easier the
application of the method itself.

In this section, some stability or instability criteria related to the direct method
are discussed.

We extensively use the notation

$$\dot{V}(\mathbf{x}) \equiv f(\mathbf{x}) \cdot \nabla V(\mathbf{x}), \tag{10.9}$$

together with the remark that along the solutions of (10.1) $\dot{V}(\mathbf{x})$ coincides with the
total derivative of $V(\mathbf{x}(t, \mathbf{x}_0))$ because

$$\dot{V}(\mathbf{x}(t, \mathbf{x}_0)) \equiv \dot{\mathbf{x}}(t, \mathbf{x}_0) \cdot \nabla V(\mathbf{x}(t, \mathbf{x}_0)) = \mathbf{f}(\mathbf{x}(t, \mathbf{x}_0)) \cdot \nabla V(\mathbf{x}(t, \mathbf{x}_0)). \tag{10.10}$$

This remark implies that the sign of $\dot{V}(\mathbf{x}(t, \mathbf{x}_0))$ along the solutions can be established
without the preliminary knowledge if the sign of $\dot{V}(\mathbf{x})$ in $\mathfrak{R}^n$.

Definition 10.1. Let $V : U_{\epsilon^*} \to \mathfrak{R}$ be a continuous function such that $V(\mathbf{0}) = 0$. V
is said to be *positive (negative) definite* on U_{ϵ^*} if $V(\mathbf{x}) > 0$ ($V(\mathbf{x}) < 0$) for $\mathbf{x} \neq 0$ or
positive (negative) semidefinite if $V(\mathbf{x}) \geq 0$ ($V(\mathbf{x}) \leq 0$).

Theorem 10.2. *Let $V : U_{\epsilon^*} \to \mathfrak{R}$ be a class C^1 function that is positive definite
in a neighborhood U_{ϵ^*} of the origin. If $\dot{V}(\mathbf{x})$ is negative semidefinite in U_{ϵ^*}, then
the origin is stable. Moreover, if $\dot{V}(\mathbf{x})$ is negative definite in U_{ϵ^*}, then the origin is
asymptotically stable.*

Proof. $\forall \epsilon \in (0, \epsilon^*]$ we denote a sphere with its center at the origin and radius ϵ. Since
$U_\epsilon \subset U_{\epsilon^*}$ and in $U_{\epsilon^*} - \{\mathbf{0}\}$ the function $V(\mathbf{x})$ is positive definite and continuous, there
exists a positive minimum of the restriction of $V(\mathbf{x})$ on ∂U_ϵ:

$$0 < V_\epsilon \equiv \min_{\mathbf{x} \in \partial U_\epsilon} V(\mathbf{x}).$$

Again for the continuity of $V(\mathbf{x})$ at the origin, there is a positive real quantity $\delta < \epsilon$
such that

$$|\mathbf{x}| < \delta \Rightarrow 0 < V(\mathbf{x}) < V_\epsilon.$$

Now we prove that any solution $\mathbf{x}(t, \mathbf{x}_0)$ of (10.1), for which $|\mathbf{x}_0| < \delta$, remains for
any $t > 0$ in the sphere U_ϵ. In fact, if this statement were false, there should be an
instant t^* in which the orbit $\mathbf{x}(t, \mathbf{x}_0)$ should cross ∂U_ϵ. But initially it is $V(\mathbf{x}_0) < V_\epsilon$,
so that we should have $V(\mathbf{x}_0) < V_\epsilon \leq V(\mathbf{x}(t^*, \mathbf{x}_0))$ against the condition $\dot{V} \leq 0$ along
the orbit $\mathbf{x}(t, \mathbf{x}_0)$. $\qquad \square$

Fig. 10.5 Asymptotically
stable origin

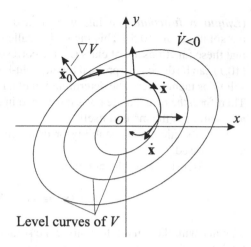

Level curves of V

The proof of the following theorem is omitted.

Theorem 10.3. *Let $V : U_{\epsilon^*} \to \Re$ be a class C^1 function in a neighborhood U_{ϵ^*} of the origin such that (1) $V(0) = 0$ and (2) for any neighborhood $U_\epsilon \subset U_{\epsilon^*}$ a point $\mathbf{x} \in U_\epsilon$ exists at which $V(\mathbf{x}) > 0$. If $\dot{V}(\mathbf{x})$ is positive definite in U_{ϵ^*}, then the origin is unstable.*

It is possible to recognize the role of the hypotheses contained in Theorems (10.2) and (10.3) by a simple geometric description in $\Re^2$. Limiting our considerations to Theorem (10.2), let c be a trajectory starting from a point $\mathbf{x}_0$ of a level curve γ of the Lyapunov function. At this point, c has a tangent vector $\dot{\mathbf{x}} = \mathbf{f}(\mathbf{x}_0)$ that points toward the internal region of γ, and the angle between ∇V and $\dot{\mathbf{x}}$ is greater than π. Consequently, c meets the more internal level curves under the same conditions. In this way, c approaches the origin (see Fig. 10.5 for a polynomial form of Liapunov's function).

Similar considerations, applied to the case $\dot{V} \leq 0$, allow us to conclude that the trajectory does not leave the internal region to the level curve of V containing x_0. However, the condition $\dot{V} \leq 0$ does not exclude the possibility that the trajectory will remain on a level curve so that it cannot approach the origin.

In particular, when $\dot{V} = 0$ along the solutions, one has the situation illustrated in Fig. 10.6 for a polynomial form of Liapunov's function.

Before stating the next theorem of Barbashin–Krasovskii, a simple physical example will be discussed. Let P be a harmonic oscillator of unit mass moving on the Ox-axis subject to an elastic force $-kx$, $k > 0$, and to a linear friction $-h\dot{x}$, $h > 0$. Its motion $x(t)$ is given by the linear system

$$\begin{cases} \dot{x} = v, \\ \dot{v} = -kx - hv. \end{cases}$$

Fig. 10.6 Stable origin

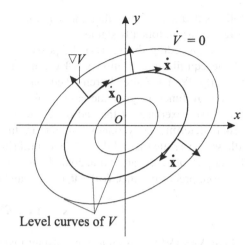

Level curves of V

It is quite evident from a physical point of view that the origin is an asymptotically stable equilibrium position. To verify this property of the origin, the total energy $V(x, v)$ is chosen as a Lyapunov function because it is positive definite and decreases along any motion for the dissipation. In fact, from the total energy

$$V(x, v) = \frac{1}{2}(v^2 + kx^2)$$

we have

$$\dot{V}(x, v) = -hv^2.$$

However, $\dot{V}(x, v)$ vanishes not only at the origin but also along the x-axis. From the previous theorems we can only conclude that the origin is stable. The following theorems allow us to establish that the origin is asymptotically stable by using the same Lyapunov function.

Theorem 10.4 (Barbashin–Krasovskii). *Let* $V : P_\gamma \to \Re$ *be a class* C^1 *function that is positive definite in a neighborhood* P_γ *of the origin. If* $\dot{V}(x)$ *is negative semidefinite in* P_γ *and if* $x = 0$ *is the only solution of (10.1) for which* $\dot{V}(x) = 0$, *then the origin is asymptotically stable.*

Theorem 10.5 (Krasovskii). *Let* $V : P_\gamma \to \Re$ *be a class* C^1 *function such that*

1. $\forall \eta \in (0, \gamma), \exists x \in P_\eta : V(x) > 0$;
2. $\dot{V}(x) \leq 0, \forall x \in P_\gamma$; *and*
3. $x = 0$ *is the only solution of (5.1) for which* $\dot{V}(x) = 0$.

Then, the origin is unstable.

Returning to the preceding example, we remark that the equation $\dot{V} = -hv^2 = 0$ implies $v = 0$ along the solutions. On the other hand, from the differential system it follows that $x = v = 0$; then the hypotheses of Theorem 10.4 are satisfied and the origin is asymptotically stable.

The preceding theorems make possible the analysis of the stability properties of the equilibrium solution $\mathbf{x} = \mathbf{0}$ given a Lyapunov function possessing suitable property. We have already remarked that a first integral of (10.1) can be used as a Lyapunov function. In some cases, polynomial Lyapunov functions can be easily found (see exercises at the end of chapter). However, in general, it is not an easy task to determine a Lyapunov function. In this regard, the following theorem, which allows us to state the stability of the equilibrium solution $\mathbf{x} = \mathbf{0}$ by an analysis of the linear part of (10.1) around $\mathbf{0}$, could be useful.[5]

More precisely, since $\mathbf{f}(\mathbf{0}) = \mathbf{0}$, (10.1) can be written as follows:

$$\dot{\mathbf{x}} = \mathbf{A}\mathbf{x} + \boldsymbol{\Phi}(\mathbf{x}), \tag{10.11}$$

where $\mathbf{A} = (\nabla \mathbf{f})_{\mathbf{x}=0}$ is an $n \times n$ constant matrix and $\boldsymbol{\Phi}(\mathbf{x}) = O(\| \mathbf{x} \|)$ when $\mathbf{x} \to \mathbf{0}$. Then, it is possible to prove the following theorem.

Theorem 10.6. *If all the eigenvalues of matrix A have negative real part, then the solution $x = 0$ of (10.1) is asymptotically stable. If among the eigenvalues of matrix A there is at least one whose real part is positive, the solution $x = 0$ of (10.1) is unstable.*

The notebook **LinStab** is based on this theorem.

10.6   Poincaré's Perturbation Method

In this section, we search for analytical approximate solutions of (10.1). It should be possible to search for such a solution as a Taylor expansion. This approach supplies a polynomial of a fixed degree r, approximating the solution of (10.1) *at least in a neighborhood of the initial value* $\mathbf{x}_0$. A better approximation is obtained by increasing the value r, that is, by considering more terms of the power expansion, or values of the independent variable t closer to the initial value $t = 0$. In contrast, Poincaré's method tries to give an approximate solution in an extended interval of the variable t, possibly for any t. On this logic, one is ready to accept a less accurate solution, provided that it approximates the solution uniformly with respect to t. In effect, Poincaré's approach does not always give a solution with this characteristic; more frequently, it gives an approximate solution whose degree of accuracy is not uniform with respect to time. This is due, as we shall see in this chapter, to the presence in Poincaré's expansion of so-called secular terms that go oscillating to infinity with t. This behavior of the approximate solution is not acceptable when the solution we are searching for is

---

[5]Readers will find in [35] many programs, written using Mathematica, that allow for the analysis of many stability problems.

periodic. However, in this case, by applying another procedure (Lindstedt, Poincaré technique), a uniform expansion with respect to t can be derived (see, for instance, [28, 29, 35]).

Poincaré's method is applicable to a differential equation, which can be written in the form

$$\dot{\mathbf{x}} = \mathbf{f}(\mathbf{x}) + \epsilon \mathbf{F}(\mathbf{x}), \tag{10.12}$$

where ϵ is a suitable "small" dimensionless parameter related to the physical system modeled by the preceding differential equation.

The method can be developed, provided that the solution of the equation

$$\dot{\mathbf{x}} = \mathbf{f}(\mathbf{x}) + \mathbf{g}(t),$$

where $\mathbf{g}(t)$ is a known function belonging to a suitable class, can be computed for a given initial condition and if the perturbation term $\mathbf{F}$ is analytic in its argument. When both conditions are fulfilled, it is possible to look for the solution in powers of ϵ. More precisely, it can be proved that the solution of (10.12), under suitable hypotheses on the functions $\mathbf{f}$ and $\mathbf{F}$, can be written as a power series of the parameter ϵ,

$$\mathbf{x}(t, \epsilon) = \mathbf{x}_0(t) + \epsilon \mathbf{x}_1(t) + \epsilon^2 \mathbf{x}_n(t) + \cdots,$$

that converges to the solution of (10.12) uniformly with respect to ϵ but usually not with respect to t. The general term $\mathbf{x}_n(t)$ of the preceding expansion is the solution of the equation

$$\dot{\mathbf{x}}_n = \mathbf{f}(\mathbf{x}_n) + \mathbf{g}_n(t),$$

where the function $\mathbf{g}_n(t)$ is completely determined since it only depends on the previous terms $\mathbf{x}_1(t), \ldots, \mathbf{x}_{n-1}(t)$.

The meaning of the attribute *small* applied to ϵ requires an explanation. The parameter ϵ is small if $\epsilon \ll 1$ and the terms $\dot{\mathbf{x}}$, $\mathbf{f}(\mathbf{x})$, and $\mathbf{F}(\mathbf{x})$ are all comparable with unity. However, it is not possible to verify this property because usually there is no knowledge of the solution. To overcome this difficulty, we can resort to a dimensionless analysis of the problem modeled by (10.12), provided that we have a sufficient knowledge of the phenomenon we are analyzing. We make clear this procedure with some examples.

We analyze in more detail the case where the function $\mathbf{f}$ in (10.12) depends linearly on $\mathbf{x}$, i.e., we consider the following Cauchy problem:

$$\begin{cases} \dot{\mathbf{x}} = \mathbf{A}\mathbf{x} + \epsilon \mathbf{F}(\mathbf{x}, \epsilon), \\ \mathbf{x}(t_0) = \bar{\mathbf{x}}, \end{cases} \tag{10.13}$$

where $\mathbf{A}$ is an $n \times n$ matrix with constant coefficients.

Poincaré proved that if $\mathbf{F}(\mathbf{x}, \epsilon)$ is an analytic function of its variables, then the solution $\mathbf{x}(t, \epsilon)$ of (10.13) is analytic with respect to ϵ; consequently, it can be expressed by an expansion

$$\mathbf{x}(t, \epsilon) = \mathbf{x}_0(t) + \mathbf{x}_1(t)\epsilon + \mathbf{x}_2(t)\epsilon^2 + \cdots \tag{10.14}$$

uniformly convergent with respect to ϵ in a neighborhood of the origin. It is evident that the functions $\mathbf{x}_i(t)$ must verify the initial data

$$\begin{cases} \mathbf{x}_0(t_0) = \bar{\mathbf{x}}, \\ \mathbf{x}_1(t_0) = \mathbf{0}, \\ \mathbf{x}_2(t_0) = \mathbf{0}, \\ \dots \end{cases} \tag{10.15}$$

To find the terms of the expansion (10.14), we expand $\mathbf{F}(\mathbf{x}, \epsilon)$ with respect to ϵ and then introduce (10.14) into (10.13)$_1$. In this way, the following sequence of Cauchy problems is derived, which determine $\mathbf{x}_0(t)$, $\mathbf{x}_1(t)$, $\mathbf{x}_2(t)$, and so on:

$$\begin{cases} \dot{\mathbf{x}}_0 = \mathbf{A}\mathbf{x}_0, \\ \mathbf{x}_0(t_0) = \bar{\mathbf{x}}, \end{cases} \tag{10.16}$$

$$\begin{cases} \dot{\mathbf{x}}_1 = \mathbf{A}\mathbf{x}_1 + \mathbf{F}(\mathbf{x}_0(t), t, 0), \\ \mathbf{x}_1(t_0) = \mathbf{0}, \end{cases} \tag{10.17}$$

$$\begin{cases} \dot{\mathbf{x}}_2 = \mathbf{A}\mathbf{x}_2 + (\mathbf{F}_\epsilon)(\mathbf{x}_0(t), t, 0) + \mathbf{x}_1(t) \cdot (\nabla_{\mathbf{x}}\mathbf{F})(\mathbf{x}_0(t), t, 0), \\ \mathbf{x}_2(t_0) = \mathbf{0}, \end{cases} \tag{10.18}$$

where $(\mathbf{F}_\epsilon)(\mathbf{x}_0(t), t, 0)$ is the derivative of $\mathbf{F}$ with respect to ϵ evaluated at $(\mathbf{x}_0(t), t, 0)$ and $(\nabla_{\mathbf{x}}\mathbf{F})(\mathbf{x}_0(t), t, 0)$ denotes the gradient of $\mathbf{F}$ with respect to the variable $\mathbf{x}$ at $(\mathbf{x}_0(t), t, 0)$.

Notice that the problems (10.16), (10.17), ... refer to the same linear differential equation, which is homogeneous at the first step and nonhomogeneous at the next steps. However, the terms appearing at the ith step are known when the previous Cauchy problems have been solved. Although the simplification reached in solving the original Cauchy problem is clear, the calculations to write and solve the different systems are very heavy and cumbersome. For these reasons, the reader can use the notebook **Poincare**.

It is also very important to recall that expansion (10.14) is usually *not uniform* with respect to time. As we shall see, in the series (10.14) some *secular terms* $\epsilon^n t^n \sin nt$, $\epsilon^n t^n \cos nt$ can appear. When this happens, one is compelled to accept into (10.14) time values verifying the inequality $t \ll a/b\epsilon$, where a and b denote the maximum values of the coefficients of, respectively, $\mathbf{x}_0(t)$ and the secular terms in $\mathbf{x}_1(t)$, to be sure that the second term of the expansion (10.14) is small with respect to the first one for time values $t \ll a/b\epsilon$.

10.7 Introducing the Small Parameter

To establish a model describing a certain system (e.g., physical, economical), we must first identify a set of variables $\mathbf{x} = (x_1, \dots, x_n)$ depending on an independent variable t that should describe, in the framework of the mathematical model, the physical state of the system we are dealing with.

Generally, independent and dependent variables of a model have dimensional values related to the physical system. This may cause problems in comparing small and large deviations of the variables. This problem can be overcome by putting all the variables in a suitable dimensionless form, relating them to convenient reference quantities. The criterion to choose these quantities must be such that the derived dimensionless quantities have values close to one. Only after this analysis it is possible to check whether the system has the form (10.13) and Poincaré's method is applicable.

This first step is a necessary step to apply Poincaré's method and, in particular, to verify whether (10.13) satisfies all the required conditions. Two mechanical examples will be considered to explain this procedure in detail. Two other examples will be considered in Chap. 14.

Example 10.3. If P is a material point of mass m subject to a nonlinear elastic force $-hx - kx^3$ and constrained to move on a straight line Ox, the Newtonian equation governing its motion yields

$$m\ddot{x} = -hx - kx^3. \tag{10.19}$$

To put (10.19) in a dimensionless form, two reference quantities L and T are introduced with, respectively, the dimensions of length and time

$$L = x_0, \quad T = \sqrt{\frac{m}{h}}, \tag{10.20}$$

where x_0 is the maximum value of the oscillations and T is equal to 2π times the period of these oscillations when only the linear part $-hx$ of the elastic force is acting. Defining the dimensionless quantities

$$x^* = \frac{x}{L}, \quad t^* = \frac{t}{T} \tag{10.21}$$

and remarking that

$$\dot{x} = \frac{L}{T}\frac{dx^*}{dt^*} = \frac{L}{T}\dot{x}^*, \quad \ddot{x} = \frac{L}{T^2}\frac{d^2x^*}{dt^{*2}} = \frac{L}{T^2}\ddot{x}^*,$$

(10.19), in view of (10.20), (10.21), and the last relations, yields

$$\ddot{x}^* = -x^* - \frac{kx_0^2}{h}x^{*3}.$$

Since this equation is equivalent to the first-order system

$$\dot{x}^* = y^*,$$
$$\dot{y}^* = -x^* - \frac{kx_0^2}{h}x^{*3},$$

the dimensionless parameter

$$\epsilon = \frac{kx_0^2}{h}$$

can be identified with the small parameter of Poincaré's method, provided that $\epsilon << 1$.

Example 10.4. Now we take another mechanical example into account: a moving point P under the action of its weight and a friction. The motion equation is written as

$$\dot{\mathbf{v}} = \mathbf{g} - hv\mathbf{v}, \tag{10.22}$$

where $\mathbf{g}$ is the gravity acceleration, $v = |\mathbf{v}|$, and h is a positive constant depending on the medium in which P is moving as well as on its form. Refer the previous vector equation to a frame Oxy whose origin O is at the initial position of P and whose Ox-axis and Oy-axis are taken to be horizontal and vertical, respectively, and such that the (vertical) plane Oxy contains the initial velocity of P. Since the motion is planar, the whole trajectory is contained in the plane Oxy. To identify the small parameter, we note that when friction is absent, the motion equation

$$\dot{\mathbf{v}} = \mathbf{g}$$

admits the solutions

$$v_x(t) = v_0\cos\alpha, \tag{10.23}$$
$$v_y(t) = v_0\sin\alpha + gt, \tag{10.24}$$

where $v_0 = |\mathbf{v}(0)|$ and α is the angle between the initial velocity $\mathbf{v}(0)$ and the horizontal Ox-axis. Since the quantities $v_x(t)$ and $v_y(t)$, in the presence of friction, are of the same order of magnitude as the previous ones, the height L at which a heavy body arrives, before inverting its motion, and the time T to reach the soil again can be taken as reference quantities of length and time, respectively. It is easy to deduce from (10.23) and (10.24) the formulae

$$L = \frac{v_0^2}{2g}, \quad T = \frac{2v_0}{g}$$

so that

$$\frac{L}{T} = \frac{v_0}{4}, \quad \frac{L}{T^2} = \frac{g}{8}.$$

With the introduction of these reference quantities, (10.22) assumes the following dimensionless form:

$$\dot{\mathbf{v}}^* = -8\mathbf{k} - Th v \mathbf{v}^*, \tag{10.25}$$

where $\mathbf{k}$ is the unit vector along the Oy-axis. If the factor

$$\epsilon = hT \ll 8,$$

then we have an equation to which the perturbation method is applicable. It is well known that in the presence of nonlinear friction, the velocity of a heavy point tends to a limit value. It has already been pointed out that Poincaré's method cannot supply correct results if the approximate solution is used in an extended time interval. This implies that the method does not result in a behavior that exhibits a limit velocity. It approximates the effective solution only if the motion lasts a sufficiently short time.

10.8 Weierstrass' Qualitative Analysis

In this section, we consider the second-order scalar differential equation

$$\ddot{x} = f(x) \tag{10.26}$$

in the unknown $x(t)$ since we will encounter it in many mechanical applications.

First, we notice that (10.26) is equivalent to the following first-order system:

$$\dot{x} = v, \tag{10.27}$$
$$\dot{v} = f(x), \tag{10.28}$$

in the unknowns $x(t)$, $v(t)$. This pair of functions defines a curve γ in the state space Γ whose structure depends on the meaning of the variable x. For instance, if x is the angle φ varying on a circumference, then Γ is a cylinder. The equilibrium positions of (10.27), (10.28) are the solutions of the system

$$f(x) = 0, \quad v = 0. \tag{10.29}$$

The purpose of Weierstrass's analysis consists in determining the qualitative behavior of the phase portrait of (10.27), (10.28).

To localize the orbits of (10.27), (10.28), we note that

$$V(x, v) = \frac{1}{2}v^2 + U(x), \tag{10.30}$$

where

$$U(x) = -\int f(x)\,dx \tag{10.31}$$

is a first integral of (10.27), (10.28) since

$$\dot{V} = v\dot{v} + U'(x)\dot{x} = v\dot{v} - f(x)\dot{x}$$

vanishes along the solution of (10.27), (10.28).

Theorem 10.7 (Dirichlet). *If $U(x)$ is a C^1 function with an effective minimum at the point x^*, then $(x^*, 0)$ is a stable equilibrium of the system (10.27), (10.28).*

Proof. If U is of class C^1 and has a minimum at x^*, then we have $U'(x^*) = f(x^*) = 0$ and $(x^*, 0)$ is an equilibrium position. Moreover, since U is defined to within a constant, we can always suppose that $U(x^*) = 0$. Consequently, we have that $V(x^*, 0) = 0$. Further, since $U(x)$ has a minimum at x^*, there exists a neighborhood of x^* in which $V(x, v) = v^2/2 + U(x) > 0$. Finally, we have already proved that $\dot{V} = 0$ along the solutions of (10.27), (10.28). In other words, $V(x, v)$ is a Lyapunov function that satisfies the hypotheses of Theorem 10.2. $\qquad\qquad\square$

We omit the proof of the following theorem, which supplies an instability criterion.

Theorem 10.8 (Chetaiev). *If the point x^* is not a minimum of $U(x)$ and the absence of the minimum can be determined by the derivatives $U^{(k)}(x^*)$ of $U(x)$ at x^*, where k is a finite integer number, then $(x^*, 0)$ is an unstable equilibrium position of the system (10.27), (10.28).*

Now we show that (10.26) and (10.30) allow us to discover the behavior of the solutions and the structure of the phase portrait applying a procedure proposed by Weierstrass. In fact, from (10.30) when we assume $V(x, v) = E$, where E is the total energy, we deduce that

$$v = \mp\sqrt{2(E - U(x))} \tag{10.32}$$

at any instant t for which $v(t) \neq 0$. This relation (recall that $v = dx/dt$) implies that

$$t = \mp \int_{x_0}^{x} \frac{ds}{\sqrt{2(E - U(x))}}. \tag{10.33}$$

Equation (10.32) is the explicit form of level curves $V = \text{cost}$. It shows that these curves are symmetric with respect to the Ox-axis. Moreover, in view of (10.27), (10.28), when $v = 0$, we also have $\dot{x} = 0$. Consequently, the tangent straight line to the level curves at the points in which they intersect Ox is orthogonal to Ox. Finally, for any fixed value of E, the admissible values of x satisfy the condition $U(x) \leq E$. Equation (10.33) gives the time needed to go from x_0 to x.

If we assign an initial datum (x_0, v_0), where $v_0 > 0$, then we must choose the $+$ sign in (10.32), (10.33) up to the instant at which $U(x) = E$. Consequently, if $U(x) < E$, $\forall x > x_0$, the function $x(t, x_0, v_0)$ goes to infinity in a finite or infinite time according to whether the function $1/(\sqrt{2(E - U(x))})$ is integrable or not.

In contrast, if on the right-hand side of x_0 there is a point $\overline{x}$ such that $U(\overline{x}) = E$, then we must consider the following two cases:

$$(a)\, U'(\overline{x}) \neq 0, \quad (b)\, U'(\overline{x}) = 0.$$

In case (a), Taylor's expansion of $U(x)$ at $\overline{x}$ is

$$U(x) = E + U'(\overline{x})(x - \overline{x}) + O(|x - \overline{x}|).) \tag{10.34}$$

Therefore, the function under the integral in (10.33) is integrable in $[x_0, \overline{x})$, and $x(t, x_0, v_0)$ reaches the point $\overline{x}$ in the finite time

$$\overline{t} = \int_{x_0}^{\overline{x}} \frac{ds}{\sqrt{2(E - U(x))}}. \tag{10.35}$$

Further, $U(x) < E$ for $x < \overline{x}$ and $U(\overline{x}) = E$ for $x = \overline{x}$. Therefore, the function $U(x)$ increases at the point $\overline{x}$ so that $f(\overline{x}) = -U'(\overline{x}) < 0$. Finally, if we take into account (10.28), then we also have $\dot{v}(\overline{t}, x_0, v_0) = \ddot{x}(\overline{t}, x_0, v_0) < 0$, whereas (10.27) and (10.32) imply that $v(\overline{t}, x_0, v_0) = \dot{x}(\overline{t}, x_0, v_0) = 0$. In conclusion, $\dot{x}(\overline{t}, x_0, v_0) < 0$ for $t < \overline{t}$ and $x(\overline{t}, x_0, v_0)$ comes back toward x_0. In particular, if to the left of x_0 is another point $\overline{\overline{x}}$ with the same characteristics of $\overline{x}$, then the function $x(\overline{t}, x_0, v_0)$ is periodic with a period

$$T = 2 \int_{\overline{x}}^{\overline{\overline{x}}} \frac{ds}{\sqrt{2(E - U(x))}}. \tag{10.36}$$

In case (b), Taylor's expansion of $U(x)$ at $\overline{x}$ is

$$U(x) = E + \frac{1}{2}U''(\overline{x})(x - \overline{x})^2 + O(|x - \overline{x}|^2),) \tag{10.37}$$

and the approach time to $\overline{x}$ is infinity since the function $1/(\sqrt{2(E - U(x))})$ is no more integrable and $x(\overline{t}, x_0, v_0)$ tends asymptotically to $\overline{x}$. We notice that, in this case, $\overline{x}$ is an unstable equilibrium position since the conditions $U(x) < E$ for $x < \overline{x}$, $U(\overline{x}) = E$ for $x = \overline{x}$, and $U'(\overline{x}) = 0$, ensure that $\overline{x}$ is not a minimum.

The analysis we have just described can be carried out by the notebook **Weierstrass**.

Example 10.5. Figure 10.7 shows the qualitative behavior of the potential energy $U(x)$ and the corresponding phase portrait of (10.27), (10.28). For $E < E_1$ there is no solution. For $E = E_1$ the level curve reduces to the equilibrium position $(x_1, 0)$. For $E_1 < E < E^*$, the level curves are closed and the orbits correspond to periodic motions. For $E^* < E < E_2$, the level curves have two connected components: the first of them corresponds to a periodic orbit, whereas the second one is an aperiodic orbit going to infinity. For $E = E_2$ we have a level curve with three connected components corresponding to a bounded open orbit, an unstable equilibrium position, and an

Fig. 10.7 Phase portrait

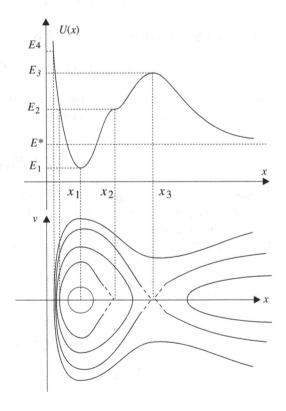

open orbit going to infinity. It is a simple task to complete the analysis of the phase portrait. We conclude by noting that the dashed lines represent orbits that tend to an unstable equilibrium position without reaching it.

10.9 Exercises

1. Verify that $V = \frac{1}{2}\left(x^2 + y^2\right)$ is a Lyapunov function for the system

$$\dot{x} = -y + ax(x^2 + y^2),$$
$$\dot{y} = x + ay(x^2 + y^2),$$

and determine the stability property of the origin upon varying the constant a.

2. Verify that $V = \frac{1}{2}\left(x^2 + y^2\right)$ is a Lyapunov function for the system

$$\dot{x} = -y + axy^2,$$
$$\dot{y} = x - yx^2,$$

and determine the stability property of the origin upon varying the constant a.

3. Determine the periodic orbits of the following system in polar coordinates (r, φ):

$$\dot{r} = r(1 - r)(2 - r),$$
$$\dot{\varphi} = -2.$$

Are there equilibrium positions? Control the results using the notebook **PolarPhase**.

4. Determine the equilibrium positions of the system

$$\dot{x} = 2xy,$$
$$\dot{y} = 1 - 3x^2 - y^2,$$

and analyze their linear stability properties. Control the obtained results using the notebook **LinStab**.

5. Determine the equilibrium positions of the system

$$\dot{x} = -x + y,$$
$$\dot{y} = -x + 2xy,$$

and analyze their linear stability properties. Control the obtained results using the notebook **LinStab**.

6. Determine the equilibrium positions of the system

$$\dot{x} = x(x^2 + y^2 - 1) - y(x^2 + y^2 + 1),$$
$$\dot{y} = y(x^2 + y^2 - 1) + x(x^2 + y^2 + 1),$$

and analyze their linear stability properties. Control the obtained results using the notebook **LinStab**.

7. Determine an approximate solution of the system

$$\dot{x} = y,$$
$$\dot{y} = -x + \epsilon(1 - x^2)y$$

using Poincaré's method, and control the obtained results using the notebook **Poincare**.

8. Determine an approximate solution of the system

$$\dot{x} = y,$$
$$\dot{y} = -x - \epsilon x^3$$

using Poincaré's method, and control the obtained results using the notebook **Poincare**.

9. Determine an approximate solution of the system

$$\dot{x} = y + \epsilon(x^2 - y^2),$$
$$\dot{y} = -x + \epsilon y^3$$

using Poincaré's method, and control the obtained results using the notebook **Poincare**.

10. Determine the equilibrium configurations of the potential energy

$$U = \frac{1}{2}x^2 + \frac{1}{4}x^4,$$

their stability properties, and the phase portrait using a Weierstrass analysis. Control the obtained results using the notebook **Weierstrass**.

11. Determine the equilibrium configurations of the potential energy

$$U(x) = -\cos(x),$$

their stability properties, and the phase portrait using a Weierstrass analysis. Control the obtained results using the notebook **Weierstrass**.

12. Determine the equilibrium configurations of the potential energy

$$U(x) = -e^{x^2}\cos(x),$$

their stability properties, and the phase portrait using a Weierstrass analysis. Control the obtained results using the notebook **Weierstrass**.

Part II
Mechanics

Chapter 11
Kinematics of a Point Particle

Abstract Kinematics analyzes the trajectories, velocities, and accelerations of the points of a moving body, the deformations of its volume elements, and the dependence of all these quantities on the frame of reference. In many cases, when such an accurate description of motion is too complex, it is convenient to substitute the real body with an ideal body for which the analysis of the preceding characteristics is simpler, provided that the kinematic description of the ideal body is sufficiently close to the behavior of the real one. For instance, when the deformations undergone by a body under the influence of the acting forces can be neglected, we adopt the *rigid body* model, which is defined by the condition that the distances among its points do not change during the motion. More particularly, if the body is contained in a sphere whose radius is much smaller than the length of its position vectors relative to a frame of reference, then the whole body is sufficiently localized by the position of any one of its points. In this case, we adopt the model of a *point particle*. In this chapter we analyze the kinematics of a point particle defining velocity, acceleration, trajectory, and compound motion.

11.1 Space–Time Frames of Reference of Classical Kinematics

To analyze physical[1] phenomena, an observer adopts a *body of reference S* at whose points he places identical *clocks*. For the present, we consider a clock as an arbitrary device defining the *local time*, i.e., an increasing continuous variable t. The set R, formed by S and the clocks placed in its points, is called a *space–time frame of reference*.

---

[1] The contents of Chapters 11–17 can also be found in [4, 12, 20, 23, 24, 26, 32, 33, 45, 46, 49, 54].

© Springer International Publishing AG, part of Springer Nature 2018   163
A. Romano and A. Marasco, *Classical Mechanics with Mathematica®*,
Modeling and Simulation in Science, Engineering and Technology,
https://doi.org/10.1007/978-3-319-77595-1_11

By a space–time frame of reference R, we can introduce a time order into the set of events happening at an arbitrary point $A \in S$. In fact, it is sufficient to label each event with the instant at which it takes place. Although we can introduce a chronology at any point A of S using the local time, at present we have no criterion to compare the chronology of the events at A with the chronology of the events at another point B of S. Only after the introduction of such a criterion can we state when an event at A is simultaneous with an event at B or when an event at A happens before or after an event at B. Already the measure of the distance $d(A, B)$ between two points of S at a given instant requires the definition of simultaneity of events happening at two different points.

We suppose that there is no problem in synchronizing two clocks at the same point A of S. To introduce a definition of synchronism between two distant clocks, we start by placing at the point A many identical and synchronous clocks. Then we postulate that these clocks remain synchronous when they are placed at different points of S. In other words, we assume the following axiom.

Axiom 11.1. *The behavior of a clock does not depend on its motion. In particular, it is not influenced by the transport.*

This postulate allows us to define a ***universal time***, i.e., a time that does not depend on a point of S. In fact, it is sufficient to place synchronous clocks at A and then to transport them to different points of S. Adopting this universal time, measured by synchronous clocks distributed at different points of S, we can define a ***rigid body*** as a body S for which the distance $d(A, B)$ between two arbitrary points A and B of S is constant in time.

Another fundamental axiom of classical kinematics follows.

Axiom 11.2. *There exist rigid bodies inside which the geometry of a three-dimensional Euclidean space $\mathcal{E}_3$ holds.*

11.2 Trajectory of a Point Particle

Let $R = \{S, t\}$ be a frame of reference formed by the rigid body S and universal time t. Henceforth, we will often identify a frame of reference with an orthonormal frame $(O, (\mathbf{e}_i))$, $i = 1, 2, 3$, of the three-dimensional Euclidean space describing the geometry of S (Axiom 11.2). In this section, we analyze the kinematic behavior of a point particle P moving with respect to $(O, (\mathbf{e}_i))$.

The ***equation of motion*** of P is a function

$$\mathbf{r} = \mathbf{r}(t), \quad t \in [t_1, t_2], \tag{11.1}$$

giving the position vector $\mathbf{r}(t)$ of P relative to the orthonormal frame $(O, (\mathbf{e}_i))$ at any instant t (Fig. 11.1). If (x_i) are the Cartesian coordinates of P in $(O, (\mathbf{e}_i))$, then (11.1)

Fig. 11.1 Trajectory of a
point particle

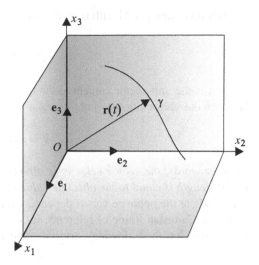

can also be written in the following form:[2]

$$\mathbf{r} = x_i(t)\mathbf{e}_i, \quad t \in [t_1, t_2], \tag{11.2}$$

where the functions $x_i(t)$ should be of class $C^2[t_1, t_2]$.

The locus γ of points (11.1) upon varying t in the interval $[t_1, t_2]$ is called the *trajectory* of P. If we denote by $s(t)$ the value of the curvilinear abscissa on γ at instant t, then the trajectory and the equation of motion can respectively be written as

$$\mathbf{r} = \hat{\mathbf{r}}(s), \tag{11.3}$$

$$\mathbf{r}(t) = \hat{\mathbf{r}}(s(t)). \tag{11.4}$$

11.3 Velocity and Acceleration

Let P be a point particle moving relative to the reference $(O, (\mathbf{e}_i))$, and let $\mathbf{r}(t)$ be the equation of motion of P. The *vector velocity* or simply *velocity* of P in $(O, (\mathbf{e}_i))$ is the vector function

$$\dot{\mathbf{r}} = \frac{d\mathbf{r}}{dt} \tag{11.5}$$

and the *scalar velocity* is the derivative $\dot{s}(t)$.

---

[2]Henceforth we intend the summation from 1 to 3 on repeated indices. We use only covariant indices since, in Cartesian orthogonal coordinates, there is no difference between covariant and contravariant components.

Differentiating (11.4) with respect to time and recalling the first Frenet formula

$$\frac{d\hat{\mathbf{r}}}{ds} = \mathbf{t},$$

where $\mathbf{t}$ is the unit vector tangent to the trajectory, we obtain the following relation between the vector velocity and the scalar velocity:

$$\dot{\mathbf{r}} = \dot{s}\,\mathbf{t}. \tag{11.6}$$

In other words, *the vector velocity is directed along the tangent to the trajectory and its length is equal to the absolute value of the scalar velocity. Finally, it has the versus of* $\mathbf{t}$ *or the opposite versus depending on the sign of* $\dot{s}$.

In the Cartesian frame of reference $(O, (\mathbf{e}_i))$, we can write $\mathbf{r}(t) = x_i(t)\mathbf{e}_i$ so that the velocity has the following Cartesian expression:

$$\dot{\mathbf{r}} = \dot{x}_i\mathbf{e}_i. \tag{11.7}$$

The **vector acceleration**, or simply the **acceleration**, of P is given by the vector

$$\ddot{\mathbf{r}} = \frac{d\dot{\mathbf{r}}}{dt} = \frac{d^2\mathbf{r}}{dt^2}, \tag{11.8}$$

which, in view of (11.7), in Cartesian coordinates has the form

$$\ddot{\mathbf{r}} = \ddot{x}_i\mathbf{e}_i. \tag{11.9}$$

Recalling the second Frenet formula

$$\frac{d\mathbf{t}}{dt} = \frac{\mathbf{n}}{R},$$

where $\mathbf{n}$ is the unit principal normal to the trajectory and R the radius of the osculating circle, we obtain another important expression of acceleration differentiating (11.6):

$$\ddot{\mathbf{r}} = \ddot{s}\,\mathbf{t} + \frac{v^2}{R}\mathbf{n}, \quad (v = |\dot{\mathbf{r}}| = |\dot{s}|). \tag{11.10}$$

The preceding formula shows that the acceleration

- Lies in the osculating plane of trajectory.
- Is directed along the tangent if the trajectory is a straight line, and
- Is normal to the trajectory when $\dot{s} = 0$ (uniform motion).

Henceforth, we will use the following definitions:

$$\ddot{s} = \text{scalar acceleration}$$
$$\ddot{s}\,\mathbf{t} = \text{tangential acceleration} \tag{11.11}$$
$$\frac{v^2}{R}\mathbf{n} = \text{centripetal acceleration}$$

We say that the motion of P is **accelerated** if v (equivalently, v^2) is an increasing function of time and **decelerated** in the opposite case. Since

$$\frac{dv^2}{dt} = \frac{d\dot{s}^2}{dt} = 2\dot{s}\ddot{s},$$

the motion is accelerated if $\dot{s}\ddot{s} > 0$ and decelerated if $\dot{s}\ddot{s} < 0$.

11.4 Velocity and Acceleration in Plane Motions

Let P be a point moving on the plane α. Introducing in α polar coordinates (r, φ), the motion of P is represented by the equations

$$r = r(t), \quad \varphi = \varphi(t). \tag{11.12}$$

The square distance ds^2 between the points (r, φ) and $(r + dr, \varphi + d\varphi)$ is

$$ds^2 = dr^2 + r^2 d\varphi^2,$$

and the vectors

$$\hat{\mathbf{e}}_r = \frac{\partial}{\partial r}, \quad \hat{\mathbf{e}}_\varphi = \frac{\partial}{\partial \varphi}$$

of the holonomic base relative to the polar coordinates verify the conditions

$$\hat{\mathbf{e}}_r \cdot \hat{\mathbf{e}}_r = 1, \ \hat{\mathbf{e}}_\varphi \cdot \hat{\mathbf{e}}_\varphi = r^2, \ \hat{\mathbf{e}}_r \cdot \hat{\mathbf{e}}_\varphi = 0.$$

We denote by $(P, \mathbf{e}_r, \mathbf{e}_\varphi)$ the orthonormal frame relative to the polar coordinates at any point $P \in \alpha$ (Fig. 11.2), where

$$\mathbf{e}_r = \hat{\mathbf{e}}_r, \quad \mathbf{e}_\varphi = \frac{1}{r}\hat{\mathbf{e}}_\varphi.$$

Then we define the **radial velocity** v_r and **transverse velocity** v_φ as the scalar quantities

$$v_r = \dot{\mathbf{r}} \cdot \mathbf{e}_r, \quad v_\varphi = \dot{\mathbf{r}} \cdot \mathbf{e}_\varphi, \tag{11.13}$$

that is, the components of $\dot{\mathbf{v}}$ relative to the base $(\mathbf{e}_r, \mathbf{e}_\varphi)$ at the point $P \in \gamma$. To determine the form of (11.13) in terms of (11.12), we note that, in polar coordinates, the position vector $\mathbf{r}$ is written as

$$\mathbf{r} = r\mathbf{e}_r(r, \varphi).$$

Differentiating this expression with respect to time, we obtain

$$\dot{\mathbf{r}} = \dot{r}\mathbf{e}_r + r\left(\frac{\partial \mathbf{e}_r}{\partial r}\dot{r} + \frac{\partial \mathbf{e}_r}{\partial \varphi}\dot{\varphi}\right).$$

On the other hand, in a Cartesian frame of reference $(O, \mathbf{i}, \mathbf{j})$ (Fig. 11.2), this results in

$$\mathbf{e}_r = \cos\varphi\,\mathbf{i} + \sin\varphi\,\mathbf{j}, \quad \mathbf{e}_\varphi = -\sin\varphi\,\mathbf{i} + \cos\varphi\,\mathbf{j},$$

so that we have

$$\frac{\partial \mathbf{e}_r}{\partial r} = \mathbf{0}, \quad \frac{\partial \mathbf{e}_r}{\partial \varphi} = \mathbf{e}_\varphi,$$

$$\frac{\partial \mathbf{e}_\varphi}{\partial r} = \mathbf{0}, \quad \frac{\partial \mathbf{e}_\varphi}{\partial \varphi} = -\mathbf{e}_r, \tag{11.14}$$

and the velocity in polar coordinates is written as

$$\dot{\mathbf{r}} = \dot{r}\mathbf{e}_r + r\dot{\varphi}\mathbf{e}_\varphi. \tag{11.15}$$

In view of (11.13) and (11.15), the radial and transverse velocities are given by the following formulae:

$$v_r = \dot{r}, \quad v_\varphi = r\dot{\varphi}. \tag{11.16}$$

Differentiating (11.15) with respect to time, we obtain the acceleration of P in polar coordinates:

$$\ddot{\mathbf{r}} = (\ddot{r} - r\dot{\varphi}^2)\mathbf{e}_r + \frac{1}{r}\frac{d}{dt}(r^2\dot{\varphi})\mathbf{e}_\varphi, \tag{11.17}$$

so that its components along $\mathbf{e}_r$ and $\mathbf{e}_\varphi$ are

$$a_r = \mathbf{a} \cdot \mathbf{e}_r = \ddot{r} - r\dot{\varphi}^2, \tag{11.18}$$

$$a_\varphi = \mathbf{a} \cdot \mathbf{e}_\varphi = \frac{1}{r}\frac{d}{dt}(r^2\dot{\varphi}). \tag{11.19}$$

Now we introduce a new definition that plays an important role in describing the motions of planets. Consider the area ΔA spanned by the radius $OP(t)$, where P belongs to the trajectory γ, in the time interval $(t, t + \Delta t)$ (Fig. 11.3). Then the limit

Fig. 11.2 Trajectory in polar coordinates

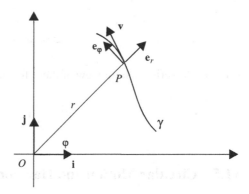

Fig. 11.3 Areal velocity

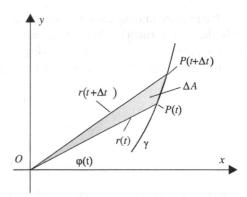

$$\dot{A} = \lim_{\Delta t \to 0} \frac{\Delta A}{\Delta t}$$

is the *areal velocity* of P. To obtain the explicit expression of $\dot{A}$ in terms of (11.12), we denote by $r_{\min}$ and $r_{\max}$ the minimum and the maximum of $r(t)$ in the interval $(t, t + \Delta t)$, respectively. Since

$$\frac{1}{2} r_{\min}^2 \Delta \varphi \le \Delta A \le \frac{1}{2} r_{\max}^2 \Delta \varphi,$$

in the limit $\Delta t \to 0$, we obtain

$$\dot{A} = \frac{1}{2} r^2 \dot{\varphi}. \tag{11.20}$$

In some cases, the Cartesian form of the areal velocity is more convenient. It can be obtained by considering the coordinate transformation from Cartesian coordinates (x, y) to polar ones

$$r = \sqrt{x^2 + y^2},$$
$$\varphi = \arctan \frac{y}{x}.$$

It is easy to verify that, in view of these formulae, (11.20) becomes

$$\dot{A} = \frac{1}{2}(x\dot{y} - \dot{x}y). \tag{11.21}$$

11.5   Circular Motion and Harmonic Motion

Let P be a point moving on a circumference γ with center O and radius r. Further, let φ be the angle formed by the position vector $\mathbf{r}(t)$ of P with the Ox-axis of an arbitrary Cartesian system of coordinate Oxy (Fig. 11.4). Since the curvilinear abscissa s is related to the angle φ and the radius r by the relation

$$s = r\varphi(t),$$

we have

$$\dot{s} = r\dot{\varphi}(t), \quad \ddot{s} = r\ddot{\varphi}(t).$$

Introducing these results into (11.6) and (11.10), we obtain the expressions of velocity and acceleration in a circular motion

$$\dot{\mathbf{r}} = r\dot{\varphi}\mathbf{t}, \tag{11.22}$$
$$\ddot{\mathbf{r}} = r\ddot{\varphi}\mathbf{t} + r\dot{\varphi}^2\mathbf{n}, \tag{11.23}$$

where $\mathbf{t}$ is the unit vector tangent to γ and $\mathbf{n}$ is the unit vector orthogonal to γ and directed toward O. In particular, if P moves uniformly on γ, then (11.23) gives

$$\ddot{\mathbf{r}} = -\omega^2\mathbf{r}, \tag{11.24}$$

where we have introduced the notation

$$\omega = |\dot{\varphi}|. \tag{11.25}$$

Let P be a material point that *uniformly* moves counterclockwise ($\dot{\varphi} > 0$) along a circumference γ with its center at O and radius r. We define **harmonic motion** with its center at O along the straight line Ox as the motion that is obtained projecting the point P along any diameter Ox of γ. In the Cartesian coordinates of Fig. 11.4, we have that the motion of P is represented by the following equation:

Fig. 11.4 Circular motion

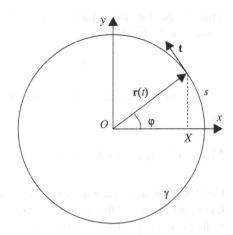

$$x(t) = r \cos \varphi(t) = r \cos(\omega t + \varphi_0). \tag{11.26}$$

In (11.26), r is called the **amplitude** of the harmonic motion and ω its **frequency**. Differentiating (11.26) twice, we obtain

$$\dot{x}(t) = -r\omega \sin(\omega t + \varphi_0), \quad \ddot{x}(t) = -r\omega^2 \cos(\omega t + \varphi_0),$$

so that in a harmonic motion the following relation between position and acceleration holds:

$$\ddot{x}(t) = -\omega^2 x(t). \tag{11.27}$$

It is simple to recognize that the homogeneous second-order linear differential equation with constant coefficients (11.27) *characterizes* the harmonic motions, i.e., its solutions are only harmonic motions. In fact, $\lambda = \mp i\omega$ are the roots of the characteristic equation of (11.27)

$$\lambda^2 + \omega^2 = 0,$$

and the general solution of (11.27) is

$$x(t) = C_1 e^{-i\omega t} + C_2 e^{i\omega t},$$

where C_1 and C_2 are arbitrary constants. Taking into account Euler's formulae, the preceding solution can also be written as

$$x(t) = A \cos \omega t + B \sin \omega t,$$

where A and B are two arbitrary constants. Finally, we obtain (11.26) when we write A and B in the form

$$A = r \cos \varphi_0, \quad B = -r \sin \varphi_0,$$

where r and φ_0 are still arbitrary constants.

11.6 Compound Motions

Let P be a moving point with respect to a frame of reference $(O, \mathbf{i}, \mathbf{j}, \mathbf{k})$, where $\mathbf{i}, \mathbf{j}$, and $\mathbf{k}$ are unit vectors along the (Ox, Oy, Oz) axes of a Cartesian system of coordinates. If the position vector $\mathbf{r}$ can be written in the form

$$\mathbf{r}(t) = \sum_{i=1}^{n} \mathbf{r}_i(t), \tag{11.28}$$

then we say that the motion of P is a **compound motion** of the n motions $\mathbf{r}_i(t)$.

In this section, we analyze some interesting compound motions. First, we consider the motions

$$\mathbf{r}_1(t) = v_0 t \mathbf{i},$$
$$\mathbf{r}_2(t) = \left(-\frac{1}{2}gt^2 + z_0\right) \mathbf{k},$$

where v_0, g, and z_0 are arbitrary positive constants. Eliminating t from the components of the preceding equations, we at once see that the trajectory of the compound motion is a parabola of the Oxz plane with the concavity downward.

Now we consider the motion obtained by compounding the following motions:

$$\mathbf{r}_1(t) = R \cos(\omega t + \varphi_0)\mathbf{i} + R \sin(\omega t + \varphi_0)\mathbf{j},$$
$$\mathbf{r}_2(t) = \dot{z}_0 t \mathbf{k}$$

The first equation represents a uniform motion along a circumference of radius R and center O lying in the coordinate plane Oxy. The second equation is relative to a uniform motion along the Oz-axis. The trajectory γ of the compound motion is a helix lying on a cylinder of radius R and axis Oz (Fig. 11.5). The helix is regular since the unit tangent vector $\mathbf{t} = \Gamma(-\omega R \sin(\omega t + \varphi_0), \omega R \cos(\omega t + \varphi_0), \dot{z}_0)$ to γ, where

$$\Gamma = \sqrt{\omega^2 R^2 + \dot{z}_0^2},$$

Fig. 11.5 Helicoidal motion

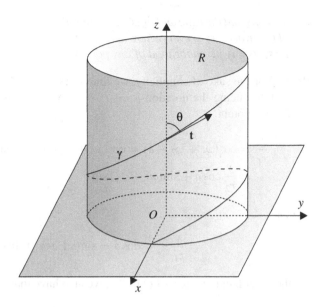

forms a constant angle θ with the directrices of the cylinder. In fact, it is

$$\cos \theta = \frac{\dot{z}_0}{\Gamma}.$$

Finally, we consider the motion of a point P obtained by compounding two harmonic motions along the Ox- and Oy-axes of a Cartesian system of coordinates:

$$\mathbf{r}_1(t) = x(t)\mathbf{i} = r_1 \cos(\omega_1 t + \varphi_1)\mathbf{i},$$
$$\mathbf{r}_2(t) = y(t)\mathbf{j} = r_2 \cos(\omega_2 t + \varphi_2)\mathbf{j}. \qquad (11.29)$$

The compound motion of P may be very complex, as is proved by the following theorem.

Theorem 11.1. *Let ABCD be the rectangle* $[-r_1, r_1] \times [-r_2, r_2]$ *with vertices* $A(-r_1, -r_2)$, $B(r_1, -r_2)$, $C(r_1, r_2)$, *and* $D(-r_1, r_2)$. *The compound motion of the harmonic motions (11.29) has the following properties:*

- *If* $\varphi_1 = \varphi_2$, *or* $\varphi_1 = \varphi_2 \mp \pi$ *and* $\omega_1 = \omega_2$, *then it is harmonic along one of the diagonals of the rectangle ABCD;*
- *For all other values of* φ_1 *and* φ_2 *and* $\omega_1 = \omega_2$, *the trajectory* γ *is an ellipsis tangent to the sides of the rectangle ABCD every time* γ *touches one of them; further, P moves along* γ *with constant areal velocity, and the motion is periodic;*
- *If* $\omega_1 \neq \omega_2$ *and the ratio* ω_1/ω_2 *is a rational number, the motion is periodic. Further, the trajectory may be either an open curve or a closed curve;*

- If $\omega_1 \neq \omega_2$ and the ratio ω_1/ω_2 is an irrational number, then the motion is aperiodic and the trajectory is everywhere dense on the rectangle ABCD (i.e., the trajectory intersects any neighborhood of any point in the rectangle).

Proof. For the sake of simplicity, we limit ourselves to proving only the first three items. Introducing the notation $\delta = \varphi_2 - \varphi_1$ and supposing $\omega_1 = \omega_2 = \omega$, from (11.29) we obtain

$$\frac{x}{r_1} = \cos(\omega t + \varphi_2 - \delta) = \cos(\omega t + \varphi_2)\cos\delta + \sin(\omega t + \varphi_2)\sin\delta$$
$$= \frac{y}{r_2}\cos\delta + \sin(\omega t + \varphi_2)\sin\delta,$$

so that

$$\frac{x}{r_1} - \frac{y}{r_2}\cos\delta = \sin(\omega t + \varphi_2)\sin\delta. \tag{11.30}$$

On the other hand, in view of $(11.29)_2$, we also have that

$$\frac{y}{r_2}\sin\delta = \cos(\omega t + \varphi_2)\sin\delta. \tag{11.31}$$

Consequently, from (11.30) and (11.31) we obtain that the trajectory of the compound motion is

$$\frac{x^2}{r_1^2} + \frac{y^2}{r_2^2} - \frac{2}{r_1 r_2}x\,y\cos\delta = \sin^2\delta. \tag{11.32}$$

If $\varphi_1 = \varphi_2$, then $\delta = 0$, whereas if $\varphi_1 = \varphi_2 \mp \pi$, then $\delta = \mp\pi$. In all these cases, (11.32) gives

$$\left(\frac{x}{r_1} \mp \frac{y}{r_2}\right)^2 = 0, \tag{11.33}$$

where the $-$ sign corresponds to $\delta = 0$, the $+$ sign corresponds to $\delta = \mp\pi$, and the trajectory coincides with one of the diagonals of the rectangle $ABCD$. To prove that the motion along the diagonal $y = r_2 x/r_1$, for instance, is harmonic, we note that the distance s of P from O along this diagonal is given by

$$s = \sqrt{x^2 + y^2} = \sqrt{r_1^2 + r_2^2}\cos(\omega t + \varphi_1).$$

When δ is different from 0 and $\mp\pi$, the trajectory (11.32) is an ellipse. In this case, the areal velocity A [see (11.21)]

$$A = \frac{1}{2}(\dot{x}y - x\dot{y}) = -\frac{1}{2}wr_1r_2 \sin \delta$$

is constant and the motion is periodic. Moreover, the ellipsis touches the sides $x = \mp r_1$ of the rectangle $ABCD$ when $\cos(\omega t + \varphi_1) = \mp 1$, that is, when $\omega t + \varphi_1 = n\pi$, where n is an arbitrary relative integer. It is plain to verify that the corresponding value of $\dot{x}$ vanishes and the ellipsis is tangent to the rectangle any time it touches the sides $x = \mp r_1$.

We now determine the conditions under which the motion is periodic when $\omega_1 \neq \omega_2$. The compound motion is periodic if and only if for any instant t there exists a value T of time such that

$$\cos(\omega_1(t + T) + \varphi_1) = \cos(\omega_1 t + \varphi_1),$$
$$\cos(\omega_2(t + T) + \varphi_2) = \cos(\omega_2 t + \varphi_2).$$

The preceding conditions can equivalently be written as

$$\omega_1 T = 2n\pi,$$
$$\omega_2 T = 2m\pi,$$

where n and m are arbitrary relative integers. In conclusion, the motion is periodic if and only if the ratio

$$\frac{\omega_1}{\omega_2} = \frac{n}{m} \qquad\qquad (11.34)$$

is a rational number.   □

Figures 11.6–11.10 show different cases of compounding two harmonic motions along orthogonal axes. They were obtained by using the notebook **CompositionP**, which allows to draw the trajectory obtained by compounding two or three motions.

Fig. 11.6 $r_1 = 2, r_2 = 1$, $\omega_1 = \omega_2 = 1, \varphi_1 = 0$, $\varphi_2 = \pi/3$

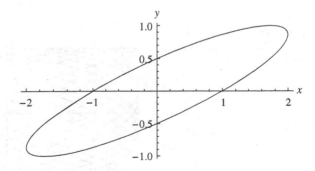

Fig. 11.7 $r_1 = 2, r_2 = 1, \omega_2$
$= 2, \omega_2 = 1, \varphi_1 = 1,$
$\varphi_2 = \pi/2$

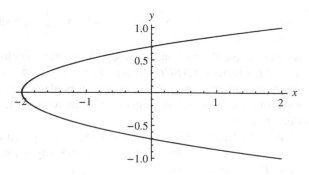

Fig. 11.8 $r_1 = 2, r_2 = 1, \omega_2$
$= 3, \omega_2 = 1, \varphi_1 = 1,$
$\varphi_2 = \pi/2$

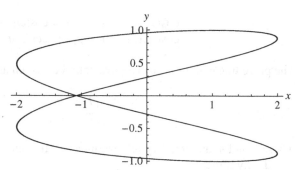

Fig. 11.9 $r_1 = 2, r_2 = 1, \omega_2$
$= 3, \omega_2 = 1, \phi_1 = 1,$
$\varphi_2 = \pi/3$

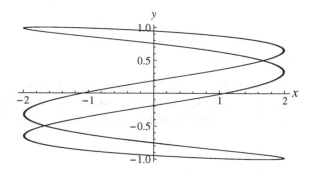

Fig. 11.10 $r_1 = 2, r_2 = 1,$
$\omega_2 = \sqrt{2}, \omega_2 = 1, \varphi_1 = 0,$
$\varphi_2 = \pi/3$

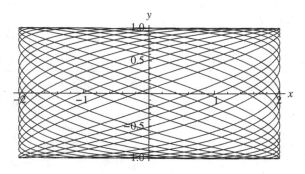

11.7 Exercises

1. Let Oxy be a plane in which we adopt polar coordinates (r, φ), and let $r = a\varphi$ be the trajectory of a point particle P, where a is a constant. Assuming that $\dot{\varphi} = \dot{\varphi}_0$ is constant, determine the radial and transverse velocity, the radial and transverse acceleration of P, and the areal velocity.

2. In the same plane of the preceding exercise, a point particle P describes the trajectory $r = r_0 e^{-\varphi}$, where r_0 and $\dot{\varphi}$ are constant. Determine the radial and transverse velocity and acceleration.

3. In the plane Oxy, the equations of motion of a point particle P are

$$x = \dot{x}_0 t, \quad y = y_0 - \beta(1 - e^{-\alpha t}),$$

where $\dot{x}_0$, y_0, α, and β are positive constant. Determine the velocity and acceleration of P, its trajectory γ, and the unit vector tangent to γ.

4. Let P be a point particle moving on a parabola γ, and let $y = x^2$ be the equation of γ in the frame Oxy. Determine the velocity, acceleration, and the curvilinear abscissa of P.
   Hint: The curvature $1/R$ of a curve $y = f(x)$ is $1/R = |f''(x)|1/(1 + f'(x)^2)^{3/2}$.

5. Let $Oxyz$ be a Cartesian frame of reference in space. A point particle P moves on the parabola $z = x^2$, which uniformly rotates with angular velocity ω about the Oz-axis. Determine the velocity and acceleration of P.

6. The acceleration and velocity of a point particle P moving on the Ox-axis are related by the equation

$$\ddot{x} = -a\dot{x}, \quad a > 0.$$

Determine the equation of motion of P, and evaluate the limit of $x(t)$ and $\dot{x}(t)$ when $t \to \infty$.

7. The acceleration and velocity of a point particle P moving on the Ox-axis are related by the equation

$$\ddot{x} = -a|\dot{x}|\dot{x}, \quad a > 0.$$

Prove that $\lim\limits_{t \to \infty} \dot{x} = 0$.
   Hint: Multiply both sides of the equation by $\dot{x}$, and take into account that

$$\dot{x}\ddot{x} = \frac{1}{2}\frac{d\dot{x}^2}{dt}.$$

8. Determine the velocity, acceleration, and trajectory of a point particle P whose equations of motion in the frame Oxy are

$$x(t) = \sin \omega t, \quad y(t) = e^{-t}.$$

Further, determine the instants at which the velocity is parallel to Oy.

9. Let P be a point particle moving on the ellipse γ whose equation in polar coordinates (r, φ) is

$$r = \frac{p}{1 + e \cos \varphi},$$

where p is a constant and e the eccentricity. If P moves on γ with constant areal velocity, prove that the velocity square has a maximum at the perihelion and a minimum at the aphelion.

Hint: Evaluate $\dot{r}$ taking into account that $r^2 \dot{\varphi} = c$, where c is a constant. Then recall that the radial velocity is $\dot{r}$ and the transverse velocity is $r\dot{\varphi}$.

10. A particle P moves along a curve $y = a \sin bx$, where a and b are constant. Determine the acceleration of P assuming that the component of its velocity along the Ox-axis is constant.

Chapter 12
Kinematics of Rigid Bodies

Abstract In this chapter a change of rigid frame of reference is considered and the general formula of velocity field in a rigid motion is given. Then, the translational, rotational, spherical, and planar motions are studied. Finally, the transformation formulae of velocity and acceleration from a rigid frame of reference to another one are determined. Exercises conclude the chapter.

12.1 Change of the Frame of Reference

An important task of kinematics consists of comparing the measures of lengths and time intervals carried out by two observers R and R' moving each with respect to the other with an arbitrary rigid motion. The solution of this problem allows us to compare the measures of velocities, accelerations, etc., carried out by the observers' co-moving with R and R'.

Let $E = (P', t')$ be an event happening at the point P' at instant t' for an observer at rest in the frame of reference R'. Let $E = (P, t)$ be the *same* event evaluated from an observer at rest in R. The aforementioned problem of kinematics can mathematically be formulated as follows:

Determine the functions

$$P = f(P', t'), \tag{12.1}$$

$$t = g(P', t'), \tag{12.2}$$

relating the space–time measures that R and R'associate with the same event.

When we recall the postulate stating that the behavior of a clock is not influenced by the transport, we can equip R and R' with synchronized clocks and be sure that they will remain synchronous. Therefore, (12.2) becomes $t = t'$. More generally, even if R and R' use clocks with the same behavior, they could choose a different time unit and a different origin of time. In this case, instead of $t = t'$, we have

© Springer International Publishing AG, part of Springer Nature 2018 179
A. Romano and A. Marasco, *Classical Mechanics with Mathematica®*,
Modeling and Simulation in Science, Engineering and Technology,
https://doi.org/10.1007/978-3-319-77595-1_12

$$t = at' + b, \tag{12.3}$$

where a and b are constants. We remark that (12.3) still implies that two simultaneous events for R are also simultaneous for R'.

Let $E_A = (A', t')$ and $E_B = (B', t')$ be two events happening in two different points A' and B' at the same instant for the observer R'. The same events are also simultaneous for R but happening at points A and B. To determine the function (12.1), classical kinematics resorts to another axiom.

Axiom 12.1. *At any instant, transformation (12.1) preserves the spatial distance d between simultaneous events, that is,*

$$d(A, B) = d(f(A', t'), f(B', t')) = d(A', B'). \tag{12.4}$$

In other words, Axiom 12.1 states that, at any instant, (12.1) is an isometry between the three-dimensional Euclidean spaces $\mathcal{E}_3$ and $\mathcal{E}_3'$ of observers R and R', respectively. Let us introduce Cartesian axes $(O', (x_i'))$ in the frame R' and Cartesian axes $(O, (x_i))$ in R. If (x_i') and (x_i) are the coordinates that R' and R, respectively, associate with the points P' and P of the same event $E = (P', t') = (P, t)$, then from linear algebra we know that transformation (12.1) is expressed by the formulae

$$x_i = x_{\Omega i} + Q_{ij}(t)x_j', \tag{12.5}$$

where Ω is the place in which occurs the arbitrary event $4E_\Omega = (\Omega, t)$, evaluated by observer R, and $(Q_{ij}(t))$ is an arbitrary time-dependent orthogonal matrix, that is, a matrix satisfying the conditions

$$Q_{ij}(Q^T)_{jh} = \delta_{ih}. \tag{12.6}$$

Remark 12.1. When $\Omega = O$, (12.5) reduces to the relation

$$x_i = Q_{ij}(t)x_j',$$

which gives the transformation formulae of the components of the position vector $\mathbf{r}'$ in passing from the frame $(O, (\mathbf{e}_i'))$ to the frame $(O, (\mathbf{e}_i))$. Consequently, the two bases $(\mathbf{e}_i)$ and $(\mathbf{e}_i')$ are related by the equations

$$\mathbf{e}_i' = (Q^{-1})_{ij}\mathbf{e}_j = (Q^T)_{ij}\mathbf{e}_j = Q_{ji}\mathbf{e}_j, \tag{12.7}$$

so that the coefficients $(Q^T)_{ji}$ are the components of the vectors $\mathbf{e}_i'$ with respect to the base $\mathbf{e}_i$.

Remark 12.2. In many textbooks, the transformation formulae (12.5) are proved by noting that from Fig. 12.1 we have that

$$\mathbf{r}_P = \mathbf{r}_\Omega + \mathbf{r}'. \tag{12.8}$$

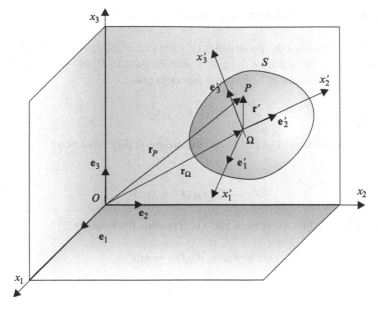

Fig. 12.1 Coordinates in a rigid motion

Therefore, referring this relation to the bases $(\mathbf{e}_i)$ and $(\mathbf{e}'_i)$, we obtain

$$x_i \mathbf{e}_i = x_{\Omega i} \mathbf{e}_i + x'_j \mathbf{e}'_j.$$

Now, representing the base $(\mathbf{e}'_j)$ with respect to the base $(\mathbf{e}_i)$ by the relations $\mathbf{e}_{j'} = Q_{ij} \mathbf{e}_i$, where (Q_{ij}) is an orthogonal matrix, we again obtain (12.5). However, this approach hides a logical mistake. In fact, in both of the preceding relations, the vector $\mathbf{r}'$ is intended to be a vector belonging to both vector spaces relative to the two observers R and R', that is, it is intended as the *same* vector. But this assumption is just equivalent to the condition that the correspondence between the spaces relative to the two observers is an isometry. In other words, we can accept (12.8) only after proving (12.5).

12.2 Velocity Field of a Rigid Motion

In this section we evaluate the velocity field and the acceleration field of any rigid body S moving relative to a rigid frame of reference R.

Let $(O, (\mathbf{e}_i))$ be a Cartesian frame in R, and denote by (x_i) the coordinates of any point of S relative to $(O, (\mathbf{e}_i))$. Introduce a Cartesian frame $(\Omega, (\mathbf{e}'_i))$ in the body S (Fig. 12.1) with the origin at an arbitrary point $\Omega \in S$. Owing to the rigidity of S, any point P of S has coordinates (x'_i) relative to $(\Omega, (\mathbf{e}'_i))$ that *do not depend on time*. For the sake of brevity, we call R the **space frame** or the **lab frame** and $(\Omega, (\mathbf{e}'_i))$ the **body frame**.

The relation between the coordinates (x'_i) and (x_i) is expressed by (12.5). We are now interested in the velocity field of S with respect to R and S is a rigid body. Therefore, we differentiate (12.5) with respect to time:

$$\dot{x}_i = \dot{x}_{\Omega i} + \dot{Q}_{ij} x'_j. \tag{12.9}$$

On the other hand, in view of (12.5) and the orthogonality condition (12.6), we have that

$$x'_j = (Q^T)_{jh}(x_h - x_{\Omega h}). \tag{12.10}$$

Inserting the preceding expression into (12.9) yields the relation

$$\dot{x}_i = \dot{x}_{\Omega i} + W_{ih}(x_h - x_{\Omega h}), \tag{12.11}$$

where

$$W_{ih} = \dot{Q}_{ij}(Q^T)_{jh}. \tag{12.12}$$

In (12.11), $\dot{x}_i$, $\dot{x}_{\Omega i}$, and $x_h - x_{\Omega h}$ are components of vectors; therefore, from the tensorial criteria of Chap. 2, we can state that W_{ih} are the components of a Euclidean 2-tensor $\mathbf{W}$. This tensor is skew-symmetric since (12.6) implies the equality

$$\dot{Q}_{ij}(Q^T)_{jh} = -Q_{ij}(\dot{Q}^T)_{jh}, \tag{12.13}$$

and consequently

$$W_{ih} = -Q_{ij}(\dot{Q}^T)_{jh} = -Q_{ij}\dot{Q}_{hj} = -\dot{Q}_{hj}(Q^T)_{ji} = -W_{hi}. \tag{12.14}$$

Introducing the pseudovector (Chap. 5) $\boldsymbol{\omega}$ such that

$$W_{ih} = \epsilon_{ijh}\omega_j, \tag{12.15}$$

where ϵ_{ijh} is the Levi–Civita symbol, (12.9) becomes

$$\dot{x}_i = \dot{x}_{\Omega i} + \epsilon_{ijh}\omega_j(x_h - x_{\Omega h}). \tag{12.16}$$

With the notations of Fig. 12.1, we can put (12.16) in the following vector form:

$$\dot{\mathbf{r}}_P = \dot{\mathbf{r}}_\Omega + \boldsymbol{\omega} \times \overrightarrow{\Omega P}, \quad \overrightarrow{\Omega P} = \mathbf{r}_P - \mathbf{r}_\Omega. \tag{12.17}$$

Differentiating the preceding formula with respect to time and noting that

$$\frac{d}{dt}\overrightarrow{\Omega P} = \frac{d}{dt}(\mathbf{r}_P - \mathbf{r}_\Omega) = \dot{\mathbf{r}}_P - \dot{\mathbf{r}}_\Omega = \boldsymbol{\omega} \times \overrightarrow{\Omega P},$$

we derive the acceleration field of a rigid motion

$$\ddot{\mathbf{r}}_P = \ddot{\mathbf{r}}_\Omega + \dot{\boldsymbol{\omega}} \times \overrightarrow{\Omega P} + \boldsymbol{\omega} \times (\boldsymbol{\omega} \times \overrightarrow{\Omega P}). \tag{12.18}$$

Exercise 12.1. Verify that (12.15) implies that

$$\omega_i = -\frac{1}{2}\epsilon_{ijh}W_{jh}. \tag{12.19}$$

Remark 12.3. Equation (12.17) shows that the velocity field of a rigid motion is completely determined by giving two vector functions of time, that is, the velocity $\dot{\mathbf{r}}_\Omega(t)$ of an arbitrary point Ω of the rigid body and the vector function $\boldsymbol{\omega}(t)$.

We now prove the important *Mozzi's theorem*, which supplies both an interesting interpretation of (12.17) and the physical meaning of the vector $\boldsymbol{\omega}(t)$.

Theorem 12.1. *At any instant t at which $\boldsymbol{\omega}(t) \neq \mathbf{0}$, the locus $\mathcal{A}(t)$ of points A such that $\dot{\mathbf{r}}_A(t) = \lambda(t)\boldsymbol{\omega}(t)$, where $\lambda(t)$ is a real function, is a straight line, parallel to $\boldsymbol{\omega}(t)$, which is called **Mozzi's axis**. Moreover, all points belonging to $\mathcal{A}(t)$ move with the same velocity $\boldsymbol{\tau}(t)$.*

Proof. The points A of S having the property stated by the theorem are characterized by the condition

$$\dot{\mathbf{r}}_A \times \boldsymbol{\omega} = \mathbf{0},$$

which, in view of (12.17), becomes

$$(\dot{\mathbf{r}}_\Omega + \boldsymbol{\omega} \times \overrightarrow{\Omega A}) \times \boldsymbol{\omega} = \mathbf{0}.$$

When we take into account the vector identity $(\mathbf{a} \times \mathbf{b}) \times \mathbf{c} = (\mathbf{a} \cdot \mathbf{c})\mathbf{b} - (\mathbf{b} \cdot \mathbf{c})\mathbf{a}$, the preceding equation can also be written as

$$\dot{\mathbf{r}}_\Omega \times \boldsymbol{\omega} = (\overrightarrow{\Omega A} \cdot \boldsymbol{\omega})\boldsymbol{\omega} - \omega^2 \overrightarrow{\Omega A}. \tag{12.20}$$

In conclusion, the locus $\mathcal{A}(t)$ coincides with the set of solutions $\overrightarrow{\Omega A}$ of (12.20). Introducing the unit vector $\mathbf{k} = \boldsymbol{\omega}/|\boldsymbol{\omega}|$, we have that

$$\overrightarrow{\Omega A} = (\overrightarrow{\Omega A} \cdot \mathbf{k})\mathbf{k} + \overrightarrow{\Omega A}_\perp,$$

where $\overrightarrow{\Omega A}_\perp \cdot \mathbf{k} = 0$, and (12.20) becomes

$$\dot{\mathbf{r}}_\Omega \times \boldsymbol{\omega} = (\overrightarrow{\Omega A} \cdot \mathbf{k})\omega^2\mathbf{k} - \omega^2[(\overrightarrow{\Omega A} \cdot \mathbf{k})\mathbf{k} + \overrightarrow{\Omega A}_\perp].$$

In the hypothesis $\boldsymbol{\omega} \neq \mathbf{0}$, the preceding equation supplies

$$\overrightarrow{\Omega A}_\perp = -\frac{\dot{\mathbf{r}}_\Omega \times \omega}{\omega^2},$$

so that the requested vectors $\overrightarrow{\Omega A}$ are given by the formula

$$\overrightarrow{\Omega A} = -\frac{\dot{\mathbf{r}}_\Omega \times \boldsymbol{\omega}}{\omega^2} + \lambda \boldsymbol{\omega}, \tag{12.21}$$

where λ is an arbitrary real number. The preceding formula proves that points A belong to a straight line $\mathcal{A}(t)$, parallel to $\boldsymbol{\omega}$. Finally, we write (12.17) for any pair $P, \Omega \in \mathcal{A}(t)$. Since in this case $\overrightarrow{\Omega P}$ is parallel to $\boldsymbol{\omega}$, we obtain

$$\dot{\mathbf{r}}_P = \dot{\mathbf{r}}_\Omega,$$

and the theorem is proved. $\square$

The existence of Mozzi's axis allows us to write (12.17) in the following form:

$$\dot{\mathbf{r}}_P = \boldsymbol{\tau} + \boldsymbol{\omega} \times \overrightarrow{AP}, \quad \boldsymbol{\tau} \parallel \boldsymbol{\omega}, \ A \in \mathcal{A}(t), \tag{12.22}$$

where $\boldsymbol{\tau}$ is the common velocity of the points $A \in \mathcal{A}(t)$.

12.3   Translational and Rotational Motions

Definition 12.1. A rigid motion is said to be *translational* relative to the frame of reference R if, at any instant t, the velocity field is uniform:

$$\dot{\mathbf{r}}_P = \boldsymbol{\tau}(t), \quad \forall P \in S. \tag{12.23}$$

Theorem 12.2. *The following statements are equivalent.*

- *A rigid motion is translational.*
- $\omega(t) = \mathbf{0}$, *at any instant.*
- *The orthogonal matrix* (Q_{ij}) *in (12.5) does not depend on time.*

Proof. The equivalence between the first two conditions is evident in view of (12.22). The equivalence between the first and third conditions follows at once from (12.12) and (12.15). $\square$

Definition 12.2. A rigid motion is a *rotational motion* about axis a if during the motion the points belonging to a are fixed. Further, a is called the *rotation axis*.

Fig. 12.2 Lab and body
frames in a rotational motion

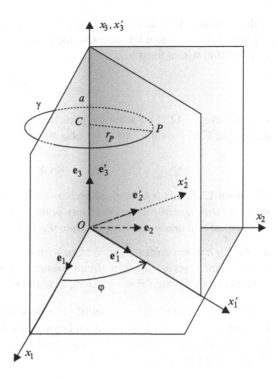

To find the particular expression of the velocity field in a rotational motion about
the fixed axis a, we start by choosing the frames of coordinates as in Fig. 12.2.
Recalling the meaning of the orthogonal matrix (Q_{ij}) [see (12.7)] and looking at
Fig. 12.2, we at once see that

$$(Q_{ij}) = \begin{pmatrix} \cos\varphi & -\sin\varphi & 0 \\ \sin\varphi & \cos\varphi & 0 \\ 0 & 0 & 1 \end{pmatrix}.$$

Taking into account (12.7), (12.12), and (12.19), by simple calculations we obtain
the following results:

$$W_{ij} = \dot\varphi \begin{pmatrix} 0 & -1 & 0 \\ 1 & 0 & 0 \\ 0 & 0 & 0 \end{pmatrix},$$

$$\omega_i = -\frac{1}{2}\epsilon_{ijh}W_{jh} = (0, 0, \dot\varphi).$$

If the arbitrary point Ω in (12.17) is chosen on the fixed axis a and the preceding expression of ω is taken into account, then we obtain the velocity field of a rotational rigid motion with a fixed axis a:

$$\dot{\mathbf{r}}_P = \dot{\varphi}\mathbf{e}_3 \times \overrightarrow{\Omega P}. \qquad (12.24)$$

In this formula, Ω is an arbitrary point of a; in particular, identifying Ω with the projection C of P on the axis a, we obtain

$$\dot{\mathbf{r}}_P = \dot{\varphi}\mathbf{e}_3 \times \overrightarrow{CP}. \qquad (12.25)$$

Theorem 12.3. *Let S be a rigid body rotating about the fixed axis a, and denote by C the orthogonal projection of any point $P \in S$ on axis a and by π_P the plane orthogonal to a and containing C. Then the following properties hold.*

- *Mozzi's axis coincides at any instant with a.*
- *Any point P moves along a circumference γ contained in the plane π_P with its center at C. Further, P moves along γ with velocity $\dot{\varphi}|CP|$*

Proof. Mozzi's axis $\mathcal{A}(t)$ is a set of points whose velocities vanish or are parallel to ω. Since a is fixed, its points must belong to $\mathcal{A}(t)$. Consequently, $\mathcal{A}(t)$ coincides with the fixed axis a. The second statement follows at once from both (12.25) and the properties of the cross product. $\square$

The vector $\omega(t)$ vanishes in a translational rigid motion, whereas in a rotational motion about a fixed axis a, it is directed along a and its component along a gives the angular velocity of all points of the rigid body.

Definition 12.3. A rigid motion is ***helicoidal*** if it is obtained by compounding a rotational motion about axis a and a translational motion along a.

In view of (12.23) and (12.25), the velocity field in such a motion yields

$$\dot{\mathbf{r}}_P = \boldsymbol{\tau} + \dot{\varphi}\mathbf{k} \times \overrightarrow{OP}, \qquad (12.26)$$

where τ is the sliding speed of a, $\mathbf{k}$ a unit vector along a, and $\dot{\varphi}$ the angular velocity about a.

Comparing the velocity field of a helicoidal motion (12.26) with (12.22), we obtain the following equivalent form of Mozzi's theorem.

Theorem 12.4. *At any instant t, the velocity field of a rigid motion M coincides with the velocity field of a helicoidal motion, called a helicoidal motion tangent to M at instant t.*

At this point, it is quite natural to call ω the ***instantaneous angular velocity*** of the rigid motion.

12.4 Spherical Rigid Motions

Definition 12.4. A rigid motion with a fixed point O is said to be a *spherical motion*.

Choosing $\Omega \equiv O$ in (12.5), we obtain

$$x_i = Q_{ij}(t)x'_j, \tag{12.27}$$

so that a spherical motion is determined by giving the matrix function $(Q_{ij}(t))$. Since this matrix is orthogonal, that is, it confirms the six conditions (12.6), we can say that a spherical motion is determined assigning three independent coefficients of the orthogonal matrix (Q_{ij}) as functions of time.

Further, the velocity field of a spherical motion is given by the equation

$$\dot{\mathbf{r}}_P = \boldsymbol{\omega}(t) \times \overrightarrow{OP}, \tag{12.28}$$

in which both the value and *direction* of the angular velocity $\boldsymbol{\omega}$ depend on time.

Instead of giving three independent coefficients of the orthogonal matrix (Q_{ij}), we can determine the orientation of the Cartesian axes $(O, (x'_i))$ with respect to the Cartesian axes $(O, (x_i))$ using three other independent parameters, called *Euler's angles* (Fig. 12.3). These angles can be defined when the planes Ox_1x_2 and $Ox'_1x'_2$ do not coincide. In fact, in this case, they intersect each other along a straight line, called a *line of nodes* or *nodal line*, and we can define the angles ψ, φ, and θ shown in Fig. 12.3. The angle ψ between the axis Ox_1 and the line of nodes is called the *angle of precession*, the angle φ between the line of nodes and the axis Ox'_1 is the *angle of proper rotation*, and the angle between the axes Ox_3 and Ox_3' is called the

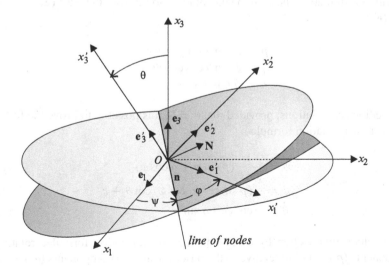

Fig. 12.3 Euler's angles

angle of nutation. To recognize the independence of these angles, it is sufficient to verify that they define three independent rotations transforming the axes (Ox_i) into the axes (Ox''^i). It is easy to understand that the following three rotations have this property: a rotation through the angle ψ about the axis Ox_3, a rotation through the angle φ about the axis Ox_3', and a rotation through the angle θ about the line of nodes.

We now determine the fundamental **Euler kinematical relations** relating the angular velocity with the Euler angles. First, we start with the following representation of ω in the base (e_i'):

$$\omega = p e_1' + q e_2' + r e_3'. \tag{12.29}$$

Owing to the independence of the Euler angles, we can write

$$\omega = \dot{\psi} e_3 + \dot{\varphi} e_3' + \dot{\theta} \mathbf{n}. \tag{12.30}$$

Denoting by $\mathbf{N}$ the unit vector in the plane (O, e_1', e_2') and orthogonal to $\mathbf{n}$ and resorting to Fig. 12.3, we can easily prove the following formulae:

$$\mathbf{N} = \sin \varphi e_1' + \cos \varphi e_2',$$
$$e_3 = \cos \theta e_3' + \sin \theta \mathbf{N}.$$

Combining these relations and recalling (12.30), we obtain

$$e_3 = \sin \theta \sin \varphi e_1' + \sin \theta \cos \varphi e_2' + \cos \theta e_3',$$
$$p = \omega \cdot e_1' = \dot{\psi} e_3 \cdot e_1' + \dot{\theta} \mathbf{n} \cdot e_1' = \dot{\psi} \sin \theta \sin \varphi + \dot{\theta} \cos \varphi. \tag{12.31}$$

Applying a similar reasoning for evaluating q and r, we obtain the following transformation formulae of the components of ω in going from the base $(e_3, e_3', \mathbf{n})$ to the base (e_i'):

$$p = \dot{\psi} \sin \theta \sin \varphi + \dot{\theta} \cos \varphi,$$
$$q = \dot{\psi} \sin \theta \cos \varphi - \dot{\theta} \sin \varphi,$$
$$r = \dot{\psi} \cos \theta + \dot{\varphi}$$

With tedious calculations, provided that $\theta \neq 0$, it is possible to prove the following inverse transformation formulae:

$$\dot{\psi} = \frac{1}{\sin \theta} (p \sin \varphi + q \cos \varphi), \tag{12.32}$$
$$\dot{\varphi} = -(p \sin \varphi + q \cos \varphi) \cot \theta + r, \tag{12.33}$$
$$\dot{\theta} = p \cos \varphi - q \sin \varphi. \tag{12.34}$$

We conclude this section by evaluating the transformation formulae relating the components (p, q, r) of ω relative to the base (e_i') and the components $(\omega_1, \omega_2, \omega_3)$ relative to the base (e_i). Owing to (12.7), applying the matrix (Q_{ij}^T) to (e_i), we obtain the base $(\mathbf{e}'_i)$. As we have already noted, we can bring the base $(\mathbf{e}_i)$ to coincide with the base $(\mathbf{e}'_i)$ by applying the following three rotations: the rotation through ψ about Ox_3, the rotation through θ about the line of nodes, and, finally, the rotation through φ about $= x'_3$. These rotations are, respectively, represented by the following matrices:

$$A = \begin{pmatrix} \cos\psi & \sin\psi & 0 \\ -\sin\psi & \cos\psi & 0 \\ 0 & 0 & 1 \end{pmatrix},$$

$$B = \begin{pmatrix} 1 & 0 & 0 \\ 0 & \cos\theta & \sin\theta \\ 0 & -\sin\theta & \cos\theta \end{pmatrix},$$

$$C = \begin{pmatrix} \cos\varphi & \sin\varphi & 0 \\ -\sin\varphi & \cos\varphi & 0 \\ 0 & 0 & 1 \end{pmatrix},$$

so that

$$Q^T_{ij} = C_{ih} B_{hk} A_{kj}.$$

On the other hand, the components of a vector are transformed according to the inverse matrix of (Q^T_{ij}), that is, according to (Q_{ij}), since (Q^T_{ij}) is orthogonal. Consequently, the matrix giving the transformation $(p, q, r) \rightarrow (\omega_i)$ is

$$Q_{ij} = A^T_{ih} B^T_{hk} C^T_{kj},$$

and by simple calculations it is possible to show that

$$\begin{aligned}
\omega_1 &= (\cos\varphi \cos\psi - \sin\varphi \sin\psi \cos\theta)p \\
&\quad -(\sin\varphi \cos\psi + \cos\varphi \sin\psi \cos\theta)q + \sin\psi \sin\theta\, r \\
\omega_2 &= (\cos\varphi \sin\psi + \sin\varphi \cos\psi \cos\theta)p \\
&\quad +(-\sin\varphi \sin\psi + \cos\varphi \cos\psi \cos\theta)q - \cos\psi \sin\theta\, r \\
\omega_3 &= \sin\varphi \sin\theta\, p + \cos\varphi \sin\theta\, q + \cos\theta\, r.
\end{aligned} \tag{12.35}$$

The program **SphereMotion** contained in the notebook **RigidMotion** supplies a space representation of the instantaneous axis of rotation during the rigid motion with a fixed point.

12.5   Rigid Planar Motions

Definition 12.5. A rigid motion is a *planar motion* if the velocity field is parallel to a fixed plane π.

It is trivial that the rigidity of the planar motion implies that all points that belong to the same straight line orthogonal to π have the same velocity. Consequently, to characterize the velocity field of a planar motion, it is sufficient to analyze the velocity field on an arbitrary plane parallel to π. We must determine the restrictions on the vectors τ and ω of (12.22) due to the hypothesis of planar motion. The answer to this question is given by the following theorem.

Theorem 12.5. *In a planar motion, we have that*

1. *If* $\tau(t) \neq \mathbf{0}$, *then necessarily* $\omega(t) = \mathbf{0}$ *and* $\tau(t)$ *is parallel to* π.
2. *If* $\omega(t) \neq \mathbf{0}$, *then necessarily* $\tau(t) = \mathbf{0}$ *and* $\omega(t)$ *is orthogonal to* π.

Proof. At instant t, the vector τ in (12.26) gives the common velocity of all points belonging to Mozzi's axis $\mathcal{A}(t)$. Consequently, $\mathcal{A}(t)$ must be parallel to π. But also $\omega(t) \times \overrightarrow{AP}$ must be parallel to π for any choice of P, and that is possible if and only if $\omega(t) = \mathbf{0}$. Inversely, if $\omega(t) \neq \mathbf{0}$, then the term $\omega(t) \times \overrightarrow{AP}$ is parallel to π for any P if and only if $\omega(t)$ is orthogonal to π. Further, $\tau(t)$ must also be parallel to both π and ω so that it vanishes. $\square$

The preceding theorem states that, at any instant t, the velocity field of a planar motion coincides either with that of a translational motion parallel to π or with the velocity field of a rotational motion about an axis orthogonal to π. In this last case, since the points of this axis are at rest, it coincides with Mozzi's axis $\mathcal{A}(t)$. The intersection point A of $\mathcal{A}(t)$ with the plane π is called the *instantaneous rotation center*. It is evident that, for any point P, A belongs to a straight line orthogonal to the velocity $\dot{\mathbf{r}}_P$ at point P.

The notebook **PlaneMotion** contained in the notebook **RigidMotion** supplies the trajectory of the instantaneous rotation center during a rigid planar motion.

12.6   Two Examples

In this section, we apply the results of the previous sections to two examples. Let S be a disk and let a be an axis orthogonal to S containing the center C of S. We suppose that S rotates with angular velocity $\omega = \dot{\psi}\mathbf{k}$ about a, where $\mathbf{k}$ is the unit vector along a. Further, we suppose that C moves with constant angular velocity $\dot{\varphi}$ on a circumference γ having its center at O and radius r_C (Fig. 12.4). We take the position vectors starting from O and denote by $\mathbf{r}_A$ the unknown position vector of the instantaneous rotation center A. Since we are considering a nontranslational planar motion, Mozzi's axis $\mathcal{A}(t)$ is orthogonal to the plane π containing S and,

Fig. 12.4 Example of planar motion

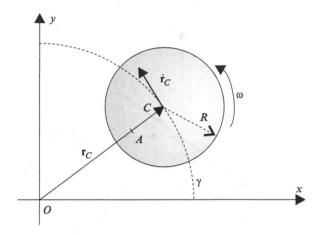

consequently, is determined by the single point A at which it intersects π. With our notations, relation (12.22) reduces to the equation

$$\mathbf{r}_A = \mathbf{r}_C - \frac{\dot{\mathbf{r}}_C \times \dot{\psi}\mathbf{k}}{\dot{\psi}^2}.$$

Introducing the unit vector $\mathbf{n}$ orthogonal to γ and directed toward O, the preceding equation can also be written as

$$\mathbf{r}_A = \mathbf{r}_C \pm R\frac{|\dot{\varphi}|}{|\dot{\psi}|}\mathbf{n},$$

where we must choose the $+$ sign when $\dot{\varphi}$ and $\dot{\psi}$ have the same sign and the $-$ sign in the opposite case. In conclusion, during motion, the instantaneous rotation center A describes a circumference with center at O and radius r_A less or greater than r_C, depending on whether or not $\dot{\varphi}$ and $\dot{\psi}$ have the same sign.

As a second example, we consider a disk S rotating with constant angular velocity $\dot{\psi}$ about an axis a contained in the plane of S (Fig. 12.5). We also suppose that the center C of S moves along a straight line Ox orthogonal to a with a velocity $\dot{\mathbf{r}}_C$. To determine the evolution of Mozzi's axis $\mathcal{A}(t)$, we denote by $\mathbf{k}$ a unit vector along a and by A the intersection point of $\mathcal{A}(t)$ with the plane Oxy. Then from (12.22) we obtain

$$\mathbf{r}_A = \mathbf{r}_C - \frac{\dot{\mathbf{r}}_C \times \dot{\psi}\mathbf{k}}{\dot{\psi}^2} = \mathbf{r}_C - \frac{|\dot{\mathbf{r}}_C|}{|\dot{\psi}|}\mathbf{j},$$

where $\mathbf{j} = \mathbf{k} \times \mathbf{i}$ and $\mathbf{i}$ is the unite vector directed along Ox.

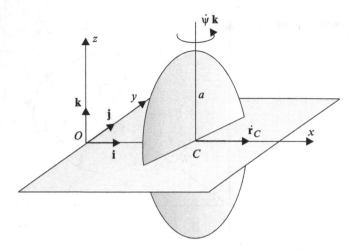

Fig. 12.5 Example of rigid motion

12.7   Relative Motions

Let R and R' be two observers moving relative to each other. Let the relative motion be rigid. If point P is moving relative to both R and R', what is the relation between the velocities and the accelerations of P measured by both observers? In this section, we answer this question.

We denote by $(O, \mathbf{e}_i)$ and $(\Omega, \mathbf{e}'_i)$ two systems of Cartesian coordinates fixed in R and R', respectively. Further, we call the motion of P relative to R' *relative motion* and the motion of P relative to R *absolute motion*.

Let $x'_i(t)$ be the parametric equation of the moving point P in the frame $(\Omega, \mathbf{e}'_i)$. To know the coordinates x_i of P relative to $(O, \mathbf{e}_i)$, we introduce the functions $x'_i(t)$ into (12.5):

$$x_i(t) = x_{\Omega_i}(t) + Q_{ij}(t)x'_j(t). \tag{12.36}$$

Differentiating this relation with respect to time, we obtain

$$\dot{x}_i(t) = \dot{x}_{\Omega_i}(t) + \dot{Q}_{ij}(t)x'_j(t) + Q_{ij}(t)\dot{x}'_j(t). \tag{12.37}$$

Now, we must recognize the physical meaning of the different terms appearing in (12.37). First, on the left-hand side are the components relative to the base $(\mathbf{e}_i)$ of the velocity $\dot{\mathbf{r}}_P$ of P evaluated by observer R, that is, of the *absolute velocity*. The first two terms on the right-hand side, which coincide with (12.9), are obtained by differentiating (12.36) with respect to time and holding constant the components $x'_j(t)$. In other words, these terms give the velocity $\mathbf{v}_\tau$ of the point *at rest* relative to $(\Omega, \mathbf{e}'_i)$ that, at instant t, is occupied by P. Since the motion of R' relative to R is rigid,

$\mathbf{v}_\tau$ is given by (12.17), i.e., by the first two terms on the right-hand side of (12.37). Finally, the last term of (12.37) is the velocity $\dot{\mathbf{r}}'_P$ of P evaluated by observer R'. In conclusion, we can give (12.37) the following form:

$$\dot{\mathbf{r}}_P = \mathbf{v}_\tau + \dot{\mathbf{r}}'_P, \qquad (12.38)$$

where

$$\mathbf{v}_\tau = \dot{\mathbf{r}}_\Omega + \boldsymbol{\omega} \times \overrightarrow{\Omega P}. \qquad (12.39)$$

Equation (12.38) is the mathematical formulation of the classical *principle of velocity composition*.

Let $\mathbf{u}$ be a vector varying in time with respect to both observers R and R'. Due to the relative motion of R' relative to R, the dependence of $\mathbf{u}$ on time is expressed by a function $\mathbf{f}(t)$ in R and by a function $\mathbf{g}(t)$ in R'. We call

$$\frac{d_a \mathbf{u}(t)}{dt} = \dot{\mathbf{f}}(t)$$

the *absolute derivative* of $\mathbf{u}(t)$ and

$$\frac{d_r \mathbf{u}(t)}{dt} = \dot{\mathbf{g}}(t)$$

the *relative derivative* of $\mathbf{u}(t)$. Here, we show that (12.38) allows us to determine the relation between these two derivatives. In fact, the following two representations of the vector $\mathbf{u}(t) \equiv \overrightarrow{AB}$ hold in R and R', respectively (Fig. 12.6):

$$\mathbf{u}(t) = \mathbf{r}_B(t) - \mathbf{r}_A(t), \quad \mathbf{u}(t) = \mathbf{r}'_B(t) - \mathbf{r}'_A(t), \qquad (12.40)$$

where $\mathbf{r}_B(t)$ and $\mathbf{r}_A(t)$ are the position vectors of points B and A relative to the frame $(O, \mathbf{e}_i)$, and $\mathbf{r}'_B(t)$ and $\mathbf{r}'_A(t)$ are the position vectors of the same points in the frame $(\Omega, \mathbf{e}'_i)$. It is evident that

$$\frac{d_a \mathbf{u}(t)}{dt} = \dot{\mathbf{r}}_B(t) - \dot{\mathbf{r}}_A(t),$$

$$\frac{d_r \mathbf{u}(t)}{dt} = \dot{\mathbf{r}}'_B(t) - \dot{\mathbf{r}}'_A(t)$$

Using (12.38) and (12.39) in the preceding expressions, we obtain the fundamental formula

$$\frac{d_a \mathbf{u}(t)}{dt} = \frac{d_r \mathbf{u}(t)}{dt} + \boldsymbol{\omega} \times \mathbf{u}(t). \qquad (12.41)$$

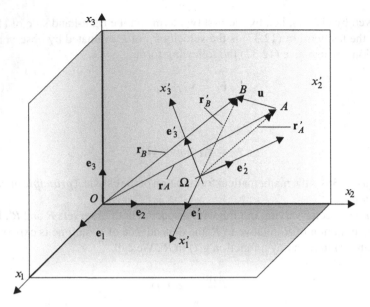

Fig. 12.6 Absolute and relative derivatives

We are interested in determining the relation between the absolute acceleration and the relative acceleration. We reach this result differentiating with respect to time either (12.37) or (12.38). Here, we adopt the second approach, which is based on (12.41).

Differentiating (12.38) and taking into account (12.41), we obtain

$$
\begin{aligned}
\ddot{\mathbf{r}}_P &= \frac{\mathrm{d}_a \dot{\mathbf{r}}_P}{\mathrm{d}t} = \frac{\mathrm{d}_a}{\mathrm{d}t}(\dot{\mathbf{r}}'_P + \mathbf{v}_\tau) \\
&= \frac{\mathrm{d}_r \dot{\mathbf{r}}'_P}{\mathrm{d}t} + \boldsymbol{\omega} \times \dot{\mathbf{r}}'_P + \frac{\mathrm{d}_a \mathbf{v}_\tau}{\mathrm{d}t} \\
&= \ddot{\mathbf{r}}'_P + \boldsymbol{\omega} \times \dot{\mathbf{r}}'_P + \frac{\mathrm{d}_a \mathbf{v}_\tau}{\mathrm{d}t}.
\end{aligned}
\tag{12.42}
$$

On the other hand, from (12.39) we have that

$$
\begin{aligned}
\frac{\mathrm{d}_a \mathbf{v}_\tau}{\mathrm{d}t} &= \ddot{\mathbf{r}}_\Omega + \dot{\boldsymbol{\omega}} \times \overrightarrow{\Omega P} + \boldsymbol{\omega} \times (\dot{\mathbf{r}}_P - \dot{\mathbf{r}}_\Omega) \\
&= \ddot{\mathbf{r}}_\Omega + \dot{\boldsymbol{\omega}} \times \overrightarrow{\Omega P} + \boldsymbol{\omega} \times (\dot{\mathbf{r}}'_P + \mathbf{v}_{\tau P} - \dot{\mathbf{r}}_\Omega) \\
&= \ddot{\mathbf{r}}_\Omega + \dot{\boldsymbol{\omega}} \times \overrightarrow{\Omega P} + \boldsymbol{\omega} \times \dot{\mathbf{r}}'_P + \boldsymbol{\omega} \times (\boldsymbol{\omega} \times \overrightarrow{\Omega P}).
\end{aligned}
\tag{12.43}
$$

Collecting (12.42) and (12.43), we can write

$$
\ddot{\mathbf{r}}_P = \ddot{\mathbf{r}}'_P + \mathbf{a}_\tau + \mathbf{a}_c,
\tag{12.44}
$$

Fig. 12.7 Example of relative motion

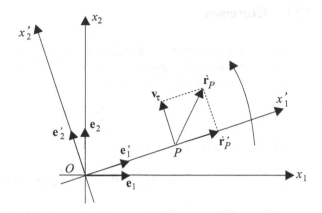

where

$$\mathbf{a}_\tau = \ddot{\mathbf{r}}_\Omega + \dot{\boldsymbol{\omega}} \times \overrightarrow{\Omega P} + \boldsymbol{\omega} \times (\boldsymbol{\omega} \times \overrightarrow{\Omega P}),$$
$$\mathbf{a}_c = 2\boldsymbol{\omega} \times \dot{\mathbf{r}}'_P.$$

In view of (12.18), the term $\mathbf{a}_\tau$ gives the acceleration of that fixed point of $(\Omega, \mathbf{e}'_i)$, which is occupied by P at instant t. The term $\mathbf{a}_c$ is called *Coriolis' acceleration*.

Example 12.1. Let P be a point moving on a straight line Ox_1 that rotates with constant angular velocity $\dot{\varphi}_0$ about a fixed axis a orthogonal to Ox_1x_2 and containing O (Fig. 12.7). Verify that

$$\dot{\mathbf{r}}' = \dot{x}'_1 \mathbf{e}'_1, \quad \mathbf{v}_\tau = x'_1 \dot{\varphi}_0 \mathbf{e}'_2,$$
$$\ddot{\mathbf{r}} = \ddot{x}'_1 \mathbf{e}'_1, \quad \mathbf{a}_\tau = -\dot{\varphi}_0^2 x'_1 \mathbf{e}'_1,$$
$$\mathbf{a}_c = 2\dot{\varphi}_0 \dot{x}'_1 \mathbf{e}'_2.$$

Later, we will use the result stated in the following theorem.

Theorem 12.6. *Let P be a point uniformly moving relative to observer R. Then the acceleration relative to another observer R' vanishes if and only if the motion of R' relative to R is translational and uniform.*

Proof. If $\ddot{\mathbf{r}} = \mathbf{0}$ and the motion of R' relative to R is translational and uniform, then $\boldsymbol{\omega} = \mathbf{a}_\tau = \mathbf{0}$. Consequently, we also have that $\mathbf{a}_c = \mathbf{0}$ and $\ddot{\mathbf{r}}' = \mathbf{0}$. In contrast, if $\ddot{\mathbf{r}} = \ddot{\mathbf{r}}' = \mathbf{0}$, then the condition

$$\mathbf{a}_\tau + 2\boldsymbol{\omega} \times \dot{\mathbf{r}}' = \mathbf{0}$$

must hold for any vector $\dot{\mathbf{r}}'$, so that $\mathbf{a}_\tau = \boldsymbol{\omega} = \mathbf{0}$, and the theorem is proved. $\square$

12.8 Exercises

1. Let ω_i be the components of the angular velocity in the lab frame (O, x_i). Determine the evolution equations of the matrix $\mathbb{Q} = (Q_{ij})$ relative to (O, x_i).
   From (12.12) and (12.15) we obtain

$$\dot{\mathbb{Q}} = \mathbb{W}\mathbb{Q}, \tag{12.45}$$

   which in components is written as

$$\dot{Q}_{ij} = W_{ih}Q_{hj} = Q_{hj}\epsilon_{ikh}\omega_k. \tag{12.46}$$

   This is a system of nine ODEs in the nine unknowns $(Q_{ij}(t))$. However, we can only accept solutions for which the matrix $\mathbb{Q}$ is orthogonal at any instant. Now we prove that this property is confirmed by *any* solution of (12.46) corresponding to initial data $\mathbb{Q}(0)$ such that

$$\mathbb{Q}(0)\mathbb{Q}^T(0) = \mathbb{I}. \tag{12.47}$$

   In fact, (12.45) can equivalently be written as

$$\dot{\mathbb{Q}}\mathbb{Q}^T = \mathbb{W} = -\mathbb{W}^T = -\mathbb{Q}\dot{\mathbb{Q}}^T,$$

   that is,

$$\dot{\mathbb{Q}}\mathbb{Q}^T + \mathbb{Q}\dot{\mathbb{Q}}^T = 0 \Leftrightarrow \frac{d}{dt}(\mathbb{Q}\mathbb{Q}^T) = 0.$$

   In other words, for any solution of (12.46), the matrix $\mathbb{Q}\,\mathbb{Q}^T$ is constant. Therefore, if we assign the initial data satisfying (12.47), the corresponding solution of (12.46) will be an orthogonal matrix.

2. Determine the time evolution of the matrix $\mathbb{Q} = (Q_{ij})$ in the lab frame when the angular velocity ω is given by the functions $(\dot{\psi}(t), \dot{\varphi}(t), \dot{\theta}(t))$.
   When the initial values $(\psi_0, \varphi_0, \theta_0)$ of Euler's angles are given, by integrating $(\dot{\psi}(t), \dot{\varphi}(t), \dot{\theta}(t))$, we obtain Euler's angles $(\psi(t), \varphi(t), \theta(t))$. Then, the time evolution of the matrix $\mathbb{Q} = (Q_{ij})$ follows from (12.32)–(12.35).

3. Let S be a rigid homogeneous bar whose end points A and B are constrained to move, respectively, along the axes Ox and Oy of a Cartesian system of coordinates. Prove that, during the motion of S, the instantaneous rotation center C describes a circumference with its center at O and radius equal to the length $2l$ of the bar (Fig. 12.8). Justify the result recalling that the instantaneous rotation center belongs to the intersection of the straight lines orthogonal to the velocities.
   Hint: Point A has coordinates $(2l \sin \varphi, 0)$, where φ is the angle between S and a downward-oriented vertical straight line. Then, the velocity of A has components

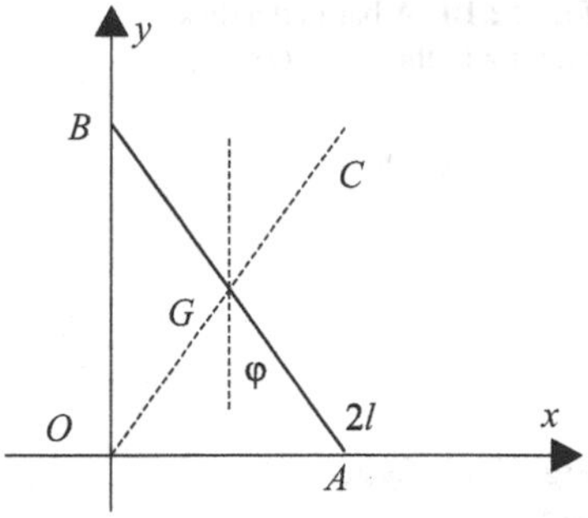

Fig. 12.8 A constrained bar

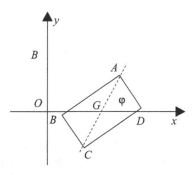

Fig. 12.9 A rectangle with center moving along Ox

$(2l\dot{\varphi}\cos\varphi, 0)$. In view of (12.21) and Theorem 12.4, we have that

$$\mathbf{r}_C = \mathbf{r}_A - \frac{\dot{\mathbf{r}}_A \times \boldsymbol{\omega}}{\omega^2}.$$

Since $\boldsymbol{\omega} = \dot{\varphi}\mathbf{k}$, where $\mathbf{k}$ is a unit vector orthogonal to the plane Oxy, we have that $\mathbf{r}_C = (2l\sin\varphi, 2l\cos\varphi)$.

4. The center G of a rectangle $ABCD$ (Fig. 12.9) moves along the Ox-axis of a Cartesian system of coordinates. Further, $ABCD$ rotates about an axis orthogonal to Oxy and containing G. Denoting by $2l$ the diagonal of the rectangle, determine the velocity of point A and the trajectory of the instantaneous rotation center.
   Hint: $\dot{\mathbf{r}}_A = \dot{x}_G\mathbf{i} + \dot{\varphi}\mathbf{k} \times \overrightarrow{GA}$. Apply (12.21) to determine C.
5. A bar OC and a disk are constrained to move in the plane Oxy (Fig. 12.10). The bar has point A moving on the Ox-axis and rotates about O. The disk rotates about C. Determine the velocity of point P of the disk.
6. A uniform circular hoop of radius r rolls on the outer rim of a fixed wheel of radius b; the hoop and the wheel are coplanar. If the angular velocity ω of the hoop is constant, find

   - The velocity and acceleration of the center of the hoop;
   - The acceleration of that point of the hoop which is at the greatest distance from the center of the wheel.

Fig. 12.10 A bar and a disk
moving in the plane Oxy

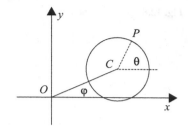

Fig. 12.11 System of two
bars

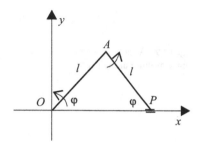

7. A point A rotates about a fixed point with angular velocity ω. A point B has a
   uniform circular motion about A with angular velocity Ω. What relation connects
   Ω and ω if the acceleration of B is always directed toward O?
8. A wheel of radius r rolls without slipping along a straight line a. If the center of
   the wheel has a uniform velocity v parallel to a, find at any instant the velocity
   and the acceleration of the two points of the rim that are at the height $h \leq r$
   above a.
9. Show that in a general motion of a rigid lamina in its plane, there is just one
   point with zero acceleration.
10. Consider a system of two bars, as shown in Fig. 12.11. Determine the velocity
    of point P in terms of the angular velocity $\dot{\varphi}$ of the bar OA about O.

Chapter 13
Principles of Dynamics

Abstract In this chapter the principle of inertia and inertial frames are introduced. In the inertial frames the axiomatic of classical forces is discussed. Then, the Newton laws are recalled together with some important consequences (momentum balance, angular momentum balance, kinetic energy, König's theorem). Dynamics in non-inertial frames both in the absence and in the presence of constraints is presented together with some exercises of point dynamics.

13.1 Introduction

Dynamics has the aim of determining the motion starting from its causes. If we take into account all the characteristics of real bodies, such as, extension and deformability, with the aim of reaching a more accurate description of the real world, then we reach such a complex mathematical model that it is usually impossible to extract concrete results to compare with experimental data. Consequently, it is convenient to start with simplified models that necessarily neglect some aspects of real bodies. These models will describe real behavior with sufficient accuracy, provided that the neglected aspects have negligible effects on the motion.

In this chapter, we start with the *Newtonian model* in which the bodies are modeled as *material points* or *point masses*. This model does not take into account the extension and form of bodies, which are characterized only by their mass. It is evident that such a model supplies a reasonable description of the real behavior of bodies only when their extension can be neglected with respect to the extension of the region in which the motion takes place. A famous example satisfying this condition is supplied by the motion of the planets around the Sun. In fact, the Newtonian model was created simply to describe the behavior of the planetary system and, more generally, of celestial bodies such as comets, asteroids, and others. The evolution of systems formed by material points is governed by the Newtonian laws of dynamics, whose predictions have been confirmed by a large body of experimental data.

© Springer International Publishing AG, part of Springer Nature 2018 199
A. Romano and A. Marasco, *Classical Mechanics with Mathematica®*,
Modeling and Simulation in Science, Engineering and Technology,
https://doi.org/10.1007/978-3-319-77595-1_13

In Chap. 15 we analyze the **Eulerian model** in which the extension of a body B becomes a relevant characteristic of the model; the mass distribution of B plays a crucial role but its deformations are neglected. In other words, the Eulerian model analyzes the motion of *extended rigid* bodies. A model to describe a system of material points or rigid bodies, free or subjected to constraints, is due to Lagrange and will be analyzed in Chap. 17.

13.2   Principle of Inertia and Inertial Frames

As was remarked in Chap. 11, the description of motion requires the introduction of a frame of reference equipped with synchronized clocks. All frames of reference having this characteristic are quite equivalent from a kinematical point of view. In dynamics we are faced with a completely different situation since the choice of a frame of reference deeply affects the connection between the motion and its causes. Then it is fundamental to check the existence of a subset $\mathcal{I}$ of the whole class of frames of reference adopted in kinematics, in which the description between motion and causes admits the simplest description.

The criterion that helps us in determining set $\mathcal{I}$ is based on a fundamental assumption: *any action on a material point P is due to the presence of other bodies B in the space around it; moreover, this action reduces to zero when the distances between P and the bodies belonging to B go to infinity.* This vague assumption on the interaction among bodies makes reasonable the following definition: a material point P is said to be **isolated** if it is so far from any other body that we can neglect their action on P. It is natural to presume that in the class $\mathcal{I}$ the motion of an isolated material point P is as simple as possible. In this respect, we have the following principle.

Axiom 13.1. *Principle of inertia (Newton's First Law) – There is at least a frame of reference I, called an **inertial frame**, in which any isolated material point is at rest or is uniformly moving on a straight line.*

An inertial frame I whose existence is stated by the preceding principle can be localized using the following procedure. First, I is chosen in such a way that *any* isolated material point moves on a straight line relative to $\mathcal{I}$. On the straight line r described by a particular isolated material point P we fix a point O and denote by x the abscissa of P with respect to O. Then I is equipped with synchronized clocks (Sect. 11.1) that associate with any position x of P the instant

$$ t = \frac{x}{v}, \tag{13.1} $$

where v is an arbitrary constant. It is evident that, with the time t, point P moves uniformly on the straight line r with constant speed v. The variable t is defined up to a linear transformation since we can change the origin of time by varying point O and the unit of time by changing the value of v. The principle of inertia states that, with this choice of time, the motion of *any isolated material point* is still rectilinear and uniform. Such a time will be called a ***dynamical time***.

In practice, we choose an arbitrary periodic device and then control whether or not the time that it defines is dynamical. Usually, we adopt the Earth as a fundamental clock, dividing its period of rotation into hours, primes, and seconds. A more accurate measure of dynamical time is obtained by fractionizing the period of revolution of the Earth about the Sun.

The existence of at least one inertial frame implies the existence of infinite inertial frames, as is proved by the following theorem, which follows from Theorem 12.5.

Theorem 13.1. *Let I be an inertial frame equipped with the dynamical time t. Then, a frame of reference I' is an inertial frame if and only if its motion relative to I is translational and uniform and it is equipped with clocks measuring the time $t' = at + b$, where a and b are arbitrary real constants.*

13.3   Dynamical Interactions and Force Laws

The dynamics of a system of material points can be based on different sets of axioms. These different axiomatic formulations, although they lead to the same physical description of dynamical phenomena, differ from each other in the more or less abstract principles on which they are based. Roughly, they can be partitioned into two main classes according to whether the force is introduced as a primitive or a derived concept. Here, we adopt a formulation of the first type for the following two reasons. First, it is a faithful mathematical transcription of the fundamental problem of dynamics that consists in finding the motion when its causes are given. Second, many forces admit a *static* measure, whereas the axiomatic formulations of the second class only lead to *dynamical* measures of forces. In any case, for completeness, at the end of this section we present an axiomatic approach of the second type proposed by Mach and Kirchhoff.

The following axiom describes the interaction between material points.

Axiom 13.2. *Let $S = \{P_1, P_2, \ldots, P_n\}$ be an **isolated** system of n material points. In an inertial frame of reference I, the dynamical action on $P_i \in S$, due to the point $P_j \in S$, is represented by a vector $\mathbf{F}(P_i, P_j)$ that is applied at P_i and is independent of the presence of the other points of S. Further, $\mathbf{F}(P_i, P_j)$ depends on the position vectors $\mathbf{r}_i$, $\mathbf{r}_j$ and the speeds $\dot{\mathbf{r}}_i$, $\dot{\mathbf{r}}_i$ relative to I of the material points P_i and P_j, that is,*

$$\mathbf{F}(P_i, P_j) = \mathbf{f}_{ij}(\mathbf{r}_i, \mathbf{r}_j, \dot{\mathbf{r}}_i, \dot{\mathbf{r}}_j). \tag{13.2}$$

The two vectors $\mathbf{F}(P_i, P_j)$ and $\mathbf{F}(P_j, P_i)$, applied at P_i and P_j, respectively, are directed along the straight line $P_i P_j$ and confirm the condition

$$\mathbf{F}(P_i, P_j) = -\mathbf{F}(P_j, P_i). \tag{13.3}$$

Finally, the action on P_i of the material points $P_i^e = S - \{P_i\}$ of S is expressed by the following vector sum:

$$\mathbf{F}(P_i, P_i^e) = \sum_{j \neq i, j=1}^{n} \mathbf{F}(P_i, P_j). \tag{13.4}$$

The vector $\mathbf{F}(P_i, P_j)$ is the *force* that P_j exerts on P_i, the function $\mathbf{f}_{ij}$ into (13.2) is called a *force law*, condition (13.3) is the mathematical version of the *action-reaction principle (Newton's third law)*, and, finally, (13.4) postulates the *parallelogram rule* for forces.

The preceding axiom lays down the laws that hold for interactions in classical mechanics. Before introducing the axiom that makes clear the connection between these actions and the motion they produce, we must deal with another important aspect of classical interactions. More precisely, we must define the behavior of forces and force laws under a change of the frame of reference. It is important to solve this problem for the following reasons. First, we can compare the dynamical descriptions obtained by two inertial observers analyzing the same phenomenon. Second, there are some interesting phenomena (for instance, dynamics relative to the Earth) in which a noninertial frame is spontaneously adopted, so that we are compelled to decide how the force law is modified by adopting this new frame of reference.

To solve the preceding problem, we start by introducing some definitions.

Definition 13.1. A tensor is *objective* if it is invariant under a rigid frame change $R = (O, x_i) \rightarrow R' = (\Omega', x_i')$ given by

$$x_i' = x_{\Omega i}(t) + Q_{ij}(t)x_j, \tag{13.5}$$

where $(Q_{ij}(t))$ is an orthogonal matrix at any instant t.

In particular, a scalar quantity is objective if it has the same value when it is evaluated by R and R'. A vector quantity $\mathbf{v}$ is objective if, denoting by (v_i) and (v_i') its components relative to R and R', respectively, the following results are obtained:

$$v_i' = Q_{ij}v_j. \tag{13.6}$$

Similarly, a 2-tensor $\mathbf{T}$ is objective if

$$T_{ij}' = Q_{ih}Q_{jk}T_{hk}. \tag{13.7}$$

Remark 13.1. To better understand the meaning of the preceding definition, we note that the tensor quantity $\mathbf{T}$ is "a priori" relative to the frame of reference R in which it is evaluated; more precisely, it belongs to the Euclidean space moving with R. Consequently, in measuring the quantity $\mathbf{T}$, the two observers R and R' could find two different tensors belonging to their own Euclidean spaces. For instance, a measure of velocity of a material point moving relative to R and R' leads to different vectors: the absolute velocity and the relative velocity (Sect. 12.7). On the contrary, the tensor $\mathbf{T}$ is objective if *it is invariant under a rigid change of frame* or, equivalently, if its components are modified according to (13.7). Bearing that in mind, we say that $\mathbf{T}$ is objective if

$$\mathbf{T}' = \mathbf{T}. \tag{13.8}$$

Definition 13.2. Let

$$\mathbf{v} = \mathbf{g}(\mathbf{u}_1, \ldots, \mathbf{u}_n) \tag{13.9}$$

be a tensor quantity depending on the tensor variables $(\mathbf{u}_1, \ldots, \mathbf{u}_n)$. The function $\mathbf{g}$ is said to be *objective* if

$$\mathbf{v} = \mathbf{g}(\mathbf{u}_1, \ldots, \mathbf{u}_n) = \mathbf{g}(\mathbf{u}'_1, \ldots, \mathbf{u}'_n) = \mathbf{v}', \tag{13.10}$$

i.e., if it is invariant under a frame change.

Now we complete the description of classical interactions by adding the following axiom.

Axiom 13.3. *The force* $\mathbf{F}(P_i, P_j)$ *acting between two material points* P_i *and* P_j *of an isolated system S, as well as the force law (13.2), is objective under a change from an inertial frame I to any rigid frame* R'.

The following theorem, which characterizes the force laws satisfying the preceding axiom, is very important.

Theorem 13.2. *The force law* $\mathbf{f}_{ij}(\mathbf{r}_i, \mathbf{r}_j, \dot{\mathbf{r}}_i, \dot{\mathbf{r}}_i)$ *is objective if and only if*

$$\mathbf{f}_{ij}(\mathbf{r}_i, \mathbf{r}_j, \dot{\mathbf{r}}_i, \dot{\mathbf{r}}_j) = \varphi_{ij}(|\mathbf{r}|, \dot{\mathbf{r}} \cdot \mathbf{k})\mathbf{k}, \tag{13.11}$$

where

$$\mathbf{r} = \mathbf{r}_i - \mathbf{r}_j, \quad \dot{\mathbf{r}} = \dot{\mathbf{r}}_i - \dot{\mathbf{r}}_j, \quad \mathbf{k} = \frac{\mathbf{r}}{|\mathbf{r}|},$$

and φ_{ij} *is an objective scalar function.*

Remark 13.2. The objectivity principle implies that the force $\mathbf{f}_{ij}(\mathbf{r}_i, \mathbf{r}_j, \dot{\mathbf{r}}_i, \dot{\mathbf{r}}_i)$ acts along the straight line joining points P_i and P_j, in accordance with the action and reaction principle.

Proof. To simplify the notations, we omit the indices i and j and set $\mathbf{r}_i = \mathbf{u}$, $\mathbf{r}_j = \mathbf{v}$. Further, we notice that the length of any vector $\mathbf{r} = \mathbf{u} - \mathbf{v}$ is invariant under transformation (13.5). If (13.11) holds, then we have that

$$f_h'(\mathbf{u}', \mathbf{v}', \dot{\mathbf{u}}', \dot{\mathbf{v}}') = \varphi\left(|\mathbf{r}'|, \frac{\dot{r}_l' r_l'}{|\mathbf{r}'|}\right) \frac{r_h'}{|\mathbf{r}'|}$$

$$= \varphi\left(|\mathbf{r}|, \frac{(Q_{lk}\dot{r}_k + \dot{Q}_{lk}r_k)Q_{lm}r_m}{|\mathbf{r}|}\right) \frac{Q_{hl}r_l}{|\mathbf{r}|}.$$

Since (Q_{ij}) is an orthogonal matrix and $\dot{Q}_{ij}Q_{il}$ is skew-symmetric, we obtain

$$\varphi\left(|\mathbf{r}'|, \frac{\dot{r}_l' r_l'}{|\mathbf{r}'|}\right) \frac{r_h'}{|\mathbf{r}'|} = Q_{hl}\varphi\left(|\mathbf{r}|, \frac{\dot{r}_m r_m}{|\mathbf{r}|}\right) \frac{r_l}{|\mathbf{r}|},$$

and the force law is objective. Conversely, if the force law is objective, i.e., if

$$f_h(\mathbf{u}', \mathbf{v}', \dot{\mathbf{u}}', \dot{\mathbf{v}}') = Q_{hk} f_k(\mathbf{u}, \mathbf{v}, \dot{\mathbf{u}}, \dot{\mathbf{v}}),$$

then we have that the condition

$$\begin{aligned} f_h\,(Q_{kl}(u_l - x_{\Omega l}), Q_{kl}(v_l - x_{\Omega l}), Q_{kl}(\dot{u}_l - \dot{x}_{\Omega l}) \\ + \dot{Q}_{kl}(u_l - x_{\Omega l}), Q_{kl}(\dot{v}_l - \dot{x}_{\Omega l}) + \dot{Q}_{kl}(v_l - x_{\Omega l})) \\ = Q_{hm} f_m(u_k, v_k, \dot{u}_k, \dot{v}_k) \end{aligned} \tag{13.12}$$

must hold for any possible choice of the quantities of Q_{kl}, $\dot{Q}_{kl}$, $x_{\Omega i}$, and $\dot{x}_{\Omega i}$. Now we choose

$$Q_{kl} = \delta_{kl}, \quad \dot{Q}_{kl} = 0, \quad x_{\Omega k} = v_k, \quad \dot{x}_{\Omega i} = \dot{v}_k.$$

This choice corresponds to taking a frame of reference R' that, at the arbitrary instant t, moves relative to I with a translational velocity equal to the velocity at instant t of point P_j, which is at rest at the origin of R'. For this choice, (13.12) becomes

$$f_h(r_k, 0, \dot{r}_k, 0) = f_h(u_k, v_k, \dot{u}_k, \dot{v}_k)$$

so that $\mathbf{f}$ must depend on $\mathbf{r} = \mathbf{u} - \mathbf{v}$ and $\dot{\mathbf{r}} = \dot{\mathbf{u}} - \dot{\mathbf{v}}$. Then, for any orthogonal matrix (Q_{kl}), we have that

$$f_h(Q_{kl}r_l, Q_{kl}\dot{r}_l + \dot{Q}_{kl}r_l) = Q_{hm} f_m(r_k, \dot{r}_k). \tag{13.13}$$

Now we take $Q_{kl} = \delta_{kl}$ and suppose that $\dot{Q}_{kl}$ is an arbitrary skew-symmetric matrix. Then (Sect. 12.2) the equation $\dot{Q}_{kl}r_l = (\boldsymbol{\omega} \times \mathbf{r})_k$ uniquely defines the vector $\boldsymbol{\omega}$. Moreover, decomposing $\dot{\mathbf{r}}$ to

$$\dot{\mathbf{r}} = \dot{\mathbf{r}}_\perp + (\dot{\mathbf{r}} \cdot \mathbf{k})\mathbf{k},$$

where $\mathbf{k} = \mathbf{r}/|\mathbf{r}|$ and $\dot{\mathbf{r}}_\perp \cdot \mathbf{k} = 0$, (13.13) can also be written as

$$f_i(r_k, \dot{r}_k) = f_i(r_k, (\dot{r}_\perp)_k + (\omega \times \mathbf{r}_\perp)_k + (\dot{\mathbf{r}} \cdot \mathbf{k})\mathbf{k}).$$

Since this condition holds for any vector ω, we have that

$$f_i(r_k, \dot{r}_k) = f_i(r_k, (\dot{\mathbf{r}} \cdot \mathbf{k})\mathbf{k}).$$

When we note that $|\mathbf{r}|$ and $\dot{\mathbf{r}} \cdot \mathbf{k}$ are invariant under transformation (13.5), we conclude that (13.13) assumes the final form

$$f_i(r_k, (\dot{\mathbf{r}} \cdot \mathbf{k})\mathbf{k}) = Q_{hi} f_h(Q_{kl} r_l, (\dot{r}_k \cdot \mathbf{k}) Q_{kl} k_l),$$

so that $\mathbf{f}(\mathbf{r}, (\dot{\mathbf{r}} \cdot \mathbf{k})\mathbf{k})$ depends isotropically on the variables $\mathbf{r}$ and $(\dot{\mathbf{r}} \cdot \mathbf{k})\mathbf{k}$. Finally, (13.11) follows from known theorems of algebra about isotropic functions.   □

13.4 Newton's Second Law

The axioms of the preceding sections describe some fundamental aspects of the classical interactions between material points. The following axiom describes the connection existing between a force acting upon a material point and a corresponding motion.

Axiom 13.4. *Let $S = \{P_1, \dots, P_n\}$ be an isolated system of n material points, and let I be an inertial system of reference. Denote by $P_i^{(e)}$ the set of all points of S different from P_i. Then, to any material point P_i it is possible to associate a positive number m_i, the* **mass** *of P_i, which only depends on the material constitution of P_i. Further, the acceleration $\ddot{\mathbf{r}}_i$ of P_i relative to I satisfies the equation [see (13.4)]*

$$m_i \ddot{\mathbf{r}}_i = \mathbf{F}_i(P_i, P_i^e). \tag{13.14}$$

In particular, if we apply (13.14) to the pair $P_i, P_j \in S$ and recall (13.3), we obtain

$$m_i \ddot{\mathbf{r}}_i + m_j \ddot{\mathbf{r}}_j = \mathbf{0}. \tag{13.15}$$

This result proves the following theorem.

Theorem 13.3. *Let $S = \{P_1, \dots, P_n\}$ be an isolated system of n material points, and let I be an inertial system of reference. The accelerations $\ddot{\mathbf{r}}_i$ and $\ddot{\mathbf{r}}_j$ of the points P_i, $P_j \in S$, produced by their mutual interaction, have the same direction and opposite versus. Moreover, the ratio of their lengths is constant and equal to the inverse ratio of their masses*

$$\frac{|\ddot{\mathbf{r}}_i|}{|\ddot{\mathbf{r}}_j|} = \frac{m_j}{m_i}. \tag{13.16}$$

This is precisely the result proved in the preceding theorem, which is postulated in the formulation due to Mach and Kirchhoff. In this approach the force $\mathbf{F}_i(P_i, P_j)$, which P_j exerts upon P_i, is defined as the product $m_i \ddot{\mathbf{r}}_i$. Postulating that the forces satisfy the parallelogram rule, the action and reaction principle is derived. We conclude by noting that (13.16) leads to an operative measure of mass by measures of accelerations.

13.5 Mechanical Determinism and Galilean Relativity Principle

In this section, we show that the Newtonian laws of dynamics allow one, at least in principle, to determine the motion relative to an inertial frame I of material points belonging to an isolated system $S = \{P_1, \ldots, P_n\}$. In fact, applying (13.14) to any material point $P_i \in S$ and taking into account (13.3), we obtain the system

$$m_i \ddot{\mathbf{r}}_i = \mathbf{f}_i(\mathbf{r}_1, \ldots, \mathbf{r}_n, \dot{\mathbf{r}}_1, \ldots \dot{\mathbf{r}}_n), \quad i = 1, \ldots, n, \tag{13.17}$$

where

$$\mathbf{f}_i(\mathbf{r}_1, \ldots, \mathbf{r}_n, \dot{\mathbf{r}}_1, \ldots \dot{\mathbf{r}}_n) = \sum_{j \neq i, j=1}^{n} \mathbf{f}_{ij}(\mathbf{r}_i, \mathbf{r}_j, \dot{\mathbf{r}}_i, \dot{\mathbf{r}}_j)$$

of n second-order ODEs in the n unknown functions $\mathbf{r}_1(t), \ldots, \mathbf{r}_n(t)$. Since the system has a normal form, under suitable hypotheses about the functions $\mathbf{f}_i$, there exists one and only one solution of (13.17) satisfying the following *initial conditions* (Cauchy's problem):

$$\mathbf{r}_i(t_0) = \mathbf{r}_i^0, \quad \dot{\mathbf{r}}_i(t_0) = \dot{\mathbf{r}}_i^0, \quad i = 1, \ldots, n. \tag{13.18}$$

In other words, when the state of the system S is assigned by giving the position and velocity of any point of S at the instant t_0, then the evolution of S is uniquely determined for $t \geq t_0$. This circumstance, characteristic of classical physics, is called *mechanical determinism*.

The explicit solution of the initial value problem (13.17) and (13.18) can only be found for some particular force laws. However, we show throughout the book that we have at our disposal many mathematical instruments to obtain meaningful information about the solution and its properties.

We are now in a position to state a fundamental principle of dynamics. Let I and I' be two inertial frames, and let $S = \{P_1, \ldots, P_n\}$ be an isolated system moving relative to them. Since the motion of I' relative to I is translational and uniform, the acceleration of any point $P_i \in S$ is the same in both inertial frames [see (12.44)]

$$\ddot{\mathbf{r}}' = \ddot{\mathbf{r}}.$$

Further, the general axioms of dynamics state that the mass of P_i is invariant, and the force $\mathbf{F}_i$ acting on it is objective, together with the force law. Consequently, in the inertial frame I', system (13.17) becomes

$$m_i\ddot{\mathbf{r}}'_i = \mathbf{f}_i(\mathbf{r}'_1, \ldots, \mathbf{r}'_n, \dot{\mathbf{r}}'_1, \ldots \dot{\mathbf{r}}'_n), \quad i = 1, \ldots, n. \tag{13.19}$$

This result proves the following theorem.

Theorem 13.4 (Galilean relativity principle). *The general laws of dynamics preserve their form (are covariant) in any inertial frame.*

The preceding theorem implies that, under the same conditions, the mechanical evolution of an isolated system is the same for any inertial observer. More precisely, since systems (13.17) and (13.19) have the same form, they will admit the same solution when we choose the same initial conditions. In other words, two inertial observers that repeat the same dynamical experiment will arrive at the same results.[1]

13.6 Balance Equations

Let $S = \{P_1, \ldots, P_n, P_{n+1}, \ldots, P_{n+m}\}$ be an isolated system of $n+m$ material points moving relative to an inertial frame I. We now propose to analyze the motion of the points of the system $S^{(i)} = \{P_1, \ldots, P_n\}$ in the presence of the remaining material points $S^{(e)} = \{P_{n+1}, \ldots, P_{n+m}\}$. The motion of any point $P_i \in S$ is governed by the equation

$$m_i\ddot{\mathbf{r}}_i = \mathbf{F}_i^{(i)} + \mathbf{F}_i^{(e)}, \quad i = 1, \ldots, n, \tag{13.20}$$

where $\mathbf{F}_i^{(i)}$ is the total force acting upon P_i, due to the other points belonging to $S^{(i)}$, and $\mathbf{F}_i^{(e)}$ is the total force due to the interaction of P_i with the material points belonging to $S^{(e)}$. By adding all the equations (13.20) and taking into account the action and reaction principle, according to which [see (13.3)]

$$\sum_{i=1}^{n} \mathbf{F}_i^{(i)} = \sum_{i=1}^{n} \sum_{j=1, j\neq i}^{n} \mathbf{F}_{ij}(P_i, P_j), \tag{13.21}$$

we obtain

$$\sum_{i=1}^{n} m_i\ddot{\mathbf{r}}_i = \sum_{i=1}^{n} \mathbf{F}_i^{(e)}. \tag{13.22}$$

---

[1] For a deeper analysis of the relativity principle, see Chap. 26.

Introducing the *linear momentum*

$$\mathbf{Q} = \sum_{i=1}^{n} m_i \dot{\mathbf{r}}_i \tag{13.23}$$

and the *external total force*

$$\mathbf{R}^{(e)} = \sum_{i=1}^{n} \mathbf{F}_i^{(e)}, \tag{13.24}$$

from (13.22) we derive the following theorem.

Theorem 13.5. *The time derivative of the linear momentum of the system $S^{(i)}$ of material points under the action of the external system $S^{(e)}$ is equal to the total external force:*

$$\dot{\mathbf{Q}} = \mathbf{R}^{(e)}. \tag{13.25}$$

Equation (13.25) is called a *balance equation of linear momentum* or *first cardinal equation of dynamics*.

Define the *center of mass G* as the point whose position vector $\mathbf{r}_G$ with respect to the axes of the inertial frame I is given by the relation

$$m\mathbf{r}_G = \sum_{i=1}^{n} m_i \mathbf{r}_i, \tag{13.26}$$

where $m = \sum_{i=1}^{n} m_i$ is the total mass of S. Then we have that

$$\mathbf{Q} = m\dot{\mathbf{r}}_G, \tag{13.27}$$

and (13.25) can be written in the equivalent form

$$m\ddot{\mathbf{r}}_G = \mathbf{R}^{(e)}, \tag{13.28}$$

which makes the following statement.

Theorem 13.6. *The center of mass of the system S of material points moves relative to the inertial frame I as a point having the total mass of the system and subject to the total force acting on S.*

In particular, when the total force vanishes, from (13.25) and (13.28) we derive that the linear momentum is constant and the center of mass moves uniformly with respect to I.

Remark 13.3. The center of mass does not depend on the choice of the origin of the axes introduced in the inertial frame I. In fact, choosing another origin O' and evaluating the position vectors with respect to this new origin, we have that

$$\sum_{i=1}^{n} m_i \mathbf{r}'_i = m\overrightarrow{O'O} + \sum_{i=1}^{n} m_i \mathbf{r}_i = m\overrightarrow{O'O} + m\mathbf{r}_G = m\mathbf{r}'_G.$$

Remark 13.4. It is important to note that (13.28) does not allow us to find the motion of G since, in general, the total force depends on the position and velocity of all material points of S. It is possible to determine the motion of G by (13.28) only if the total force only depends on the position and velocity of G.

Let $\mathbf{r}_i$ be the position vector of $P_i \in S$ relative to the origin O of an inertial frame I. Taking into account the action and reaction principle, after simple calculations, from (13.20) we derive the condition

$$\sum_{i=1}^{n} \mathbf{r}_i \times m_i \ddot{\mathbf{r}}_i = \sum_{i=1}^{n} \mathbf{r}_i \times \mathbf{F}_i^{(e)}. \tag{13.29}$$

If we introduce the **angular momentum with respect to the pole O**

$$\mathbf{K}_O = \sum_{i=1}^{n} \mathbf{r}_i \times m_i \dot{\mathbf{r}}_i \tag{13.30}$$

and the **total torque with respect to the pole O**

$$\mathbf{M}_O^{(e)} = \sum_{i=1}^{n} \mathbf{r}_i \times \mathbf{F}_i^{(e)}, \tag{13.31}$$

then from (13.29) we obtain the following theorem.

Theorem 13.7. *The time derivative of the angular momentum with respect to pole O of the system S is equal to the total torque with respect to the same pole:*

$$\dot{\mathbf{K}}_O = \mathbf{M}_O^{(e)}. \tag{13.32}$$

Equation (13.32) is called a **balance equation of angular momentum** or **second cardinal equation of dynamics**.

Remark 13.5. It is possible to extend (13.32) to the case where the angular momentum and the torque are evaluated with respect to an arbitrary moving pole A. In fact, we have that

$$\mathbf{K}_A = \sum_{i=1}^{n} \overrightarrow{AP}_i \times m_i \dot{\mathbf{r}}_i = \sum_{i=1}^{n} (\mathbf{r}_i - \mathbf{r}_A) \times m_i \dot{\mathbf{r}}_i = \mathbf{M}_A^{(e)}.$$

Differentiating with respect to time and recalling (13.25), (13.27), and (13.30), instead of (13.32), we obtain

$$\dot{\mathbf{K}}_A + \dot{\mathbf{r}}_A \times m \dot{\mathbf{r}}_G = \mathbf{M}_A^{(e)}. \tag{13.33}$$

This generalization of the balance equation of angular momentum shows that (13.32) holds either if A is fixed or if it coincides with the center of mass G.

Remark 13.6. The balance equations of dynamics do not determine the motion of a system S since they do not contain the internal forces, which are fundamental in determining the evolution of S. However, they will play a fundamental role in describing the motion of a rigid system in which the internal interactions do not affect its evolution.

13.7 Kinetic Energy and König's Theorem

Let $S = \{P_1, \ldots, P_n\}$ be a system of n material points, and let R be an *arbitrary* rigid frame of reference. Denoting by $\dot{\mathbf{r}}_i$ the velocity of $P_i \in S$, we define the **kinetic energy** of S relative to R the (nonobjective) scalar quantity

$$T = \frac{1}{2} \sum_{i=1}^{n} m_i \dot{r}_i^2, \quad \dot{r}_i = |\dot{\mathbf{r}}_i|. \tag{13.34}$$

Introduce the frame of reference R_G whose motion relative to R is translational with velocity $\dot{\mathbf{r}}_G$. The motion of S relative to R_G is called **motion about the center of mass**. The following **König's theorem** holds.

Theorem 13.8. *If T' denotes the kinetic energy relative to the motion about the center of mass, then the kinetic energy T of S relative to R is given by the formula*

$$T = T' + \frac{1}{2} m \dot{r}_G^2. \tag{13.35}$$

Proof. Denote by $\dot{\mathbf{r}}_i'$ the velocity of P_i relative to R_G. Since the motion of R_G relative to R is translational with velocity $\dot{\mathbf{r}}_G$, we have that

$$\dot{\mathbf{r}}_i = \dot{\mathbf{r}}_i' + \dot{\mathbf{r}}_G.$$

Consequently, we also have

$$T = \tfrac{1}{2} \sum_{i=1}^{n} m_i \dot{r}_i^2 = \tfrac{1}{2} \sum_{i=1}^{n} m_i (\dot{\mathbf{r}}_i' + \dot{\mathbf{r}}_G)^2$$
$$= T' + \tfrac{1}{2} m \dot{r}_G^2 + \dot{\mathbf{r}}_G \cdot \sum_{i=1}^{n} m_i \dot{\mathbf{r}}_i'.$$

In the frame R_G the center of mass is given by the formula

$$m \mathbf{r}_G' = \sum_{i=1}^{n} m_i \mathbf{r}_i'.$$

But in R_G the center of mass is at rest; therefore, also

$$\sum_{i=1}^{n} m_i \dot{\mathbf{r}}_i' = \mathbf{0}, \tag{13.36}$$

and the theorem is proved. $\qquad\square$

Another important result is stated by the following theorem.

Theorem 13.9. *The angular momentum* $\mathbf{K}'$, *relative to the motion about the center of mass, is independent of the pole. Further, if* $\mathbf{K}_G$ *is the angular momentum relative to pole G in any frame of reference, it results that*

$$\mathbf{K}_G = \mathbf{K}'. \tag{13.37}$$

Proof. Let O and O' be two arbitrary points moving or at rest in R_G. Then, taking into account (13.36), we have

$$\mathbf{K}_{O'}' = \sum_{i=1}^{n} \overrightarrow{O'P_i} \times m_i \dot{\mathbf{r}}_i' = \sum_{i=1}^{n} \overrightarrow{O'O} \times m_i \dot{\mathbf{r}}_i' + \sum_{i=1}^{n} \overrightarrow{OP_i} \times m_i \dot{\mathbf{r}}_i' = \mathbf{K}_O'.$$

To complete the proof of the theorem, we note that

$$\mathbf{K}_G = \sum_{i=1}^{n} \overrightarrow{GP_i} \times m_i \dot{\mathbf{r}}_i = \sum_{i=1}^{n} \overrightarrow{GP_i} \times m_i (\dot{\mathbf{r}}_i' + \dot{\mathbf{r}}_G)$$
$$= \mathbf{K}' + \left(\sum_{i=1}^{n} m_i \overrightarrow{GP_i} \right) \times \dot{\mathbf{r}}_G,$$

where, recalling the definition of center of mass, we obtain

$$\sum_{i=1}^{n} m_i \overrightarrow{GP_i} = m \overrightarrow{GG} = \mathbf{0}.$$

$\qquad\square$

13.8   Kinetic Energy Theorem

Let $S = \{P_1, \ldots, P_n\}$ be a system of material points. We denote by $\mathbf{r}_1(t), \ldots, \mathbf{r}_n(t)$ the trajectories of the points of S moving relative to the frame of reference R under the action of the forces $\Sigma = \{(\mathbf{r}_1, \mathbf{F}_1), \ldots, (\mathbf{r}_n, \mathbf{F}_n)\}$. The *work of the forces* in the time interval $[t_0, t_1]$ relative to the motion $\mathbf{r}_1(t), \ldots, \mathbf{r}_n(t)$ is given by

$$W_{[t_0,t_1]} = \sum_{i=1}^{n} \int_{t_0}^{t_1} \mathbf{F}_i \cdot \dot{\mathbf{r}}_i \mathrm{d}t. \tag{13.38}$$

From (13.20) we have the equation

$$\sum_{i=1}^{n} m_i \ddot{\mathbf{r}}_i \cdot \dot{\mathbf{r}}_i = \sum_{i=1}^{n} \left(\mathbf{F}_i^{(i)} + \mathbf{F}_i^{(e)} \right) \cdot \dot{\mathbf{r}}_i,$$

which, taking into account (13.34), can be written as

$$\dot{T} = \sum_{i=1}^{n} \left(\mathbf{F}_i^{(i)} + \mathbf{F}_i^{(e)} \right) \cdot \dot{\mathbf{r}}_i. \tag{13.39}$$

Integrating (13.38) in the time interval $[t_0, t_1]$, we obtain that *the variation of kinetic energy is equal to the work of internal and external forces acting upon the system S*:

$$T(t_1) - T(t_0) = W_{[t_0,t_1]}^{(i)} + W_{[t_0,t_1]}^{(e)}. \tag{13.40}$$

The forces acting on S are said to be *positional* if they depend only on the position vectors of the material points belonging to $S_{(i)}$ and $S^{(e)}$ (Sect. 13.6). In this case, the work of these forces in the time interval $[t_0, t_1]$ only depends on the trajectories of the points of $S_{(i)}$ and $S^{(e)}$ and is independent of their velocities. More particularly, a system of positional forces $\Sigma = \{(\mathbf{r}_1, \mathbf{F}_1), \ldots, (\mathbf{r}_1, \mathbf{F}_1)\}$ acting upon a system of material points $\{P_1, \ldots, P_n\}$ is *conservative* if there exists a C^1 function $U(\mathbf{r}_1, \ldots, \mathbf{r}_n)$, called *potential energy*, such that

$$\mathbf{F}_i = -\nabla_{\mathbf{r}_i} U. \tag{13.41}$$

The work of conservative forces in the time interval $[t_0, t_1]$ assumes the form

$$W_{[t_0,t_1]} = - \left(U(\mathbf{r}_1(t_1), \ldots, \mathbf{r}_n(t_1)) - U(\mathbf{r}_1(t_0), \ldots, \mathbf{r}_n(t_0)) \right), \tag{13.42}$$

that is, the work depends not on the trajectories and velocities of the points of S but only on their initial and final positions.

We conclude this section with the following theorem, which follows immediately from (13.40) and (13.42).

Theorem 13.10 (**Conservation of mechanical energy**). *If the forces acting on a system S of material points are conservative and U is the corresponding potential energy, during the motion the **total energy** is constant:*

$$T + U = \text{const.} \tag{13.43}$$

13.9 Dynamics in Noninertial Frames

In the preceding sections, the fundamental laws of dynamics were stated with respect to inertial frames. However, in some cases it is necessary to analyze the motion of a system $S = \{P_1, \ldots, P_n\}$ in an arbitrary frame of reference. For instance, when we study the motion of a material point with respect to the Earth, we adopt a terrestrial frame of reference that is not inertial.

Let I and R' be, respectively, an inertial frame and an arbitrary rigid frame of reference. To find the form that the Eqs. (13.17) assume in R', we first recall that the masses and the force laws are invariant under an *arbitrary* change in the rigid frame of reference. Then, taking into account (13.36), we can give (13.17) the form

$$m_i \ddot{\mathbf{r}}'_i = \mathbf{f}_i(\mathbf{r}'_1, \ldots, \mathbf{r}'_n, \dot{\mathbf{r}}'_1, \ldots, \dot{\mathbf{r}}'_n) - m_i \mathbf{a}_{\tau i} - 2m_i \boldsymbol{\omega} \times \dot{\mathbf{r}}'_i, \tag{13.44}$$

where $\mathbf{a}_{\tau i}$ is the acceleration of point P_i due to the relative motion of R' with respect to I (Sect. 12.7). Relation (13.14) shows that, to describe the evolution of S in the frame R', besides the forces $\mathbf{f}_i, i = 1, \ldots, n$, we must consider the forces $- m_i \mathbf{a}_{\tau i}$ and $-2m_i \boldsymbol{\omega} \times \dot{\mathbf{r}}'_i$, which are called *inertial or fictitious forces*. More precisely, the first of these is the *drag force*, whereas the second one is the *Coriolis force*. The name we have given these forces is due to the fact that the observer R finds no physical justification of their presence, that is, he cannot find bodies that, interacting with S, can explain the existence of those forces. For these reasons, the forces due to the presence of bodies are also called *real forces*.

We conclude by remarking that all the results we proved for inertial frames can be extended to noninertial frames, provided we consider both real and fictitious forces.

13.10 Constrained Dynamics

In all the preceding sections, we considered systems $S = \{P_1, \ldots, P_n\}$ of material points in which, a priori, there is no limitation to the possible positions of the points P_i. Actually, among all the possible configurations, the points assume only those that are compatible with the applied forces and the equations of motion. However, in an application, it may happen that some obstacles or **constraints** can limit the mobility of the points of S. It is evident that the presence of constraints reduces the number of coordinates we must find by the equations of motion to determine the trajectories of the points P_i. On the other hand, the presence of constraints introduces as new unknowns the *reactive forces* that the constraints exert on the points during motion. To understand why these forces are unknown, we must remember that the constraints are due to the presence of suitable *rigid* mechanical devices. The reactive forces are a consequence of very small deformations undergone by these devices during motion. Since we have erased these deformations assuming the rigidity of the obstacles, we cannot evaluate the reactive forces and, consequently, cannot assign their force laws. For instance, a body B on a table, provided it is not too heavy, deforms the table, which, to return to its undeformed state, balances the weight of B with a reactive force. To determine this force, one should resort to the theory of elasticity relating the reactive force to the deformation of the table. However, such a problem is very difficult to solve, and we prefer to assume that the table is rigid. This hypothesis makes it impossible to determine the reactive forces that become unknowns.

Although we are not able to assign the reactive forces, it is possible by experimentation to determine some of their important characteristics. In particular, for smooth constraints, the reactive force $\boldsymbol{\Phi}$ acting on a material point P_i of S during motion is orthogonal to the constraint at the point occupied by P_i.

At this point we must verify if the laws of dynamics are able to determine the motion of S in the presence of constraints. In this chapter, we limit ourselves to proving that the motion can still be determined in particular cases, deferring to a later chapter the general analysis of this problem.

Let P be a material point constrained to move on the surface $f(\mathbf{r}) = 0$. If the surface is smooth, then the reactive force acting on P can be written as $\boldsymbol{\Phi} = \lambda \nabla f$, where $\lambda(\mathbf{r}(t))$ is an unknown scalar function. Consequently, with the usual meaning of the symbols, the equation of motion

$$m\ddot{\mathbf{r}} = \mathbf{F} + \lambda \nabla f, \tag{13.45}$$

together with the equation of the surface, at least in principle, makes it possible for us to determine the unknowns $\mathbf{r}(t)$ and $\lambda(\mathbf{r}(t))$.

Let P be a material point moving along the curve γ obtained by intersecting the surfaces

$$f_1(\mathbf{r}) = 0, \quad f_2(\mathbf{r}) = 0 \tag{13.46}$$

such that the rank of the Jacobian matrix of the functions f_1 and f_2 is equal to 2. If the surfaces are smooth, then the reactive force of the constraint is a linear combination of two reactive forces orthogonal to each surface. Consequently, the equation of motion can be written as

$$m\ddot{\mathbf{r}} = \mathbf{F} + \lambda_1 \nabla f_1 + \lambda_2 \nabla f_2. \tag{13.47}$$

Equations (13.46) and (13.47) allow, at least in principle, to determine the unknowns $\mathbf{r}(t)$, $\lambda_1(\mathbf{r}(t))$, and $\lambda_2(\mathbf{r}(t))$.

If the curve γ is given in parametric form

$$\mathbf{r} = \mathbf{r}(s) \tag{13.48}$$

as a function of the curvilinear abscissa s, then we have that [see (11.9)]

$$\ddot{\mathbf{r}} = \ddot{s}\mathbf{t} + \frac{\dot{s}^2}{r}\mathbf{n}, \tag{13.49}$$

where $\mathbf{t}$ is the unit vector tangent to γ, $\mathbf{n}$ the principal unit vector, and r the curvature radius of γ. The equation of motion

$$m\left(\ddot{s}\mathbf{t} + \frac{\dot{s}^2}{r}\mathbf{n}\right) = \mathbf{F} + \mathbf{\Phi}, \tag{13.50}$$

when it is projected along $\mathbf{t}$ and the condition $\mathbf{\Phi} \cdot \mathbf{t} = 0$ is taken into account, gives the scalar differential equation

$$m\ddot{s} = F_t(s, \dot{s}) \equiv \mathbf{F} \cdot \mathbf{t} \tag{13.51}$$

in the unknown $s(t)$.

13.11 Exercises

1. Let P_1 and P_2 be two point particles of masses m_1 and m_2, respectively. Suppose that P_1 is constrained to moving along the Ox-axis and P_2 along the Oy-axis of a Cartesian system of coordinates Oxy under the action of the elastic force $-k\overrightarrow{P_1 P_2}$. If at the time $t = 0$ the positions and velocities of P_1 and P_2 are $x_1(0) = x_1^0$, $x_2(0) = x_2^0$, $\dot{x}_1(0) = \dot{x}_1^0$, and $\dot{x}_2(0) = \dot{x}_2^0$, determine the trajectory of the center of mass of the system P_1, P_2.
2. At a certain instant t, a particle of mass m, moving freely in a vertical plane under the action of gravity, is at a height h above the ground and has a speed v. Determine when it strikes the ground using energy conservation.
3. A heavy particle rests on top of a smooth fixed sphere. If it is slightly displaced, find the angular distance from the top at which it leaves the surface.
4. Show that, if an airplane of mass M in horizontal flight drops a bomb of mass m, the airplane experiences an upward acceleration mg/M, where g is the gravity acceleration.
5. Let S be a disk with a vertical fixed axis a, intersecting S at its center. A particle P of mass m can move along a radius of S. If initially S rotates with angular velocity $\omega(0)$ about a and P is at rest, determine at any instant t the relation between the angular velocity $\omega(t)$ of S and the velocity v of P along the radius of S.

6. A particle P moves along a vertical parabola $y = x^2$ under the action of its weight starting from the point $(2,4)$. Determine the velocity of P when it arrives at the point $(0,0)$ and evaluate where it arrives before inverting its motion.

7. Let P be a particle moving along the Ox-axis under the influence of the force $F = -kx - h\dot{x}$. Determine the work done by F in the time interval $[0,t]$.

Chapter 14
Dynamics of a Material Point

Abstract In this chapter we analyze the motion of a material point acted upon by a Newtonian force and derive Kepler's laws. The motion in the air of a material point subjected to its weight is described. Then, terrestrial dynamics is formulated and the motions of a simple pendulum, rotating simple pendulum, and Foucault's pendulum are studied.

14.1 Central Forces

Definition 14.1. A positional force is said to be *central* with center O if its force law is

$$\mathbf{F} = f(r)\frac{\mathbf{r}}{r}, \quad r = |\mathbf{r}|, \tag{14.1}$$

where $\mathbf{r}$ is the position vector relative to O.

The elementary work dW of $\mathbf{F}$ in the elementary displacement $d\mathbf{r}$ is

$$dW = \mathbf{F} \cdot d\mathbf{r} = f(r)\frac{\mathbf{r}}{r} \cdot d\mathbf{r} = f(r)dr.$$

Then, taking into account (13.41), we can state that a central force is conservative and its potential energy is given by

$$U = -\int f(r)dr. \tag{14.2}$$

A first example of central force is the **elastic force** for which (14.1) and (14.2) become

$$\mathbf{F} = -kr\frac{\mathbf{r}}{r}, \quad U = \frac{1}{2}kr^2,\tag{14.3}$$

respectively. Another important example of central force is the **Newtonian force** for which

$$\mathbf{F} = -\frac{k}{r^2}\frac{\mathbf{r}}{r}, \quad U = -\frac{k}{r}.\tag{14.4}$$

A Newtonian force is said to be *attractive* if $k > 0$, *repulsive* if $k < 0$.

The torque with respect to the pole O of any central force vanishes. Consequently, the angular momentum $\mathbf{K}_O$ is constant:

$$\mathbf{K}_O = \mathbf{K}_O(0),\tag{14.5}$$

where $\mathbf{K}_O(0)$ is the initial value of $\mathbf{K}_O$. On the basis of this result we can prove the following theorem.

Theorem 14.1. *The trajectory of a material point P subject to a central force lies in a plane containing the initial position and velocity of P and orthogonal to the initial angular momentum $\mathbf{K}_O(0)$ (Laplace's plane). Further, in this plane the areal velocity of P is constant (Kepler's second law).*

Proof. Since $\mathbf{K}_O$ is constant, we have that

$$\mathbf{K}_O = \mathbf{r} \times m\dot{\mathbf{r}} = \mathbf{r}(0) \times m\dot{\mathbf{r}}(0) = \mathbf{K}_O(0).\tag{14.6}$$

Consequently, we can state that $\mathbf{r}$ and $\dot{\mathbf{r}}$ belong to the plane π orthogonal to $\mathbf{K}_O$ and containing $\mathbf{r}(0)$ and $\dot{\mathbf{r}}(0)$. Let $Oxyz$ be a Cartesian system of coordinates such that $Oxy \equiv \pi$. In these coordinates, we have that $\mathbf{r} = (x, y, 0)$, $\dot{\mathbf{r}} = (\dot{x}, \dot{y}, 0)$, and $\mathbf{K}_O(0) = (0, 0, K_z(0))$. Then (14.6) implies that

$$x\dot{y} - \dot{x}y = K_z(0)/m,\tag{14.7}$$

and the theorem is proved when we recall (11.21).   □

The dynamic equation of a point mass P subject to a central force is

$$m\ddot{\mathbf{r}} = f(r)\frac{\mathbf{r}}{r}.\tag{14.8}$$

If we introduce polar coordinates (r, φ) with center O into the Laplace plane π and we recall formulae (11.7), which give the components of the acceleration in polar coordinates, from (14.8) we obtain the following system of ordinary differential equations:

$$m(\ddot{r} - r\dot{\varphi}^2) = f(r), \tag{14.9}$$

$$\frac{d}{dt}(r^2\dot{\varphi}) = 0 \tag{14.10}$$

in the unknowns $r(t)$ and $\varphi(t)$. Equation (14.10) again states that the areal velocity is constant:

$$\frac{1}{2}r^2\dot{\varphi} = \frac{1}{2m}K_z(0) \equiv \frac{1}{2}c. \tag{14.11}$$

Introducing (14.11) into (14.9), we obtain the following equation in the unknown function $r(t)$:

$$\ddot{r} = \frac{1}{m}\left(f(r) + \frac{mc^2}{r^3}\right). \tag{14.12}$$

Finally, as the central forces are conservative, the total energy [see (13.43)]

$$\frac{1}{2}m\dot{\mathbf{r}}^2 + U(r) = E \tag{14.13}$$

assumes a constant value E depending on the initial data. In polar coordinates, we have that $\dot{\mathbf{r}}^2 = \dot{r}^2 + r^2\dot{\varphi}^2$ [see (11.15)] and taking into account (14.11), (14.13) becomes

$$\frac{1}{2}m\left(\dot{r}^2 + \frac{c^2}{r^2}\right) + U(r) = E. \tag{14.14}$$

In view of (14.11), (14.13) gives

$$\dot{r}^2 = \frac{2}{m}(E - U_{\text{eff}}(r)), \tag{14.15}$$

where we have introduced the **effective potential**

$$U_{\text{eff}}(r) = U(r) + \frac{mc^2}{2r^2}. \tag{14.16}$$

Equation (14.11) shows that, if the initial data are chosen in such a way that $c = 0$, then $\varphi(t) = \varphi(0)$ for any $t > 0$ and the trajectory lies on a straight line containing O. When $c \neq 0$, the function $\dot{\varphi}(t)$ has a constant sign, so that $\varphi(t)$ is always increasing or decreasing, according to the initial data. Consequently, there exists the inverse function

$$t = t(\varphi), \tag{14.17}$$

and the trajectory, instead of being expressed by the parametric equations $r(t)$, $\varphi(t)$, can be put in the polar form $r(\varphi)$. To find a differential equation giving the function $r(\varphi)$, it is sufficient to note that the formulae

$$\dot{r} = \frac{dr}{d\varphi}\dot{\varphi} = \frac{c}{r^2}\frac{dr}{d\varphi} = -c\frac{d}{d\varphi}\left(\frac{1}{r}\right), \tag{14.18}$$

$$\ddot{r} = \frac{d\dot{r}}{d\varphi}\dot{\varphi} = \frac{c}{r^2}\frac{d\dot{r}}{d\varphi} = -\frac{c^2}{r^2}\frac{d^2}{d\varphi^2}\left(\frac{1}{r}\right), \tag{14.19}$$

allow us to write (14.9) and (14.14), respectively, in the form (*Binet*)

$$\frac{d^2u}{d\varphi^2} + u = -\frac{1}{mc^2u^2}f\left(\frac{1}{u}\right), \tag{14.20}$$

$$\frac{1}{2}\left(\frac{du}{d\varphi}\right)^2 = \frac{E}{mc^2} - \frac{1}{2}u^2 - \frac{1}{mc^2}U\left(\frac{1}{u}\right), \tag{14.21}$$

where

$$u = \frac{1}{r(\varphi)}.$$

We highlight that (14.12) and (14.20) contain the constant quantity c. In other words, these equations require a knowledge of the areal velocity or, equivalently, of the angular momentum about an axis orthogonal to the plane in which the orbit lies [see (14.11)] and containing the center of the force. This plane is obtained by giving the initial data $r(0)$ and $\dot{\varphi}(0)$. We can always assume that the initial datum $\varphi(0)$ is equal to zero by a convenient choice of the polar axis. On the other hand, to integrate (14.12), we must also assign the initial datum $\dot{r}(0)$. The data $r(0)$, $\dot{\varphi}(0)$, and $\dot{r}(0)$ are also necessary for integrating (14.20) since $(dr/d\varphi)(0) = \dot{r}(0)/\dot{\varphi}(0)$ [see (14.18)].

Weierstrass' analysis (Chap. 10) can be applied to (14.12) and (14.14) to determine the qualitative behavior of the function $r(t)$ as well as the phase portrait in the space $(r, \dot{r})$. In contrast, if we are interested in determining the qualitative behavior of the function $r(\varphi)$ and the corresponding phase portrait in the space $(r, dr / d\varphi)$, then we apply Weierstrass's analysis to equations (14.20) and (14.21) since (14.21) is a first integral of (14.20). We explicitly note that the preceding phase portrait depends on the choice of the constant c. The different curves of the phase portrait in the space $(r, \dot{r})$ are obtained by assigning the initial data $r(0)$, $\dot{r}(0)$, whereas the curve of the phase portrait in the space $(r, dr / d\varphi)$ is determined by the initial data $(r(0), (dr / d\varphi)(0))$. The notebook **Weierstrass** provides examples of these phase portraits.

Regarding the central forces we recall the following important theorem.

Theorem 14.2 (Bertrand). *The elastic forces and the Newtonian forces are the only central forces for which **all** bounded orbits are closed.*

14.2 Newtonian Forces

Before analyzing the motion of a material point subject to a Newtonian force, we remember one possible way to define a conic. In the plane π we fix a point O at a distance D from a straight line a. Finally, we denote by OP and d the distances of any point $P \in \pi$ from O and a, respectively. Then, a conic is the locus Γ of the points P such that the **eccentricity** of Γ, defined by the ratio

$$e = \frac{OP}{d},$$

is constant. The point O and the straight line a are respectively called the *focus* and the **directrix** of the conic Γ. Applying the preceding definition of a conic to both cases of Fig. 14.1 and using polar coordinates of center O in the plane π, we obtain the relations

$$e = \frac{r}{D - r\cos(\varphi - \varphi_0)}, \tag{14.22}$$

$$e = \frac{r}{r\cos(\varphi - \varphi_0) - D}. \tag{14.23}$$

In particular, left panel of Fig. 14.1 refers to a conic with the focus O near to the perihelion and leads to Eq. (14.22). Differently, Eq. (14.23) which refers to the right panel of Fig. 14.1 describes a conic (only parabola or hyperbola) when the origin of polar coordinates coincides with the focus O exterior to the branch Γ of the conic.

From Eqs. (14.22)–(14.23) we obtain the following equations of a conic:

$$r = \frac{eD}{1 + e\cos(\varphi - \varphi_0)}, \tag{14.24}$$

$$r = \frac{eD}{-1 + e\cos(\varphi - \varphi_0)}. \tag{14.25}$$

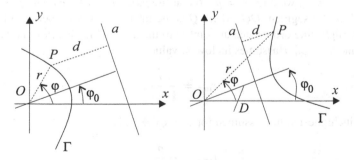

Fig. 14.1 In the left panel the focus is near to perihelion; in the right panel the focus is exterior to the branch of parabola or hyperbola

For a material point subject to a Newtonian force (14.4), (14.20) yields

$$\frac{d^2 u}{d\varphi^2} + u = \frac{1}{p},$$ (14.26)

where

$$p = \frac{mc^2}{k}.$$ (14.27)

The general integral of the linear equation (14.26), with constant coefficients, is

$$u = \frac{1}{p} + R \cos(\varphi - \varphi_0),$$ (14.28)

where $R > 0$ and φ_0 are arbitrary constants depending on the initial data. Going back to the variable r and introducing the quantity

$$e = |p|R,$$ (14.29)

we obtain the equation of the orbit

$$r = \frac{p}{1 + e \cos(\varphi - \varphi_0)}, \text{ if } k > 0,$$ (14.30)

$$r = \frac{p}{-1 + e \cos(\varphi - \varphi_0)}, \text{ if } k < 0,$$ (14.31)

for attractive ($k > 0$) or repulsive ($k < 0$) Newtonian forces. In view of (14.24) and (14.25), we conclude that *a material point under the influence of a Newtonian force moves, with constant areal velocity, along a conic having a focus at the center O.*

To recognize the meaning of the constants p and φ_0 in (14.30) and (14.31), we begin by noting that these relations do not depend on the sign of the difference $(\varphi - \varphi_0)$. Consequently, the axis $\varphi = \varphi_0$ is a symmetry axis s of the conic. Further, for $\varphi = \varphi_0 \pm \pi/2$, we have $r = p$, so that the *parameter of the conic p* is equal to the length of the segment OQ, where Q is the intersection point between the conic and the straight line orthogonal at point O to the axis s (Fig. 14.2). Finally, when $0 \leq e < 1$, and $\varphi = \varphi_0$, r assumes its lowest value

$$r_{\min} = \frac{p}{1 + e},$$ (14.32)

whereas its highest value, assumed for $\varphi = \varphi_0 \pm \pi$, is

$$r_{\max} = \frac{p}{1 - e}.$$ (14.33)

Fig. 14.2 Axis and
parameter of the conic

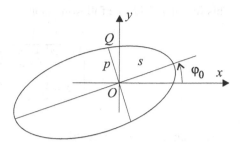

The values (14.32) and (14.33) correspond to the **perihelion** and **aphelion**, respectively. In conclusion, when $0 \leq e < 1$, we have that $r_{min} \leq r \leq r_{max}$, and the conic is an ellipse. In particular, it becomes a circumference when $e = 0$.

If $e = 1$, then we have

$$r_{min} = \frac{p}{2}, \quad r_{max} = \lim_{\varphi \to \varphi_0 + \pi} r = \infty, \tag{14.34}$$

and the conic is a parabola.

Finally, when $e > 1$, the perihelion is still given by (14.32) but there exist two values $\varphi = \varphi_0 \pm \alpha$ for which $r = \infty$. The two straight lines with these angular coefficients are asymptotes of the conic, which is a branch of a hyperbola with its focus at O.

We can summarize the preceding considerations as follows: *a material point subject to an attractive Newtonian central force moves along a conic that may be an ellipse, a parabola, or a branch of a hyperbola. In this last case, the focus O is internal to the branch of the hyperbola. If the force is repulsive, then the trajectory is still a branch of the hyperbola and O is an external focus. In any case, during the motion the areal velocity is constant.*

It is important to relate the eccentricity of the conic to the total energy E. In view of (14.18), the conservation of energy (14.14) assumes the form

$$\frac{1}{2}mc^2 \left[\left(\frac{d}{d\varphi} \left(\frac{1}{r} \right) \right)^2 + \frac{1}{r^2} \right] - \frac{k}{r} = E.$$

Taking into account (14.27)–(14.29), after some calculations, we can prove the result

$$E = \frac{k}{2p}(e^2 - 1) = \frac{k^2}{2mc^2}(e^2 - 1),$$

which implies that

$$e = \sqrt{1 + \frac{2mc^2}{k^2}E}. \tag{14.35}$$

This formula yields the following table:

Energy	Eccentricity	Orbit
$E < 0$	$0 \le e < 1$	Ellipse
$E = 0$	$e = 1$	Parabola
$E > 0$	$e > 1$	Hyperbola

relating energy and eccentricity.

We conclude this section with two important remarks.

Remark 14.1. After determining the orbit $r(\varphi)$ of a material point subject to a New-tonian force, we must know $\varphi(t)$ to complete the analysis of the motion. We limit ourselves to the case of closed orbits for which $0 \le e \le 1$. Since the areal velocity is constant, the function $\varphi(t)$ is a solution of the equation [see (14.11) and (14.30)]

$$\dot{\varphi} = \frac{c}{r^2(\varphi)} = \frac{c}{p^2}(1 + e\cos(\varphi - \varphi_0))^2. \tag{14.36}$$

By changing the polar axis, we can always suppose $\varphi(0) = \varphi_0 = 0$. Then, separating the variables in the preceding equation, we obtain the implicit form of the solution

$$\int_0^{\varphi} \frac{d\varphi}{(1 + e\cos\varphi)^2} = \frac{c}{p^2}t. \tag{14.37}$$

To find for elliptic orbits an approximate solution of (14.36), we note that if, in particular, the orbit is a circumference ($e = 0$), then the angular velocity has the constant value $a \equiv c / p^2$ since r is constant. For elliptic orbits, which are similar to circular orbits, it is reasonable to search for an approximate solution of (14.36) in the form

$$\varphi(t) = at + e\varphi_1(t) + e^2\varphi_2(t) + \cdots . \tag{14.38}$$

Introducing (14.38) into (14.36) we obtain

$$a + e\dot{\varphi}_1 + e^2\dot{\varphi}_2 + \cdots = a(1 + e\cos(a + e\dot{\varphi}_1 + e^2\dot{\varphi}_2 + \cdots))^2. \tag{14.39}$$

Applying Taylor's expansion to the right-hand side of (14.39) and collecting the terms that multiply the same power of the eccentricity, we obtain the following sequence of equations:

$$\dot{\varphi}_1 = 2a\cos(at), \tag{14.40}$$

$$\dot{\varphi}_2 = a\cos^2(at) - 2a\sin(at)\varphi_1(t), \tag{14.41}$$

$$\cdots \cdots$$

Finally, solving the preceding equations with the initial data $\varphi_i(0) = 0$, $i = 0, 1, 2, \ldots$, we obtain

$$\varphi(t) = at + 2e\sin(at) + e^2\left(-\frac{3}{2}at + \frac{5}{4}\sin(2at)\right) + \cdots. \tag{14.42}$$

Remark 14.2. A material point P with mass m, subject to an attractive Newtonian force, initially occupies the position (r_0, φ_0). We want to determine for which initial velocity v_0 of P the orbit is a parabola or a hyperbola, i.e., is open. The lowest value of these velocities is called the *escape velocity*. We have already shown that open orbits are possible if and only if the total energy verifies the condition

$$E = \frac{1}{2}mv_0^2 - \frac{k}{r} \geq 0.$$

Consequently, the escape velocity is given by

$$v_0 = \sqrt{\frac{2k}{mr_0}}. \tag{14.43}$$

14.3 Two-Body Problem

Let S be an isolated system of two material points P_1 and P_2 subject to the action of the mutual gravitational force. If m_1 and m_2 denote their masses, the motion of S is described by the equations

$$m_1\ddot{\mathbf{r}}_1 = -\mathfrak{h}\frac{m_1m_2}{|\mathbf{r}_1 - \mathbf{r}_2|^3}(\mathbf{r}_1 - \mathbf{r}_2), \tag{14.44}$$

$$m_2\ddot{\mathbf{r}}_2 = -\mathfrak{h}\frac{m_1m_2}{|\mathbf{r}_2 - \mathbf{r}_1|^3}(\mathbf{r}_2 - \mathbf{r}_1), \tag{14.45}$$

where the position vector $\mathbf{r}_i$, $i = 1, 2$, is relative to the origin O of an inertial frame I (Fig. 14.3). Further, the position vector $\mathbf{r}_G$ relative to O of the center of mass G of S yields

$$(m_1 + m_2)\mathbf{r}_G = m_1\mathbf{r}_1 + m_2\mathbf{r}_2, \tag{14.46}$$

whereas the position vector $\mathbf{r}$ of P_2 relative to P_1 is given by

$$\mathbf{r} = \mathbf{r}_2 - \mathbf{r}_1. \tag{14.47}$$

Solving (14.46) and (14.47) with respect to $\mathbf{r}_1$ and $\mathbf{r}_2$, we obtain

Fig. 14.3 Two-body system

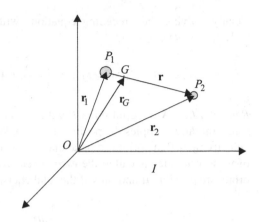

$$\mathbf{r}_1 = \mathbf{r}_G - \frac{m_2}{m_1 + m_2}\mathbf{r}, \tag{14.48}$$

$$\mathbf{r}_2 = \mathbf{r}_G + \frac{m_1}{m_1 + m_2}\mathbf{r}. \tag{14.49}$$

These equations show that the two-body problem can be solved by determining the vector functions $\mathbf{r}_G(t)$ and $\mathbf{r}(t)$. Since the system S is isolated, we can state that the center of mass moves uniformly on a straight line; therefore, the function $\mathbf{r}_G(t)$ is known when the initial data $\mathbf{r}_1(0)$, $\dot{\mathbf{r}}_1(0)$, $\mathbf{r}_2(0)$, $\dot{\mathbf{r}}_2(0)$ are given [see (14.46)]. Regarding the function $\mathbf{r}(t)$, we note that, dividing (14.44) by m_1 and (14.44) by m_2 and subtracting the second equation from the first one, we obtain the equation

$$\ddot{\mathbf{r}} = -\hbar\frac{(m_1 + m_2)}{|\mathbf{r}|^3}\mathbf{r}, \tag{14.50}$$

which can equivalently be written in the form

$$\mu\ddot{\mathbf{r}} = -\hbar\frac{m_1 m_2}{|\mathbf{r}|^3}\mathbf{r}, \tag{14.51}$$

where we have introduced the **reduced mass**

$$\mu = \frac{m_1 m_2}{m_1 + m_2}. \tag{14.52}$$

Both Eqs. (14.50) and (14.51) show that $\mathbf{r}(t)$ can be determined by solving a problem of a material point subject to a Newtonian force. In other words, if we introduce the variables $\mathbf{r}_G$ and $\mathbf{r}$, the two-body problem is reduced to the problem of a material point moving under the action of a Newtonian force.

Example 14.1. Let $P_1 x_1' x_2' x_3'$ be a noninertial frame of reference with the origin at the point P_1 and axes parallel to the axes of an inertial frame $O x_1 x_2 x_3$. Prove that Newton's equation relative to the material point P_2 in the frame $P_1 x_1' x_2' x_3'$ coincides with (14.50), provided that we take into account the fictitious forces.

14.4 Kepler's Laws

In this section, we study the motion of the planets about the Sun. First, we note that the dimension of the planets is much smaller compared with their distances from the Sun. Consequently, it is reasonable to model the planetary system with a system of material points moving under the influence of the mutual gravitational interaction. Such a problem is too complex to be solved. However, the solar mass is much greater than the mass of any planet. This circumstance suggests adopting a simplified model to analyze the motion of the planets. This model is justified by the following considerations.

The motion of a planet P_i is determined by the gravitational action of the Sun and the other planets. Consequently, in an inertial frame I, the motion of the planets P_i of mass m_i is a solution of the equation

$$m_i \ddot{\mathbf{r}}_i = -\mathfrak{h} \frac{M m_i}{|\mathbf{r}_i|^3} \mathbf{r}_i - \sum_{j \neq i} \mathfrak{h} \frac{m_i m_j}{|\mathbf{r}|_{ij}^3} \mathbf{r}_{ij}, \tag{14.53}$$

where $\mathfrak{h}$ is the gravitational constant, M is the solar mass, $\mathbf{r}_i$ is the position vector of the planet P_i with respect to the Sun, and $\mathbf{r}_{ij} = \mathbf{r}_i - \mathbf{r}_j$ is the position vector of P_i relative to the planet P_j. Since the solar mass $M \gg m_i$, for any planet, while the distances $|\mathbf{r}_i|$ and $|\mathbf{r}_{ij}|$ are of the same order of magnitude, in a first approximation, we can neglect the action of the other planets on P_i with respect to the prevailing action of the Sun. In this approximation we are faced with a two-body problem: the Sun and the single planet. Consequently, to describe the Sun–planet system, we can resort to (14.53), which, owing to the condition $M \gg m_i$, reduces to the equation

$$m_i \ddot{\mathbf{r}}_i = -\mathfrak{h} \frac{m_i M}{|\mathbf{r}_i|^3} \mathbf{r}_i, \tag{14.54}$$

according to which, in a first approximation, the planet moves about the Sun as a material point subject to a Newtonian force. We are in a position to prove the following *Kepler's laws*:

1. *The orbit of a planet is an ellipse having a focus occupied by the center of the Sun.*
2. *The orbit is described by a constant areal velocity.*
3. *The ratio between the square of the revolution period T_i and the cube of the greatest semiaxis a_i of the ellipse is given by*

$$\frac{T_i^2}{a_i^3} = \frac{4\pi^2}{\mathfrak{h}M},$$ (14.55)

i.e., it does not depend on the planet.

The first two laws are contained in the results we have already proved for the motion of a point under the influence of a Newtonian force. To prove the third law, we start by recalling that the area bounded by the ellipse described by the planet is given by the formula $\pi a_i b_i$, where b_i is the smallest semiaxis of the ellipse. By the second law, we have that [see (14.11)]

$$\pi a_i b_i = \frac{1}{2}c_i T_i.$$

On the other hand, it can be proved that $p_i = b_i^2/a_i$, so that

$$p_i = \frac{c_i^2 T_i^2}{4\pi^2 a_i^3}.$$

Comparing this result with (14.27), we obtain the formula

$$\frac{T_i^2}{a_i^3} = \frac{4\pi^2 m_i}{k_i}$$

which, recalling that for Newtonian forces $k_i = \mathfrak{h}m_i M$, implies the Kepler's third law.

The preceding considerations show that Kepler's laws are not exact laws since they are obtained by neglecting the mutual actions of the other planets on the motion of a single planet. Taking into account this influence is a very difficult task since it requires a more sophisticated formulation of the problem. This more accurate analysis shows that the orbits of the planets are more complex since they are open curves that, roughly, can be regarded as ellipses slowly rotating about the focus occupied by the Sun (see Chap. 24).

14.5   Scattering by Newtonian Forces

Let P be a material point of mass m moving under the action of a repulsive Newtonian force with center at O. If we suppose its energy E to be positive, the trajectory of P will be a branch of a hyperbola γ that does not contain the center O. We denote by ψ the angle between the two asymptotes of γ and define as a *scattering angle* the angle $\alpha = \pi - \psi$. This is the angle formed by the velocity $\mathbf{v}_\infty^{(i)}$ of P when it comes from infinity and the velocity $\mathbf{v}_\infty^{(e)}$ when P comes back to infinity (Fig. 14.4). Finally,

Fig. 14.4 Scattering of a particle

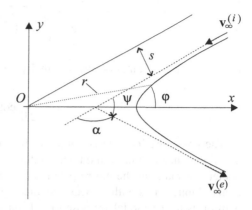

we call a **shock parameter** s the distance of O from the asymptote corresponding to the velocity $\mathbf{v}_\infty^{(i)}$.

In the polar coordinates (r, φ), the equation of γ is [see (14.31)]

$$r = \frac{|p|}{e \cos \varphi - 1},$$

and it gives for the angle ψ the value

$$\cos \frac{\psi}{2} = \frac{1}{e}.$$

Further, we also have that

$$\sin \frac{\alpha}{2} = \sin \left(\frac{\pi}{2} - \frac{\psi}{2} \right) = \cos \frac{\psi}{2} = \frac{1}{e}$$

and

$$\tan \frac{\alpha}{2} = \frac{1}{\sqrt{e^2 - 1}}. \tag{14.56}$$

On the other hand, the conservation of total energy and angular momentum is expressed by the following formulae:

$$E = \frac{1}{2} v^2 - \frac{k}{r} = \frac{1}{2} m \left(v_\infty^{(i)} \right)^2, \tag{14.57}$$

$$|\mathbf{K}_O| = m s v_\infty^{(i)} = s \sqrt{2mE}, \tag{14.58}$$

which allow us to give (14.35) the equivalent form

$$e = \sqrt{1 + \left(\frac{2sE}{k}\right)^2}. \tag{14.59}$$

Using this relation and (14.56), we finally obtain that

$$s = \frac{k}{2E} \cot \frac{\alpha}{2}. \tag{14.60}$$

The preceding formula is important for the following reasons. Rutherford proposed the nuclear atomic model according to which an atom is formed by a small nucleus, containing the entire positive charge of the atom, and electrons rotating along elliptic orbits with a focus occupied by the nucleus. This model was in competition with the model proposed by Thomson in which the positive charge was continuously distributed in a cloud having the dimension of the atom, whereas the electrons were contained in fixed positions inside this cloud. Rutherford understood that, to resolve the dispute regarding the correct model, it was sufficient to throw charged particles (α particles) against the atoms contained in a thin gold leaf and to evaluate the scattering angles of those particles. In fact, Thomson's model, owing to the assumed charge distribution inside the atom, implied small scattering angles. In contrast, the concentrated positive charge of Rutherford's model implied high scattering angles, especially for small shock parameters. The experimental results of Rutherford's experience were in full agreement with Eq. (14.60), which, in particular, implies a back scattering for very small shock parameters.

14.6   Vertical Motion of a Heavy Particle in Air

Let P be a mass point acted upon by its weight $m\mathbf{g}$ and the air resistance

$$\mathbf{F} = -R(v)\frac{\mathbf{v}}{v}, \tag{14.61}$$

where m is the mass of P, $\mathbf{g}$ the gravitational acceleration, and v the length of the velocity $\mathbf{v}$. To suggest a reasonable form of the function $R(v)$, we recall that a mass point is a schematic model of a small body B. When B is falling in the air, the motion is influenced by the form of B, even if B is small. Consequently, the air resistance must include something resembling those characteristics of B that we discarded in adopting the model of a point particle. A good description of the phenomenon we are considering can be obtained assuming that the air resistance has the following form:

$$R(v) = \mu A \alpha f(v), \tag{14.62}$$

where μ is the mass density of the air, A is the area of the projection of B on a plane orthogonal to $\mathbf{v}$, and α is a *form coefficient* that depends on the profile of B. Finally, $f(v)$ is an increasing function confirming the conditions

$$f(0) = 0, \quad \lim_{v \to \infty} f(v) = \infty. \tag{14.63}$$

Taking into account the preceding considerations, we can state that the equation governing the motion of P is

$$m\dot{\mathbf{v}} = m\mathbf{g} - R(v)\frac{\mathbf{v}}{v}. \tag{14.64}$$

Since in this section we limit our attention to falls along the vertical a, we begin by proving that *if the initial velocity* $\mathbf{v}_0$ *is directed along* a, *then the whole trajectory lies on a vertical straight line*. Let $Oxyz$ be a frame of reference with the origin O at the initial position of P and the axis Oz vertical and downward directed. In this frame, the components along Ox and Oy of (14.64) are written as

$$\ddot{x} = -\frac{R(v)}{mv}\dot{x}, \tag{14.65}$$

$$\ddot{y} = -\frac{R(v)}{mv}\dot{y}. \tag{14.66}$$

Multiplying (14.65) by $\dot{x}$ and (14.66) by $\dot{y}$, we obtain

$$\frac{d}{dt}\dot{x}^2 = -2\frac{R(v)}{mv}\dot{x}^2 \le 0, \tag{14.67}$$

$$\frac{d}{dt}\dot{y}^2 = -2\frac{R(v)}{mv}\dot{y}^2 \le 0, \tag{14.68}$$

so that $\dot{x}^2(t)$ and $\dot{y}^2(t)$ are not increasing functions of time. Since initially $\dot{x}(0) = \dot{y}(0) = 0$, we can state that $\dot{x}(t)$ and $\dot{y}(t)$ vanish identically. But in addition, $x(0) = y(0) = 0$, and then $x(t) = y(t) = 0$, for any value of time, and P moves vertically. It remains to analyze the motion along the vertical axis Oz. To this end, we denote by $\mathbf{k}$ the vertical downward-oriented unit vector and note that

$$-\frac{\mathbf{v}}{v} = -\frac{\dot{z}}{|\dot{z}|}\mathbf{k} = \mp\mathbf{k}, \tag{14.69}$$

where we must take the $-$ sign if $\dot{z} > 0$ (downward motion) and the $+$ sign if $\dot{z} < 0$ (upward motion). Finally, to evaluate the vertical motion of P, we must integrate the equation

$$\ddot{z} = mg\left(1 \mp \frac{R(|\dot{z}|)}{mg}\right) \tag{14.70}$$

with the initial conditions

$$z(0) = 0, \quad \dot{z}(0) = \dot{z}_0. \tag{14.71}$$

We must consider the following possibilities:

$$(a) \ \dot{z}_0 < 0, \quad (b) \ \dot{z}_0 \geq 0. \tag{14.72}$$

In case (a), we must take the $+$ sign in (14.70) at least up to the instant t^* for which $\dot{z}(t^*) = 0$. In the time interval $[0, t^*]$, we have that $\ddot{z} > 0$, $\dot{z} < 0$, so that

$$\frac{d\dot{z}^2}{dt} = 2\dot{z}\ddot{z} < 0,$$

and the motion is decelerated. Integrating (14.70) in the interval $[0, t^*]$ we obtain the formula

$$\int_{\dot{z}_0}^{0} \frac{d\dot{z}}{1 + \dfrac{R(|\dot{z}|)}{mg}} = gt^*, \tag{14.73}$$

and t^* is finite since the function under the integral is bounded in the interval $[\dot{z}_0, 0]$.

In case (b), the motion is initially progressive, so that we take the $-$ sign in (14.70) at least up to an eventual instant t_* in which $\dot{z}(t_*) = 0$:

$$\ddot{z} = mg \left(1 - \frac{R(\dot{z})}{mg} \right). \tag{14.74}$$

On the other hand, it is evident that, under hypotheses (14.63) on $f(v)$, there is one and only one value V such that

$$mg = R(V). \tag{14.75}$$

Consider the following three cases:

$$(i) \ \dot{z}(0) = V, \quad (ii) \ \dot{z}(0) > V, \quad (iii) \ \dot{z}(0) < V. \tag{14.76}$$

It is plain to prove that $\dot{z}(t) = V$ is a solution of (14.74) verifying the initial condition (i). Owing to the uniqueness theorem, this is the only solution satisfying datum (i). In case (ii) we have that $\dot{z} > V$ for any value of t. In fact, if there is an instant t_* in which $\dot{z}(t_*) = V$, then, taking this instant as the initial time, we should always have $\dot{z}(t) = V$, against the hypothesis $\dot{z}_0 > V$. Further, if $\dot{z}(t) > V$ for any t, then from (14.74) we have $\ddot{z}(t) < 0$; consequently, $\dot{z}(t)\ddot{z}(t) < 0$ and $\dot{z}^2(t)$ is always decreasing. In conclusion, the motion is progressive and decelerated, and

$$\lim_{t \to \infty} \dot{z}(t) = V. \tag{14.77}$$

Reasoning in the same way, we conclude that in case (*iii*) the motion is progressive and accelerated, and (14.77) still holds.

As an application of the previous analysis, let us consider the case of a parachutist of mass m for which we require a prefixed limit velocity V, i.e., an impact velocity with the ground that does not cause damage to the parachutist. From (14.62) we obtain the cross section A of the parachute in order to attain the requested limit velocity:

$$A = \frac{mg}{\alpha \mu f(V)}.$$

14.7 Curvilinear Motion of a Heavy Particle in Air

In this section, a qualitative analysis of Eq. (14.64) is presented in the general case in which the direction of the initial velocity $\mathbf{v}_0$ is arbitrary. Let $Oxyz$ be a frame of reference with the origin O at the initial position of P and an upward directed vertical axis Oz. Further, we suppose that the coordinate plane Oxz contains $\mathbf{v}_0$, and Ox is directed in such a way as to have $\dot{x}_0 > 0$. Since in this frame $\dot{y}_0 = 0$, and the component of Eq. (14.64) along Oy is still (14.66), we have that $\dot{y}(t) = 0$, at any instant, and we can state that *the trajectory lies in the vertical plane containing the initial position and velocity.*

From (14.65) and the initial condition $\dot{x}(0) = \dot{x}_0 > 0$ it follows that the function $|\dot{x}(t)|$ is not increasing [see (14.67)]. Moreover, there is no instant t^* such that $\dot{x}(t^*) = 0$. In fact, if such an instant existed, we could consider the initial value problem given by (14.65) and the initial datum $\dot{x}(t^*) = 0$. Since the only solution of this problem is given by $\dot{x}(t) = 0$, the motion should take place along the vertical axis Oz, and this conclusion should be in contradiction with the condition $\dot{x}(0) = \dot{x}_0 > 0$. In conclusion, $\dot{x}(t)$ is a positive nonincreasing function for any $t \geq 0$ so that

$$\lim_{t \to \infty} \dot{x}(t) = \dot{x}_\infty < \dot{x}_0, \tag{14.78}$$

$$\lim_{t \to \infty} \ddot{x}(t) = 0. \tag{14.79}$$

In the limit $t \to \infty$, (14.65) and (14.79) lead us to the result

$$\lim_{t \to \infty} \dot{x}(t) = 0. \tag{14.80}$$

We have already proved that the trajectory γ of the material point P is contained in the vertical plane Oxz. To discover important properties of γ, we introduce the unit vector $\mathbf{t} = \mathbf{v}/v$ tangent to γ and denote by α the angle that $\mathbf{t}$ forms with the Ox-axis. Then, projecting (14.64) onto the Ox-axis and along $\mathbf{t}$, we have the system

$$m\frac{d}{dt}(v\cos\alpha) = -R(v)\cos\alpha, \tag{14.81}$$

$$m\dot{v} = -mg\sin\alpha - R(v) \tag{14.82}$$

in the unknowns $\alpha(t)$ and $v(t)$. Introducing into (14.81) the value of $\dot{v}$ deduced from (14.82), we obtain the equation

$$\frac{d}{dt}(\cos\alpha) = \frac{g}{2v}\sin 2\alpha, \tag{14.83}$$

which can also be written as

$$\dot{\alpha}(t) = -\frac{g\cos\alpha}{v}. \tag{14.84}$$

Since we refer to the case $\dot{x}(0) > 0$, at the initial instant $t = 0$ the angle α satisfies the condition

$$-\frac{\pi}{2} < \alpha_0 < \frac{\pi}{2}.$$

Then from (14.84) we have that

$$\dot{\alpha}(0) < 0. \tag{14.85}$$

Consequently, we can state that $\alpha(t)$, starting from any initial value belonging to the interval $(-\pi/2, \pi/2)$, decreases at least up to an eventual instant t^* in which the right-hand side of (14.84) vanishes, that is, when (see Fig. 14.5)

$$\alpha(t^*) = \mp\frac{\pi}{2}.$$

The value t^* is infinite since, in the opposite case, if we adopt the initial datum $\dot{x}(t^*) = v(t^*)\cos\alpha(t^*) = 0$, the corresponding motion should always take place along a vertical line, against the hypothesis that for $t = 0$ the velocity has a component along the Ox-axis.

In conclusion, we can state that *the trajectory is downward concave ($\alpha(t)$ decreases) and has a vertical asymptote* (Fig. 14.6).

Finally, we want to prove that, if $v(t)$ is a regular function of t at infinity, then

$$\lim_{t\to\infty} v(t) = V, \tag{14.86}$$

where V is the unique solution of (14.75). In fact, the regularity of the *positive* function $v(t)$ implies one of the following possibilities:

$$\lim_{t\to\infty} v(t) = +\infty, \quad \lim_{t\to\infty} v(t) = v_\infty < +\infty. \tag{14.87}$$

Fig. 14.5 Behavior of
function $\alpha(t)$

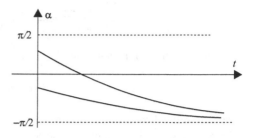

Fig. 14.6 Trajectories in air

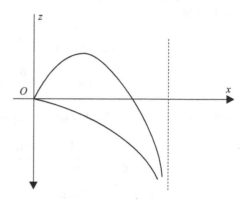

In the first case, since $v(t)$ is positive, either of the following conditions holds:

$$\lim_{t\to\infty} \dot{v}(t) = +\infty, \quad \lim_{t\to\infty} \dot{v}(t) = a > 0.$$

On the other hand, when $\lim_{t\to\infty} v(t) = +\infty$, from (14.82) we obtain $\lim_{t\to\infty} \dot{v}(t) = -\infty$, and then $(14.87)_1$ is false.

When $(14.87)_2$ is verified, we have

$$\lim_{t\to\infty} v(\dot{t}) = 0,$$

so that, by (14.82), we prove (14.86).

We wish to conclude this section with some considerations about the mathematical problem we are faced with. Equations (14.84) and (14.82) can also be written in the form

$$\dot{\alpha}(t) = -\frac{g\cos\alpha}{v}, \tag{14.88}$$

$$\frac{dv}{d\alpha}\dot{\alpha}(t) = -g\sin\alpha - \frac{R(v)}{m}, \tag{14.89}$$

from which we deduce the equation

$$\frac{dv}{d\alpha} = v \tan \alpha + \frac{vR(v)}{mg \cos \alpha} \tag{14.90}$$

in the unknown $v(\alpha)$. Its solution,

$$v = F(\alpha, \alpha_0, v_0), \tag{14.91}$$

for given initial data α_0 and v_0, defines a curve in the plane (α, v). When this function is known, it is possible to integrate the equation of motion. In fact, in view of (14.88), we have that

$$v \cos \alpha = \dot{x} = \dot{\alpha}\frac{dx}{d\alpha} = -\frac{g \cos \alpha}{v}\frac{dx}{d\alpha},$$
$$v \sin \alpha = \dot{z} = \dot{\alpha}\frac{dz}{d\alpha} = -\frac{g \cos \alpha}{v}\frac{dz}{d\alpha}.$$

In turn, in view of (14.88), this system implies

$$\frac{dx}{d\alpha} = -\frac{v^2}{g} = -\frac{F^2(\alpha, \alpha_0, v_0)}{g},$$
$$\frac{dz}{d\alpha} = -\frac{v^2}{g}\tan \alpha = -\frac{F^2(\alpha, \alpha_0, v_0)}{g}\tan \alpha,$$

so that

$$x = -\frac{1}{g}\int_{\alpha_0}^{\alpha} F^2(\xi, \alpha_0, v_0)d\xi \equiv G(\alpha, \alpha_0, v_0), \tag{14.92}$$

$$z = -\frac{1}{g}\int_{\alpha_0}^{\alpha} F^2(\xi, \alpha_0, v_0) \tan \alpha \, d\xi \equiv H(\alpha, \alpha_0, v_0). \tag{14.93}$$

These relations, which give the parametric equations of the trajectory for any choice of the initial data, represent the solution of the *fundamental ballistic problem* since they make it possible to determine the initial angle α_0 and the initial velocity v_0 to hit a given target.

14.8   Terrestrial Dynamics

By *terrestrial dynamics* we mean the dynamics relative to a frame of reference $R_e = (O', x_1, x_2, x_3)$ at rest with respect to the Earth. Before writing the equation of motion of a material point in the frame $R_e = (O', x_1, x_2, x_3)$, we must determine how $R_e = (O', x_1, x_2, x_3)$ moves relative to an inertial frame $I = (O\xi_1, \xi_2, \xi_3)$. We take the frame $I = (O\xi_1, \xi_2, \xi_3)$ with its origin at the center O of the Sun, axes oriented toward fixed stars, and containing the terrestrial orbit in the coordinate plane $O\xi_1, \xi_2$ (Fig. 14.7).

Fig. 14.7 Terrestrial frame

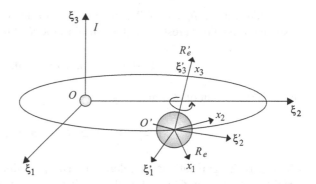

It is well known that the motion of the Earth relative to I is very complex. However, we reach a sufficiently accurate description of terrestrial dynamics supposing that the Earth describes an elliptic orbit about the Sun while it rotates uniformly about the terrestrial axis a.

Then we consider the frame $R'_e = (O', \xi'_1, \xi'_2, \xi'_3)$ with its origin at the center O' of the Earth, axes $O\xi'_3 \equiv a$, and the other axes with fixed orientation relative to the axes of I. It is evident that the motion of R'_e relative to I is translational but not uniform. Finally, we consider a frame $R_e = (O', x_1, x_2, x_3)$ with $Ox_3 \equiv a$ and the other two axes at rest with respect to the Earth. Because the dynamical phenomena we want to describe take place in smaller time intervals compared with the solar year, we can assume that the rigid motion of R'_e relative to I is translational and uniform. In other words, R'_e can be supposed to be inertial, at least for time intervals much shorter than the solar year. Finally, the motion of R_e relative to R'_e is a uniform rotation about the terrestrial axis a. Now we are in a position to determine the fictitious forces acting on a material point P moving relative to the frame R_e:

$$-m\mathbf{a}_\tau = m\omega_T^2 \, \overrightarrow{QP}, \quad -m\mathbf{a}_c = -2m\boldsymbol{\omega}_T \times \dot{\mathbf{r}}', \tag{14.94}$$

where m is the mass of P, $\boldsymbol{\omega}_T$ the terrestrial angular velocity, Q the projection of P onto the rotation axis a, and $\dot{\mathbf{r}}'$ the velocity of P relative to R_e. Consequently, the equation governing the motion of P relative to the frame R_e is

$$m\ddot{\mathbf{r}}' = \mathbf{F}^* + m\mathbf{A} + m\omega_T^2 \, \overrightarrow{QP} - 2m\boldsymbol{\omega}_T \times \dot{\mathbf{r}}', \tag{14.95}$$

where

$$m\mathbf{A} = -\mathfrak{h} \frac{M_T m}{|\mathbf{r}|^{/3}} \mathbf{r}' \tag{14.96}$$

is the gravitational force acting on P due to the Earth and $\mathbf{F}^*$ is the nongravitational force.

We define the **weight** **p** of P as the opposite of the force $\mathbf{F}_e$ we need to apply to P, so that it remains at rest in the terrestrial frame R_e. From (14.95) we have that

$$\mathbf{F}_e + m\mathbf{A} + m\omega_T^2 \overrightarrow{QP} = \mathbf{0},$$

and then the weight of P is given by

$$\mathbf{p} = m\left(\mathbf{A} + \omega_T^2 \overrightarrow{QP}\right) \equiv m\mathbf{g}, \tag{14.97}$$

where **g** is the gravitational acceleration. The preceding equation shows that the weight of P is obtained by adding the gravitational force $m\mathbf{A}$, which is exerted on P by the Earth, and the centrifugal force $m\omega_T^2 \overrightarrow{QP}$. Supposing that the Earth is spherical and recalling that the first force is always directed toward the center O' of the Earth, whereas the second force is orthogonal to the rotation axis of the Earth, we can state that the weight reaches its minimum at the equator and its maximum at the poles. Further, at the poles and at the equator, the direction of **p** coincides with the vertical to the surface of the Earth. Finally, we remark that the direction of **g** *does not depend on the mass of* P. This fundamental property follows from the identification of the masses m appearing in (14.96) and in the centrifugal force $m\omega_T^2 \overrightarrow{QP}$. If we define as **gravitational mass** m_g the mass appearing in the gravitational force (14.96) and as **inertial mass** the mass m_i appearing in Newton's equation of motion, instead of (14.97), we obtain

$$\mathbf{g} = \mathbf{A} + \frac{m_i}{m_g}\omega_T^2 \overrightarrow{QP}. \tag{14.98}$$

In other words, if we distinguish the two masses respectively appearing in the universal attraction law and in Newton's law of motion, then, due to the presence of the ratio m_i/m_g, the acceleration **g** depends on the nature of the body. However, Eötvös and Zeeman showed, with very high experimental accuracy, that **g** does not depend on the ratio m_i/m_g. Consequently, if we adopt the same unit of measure for m_i and m_g, we can assume that $m_i/m_g = 1$. This result plays a fundamental role in general relativity.

With the introduction of the weight (14.97) and in the absence of other forces, (14.95) becomes

$$\ddot{\mathbf{r}}' = \mathbf{g} - 2\boldsymbol{\omega}_T \times \dot{\mathbf{r}}'. \tag{14.99}$$

To consider a first application of (14.99), we choose a frame of reference R_e as in Fig. 14.8 and write (14.99) in a convenient nondimensional form.

Let us introduce a reference length L and a reference time interval T. Since the rotation of the Earth produces much smaller effects on the falls of heavy bodies than does the weight, we can take L and T in such a way that $L/T^2 = g$. On the other hand, we cannot use the same time interval T as a measure of ω since a turn of the Earth

Fig. 14.8 Terrestrial frame

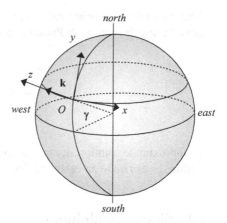

about its axis requires 24 h, whereas a heavy body falls in a few minutes or seconds depending on the throw height. Consequently, we introduce a new reference time interval, T_T. If we use the same letters to denote dimensional and nondimensional quantities, (14.99) can also be written as

$$\ddot{\mathbf{r}}' = -\mathbf{k} + 2\epsilon\omega_T \times \dot{\mathbf{r}}', \tag{14.100}$$

where $\mathbf{k}$ is the unit vector along Oz (Fig. 14.8) and ϵ is the nondimensional parameter

$$\epsilon = \frac{T}{T_T}.$$

For a falling time $T = 60$ s and $T = 86,400$ s (24 h) we have $\epsilon \simeq 0.0007$. In other words, we are faced with a problem to which we can apply Poincaré's method. Since in the frame R_e it is $\mathbf{k} = (0, 0, 1)$ and $\boldsymbol{\omega}_T = (0, \omega_T \cos \gamma, \omega_T \sin\gamma)$, (14.100) gives the system

$$\begin{aligned}
\ddot{x} &= -2\epsilon\omega_T (\dot{z}\cos\gamma - \dot{y}\sin\gamma), \\
\ddot{y} &= -2\epsilon\omega_T \dot{x}\sin\gamma, \\
\ddot{z} &= -1 + 2\epsilon\omega_T \dot{x}\cos\gamma,
\end{aligned}$$

which is equivalent to the first-order system

$$\begin{aligned}
\dot{x} &= u, \\
\dot{y} &= v, \\
\dot{z} &= w, \\
\dot{u} &= -2\epsilon\omega_T (w\cos\gamma - v\sin\gamma), \\
\dot{v} &= -2\epsilon\omega_T u\sin\gamma, \\
\dot{w} &= -1 + 2\epsilon\omega_T u\cos\gamma.
\end{aligned}$$

The first-order Poincaré expansion of the solution of the preceding system, obtained using the notebook **Poincare** and referred to by $x(t)$, $y(t)$, and $z(t)$, is

$$x = \frac{1}{3}\epsilon t^3 \omega \cos \gamma,$$
$$y = 0,$$
$$z = 1 - \frac{t^2}{2}.$$

This approximate solution shows that the rotation of the Earth, while a heavy body is falling, gives rise to an eastward deviation.

14.9 Simple Pendulum

We call a heavy particle P with mass m, constrained to moving along a *smooth* circumference γ lying in a vertical plane π (Fig. 14.9), a **simple pendulum**.

The equation of motion of P is

$$m\ddot{\mathbf{r}} = m\mathbf{g} + \boldsymbol{\Phi}, \tag{14.101}$$

where $\mathbf{g}$ is the gravitational acceleration and $\boldsymbol{\Phi}$ is the reactive force exerted by the constraint. Projecting (14.101) onto the unit vector $\mathbf{t}$ tangent to γ and recalling that the curvilinear abscissa $s = l\varphi$, where l is the radius of γ, and $\boldsymbol{\Phi} \cdot \mathbf{t} = 0$ [see (13.51)], we obtain a second-order differential equation

$$\ddot{\varphi} = -\frac{g}{l} \sin \varphi \tag{14.102}$$

Fig. 14.9 Simple pendulum

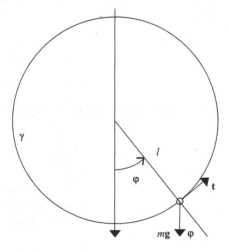

Fig. 14.10 Phase portrait of
simple pendulum

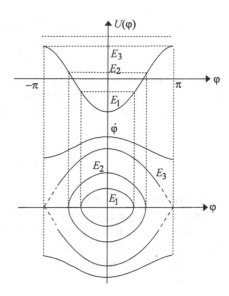

in the unknown $\varphi(t)$. This equation admits one and only one solution when the
initial data $\varphi(0)$ and $\dot\varphi(0)$ are given. Since there is no solution in a finite form of the
preceding equation, we resort to a qualitative analysis of (14.102). Due to the form
of (14.102), we can resort to the method discussed in Sect. 10.8 to obtain a phase
portrait (Fig. 14.10).

The conservation of the total mechanical energy is written as

$$\frac{1}{2}ml^2\dot\varphi^2 - mgl\cos\varphi = E,$$

so that we have

$$\dot\varphi^2 = \frac{2}{ml^2}(E + mgl\cos\varphi). \tag{14.103}$$

Since the potential energy $U(\varphi) = -mgl\cos\varphi$ is periodic, it is sufficient to analyze
the motion for $\varphi \in [-\pi, \pi]$.

The motion is possible if $E + mgl\cos\varphi \geq 0$. When $E = -mgl$, point P occupies
the position $\varphi = 0$ with velocity equal to zero [see (14.103)]. This position is a stable
equilibrium position since the potential energy has a minimum at $\varphi = 0$. For $-mgl
< E < mgl$ the motion is periodic between the two simple zeros of the equation

$$E + mgl\cos\varphi = 0,$$

and the corresponding period of the motion is given by the formula

$$T = 2 \int_{\varphi_1(E)}^{\varphi_2(E)} \frac{\mathrm{d}\varphi}{\sqrt{\frac{2}{ml^2}(E + mgl \cos \varphi)}}, \tag{14.104}$$

which is easily obtained by (14.103). For $E = mgl$ we have again an equilibrium position $\varphi = \mp\pi$ that is unstable since in this position the potential energy has a maximum. In view of (14.103), the level curve corresponding to this value of the energy is

$$\dot{\varphi}^2 = 2\frac{g}{l}(1 + \cos \varphi). \tag{14.105}$$

This level curve contains three trajectories. One of them corresponds to the already mentioned unstable equilibrium position. Since in this position the first derivative of potential energy vanishes, any motion with initial condition $-\pi < \varphi(0) < \pi$ and velocity such that the total energy assumes the value $E = mgl$ tends to the position $\pm\pi$ without reaching it. This circumstance is highlighted by the dashed part of the level curve.

14.10   Rotating Simple Pendulum

In this section we consider a simple pendulum P moving along a smooth vertical circumference γ that uniformly rotates about a vertical diameter a with angular velocity ω. We study the motion of P relative to a frame of reference $R = Cxyz$, having its origin at the center C of γ, $Cz \equiv a$, and the plane Cyz containing γ (Fig. 14.11). Since R is not an inertial frame, the equation governing the motion of P is

$$m\ddot{\mathbf{r}}' = m\mathbf{g} + m\omega^2 \overrightarrow{QP} - 2m\boldsymbol{\omega} \times \dot{\mathbf{r}}' + \boldsymbol{\Phi}, \tag{14.106}$$

where Q is the orthogonal projection of P on a, $\mathbf{t}$ the unit vector tangent to γ, and $\boldsymbol{\Phi}$ the reactive force satisfying the condition $\boldsymbol{\Phi} \cdot \mathbf{t} = 0$. Denoting by l the radius of γ, recalling the relation $s = l\varphi$, where s is the curvilinear abscissa on γ, and projecting (14.106) along $\mathbf{t}$, we obtain the second-order differential equation

$$m\ddot{\varphi} = \sin \varphi \left(\omega^2 \cos \varphi - \frac{g}{l}\right). \tag{14.107}$$

There is no closed solution of this equation, and then we again resort to the qualitative analysis of Sect. 10.8.

Multiplying (14.107) by $\dot{\varphi}$, we obtain

$$\dot{\varphi}^2 = \frac{2}{ml^2} \left(E + mgl \cos \varphi - \frac{m\omega^2 l^2}{2} \cos^2 \varphi\right). \tag{14.108}$$

Fig. 14.11 Rotating simple pendulum

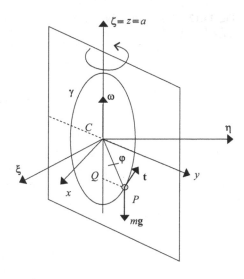

To determine the behavior of the solutions of (14.107), we analyze the potential

$$U(\varphi) = -mgl \cos \varphi + \frac{m\omega^2 l^2}{2} \cos^2 \varphi \qquad (14.109)$$

in the interval $[-\pi, \pi]$. First, we have

$$U(0) = ml \left(\frac{\omega^2 l}{2} - g \right), \qquad (14.110)$$

and

$$U'(\varphi) = ml \sin \varphi (g - \omega^2 l \cos \varphi). \qquad (14.111)$$

Therefore, if

$$\frac{g}{\omega^2 l} > 1,$$

then $U'(\varphi) = 0$ if and only if $\varphi = 0, \pm \pi$. In this hypothesis, we can state that the sign of $U'(\varphi)$ [see (14.111)] depends on the sign of $\sin \varphi$. Consequently, $U(\varphi)$ decreases for $\varphi < 0$, increases for $\varphi > 0$, and has a minimum when $\varphi = 0$ and a maximum for $\varphi = \pm \pi$ (Fig. 14.12).

In contrast, if

$$\frac{g}{\omega^2 l} < 1,$$

Fig. 14.12 Phase portrait for $(g/\omega^2 l) > 1$

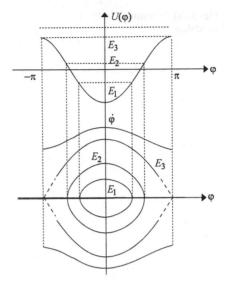

Fig. 14.13 Phase portrait for $(g/\omega^2 l) < 1$

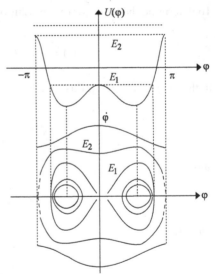

then $U'(\varphi) = 0$ when $\varphi = 0, \pm\pi, \pm\arccos\left(\frac{g}{\omega^2 l}\right)$. Moreover, $\varphi = 0, \pm\pi$ are maxima of $U(\varphi)$, whereas the angles $\pm\arccos(g/\omega^2 l)$ correspond to minima. The relative phase portrait is shown in Fig. 14.13. This analysis shows that the phase portrait undergoes a profound change when the ratio $g/\omega^2 l$ changes in a small neighborhood of 1. For this reason it is called a *bifurcation value*.

Fig. 14.14 Spherical
pendulum

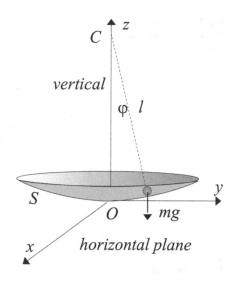

Fig. 14.14 Spherical
pendulum

14.11 Foucault's Pendulum

Let P be a heavy point moving on the surface S of a smooth sphere with radius l
whose center C is fixed with respect to a terrestrial frame of reference R. We wish
to study the small oscillations of this *spherical pendulum* relative to a terrestrial
observer $R = Oxyz$. The axis Oz is chosen to coincide with the vertical of the place
containing the center C of S, and the axes Ox and Oy are horizontal (Figs. 14.14 and
14.15). The equation governing the motion of P relative to a terrestrial observer $R =
Oxyz$ is

$$\ddot{\mathbf{r}} = \mathbf{g} - 2\boldsymbol{\omega}_T \times \dot{\mathbf{r}} + \boldsymbol{\Phi}, \tag{14.112}$$

where $\boldsymbol{\omega}_T$ is the angular velocity of the Earth and $\boldsymbol{\Phi}$ is the reactive force exerted on
P by the spherical constraint S with the equation

$$f(x, y, z) \equiv x^2 + y^2 + (z - l)^2 - l^2 = 0. \tag{14.113}$$

The sphere S is smooth so that $\boldsymbol{\Phi} = \lambda \nabla f$, where λ is an unknown function of
(x, y, z). Since also $\boldsymbol{\omega}_T = (\omega_T \sin \alpha, 0, \omega_T \cos \alpha)$, the components of (14.112) along
the axes of R are

$$\ddot{x} = 2\omega_T \dot{y} \cos \alpha + 2x\lambda, \tag{14.114}$$

$$\ddot{y} = 2\omega_T \dot{z} \sin \alpha - 2\omega_T \dot{x} \cos \alpha + 2y\lambda, \tag{14.115}$$

$$\ddot{z} = -g - 2\omega_T \dot{y} \sin \alpha - 2(l - z)\lambda. \tag{14.116}$$

Fig. 14.15 Spherical
pendulum and terrestrial
frame

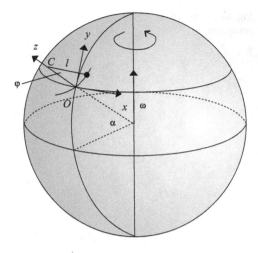

Fig. 14.16 Order of
magnitude of x, y, and z

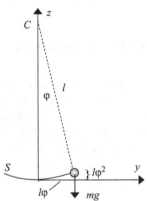

We are interested in the small oscillations of the spherical pendulum about the position $\varphi = 0$. This circumstance simplifies the study of Eqs. (14.114)–(14.116), as we can see by a nondimensional analysis. Introduce as the length of reference the radius l of the sphere S, and denote by $\overline{\varphi}$ the angle corresponding to the largest oscillation; then we have that (Fig. 14.16)

$$x, y \simeq l\overline{\varphi}, \quad z \simeq l\overline{\varphi}^2. \tag{14.117}$$

Now we introduce the two reference times

$$T = \sqrt{\frac{l}{g}}, \quad T_T, \tag{14.118}$$

where T is approximately equal to the oscillation period of a simple pendulum and T_T is equal to the number of seconds in a day. Writing (14.116) in nondimensional form and using the same symbols for nondimensional and dimensional quantities, we have

$$\frac{l\overline{\varphi}^2}{T^2}\ddot{z} = -g - 2\frac{l\overline{\varphi}}{T\,T_T}\omega_T\dot{y}\sin\alpha - 2l\lambda + l\overline{\varphi}^2 z\lambda,$$

that is,

$$\lambda \simeq -\frac{g}{2l}. \tag{14.119}$$

Operating in the same way with (14.116) and taking into account (14.119), we obtain the approximate system

$$\ddot{x} = 2\omega_T\dot{y}\cos\alpha - \frac{g}{l}x, \tag{14.120}$$

$$\ddot{y} = -2\omega_T\dot{x}\cos\alpha - \frac{g}{l}y, \tag{14.121}$$

which in vector form is written as

$$\ddot{\mathbf{r}}_\perp = -\frac{g}{l}\mathbf{r}_\perp - 2\boldsymbol{\omega}^* \times \dot{\mathbf{r}}_\perp, \tag{14.122}$$

where $\mathbf{r}_\perp$ is the vector obtained projecting the position vector $\mathbf{r}$ of P onto the horizontal plane Oxy and $\boldsymbol{\omega}^* = (0,0,\omega_T\cos\alpha)$.

To solve (14.122), we introduce a new frame of reference $R' = (Ox'y'z')$ with the same origin of R, $Oz' \equiv Oz$, and rotating about Oz with angular velocity $-\boldsymbol{\omega}^*$. In the transformation $R \to R'$ we have [see (12.38), (12.39), and (12.44)]

$$\mathbf{r} = \mathbf{r}',$$
$$\dot{\mathbf{r}} = \dot{\mathbf{r}}' - \boldsymbol{\omega}^* \times \mathbf{r}',$$
$$\ddot{\mathbf{r}} = \ddot{\mathbf{r}}' + \mathbf{a}_\tau - 2\boldsymbol{\omega}^* \times \dot{\mathbf{r}}' = \ddot{\mathbf{r}}' - \omega^{*2}\mathbf{r}' - 2\boldsymbol{\omega}^* \times \dot{\mathbf{r}}'.$$

Introducing the preceding relations into (14.122) and noting that

$$2\boldsymbol{\omega}^* \times (\boldsymbol{\omega}^* \times \mathbf{r}') = -2\omega^{*2}\mathbf{r}' + 2(\boldsymbol{\omega}^* \cdot \mathbf{r}')\boldsymbol{\omega}^* = -2\omega^{*2}\mathbf{r}'$$

since $\boldsymbol{\omega}^* \cdot \mathbf{r}' = 0$, (14.122) becomes

$$\ddot{\mathbf{r}}' = -\left(\frac{g}{l} + \omega^{*2}\right)\mathbf{r}'. \tag{14.123}$$

This equation shows that in the frame R', the pendulum oscillates with a period $2\pi/\sqrt{\frac{g}{l}+\omega^{*2}}$, whereas its plane of oscillation rotates with angular velocity $-\omega^*$ about the vertical axis Oz.

14.12 Exercises

The notebooks **CentralForces**, **CurveMot**, and **Motions** can help the reader to solve some dynamical problems of a material point.

1. Neglecting the resistance of the air, prove that the velocity of a point particle projected upward in a direction inclined at 60 to the horizontal is half of its initial velocity when it arrives at its greatest height.
2. Neglecting the resistance of the air, find the greatest distance that a stone can be thrown with initial velocity v_0.
3. A simple pendulum executes small oscillations in a resisting medium with damping $-h\dot{\varphi}$. Determine the qualitative behavior of the motion and the loss of energy after n oscillations.
4. Supposing the Earth to be spherical, with what velocity must a projectile be fired from the Earth's surface for its trajectory to be an ellipse with major axis a?
5. A bomb is dropped from an airplane flying horizontally at height h and with speed v. Assuming a linear law of resistance $-\,mg k\mathbf{v}$, show that, if k is small, the fall time is approximately

$$\sqrt{\frac{2h}{g}}\left(1+\frac{1}{6}gk\sqrt{\frac{2h}{g}}\right).$$

6. A particle moves in a plane under the influence of the central force

$$f(r) = \frac{b}{r^2} + \frac{c}{r^4},$$

where a and b are positive constants and r is the distance from the center. Analyze qualitatively the motion, prove the existence of a circular orbit, and determine its radius.
7. Repeat the analysis required in the preceding exercise for a particle moving under the action of the central force

$$f(r) = kr^{-2}e^{-r^2}.$$

8. A point moves under the action of a Newtonian attractive force. Resorting to the stability theorems of Chap. 10 relative to the equilibrium stability, verify that the elliptic orbits are stable. Are the motions along these orbits stable?

9. A particle moves in a plane attracted to a fixed center by a central force

$$f(r) = -\frac{k}{r^3},$$

where k is a positive constant. Find the equation of the orbits.

10. Let γ be a vertical parabola with equation $y = x^2$. Determine the motion of a particle P of mass m constrained to moving smoothly on γ under the influence of its weight and of an elastic force $\mathbf{F} = -k\overrightarrow{QP}$, where k is a positive constant and Q is the orthogonal projection of P onto the axis of the parabola.

Chapter 15
General Principles of Rigid Body Dynamics

Abstract In this chapter the dynamical description of real bodies is improved by adopting the rigid body model. After defining the center of mass, linear momentum, angular momentum, and kinetic energy of a rigid body, the tensor of inertia is introduced and the fundamental properties of this tensor are discussed. Then, active and reactive forces are described and the balance equations of a rigid body are formulated.

15.1 Mass Density and Center of Mass

To improve the dynamical description of real bodies, we resort to a model of rigid bodies that takes into account both the extension and the mass distribution in the spatial region occupied by the body.

Let $\mathcal{B}$ be a rigid body, and Ω an arbitrary point of $\mathcal{B}$. Henceforth we denote by $R = (Ox_1 x_2 x_3)$ and $R' = (\Omega x_1' x_2' x_3')$ the lab frame and the body frame, respectively. Further, we denote by $\mathbf{e}_i$ and $\mathbf{e}_i'$, $i = 1, 2, 3$, the unit vector directed along the orthogonal axes of R and R', respectively. Finally, C is the *fixed* region occupied by the rigid body $\mathcal{B}$ in the body frame R', and $\mathbf{r}'$ is the position vector of any point of C relative to the origin Ω of R'.

We suppose that there exists a positive and Riemann-integrable function $\rho(\mathbf{r}')$: $C \to R^+$ such that the integral

$$m(c) = \int_c \rho(\mathbf{r}') \, dc \tag{15.1}$$

gives the mass of the arbitrary measurable region $c \subset C$. The function $\rho(\mathbf{r}')$, which is called the ***mass density*** of the body $\mathcal{B}$, could exhibit finite discontinuity across some surfaces contained in $\mathcal{B}$.

We call the point $G \in C$ whose position vector $\mathbf{r}_G$ relative to R is defined by the formula

© Springer International Publishing AG, part of Springer Nature 2018 251
A. Romano and A. Marasco, *Classical Mechanics with Mathematica®*,
Modeling and Simulation in Science, Engineering and Technology,
https://doi.org/10.1007/978-3-319-77595-1_15

$$m(C)\mathbf{r}_G = \int_C \rho(\mathbf{r}')\mathbf{r}\, dc \tag{15.2}$$

the *center of mass*.

Exercise 15.1. Prove that the center of mass is a characteristic point of the rigid body $\mathcal{B}$, i.e., it is independent of the origin O of the lab frame.

15.2   Linear Momentum, Angular Momentum, and Kinetic Energy of a Rigid Body

The velocity field, which in a rigid motion has the form (12.17)

$$\dot{\mathbf{r}} = \dot{\mathbf{r}}_\Omega(t) + \boldsymbol{\omega}(t) \times (\mathbf{r} - \mathbf{r}_\Omega) = \dot{\mathbf{r}}_\Omega(t) + \boldsymbol{\omega}(t) \times \mathbf{r}',$$

can be considered as depending on the time t and the position vector $\mathbf{r}$ in the lab frame or as a function of t and the position vector $\mathbf{r}'$ in the body frame. In all of the following formulae, it is supposed to be a function of t, $\mathbf{r}'$.

In this section, we extend to a rigid body the results of Sects. 13.6 and 13.7 relative to systems of material points. To this end, besides the lab and body frames R and R', we need to introduce a third frame of reference $R_G = (G\overline{x}_1\overline{x}_2\overline{x}_3)$ having a translational motion relative to R with velocity $\dot{\mathbf{r}}_G$. As in Sect. 13.7, we call the motion of $\mathcal{B}$ relative to R_G the *motion about the center of mass*. Any quantity a related to R_G will be denoted by $\overline{a}$.

We call *linear momentum* the vector

$$\mathbf{Q} = \int_C \rho\dot{\mathbf{r}}\, dc \tag{15.3}$$

and *angular momentum with respect to the pole O* the vector

$$\mathbf{K}_O = \int_C \rho\overrightarrow{OP} \times \dot{\mathbf{r}}\, dc. \tag{15.4}$$

Since the region C does not depend on time owing to the rigidity of $\mathcal{B}$, differentiating (15.2) with respect to time, we obtain

$$\mathbf{Q} = m\dot{\mathbf{r}}_G. \tag{15.5}$$

We can comment on this result as follows. *The momentum of $\mathcal{B}$ is equal to the momentum of a mass point having the total mass m of $\mathcal{B}$ and moving with the velocity of the center of mass $\mathbf{r}_G$.*

Since in the frame R_G the center of mass is at rest, we have that

$$\overline{\mathbf{Q}} = \mathbf{0}. \tag{15.6}$$

We now prove the following important properties of $\mathbf{K}_O$.

Theorem 15.1. *The angular momentum with respect to G in the frame R is equal to the angular momentum with respect to G in the frame R_G; that is,*

$$\mathbf{K}_G = \overline{\mathbf{K}}_G. \tag{15.7}$$

The angular momentum relative to R_G does not depend on the pole; that is,

$$\overline{\mathbf{K}}_O = \overline{\mathbf{K}}_G \tag{15.8}$$

for any pole O.

Proof. Since the motion of R_G relative to R is translational, it is $\dot{\mathbf{r}} = \dot{\overline{\mathbf{r}}} + \dot{\mathbf{r}}_G$. Consequently, we have that

$$\mathbf{K}_O = \int_C \overrightarrow{OP} \times \rho(\dot{\overline{\mathbf{r}}} + \dot{\mathbf{r}}_G)\,\mathrm{d}c$$
$$= \overline{\mathbf{K}}_O + \left(\int_C \rho\overrightarrow{OP}\,\mathrm{d}c\right) \times \dot{\mathbf{r}}_G,$$

and, taking into account (15.2), the following formula is proved:

$$\mathbf{K}_O = \overline{\mathbf{K}}_O + m\overrightarrow{OG} \times \dot{\mathbf{r}}_G. \tag{15.9}$$

Choosing $O \equiv G$ in this formula, we obtain (15.7). Further, the definition of angular momentum allows us to write that

$$\overline{\mathbf{K}}_O = \int_C (\overrightarrow{OG} + \overrightarrow{GP}) \times \rho\dot{\overline{\mathbf{r}}}\,\mathrm{d}c = \overrightarrow{OG} \times \int_C \rho\dot{\overline{\mathbf{r}}}\,\mathrm{d}c + \overline{\mathbf{K}}_G.$$

In view of (15.6), the integral on the right-hand side of the preceding equation vanishes and the theorem is proved. $\square$

We conclude this section defining the **kinetic energy** of the body $\mathcal{B}$ as follows:

$$T = \frac{1}{2} \int_C \rho\dot{\mathbf{r}}^2\,\mathrm{d}c. \tag{15.10}$$

It is trivial to extend the König theorem (13.35) to the case of a rigid body so that

$$T = \overline{T} + \frac{1}{2}m\dot{\mathbf{r}}_G^2. \tag{15.11}$$

15.3   Tensor of Inertia

The quantities we introduced in the preceding section refer to any possible rigid motion, but we have not yet used the explicit expression of the rigid velocity field. In this section, we determine useful expressions of angular momentum and kinetic energy that explicitly take into account the relation existing in a rigid motion between the angular velocity and the velocity of any point.

Let $\mathcal{B}$ be a rigid body with a fixed point O in the frame of reference R. It is always possible to take O coinciding both with the origin of the lab frame R and the origin Ω of the body frame R'. With this choice of the frames of reference, $\overrightarrow{OP} = \mathbf{r}$ and the velocity field is written as $\dot{\mathbf{r}} = \omega \times \mathbf{r}$. Consequently, (15.4) gives

$$
\begin{aligned}
\mathbf{K}_O &= \int_C \rho \mathbf{r} \times (\omega \times \mathbf{r}) \, dc \\
&= \int_C \rho(|\mathbf{r}|^2 \omega - (\mathbf{r} \cdot \omega)\mathbf{r}) \, dc.
\end{aligned}
$$

In conclusion, we can state that

$$
\mathbf{K}_O = \mathbf{I}_O \cdot \omega, \tag{15.12}
$$

where the 2-tensor

$$
\mathbf{I}_O = \int_C \rho(|\mathbf{r}|^2 \mathbf{1} - \mathbf{r} \otimes \mathbf{r}) \, dc \tag{15.13}
$$

is said to be the **tensor of inertia** of the body $\mathcal{B}$ relative to point O. Formula (15.12) shows that $\mathbf{K}_O$ depends linearly on ω. The components I_{Oii} of $\mathbf{I}_O$ are called **moments of inertia** with respect to the axes $\mathbf{e}_i$, whereas the components I_{Oij}, $i \neq j$, are the **products of inertia**. It is plain to verify that

$$
I_{O11} = \int_C \rho(x_2^2 + x_3^2) dc, \quad I_{O22} = \int_C \rho(x_1^2 + x_3^2) dc, \tag{15.14}
$$

$$
I_{O33} = \int_C \rho(x_1^2 + x_2^2) dc, \quad I_{Oij} = -\int_C \rho x_i x_j \, dc, \, (i \neq j), \tag{15.15}
$$

so that $\mathbf{I}_O$ is *symmetric*.

When a rigid motion with a fixed point O and its angular velocity ω are given, it is possible to evaluate not only the angular momentum $\mathbf{K}_O$ by (15.12) but also the kinetic energy. In fact, introducing the velocity $\dot{\mathbf{r}} = \omega \times \mathbf{r}$ into (15.10), we obtain

$$
T = \frac{1}{2} \int_C \rho \omega \times \mathbf{r} \cdot \dot{\mathbf{r}} \, dc = \frac{1}{2} \int_C \rho \mathbf{r} \times \dot{\mathbf{r}} \cdot \omega \, dc.
$$

If we note that the angular velocity does not depend on the point of C and we take into account (15.4) and (15.12), then the preceding relation can also be written as

$$T = \frac{1}{2}\mathbf{K}_O \cdot \omega = \frac{1}{2}\omega \cdot \mathbf{I}_O\omega. \tag{15.16}$$

The kinetic energy is a positive quantity that vanishes if and only if the velocity $\dot{\mathbf{r}} = \omega \times \mathbf{r}$ vanishes identically. In turn, this condition is equivalent to requiring that $\omega = \mathbf{0}$. In conclusion, in view of (15.16), we can state that the tensor of inertia is *symmetric and positive definite*.

The quantities $\mathbf{K}_O$ and ω in relations (15.12) and (15.16) are relative to the lab frame R. However, since they are vector quantities, we can evaluate their components in any frame of reference. Consistent advantages are obtained if (15.12) and (15.16) are projected along the axes of the body frame R'. In fact, with respect to the basis $(\mathbf{e}'_i)$ of the unit vectors directed along the axes of R', we have

$$K'_{Oi} = I'_{ij}\omega'_{ij}, \quad T = \frac{1}{2}I'_{ij}\omega'_i\omega'_j, \tag{15.17}$$

where the components of the tensor of inertia are

$$I'_{ij} = \int_C \rho(r'^2\delta_{ij} - x'_i x'_j)\,dc. \tag{15.18}$$

In the body frame R' the components I'_{ij} are *independent of time* since they only depend on the geometry and mass distribution of the rigid body $\mathcal{B}$. On the other hand, we have already noted that the 2-tensor $\mathbf{I}_O$ is symmetric and positive definite. Consequently, the eigenvalues A, B, and C of $\mathbf{I}_O$ are positive, and there exists at least an orthonormal basis $(\mathbf{e}'_i)$ formed by eigenvectors of $\mathbf{I}_O$. These vectors are at rest in R', i.e., relative to the body $\mathcal{B}$. Any frame of reference that is formed by an arbitrary point O and axes parallel to eigenvectors of $\mathbf{I}_O$ is called a ***principal frame of inertia*** relative to O, whereas the eigenvalues A, B, and C of $\mathbf{I}_O$ are the ***principal moments of inertia*** relative to O. In particular, if O is the center of mass of S, then we substitute the attribute principal with ***central***.

In a principal frame of inertia, the tensor of inertia is represented by the following diagonal matrix:

$$\begin{pmatrix} A & 0 & 0 \\ 0 & B & 0 \\ 0 & 0 & C \end{pmatrix}, \tag{15.19}$$

so that (15.12) and (15.16) become

$$\mathbf{K}_O = Ap\mathbf{e}'_1 + Bq\mathbf{e}'_2 + Cr\mathbf{e}'_3, \tag{15.20}$$

$$T = \frac{1}{2}\left(Ap^2 + Bq^2 + Cr^2\right),\tag{15.21}$$

where the eigenvalues A, B, C are constant, and p, q, r are the components of the angular velocity ω relative to R but projected along the axes of the body frame.

We now determine the expressions of $\mathbf{K}_O$ nd T when the solid B has a fixed axis a or is free from any constraint.

(a) In the first case, provided that we choose the lab frame R and the body frame R' as in Fig. 12.2, we have that $\omega = \dot{\varphi}\mathbf{e}_3'$. Consequently, for any $O \in a$, (15.17) gives

$$\mathbf{K}_O = \dot{\varphi}(I_{13}'\mathbf{e}_1' + I_{23}'\mathbf{e}_2' + I_{33}'\mathbf{e}_3'), \quad T = \frac{1}{2}I_{33}'\dot{\varphi}^2.\tag{15.22}$$

In particular, if $\mathbf{e}_3'$ is an eigenvector of $\mathbf{I}_O$ belonging to the eigenvalue C, then

$$\mathbf{I}_O\mathbf{e}_3' = C\mathbf{e}_3'$$

or, equivalently,

$$\begin{pmatrix} I_{11}' & I_{12}' & I_{13}' \\ I_{12}' & I_{22}' & I_{23}' \\ I_{13}' & I_{23}' & C \end{pmatrix}\begin{pmatrix} 0 \\ 0 \\ 1 \end{pmatrix} = C\begin{pmatrix} 0 \\ 0 \\ 1 \end{pmatrix}.$$

This equation implies that

$$I_{13}' = I_{23}' = 0,$$

and finally we can write (15.22) in the following simple form:

$$\mathbf{K}_O = C\dot{\varphi}\mathbf{e}_3',\tag{15.23}$$

which holds if the rotation axis is a principal axis of inertia.

(b) Let B be a rigid body freely moving relative to the lab frame $R = (O, \mathbf{e}_i)$. Then, in view of (15.4) and (15.7), the angular momentum $\mathbf{K}_O$ relative to R is

$$\mathbf{K}_O = \overline{\mathbf{K}}_G + \overrightarrow{OG} \times m\dot{\mathbf{r}}_G,\tag{15.24}$$

where $\overline{\mathbf{K}}_G$ is the angular momentum of S relative to the motion about the center of mass, i.e., relative to the frame $R_G = (G, \overline{\mathbf{e}}_i)$. Consequently, in the motion of S with respect to R_G the center of mass is at rest and the angular momentum $\overline{\mathbf{K}}_G$ has the form (15.20), provided that we denote by A, B, and C the eigenvalues of $\mathbf{I}_G$ and we choose the basis $(\mathbf{e}_i')$ moving with B as a *central* base of inertia. Bearing in mind the preceding remarks, we can write

$$\mathbf{K}_O = Ap\mathbf{e}'_1 + Bq\mathbf{e}'_2 + Cr\mathbf{e}'_3 + \overrightarrow{OG} \times m\dot{\mathbf{r}}_G, \tag{15.25}$$

$$T = \frac{1}{2}\left(Ap^2 + Bq^2 + Cr^2\right) + \frac{1}{2}m\dot{\mathbf{r}}_G^2. \tag{15.26}$$

15.4   Some Properties of the Tensor of Inertia

At first sight, it seems that we have to evaluate the tensor of inertia $\mathbf{I}_O$ every time we change the fixed point O. In this section we show that there exists a simple relation between the tensor $\mathbf{I}_G$ and the tensor of inertia $\mathbf{I}_O$ relative to any point O. Further, the evaluation of $\mathbf{I}_G$ becomes simpler when the rigid body $\mathcal{B}$ possesses some symmetries.

Denoting by $\mathbf{r}$ the position vector of any point P of $\mathcal{B}$ in the lab frame R_O with the origin at O, by $\mathbf{r}_G$ the position vector of the center of mass G of $\mathcal{B}$ in R_O, and by $\mathbf{r}'$ the position vector of P in the body frame R'_G with the origin at G, we have that $\mathbf{r} = \mathbf{r}_G + \mathbf{r}'$, and relation (15.13) can also be written as follows:

$$\mathbf{I}_O = \int_C \rho[(|\mathbf{r}_G|^2 + |\mathbf{r}'|^2 + 2\mathbf{r}_G \cdot \mathbf{r}')\mathbf{1} - (\mathbf{r}_G + \mathbf{r}') \otimes (\mathbf{r}_G + \mathbf{r}')]\, dc.$$

Since

$$\mathbf{I}_G = \int_C \rho(|\mathbf{r}_G|^2\mathbf{1} - \mathbf{r}' \otimes \mathbf{r}')\, dc, \quad m\mathbf{r}'_G = \int_C \rho\mathbf{r}'\, dc = \mathbf{0},$$

we obtain the following formula:

$$\mathbf{I}_O = \mathbf{I}_G + m(|\mathbf{r}_G|^2\mathbf{1} - \mathbf{r}_G \otimes \mathbf{r}_G), \tag{15.27}$$

which relates $\mathbf{I}_G$ to the tensor of inertia $\mathbf{I}_O$ relative to any point O.

Formula (15.27) allows us to prove the following theorem.

Theorem 15.2. *Choose frames of reference* (O, x_i) *and* $(G, \overline{x}_i)$ *in* R_O *and* R_G, *respectively, in such a way that the corresponding axes are parallel. The moments of inertia relative to the axes are modified according to the formula*

$$I_{Oii} = \overline{I}_{Gii} + m\delta_i^2, \tag{15.28}$$

where δ_i *is the distance between the axes* Ox_i *and* $G\overline{x}_i$; *moreover, the products of inertia are modified according to the rule*

$$I_{Oij} = \overline{I}_{Gij} - mx_{Gi}x_{Gj}, \quad i \neq j. \tag{15.29}$$

If the center of mass belongs to the axis Ox_i, *then*

$$I_{Oij} = \overline{I}_{Gij}, \quad i \neq j. \tag{15.30}$$

Finally, if $(G, \overline{x}_i)$ is a central frame of inertia, then any frame (O, x_i), where $O \in \overline{x}_i$ and the axes are parallel to the corresponding axes of $(G, \overline{x}_i)$, is a principal frame of inertia.

Proof. Relations (15.28) and (15.29) follow at once from (15.27). Equation (15.30) follows from (15.29) when we note that two coordinates of G vanish when one of the axes Ox_i contains G. Finally, from (15.30) it follows that if $(G, \overline{x}_i)$ is a central frame of inertia, then $\overline{I}_{Gij} = 0$ for $i \neq j$ and $I_{Oij} = 0$ for $i \neq j$. $\square$

To determine $\mathbf{I}_G$, we arbitrarily fix a frame of reference $(G, \overline{x}_i)$ and we evaluate the components of $\mathbf{I}_G$ in this frame by (15.14) and (15.15). Then we solve the eigenvalue equation

$$\overline{I}_{Gij} u_j = \lambda u_i = \lambda \delta_{ij} u_j$$

to find the eigenvalues and eigenvectors of $\mathbf{I}_G$, i.e., the central frames of inertia.

The following theorem shows that the last problem can be simplified if the body $\mathcal{B}$ exhibits some symmetries with respect to the center of mass.

Theorem 15.3. *If the plane π is a symmetry plane of $\mathcal{B}$, then π is a principal plane of inertia for any point of it.*

Proof. Let (O, x_i) be a frame of reference whose coordinate plane Ox_1x_2 coincides with π. Since π is a symmetry plane of $\mathcal{B}$, for any point $\mathbf{r} = (x_1, x_2, x_3)$ of $\mathcal{B}$ there is a symmetric point $\mathbf{r}' = (x_1, x_2, -x_3)$ such that $\rho(\mathbf{r}) = \rho(\mathbf{r}')$. Consequently,

$$I_{O13} = \int_C \rho x_1 x_3 \, dc = 0, \quad I_{O23} = \int_C \rho x_2 x_3 \, dc = 0,$$

and the matrix of the components of $\mathbf{I}_O$ becomes

$$\begin{pmatrix} I_{11} & I_{12} & 0 \\ I_{12} & I_{22} & 0 \\ 0 & 0 & I_{33} \end{pmatrix}.$$

It is a simple exercise to verify that an eigenvector of $\mathbf{I}_O$ is orthogonal to π so that the other eigenvectors belong to π. $\square$

The preceding theorem yields the following result.

Theorem 15.4. *If the body $\mathcal{B}$ has two orthogonal planes of symmetry π_1 and π_2, then they intersect along a straight line r, which is a principal axis of inertia for any $O \in r$. Further, a principal frame of inertia relative to O is obtained by intersecting π_1 and π_2 with a third plane π_3 that is orthogonal to the first two planes and contains O.*

15.5 Ellipsoid of Inertia

We define the ***moment of inertia*** of a rigid body $\mathcal{B}$ relative to the straight line a by the quantity

$$\mathcal{I}_a = \int_C \rho \delta^2(\mathbf{r}) \, dc, \tag{15.31}$$

where δ denotes the distance of an arbitrary point $\mathbf{r}$ of $\mathcal{B}$ from the straight line a (Fig. 15.1). Denoting by $\mathbf{u}$ a unit vector directed along a, we have that

$$\delta^2 = |\mathbf{r}|^2 - |\mathbf{r} \cdot \mathbf{u}|^2 = |\mathbf{r}|^2 \delta_{ij} u_i u_j - x_i x_j u_i u_j$$

and recalling (15.18), relation (15.31) can also be written as follows:

$$\mathcal{I}_a = \mathbf{u} \cdot \mathbf{I}_O \mathbf{u}. \tag{15.32}$$

Finally, introducing the notation

$$\boldsymbol{\nu} = \frac{\mathbf{u}}{\sqrt{\mathcal{I}_a}}, \tag{15.33}$$

(15.32) assumes the form

$$\boldsymbol{\nu} \cdot \mathbf{I}_O \boldsymbol{\nu} = I_{Oij} \nu_i \nu_j = 1. \tag{15.34}$$

The positive definite character of the tensor of inertia $\mathbf{I}_O$ implies that (15.34) defines an ellipsoid E, which is called the ***ellipsoid of inertia relative to*** O. In particular, if $O \equiv G$, then E is called the ***central ellipsoid of inertia***. In view of (15.33), the moment of inertia of $\mathcal{B}$ relative to any axis a containing the center O of the ellipsoid is given by the formula

Fig. 15.1 Ellipsoid of inertia

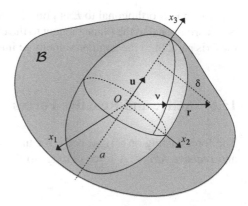

$$\mathcal{I}_a = \frac{1}{|\boldsymbol{\nu}|^2}.$$ (15.35)

In other words, the ellipsoid of inertia relative to O allows one to determine geometrically the moment of inertia of $\mathcal{B}$ with respect to any axis containing O. Recalling the theorems we have proved regarding the dependence on O of the tensor $\mathbf{I}_O$, we understand that knowledge of the central ellipsoid of inertia is sufficient to determine the ellipsoid of inertia relative to any other point O. If the unit vectors $(\mathbf{e}_i)$ along the axes of the frame of reference (O, x_i) are eigenvectors of $\mathbf{I}_O$, then (15.34) becomes

$$Ax_1^2 + Bx_2^2 + Cx_3^2 = 1,$$ (15.36)

where A, B, and C are the corresponding eigenvalues. When $A = B$, the ellipsoid E is an ellipsoid of revolution about Ox_3; finally, it reduces to a sphere when $A = B = C$.

We conclude this section by proving an important relation existing between the ellipsoid of inertia $\mathbf{I}_O$ relative to O and the angular momentum $\mathbf{K}_O$. If we introduce the notation $\boldsymbol{\omega} = |\omega|\mathbf{u}$, where $\mathbf{u}$ is a unit vector, then we can write [see (15.12)]

$$\mathbf{K}_O = \mathbf{I}_O \cdot \boldsymbol{\omega} = |\omega|\mathbf{I}_O \cdot \mathbf{u} = \sqrt{\mathcal{I}_a}|\omega|\mathbf{I}_O \cdot \frac{\mathbf{u}}{\sqrt{\mathcal{I}_a}},$$

where a is the axis containing O and parallel to $\boldsymbol{\omega}$. In view of (15.33), the vector $\boldsymbol{\nu} = \frac{\mathbf{u}}{\sqrt{\mathcal{I}_a}}$ is the position vector of the intersection point P of the ellipsoid E with axis a; therefore, we have that

$$\mathbf{K}_O = \sqrt{\mathcal{I}_a}|\omega|\mathbf{I}_O \cdot \boldsymbol{\nu}.$$

On the other hand, from the equation $f(\boldsymbol{\nu}) \equiv \boldsymbol{\nu} \cdot \mathbf{I}_O \cdot \boldsymbol{\nu} - 1 = 0$ of E [see (15.34)] we obtain $(\nabla f)_P = 2\mathbf{I}_O \cdot \boldsymbol{\nu}$, and the preceding equation assumes the following final form:

$$\mathbf{K}_O = \frac{1}{2}\sqrt{\mathcal{I}_a}|\omega|(\nabla f)_P.$$ (15.37)

Since $(\nabla f)_P$ is orthogonal to E at point P, we can state that the angular momentum $\mathbf{K}_O$ is orthogonal to the plane tangent to the ellipsoid of inertia E at point P in which the axis parallel to the angular velocity $\boldsymbol{\omega}$ intersects the ellipsoid.

15.6   Active and Reactive Forces

The forces acting on a rigid body $\mathcal{B}$ can be divided into two classes: *active forces* and *reactive forces*. A force belongs to the first class if we can assign a priori the force law, i.e., the way in which the force depends on the position of $\mathcal{B}$, its velocity field, and time. These forces can be **concentrated forces** if they are applied at a point of the body $\mathcal{B}$ and **mass forces** if they act on the whole volume occupied by $\mathcal{B}$. Consequently, the total force and the total torque relative to a pole O of the active forces can be written as

$$\mathbf{R}^{(a)} = \int_C \rho \mathbf{b} \, dc + \sum_{i=1}^{p} \mathbf{F}_i^{(a)}, \tag{15.38}$$

$$\mathbf{M}_O^{(a)} = \int_C \rho \overrightarrow{OP} \times \mathbf{b} \, dc + \sum_{i=1}^{p} \overrightarrow{OP}_i \times \mathbf{F}_i^{(a)}, \tag{15.39}$$

where ρ is the mass density of $\mathcal{B}$, $\mathbf{b}$ is the force per unit mass, and $\mathbf{F}_i^{(a)}$, $i = 1, \ldots, p$, are the concentrated forces acting at the points $P_1, \ldots, P_p$ of $\mathcal{B}$.

The reactive forces are due to the contact between $\mathcal{B}$ and the external obstacles (constraints). In turn, these forces are determined by the very small deformations that both $\mathcal{B}$ and the constraints undergo in very small regions around the contact surfaces. The hypothesis that both $\mathcal{B}$ and the constraints are rigid eliminates the preceding small deformations so that it is impossible to determine the force law of the reactive forces. We suppose that the reactive forces can be distributed on the surfaces $\sigma_1, \ldots, \sigma_m$ of $\mathcal{B}$ as well as on very small areas around the points $Q_1, \ldots Q_s$ of contact between $\mathcal{B}$ and the constraints. Then we assume that the reactive forces acting on these small areas can be described by a total force $\boldsymbol{\Phi}_i$ and a couple with a moment $\boldsymbol{\Gamma}_i$, which is called **rolling friction**. As a consequence of these assumptions, we have that the total reactive force and torque can be written as follows:

$$\mathbf{R}^{(r)} = \sum_{i=1}^{m} \int_{\sigma_i} \mathbf{t} \, d\sigma + \sum_{i=1}^{s} \boldsymbol{\Phi}_i, \tag{15.40}$$

$$\mathbf{M}_O^{(r)} = \sum_{1}^{m} \int_{\sigma_i} \overrightarrow{OP} \times \mathbf{t} \, dc + \sum_{i=1}^{s} (\overrightarrow{OQ}_i \times \boldsymbol{\Phi}_i + \boldsymbol{\Gamma}_i), \tag{15.41}$$

where $\mathbf{t}$ is the force per unit surface.

The vectors $\boldsymbol{\Phi}_i$ and $\boldsymbol{\Gamma}_i$ are not arbitrary since, although their force laws cannot be assigned, they must satisfy some general phenomenological conditions. In this regard, it is useful to distinguish between contact with friction and frictionless contacts. In the latter case, starting from experimental results the following definition is induced.

Definition 15.1. A constraint acting on a rigid body $\mathcal{B}$ is called **ideal** or **smooth** if at the contact point Q_i of $\mathcal{B}$ with the constraint surface Σ, $\boldsymbol{\Gamma}_i = \mathbf{0}$ and the reactive force $\boldsymbol{\Phi}_i$ exerted by Σ is normal to the boundary and directed toward the interior of $\mathcal{B}$. Moreover, $\mathbf{t}$ is orthogonal to Σ at any contact point.

If the constraints are **rough**, then friction is present and the preceding conditions must be replaced by more complex ones. Moreover, they depend on whether B is moving or at rest relative to the constraints. The conditions that are satisfied by the reactive forces are called **Coulomb's friction laws**.

For the sake of simplicity, we refer these empirical laws to the case of concentrated reactive forces since its extension to surface reactive forces is obvious. It is convenient to distinguish in $\mathbf{\Phi}$ a component $\mathbf{\Phi}_\perp$ normal to the contact surface Σ, which is always present and corresponds to the frictionless situation, and a component $\mathbf{\Phi}_{||}$ tangential to Σ. With similar meanings, we set $\mathbf{\Gamma} = \mathbf{\Gamma}_\perp + \mathbf{\Gamma}_{||}$, where the components $\mathbf{\Gamma}_\perp$ and $\mathbf{\Gamma}_{||}$ are respectively called **spin** and **rolling friction**.

All the aforementioned reactive forces are modeled by the following empirical laws.

Definition 15.2. Rough constraints in static conditions are such that

$$|\mathbf{\Phi}_{||}| \le \mu_s |\mathbf{\Phi}_\perp|, \tag{15.42}$$

$$|\mathbf{\Gamma}_{||}| \le h_s |\mathbf{\Phi}_\perp|, \tag{15.43}$$

$$|\mathbf{\Gamma}_\perp| \le k_s |\mathbf{\Phi}_\perp|, \tag{15.44}$$

where the constants $0 < \mu_s, h_s, k_s < 1$, depending on the contact between the body and the constraints, are called **static friction coefficients**.

Definition 15.3. Rough constraints in dynamic conditions are such that

$$|\mathbf{\Phi}_{||}| = -\mu_d |\mathbf{\Phi}_\perp| \frac{\dot{\mathbf{r}}}{|\dot{\mathbf{r}}|}, \tag{15.45}$$

$$|\mathbf{\Gamma}_{||}| = -h_d |\mathbf{\Phi}_\perp| \frac{\boldsymbol{\omega}_{||}}{|\boldsymbol{\omega}_{||}|}, \tag{15.46}$$

$$|\mathbf{\Gamma}_\perp| = -k_d |\mathbf{\Phi}_\perp| \frac{\boldsymbol{\omega}_\perp}{|\boldsymbol{\omega}_\perp|}, \tag{15.47}$$

where $\dot{\mathbf{r}}$ is the relative velocity of the contact point of B with respect to the constraint, and $\boldsymbol{\omega}_{||}$ and $\boldsymbol{\omega}_\perp$ denote respectively the tangential and normal components of the relative angular velocity $\boldsymbol{\omega}$ of B with respect to the constraint. The coefficients $0 < \mu_d < \mu_s < 1$, h_d, and k_d have the same meaning as in the static case but are now called the **dynamic friction coefficients**.

15.7 Balance Equations for a Rigid Body: Theorem of Kinetic Energy

In the preceding section, the characteristics of the forces acting on a rigid body B were given. In this section, we introduce the balance equations governing its dynamics.

It is reasonable to assume that, owing to the hypothesis of rigidity, the internal forces, acting between parts of B, do not influence the motion. Consequently, they must not appear in the dynamic equations describing the effect of the acting forces on the motion of B. Now, the equations possessing this characteristic are the balance equations of linear and angular momentum, (13.28) and (13.32). Although these equations were proved for a system of mass points, it is quite natural to require their validity for a rigid body that is a collection of the elementary masses $\rho \, dc$. Therefore, we postulate that the dynamical behavior of a rigid body B is governed by the following **balance equations**:

$$m\ddot{\mathbf{r}}_G = \mathbf{R}^{(a)} + \mathbf{R}^{(r)}, \tag{15.48}$$

$$\dot{\mathbf{K}}_O = \mathbf{M}_O^{(a)} + \mathbf{M}_O^{(r)}, \tag{15.49}$$

where m is the total mass of B and G its center of mass, O coincides with G or with an arbitrary fixed pole, and the angular momentum is expressed by (15.25). In the next chapter, we prove that the preceding balance equations allow us to determine the motion of a rigid body B when it has a fixed axis or a fixed point or is free of constraints.

We conclude this section by proving an important theorem that holds during the motion of any rigid body regardless of the constraints to which it is subjected.

Let us suppose that B moves relative to the frame of reference (O, x_i) under the action of the forces $\mathbf{F}_1, \ldots, \mathbf{F}_s$ applied at the points $P_1, \ldots, P_s$ of B. Then, taking into account that the motion of B is rigid, the power W of these forces is

$$W = \sum_{i=1}^{s} \mathbf{F}_i \cdot \dot{\mathbf{r}}_i = \sum_{i=1}^{s} \mathbf{F}_i \cdot (\dot{\mathbf{r}}_\Omega + \boldsymbol{\omega} \times \overrightarrow{\Omega P_i})$$

$$= \left(\sum_{i=1}^{s} \mathbf{F}_i \right) \cdot \dot{\mathbf{r}}_\Omega + \left(\sum_{i=1}^{s} \overrightarrow{\Omega P_i} \times \mathbf{F}_i \right) \cdot \boldsymbol{\omega},$$

where Ω is an arbitrary point of B and $\boldsymbol{\omega}$ is the angular velocity of the rigid motion. Introducing the total force $\mathbf{R}$ and torque $\mathbf{M}_O$ acting on B, we can put the preceding relation in the form

$$W = \mathbf{R} \cdot \dot{\mathbf{r}}_\Omega + \mathbf{M}_O \cdot \boldsymbol{\omega}, \tag{15.50}$$

which shows that *the power of the forces acting on B depends only on their total force and torque.*

Denoting by T the kinetic energy of B, we prove the following formula, which expresses the **theorem of kinetic energy**:

$$\dot{T} = W. \tag{15.51}$$

First, we choose $O \equiv G$ in (15.50). Then, by adding the scalar product of (15.48) by $\dot{\mathbf{r}}_\Omega$ and the scalar product of (15.49) by ω and taking into account (15.50), we obtain

$$\frac{d}{dt}\left(\frac{1}{2}m|\dot{\mathbf{r}}|_G^2\right) + \dot{\mathbf{K}}_G \cdot \omega = W^{(a)} + W^{(r)}. \tag{15.52}$$

On the other hand, in a central frame of inertia $(G, \mathbf{e}_i')$, $\omega = p\mathbf{e}_1' + q\mathbf{e}_2' + r\mathbf{e}_3'$, and, in view of (15.25), we have that

$$\dot{\mathbf{K}}_G = A\dot{p}\mathbf{e}_1' + B\dot{q}\mathbf{e}_2' + C\dot{r}\mathbf{e}_3' + \omega \times \mathbf{K}_G.$$

Recalling (15.26), we finally obtain

$$\dot{\mathbf{K}}_G \cdot \omega = Ap\dot{p} + Bq\dot{q} + Cr\dot{r} = \frac{dT}{dt},$$

and (15.51) is proved.

15.8 Exercises

1. A cannon resting on a rough horizontal plane is fired, and the muzzle velocity of the projectile with respect to the cannon is v. If m_1 is the mass of the cannon and m_2 the mass of the projectile, the mass of the powder being negligible, and μ is the coefficient of friction, show that the distance of recoil of the cannon is

$$\left(\frac{m_2 v}{m_1 + m_2}\right)^2 \frac{1}{2\mu g},$$

where g is the gravity acceleration.

2. Two men, each of mass m_2, are standing at the center of a homogeneous horizontal beam of mass m_1 that is rotating with uniform angular velocity ω about a vertical axis through its center. If the two men move to the ends of the beam and ω_1 is then the angular velocity, show that

$$\omega_1 = \frac{m_1}{m_1 + 6m_2}\omega.$$

3. If a shell at rest explodes and breaks into two fragments, show that the two fragments move in opposite directions along the same straight line with speeds that are inversely proportional to their masses.

4. Prove that, if μ is the coefficient of friction, the couple necessary to set into rotation a right circular cylinder of radius r and weight mg that is standing on its base on a rough horizontal plane is $\frac{2}{3}mgr\mu$.

5. A uniform disk of radius a that is spinning with angular speed ω about a vertical axis is placed upon a horizontal table. If the coefficient of the friction between the surface of the disk and the plane is μ, determine when the disk stops.

6. A heavy bar's end points A and B are moving, respectively, along the axes Ox and Oy of a vertical plane π. Neglecting the friction of the constraints, determine the equations of motion of the bar and the reactive forces at A and B using the balance equations.

Chapter 16
Dynamics of a Rigid Body

Abstract In this chapter the motion of a solid with a smooth fixed axis is analyzed. Then, some rigid motions of a solid with a (smooth) fixed point when the moment of forces vanishes are studied: Poinsot's motions, uniform rotations, and inertial motions of a gyroscope. The different properties of motion of a heavy gyroscope and the gyroscopic effect are analyzed. Finally, the motions of Foucault's gyroscope, gyroscopic compass, and free heavy solid are taken into account.

16.1 Rigid Body with a Smooth Fixed Axis

Consider a rigid body $\mathcal{B}$ with a fixed smooth axis. Figure 16.1 shows two devices forming a fixed axis r: two spherical hinges and a spherical hinge coupled with a cylindrical one. If the contact surfaces between the moving parts of these devices are smooth, then the reactive forces are orthogonal to them and, at least in the devices shown, their application straight lines intersect the fixed axis r. Consequently, the component $M_3^{(r)}$ of the total reactive force vanishes:

$$M_3^{(r)} = 0. \tag{16.1}$$

Henceforth we will assume this condition as being characteristic of any smooth device that gives rise to a fixed axis.

The lab and body frames are chosen as shown in Fig. 16.2. It is evident that the motion of $\mathcal{B}$ is determined by a knowledge of the function $\varphi(t)$, which gives the angle between the fixed plane Ox_1x_3 and the plane $Ox_1'x_3'$ moving with the body.

To find an equation containing the function $\varphi(t)$, we note that, since the axis r could not be a principal axis of inertia, we cannot assume that the body frame is a principal frame of inertia. Therefore, the angular momentum $\mathbf{K}_O$ is given by (15.22). Introducing this expression into the balance of angular momentum (15.49), projecting this equation on r, and taking into account (16.1), we obtain that

$$I_{33}\ddot{\varphi} = M_3^{(a)}, \tag{16.2}$$

Fig. 16.1 Devices for a
fixed axis

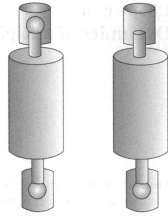

Fig. 16.2 Solid rotating
about a fixed axis

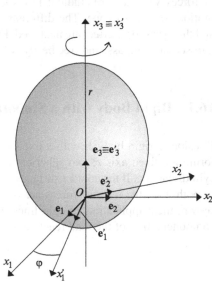

where I_{33} is the moment of inertia of $\mathcal{B}$ relative to the axis r. It is possible to determine, at least in principle, the function $\varphi(t)$ by (16.2), provided that we recall that active forces depend on the position of $\mathcal{B}$, its velocity field, and time. But the position of $\mathcal{B}$ is determined by the angle φ, and the velocity field is given by the relation $\dot{\mathbf{r}} = \dot{\varphi}\mathbf{e}_3 \times \mathbf{r}$, where $\mathbf{r} = \overrightarrow{OP}$. Consequently, (16.2) assumes the final form

$$I_{33}\ddot{\varphi} = M_3^{(a)}(\varphi, \dot{\varphi}, t), \qquad (16.3)$$

which determines one and only one motion satisfying the initial conditions ($t = 0$)

$$\varphi(0) = \varphi_0, \quad \dot{\varphi}(0) = \dot{\varphi}_0. \qquad (16.4)$$

16.2 Compound Pendulum

A *compound pendulum* is a heavy rigid body $\mathcal{B}$ rotating about a smooth horizontal axis Ox_3, which is called the *suspension axis* (Fig. 16.3). Let O be the orthogonal projection of the center of mass G of $\mathcal{B}$ on the suspension axis, and let φ be the angle between OG and the vertical straight line containing O. Finally, we denote by h the length of OG, by m the mass of $\mathcal{B}$, and by $\mathbf{g}$ the gravity acceleration. Since the active force reduces to the weight, which is equivalent to the single force $m\mathbf{g}$ applied at the center of mass, we have that the component of the active torque becomes $M_3 = -mgh \sin \varphi$, and (16.3) assumes the following form:

$$\ddot{\varphi} = -\frac{g}{l} \sin \varphi, \qquad (16.5)$$

where

$$l = \frac{I_{33}}{mh} \qquad (16.6)$$

is called the *reduced length* of the pendulum. Equation (16.5) shows that a compound pendulum moves as a simple pendulum whose length is equal to the reduced length of the compound pendulum.

16.3 Solid with a Fixed Point: Euler's Equations

Let $\mathcal{B}$ be a rigid body constrained to moving around a smooth fixed point O. If the constraint is realized with a spherical smooth hinge, then the reactive forces are

Fig. 16.3 Compound pendulum

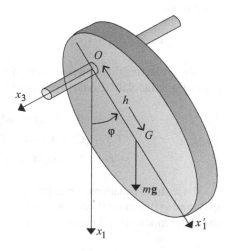

orthogonal to the contact spherical surface. Consequently, their application straight lines contain point O and the total reactive torque relative to O vanishes:

$$\mathbf{M}_O^{(r)} = \mathbf{0}. \tag{16.7}$$

We use the preceding condition as a definition of a smooth fixed point, regardless of the device realizing the fixed point O.

The position of $\mathcal{B}$ is determined by the Euler angles ψ, φ, and θ formed by the body frame $(O, \mathbf{e}_i')$ with the lab frame $(O, \mathbf{e}_i)$. Consequently, we need to find three differential equations that can determine these variables as functions of time. In view of (16.7), the balance of angular momentum assumes the form

$$\dot{\mathbf{K}}_O = \mathbf{M}_O^{(a)}. \tag{16.8}$$

The torque $\mathbf{M}_O^{(a)}$ of the active forces must depend on the position of $\mathcal{B}$, its velocity field, and time. Since the velocity of any point $\mathbf{r} = \overrightarrow{OP}$ is given by the formula $\dot{\mathbf{r}} = \omega \times \mathbf{r}$, where ω is the angular velocity of the rigid motion, we can state that

$$\mathbf{M}_O^{(a)} = \mathbf{M}_O^{(a)}(\psi, \varphi, \theta, p, q, r, t), \tag{16.9}$$

where p, q, and r are the components of ω in the body frame. On the other hand, we have proved that the angular momentum relative to the lab frame of a solid with a fixed point, when it is projected onto the body frame, is given by (15.20). The time derivative of $\mathbf{K}_O$, in view of (12.42), is given by

$$\dot{\mathbf{K}}_O = A\dot{p}\mathbf{e}_1' + B\dot{q}\mathbf{e}_2' + C\dot{r}\mathbf{e}_3' + \omega \times \mathbf{K}_O. \tag{16.10}$$

Introducing expressions (16.9) and (16.10) of $\mathbf{M}_O^{(a)}$ and $\dot{\mathbf{K}}_O$ into the balance equation of angular momentum, we obtain the following equation for a solid with a smooth fixed point:

$$A\dot{p}\mathbf{e}_1' + B\dot{q}\mathbf{e}_2' + C\dot{r}\mathbf{e}_3' + \omega \times \mathbf{K}_O = \mathbf{M}_O^{(a)}(\psi, \varphi, \theta, p, q, r, t). \tag{16.11}$$

Finally, projecting this equation along the axes of the body frame (which are principal axes of inertia relative to the fixed point O), we obtain the **Euler equations**

$$A\dot{p} - (B - C)qr = M_{Ox_1}(\psi, \varphi, \theta, p, q, r, t), \tag{16.12}$$

$$B\dot{q} - (C - A)rp = M_{Ox_2}(\psi, \varphi, \theta, p, q, r, t), \tag{16.13}$$

$$C\dot{r} - (A - B)pq = M_{Ox_31}(\psi, \varphi, \theta, p, q, r, t). \tag{16.14}$$

These three equations, since they contain the six unknowns $\psi(t)$, $\varphi(t)$, $\theta(t)$, $p(t)$, $q(t)$, and $r(t)$, cannot determine the motion of $\mathcal{B}$. However, if we take into account the Euler kinematic relations (12.32)–(12.34),

$$\dot{\psi} = \frac{1}{\sin \theta}(p \sin \varphi + q \cos \varphi), \tag{16.15}$$

$$\dot{\varphi} = -(p \sin \varphi + q \cos \varphi) \cot \theta + r, \tag{16.16}$$

$$\dot{\theta} = p \cos \varphi - q \sin \varphi, \tag{16.17}$$

we obtain a system that, under general hypotheses of regularity of the functions appearing on the right-hand side of (16.12)–(16.14), admits one and only one solution satisfying the initial data

$$\psi(0) = \psi_0, \quad \varphi(0) = \varphi_0, \quad \theta(0) = \theta_0, \tag{16.18}$$

$$p(0) = p_0, \quad q(0) = q_0, \quad r(0) = r_0, \tag{16.19}$$

which correspond to giving the initial position and velocity field of B.

The *nonlinear* equations (16.12)–(16.17) exhibit a very complex form. It is utopian to search for explicit solutions of them, even in simple cases. In the following sections, we analyze some simple but interesting situations in which we do not succeed in finding explicit solutions, but, by a qualitative analysis, we obtain meaningful information about the behavior of the motion.

16.4 Poinsot's Motions

We define *free rotations* or *Poinsot's motions* all the motions of a solid B with a smooth fixed point O in the absence of the total active torque $\mathbf{M}_O^{(a)}$:

$$\mathbf{M}_O^{(a)} = \mathbf{0}. \tag{16.20}$$

In other words, these motions correspond to the solutions of Euler's equations when their right-hand sides vanish. Also in this case, it is not possible to find the closed form of solutions. However, Poinsot (1851) proposed a very interesting geometric qualitative analysis of free rotations. As a result of this analysis, we are able to determine the *configurations that B assumes during the motion, although we are not able to give the instant at which B occupies one of them.*

This analysis is based on the following conservation laws of angular momentum and kinetic energy:

$$\mathbf{K}_O = \mathbf{K}_O(0), \quad T = T(0), \tag{16.21}$$

which follow at once from (15.49) and (15.51) and hypothesis (16.20).

Let E_O be the ellipsoid of inertia of B relative to point O, and denote by $r(t)$ the instantaneous rotation axis, that is, the straight line parallel to the angular velocity $\omega \equiv |\omega|\mathbf{u}$ and containing O. Finally, we denote by $P(t)$ the intersection point of $r(t)$ and E (Fig. 16.4).

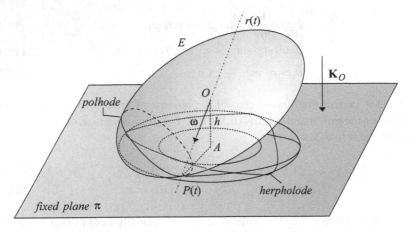

Fig. 16.4 Polhode and herpolhode

We have proved [see (15.49)] that the angular momentum $\mathbf{K}_O$ is orthogonal to the plane π tangent to the ellipsoid of inertia at the point $P(t)$. Since, in view of (16.21), $\mathbf{K}_O$ is a constant vector, we can state that during motion this tangent plane remains parallel to itself. Now we prove that the conservation of kinetic energy implies that the distance h of π from O is constant during the motion. First, we note that

$$h = \overrightarrow{OP} \cdot \frac{\mathbf{K}_O}{|\mathbf{K}_O|}. \tag{16.22}$$

Further, the point $P(t) \in E$ confirms the equation of the ellipsoid so that, in view of (15.33), we can write

$$h = \frac{\mathbf{u}}{\sqrt{\mathcal{I}_r}} \cdot \frac{\mathbf{K}_O}{|\mathbf{K}_O|} = \frac{\boldsymbol{\omega} \cdot \mathbf{K}_O}{|\boldsymbol{\omega}|\sqrt{\mathcal{I}_r}|\mathbf{K}_O|} = \frac{2T}{|\boldsymbol{\omega}|\sqrt{\mathcal{I}_r}|\mathbf{K}_O|}. \tag{16.23}$$

On the other hand, we have

$$2T = \boldsymbol{\omega} \cdot \mathbf{I}_O \cdot \boldsymbol{\omega} = |\boldsymbol{\omega}|^2 \mathbf{u} \cdot \mathbf{I}_O \cdot \mathbf{u} = |\boldsymbol{\omega}|^2 \mathcal{I}_r,$$

and (16.23) can be written in the form

$$h = \frac{\sqrt{2T(0)}}{|\mathbf{K}_O(0)|}, \tag{16.24}$$

which shows that h is constant.

In conclusion, during the motion of $\mathcal{B}$, the plane π tangent to the ellipsoid of inertia E at point P, where P is the intersection of the instantaneous axis of rotation with E, does not change its position. To understand the consequences of this result on the motion of $\mathcal{B}$, we assign the initial angular velocity $\omega(0)$. This vector determines the initial values of $\mathbf{K}_O$ and T_O and, consequently, owing to (16.24), the fixed position of the plane π relative to O and the initial position $P(0)$ of $P(t)$. Since $P(t) \in E \cap r(t)$, its velocity is equal to zero. Finally, we can state that *during a Poinsot motion of $\mathcal{B}$ about O, the ellipsoid of inertia E rolls, without slipping, on the fixed plane π.*

During motion, the point $P(t)$ describes a curve on E, which is called a **polhode**, and a curve on π, known as a **herpolhode** (Fig. 16.4). It is possible to prove that the polhode is a closed curve on E, which may reduce to a point. In contrast, the herpolhode could be an open curve. More precisely, consider the arc γ described by $P(t)$ during a complete turn on the ellipsoid E, and let α be the angle subtended by γ having the vertex in the orthogonal projection A of O on π. If $\alpha = 2\pi(m/n)$, where m and n are integers, then the herpolhode is closed; otherwise it is open. It is also possible to prove that the herpolhode is contained in an annulus σ whose center is A, is always concave toward A, and, when it is open, is everywhere dense in σ. In particular, if the initial angular velocity $\omega(0)$ is parallel to a principal axis of inertia, then the angular momentum $\mathbf{K}_O$ is always parallel to $\omega(0)$ and h is equal to the half-length of the axis of E. Consequently, the polhode and the herpolhode reduce to a point. If the ellipsoid E has a symmetry of revolution around an axis a, then the polhode and the herpolhode become two circles. Finally, if E reduces to a sphere, the polhode and the herpolhode reduce to a point and the motion is always a uniform rotation. With the notebook **MotiPoin** it is possible to simulate all the preceding cases. Figure 16.5 shows a herpolhode obtained with this notebook corresponding to the data $A = 1$, $B = 1.5$, $C = 0.5$, $r_0 = 3$, and $\theta(0) = \pi/4$ in an interval of motion of 15 s.

Fig. 16.5 Herpolhode

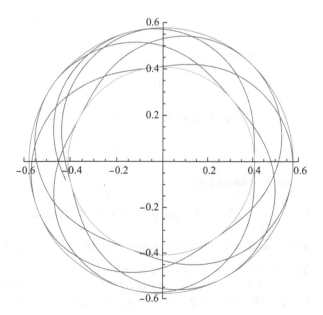

16.5 Uniform Rotations

Let $\mathcal{B}$ be a solid with a smooth fixed point O. In this section, we prove that, among all the possible Poinsot motions, there are *uniform rotations* about certain fixed axes containing O. First we note that the condition $\omega = const.$ must be considered in the lab frame in which we are searching for uniform rotations. However, in view of (12.41), we have

$$\frac{d_a \omega}{dt} = \frac{d_r \omega}{dt} + \omega \times \omega = \frac{d_r \omega}{dt}. \tag{16.25}$$

In other words, if a uniform rotation is possible in the lab frame, then this rotation is also uniform in the body frame, i.e., the components p, q, and r of ω along the axes of a principal frame of inertia in the body are constant. In view of Euler's equations (16.12)–(16.14), these components must satisfy the following *algebraic system*:

$$(B - C)qr = 0, \tag{16.26}$$
$$(C - A)rp = 0, \tag{16.27}$$
$$(A - B)pq = 0. \tag{16.28}$$

To find the solutions of this system, we distinguish three cases:

- If the eigenvalues A, B, and C differ from each other, then system (16.26)–(16.28) becomes

$$qr = pr = pq = 0,$$

and its solutions are

$$\{p = p_0, q = r = 0\}, \ \{p = r = 0, q = q_0\}, \ \{p = q = 0, r = r_0\}.$$

- If $A = B \neq C$, then system (16.26)–(16.28) becomes

$$pr = qr = 0,$$

and its solutions are

$$\{p = p_0, q = q_0, r = 0\}, \ \{p = q = 0, r = r_0\}.$$

- If $A = B = C$, then the solutions are given by arbitrary values of p, q, and r.

We note that in the first case, there is a single body frame formed by principal axes of inertia; in the second case, any body frame formed by the eigenvectors belonging to the axis Ox_3' and any pair of orthogonal axes that are also orthogonal to Ox_3' form a principal frame of inertia. Finally, if $A = B = C$, then any body frame is a principal frame. We can summarize the preceding results as follows: *if a uniform rotation is realized, then its axis of rotation coincides necessarily with a principal axis of inertia of the tensor of inertia* $\mathbf{I}_O$.

One of these rotations can be obtained by assigning the *initial* angular velocity $\omega(0)$ directed along a principal axis of inertia. Thus, we have just proved that a uniform rotation $\omega(t) = \omega(0)$ is possible about this axis. On the other hand, we know that there is only one solution satisfying the assigned initial data. Therefore, this uniform rotation is the only possible motion satisfying the initial data.

We conclude this section by recalling a fundamental result relative to uniform rotations about principal axes of inertia. If the eigenvalues A, B, C of the tensor of inertia $\mathbf{I}_O$ relative to the smooth fixed point O are assumed to satisfy the condition $A < B < C$, then the uniform rotations about the principal axis of inertia corresponding to A and C are stable, whereas the uniform rotations about the principal axis corresponding to the eigenvalue B are unstable. The axes corresponding to A and C are said to be ***principal steady axes***.

16.6 Poinsot's Motions in a Gyroscope

A ***gyroscope*** $\mathcal{B}$ is a rigid body whose central tensor of inertia $\mathbf{I}_G$ has a double eigenvalue $(A = B)$, which differs from the remaining simple eigenvalue C. The one-dimensional eigenspace a corresponding to the eigenvalue C is called a ***gyroscopic axis***. Figure 16.6 shows a gyroscope whose fixed point O is obtained with a complex structure (*cardan joint*) that allows for the complete mobility of $\mathcal{B}$ about O.

From the properties we proved about the tensor of inertia in Chap. 15 we know that a body frame whose axes are principal axes of inertia can be obtained by adding to the axis $Ox_3' = a$ any arbitrary pair of axes Ox_1' and Ox_2' that are orthogonal to each other and to Ox_3'. Moreover, any other frame $(\Omega x_1'' x_2'' x_3'')$, where $\Omega \in a$ and Ox_1' is parallel to $\Omega x_1''$, Ox_2' is parallel to $\Omega x_2''$, and $Ox_3' \equiv \Omega x_3''$, is again a principal frame of inertia. Finally, since $A = B$ for the central tensor of inertia, the corresponding ellipsoid of inertia E has a symmetry of revolution about a, *and these properties hold for all the tensors of inertia and corresponding ellipsoids relative to any other point of* a.

After recalling these properties, we are now in a position to evaluate Poinsot's motions of a gyroscope with a smooth fixed point O belonging to the gyroscopic axis a. Since we have $A = B$ and $\mathbf{M}_O^{(a)} = \mathbf{0}$, from the third Euler equation (16.14) we derive the following conservation law:

$$r = r_0, \tag{16.29}$$

which states that *in a free rotation of a gyroscope the component of the angular velocity along the gyroscopic axis is constant*. Further, the angular momentum is constant. Collecting these results, we can write that

Fig. 16.6 Gyroscope with
its suspension device

$$\mathbf{K}_O = A\left(p\mathbf{e}_1' + q\,\mathbf{e}_2'\right) + Cr_0\mathbf{e}_3' = \mathbf{K}_O(0), \tag{16.30}$$

where the unit vectors $\mathbf{e}_i'$ are directed along the principal axes of the body frame. By the preceding relation, we can write the angular velocity $\boldsymbol{\omega} = p\mathbf{e}_1' + q\mathbf{e}_2' + r\mathbf{e}_3'$ in the following form:

$$\boldsymbol{\omega} = \boldsymbol{\omega}_1 + \boldsymbol{\omega}_2, \tag{16.31}$$

where

$$\boldsymbol{\omega}_1 = \left(1 - \frac{C}{A}\right)r_0\mathbf{e}_3', \quad \boldsymbol{\omega}_2 = \frac{1}{A}\mathbf{K}_O(0). \tag{16.32}$$

The preceding relations show that any Poinsot motion of a gyroscope is a spherical motion obtained by composing a uniform rotation about a fixed axis containing O and parallel to $\mathbf{K}_O$, and a uniform rotation about the gyroscopic axis.

16.7   Heavy Gyroscope

In this section we consider a gyroscope $\mathcal{B}$ with a fixed smooth point O, $O \neq G$, of its gyroscopic axis that is acted upon by its weight $\mathbf{p}$ (Fig. 16.7). The balance of angular momentum yields

$$\dot{\mathbf{K}}_O = h_G\mathbf{e}_3' \times \mathbf{p}, \tag{16.33}$$

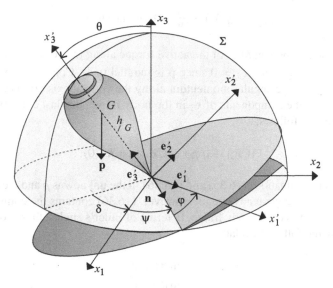

Fig. 16.7 Heavy gyroscope

where h_G is the abscissa of the center of mass G along the gyroscopic axis Ox'_3. We now prove that the three Euler equations can be substituted, to great advantage, by three conservation laws.

First, since $A = B$ and the component $M_{x'_3} = h_G e'_3 \times \mathbf{p} \cdot e'_3 = 0$, the third Euler equation still implies the first conservation law $r = r_0$.

Second, the total energy is constant. In fact, in view of (15.50), we have that the power of the reactive forces vanishes,

$$W^{(r)} = \mathbf{R}^{(r)} \cdot \dot{\mathbf{r}}_0 + \mathbf{M}_O^{(r)} \cdot \boldsymbol{\omega} = 0,$$

since $\dot{\mathbf{r}}_0 = \mathbf{0}$ and $\mathbf{M}_O^{(r)} = \mathbf{0}$. On the other hand, the power $W^{(a)}$ of the active forces is given by

$$\mathbf{p} \cdot \dot{\mathbf{r}}_G = \frac{d}{dt}(\mathbf{p} \cdot \mathbf{r}_G) = -\frac{d}{dt}(ph_G \cos\theta).$$

Taking into account the theorem of kinetic energy (15.51), we obtain conservation of the total energy E:

$$T + ph_G \cos\theta = E, \tag{16.34}$$

which, by expression (15.21) of the kinetic energy of a solid with a fixed point and the conditions $A = B$ and $r = r_0$, assumes the form

$$A \left(p^2 + q^2 \right) + Cr_0^2 + 2 p h_G \cos \theta = 2E. \tag{16.35}$$

Finally, the component $M_3^{(a)}$ of the active torque along the vertical axis Ox_3 vanishes $M_3^{(a)} = h_G \mathbf{e}_3' \times \mathbf{p} \cdot \mathbf{e}_3 = 0$ since $\mathbf{p}$ is parallel to $\mathbf{e}_3$. Consequently, the component $\mathbf{K}_O \cdot \mathbf{e}_3$ of the angular momentum along the vertical axis Ox_3 is constant. If we denote by γ_i the components of $\mathbf{e}_3$ in the body frame, this last conservation law can be written as follows:

$$A(p\gamma_1 + q\gamma_2) + Cr_0\gamma_3 = K_3(0). \tag{16.36}$$

The algebraic equations (16.35) and (16.36) in the unknowns p and q can replace Euler's equations to great advantage. However, (16.36) contains the components γ_i of $\mathbf{e}_3$ along the axes of the body frame. In terms of Euler's angles, these components are given by the following relations:

$$\gamma_1 = \sin \theta \sin \varphi, \tag{16.37}$$
$$\gamma_2 = \sin \theta \cos \varphi, \tag{16.38}$$
$$\gamma_3 = \cos \theta. \tag{16.39}$$

The presence of Euler's angles in these relations requires the use of the Euler kinematic relations (Sect. 12.4):

$$p = \dot{\psi} \sin \theta \sin \varphi + \dot{\theta} \cos \varphi, \tag{16.40}$$
$$q = \dot{\psi} \sin \theta \cos \varphi - \dot{\theta} \sin \varphi, \tag{16.41}$$
$$r_0 = \dot{\psi} \cos \theta + \dot{\varphi}. \tag{16.42}$$

In conclusion, the motion of a heavy gyroscope $\mathcal{B}$ can be determined by solving the system of eight equations (16.35)–(16.42) in the unknowns $p(t)$, $q(t)$, $\gamma_1(t)$, $\gamma_2(t)$, $\gamma_3(t)$, $\psi(t)$, $\varphi(t)$, and $\theta(t)$.

From (16.40) and (16.41) we obtain that

$$p^2 + q^2 = \dot{\psi}^2 \sin^2 \theta + \dot{\theta}^2. \tag{16.43}$$

Introducing this equation into (16.35) and relations (16.37)–(16.41) into (16.36) and taking into account (16.42), we can write the following system:

$$\dot{\psi}^2 \sin^2 \theta + \dot{\theta}^2 = \alpha - a \cos \theta, \tag{16.44}$$
$$\dot{\psi} \sin^2 \theta = \beta - br_0 \cos \theta, \tag{16.45}$$
$$\dot{\psi} \cos \theta + \dot{\varphi} = r_0, \tag{16.46}$$

where

$$\alpha = \frac{2E - Cr_0^2}{A}, \quad a = \frac{2ph_G}{A}, \tag{16.47}$$

$$\beta = \frac{K_3(0)}{A}, \quad b = \frac{C}{A} > 0. \tag{16.48}$$

From (16.45) we obtain

$$\dot{\psi} = \frac{\beta - br_0 \cos\theta}{\sin^2\theta}, \tag{16.49}$$

and substituting this result into (16.44) we arrive at the system

$$(\beta - br_0 \cos\theta)^2 + \dot{\theta}^2 \sin^2\theta = (\alpha - a\cos\theta)\sin^2\theta, \tag{16.50}$$

$$\dot{\psi} = \frac{\beta - br_0 \cos\theta}{\sin^2\theta}, \tag{16.51}$$

$$\dot{\varphi} = r_0 - \dot{\psi}\cos\theta. \tag{16.52}$$

The first of the preceding equations can be written in a more convenient form by introducing the new variable $u = \cos\theta$. In fact, since

$$\dot{u} = -\dot{\theta}\sin\theta, \quad \sin^2\theta = 1 - u^2, \tag{16.53}$$

system (16.50)–(16.52) assumes the final form

$$\dot{u}^2 = (\alpha - au)(1 - u^2) - (\beta - br_0 u)^2 \equiv f(u), \tag{16.54}$$

$$\dot{\psi} = \frac{\beta - br_0 u}{1 - u^2}, \tag{16.55}$$

$$\dot{\varphi} = r_0 - \frac{\beta - br_0 u}{1 - u^2}u. \tag{16.56}$$

A complete integration of the preceding system is impossible without resorting to numerical methods. However, a qualitative analysis of the solution can be developed.

Consider the unit sphere Σ (Fig. 16.7) with its center at O and the curve Γ that the gyroscopic axis Ox_3' draws on Σ during the motion. When we note that both axes Ox_3 and Ox_3' are orthogonal to the line of nodes, identified in Fig. 16.7 by the unit vector $\mathbf{n}$, we deduce that the angle between the plane Ox_3x_3' and the fixed axis Ox_1 results in $\delta = \psi - \pi/2$. Consequently, in spherical coordinates, the curve Γ can be described in terms of the angles δ and θ. Remembering that $u = \cos\theta$, (16.54) and (16.55) represent a system of differential equations in the unknown parametric equations of the curve Γ, i.e., in the functions $\delta(t)$ and $\theta(t)$. Finally, (16.56) describes how the gyroscope rotates about its axis.

We do not prove that the relations existing among the quantities α, β, a, and b imply that the three roots of the equation $f(u) = 0$ are real. In view of (16.54) and the meaning of u, we must have

Fig. 16.8 Behavior of
function $f(u)$

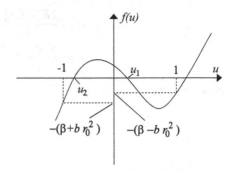

$$f(u) \geq 0, \quad -1 \geq u \geq 1. \tag{16.57}$$

On the other hand, since $a > 0$, we also have (Fig. 16.8)

$$\lim_{u \to -\infty} f(u) = -\infty, \quad \lim_{u \to \infty} f(u) = \infty; \tag{16.58}$$

$$f(-1) = -(\beta + br_0^2), \quad f(1) = -(\beta - br_0^2). \tag{16.59}$$

The preceding conditions imply that the two acceptable roots u_1 and u_2 of the equation $f(u) = 0$ belong to the interval $[-1, 1]$, that is,

$$-1 \leq u_2 \leq u_1 \leq 1. \tag{16.60}$$

Equivalently, since $u = \cos\theta$, we can state that during the motion of the gyroscope the angle θ satisfies the condition

$$\theta_1 \leq \theta \leq \theta_2. \tag{16.61}$$

This shows that the curve Γ described on the unit sphere Σ is always contained between the parallel $\theta = \theta_1$ and $\theta = \theta_2$. In particular, if the initial data are such that $\theta_1 = \theta_2$, then Γ reduces to the parallel $\theta = \theta_1 = \theta_2$ and, in view of (16.55) and (16.56), both $\dot{\psi}$ and $\dot{\varphi}$ are constant. In other words, if $\theta_1 = \theta_2$, the gyroscopic axis moves around the vertical with constant slope θ_1, constant precession velocity $\dot{\psi}$, and proper angular velocity $\dot{\varphi}$.

If $\theta_1 \neq \theta_2$, then the motion of the gyroscopic axis is called **nutation**. Denoting

$$\bar{u} = \frac{\beta}{br_0}, \quad \bar{\theta} = \arcsin\bar{u}, \tag{16.62}$$

we have the following three possibilities:

- $\bar{\theta} \notin [\theta_1, \theta_2]$;
  then (16.55) shows that $\dot{\psi}$ always has the same sign. Moreover, for $\theta = \theta_1$ and $\theta = \theta_2$, $\dot{\theta} = 0$ and Γ is tangent to the aforementioned parallels.

Fig. 16.9 $A = B = 1$,
$C = 0.5$, $p(0) = -0.3$,
$q(0) = 0.5$, $r(0) = 3.5$,
$\theta(0) = \pi/4$

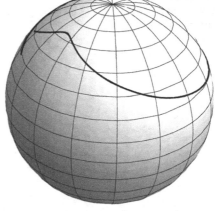

Fig. 16.10 $A = B = 1$,
$C = 1.2$, $p(0) = -0.3$,
$q(0) = 0.5$, $r(0) = 3.5$,
$\theta(0) = \pi/4$

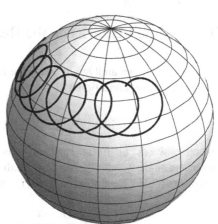

- $\overline{\theta} \in (\theta_1, \theta_2)$;
  then $\dot{\psi} = 0$ when $\theta = \overline{\theta}$ and the curve Γ is perpendicular to the parallel $\theta = \overline{\theta}$.
  Moreover, $\dot{\psi}$ has a definite sign when $\theta \in (\theta_1, \overline{\theta})$ and an opposite but still definite
  sign if $\theta \in (\overline{\theta}, \theta_2)$. This means that $\psi(t)$ increases in the first interval and decreases
  in the second one or vice versa. Moreover, for $\theta = \theta_1$ and $\theta = \theta_2$, we have $\dot{\theta} = 0$
  and the curve Γ is tangent to the parallels $\theta = \theta_1$ and $\theta = \theta_2$.
- $\theta = \overline{\theta}$; < for $\theta = \theta_2$ both $\dot{\theta}$ and $\dot{\psi}$ vanish, but $\dot{\psi}$ has a constant sign at any other
  point. It is possible to prove that Γ exhibits cusps when $\theta = \overline{\theta} = \theta_1$.

We omit proving that the case $\theta_2 = \overline{\theta}$ is impossible.

The previously listed cases are shown in Figs. 16.9–16.11, which were obtained
by the notebook **Solid** and refer to a gyroscope for which $p = h_G = 1$. Their captions
contain the relative initial data.

Fig. 16.11 $A = B = 1$,
$C = 1.2$, $p(0) = q(0) = 0$,
$r(0) = 2$, $\theta(0) = \pi/4$

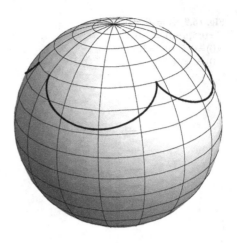

16.8   Some Remarks on the Heavy Gyroscope

In this section, we analyze in detail the motion of a gyroscope corresponding to the following initial conditions:

$$0 < \theta_0 < \pi, \;\; p_0 = q_0 = 0, \;\; r_0 > 0. \tag{16.63}$$

In other words, an initial rotation r_0 is impressed about the gyroscopic axis that forms an angle θ_0 with the vertical. With the preceding initial conditions, taking into account (16.35), the constants (16.47) and (16.48) become

$$\alpha = \frac{2E - Cr_0^2}{A} = \frac{2Ph_G}{A}\cos\theta_0 = au_0, \tag{16.64}$$

$$\beta = br_0\cos\theta_0 = br_0u_0. \tag{16.65}$$

Introducing these values of α and β into (16.54), we obtain

$$\dot{u}^2 = (u_0 - u)\left[a(1 - u^2) - b^2r_0^2(u_0 - u)\right] \equiv f(u). \tag{16.66}$$

Since the polynomial $f(u)$ admits the root

$$u_1 = u_0 \Rightarrow \cos\theta_1 = \cos\theta_0, \tag{16.67}$$

the motion of the gyroscope corresponds to case (c) of the previous section. The other solution u_2 in the interval $[-1, 1]$ is the root of the equation

$$a(1 - u_2^2) = b^2r_0^2(u_0 - u_2), \tag{16.68}$$

from which, in view of (16.67), we derive the following expression for the difference of the two roots:

$$u_1 - u_2 = \frac{a(1 - u_2^2)}{b^2 r_0^2}.$$

(16.69)

The preceding difference is positive since u_1, $u_2 \in [-1, 1]$, so that we have $u_2 < u_1$. But during the motion, $u_1 \leq u \leq u_2$, and consequently $u_1 - u \leq u_1 - u_2$. Bearing in mind these remarks, (16.55) supplies

$$0 \leq \dot{\psi} = b r_0 \frac{u_1 - u}{1 - u^2} \leq b r_0 \frac{u_1 - u_2}{1 - u^2}.$$

Recalling (16.69), from the preceding condition we obtain that

$$0 \leq \dot{\psi} \simeq \frac{a}{b r_0}.$$

(16.70)

The results (16.69) and (16.70) illustrate the following points.

- The nutation angle is proportional to $1/r_0^2$; in particular, $\theta \to 0$ when $r_0 \to \infty$.
- The velocity of precession is proportional to $1/r_0$.

We do not prove that $|\dot{\theta}|$ is proportional to r_0.
Finally, the balance of energy (16.35) can be written as

$$A(p^2 + q^2) + C r_0^2 + 2 p h_G u = C r_0^2 + 2 p h_G u_0,$$

and taking into account (16.69), we attain the result

$$A(p^2 + q^2) = 2 p h_G (u_0 - u) < 2 p h_G (u_1 - u_2) \leq \frac{a(1 - u_2^2)}{b^2 r_0^2}.$$

This inequality allows us to state that

$$\mathbf{K}_O = C r_0 \mathbf{e}_3' + O\left(\frac{1}{r_0}\right).$$

(16.71)

The preceding remarks prove that, *for the initial conditions (16.63) and high values of r_0, the motion of a gyroscope is approximately a slow regular precession around the vertical straight line containing O. Since $|\dot{\theta}|$ increases with r_0, during the low precession, the gyroscopic axis vibrates with high frequency* (Fig. 16.12).

This result can also be obtained supposing that the motion of the heavy gyroscope, corresponding to the initial data (16.63) and high values of r_0, is described by the equation [see (16.71)]

$$C r_0 \dot{\mathbf{e}}_3' = h_G \mathbf{e}_3' \times \mathbf{p}.$$

(16.72)

Fig. 16.12 $A = B = 1$,
$C = 0.5$, $p(0) = 0.1$, $q(0) = 0$,
$r(0) = 12$, $\theta(0) = \pi/4$

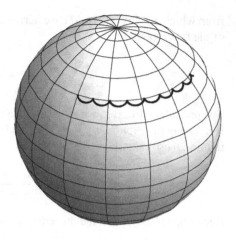

In fact, denoting by $\boldsymbol{\omega}$ the angular velocity of the gyroscope and recalling (12.41), the preceding equation can also be written as

$$\left(\boldsymbol{\omega} - \frac{h_G}{Cr_0}\mathbf{p}\right) \times \mathbf{e}_3' = \mathbf{0},$$

so that

$$\boldsymbol{\omega} = \lambda \mathbf{e}_3' + \frac{h_G}{Cr_0}\mathbf{p},$$

where the approximate value of λ is

$$\lambda = r_0 + \frac{h_G}{Cr_0}\mathbf{p} \cdot \mathbf{e}_3' \simeq r_0.$$

Finally, we obtain that the angular velocity is given by

$$\boldsymbol{\omega} = r_0 \mathbf{e}_3' - \frac{h_G}{Cr_0}\mathbf{p}. \tag{16.73}$$

In conclusion, substituting the balance equation of momentum with (16.72), we determine with sufficient approximation the position of the gyroscopic axis. However, the velocity field is not well approximated since the fast vibrations of the gyroscopic axis are neglected.

We say that for *any solid* $\mathcal{B}$ with a smooth fixed point O, the ***principle of the gyroscopic effect*** holds if, substituting the balance of angular momentum with the equation

$$Cr_0 \dot{\mathbf{e}}_3' = \mathbf{M}_O^{(a)}, \tag{16.74}$$

we determine, with sufficient approximation, the *configurations* that $\mathcal{B}$ assumes during its motion. The preceding considerations prove that such a principle holds for a heavy gyroscope.

Many references on this subject can be found in [34] where many interesting cases are analyzed in which the principle of the gyroscopic effect holds, at least when the initial rotation is impressed about a steady principal axis of inertia (Sect. 16.5) and the condition

$$\mathbf{M}_O^{(a)} \cdot \mathbf{e}_3' = 0$$

is satisfied. The reader interested in analyzing the behavior of any body under the preceding conditions can resort to the notebook **Solid**.

16.9 Torque of Terrestrial Fictitious Forces

In the previous sections, we considered the motion of a solid $\mathcal{B}$ with a smooth fixed point when the torque $\mathbf{M}_O$ relative to O of the acting forces vanishes or is equal to the torque of the weight. It is evident that the lab frame $Ox_1x_2x_3$, in which we analyze the motion of $\mathcal{B}$, is at rest with respect to the Earth, i.e., $Ox_1x_2x_3$ is a terrestrial frame. This implies that, in the torque of the acting forces, we should also include the torque of the fictitious forces due to the noninertial character of the terrestrial frame.

In this section, we evaluate the torque of these forces with respect to the center of mass G of a gyroscope $\mathcal{B}$. Let $Ox_1x_2x_3$ be the lab frame, which is at rest relative to the Earth. If we analyze the motion of $\mathcal{B}$ during a time interval which is much shorter than a month, we can suppose that the motion of the lab frame with respect to an inertial frame is a uniform rotation with terrestrial angular velocity $\boldsymbol{\omega}_T$. Then, if C is the region occupied by $\mathcal{B}$ and ρ the mass density of $\mathcal{B}$, then the torque relative to G of the fictitious forces due to the Earth's rotation is

$$\mathbf{M}_G^{(t)} = -\omega_T^2 \int_C \rho \, \overrightarrow{GP} \times \overrightarrow{QP} \, dc, \tag{16.75}$$

where Q is the orthogonal projection of the point $P \in \mathcal{B}$ on the rotation axis a of the Earth. Denoting by Q_* the orthogonal projection of G on a, we have $\overrightarrow{QP} = \overrightarrow{QQ_*} + \overrightarrow{Q_*G} + \overrightarrow{GP}$. But, $\forall P \in C$, we have that $|\overrightarrow{QQ_*}| \ll |\overrightarrow{Q_*G}|$, and (16.75) yields

$$\mathbf{M}_G^{(t)} = \omega_T^2 \overrightarrow{Q_*G} \times \int_C \rho \, \overrightarrow{GP} \, dc = \mathbf{0}, \tag{16.76}$$

owing to the definition of center of mass. It remains to evaluate the torque relative to G of Coriolis' forces:

$$\mathbf{M}_G^{(c)} = -2 \int_C \rho \overrightarrow{GP} \times (\boldsymbol{\omega}_T \times \dot{\mathbf{r}}_P) \, dc. \tag{16.77}$$

For the next applications, it will be sufficient to consider the center of mass at rest in a lab frame. In this case, $\dot{\mathbf{r}}_P = \boldsymbol{\omega} \times \overrightarrow{GP}$ and (16.77) can also be written as follows:

$$\mathbf{M}_G^{(c)} = -2 \int_C \rho \overrightarrow{GP} \times \left[\boldsymbol{\omega}_T \times (\boldsymbol{\omega} \times \overrightarrow{GP}) \right] dc. \tag{16.78}$$

On the other hand, this results in

$$\boldsymbol{\omega}_T \times \left(\boldsymbol{\omega} \times \overrightarrow{GP} \right) = \left(\boldsymbol{\omega}_T \cdot \overrightarrow{GP} \right) \boldsymbol{\omega} - (\boldsymbol{\omega}_T \cdot \boldsymbol{\omega}) \overrightarrow{GP},$$

so that

$$\overrightarrow{GP} \times \left[\boldsymbol{\omega}_T \times \left(\boldsymbol{\omega} \times \overrightarrow{GP} \right) \right] = - \left(\boldsymbol{\omega}_T \cdot \overrightarrow{GP} \right) \boldsymbol{\omega} \times \overrightarrow{GP}.$$

Taking into account this result, (16.78) becomes

$$\mathbf{M}_G^{(c)} = 2\boldsymbol{\omega} \times \int_C \rho \left(\boldsymbol{\omega}_T \cdot \overrightarrow{GP} \right) \overrightarrow{GP} \, dc. \tag{16.79}$$

Denoting by ω'_{Ti} the components of $\boldsymbol{\omega}_T$ in a central frame of inertia $Gx'_1 x'_2 x'_3$, we have that

$$\int_C \rho \left(\boldsymbol{\omega}_T \cdot \overrightarrow{GP} \right) \overrightarrow{GP} \, dc = \int_C \rho \left(\omega'_{T1} x'_1 + \omega'_{T2} x'_2 + \omega'_{T3} x'_3 \right) \left(x'_1 \mathbf{e}'_1 + x'_2 \mathbf{e}'_2 + x'_3 \mathbf{e}'_3 \right) dc.$$

Since the body frame $Gx'_1 x'_2 x'_3$ is a central frame of inertia, the products of inertia vanish and the preceding relation reduces to the following one:

$$\int_C \rho \left(\boldsymbol{\omega}_T \cdot \overrightarrow{GP} \right) \overrightarrow{GP} \, dc = \omega'_{T1} \mathbf{e}'_1 \int_C \rho x_1'^2 \, dc + \omega'_{T2} \mathbf{e}'_2 \int_C \rho x_2'^2 \, dc$$

$$+ \omega'_{T3} \mathbf{e}'_3 \int_C \rho x_3'^2 \, dc. \tag{16.80}$$

Since $\mathcal{B}$ is a gyroscope, we have

$$A = \int_C \rho \left(x_2'^2 + x_3'^2 \right) dc = \int_C \rho \left(x_1'^2 + x_3'^2 \right) dc = B \Rightarrow \int_C \rho x_2'^2 dc = \int_C \rho x_1'^2 dc.$$

On the other hand, we also have

$$C = \int_C \rho \left(x_1'^2 + x_2'^2 \right) dc = 2 \int_C \rho x_1'^2 dc,$$

so that

$$\int_C \rho x_1'^2 dc = \int_C \rho x_2'^2 dc = \frac{C}{2}. \tag{16.81}$$

$$\int_C \rho x_3'^2 dc = A - \frac{C}{2}. \tag{16.82}$$

In view of (16.81) and (16.82), Eq. (16.80) assumes the final form

$$2 \int_C \rho \left(\boldsymbol{\omega}_T \cdot \overrightarrow{GP} \right) \overrightarrow{GP} \, dc = C \left(\omega_{T1}' \mathbf{e}_1' + \omega_{T2}' \mathbf{e}_2' \right) + \left(A - \frac{C}{2} \right) \omega_{T3}' \mathbf{e}_3'. \tag{16.83}$$

Introducing this expression into (16.79), we finally obtain

$$\mathbf{M}_G^{(c)} = \boldsymbol{\omega} \times \left(C\boldsymbol{\omega}_T + \left(A - \frac{C}{2} \right) (\boldsymbol{\omega} \cdot \mathbf{e}_3') \mathbf{e}_3' \right). \tag{16.84}$$

16.10 Foucault's Gyroscope and Gyroscope Compass

In this section, we show two interesting applications of formulae (16.76) and (16.84).

A *Foucault gyroscope* $\mathcal{B}$ is a heavy gyroscope whose center of mass G is fixed with respect to the Earth. Let $\Omega x_1 x_2 x_3$ be an inertial frame of reference with the origin at the center of the Sun and the axes oriented toward fixed stars. Consider another frame $G x_1 x_2 x_3$ whose axes are parallel to the axes of the inertial frame $\Omega x_1 x_2 x_3$. The frame $G x_1 x_2 x_3$ is not an inertial frame since its origin is at rest with respect to the Earth. The forces acting on $\mathcal{B}$ are the gravitational attraction of the Earth and the fictitious forces. The latter is applied at G so that its torque relative to G vanishes. Further, the fictitious forces reduce to the drag forces since the motion of the frame $G x_1 x_2 x_3$ relative to $\Omega x_1 x_2 x_3$ is translational. In view of (16.76), the torque of these forces vanishes. Therefore, the balance equation of momentum in the frame $G x_1 x_2 x_3$ is

$$\dot{\mathbf{K}}_G = 0, \tag{16.85}$$

and the motion of $\mathcal{B}$ reduces to a regular precession. Consequently, if initially the angular velocity $\boldsymbol{\omega}$ is directed along the gyroscopic axis a, then the motion of $\mathcal{B}$ reduces to a uniform rotation about a. In other words, the gyroscopic axis is constantly directed toward a fixed point of the celestial sphere. An observer in a frame of reference $G \hat{x}_1 \hat{x}_2 \hat{x}_3$ at rest relative to the Earth will see the gyroscopic axis moving with the terrestrial angular velocity $\boldsymbol{\omega}_T$.

The *gyroscopic compass* is a heavy gyroscope $\mathcal{B}$ having a center of mass at rest in a terrestrial frame $G x_1 x_2 x_3$ and the gyroscopic axis a constrained to move on a fixed plane π. Suppose that an initial rotation is applied to $\mathcal{B}$ about the axis a. We

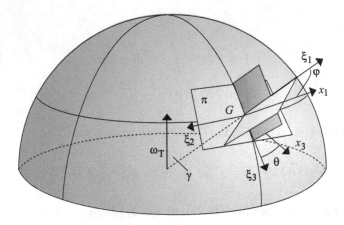

Fig. 16.13 Gyroscopic compass

prove that this axis remains at rest on π if and only if it coincides with the projection of the terrestrial axis. Consequently, if π is horizontal (Fig. 16.13), then $\mathcal{B}$ serves as a declination compass since a is directed toward the pole; if π is a meridian plane, then $\mathcal{B}$ gives the terrestrial axial tilt.

Let $G\xi_1\xi_2\xi_3$ be a terrestrial frame with axes oriented as in Fig. 16.13, and let $Gx_1x_2x_3$ be a central frame of inertia with an axis $Gx_3 \equiv a$. If ψ, φ, and θ denote the Euler angles, the condition that a lies on π implies that $\psi = 0$ since the nodal line coincides with the vertical axis Gx_1. Bearing this in mind, we can write (12.30) as

$$\boldsymbol{\omega} = \dot{\varphi}\mathbf{e}'_3 + \dot{\theta}\mathbf{e}_1, \tag{16.86}$$

where $\mathbf{e}'_3$ is the unit vector along the gyroscopic axis Gx_3 and $\mathbf{e}_1$ is the unit vector along the lab axis Gx_1. Moreover (Sect. 12.4)

$$p = \dot{\theta}\cos\varphi, \quad q = -\dot{\theta}\sin\varphi, \quad r = \dot{\varphi}, \tag{16.87}$$

and the angular momentum of $\mathcal{B}$ is given by

$$\mathbf{K}_G = A\dot{\theta}(\cos\varphi\,\mathbf{e}'_1 - \sin\varphi\,\mathbf{e}'_2) + C\dot{\varphi}\mathbf{e}'_3. \tag{16.88}$$

Finally, the balance equation of angular momentum becomes

$$\dot{\mathbf{K}}_G = \mathbf{M}_G^{(c)} + \mathbf{M}_G^{(v)}, \tag{16.89}$$

where $\mathbf{M}_G^{(c)}$ and $\mathbf{M}_G^{(v)}$ denote the torques of fictitious forces and reactive forces, respectively. It is possible to prove (see next chapter) that, if the constraints are smooth, then $\mathbf{M}_G^{(v)}$ is orthogonal to the plane $G\xi_1x_3$, that is, to the unit vectors $\mathbf{e}_1$ and $\mathbf{e}'_3$. Consequently, if we consider the projections of (16.89) along these vectors, then we obtain two differential equations without reactive forces. Now, since $\mathbf{e}_1$ is fixed and $\mathbf{e}_1' \cdot \mathbf{e}_1 = \cos \varphi$, $\mathbf{e}_2' \cdot \mathbf{e}_1 - \sin\varphi$, we have that

$$\dot{\mathbf{K}}_G \cdot \mathbf{e}_1 = \frac{d}{dt}(\mathbf{K}_G \cdot \mathbf{e}_1) = A\ddot{\theta}. \tag{16.90}$$

On the other hand, in view of (16.86) and (16.88), we also have

$$\dot{\mathbf{K}}_G \cdot \mathbf{e}_3' = \frac{d}{dt}\left(\mathbf{K}_G \cdot \mathbf{e}_3'\right) - \mathbf{K}_G \cdot \boldsymbol{\omega} \times \mathbf{e}_3'$$
$$= \frac{d}{dt}\left(\mathbf{K}_G \cdot \mathbf{e}_3'\right) + \mathbf{K}_G \times \boldsymbol{\omega} = C\ddot{\varphi}. \tag{16.91}$$

It remains to evaluate the components of $\mathbf{M}_G^{(c)}$ along $\mathbf{e}_1$ and $\mathbf{e}_3'$. First, from (16.84) we obtain

$$\mathbf{M}_G^{(c)} \cdot \mathbf{e}_3' = C\boldsymbol{\omega} \times \boldsymbol{\omega}_T \cdot \mathbf{e}_3'. \tag{16.92}$$

On the other hand, the terrestrial angular velocity is contained in the meridian plane $G\xi_1\xi_3$ (Fig. 16.13), and we can write

$$\boldsymbol{\omega}_T = \omega_T \cos\gamma \mathbf{e}_1 - \omega_T \sin\gamma \mathbf{e}_3, \tag{16.93}$$

where γ is the colatitude of G. In view of (16.86), we give (16.92) the following form:

$$\mathbf{M}_G^{(c)} \cdot \mathbf{e}_3' = C\left(\dot{\varphi}\mathbf{e}_3' + \dot{\theta}\mathbf{e}_1\right) \times (\omega_T \cos\gamma \mathbf{e}_1 - \omega_T \sin\gamma \mathbf{e}_3) \cdot \mathbf{e}_3'$$
$$= -C\dot{\theta}\omega_T \sin\gamma \mathbf{e}_2 \cdot \mathbf{e}_3' = C\omega_T \sin\gamma \dot{\theta} \sin\theta. \tag{16.94}$$

We explicitly remark that (16.94) shows that a fundamental sufficient condition to apply the gyroscopic effect is not satisfied (see end of Sect. 16.9).

Taking into account (16.84), (16.86), and (16.94), we deduce the following expression of $\mathbf{M}_G^{(c)} \cdot \mathbf{e}_1$:

$$\mathbf{M}_G^{(c)} \cdot \mathbf{e}_1 = -C\omega_T \sin\gamma\dot{\varphi} \sin\theta. \tag{16.95}$$

Collecting the preceding results we obtain the equations of motion

$$\ddot{\varphi} = \omega_T \sin\gamma\dot{\theta} \sin\theta, \tag{16.96}$$

$$A\ddot{\theta} = -C\omega_T \sin\gamma\dot{\varphi} \sin\theta, \tag{16.97}$$

to which we associate the following initial data:

$$\theta(0) = \theta_0, \ \dot{\theta}(0) = 0, \ \varphi(0) = 0, \ \dot{\varphi}(0) = r_0. \tag{16.98}$$

Since ω_T and $\sin\gamma$ are constant and $\dot\theta\sin\theta = -\mathrm{d}(\cos\theta)/\mathrm{d}t$, when we recall (16.98), from (16.96) we derive the following result:

$$\dot\varphi = r_0 + \omega_T\sin\gamma(\cos\theta - \cos\theta_0) = r_0\left(1 + \frac{\omega_T}{r_0}\sin\gamma(\cos\theta - \cos\theta_0)\right). \quad (16.99)$$

Introducing (16.99) into (16.97), we obtain

$$A\ddot\theta = -C\omega_T\sin\gamma r_0\left(1 + \frac{\omega_T}{r_0}\sin\gamma(\cos\theta - \cos\theta_0)\right)\sin\theta. \quad (16.100)$$

When we suppose that $r_0 \gg \omega_T$, (16.99) and (16.100) become

$$\dot\varphi = r_0,$$
$$A\ddot\theta = -C\omega_T\sin\gamma r_0\sin\theta.$$

In conclusion, the gyroscopic compass rotates uniformly about its axis, which oscillates around the initial position. For the presence of friction, after a few oscillations, the axis stops in the position $\theta = 0$, i.e., it is directed along the meridian.

16.11 Free Heavy Solid

We conclude this chapter considering a free solid $\mathcal{B}$ for which we can write the balance equations of momentum and angular momentum in the form

$$m\ddot{\mathbf{r}}_G = \mathbf{R}^{(a)}, \quad (16.101)$$

$$\dot{\mathbf{K}}_G = \mathbf{M}_G^{(a)}, \quad (16.102)$$

where m is the mass of $\mathcal{B}$ and $\mathbf{R}^{(a)}, \mathbf{M}_G^{(a)}$ denote the active total force and total torque relative to the center of mass G of the active forces acting on $\mathcal{B}$.

We denote by $Ox_1x_2x_3$ a lab frame and by $G\bar{x}_1\bar{x}_2\bar{x}_3$ a frame of reference with its origin at G and axes parallel to the corresponding axes of the lab frame. Finally, $Gx_1'x_2'x_3'$ is the body frame, which is a principal frame of inertia. In view of (15.7) and taking into account that the motion of $G\bar{x}_1\bar{x}_2\bar{x}_3$ relative to the lab frame is translational, we have

$$\mathbf{K}_G = \bar{\mathbf{K}}_G, \quad \dot{\mathbf{K}}_G = \frac{\mathrm{d}_a}{\mathrm{d}t}\mathbf{K}_G = \frac{\mathrm{d}_r}{\mathrm{d}t}\bar{\mathbf{K}}_G, \quad (16.103)$$

where $\mathrm{d}_r/\mathrm{d}t$ denote a derivative in the frame $G\bar{x}_1\bar{x}_2\bar{x}_3$. In other words, we can substitute (16.101) and (16.102) with others, as follows:

$$m\ddot{\mathbf{r}}_G = \mathbf{R}^{(a)}, \tag{16.104}$$

$$\frac{d_r \overline{\mathbf{K}}_G}{dt} = \mathbf{M}_G^{(a)}. \tag{16.105}$$

We remark that in the frame $G\overline{x}_1\overline{x}_2\overline{x}_3$ the center of mass G is fixed so that the components of (16.105) along the body frame are the Euler equations. If we add to (16.104) and (16.105) Euler's kinematic relations, then we obtain a system of nine scalar differential equations whose unknowns are Euler angles $\psi(t)$, $\varphi(t)$, $\theta(t)$, the components $p(t)$, $q(t)$, and $r(t)$ of the angular velocity in the body frame, and the coordinates of the center of mass x_{1G}, x_{2G}, and x_{3G}. This conclusion is also due to the fact that the acting forces, since they depend on the position of $\mathcal{B}$, its velocity field, and time, are functions of the aforementioned variables.

We explicitly remark that the presence of all the unknowns in all the equations makes it impossible to analyze the motion of the center of mass independently of the motion about the center of mass. This is possible, for instance, when $\mathcal{B}$ is only acted upon by its weight $m\mathbf{g}$. In such a case, (16.104) and (16.105) become

$$\ddot{\mathbf{r}}_G = \mathbf{g}^{(a)}, \tag{16.106}$$

$$\frac{d_r \overline{\mathbf{K}}_G}{dt} = \mathbf{0}. \tag{16.107}$$

These equations imply that the center of mass G moves along a parabola, whereas the motion about the center of mass is a free rotation (Sect. 16.4).

Chapter 17
Lagrangian Dynamics

Abstract In this chapter, Lagrangian dynamics is described. After introducing the configuration space for a system of constrained rigid bodies, the principle of virtual power and its equivalence to Lagrange's equations is shown. Then, these equations are formulated in the case of conservative forces and forces deriving from a generalized potential energy. The fundamental relation between conservation laws (first integrals) and the symmetries of the Lagrangian function is proved (Noether's theorem). Further, Lagrange's equations for linear nonholonomic constraints are formulated. The chapter contains also the analysis of small oscillations about a stable equilibrium configuration (normal modes), and an introduction to variational calculus including Hamilton's principle. Finally, two geometric formulations of Lagrangian's dynamics are presented together with Legendre's transformation that allows to transform Lagrange's equations into Hamilton's equations.

17.1 Introduction

In previous chapters, we analyzed some fundamental aspects both of Newton's model, describing a system of material points, and of Euler's model, referring to a single, free or constrained, rigid body. If we attempt to apply the latter model to a *system S* of N constrained rigid bodies, we face great difficulties. In fact, it is not an easy task either to express analytically the constraints to which S is subject or to formulate mathematically the restrictions on the reactive forces exerted by smooth constraints.

D'Alembert and Lagrange proposed an elegant and efficient method to analyze free or constrained mechanical systems. This method exhibits the advantage of supplying differential equations of motion that

- Do not contain unknown reactive forces and
- Contain the *essential* functions determining the motion of S.

© Springer International Publishing AG, part of Springer Nature 2018   293
A. Romano and A. Marasco, *Classical Mechanics with Mathematica®*,
Modeling and Simulation in Science, Engineering and Technology,
https://doi.org/10.1007/978-3-319-77595-1_17

These equations, called *Lagrange's equations*, reduce any dynamical problem to the integration of a system of ordinary differential equations. This reduction of mechanics to analysis, of which Lagrange was very proud, justifies the name *analytical mechanics* of this branch of dynamics. It must again be noticed that the effective integration of Lagrange's equation is impossible in almost all cases, so that we are compelled to resort to other strategies to obtain meaningful information about unknown solutions. One of the most important of these approaches is given by Noether's theorem, relating symmetries of mechanical systems to conservation laws.

Another important aspect of the Lagrangian formulation of dynamics is represented by the fact that Lagrange's equations coincide with the equations defining the extremals of a functional. This remark suggests an integral formulation of motion (Hamilton's principle) opening new perspectives on the development of dynamics.

Hamilton proposed a set of equations that, though equivalent to the Lagrange equations, are more convenient in describing the behavior of dynamical systems. In this new formulation of dynamics, it is possible to prove very important theorems, shedding a new light on the behavior of dynamical systems. Further, the Hamiltonian formalism has been applied to statistical mechanics and quantum mechanics.

At the beginning of the last century, a geometric formulation of dynamics made it possible to prove many meaningful theorems. Finally, using this new approach we can formulate the laws of dynamics in an intrinsic way, i.e., independently of Lagrangian coordinates.

17.2 Dynamics of Constrained Systems

In this section, we formulate the fundamental problem of dynamics of a system S of N rigid bodies $\mathcal{B}_1, \ldots, \mathcal{B}_N$, subject to constraints and acted upon by active forces. We can apply the balance laws (15.48), (15.49) of linear momentum and angular momentum to each body of S:

$$m\ddot{\mathbf{r}}_{G_i} = \mathbf{R}_i^{(a)} + \mathbf{R}_i^{(r)}, \tag{17.1}$$

$$\dot{\mathbf{K}}_{G_i} = \mathbf{M}_{G_i}^{(a)} + \mathbf{M}_{G_i}^{(r)}, \tag{17.2}$$

where $i = 1, \ldots, N$. Choosing a central frame of inertia $G_i x_1' x_2' x_3'$ as a body frame for the body $\mathcal{B}_i$, we can state that the motion of any body is known when we determine the three Euler angles $\psi_i, \varphi_i, \theta_i$ that $G_i x_1' x_2' x_3'$ forms with the axes of the lab frame as well as the three coordinates $x_{G_i 1}, x_{G_i 2}, x_{G_i 3}$ of the center of mass of $\mathcal{B}_i$ relative to the lab frame. Since the six components of the vectors $\mathbf{R}_i^{(r)}$ and $\mathbf{M}_{G_i}^{(r)}$ are unknowns, the $6N$ differential equations, obtained by projecting (17.1) and (17.2) along the axes of the lab frame, contain $12N$ unknowns. In other words, the model we are proposing seems unable to describe the motion of S. However, we must not give up hope since we have not yet considered the conditions due to the presence of the constraints or the restrictions on the vectors $\mathbf{R}_i^{(r)}$ and $\mathbf{M}_{G_i}^{(r)}$, deriving from the hypothesis that the constraints are smooth. For instance, it is evident that the constraints, reducing the mobility of the system S, lead to some relations among the $6N$ parameters necessary to determine the configuration of S. On the other hand, we have not yet taken into account that the reactive forces are orthogonal to the contact surfaces between bodies and constraints.

D'Alembert and Lagrange proved that system (17.1), (17.2), together with the restrictions on the mobility of S and the characteristics of reactive forces due to smooth constraints, leads to a system of equations that completely determine the motion of S. In the following sections, we present this approach and analyze some important consequences of it.

17.3 Configuration Space

Let S be a system of N rigid bodies $\mathcal{B}_1, \ldots, \mathcal{B}_N$. We define a **constraint** as any restriction a priori imposed on the mobility of S, that is, both on the positions of the bodies of S and their velocity fields. To describe a constraint mathematically, we introduce the system S_L obtained by eliminating the constraints of S. Since a configuration of S_L is determined by assigning the $6N$ numbers $\boldsymbol{\lambda} = (\lambda^1, \ldots, \lambda^{6N})$ given by $x_{G_i1}, x_{G_i2}, x_{G_i3}, \psi_i, \varphi_i, \theta_i, i = 1, \ldots, N$, we can state that the configurations of S_L are in a one-to-one correspondence with the points of the manifold $V_{6N} = \mathfrak{R}^{3N} \times O(3)^N$, where $O(3)$ is the space of the orthogonal matrices (Sect. 12.4).[1] The introduction of this $6N$-dimensional manifold supplies a geometric representation of the motion of S. In fact, while S is moving in the three-dimensional Euclidean space E_3, the point $\boldsymbol{\lambda}$ describes a curve γ in V_{6N}, which is called the **dynamical trajectory** of S. In the absence of constraints, this curve can invade any region of V_{6N} and its tangent vector $\mathbf{t}$, with components $\dot{x}_{G_i1}, \dot{x}_{G_i2}, \dot{x}_{G_i3}, \dot{\psi}_i, \dot{\varphi}_i, \dot{\theta}_i, i = 1, \ldots, N$, can have any direction for the arbitrariness of the velocity field of S.

Adopting this geometric interpretation, we say that S is subject to **position constraints independent of time** or to **holonomic constraints independent of time**, if the dynamical trajectory γ, corresponding to any possible motion of S, lies on a submanifold V_n of V_{6N}, where $n < 6N$. The manifold V_n is called the **configuration space** of S, and any set $(q^1, \ldots, q^n)$ of local coordinates on V_n is called a set of **Lagrangian coordinates** for S. Finally, the circumstance that V_n is n-dimensional is also expressed by saying that the system S has n **degrees of freedom**. Recalling the contents of Chap. 6, we can say that the immersion map $i : V_n \to V_{6N}$ can be expressed locally in implicit form by a system

---

[1]More precisely, V_{6N} is an open $6N$-dimensional submanifold of $\mathfrak{R}^{3N} \times O(3)^N$ since we must exclude those values of $x_{G_i1}, x_{G_i2}, x_{G_i3}$, and $\psi_i, \varphi_i, \theta_i, i = 1, \ldots, N$, for which parts of two or more bodies of S occupy the same region of the three-dimensional space.

$$f_1(\boldsymbol{\lambda}) = 0,$$
$$\cdots \tag{17.3}$$
$$f_m(\boldsymbol{\lambda}) = 0,$$

where the functions $f_1, \ldots, f_m$ are of class C^1 and satisfy the condition

$$\text{rank}\left(\frac{\partial f_i}{\partial \lambda_j}\right) = m$$

in an open region Ω of V_{6N}. In fact, under these hypotheses, system (17.3) defines in Ω a submanifold of class C^1 and dimension $n = 6N - m$. Equivalently, if $q^1, \ldots, q^n$ are arbitrary local coordinates on V_n, i.e., arbitrary Lagrangian coordinates of S, this manifold can be represented in the parametric form

$$\boldsymbol{\lambda} = \boldsymbol{\lambda}(q^1, \ldots, q^n). \tag{17.4}$$

We conclude this section by exhibiting the configuration spaces of some dynamical systems with time-independent holonomic constraints.

- Let S be a system formed by a single material point P constrained to remain on a curve γ. Then the system S_L reduces to P freely moving in the three-dimensional space $\mathcal{E}_3$ and V_n coincides with γ. The immersion map $i : \gamma \to \mathcal{E}_3$ can be written in implicit form

$$f_1(x_1, x_2, x_3) = 0,$$
$$f_2(x_1, x_2, x_3) = 0,$$

  or in parametric form

$$\mathbf{r} = \mathbf{r}(s),$$

  where x_1, x_2, and x_3 are the Cartesian coordinates of P, $\mathbf{r} = (x_1, x_2, x_3)$ the position vector of P, $f_1 = 0$ and $f_2 = 0$ the implicit equations of two surfaces intersecting each other along γ, and s the curvilinear abscissa along γ. A Lagrangian coordinate is s, and S has one degree of freedom.
- Let S be a rigid body with a fixed axis a. The manifold V_L associated with S_L is $\mathfrak{R}^3 \times O(3)$. The configuration space V_1 is a circumference C. In fact, with the notations of Fig. 16.2, the configurations of S are given assigning the angle φ between the fixed plane Ox_1x_3 and the body plane $Ox_1'x_3'$. In turn, this angle defines a point belonging to C. φ is a Lagrangian coordinate of S that has one degree of freedom.
- When S is a rigid body with a fixed point, we have $V_L = \mathfrak{R}^3 \times O(3)$ and $V_n = O(3)$. Euler's angles are Lagrangian coordinates, and S has three degrees of freedom.
- Let S be a double pendulum. Then the configuration space is the torus $V_n = S^1 \times S^1$.

- The bar AB of Fig. 17.1 is contained in a plane Oxy, has point A constrained to move on the straight line Ox, and can rotate about an axis containing A and orthogonal to Oxy. The configuration space is a cylinder $\Re \times S^1$, and x and φ are Lagrangian coordinates. The system has two degrees of freedom.

Other examples can be found in the exercises at the end of the chapter.

17.4 Virtual Velocities and Nonholonomic Constraints

The existence of a one-to-one map between the configurations of a system S of N rigid bodies $\mathcal{B}_1, \ldots, \mathcal{B}_N$ and the points of the configuration space V_n is equivalent to stating that the position vector $\mathbf{r}$ of *any* point P of $\mathcal{B}_i$, $i = 1, \ldots, N$, is determined by giving the values of the Lagrangian coordinates $q^1, \ldots, q^n$. In other words, there *exists* a function

$$\mathbf{r} = \mathbf{r}(\mathbf{r}', q^1, \ldots, q^n), \tag{17.5}$$

where $\mathbf{r}'$ is the *constant* position vector of $P \in \mathcal{B}_i$ in the body frame of $\mathcal{B}_i$.

For instance, in the example of Fig. 17.1, the position vector $\mathbf{r}$ can be written as follows:

$$\mathbf{r} = x\mathbf{e}_1 + |\mathbf{r}'|(\cos \varphi \mathbf{e}_1 + \sin \varphi \mathbf{e}_2) \equiv \mathbf{r}(\mathbf{r}', x, \varphi),$$

where $\mathbf{e}_1$ and $\mathbf{e}_2$ are the unit vectors along the Ox- and Oy-axes, respectively. It could be very difficult to determine the explicit form of the function (17.5). In any case, *we are interested not in its form but in its existence.*

The sequence of configurations that S assumes during its effective motion is represented by the dynamical curve γ of the configuration space V_n. If $q^1(t), \ldots, q^n(t)$ are the parametric equations of this curve, then the spatial trajectory of the point $\mathbf{r}' \in S$ is given by

$$\mathbf{r}(t) = \mathbf{r}(\mathbf{r}', q^1(t), \ldots, q^n(t)). \tag{17.6}$$

Fig. 17.1 Constrained bar

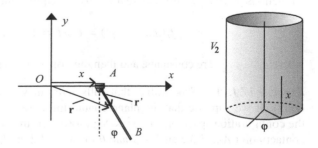

We recall that the position vector $\mathbf{r}'$ is constant since it is evaluated in a body frame. Therefore, differentiating (17.6) with respect to time, we obtain the velocity of the point $\mathbf{r}'$ in terms of the Lagrangian coordinates and the *Lagrangian velocities* $\dot{q}^1, \ldots, \dot{q}^n$

$$\dot{\mathbf{r}} = \frac{\partial \mathbf{r}}{\partial q^h} \dot{q}^h. \tag{17.7}$$

We remark that $(\dot{q}^1, \ldots, \dot{q}^n)$ is tangent to the dynamical trajectory γ, whereas the velocity vector $\dot{\mathbf{r}}$ is tangent to the trajectory of the point $\mathbf{r}'$ in three-dimensional space. These vectors are related to each other as it is shown by (17.7).

We define *virtual velocity* $\mathbf{v}$ of $P \in S$ at the instant t as any *possible velocity compatible with the constraints of S at the instant t*. It is evident that a virtual velocity is obtained by (17.7) by substituting the tangent vector $(\dot{q}^1, \ldots, \dot{q}^n)$ to the dynamical trajectory γ of V_n with an *arbitrary* vector $\mathbf{w} = (w^1, \ldots, w^n)$ tangent to V_n

$$\mathbf{v} = \frac{\partial \mathbf{r}}{\partial q^h} w^h. \tag{17.8}$$

We say that the system S, in addition to the holonomic constraints, is subject to **nonholonomic constraints** if, at any instant t, its velocity field is a priori subject to some restrictions. To give a mathematical definition of nonholonomic constraints, we consider the tangent space $T_{\mathbf{q}}(V_n)$ at the point $\mathbf{q} \equiv (q^1, \ldots, q^n)$ of V_n. In the absence of nonholonomic constraints, by (17.8), to any vector $\mathbf{w} = (w^1, \ldots, w^n) \in T_{\mathbf{q}}(V_n)$ corresponds a possible velocity $\mathbf{v}$ at the instant t. We say that S is subject to nonholonomic constraints if the possible Lagrangian velocities $(w^1, \ldots, w^n)$ belong to an r-dimensional subspace $U_{\mathbf{q}}(V_n)$ of $T_{\mathbf{q}}(V_n)$, where $r < n$. We define $U_{\mathbf{q}}(V_n)$ at any point $\mathbf{q} \in V_n$ by r equations

$$\begin{aligned} \omega_i(\mathbf{w}) &= a_{ij}(q^1, \ldots, q^n)\mathbf{d}q^j(\mathbf{w}) \\ &= a_{ij}(q^1, \ldots, q^n)w^j = 0, \end{aligned} \tag{17.9}$$

where $\omega_i = a_{ij}(q^1, \ldots, q^n)\mathbf{d}q^j$, $i = 1, \ldots, r$, are r differential forms on V_n. The differential forms ω_i must be nonintegrable. In fact, if they are integrable, there exist r functions $f_i(q^1, \ldots, q^n)$ such that (17.9) are equivalent to the system

$$f_i(q^1, \ldots, q^n) = c_i, \quad i = 1, \ldots, r,$$

where $c_1, \ldots, c_r$ are constants, and then the constraints (17.9) are holonomic.

Example 17.1. Let S be a disk that rolls without slipping on a horizontal plane Ox_1x_2. If we suppose that the plane containing S remains vertical during motion, the configuration space of S is $\Re^2 \times S^2$; moreover, the coordinates x_1 and x_2 of the contact point A and the angles φ and θ in Fig. 17.2 are Lagrangian coordinates for

Fig. 17.2 Disk rolling on a smooth surface

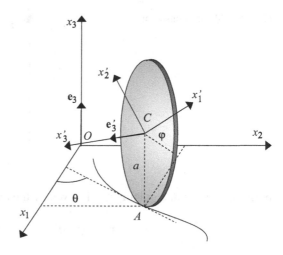

S. Finally, the angular velocity ω of S about a vertical axis containing C is

$$\omega = \dot\varphi \mathbf{e}_3' + \dot\theta \mathbf{e}_3, \tag{17.10}$$

and the condition of rolling without slipping becomes

$$\dot{\mathbf{r}}_A = \dot{\mathbf{r}}_C + \omega \times \overrightarrow{CA} = \mathbf{0}.$$

Noting that $\dot{\mathbf{r}}_C = \dot{x}_1 \mathbf{e}_1 + \dot{x}_2 \mathbf{e}_2$, $\overrightarrow{CA} = -a\mathbf{e}_3$, $\mathbf{e}_3' = \sin\theta \mathbf{e}_1 - \cos\theta \mathbf{e}_2$, and taking into account (17.10), the condition of rolling without slipping leads to the system

$$\dot{x}_1 - a\dot\varphi \cos\theta = 0,$$
$$\dot{x}_2 + a\dot\varphi \sin\theta = 0,$$

which can also be written in terms of differential forms:

$$dx_1 - a\cos\theta d\varphi = 0,$$
$$dx_2 + a\sin\theta d\varphi = 0.$$

It is evident that these differential forms are nonintegrable, so that they define a nonholonomic constraint.

17.5 Configuration Space–Time

We say that the system S of N rigid bodies $\mathcal{B}_1, \ldots, \mathcal{B}_N, i = 1, \ldots, N$, is subject to *constraints depending on time* if there exists a submanifold V_n of V_{6N} such that the

Fig. 17.3 Moving constraint

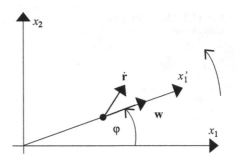

position vector $\mathbf{r}$ of a point $\mathbf{r}' \in S$ is given by the relation [see (17.5)]

$$\mathbf{r} = \mathbf{r}(\mathbf{r}', q^1, \ldots, q^n, t). \tag{17.11}$$

The manifold $V_n \times \Re$ is called the **configuration space–time**. In these conditions, the spatial trajectory of $\mathbf{r}'$ is given by

$$\mathbf{r}(t) = \mathbf{r}\left(\mathbf{r}', q^1(t), \ldots, q^n(t), t\right), \tag{17.12}$$

and its velocity yields

$$\dot{\mathbf{r}} = \frac{\partial \mathbf{r}}{\partial q^h}\dot{q}^h + \frac{\partial \mathbf{r}}{\partial t}. \tag{17.13}$$

For holonomic constraints depending on time, we define **virtual velocity** at the instant t as a possible velocity that is compatible with the constraints at that instant. Consequently, the virtual velocity is still given by (17.8).

Example 17.2. Let S be a material point constrained to move on the axis Ox_1'. Suppose that this axis rotates about a fixed axis a, orthogonal at O to Ox_1', with known angular velocity $\dot{\varphi}$ (Fig. 17.3). $\dot{\mathbf{r}}$ and $\mathbf{w}$ denote the effective velocity and a possible virtual velocity, respectively.

Example 17.3. Let S be a rotating simple pendulum (Sect. 14.10). In this case, V_{6N} is the three-dimensional Euclidean space and V_1 the rotating circumference on which the pendulum is constrained to remain.

17.6   Simple Example

Before introducing the main ideas of the new approach due to D'Alembert and Lagrange, we start with a simple example that contains all the ingredients of this theory.

Let S be a material point constrained to move on a smooth surface Σ having parametric equations

$$\mathbf{r} = \mathbf{r}(q^1, q^2), \tag{17.14}$$

where q^1 and q^2 are surface parameters. It is evident that the configuration space of S is Σ and q^1 and q^2 are possible Lagrangian coordinates. If we denote by $\boldsymbol{\Phi}$ the reactive force, by $\mathbf{F}(\mathbf{r}, \dot{\mathbf{r}})$ the active force acting on S, and by m the mass of S, then the motion of S on Σ is governed by the fundamental equation of dynamics

$$m\ddot{\mathbf{r}} = \mathbf{F}(\mathbf{r}, \dot{\mathbf{r}}) + \boldsymbol{\Phi}. \tag{17.15}$$

Let $q^1(t)$ and $q^2(t)$ be the parametric equations of the trajectory γ of S. Starting from the position occupied by S on γ at a given instant t, we consider an arbitrary virtual velocity $\mathbf{w}$ (Fig. 17.4). From (17.15), we obtain that the *virtual power* of the force of inertia $-m\ddot{\mathbf{r}}$ and active force vanishes,

$$(-m\ddot{\mathbf{r}} + \mathbf{F}) \cdot \mathbf{w} = 0, \quad \forall \mathbf{w} \in T_{\mathbf{r}}(\Sigma), \tag{17.16}$$

since the hypothesis that Σ is smooth implies $\boldsymbol{\Phi} \cdot \mathbf{w} = 0$. In (17.16) $T_{\mathbf{r}}(\Sigma)$ denotes the tangent space to Σ at the point $\mathbf{r}$ occupied by S at the considered instant. Introducing the expression (17.8) of the virtual velocity into (17.16), we obtain the condition

$$(-m\ddot{\mathbf{r}} + \mathbf{F}) \cdot \frac{\partial \mathbf{r}}{\partial q^h} w^h = 0, \quad \forall(w^1, w^2). \tag{17.17}$$

On the other hand, the vectors $\mathbf{e}_h = \partial \mathbf{r} / \partial q^h$ are tangent to the coordinate curves, and therefore $\ddot{r}_h = \ddot{\mathbf{r}} \cdot \mathbf{e}_h$ and $F_h = \mathbf{F} \cdot \mathbf{e}_h$ denote the (covariant) components of $\ddot{\mathbf{r}}$ and $\mathbf{F}$ along the vector $\mathbf{e}_h$. Finally, from (17.17), taking into account the arbitrariness of (w^1, w^2), we derive the following two scalar equations:

Fig. 17.4 Point moving on a smooth surface

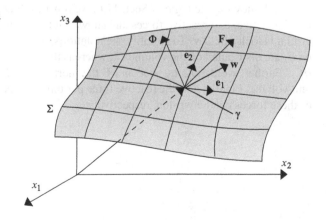

$$m\ddot{r}_h = F_h, \ h = 1, 2. \tag{17.18}$$

Since from (17.14) we obtain

$$\dot{\mathbf{r}} = \frac{\partial \mathbf{r}}{\partial q^h}\dot{q}^h, \tag{17.19}$$

$$\ddot{\mathbf{r}} = \frac{\partial \mathbf{r}}{\partial q^h}\ddot{q}^h + \frac{\partial^2 \mathbf{r}}{\partial q^h q^k}\dot{q}^h\dot{q}^k, \tag{17.20}$$

we can say that F_h are functions of q^h and $\dot{q}^h$, whereas $\ddot{r}_h$ depend on $\ddot{q}_h$, $\dot{q}^h$, and q_h. In other words, Eqs. (17.18) are a system of two second-order differential equations in the unknown functions $q^1(t)$ and $q^2(t)$ that completely determine the motion of S on the surface Σ.

We can summarize the preceding results as follows.

1. Equation (17.15) implies (17.17).
2. Equation (17.17) does not contain the unknown reactive force.
3. Equation (17.17) leads to equations that contain the unknowns determining the motion of S.

In the following sections we prove that the foregoing procedure can be extended to an arbitrary system S of N constrained rigid bodies.

17.7   Principle of Virtual Power

Let S be a system of N constrained rigid bodies $\mathcal{B}_1, \ldots, \mathcal{B}_N$, and denote by G_i the center of mass of $\mathcal{B}_i$. Finally, let C_i be the fixed region occupied by $\mathcal{B}_i$ in its body frame $G_i x_1' x_2' x_3'$. The motion of S is governed by the balance equations of linear momentum (17.1) and the balance equations of angular momentum (17.2), which in this more general situation replace (17.15) in the example considered in the previous section. Following the logic of Sect. 17.6, we must evaluate the virtual power of the inertial, active, and reactive forces and show that the virtual power vanishes for any virtual velocity field $\mathbf{v}$. To simplify the notations, we suppose that the active forces on $\mathcal{B}_i$ are distributed on the volume C_i, whereas the reactive forces act on the surface ∂C_i. Further, we denote by $-\rho\ddot{\mathbf{r}}$ and $\mathbf{F}$ the inertia forces and the active forces per unit volume. Finally, $\mathbf{\Phi}$ is the reactive force per unit surface. Then the virtual powers of these forces can be written, respectively, as

$$P^{(m)} = -\sum_{i=1}^{N} \int_{C_i} \rho \ddot{\mathbf{r}} \cdot \mathbf{v} \, dc, \qquad (17.21)$$

$$P^{(a)} = \sum_{i=1}^{N} \int_{C_i} \mathbf{F} \cdot \mathbf{v} \, dc, \qquad (17.22)$$

$$P^{(r)} = \sum_{i=1}^{N} \int_{\partial C_i} \Phi \cdot \mathbf{v} \, dc. \qquad (17.23)$$

We now prove the following theorem.

Theorem 17.1 (D'Alembert, Lagrange). *The equations of balance (17.1) and (17.2) imply the condition*

$$P^{(m)} + P^{(a)} + P^{(r)} = 0, \ \forall \mathbf{v}, \qquad (17.24)$$

where $\mathbf{v}$ *is an arbitrary virtual velocity field, i.e., compatible with the constraints to which the system is subject at the instant t.*

Proof. The virtual velocity field $\mathbf{v}$ is compatible with the constraints. In particular, it must be rigid for each body $\mathcal{B}_i$ of S. Consequently, the restriction of the velocity field $\mathbf{v}$ to the region C_i has the form

$$\mathbf{v}_i = \mathbf{v}_{G_i} + \boldsymbol{\omega}_i \times \overrightarrow{G_i P}, \qquad (17.25)$$

where $\boldsymbol{\omega}_i$ is the virtual angular velocity of $\mathcal{B}_i$. Introducing (17.25) into (17.22), we obtain the equation

$$P^{(a)} = \sum_{i=1}^{N} \mathbf{v}_{G_i} \cdot \int_{C_i} \mathbf{F} \, dc + \sum_{i=1}^{N} \boldsymbol{\omega}_i \cdot \int_{C_i} \overrightarrow{G_i P} \times \mathbf{F} \, dc$$

which can also be written in the form

$$P^{(a)} = \sum_{i=1}^{N} \mathbf{v}_{G_i} \cdot \mathbf{R}_i^{(a)} + \sum_{i=1}^{N} \boldsymbol{\omega}_i \cdot \mathbf{M}_{G_i}^{(a)}. \qquad (17.26)$$

Similarly, we have

$$P^{(r)} = \sum_{i=1}^{N} \mathbf{v}_{G_i} \cdot \mathbf{R}_i^{(r)} + \sum_{i=1}^{N} \boldsymbol{\omega}_i \cdot \mathbf{M}_{G_i}^{(r)} \qquad (17.27)$$

and

$$P^{(m)} = -\sum_{i=1}^{N} \mathbf{v}_{G_i} \cdot \int_{C_i} \rho \ddot{\mathbf{r}} \, dc - \sum_{i=1}^{N} \boldsymbol{\omega}_i \cdot \int_{C_i} \rho \overrightarrow{G_i P} \times \ddot{\mathbf{r}} \, dc.$$

This last relation assumes the final form

$$P^{(m)} = -\sum_{i=1}^{N} \mathbf{v}_{G_i} \cdot \dot{\mathbf{Q}}_i - \sum_{i=1}^{N} \boldsymbol{\omega}_i \cdot \dot{\mathbf{K}}_{G_i} \tag{17.28}$$

since

$$\int_{C_i} \rho \ddot{\mathbf{r}} \, dc = \frac{d}{dt} \int_{C_i} \rho \dot{\mathbf{r}} \, dc = \dot{\mathbf{Q}}_i,$$

$$\int_{C_i} \rho \overrightarrow{G_i P} \times \ddot{\mathbf{r}} \, dc = \frac{d}{dt} \int_{C_i} \rho \overrightarrow{G_i P} \times \dot{\mathbf{r}} \, dc - \dot{\mathbf{r}}_{G_i} \times \mathbf{Q}_i = \dot{\mathbf{K}}_{G_i},$$

and $\dot{\mathbf{r}}_{G_i} \times \mathbf{Q}_i = \dot{\mathbf{r}}_{G_i} \times m \dot{\mathbf{r}}_{G_i} = \mathbf{0}$. Collecting (17.26)–(17.28), we obtain

$$P^{(m)} + P^{(a)} + P^{(r)} = \sum_{i=1}^{N} \left[\mathbf{v}_{G-i} \cdot \left(-\dot{\mathbf{Q}}_i + \mathbf{R}_i^{(a)} + \mathbf{R}_i^{(r)} \right) \right.$$
$$\left. + \boldsymbol{\omega}_i \cdot \left(-\dot{\mathbf{K}}_{G_i} + \mathbf{M}_{G_i}^{(a)} + \mathbf{M}_{G_i}^{(r)} \right) \right],$$

and the theorem is proved.   □

By this theorem we can state that property 1, proved at the end of Sect. 17.6, holds for any constrained system S of rigid bodies. The following theorem, stated without a proof, extends property 2 to any system S.[2]

Theorem 17.2. *For any system S of rigid bodies subject to smooth constraints, we have*

$$P^{(r)} = 0 \tag{17.29}$$

for all virtual velocity fields.

Collecting the results (17.24) and (17.29), the following condition is satisfied:

$$P^{(m)} + P^{(a)} = 0 \tag{17.30}$$

for any virtual velocity field.

---

[2]In some textbooks (17.29) is given as a definition of smooth constraints.

17.8 Lagrange's Equations

In this section we prove that, for any system of constrained rigid bodies with n degrees of freedom, (17.30) leads to a system of n second-order differential equations in the unknown functions $q^1(t), \ldots, q^n(t)$.

In view of (17.21) and (17.22), we write (17.30) in the following form:

$$\sum_{i=1}^{N} \int_{C_i} (\mathbf{F}^{(a)} - \rho \ddot{\mathbf{r}}) \cdot \mathbf{w} \, dc = 0, \qquad (17.31)$$

which, resorting to (17.8) and (17.13), can also be written as

$$(Q_h(\mathbf{q}, \dot{\mathbf{q}}, t) - \tau_h(\mathbf{q}, \ddot{\mathbf{q}}, t)) w^h = 0, \quad \forall (w^1, \ldots, w^n) \in \mathfrak{R}^n, \qquad (17.32)$$

where

$$Q_h(\mathbf{q}, \dot{\mathbf{q}}, t) = \sum_{i=1}^{N} \int_{C_i} \mathbf{F}^{(a)} \cdot \frac{\partial \mathbf{r}}{\partial q^h} \, dc, \qquad (17.33)$$

$$\tau_h(\mathbf{q}, \dot{\mathbf{q}}, \ddot{\mathbf{q}}, t) = \sum_{i=1}^{N} \int_{C_i} \rho \ddot{\mathbf{r}} \cdot \frac{\partial \mathbf{r}}{\partial q^h} \, dc. \qquad (17.34)$$

For the arbitrariness of $(w^1, \ldots, w^n)$ in (17.32), the functions $q^1(t), \ldots, q^n(t)$ satisfy the following system of second-order differential equations:

$$\tau_h(\mathbf{q}, \dot{\mathbf{q}}, \ddot{\mathbf{q}}, t) = Q_h(\mathbf{q}, \dot{\mathbf{q}}, t), \quad h = 1, \ldots, n. \qquad (17.35)$$

To write (17.35) in a more expressive form, we introduce the kinetic energy of S

$$T = \sum_{i=1}^{N} \int_{C_i} \rho |\dot{\mathbf{r}}| \cdot |\dot{\mathbf{r}}| \, dc,$$

which, recalling the Lagrangian expression (17.13) of the velocity field, becomes

$$T = \frac{1}{2} a_{hk}(\mathbf{q}, t) \dot{q}^h \dot{q}^k + b_h(\mathbf{q}, t) \dot{q}^h + T_0(\mathbf{q}, t), \qquad (17.36)$$

where

$$a_{hk}(\mathbf{q}, t) = \sum_{i=1}^{N} \int_{C_i} \rho \frac{\partial \mathbf{r}}{\partial q^h} \cdot \frac{\partial \mathbf{r}}{\partial q^h} \, dc = a_{kh}(\mathbf{q}, t), \tag{17.37}$$

$$b_h(\mathbf{q}, t) = \sum_{i=1}^{N} \int_{C_i} \rho \frac{\partial \mathbf{r}}{\partial q^h} \cdot \frac{\partial \mathbf{r}}{\partial t} \, dc, \tag{17.38}$$

$$T_0 = \frac{1}{2} \sum_{i=1}^{N} \int_{C_i} \rho \left(\frac{\partial \mathbf{r}}{\partial t} \right)^2 dc. \tag{17.39}$$

Remark 17.1. When the constraints do not depend on time, we must use (17.5) instead of (17.12). Consequently, $\partial \mathbf{r}/\partial t = \mathbf{0}$, and the kinetic energy reduces to the quadratic form

$$T = \frac{1}{2} a_{hk}(\mathbf{q}) \dot{q}^h \dot{q}^k \equiv T_2. \tag{17.40}$$

The kinetic energy is a positive quantity if the velocity field $\dot{\mathbf{r}}$ does not vanish, and it is equal to zero if and only if $\dot{\mathbf{r}} = \mathbf{0}$. It is also possible to prove that $T = 0$ if and only if $\dot{q}^1 = \cdots \dot{q}^n = 0$. In other words, (17.40) is a positive definite quadratic form.

Remark 17.2. For holonomic constraints depending on time the quadratic form

$$T_2 = \frac{1}{2} a_{hk}(\mathbf{q}, t) \dot{q}^h \dot{q}^k \tag{17.41}$$

is still positive definite. In fact, (17.41) can be written as follows:

$$T_2 = \frac{1}{2} \sum_{i=1}^{N} \int_{C_i} \rho \left(\frac{\partial \mathbf{r}}{\partial q^h} \dot{q}^h \right)^2 dc \geq 0, \tag{17.42}$$

and it is equal to zero if and only if

$$\frac{\partial \mathbf{r}}{\partial q^h} \dot{q}^h = 0. \tag{17.43}$$

On the other hand, instead of (17.4) we have $\boldsymbol{\lambda} = \boldsymbol{\lambda}(q^1, \cdots q^n, t)$ and the matrix

$$\left(\frac{\partial \boldsymbol{\lambda}}{\partial q^h} \right) \tag{17.44}$$

of the homogeneous system (17.43) has rank n. Then, (17.43) is satisfied if and only if $\dot{q}^1 = \cdots \dot{q}^n = 0$.

To give (17.35) a more convenient form, we start by noting that from (17.13) there follows

$$\frac{\partial \dot{\mathbf{r}}}{\partial \dot{q}^h} = \frac{\partial \mathbf{r}}{\partial q^h}, \tag{17.45}$$

$$\frac{\partial \dot{\mathbf{r}}}{\partial q^h} = \frac{\partial}{\partial q^h}\left(\frac{\partial \mathbf{r}}{\partial q^h}\right)\dot{q}^h + \frac{\partial}{\partial t}\left(\frac{\partial \mathbf{r}}{\partial q^h}\right) = \frac{d}{dt}\frac{\partial \mathbf{r}}{\partial q^h}, \tag{17.46}$$

so that also

$$\ddot{\mathbf{r}} \cdot \frac{\partial \mathbf{r}}{\partial q^h} = \frac{d}{dt}\left(\dot{\mathbf{r}} \cdot \frac{\partial \mathbf{r}}{\partial q^h}\right) - \dot{\mathbf{r}} \cdot \frac{d}{dt}\frac{\partial \mathbf{r}}{\partial q^h}$$

$$= \frac{d}{dt}\left(\dot{\mathbf{r}} \cdot \frac{\partial \dot{\mathbf{r}}}{\partial \dot{q}^h}\right) - \dot{\mathbf{r}} \cdot \frac{\partial \dot{\mathbf{r}}}{\partial q^h}.$$

The preceding result implies that

$$T_h = \sum_{i=1}^{N} \int_{C_i} \rho \ddot{\mathbf{r}} \cdot \frac{\partial \mathbf{r}}{\partial q^h}\, dc$$

$$= \sum_{i=1}^{N} \frac{d}{dt} \int_{C_i} \rho \dot{\mathbf{r}} \cdot \frac{\partial \dot{\mathbf{r}}}{\partial \dot{q}^h} - \sum_{i=1}^{N} \int_{C_i} \rho \dot{\mathbf{r}} \cdot \frac{\partial \dot{\mathbf{r}}}{\partial q^h}\, dc$$

and we obtain the following expression for the quantities T_h:

$$T_h = \frac{d}{dt}\frac{\partial T}{\partial \dot{q}^h} - \frac{\partial T}{\partial q^h}. \tag{17.47}$$

From (17.35) and (17.47) we finally deduce the **Lagrange equations**:

$$\frac{d}{dt}\frac{\partial T}{\partial \dot{q}^h} - \frac{\partial T}{\partial q^h} = Q_h(\mathbf{q}, \dot{\mathbf{q}}, t), \quad h = 1, \ldots, n. \tag{17.48}$$

The quantities $Q_h(\mathbf{q}, \dot{\mathbf{q}}, t)$ are called **Lagrangian components of the active forces**. Taking into account expression (17.36) of the kinetic energy, we can write (17.48) as follows:

$$a_{hk}\ddot{q}^k = Q_h(\mathbf{q}, \dot{\mathbf{q}}, t) - C_h(\mathbf{q}, \dot{\mathbf{q}}, t) \equiv K_h(\mathbf{q}, \dot{\mathbf{q}}, t),$$

where the functions $C_h(\mathbf{q}, \dot{\mathbf{q}}, t)$ are given. On the other hand, the positive definite character of the quadratic form $a_{hk}\dot{q}^h\dot{q}^k$ implies that, at any configuration $\mathbf{q}$, all the principal minors of the matrix $(a_{hk}(\mathbf{q},t))$ have positive determinants; in particular, $\det(a_{hk}) \neq 0$. Consequently, the preceding system can be solved with respect to the quantities $\ddot{q}^h$, that is, it can be written in the normal form

$$\ddot{q}^h = g^h(\mathbf{q}, \dot{\mathbf{q}}, t). \tag{17.49}$$

This result allows us to state that, under suitable regularity hypotheses on the functions $T(\mathbf{q}, \dot{\mathbf{q}}, t)$ and $Q_h(\mathbf{q}, \dot{\mathbf{q}}, t)$, the system of Lagrange equations has one and only one solution satisfying the initial data

$$q^h(0) = q_0^h, \quad \dot{q}^h(0) = \dot{q}_0^h \tag{17.50}$$

that determine the initial position and velocity field of S.

17.9 Conservative Forces

We say that the *positional* forces acting on the system S of rigid bodies, with constraints depending on time, are ***conservative*** if there exists a C^1 function $U(\mathbf{q}, t)$, called ***potential energy***, such that

$$Q_h(\mathbf{q}, t) = -\frac{\partial U}{\partial q^h}. \tag{17.51}$$

Note that in (17.51) the time t appears as a parameter. If the active forces are conservative, then $\partial U/\partial \dot{q}^h = 0$. Consequently, introducing the ***Lagrange function***

$$L(\mathbf{q}, \dot{\mathbf{q}}, t) = T(\mathbf{q}, \dot{\mathbf{q}}, t) - U(\mathbf{q}, t), \tag{17.52}$$

the Lagrange equations (17.48) assume the following form:

$$\frac{\mathrm{d}}{\mathrm{d}t}\frac{\partial L}{\partial \dot{q}^h} - \frac{\partial L}{\partial q^h} = 0, \quad h = 1, \dots, n. \tag{17.53}$$

These equations show the fundamental importance of a Lagrangian description of dynamical phenomena. In fact, they prove that *the whole dynamical behavior of a mechanical system subject to smooth constraints and conservative forces is contained in a Lagrangian function*.

The following theorem is very important in many applications of (17.53).

Theorem 17.3. *Let S be a system of N rigid bodies subject to smooth holonomic constraints. Further, denote by S_L the system obtained by eliminating all the constraints of S and by $\boldsymbol{\lambda} \equiv (\lambda^1, \dots, \lambda^L)$ the Lagrangian coordinates of S_L. If S_L is acted upon by conservative forces with potential energy $\mathcal{U}(\boldsymbol{\lambda}, t)$, then S is subjected to conservative forces with potential energy $U(\mathbf{q}, t) = \mathcal{U}(\boldsymbol{\lambda}(\mathbf{q}, t), t)$, where $\lambda^1 = \lambda^1(\mathbf{q}, t), \dots, \lambda^L = \lambda^L(\mathbf{q}, t)$ are the parametric equations of $V_n \times \mathfrak{R}$.*

Proof. For the sake of simplicity, we suppose that S is formed by a single constrained rigid body. Formula (17.11) holds for S, and we denote by $\mathbf{r} = \hat{\mathbf{r}}(\mathbf{r}', \lambda^1, \dots, \lambda^L, t)$

the corresponding equation for S_L. If S_L is acted upon by conservative forces, then there exists a potential energy $\mathcal{U}(\boldsymbol{\lambda}, t)$ such that the Lagrangian components $\mathcal{Q}_i$, $i = 1, \ldots, 6N$, of the active forces are given by [see (17.33)]

$$\mathcal{Q}_i = \int_{C_i} \mathbf{F}^{(a)} \cdot \frac{\partial \hat{\mathbf{r}}}{\partial \lambda_i} \, dc = -\frac{\partial \mathcal{U}}{\partial \lambda_i}.$$

On the other hand, for the system S, we have that

$$Q_h = \int_{C_i} \mathbf{F}^{(a)} \cdot \frac{\partial \mathbf{r}}{\partial q^h} \, dc = \int_{C_i} \mathbf{F}^{(a)} \cdot \frac{\partial \hat{\mathbf{r}}}{\partial \lambda^i} \frac{\partial \lambda_i}{\partial q^h} \, dc$$

$$= -\frac{\partial \mathcal{U}}{\partial \lambda_i} \frac{\partial \lambda^i}{\partial q^h} = -\frac{\partial U}{\partial q_h},$$

and the theorem is proved.   □

We say that the C^1 function $U(\mathbf{q}, \dot{\mathbf{q}}, t)$ is the **generalized potential energy** of the forces acting on S if

$$Q_h(\mathbf{q}, \dot{\mathbf{q}}, t) = \frac{d}{dt} \frac{\partial U}{\partial \dot{q}^h} - \frac{\partial U}{\partial q^h}. \tag{17.54}$$

For forces deriving from a generalized potential energy, the Lagrange equations still assume the form (17.52), with the Lagrangian function given by

$$L(\mathbf{q}, \dot{\mathbf{q}}, t) = T(\mathbf{q}, \dot{\mathbf{q}}, t) - U(\mathbf{q}, \dot{\mathbf{q}}, t). \tag{17.55}$$

To recognize the severe restrictions that (17.54) imposes on the function $U(\mathbf{q}, \dot{\mathbf{q}}, t)$, we start by noting that it can be put in the form

$$Q_h(\mathbf{q}, \dot{\mathbf{q}}, t) = \frac{\partial^2 U}{\partial \dot{q}^h \partial \dot{q}^k} \ddot{q}^k + u_h(\mathbf{q}, \dot{\mathbf{q}}, t),$$

where $u_h(\mathbf{q}, \dot{\mathbf{q}}, t)$ are suitable functions independent of $\ddot{q}^h$. Since the left-hand sides of the preceding equations do not depend on $\ddot{q}^h$, the function $U(\mathbf{q}, \dot{\mathbf{q}}, t)$ must depend linearly on $\dot{q}^h$, i.e.,

$$\frac{\partial^2 U}{\partial \dot{q}^h \partial \dot{q}^k} = 0,$$

and we can state that

$$U(\mathbf{q}, \dot{\mathbf{q}}, t) = \hat{U}(\mathbf{q}, t) + \Pi_h(\mathbf{q}, t)\dot{q}^h. \tag{17.56}$$

Introducing (17.56) into (17.54) we are led to the equation

$$Q_h(\mathbf{q}, \dot{\mathbf{q}}, t) = -\frac{\partial \hat{U}}{\partial q^h} + \frac{d\Pi_h}{dt} - \frac{\partial \Pi_k}{\partial q^h}\dot{q}^k,$$

which can also be written as follows:

$$Q_h(\mathbf{q}, \dot{\mathbf{q}}, t) = -\frac{\partial \hat{U}}{\partial q^h} + \left(\frac{\partial \Pi_h}{\partial q^k} - \frac{\partial \Pi_k}{\partial q^h}\right)\dot{q}^k + \frac{\partial \Pi_h}{\partial t}. \tag{17.57}$$

This formula shows that the forces having a generalized potential energy, apart from the term $\partial \Pi_h/\partial t$, are formed by conservative forces with a usual potential energy of $\hat{U}(\mathbf{q}, t)$ and by **gyroscopic forces** whose Lagrangian components are

$$Q_h^*(\mathbf{q}, \dot{\mathbf{q}}, t) = \left(\frac{\partial \Pi_h}{\partial q^k} - \frac{\partial \Pi_k}{\partial q^h}\right)\dot{q}^k \equiv \gamma_{hk}\dot{q}^k. \tag{17.58}$$

These forces have the remarkable property that their power vanishes

$$Q_h^*\dot{q}^h = \gamma_{hk}\dot{q}^h\dot{q}^k = 0 \tag{17.59}$$

since γ_{hk} is skew-symmetric and $\dot{q}^h\dot{q}^k$ is symmetric.

We conclude this section with two important examples of forces having a generalized potential.

(a) *Charge moving under the action of an electromagnetic field.*

Let P be a material point freely moving relative to the Cartesian frame $Ox^1x^2x^3$, and denote by $\mathbf{r}$ a point of this frame and by $\mathbf{v}$ the velocity of P relative to $Ox^1x^2x^3$. Since the Lagrangian coordinates coincide with the Cartesian coordinates and the Lagrangian components of the active force $\mathbf{F}$ acting on P are the Cartesian components of $\mathbf{F}$ relative to $Ox^1x^2x^3$, we can write (17.56) and (17.57) in vector form:

$$U(\mathbf{r}, \mathbf{v}, t) = \hat{U}(\mathbf{r}, t) + \mathbf{\Pi} \cdot \mathbf{v}, \tag{17.60}$$

$$\mathbf{F}(\mathbf{r}, \mathbf{v}, t) = -\nabla\hat{U} + \nabla \times \mathbf{\Pi} \times \mathbf{v} + \frac{\partial \mathbf{\Pi}}{\partial t}. \tag{17.61}$$

Further, the electromagnetic force acting on a moving charge is given by Lorentz's force

$$\mathbf{F} = q(\mathbf{E} + \mathbf{v} \times \mathbf{B}) = q(\mathbf{E} - \mathbf{B} \times \mathbf{v}), \tag{17.62}$$

where $\mathbf{E}$ is the electric field and $\mathbf{B}$ the magnetic induction. Owing to Maxwell's equations, these fields can be obtained by a scalar potential φ and a vector potential $\mathbf{A}$ by the formulae

$$E = -\nabla\varphi - \frac{\partial A}{\partial t},$$

$$B = \nabla \times A.$$

Taking into account these relations, Lorentz's force (17.62) can be written as follows:

$$F = -\nabla(q\varphi) + \nabla \times (-qA) \times v + \frac{\partial}{\partial t}(-qA). \qquad (17.63)$$

Comparing (17.63) with (17.61), we conclude that Lorentz's force has the following generalized potential energy:

$$U(r, v, t) = q\varphi - qA \cdot v. \qquad (17.64)$$

(b) *Fictitious forces*

Let $Ox_1x_2x_3$ be a frame of reference whose origin O is at rest with respect to an inertial frame $O\xi\eta\zeta$. If m denotes the mass of a material point P moving relative to $Ox_1x_2x_3$ with velocity v and r is the position vector of the point of $Oxyz$ occupied by P at the instant t, then the fictitious force acting on P is [see (12.18)]

$$\begin{aligned} F &= -m[\dot{\omega} \times r + \omega \times (\omega \times r) + 2\omega \times v] \\ &= -m[\dot{\omega} \times r + (\omega \otimes \omega - \omega^2 I) \cdot r + 2\omega \times v]. \end{aligned} \qquad (17.65)$$

Bearing in mind (17.61), it is possible to verify (see exercises 15 and 16) that the generalized potential energy of F is given by

$$U(r, v, t) = \hat{U}(r, t) + \Pi(r, t) \cdot v, \qquad (17.66)$$

where

$$\hat{U} = \frac{1}{2}mr \cdot (\omega \otimes \omega - \omega^2 I) \cdot r, \qquad (17.67)$$

$$\Pi = -m\omega \times r. \qquad (17.68)$$

17.10 First Integrals

Let S be a dynamical system with n degrees of freedom described by a Lagrangian function $L(q, \dot{q}, t)$, and denote by $V_n \times \mathfrak{R}$ the configuration space–time of S. A function $f : T(V_n) \times \mathfrak{R} \to \mathfrak{R}$ is a *first integral* of the Lagrange equations if it is constant along their solutions:

$$f(\mathbf{q}(t), \dot{\mathbf{q}}(t), t) = f(\mathbf{q}_0, \dot{\mathbf{q}}_0, 0), \tag{17.69}$$

where $\mathbf{q}(t)$ is a solution satisfying the initial data $q^h(0) = q_0^h, \dot{q}^h(0) = \dot{q}_0^h, i = 1, \ldots, n$.

In other words, a first integral supplies a conservation law that holds in any motion of S. From a mathematical point of view, it is a first-order differential equation that may substitute one of the Lagrange equations, simplifying their integration. Both these remarks show that searching for first integrals is a fundamental problem of Lagrangian dynamics.

We say that a Lagrangian coordinate q^h is *cyclic* or *ignorable* if it does not appear in the Lagrangian function L. If we introduce the *generalized momentum* conjugate to q^h

$$p_h = \frac{\partial L}{\partial \dot{q}^h}, \tag{17.70}$$

from Lagrange's equation we obtain that p_h is a first integral when L is independent of q^h:

$$p_h(\mathbf{q}, \dot{\mathbf{q}}, t) = p_h(\mathbf{q}_0, \dot{\mathbf{q}}_0, 0). \tag{17.71}$$

This result shows how important is the choice of a suitable system of Lagrangian coordinates in analyzing the motion of a dynamical system. Now we prove the existence of a fundamental first integral.

Theorem 17.4. *Let S be a system of rigid bodies with smooth and fixed holonomic constraints. If the active forces derive from a generalized potential independent of time [see (17.56)]*

$$U(\mathbf{q}, \dot{\mathbf{q}}) = \hat{U}(\mathbf{q}) + \Pi_h(\mathbf{q})\dot{q}^h,$$

then the total energy

$$E = T + \hat{U} \tag{17.72}$$

is constant during any motion of S.

Proof. Along any motion of S we have that

$$\frac{dL}{dt} = \frac{\partial L}{\partial q^h}\dot{q}^h + \frac{\partial L}{\partial \dot{q}^h}\ddot{q}^h$$

$$= \frac{\partial L}{\partial q^h}\dot{q}^h + \frac{d}{dt}\left(\frac{\partial L}{\partial \dot{q}^h}\dot{q}^h\right) - \frac{d}{dt}\left(\frac{\partial L}{\partial \dot{q}^h}\right)\dot{q}^h.$$

Taking into account the Lagrange equations, the preceding relation reduces to the condition

$$\frac{\mathrm{d}L}{\mathrm{d}t} = \frac{\mathrm{d}}{\mathrm{d}t}\left(\frac{\partial L}{\partial \dot{q}^h}\dot{q}^h\right),$$

which can be written equivalently as

$$\frac{\partial L}{\partial \dot{q}^h}\dot{q}^h - L = \text{const.} \tag{17.73}$$

Further, if the constraints do not depend on time, formula (17.40) of the kinetic energy reduces to the quadratic form T_2, and we have that

$$L = T - U = T_2 - \hat{U}(\mathbf{q}) - \Pi_h(\mathbf{q})\dot{q}^h.$$

In conclusion, (17.73) becomes

$$\frac{\partial L}{\partial \dot{q}^h}\dot{q}^h - L = a_{hk}\dot{q}^h\dot{q}^k - \Pi_h\dot{q}^h - T_2 + \hat{U} + \Pi_h\dot{q}^h = T + \hat{U},$$

and the theorem is proved. $\qquad\square$

Remark 17.3. The term $\Pi_h\dot{q}^h$ does not appear in the total energy E since the power of the gyroscopic forces vanishes.

17.11 Symmetries and First Integrals

In this section we prove a fundamental theorem that relates the first integrals of Lagrange's equations to the symmetries of a Lagrangian function.

As usual, we consider a dynamical system S with smooth, bilateral, and *fixed* holonomic constraints and acted upon by conservative forces. We denote by V_n the configuration space and by $T(V_n)$ its tangent fiber bundle, which is a $2n$-dimensional manifold called the **velocity space** of S. We recall that any point of $T(V_n)$ is a pair $(\mathbf{q}, \mathbf{v})$ of a point $\mathbf{q} \in V_n$ and a tangent vector $\mathbf{v}$ belonging to the tangent space $T_\mathbf{q}(V_n)$ at $\mathbf{q}$ (Chap. 6). The dynamics of S is completely described by a Lagrangian function L which can be regarded as a C^1 real-valued function on $T(V_n)$.

Let $\varphi_s(\mathbf{q}) : \Re \times V_n \to V_n$ be a one-parameter global transformation group of V_n, and let $\mathbf{X}$ be its infinitesimal generator (Chap. 7). The group $\varphi_s(\mathbf{q})$ is said to be a **symmetry group** of L if

$$L(\mathbf{q}, \mathbf{v}) = L(\varphi_s(\mathbf{q}), (\varphi_s)_*(\mathbf{q})\mathbf{v}), \tag{17.74}$$

$\forall(\mathbf{q}, \mathbf{v}) \in T(V_n), \forall s \in \Re.$

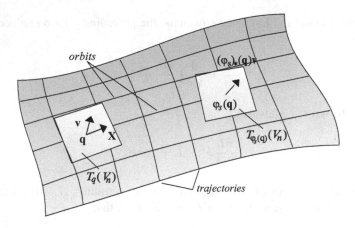

Fig. 17.5 Noether's theorem

In (17.74), $(\varphi_s)_*(\mathbf{q})$ is the differential at the point $\mathbf{q} \in V_n$ of the map $\varphi_s(\mathbf{q})$ and $(\varphi_s)_*(\mathbf{q})\mathbf{v}$ is the vector of $T_{\varphi_s(\mathbf{q})}$ corresponding to $\mathbf{v} \in T_{\mathbf{q}}(V_n)$ (Chap. 7 and Fig. 17.5). If we set $q_s''^h(\mathbf{q}) = (\varphi_s(\mathbf{q}))^h$, then the linear map $(\varphi_s)_*(\mathbf{q})$ is represented by the Jacobian matrix $(\partial q_s''^h / \partial q^k)$ and (17.74) becomes

$$L(q^h, v^h) = L\left(q_s''^h(\mathbf{q}), \frac{\partial q_s''^h}{\partial q^k} v^k\right). \tag{17.75}$$

Theorem 17.5 (E. Noether). *Let S be a system of rigid bodies with smooth, bilateral, fixed holonomic constraints and subject to conservative forces. If $\varphi_s(\mathbf{q}) : \Re \times V_n \to V_n$ is a symmetry group of the Lagrangian function $L : T(V_n) \to \Re$, then the function $f : T(V_n) \to \Re$ given by*

$$f(\mathbf{q}, \mathbf{v}) \equiv \frac{\partial L}{\partial v^h} X^h \tag{17.76}$$

is a first integral of the Lagrangian equations, that is,

$$f(\mathbf{q}(t), \dot{\mathbf{q}}(t)) = f(\mathbf{q}(0), \dot{\mathbf{q}}(0)),$$

along any solution of Lagrange's equations.

Proof. First, we evaluate (17.75) along a solution $q^h(t)$ of Lagrange's equations:

$$L(q^h(t), \dot{q}^h(t)) = L(q_s''^h(\mathbf{q}(t)), \frac{\partial q_s''^h}{\partial q^k} \dot{q}^k(t)). \tag{17.77}$$

Differentiating (17.77) with respect to s and evaluating the result at $s = 0$, we obtain

$$0 = \left(\frac{\partial L}{\partial q^h} X^h + \frac{\partial L}{\partial \dot{q}^h} \frac{\partial X^h}{\partial q^h} \dot{q}^h \right)_{(q(t),\dot{q}(t))}$$

$$= \left(\frac{\partial L}{\partial q^h} X^h + \frac{\partial L}{\partial \dot{q}^h} \frac{dX^h}{dt} \right)_{(q(t),\dot{q}(t))}$$

$$= \left[\left(\frac{\partial L}{\partial q^h} - \frac{d}{dt} \frac{\partial L}{\partial \dot{q}^h} \right) X^h + \frac{d}{dt} \left(\frac{\partial L}{\partial \dot{q}^h} X^h \right) \right]_{(q(t),\dot{q}(t))}.$$

The theorem is proved since the functions $q^h(t)$ satisfy Lagrange's equations. □

Example 17.4. Consider a dynamical system whose Lagrangian function L does not depend on the coordinate q^1. Then, the one-parameter group of transformations

$$q'^1 = q^1 + s, \ q'^2 = q^2, \ \ldots, q'^n = q^n$$

is a symmetry group of L. The infinitesimal generator of this group is the vector field $\mathbf{X} = (1, 0, \ldots, 0)$ tangent to the configuration space V_n. Consequently, the first integral associated with this symmetry group is

$$f(\mathbf{q}, \dot{\mathbf{q}}) = \frac{\partial L}{\partial \dot{q}^1} = p_1,$$

that is, the corresponding conjugate generalized momentum.

Example 17.5. Let P be a material point, constrained to move on a smooth plane π in which Cartesian coordinates (x, y) are introduced. Suppose that in these coordinates the Lagrangian function is

$$L = \frac{1}{2} m(\dot{x}^2 + \dot{y}^2) + (\alpha x + \beta y),$$

where α and β are two constants. The one-parameter group of transformations

$$x' = x + \beta s, \ y' = y - \alpha s$$

is a symmetry group of L whose infinitesimal generator is $\mathbf{X} = (\beta, -\alpha)$. Consequently, the corresponding first integral is

$$f = \frac{\partial L}{\partial \dot{x}} X^1 + \frac{\partial L}{\partial \dot{y}} X^2 = m(\beta \dot{x} - \alpha \dot{y}).$$

Example 17.6. Let S be a system of N material points $\mathbf{r}_1, \ldots, \mathbf{r}_N$ with masses $m_1, \ldots, m_N$. We suppose that, in cylindrical coordinates (r, θ, z), the forces acting on S are conservative with a potential energy $U(r_1, \ldots, r_N, z_1, \ldots, z_n)$. Then, $n = 3N$ and $V_n = \Re^{3N}$, and the Lagrangian function is given by

$$L = \frac{1}{2} \sum_{i=1}^{N} m_i \left(\dot{r}_i^2 + r_i^2 \dot{\theta}_i^2 + \dot{z}_i^2 \right) - U(r_1, \ldots, r_N, z_1, \ldots, z_n).$$

Since all the coordinates $\theta_1, \ldots, \theta_n$ are cyclic, the one-parameter group of transformations

$$\begin{aligned} r'_i &= r_i, \\ \theta'_i &= \theta_i + s, \quad i = 1, \ldots, N, \\ z'_i &= z_i, \end{aligned}$$

is a symmetry group of L. The infinitesimal generator of this group is $\mathbf{X} = (0, \ldots, 0, 1, \ldots, 1, 0, \ldots, 0)$. Consequently, the corresponding first integral of Lagrange's equations is

$$f = \sum_{i=1}^{N} m_i r_i^2 \dot{\theta}_i.$$

It is easy to verify that this first integral coincides with components K_z of the total angular momentum along the Oz-axis:

$$K_z = \sum_{i=1}^{N} m_i (x_i \dot{y}_i - y_i \dot{x}_i).$$

Remark 17.4. We underline that the preceding theorem states that to any symmetry group of L there corresponds a first integral of Lagrange's equations. It is possible to prove that also the inverse implication holds provided that we generalize the definition of symmetry group. We face this problem in the Hamiltonian formalism. It is also important to note that, although we can prove the equivalence between symmetry groups and first integrals by extending the definition of symmetry group, we are no more able, in general, to obtain the very simple expression (17.76) of the first integral corresponding to this more general symmetry group.

The extension of Noether's theorem to the case of a Lagrangian function depending on time can be obtained by including the time among the Lagrangian coordinates in the following way. Let $L(\mathbf{q}, \mathbf{v}, t) : T(V_n) \times \Re \to \Re$ be the Lagrangian function of a dynamical system S over the $(2n+1)$-manifold $T(V_n) \times \Re$. Consider the extension $\hat{L}$ of L to the $(2n+2)$ manifold $T(V_n \times \Re)$ such that

$$\hat{L}(\mathbf{q}, t, \mathbf{v}, u) = L\left(\mathbf{q}, t, \frac{\mathbf{v}}{u}\right) u. \tag{17.78}$$

The following identities can be easily proved:

$$\frac{\partial \hat{L}}{\partial q^h} = \frac{\partial L}{\partial q^h} u, \quad \frac{\partial \hat{L}}{\partial v^h} = \frac{\partial L}{\partial v^h}, \tag{17.79}$$

$$\frac{\partial \hat{L}}{\partial t} = \frac{\partial L}{\partial t} u, \quad \frac{\partial \hat{L}}{\partial u} = L - \frac{\partial L}{\partial v^h} \frac{v^h}{u}. \tag{17.80}$$

We now verify that if the functions $(q^h(t))$ are a solution of the Lagrangian equations relative to L, then $(q^h(t), t)$ are a solution of the Lagrangian equations relative to $\hat{L}$. In fact, in view of (17.79) and (17.80), regarding the first n equations relative to $\hat{L}$ we can say that

$$\frac{\mathrm{d}}{\mathrm{d}t} \frac{\partial \hat{L}}{\partial \dot{q}^h} - \frac{\partial \hat{L}}{\partial q^h} = \frac{\mathrm{d}}{\mathrm{d}t} \frac{\partial L}{\partial \dot{q}^h} - \frac{\partial L}{\partial q^h} = 0.$$

Further, when we note that $u = \dot{t} = 1$, the $(n+1)$th equation is

$$\frac{\mathrm{d}}{\mathrm{d}t} \left(\frac{\partial \hat{L}}{\partial \dot{q}^h} \right)_{u=1} - \frac{\partial \hat{L}}{\partial t} = \frac{\mathrm{d}}{\mathrm{d}t} \left(L - \frac{\partial L}{\partial \dot{q}^h} \dot{q}^h \right) - \frac{\partial L}{\partial t}$$

$$= \frac{\mathrm{d}L}{\mathrm{d}t} - \frac{\mathrm{d}}{\mathrm{d}t} \left(\frac{\partial L}{\partial \dot{q}^h} \dot{q}^h \right) - \frac{\partial L}{\partial t}$$

$$= \left(\frac{\partial L}{\partial \dot{q}^h} \ddot{q}^h + \frac{\partial L}{\partial q^h} \dot{q}^h \right) - \frac{\mathrm{d}}{\mathrm{d}t} \left(\frac{\partial L}{\partial \dot{q}^h} \dot{q}^h \right)$$

$$= - \left(\frac{\mathrm{d}}{\mathrm{d}t} \frac{\partial L}{\partial \dot{q}^h} - \frac{\partial L}{\partial q^h} \right) \dot{q}^h = 0$$

since the first n Lagrange equations are verified by the functions $(q^h(t))$.

We say that the one-parameter group of transformation $\varphi_s : \Re \times (V_n \times \Re) \to (V_n \times \Re)$ on the space–time of configurations $(V_n \times \Re)$ is a symmetry group of L if it is a symmetry group of $\hat{L}$, that is, if

$$\hat{L}(q^\alpha, v^\alpha) = \hat{L} \left(\varphi_s^\alpha(q^\beta), \frac{\partial \varphi_s^\alpha}{\partial v^\beta} v^\beta \right), \tag{17.81}$$

where $\alpha, \beta = 1, \ldots, n+1$, $(q^\alpha) = (\mathbf{q}, t)$, and $(v^\alpha) = (\mathbf{v}, u)$. Applying Noether's theorem to $\hat{L}$, we determine the following first integral of the Lagrange equations relative to $\hat{L}$:

$$f = \frac{\partial \hat{L}}{\partial v^\alpha} X^\alpha, \tag{17.82}$$

with $\mathbf{X} = (X^\alpha)$ the infinitesimal generator of $\varphi_s(q^\alpha)$. Taking into account (17.79) and (17.80), we can give (17.82) the alternative form

$$f = \frac{\partial L}{\partial \dot{q}^h} X^h + \left(L - \frac{\partial L}{\partial \dot{q}^h} \dot{q}^h\right) X^{n+1} \tag{17.83}$$

along the curve $(q^h(t), t)$.

In particular, if L does not depend on time, then a symmetry group of L is given by the following one-parameter group of transformations:

$$q'^h = q^h, \; t' = t + s.$$

Since the infinitesimal generator of this group is the vector field $\mathbf{X} = (0, \ldots, 0, 1)$, we obtain from (17.83) the conservation of the total mechanical energy

$$E = \frac{\partial L}{\partial \dot{q}^h} \dot{q}^h - L = const.$$

The reader can verify the results of this section by resorting to the notebook **Noether**.

17.12 Lagrange's Equations for Nonholonomic Constraints

Let S be a dynamical system with both holonomic and nonholonomic constraints, and let V_n be the configuration space of the system S when we take into account only the holonomic constraints. In Sect. 17.4, we assigned nonholonomic constraints by $r \leq n$ linear relations (17.9), which define, at any point $\mathbf{q} \in V_n$, a subspace $\mathcal{V}_{\mathbf{q}}$ of the tangent space $T_{\mathbf{q}}(V_n)$. The fundamental consequence of the presence of nonholonomic constraints is that in relation (17.32) the components $(w^1, \ldots, w^n)$ of the virtual velocity are not arbitrary since they must belong to $\mathcal{V}_{\mathbf{q}}$. In other words, the condition

$$(Q_h(\mathbf{q}, \dot{\mathbf{q}}, t) - \tau_h(\mathbf{q}, \dot{\mathbf{q}}, \ddot{\mathbf{q}}, t))w^h = 0 \tag{17.84}$$

must be verified for any $(w^1, \ldots, w^n) \in \mathcal{V}_{\mathbf{q}}$, that is, for all the vectors $(w^1, \ldots, w^n)$ confirming the r linear conditions

$$a_{ih}w^h = 0, \quad i = 1, \ldots, r \leq n, \tag{17.85}$$

expressing the nonholonomic constraints. Consequently, the coefficients of w^h in (17.84) do not vanish but are linear combinations of coefficients of (17.85):

$$Q_h(\mathbf{q}, \dot{\mathbf{q}}, t) - \tau_h(\mathbf{q}, \dot{\mathbf{q}}, \ddot{\mathbf{q}}, t) = \lambda_1 a_{1h} + \cdots + \lambda_r a_{rh}, \quad h = 1, \ldots, n. \tag{17.86}$$

The coefficients $\lambda_1, \ldots, \lambda_r$, which can depend on q^h, $\dot{q}^h$, $\ddot{q}^h$, and t, are called *Lagrange's multipliers*. In conclusion, the Lagrange equations, in the presence of nonholonomic constraints, are

$$\frac{d}{dt}\frac{\partial T}{\partial \dot{q}^h} - \frac{\partial T}{\partial q^h} = Q_h - \sum_{i=1}^{r} \lambda_i a_{ih}, \quad h = 1, \ldots, n. \tag{17.87}$$

These equations, together with the conditions due to the nonholonomic constraints,

$$a_{ih}\dot{q}^h = 0, \quad i = 1, \ldots, r, \tag{17.88}$$

define a system of $n+r$ differential equations in the $n+r$ unknowns $q^1(t), \ldots, q^n(t)$, $\lambda_1, \ldots, \lambda_r$.

Example 17.7. Consider the homogeneous disk S rolling without slipping on a plane already considered in Fig. 17.2. Since S is supposed to remain in a vertical plane, the weight of S has no effect on the motion. Therefore, the Lagrange function L of S reduces to the kinetic energy, which can be written as

$$L = \frac{1}{2}\left(\dot{x}_1^2 + \dot{x}_2^2\right) + \frac{1}{2}\left(A(p^2 + q^2) + Cr^2\right),$$

where A and C are the momenta of inertia of S, and p, q, and r are the components of the angular velocity ω in the body frame $Cx_1'x_2'x_3'$ (Fig. 17.2). Since $\mathbf{e}_3 = \sin \varphi \mathbf{e}_1' + \cos \varphi \mathbf{e}_2'$, the angular velocity can be written as

$$\omega = \dot{\varphi}\mathbf{e}_3' + \dot{\theta}\mathbf{e}_3 = \dot{\theta} \sin \varphi \mathbf{e}_1' + \dot{\theta} \cos \varphi \mathbf{e}_2' + \dot{\varphi}\mathbf{e}_3',$$

and the Lagrange function becomes

$$L = \frac{1}{2}\left(\dot{x}_1^2 + \dot{x}_2^2\right) + \frac{1}{2}\left(A\dot{\theta}^2 + C\dot{\varphi}^2\right).$$

From (17.87) and the conditions determined at the end of Sect. 17.4, we obtain the following system:

$$m\ddot{x}_1 = -\lambda_1, \tag{17.89}$$
$$m\ddot{x}_2 = -\lambda_2, \tag{17.90}$$
$$C\ddot{\varphi} = a(\lambda_1 \cos \theta - \lambda_2 \sin \theta), \tag{17.91}$$
$$A\ddot{\theta} = 0, \tag{17.92}$$
$$\dot{x}_1 = a\dot{\varphi} \cos \theta, \tag{17.93}$$
$$\dot{x}_2 = -a\dot{\varphi} \sin \theta. \tag{17.94}$$

Equation (17.92) implies that

$$\theta(t) = \dot{\theta}_0 t + \theta_0, \tag{17.95}$$

where the meaning of the constants $\dot{\theta}_0$ and θ_0 is evident. Introducing (17.89) and (17.90) into (17.91), we obtain the equation

$$C\ddot{\varphi} = -ma(\ddot{x}_1 \cos\theta - \ddot{x}_2 \sin\theta)$$

which, in view of (17.93) and (17.94), can also be written as

$$(C + ma^2)\ddot{\varphi} = 0,$$

so that

$$\varphi(t) = \dot{\varphi}_0 t + \varphi_0, \tag{17.96}$$

where again the meaning of $\dot{\varphi}_0$ and φ_0 is evident. Finally, from (17.95), (17.96), (17.93), and (17.94), if $\dot{\theta}_0 \neq 0$ we have that

$$x_1(t) = \frac{a\dot{\varphi}_0}{\dot{\theta}_0} \sin(\dot{\theta}_0 t + \theta_0) + x_1^0,$$

$$x_2(t) = \frac{a\dot{\varphi}_0}{\dot{\theta}_0} \cos(\dot{\theta}_0 t + \theta_0) + x_2^0,$$

whereas if $\dot{\theta}_0 = 0$ we obtain

$$x_1(t) = \dot{\varphi}_0 t \cos\theta_0 + x_1^0,$$
$$x_2(t) = \dot{\varphi}_0 t \sin\theta_0 + x_2^0.$$

In conclusion, the disk rotates uniformly about the vertical axis containing its center C, rolls uniformly about the horizontal axis containing C, and the contact point A moves uniformly along a straight line if $\dot{\theta}_0 = 0$ and moves uniformly on a circumference of radius $|a\dot{\varphi}_0/\dot{\theta}_0|$ if $\dot{\theta}_0 \neq 0$.

17.13 Small Oscillations

Let S be a dynamical system with smooth and fixed holonomic constraints. We say that a configuration C_* of S is an **equilibrium configuration** if, placing S at rest in the configuration C_* at the instant $t = 0$ it remains in this configuration at any time. Let V_n be the configuration space and denote by $\mathbf{q}_*$ the point of V_n representative of the configuration C_*. Then, $\mathbf{q}_*$ is an equilibrium configuration if

$$q^h(0) = q_*^h, \quad \dot{q}^h(0) = 0 \quad \Longrightarrow \quad q^h(t) = q_*^h, \quad \forall t \geq 0. \tag{17.97}$$

Theorem 17.6. *The configuration* $\mathbf{q}_*$ *is an equilibrium configuration of S if and only if*

$$Q_h(\mathbf{q}_*, \mathbf{0}) = 0, \quad h = 1, \ldots, n. \tag{17.98}$$

Proof. If $\mathbf{q}_*$ is an equilibrium configuration, i.e., if (17.97) holds, then the functions $q^h(t) = q_*^h, h = 1, \ldots, n$, must be a solution of Lagrange's equations. Introducing the functions $q^h(t) = q_*^h$ into Lagrange's equations, we obtain (17.98). Conversely, if this last condition is satisfied, then the Lagrange equations admit $q^h(t) = q_*^h, h = 1, \ldots, n$, as a solution. But this solution satisfies the initial data $q^h(0) = q_*^h, \dot{q}^h(0) = 0, h = 1, \ldots, n$, and, consequently, it is the only solution of Lagrange's equations satisfying the initial data $q^h(0) = q_*^h, \dot{q}^h(0) = 0, h = 1, \ldots, n$. $\qquad \square$

To apply the results of Chap. 10, we reduce Lagrange's equations to an equivalent system of $2n$ first-order differential equations. If we set $\dot{q}^h = v^h$, then (17.49) can be written in the form

$$\dot{q}^h = v^h, \tag{17.99}$$

$$\dot{v}^h = g^h(\mathbf{q}, \mathbf{v}). \tag{17.100}$$

We say that the configuration C_* of S is a **stable equilibrium configuration** if $(\mathbf{q}_*, \mathbf{0})$ is a stable equilibrium configuration of system (17.99), (17.100). This means that for any neighborhood $I(\mathbf{q}_*, \mathbf{0})$ of $(\mathbf{q}_*, \mathbf{0}) \in T(V_n)$, there is a neighborhood $I_0(\mathbf{q}_*, \mathbf{0}) \subset I(\mathbf{q}_*, \mathbf{0})$ such that

$$\forall (\mathbf{q}_0, \mathbf{v}_0) \in I_0(\mathbf{q}_*, \mathbf{0}) \Rightarrow (\mathbf{q}(t, \mathbf{q}_0, \mathbf{v}_0), \mathbf{v}(t, \mathbf{q}_0, \mathbf{v}_0)) \in I(\mathbf{q}_*, \mathbf{0}). \tag{17.101}$$

In Chap. 10, we remarked that checking the stability of an equilibrium configuration requires a knowledge of the solutions. Since, in general, it is impossible to exhibit the solutions of a given system of first-order differential equations, it is fundamental to supply criteria of stability. In the following theorem, *Dirichlet's stability criterion* is proved.

Theorem 17.7. *Let S be a mechanical system with smooth and fixed holonomic constraints subject to active conservative forces with a potential energy* $U(\mathbf{q})$. *If* $\mathbf{q}_*$ *is an effective minimum of* $U(\mathbf{q})$, *then* $\mathbf{q}_*$ *is a stable equilibrium position of S.*

Proof. First, if $U(\mathbf{q})$ has a minimum at $\mathbf{q}_*$, then

$$\left(\frac{\partial U}{\partial q^h} \right)_{\mathbf{q}_*} = Q_h(\mathbf{q}_*) = 0, \tag{17.102}$$

and $\mathbf{q}_*$ is an equilibrium configuration (Theorem 17.6). Without loss of generality, we can suppose that $q_*^h = 0, h = 1, \ldots, n$. Now we prove that the total energy

$$V(\mathbf{q}, \mathbf{v}) = T(\mathbf{q}, \mathbf{v}) + U(\mathbf{q}) \tag{17.103}$$

confirms the following conditions:

1. $V(\mathbf{0},\mathbf{0})=0$;
2. $V(\mathbf{q},\mathbf{v})>0$ in a neighborhood of $(\mathbf{0},\mathbf{0})$;
3. $\dot{V}=0$ along any solution $(q^h(t),v^h(t))$ of (17.99), (17.100),

i.e., $V(\mathbf{q},\mathbf{v})$ is a Lyapunov function for system (17.99), (17.100). First, we have that $V(\mathbf{0},\mathbf{0})=U(\mathbf{0})=0$ since the potential energy is defined up to an additive constant. Further, the kinetic energy is a positive definite quadratic form, and the hypothesis that $U(\mathbf{q})$ has an effective minimum at $\mathbf{0}$ implies that $U(\mathbf{q})>0$ in a neighborhood of $\mathbf{0}$, and the second property is proved. Finally, the third property is confirmed since the total energy is a first integral.   □

Let $q_*^h=0$, $h=1,\ldots,n$, be a configuration corresponding to a minimum of the potential energy $U(\mathbf{q})$. For the preceding theorem it is a stable equilibrium configuration. Consequently, for initial data $(\mathbf{q}_0,\mathbf{v}_0)$ in a suitable neighborhood I_0 of $(\mathbf{0},\mathbf{0})$, the corresponding solutions remain in an assigned neighborhood I containing I_0. We suppose that I is such that we can neglect in the following Taylor expansions the terms of order higher than two in the variables q^h, $\dot{q}^h$:

$$T^*=\frac{1}{2}a_{hk}^*\dot{q}^h\dot{q}^k,$$

$$U^*=U(\mathbf{0})+\left(\frac{\partial U}{\partial q^h}\right)_0 q^h+\frac{1}{2}\left(\frac{\partial^2 U}{\partial q^h\partial q^k}\right)_0 q^h q^k,$$

where $a^*=a_{hk}(\mathbf{0})$.

Since $U(\mathbf{q})$ is defined up to an arbitrary constant, we can suppose $U(\mathbf{0})=0$. Further, $U(\mathbf{q})$ has a minimum at $\mathbf{0}$, and then $(\partial U/\partial q^h)_0=0$. Consequently, the preceding expansions become

$$T^*=\frac{1}{2}a_{hk}^*\dot{q}^h\dot{q}^k, \tag{17.104}$$

$$U^*=\frac{1}{2}b_{hk}^*q^h q^k, \tag{17.105}$$

where

$$b_{hk}^*=\left(\frac{\partial^2 U}{\partial q^h\partial q^k}\right)_0. \tag{17.106}$$

The quadratic form (17.104) is positive definite since all the principal minors of the matrix (a_{hk}^*) have positive determinants (Sect. 17.8). Also, the quadratic form (17.105) is positive definite since $U(\mathbf{q})$ has an effective minimum at $\mathbf{0}$.

The motions corresponding to the approximate Lagrangian function

$$L^*(\mathbf{q},\dot{\mathbf{q}})=\frac{1}{2}a_{hk}^*\dot{q}^h\dot{q}^k-\frac{1}{2}b_{hk}^*q^h q^k, \tag{17.107}$$

which are called *small oscillations* about the stable equilibrium configuration **0**, are solutions of the following linear differential equations:

$$a_{hk}^* \dot{q}^k = -b_{hk}^* q^k. \tag{17.108}$$

The analysis of small oscillations becomes trivial by the following considerations. Let $\mathcal{E}_n$ be a Euclidean vector space equipped with the scalar product

$$\mathbf{u} \cdot \mathbf{v} = a_{hk}^* u^h v^k, \tag{17.109}$$

where $\mathbf{u}, \mathbf{v} \in \mathcal{E}_n$ and u^h, v^h denote their components in a basis $(\mathbf{e}_h)$ of $\mathcal{E}_n$. Let us consider the eigenvalue equation

$$b_{hk}^* u^k = \lambda a_{hk}^* u^k. \tag{17.110}$$

Since the matrix (b_{hk}^*) is symmetric and the quadratic form (17.10) is positive definite, all its eigenvalues $\lambda_1, \ldots, \lambda_n$ are positive, and it is possible to find an orthonormal basis $(\mathbf{u}_h)$ of $\mathcal{E}_n$, formed by eigenvectors of (b_{hk}^*). In this basis, the matrices (a_{hk}^*) and (b_{hk}^*) have a diagonal form

$$(\overline{b}_{hk}^*) = \begin{pmatrix} \lambda_1 & \cdots & 0 \\ \cdots & \cdots & \cdots \\ 0 & \cdots & \lambda_n \end{pmatrix},$$

$$(\overline{a}_{hk}^*) = \begin{pmatrix} 1 & \cdots & 0 \\ \cdots & \cdots & \cdots \\ 0 & \cdots & 1 \end{pmatrix},$$

the Lagrangian function (17.107) becomes

$$L^*(\overline{\mathbf{q}}, \dot{\overline{\mathbf{q}}}) = \frac{1}{2} \sum_{i=1}^n \left(\dot{\overline{q}}^h \right)^2 - \frac{1}{2} \sum_{i=1}^n \lambda_h \left(\overline{q}^h \right)^2, \tag{17.111}$$

and the corresponding Lagrange equations assume the simple form

$$\ddot{\overline{q}}^h = -\sqrt{\lambda_h} \overline{q}^h. \tag{17.112}$$

In conclusion, we have stated that *small oscillations about a stable equilibrium configuration, corresponding to an effective minimum of the potential energy, are obtained by combining n harmonic motions along the eigenvectors of (b_{hk}^*) with frequencies $\sqrt{\lambda_h}$*. These frequencies are called *normal frequencies* and harmonic motions *normal modes*. The calculations presented in this section can be carried out using the notebook **SmallOscill**.

17.14   Hamilton's Principle

Newton's laws, Euler's equations, and Lagrange's equations, which, respectively, govern the dynamics of systems of material points of a single rigid body and of systems of constrained rigid bodies, have a common aspect: they describe the dynamical evolution by differential equations, that is, by relations between the unknown functions describing the motion and their first and second time derivatives.

On the other hand, in physics there are many situations in which the problem we confront does not spontaneously lead to a system of differential equations but to the minimum value of an integral relation. In discussing some historical examples, we discover a new formulation of mechanical laws.

- We start with the problem of the *brachistochrone* posed by Johann Bernoulli and solved by Newton and Jakob Bernoulli. A brachistochrone curve, or a curve of fastest descent, is the curve γ between two points A and B that is covered in the least time by a body that starts from the first point A with zero speed and is constrained to move along the curve to the second point B, with A no lower than B, under the action of constant gravity and assuming no friction. In the vertical plane π containing γ, we introduce a frame of reference Oxy as shown in Fig. 17.6. Let $y(x)$ be the equation of the curve γ and denote by s the curvilinear abscissa from the origin $O \equiv A$. With our choice of the origin of coordinates and the condition $\dot{s} = 0$ at the initial time, the conservation of energy reduces to the relation

$$\frac{1}{2}m\dot{s}^2 - mgy = 0.$$

Further, the length ds of the element of arc along γ is given by

Fig. 17.6 Brachistochrone

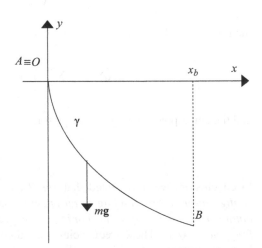

$$ds = \sqrt{1 + y'^2}dx.$$

Bearing in mind the preceding two relations, we obtain the time $\mathbb{T}$ to cover γ from A to B:

$$\mathbb{T}[y(x)] = \int_0^{x_b} \frac{\sqrt{1 + y'^2(x)}}{\sqrt{2gy(x)}} \, dx. \tag{17.113}$$

By (17.113), a real number $\mathbb{T}$ corresponds to any $C^1(0, x_b)$ function $y(x)$, assuming the values $y(0) = 0$ and $y(x_b) = y_b$, provided that the integral on the right-hand side of (17.113) exists. Our problem will have a solution if we find a function with the preceding properties that minimizes the value of $\mathbb{T}$.

- As a second example we consider *Fermat's principle*. Let S be an optical medium with refractive index $n(\mathbf{r})$. This principle states that the path γ taken between two points A and B by a ray of light is the path that can be traversed in the least time. Denoting by $(x_1(t), x_2(t), x_3(t))$ the parametric equations of γ, the time spent to cover γ with speed $v = c/n$ is given by

$$T = \frac{1}{c} \int_{t_a}^{t_b} n(x_1(t), x_2(t), x_3(t)) \sqrt{\dot{x}_1{}^2 + \dot{x}_2{}^2 + \dot{x}_3{}^2} dt. \tag{17.114}$$

It is well known that this principle includes in a very compact form all the laws of geometric optics (e.g., reflection, refraction). Once again the path of a ray minimizes the integral relation (17.114).

- Let γ be a curve given by the function $y(x)$, $a \leq x \leq b$. Denote by S the surface of revolution obtained rotating γ about the Ox-axis. Determine the curve γ for which the area of S, given by

$$S[y(x)] = \int_a^b 2\pi y(x)\sqrt{1 + y'^2} \, dx, \tag{17.115}$$

assumes the minimum value.

We define as *functional* any map

$$F : \mathfrak{F} \to \mathfrak{R},$$

where $\mathfrak{F}$ is a normed space of functions. Formulae (17.113)–(17.115) are examples of functionals. Searching for a minimum of a functional is called a *variational problem*. The elements of $\mathfrak{F}$ at which F attains a minimum are said to be *extremals* of the functional F.[3]

Many situations can be described by a variational principle. For example, a soap ball is spherical to minimize its surface for given internal and external pressures, and a

---

[3] A wider introduction to variational calculus can be found in Chapter 27.

thin elastic membrane with a fixed boundary assumes a configuration corresponding
to a minimum of its area.

In view of the preceding examples, we are led to search for the extremals of a
functional of the form

$$I = \int_{t_a}^{t_b} L(\mathbf{q}(t), \dot{\mathbf{q}}(t)) \, dt, \tag{17.116}$$

where $\mathbf{q}(t) = (q^1(t), \dots, q^n(t))$ and the functions $q^h(t)$ are of class $C^1[t_a, t_b]$ and
satisfy the boundary conditions $\mathbf{q}(t_a) = \mathbf{q}_a = (q_a^1, \dots, q_a^n)$ and $\mathbf{q}(t_b) = \mathbf{q}_b =
(q_b^1, \dots, q_b^n)$. In what follows, we consider any function $\mathbf{q}(t)$ as a curve of $\mathfrak{R}^n$. We do
not introduce the wide mathematical apparatus that has been developed to analyze the
problem of the existence of extremals. We prefer to follow a very simple approach,
proposed by Euler and Lagrange, to obtain differential equations whose solutions
are the extremals we are searching for.

Let $\mathbf{f} : (s, t) \in [-\epsilon, \epsilon] \times [t_a, t_b]$ be a family of curves, depending on the parameter
s, such that the parametric equations $(f^1(s, t), \dots, f^n(s, t))$ belong to $C^1[t_a, t_b]$.
Further, we suppose that $\mathbf{f}(s, t)$ contains the function $\mathbf{q}(t)$ for $s = 0$. More explicitly,
we suppose that

$$\mathbf{f}(s, t_a) = \mathbf{q}_a, \ \mathbf{f}(s, t_b) = \mathbf{q}_b, \ \forall s \in [-\epsilon, \epsilon]; \tag{17.117}$$

$$\mathbf{f}(0, t) = \mathbf{q}(t). \tag{17.118}$$

In other words, we are considering a family $\mathbf{f}(s, t)$ of curves of $\mathfrak{R}^n$, depending on
the parameter s, starting from the point $\mathbf{q}_a$ and ending at the point $\mathbf{q}_b$. This family
reduces to the curve $\mathbf{q}(t)$ for $s = 0$ (Fig. 17.7).

Now, introducing the family $\mathbf{f}(s, t)$ into functional (17.116), we obtain a function
of the parameter s:

Fig. 17.7 Family of curves
$\mathbf{f}(s, t)$

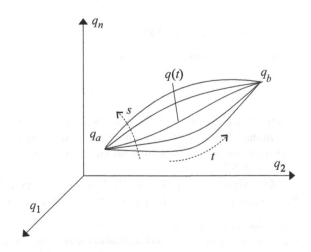

$$I(s) = \int_{t_a}^{t_b} L\left(\mathbf{f}(s,t), \dot{\mathbf{f}}(s,t)\right) dt, \tag{17.119}$$

and we can state that $\mathbf{q}(t)$ is an extremal of functional (17.116) if and only if

$$\left(\frac{dI}{ds}\right)_{s=0} = 0 \tag{17.120}$$

for any choice of the family $\mathbf{f}(s,t)$ of curves satisfying properties (17.117) and (17.118).

Since

$$\frac{\partial \dot{\mathbf{f}}}{\partial s} = \frac{\partial}{\partial s}\frac{\partial \mathbf{f}}{\partial t} = \frac{\partial}{\partial t}\frac{\partial \mathbf{f}}{\partial s},$$

we can write

$$\frac{dI}{ds} = \int_{t_a}^{t_b} \left[\frac{\partial L}{\partial \mathbf{f}}\frac{\partial \mathbf{f}}{\partial s} + \frac{\partial L}{\partial \dot{\mathbf{f}}}\frac{\partial \dot{\mathbf{f}}}{\partial s}\right] dt$$

$$= \left[\frac{\partial L}{\partial \dot{\mathbf{f}}}\frac{\partial \mathbf{f}}{\partial s}\right]_{t_a}^{t_b} + \int_{t_a}^{t_b} \left[\frac{\partial L}{\partial \mathbf{f}} - \frac{\partial}{\partial t}\frac{\partial L}{\partial \dot{\mathbf{f}}}\right]\frac{\partial \mathbf{f}}{\partial s} dt. \tag{17.121}$$

But, in view of (17.117), we have

$$\left[\frac{\partial L}{\partial \dot{\mathbf{f}}}\frac{\partial \mathbf{f}}{\partial s}\right]_{t_a}^{t_b} = 0,$$

and (17.121), for $s = 0$, gives

$$\left(\frac{dI}{ds}\right)_{s=0} = \int_{t_a}^{t_b} \left[\frac{\partial L}{\partial \mathbf{q}} - \frac{d}{dt}\frac{\partial L}{\partial \dot{\mathbf{q}}}\right]\left(\frac{\partial \mathbf{f}}{\partial s}\right)_{s=0} dt. \tag{17.122}$$

To obtain the preceding relation, we substituted $\partial/\partial t$ with d/dt since, for $s = 0$, all the functions under the integral depend only on t; further, in view of (17.118), we have that $\mathbf{f}(0,t) = \mathbf{q}(t)$, $\dot{\mathbf{f}}(0,t) = \dot{\mathbf{q}}(t)$. Comparing (17.120) and (17.122), we see that a solution $\mathbf{q}(t)$ of the Lagrange equations is an extremal of functional (17.116). Conversely, if $\mathbf{q}(t)$ is an extremal of functional (17.116), then (17.122) is satisfied for every $\mathbf{f}(s,t)$. Consequently, the right-hand side of (17.122) vanishes for every $\mathbf{f}(s,t)$, and it is possible to prove that the function $\mathbf{q}(t)$ is a solution of the Lagrange equations.

In conclusion, the extremals of (17.116) are the solutions of the following boundary value problem:

$$\frac{\partial L}{\partial \mathbf{q}} - \frac{\mathrm{d}}{\mathrm{d}t}\frac{\partial L}{\partial \dot{\mathbf{q}}} = 0, \ \mathbf{q}(t_a) = \mathbf{q}_a, \mathbf{q}(t_b) = \mathbf{q}_b. \tag{17.123}$$

No general theorem of existence and uniqueness exists for the preceding boundary value problem.

The preceding results prove the following theorem.

Theorem 17.8 (Hamilton's principle). *Let S be a dynamical system with Lagrangian function L. Then, an effective motion for which* $\mathbf{q}(t_a) = \mathbf{q}_a$ *and* $\mathbf{q}(t_b) = \mathbf{q}_b$ *is an extremal of functional (17.116).*

Functional (17.116) is called an **action functional.**

17.15   Geometric Formulations of Lagrangian Dynamics

Up to now, our formulation of Lagrangian dynamics has used in a marginal way the structures of differential geometry. Essentially, we have associated a differentiable manifold V_n, the configuration space, with any dynamical system S. The dimension n of V_n gives the degrees of freedom of S, i.e., the number of independent parameters defining the configuration of S. Finally, the motion of S is represented by a curve $\mathbf{q}(t)$ of V_n, called the dynamical trajectory of S, satisfying the Lagrange equations. This geometric formulation of Lagrangian dynamics is *local* since it uses local coordinates (q^h) on V_n; in other words, there is no *intrinsic* formulation of this theory. In this section, we present two interesting geometric versions of Lagrangian dynamics in which any mechanical aspect assumes a geometric form.

Let S be a dynamical system with smooth and fixed holonomic constraints acted upon by conservative active forces so that the dynamics of S is described by a Lagrangian function $L(\mathbf{q}, \dot{\mathbf{q}})$. We denote by V_n the configuration space of S. In local coordinates $(q^1, \ldots, q^h)$ of V_n, the kinetic energy of S is

$$T = \frac{1}{2}a_{hk}(\mathbf{q})\dot{q}^h\dot{q}^k, \tag{17.124}$$

where the quadratic form on the right-hand side of (17.124) is positive definite and the coefficients a_{hk} are symmetric. Further, the functions $a_{hk}(\mathbf{q})$ define a $(0,2)$-tensor over V_n since the kinetic energy is an invariant scalar with respect to the coordinate transformations of V_n and $\dot{q}^h$ are the contravariant components of the Lagrangian velocity relative to a coordinate basis $(\partial/\partial q^h)$. These properties allow us to equip V_n with the following Riemannian metric (Chap. 6):

$$\mathrm{d}s^2 = a_{hk}(\mathbf{q})\mathrm{d}q^h\mathrm{d}q^k. \tag{17.125}$$

With the introduction of this metric, a scalar product is introduced in the tangent space $T_{\mathbf{q}}(V_n)$ at any point $\mathbf{q} \in V_n$; moreover, for any vector $\mathbf{v} \in T_q(V_n)$ the following relations between contravariant and covariant components hold:

$$v_h = a_{hk} v^k, \quad v^h = (a^{-1})^{hk} v_k, \tag{17.126}$$

where $\left((a^{-1})^{hk}\right)$ is the inverse matrix of (a_{hk}). Consequently, the generalized momenta

$$p_h = \frac{\partial L}{\partial \dot{q}^h} = a_{hk} \dot{q}^k \tag{17.127}$$

can be regarded as the covariant components of the Lagrangian velocity, that is, of the vector $\mathbf{v}$ tangent to the dynamical trajectory of S. In formulae, we write

$$v^h = \dot{q}^h, \quad v_h = a_{hk} v^k = p_h. \tag{17.128}$$

With these notations, the Lagrange equations (17.48) assume the following form:

$$\dot{v}_h - \frac{1}{2} \frac{\partial a_{jk}}{\partial q^h} v^j v^k = Q_h(\mathbf{q}, \mathbf{v}). \tag{17.129}$$

In view of (9.60) the Christhoffel symbols Γ_{ij}^h and the metric coefficients a_{hk} of a Riemannian manifold are related by the equation

$$\frac{1}{2} \frac{\partial a_{jk}}{\partial q^h} v^j v^k = a_{pj} \Gamma_{hk}^p v^j v^k = \Gamma_{hk}^p v_p v^k, \tag{17.130}$$

in view of which we can write (17.129) in the final form

$$\frac{\nabla v_h}{dt} \equiv \frac{dv_h}{dt} - \Gamma_{hk}^p v_p v^k = Q_h(\mathbf{q}, \mathbf{v}). \tag{17.131}$$

Taking into account that on the left-hand side of (17.131) are the covariant components of acceleration in Lagrangian coordinates (q^h), we conclude that *the acceleration of the point $\mathbf{q}(t)$, representing in the configuration space V_n the motion of the material system S, is equal to the Lagrangian force (Q_h). In other words, the motion of $\mathbf{q}(t)$ in the configuration space V_n is governed by Newton's law for a material point with unit mass.*

In particular, *in the absence of forces, the dynamical trajectory $\mathbf{q}(t)$ is a geodesic of the metric (17.125).*

In the preceding considerations, the constraints to which the dynamic system is subject and the reactive forces exerted by them were geometrized by introducing a configuration space equipped with the Riemannian metric (17.125). Now we want to prove that it is possible to geometrize the active forces, provided that they derive from a potential energy $U(\mathbf{q})$.

To prove this statement, we note that, for conservative forces, the total energy E is constant during motion, i.e, $T(\mathbf{q}, \dot{\mathbf{q}}) + U(\mathbf{q}) = E$. Consequently, the dynamical trajectory, corresponding to a fixed value of the total energy, is contained in the region $W_n \subset V_n$ defined by the inequality

$$T = E - U(\mathbf{q}) \geq 0 \Rightarrow U(\mathbf{q}) \leq E. \tag{17.132}$$

Now we can prove the following theorem.

Theorem 17.9 (Maupertuis' principle). *Let S be a mechanical system with smooth and fixed holonomic constraints, subject to conservative forces with potential energy $U(\mathbf{q})$. Then the dynamical trajectories of S corresponding to a given total energy E are geodesics of the configuration space V_n, equipped with the Riemannian metric*

$$d\sigma^2 = 2(E - U)a_{hk}dq^h dq^k \equiv Fa_{hk}dq^h dq^k. \tag{17.133}$$

These geodesics are covered with the following velocity:

$$\frac{d\sigma}{dt} = 2(E - U) \equiv F. \tag{17.134}$$

Proof. The geodesic of metric (17.133), connecting two points $\mathbf{q}_0$ and $\mathbf{q}_1$ of V_n, is the extremal of the functional (Chap. 9)

$$I[\mathbf{q}(s)] = \int_{\mathbf{q}_0}^{\mathbf{q}_1} \sqrt{Fa_{hk}\frac{dq^h}{d\sigma}\frac{dq^k}{d\sigma}}d\sigma, \tag{17.135}$$

where the parameter σ along the curves is the curvilinear abscissa relative to metric (17.133). We recall (Chap. 9) that, owing to the form of the preceding functional, the extremals are independent on the parameterization. In other words, the geodesics are defined as loci of points. The Lagrange equations corresponding to the functional (17.135) are [see (6.93), (6.91)]

$$\frac{d}{d\sigma}(Fa_{hk}q'^k) - \frac{1}{2}\frac{\partial}{\partial q^h}(Fa_{jk})q'^j q'^k = 0, \quad q'^j = \frac{dq^j}{d\sigma}.$$

These equations can also be written as follows:

$$\frac{d}{d\sigma}(Fa_{hk}q'^k) - \frac{1}{2}\frac{\partial a_{jk}}{\partial q^h} = \frac{1}{2F}\frac{\partial F}{\partial q^h} \tag{17.136}$$

since along the curves $(q^h(\sigma))$ that solve the above equations it is

$$Fa_{jk}q'^jq'^k = 1. \tag{17.137}$$

On the other hand, from the conservation of the total energy we have that

$$2T = a_{ij}\dot{q}^j\dot{q}^k = 2(E - U)$$

or, equivalently,

$$a_{jk}q'^jq'^k\left(\frac{d\sigma}{dt}\right)^2 = 2(E - U).$$

In view of (17.133) and (17.137), the previous equation becomes

$$\left(\frac{d\sigma}{dt}\right)^2 = 4(E - U)^2,$$

and we obtain (17.134) when we recall that $T = (E - U) \geq 0$. Now we show that the equations of geodesics (17.136) coincide with the Lagrange equations. In fact, these equations,

$$\frac{d}{dt}\frac{\partial T}{\partial \dot{q}^h} - \frac{\partial T}{\partial q^h} = -\frac{\partial U}{\partial q^h},$$

can also be written as follows:

$$\frac{d}{dt}(a_{hk}\dot{q}^k) - \frac{1}{2}\frac{\partial a_{jk}}{\partial q^h}\dot{q}^j\dot{q}^k = -\frac{\partial U}{\partial q^h}.$$

Introducing the parameter $\sigma(t)$ and taking into account (17.134), we obtain the equations

$$F\frac{d}{d\sigma}(Fa_{hk}q'^k) - \frac{1}{2}\frac{\partial a_{jk}}{\partial q^h}F^2q'^jq'^k = -\frac{\partial U}{\partial q^h},$$

which can be put in the final form

$$\frac{d}{d\sigma}(Fa_{hk}q'^k) - \frac{F}{2}\frac{\partial a_{jk}}{\partial q^h}q'^jq'^k = -\frac{1}{F}\frac{\partial U}{\partial q^h} = \frac{1}{2F}\frac{\partial F}{\partial q^h}. \tag{17.138}$$

The theorem is proved when we take into account (17.137) and compare (17.138) and (17.136). □

Remark 17.5. Given the mechanical system S and a value E of the total energy, metric (17.133) is determined. Assigning the initial data $q^h(0) = q_0^h$ and $q'^h(0) = q_0'^h$, that is, the initial position and *direction* of the initial velocity, we determine by (17.136) the geodesic starting from this initial position in the given direction. In this way, we know the dynamical trajectory but we do not know how it is covered. The condition that during the motion the energy has a given value implies that the geodesic must be covered with the law (17.134).

17.16   Legendre's Transformation

In this section, we analyze the path leading from the Lagrangian formulation of mechanics to the Hamiltonian one. This path, in the case of constraints independent of time, consists of the following fundamental steps.

1. First, we transform the Lagrange equations into an equivalent system of $2n$ first-order differential equations by considering the Lagrangian velocities $\dot{\mathbf{q}}$ as auxiliary unknowns. A solution $\mathbf{x}_L(t)$ of this first-order system is a curve of the *velocity space*, i.e., of the fiber bundle TV_n of the configuration space. In this space, the Lagrange equations assume the form

$$\dot{\mathbf{x}}_L = \mathbf{X}_L(\mathbf{x}_L), \qquad (17.139)$$

   where $\mathbf{X}_L(x)$ is a vector field on TV_n.
2. Then we define the *Legendre transformation*, i.e., a diffeomorphism $\mathcal{L}: TV_n \to T^*V_n$ between the tangent fiber bundle TV_n and the cotangent fiber bundle T^*V_n.
3. Finally, by the differential

$$\mathcal{L}_* : T_{(\mathbf{q},\mathbf{v})}(TV_n) \to T_{\mathcal{L}(\mathbf{q},\mathbf{v})}(T^*V_n), \qquad (17.140)$$

we obtain a new system of first-order differential equations,

$$\dot{\mathbf{x}}_H = \mathbf{X}_H(\mathbf{x}_H), \qquad (17.141)$$

where $\mathbf{X}_H = \mathcal{L}_*(\mathbf{q}, \mathbf{v}) \in T^*V_n$. It is evident that the solutions of this new system are curves of T^*V_n (Fig. 17.8).

To extend the foregoing procedure to constraints depending on time, it is sufficient to substitute V_n with $V_n \times \Re$.

Let S be a dynamical system described by a Lagrangian function $L(\mathbf{q}, \dot{\mathbf{q}}, t)$. The Lagrangian equations

$$\frac{\mathrm{d}}{\mathrm{d}t} \frac{\partial L}{\partial \dot{q}^h} - \frac{\partial L}{\partial q^h} = 0, \ h = 1, \ldots, n,$$

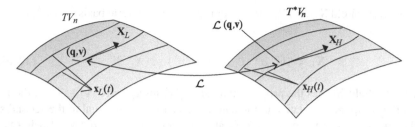

Fig. 17.8 Legendre transformation

are explicitly written as

$$L_{\dot{q}^h q^k}\dot{q}^k + L_{\dot{q}^h \dot{q}^k}\ddot{q}^k + L_{tq^h} - L_{q^h} = 0, \quad h = 1, \ldots, n, \qquad (17.142)$$

where we use the notation $\partial f/\partial x = f_x$. From $L(\mathbf{q}, \dot{\mathbf{q}}, t) = T(\mathbf{q}, \dot{\mathbf{q}}, t) - U(\mathbf{q}, t)$ and (17.36) we have that

$$L_{\dot{q}^h \dot{q}^k} = a_{hk}(\mathbf{q}, \dot{\mathbf{q}}, t),$$

where $\det(a_{hk}) \neq 0$; consequently, there exists the inverse matrix $(a^{hk}) = ((a^1)_{hk})$. Multiplying (17.142) by (a^{hl}), we obtain the Lagrangian equations in normal form:

$$\ddot{q}^h = a^{hl}(L_{q^l} - L_{t\dot{q}^l} - L_{\dot{q}^l q^k}\dot{q}^k), \quad h = 1, \ldots, n. \qquad (17.143)$$

Remark 17.6. The kinetic energy is a scalar quantity, i.e., it is invariant with respect to changes in the Lagrangian coordinates. Consequently, the quadratic form $\frac{1}{2}a_{hk}\dot{q}^h\dot{q}^k$ and the linear form $b_h\dot{q}^h$ of (17.36) are invariant with respect to these transformations. Recalling the criteria that give the tensorial character of a quantity and noting that $(\dot{q}^h)$ is a vector of V_n, we can state that (a_{hk}) and (b_h) are, respectively, the components of a *covariant* 2-tensor and a *covariant* vector of V_n. Then the generalized momenta

$$p_h = L_{\dot{q}^h} = a_{hk}\dot{q}^k + b_h$$

are the components of a *covector* of V_n.

Lagrangian function independent of time. Introducing the auxiliary unknowns $\dot{q}^h = v^h$, Eqs. (17.143) are equivalent to the first-order system

$$\dot{q}^h = v^h, \qquad (17.144)$$

$$\dot{v}^h = a^{hl}(L_{q^l} - L_{v^l q^k}v^k), \qquad (17.145)$$

in the $2n$ unknowns $q^h(t)$, $v^h(t)$, $h = 1, \ldots, n$. These unknowns are the parametric equations of a curve γ belonging to TV_n, whereas the right-hand sides of (17.145) define a vector field $\mathbf{X}_L$ on TV_n whose components in a natural basis $(\partial/\partial q^h, \partial/\partial v^h)$ of TV_n are

$$\mathbf{X}_L = (v^h, a^{hl}(L_{q^l} - L_{v^l q^k} v^k) \equiv (X^h, Y^h). \tag{17.146}$$

The vector field $\mathbf{X}_L$ has the property that, at any point $\mathbf{x}_L \equiv (\mathbf{q}, \mathbf{v}) \in TV_n$, the first set of the components relative to the natural basis of TV_n is equal to the second set of the coordinates of the point of TV_n at which $\mathbf{x}_L$ is tangent to TV_n. Such a field is called a **semispray vector field**.

Independently of the adopted coordinates, system (17.144), (17.145) can be written as

$$\dot{\mathbf{x}}_L = \mathbf{X}_L(\mathbf{x}). \tag{17.147}$$

To simplify the notations, we set $M = TV_n$ and $M^* = T^*V_n$ and adopt on both fiber bundles natural coordinates (q^h, v^h) and (q^h, p_h), respectively. Then we introduce the **Legendre transformation**

$$\mathcal{L} : M \to M^* \tag{17.148}$$

that, in natural coordinates, is defined as follows (Remark 17.6):

$$(q^h, p_h) = \mathcal{L}(q^h, v^h) = (q^h, L_{v^h}(\mathbf{q}, \mathbf{v})). \tag{17.149}$$

The differential of this mapping

$$\mathcal{L}_* : (X^h, Y^h) \in T_{(q,v)}(M) \to (X^{*h}, Y_h^*) \in T_{\mathcal{L}(q,v)}(M^*) \tag{17.150}$$

is such that

$$\begin{pmatrix} X^{*h} \\ Y_h^* \end{pmatrix} = \begin{pmatrix} \delta_k^h & 0 \\ L_{v^h q^k} & L_{v^h v^k} \end{pmatrix} \begin{pmatrix} X^k \\ Y^k \end{pmatrix}$$
$$= (X^h, L_{v^h q^k} X^k + L_{v^h v^k} Y^k). \tag{17.151}$$

Taking into account (17.149) and (17.151), we conclude that the Legendre transformation $\mathcal{L}$ associates to a point $\mathbf{x}_L = (q^h, v^h) \in M$ a point $\mathbf{x}^* = (q^h, p_h) \in M^*$ and to a tangent vector $\mathbf{X} = (X^h, Y^h) \in T_\mathbf{x} M$ a tangent vector $\mathbf{X}^* \in T_{\mathbf{x}^*} M^*$. If this correspondence is applied to (17.147), we obtain the transformed equation in M^*

$$\dot{\mathbf{x}}^* = \mathbf{X}^*(\mathbf{x}^*), \tag{17.152}$$

where $\mathbf{x}^* = (q^h, p_h)$. Further, recalling that $L_{v^h v^k} = a_{hk}$ and taking into account (17.146), we obtain

$$\mathbf{X}^* = \left(v^h, L_{v^h q^k} v^k + a_{hk} a^{kl} (L_{q^l} - L_{v^l q^m} v^m)\right) = (v^h, L_{q^h}). \tag{17.153}$$

Consequently, in natural coordinates, (17.152) leads to the system

$$\dot{q}^h = v^h, \tag{17.154}$$

$$\dot{p}_h = L_{q^h}. \tag{17.155}$$

Introducing the **Hamiltonian function**

$$H(\mathbf{q}, \mathbf{p}) = \left[p_k v^k - L(\mathbf{q}, \mathbf{v})\right]_{\mathbf{v}=\mathbf{v}(\mathbf{q}, \mathbf{p})}, \tag{17.156}$$

where $v^h = v^h(\mathbf{q}, \mathbf{p})$ is the inverse function of $p_h = L_{v^h} = a_{hk} v^k$, we have that

$$H_{q^h} = p_k \frac{\partial v^k}{\partial q^h} - L_{q^h} - L_{v^k} \frac{\partial v^k}{\partial q^h} = -L_{q^h},$$

$$H_{p_h} = v^h + p_k \frac{\partial v^k}{\partial p^h} - L_{v^k} \frac{\partial v^k}{\partial p^h} = v^h,$$

and system (17.154), (17.155) reduces to the equations

$$\dot{q}^h = \frac{\partial H}{\partial p_h}, \tag{17.157}$$

$$\dot{p}_h = -\frac{\partial H}{\partial q^h}, \tag{17.158}$$

which are called **canonical Hamiltonian equations**.

Lagrangian Function Dependent on Time. In this case, the Lagrangian function $L(\mathbf{q}, \dot{\mathbf{q}}, t)$ is defined on $M \times \Re$. Consequently, we again introduce the auxiliary unknowns $v^h = \dot{q}^h$ but write system (17.143) in the form

$$\dot{q}^h = v^h, \tag{17.159}$$

$$\dot{v}^h = a^{hk} \left(L_{q^k} - L_{tv^k} - L_{v^k q^l} v^l\right), \tag{17.160}$$

$$\dot{t} = 1. \tag{17.161}$$

The solutions $\mathbf{x}(t) = (\mathbf{q}(t), \mathbf{v}(t), t)$ of this system are curves of $M \times \Re$ that are the integral curves of the vector field

$$\mathbf{X}_L = (X^h, Y^h, 1) = \left(v^h, a^{hk} \left(L_{q^k} - L_{tv^k} - L_{v^k q^l} q^l\right), 1\right). \tag{17.162}$$

Using these notations, system (17.159)–(17.161) can be written in the form

$$\dot{\mathbf{x}} = \mathbf{X}_L(\mathbf{x}, t). \tag{17.163}$$

In natural coordinates of M and M^*, the Legendre transformation

$$\mathcal{L} : M \times \mathfrak{R} \to M^* \times \mathfrak{R} \tag{17.164}$$

is defined as

$$\mathcal{L}(q^h, v^h, t) = \left(q^h, L_{v^h}(q, v, t), t\right), \tag{17.165}$$

and its differential is

$$\begin{aligned}
\mathcal{L}_*(X) &= \begin{pmatrix} \delta_k^h & 0 & 0 \\ L_{v^h q^k} & L_{v^h v^k} & L_{v^h t} \\ 0 & 0 & 1 \end{pmatrix} \begin{pmatrix} X^k \\ Y^k \\ X \end{pmatrix} \\
&= \left(X^h, L_{v^h q^k} X^k + L_{v^h v^k} Y^k + L_{v^h t} X, X\right).
\end{aligned} \tag{17.166}$$

It is a simple exercise to verify that the Legendre transformation reduces system (17.159)–(17.161) to the form

$$\dot{q}^h = \frac{\partial H}{\partial p_h}, \tag{17.167}$$

$$\dot{p}_h = -\frac{\partial H}{\partial q^h} \tag{17.168}$$

$$\dot{t} = 1, \tag{17.169}$$

where now $H = H(\mathbf{q}, \mathbf{p}, t)$.

17.17 Exercises

In solving the following exercises the reader can be helped by the notebook **LagrEq** that gives the Lagrange equations when the Lagrangian function is given.

1. Let P be a material point subject to a Newtonian force having its center at point O. Introducing spherical coordinates (r, φ, θ) with the origin at O, prove that the kinetic energy T and the potential energy are

$$T = \frac{1}{2}m \left(\dot{r}^2 + r^2\dot{\varphi}^2 + r^2\dot{\theta}^2 \sin^2 \theta\right),$$

$$U = -\frac{h}{r},$$

where h is a constant. Write the corresponding Lagrange equations and recognize the meaning of the first integral corresponding to the cyclic variable φ. Which is the configuration space?

Hint: The spherical and Cartesian coordinates are related by the relations

$$x = r \cos \varphi \sin \theta,$$
$$y = r \sin \varphi \sin \theta,$$
$$z = r \cos \theta$$

Further, the square of velocity of P is $\dot{x}^2 + \dot{y}^2 + \dot{z}^2$.

2. Prove that the kinetic energy and the potential energy of the spherical pendulum P are

$$T = \frac{1}{2} mr^2 \left(\dot{\varphi}^2 + \dot{\theta}^2 \sin^2 \theta \right),$$
$$U = -mgr \cos \theta,$$

where g is the gravity acceleration and r the radius of the sphere S on which P is constrained to move. Write the Lagrange equations and determine the configuration space.

3. Using the notations of Sect. 16.2, prove that the kinetic energy and the potential energy of a compound pendulum are

$$T = \frac{1}{2} C \dot{\varphi}^2,$$
$$U = -mgh \cos \varphi,$$

where h is the distance between the oscillation point and the center of mass. Determine the Lagrange equation and the configuration space.

4. Let S be a heavy gyroscope (Sect. 16.7). Denoting by ψ, φ, and θ Euler's angles, prove that the kinetic energy and the potential energy are

$$T = \frac{1}{2} A \left(\dot{\psi}^2 \sin^2 \theta + \dot{\theta}^2 \right) + C (\dot{\varphi} + \dot{\psi} \cos \theta)^2,$$
$$U = -mgh_G \cos \theta.$$

Write the Lagrange equations and the first integrals related to the cyclic coordinates.

Hint: It is sufficient to recall (15.21)

$$T = \frac{1}{2} \left(Ap^2 + Aq^2 + Cr^2 \right)$$

and the Euler relations of Sect. 12.4.

Fig. 17.9 Constrained
heavy bar

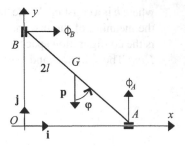

$$p = \dot\psi \sin\theta \sin\varphi + \dot\theta \cos\varphi,$$
$$q = \dot\psi \sin\theta \cos\varphi - \dot\theta \sin\varphi,$$
$$r = \dot\psi \cos\theta + \dot\varphi.$$

5. Let S be a homogeneous heavy bar with points A and B constrained to move without friction on the Ox- and Oy-axes, respectively (Fig. 17.9). Let $2l$ be the length of S and p its weight. Determine the motion of S by the balance equations. Then, noting that S^1 is the configuration space of S, determine the Lagrange equation governing the motion of S.

Hint: The momenta of $\boldsymbol\Phi_A$ and $\boldsymbol\Phi_B$ relative to center of mass G are

$$\overrightarrow{GA} \times \boldsymbol\Phi_A\mathbf{j} = l(\sin\varphi\mathbf{i} - \cos\varphi\mathbf{j}) \times \boldsymbol\Phi_A\mathbf{j} = l\Phi_A \sin\varphi\mathbf{k},$$
$$\overrightarrow{GB} \times \boldsymbol\Phi_B\mathbf{i} = l(-\sin\varphi\mathbf{i} + \cos\varphi\mathbf{j}) \times \boldsymbol\Phi_B\mathbf{i} = -l\Phi_B \cos\varphi\mathbf{k},$$

respectively, where $\mathbf{k} = \mathbf{i} \times \mathbf{j}$. Then the components along the axes Ox, Oy, and Oz (with Oz orthogonal to the plane Oxy) of the balance equations are

$$m\ddot{x}_G = \Phi_B,$$
$$m\ddot{y}_G = -p + \Phi_A,$$
$$I\ddot\varphi = l(\Phi_A \sin\varphi - \Phi_B \cos\varphi),$$

where I is the momentum of inertia relative to an axis orthogonal to Oxy and containing G. Introduce into the third equation the expressions of Φ_A and Φ_B deduced from the first two equations. Finally, the Lagrange equation is obtained by proving that

$$T = \frac{1}{2}\left(I + ml^2\right)\dot\varphi^2,$$
$$U = pl \cos\varphi.$$

6. A homogeneous bar AB is constrained to move on a smooth plane Oxy (Fig. 17.10). At points A and B two elastic forces are applied with centers O and Q, respectively. Further, $2l$ is the length of AB, M its mass, G the center of

Fig. 17.10 Bar moving in a plane

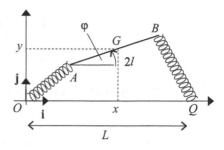

Fig. 17.11 Two constrained bars

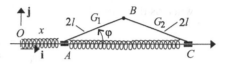

mass, and C the momentum of inertia relative to G. Show that the configuration space is $\mathfrak{R}^2 \times S^1$ and the kinetic and potential energies are

$$T = \frac{1}{2}M(\dot{x}^2 + \dot{y}^2) + \frac{1}{2}C\dot{\varphi}^2,$$

$$U = \frac{1}{2}k_1 \left[(x - l\cos\varphi)^2 + (y - l\sin\varphi)^2\right]$$
$$+ \frac{1}{2}k_2 \left[(x + l\cos\varphi - L)^2 + (y + l\sin\varphi)^2\right], \qquad (17.170)$$

where k_1 and k_2 are the elastic constants of the two springs.
Hint: The velocity of G is given by

$$\dot{\mathbf{r}}_G = \dot{x}\mathbf{i} + \dot{y}\mathbf{j},$$

and applying Koenig's theorem the expression of T is obtained. It is a simple exercise to obtain the preceding expression of the potential energy.

7. Let S be the plane system formed by two homogeneous bars of equal length (Fig. 17.11). Prove that the configuration space is $\mathfrak{R} \times S^1$ (a cylinder), and verify that the kinetic energy and the potential energy are given by

$$T = T_1 + T_2,$$
$$U = U_1 + U_2,$$

with

$$T_1 = \frac{1}{2}(C_1 + ml^2)\dot{\varphi}^2 + \frac{1}{2}m\dot{x}^2 - ml\dot{x}\dot{\varphi}\sin\varphi,$$

$$T_2 = \frac{1}{2}\left(C + ml^2(9\sin^2\varphi + \cos^2\varphi)\right)\dot{\varphi}^2 + \frac{1}{2}m\dot{x}^2 - 3ml\dot{x}\dot{\varphi}\sin\varphi,$$

$$U_1 = \frac{k}{2}x^2,$$

$$U_2 = \frac{k}{2}(4l\cos\varphi - x)^2$$

Hint: The center of mass G_1 has the following coordinates:

$$x_{G_1} = x + l\cos\varphi,$$
$$y_{G_1} = l\sin\varphi,$$

whereas the coordinates of G_2 are

$$x_{G_2} = x + 3l\cos\varphi,$$
$$y_{G_2} = l\sin\varphi.$$

Apply Koenig's theorem to each bar.

8. Let AB be a homogeneous heavy bar rotating about an axis a orthogonal to the plane Oxy and containing point A, which, in turn, is constrained to move along the Ox-axis (Fig. 17.12). Prove that the configuration space is $\Re \times S^1$ (cylinder) and the kinetic energy and the potential energy are

$$T = \frac{1}{2}(C + ml^2)\dot{\varphi}^2 + \frac{1}{2}m\dot{x}^2 + ml\dot{x}\dot{\varphi}\sin\varphi,$$
$$U = \frac{k}{2}x^2 - pl\cos\varphi.$$

Write the Lagrange equations and determine the equilibrium configurations.
Hint: The coordinates of the center of mass are $x_G = x + l\cos\varphi$, $y_G = -l\cos\varphi$.
Use Koenig's theorem.

9. Let P_1 and P_2 be two heavy material points, with masses m_1 and m_2, respectively, moving without friction on a circumference of radius r. Between the two points acts a spring as shown in Fig. 17.13. Prove that the configuration space is a torus $S^1 \times S^1$ and the kinetic energy T and the potential energy U are

Fig. 17.12 Pendulum with point A moving along Ox

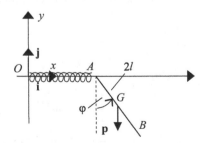

Fig. 17.13 Two heavy mass points on a circumference

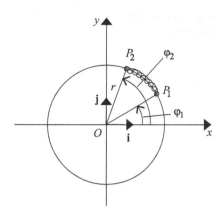

$$T = \frac{1}{2}m_1 r^2 \dot{\varphi}_1^2 + \frac{1}{2}m_2 r^2 \dot{\varphi}_2^2,$$

$$U = \frac{k}{2}r^2(\varphi_2 - \varphi_1)^2 + m_1 gr \sin \varphi_2 + m_2 r \sin \varphi_2. \qquad (17.171)$$

Determine the Lagrange equations and the equilibrium positions.

10. A homogeneous rigid disk S rotates about an axis orthogonal to the plane of S and containing its center O. Along a diameter of S a material point P moves without friction (Fig. 17.14). Denoting by M and m the masses of S and P, respectively, prove that the kinetic energy and the potential energy are

$$T = \frac{1}{2}m\dot{s}^2 + \frac{1}{2}(C + ms^2)\dot{\varphi}^2, \qquad (17.172)$$

$$U = \frac{k}{2}s^2. \qquad (17.173)$$

Determine Lagrange's equations and find a first integral of them.

Fig. 17.14 Disk and a material point

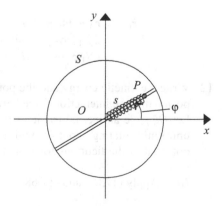

Fig. 17.15 Double
pendulum

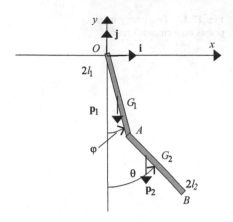

11. Prove that the potential energy of a double pendulum (Fig. 17.15) is

$$U = -p_1 l_1 \cos \varphi - p_2(2l_1 \cos \varphi + l_2 \cos \theta),$$

where $p_i = m_i g$, $i = 1, 2$. Determine the configuration space, the kinetic energy, the equilibrium configurations, and the small oscillations around the stable equilibrium configurations.

Hint: Applying Koenig's theorem to both bodies of the double pendulum with masses m_1 and m_2, we obtain

$$T = \frac{1}{2}m_1\dot{r}_{G_1}^2 + \frac{1}{2}C_1\dot{\varphi}^2 + \frac{1}{2}m_2\dot{r}_{G_2}^2 + \frac{1}{2}C_2\dot{\theta}^2.$$

On the other hand, denoting $\mathbf{k} = \mathbf{i} \times \mathbf{j}$, we have that

$$\dot{\mathbf{r}}_{G_1} = \dot{\varphi}\mathbf{k} \times l_1(\sin \varphi \mathbf{i} - \cos \varphi \mathbf{j}) = l_1\dot{\varphi}(\cos \varphi \mathbf{i} + \sin \varphi \mathbf{j})$$

and

$$\begin{aligned}
\dot{\mathbf{r}}_{G_2} &= \dot{\mathbf{r}}_A + \dot{\theta}\mathbf{k} \times \overrightarrow{AG_2} \\
&= 2\dot{\varphi}l_1(\cos \varphi \mathbf{i} + \sin \varphi \mathbf{j}) + \dot{\theta}\mathbf{k} \times (\sin \theta \mathbf{i} - \cos \theta \mathbf{j}) \\
&= (2l_1\dot{\varphi}\cos \varphi + l_2\dot{\theta}\cos \theta)\mathbf{i} + (2l_1\dot{\varphi}\sin \varphi + l_2\dot{\theta}\sin \theta)\mathbf{j}.
\end{aligned}$$

12. Write the kinetic energy and the potential energy of a bi-atomic molecule supposing that the interaction force between the two atoms is elastic.

13. Let $O\xi\eta z$ be an inertial frame of reference and denote by $Oxyz$ another frame uniformly rotating about Oz with respect to $O\xi\eta z$. Determine the generalized potential of the fictitious forces acting on a mass point P moving in the plane Oxy.

Hint: Apply (17.67) and (17.68).

14. Verify that in the Euclidean space E_3 hold the identities

$$\epsilon_{ikl}\epsilon_{imn} = \delta_{km}\delta_{ln} - \delta_{kn}\delta_{lm}, \qquad (17.174)$$

$$\epsilon_{ikl}\epsilon_{ikm} = 2\delta_{lm}. \qquad (17.175)$$

15. Let P be a material point and denote by x_i, $i = 1, 2, 3$ its Cartesian coordinates. Verify that

$$\left(\frac{\partial \Pi_k}{\partial x_h} - \frac{\partial \Pi_h}{\partial x_k}\right) v_k = ((\nabla \times \boldsymbol{\Pi}) \times \mathbf{v})_h . \qquad (17.176)$$

Hint:

$$((\nabla \times \boldsymbol{\Pi}) \times \mathbf{v})_h = \epsilon_{hik}\left(\epsilon_{ilj}\frac{\partial \Pi_j}{\partial x_l}\right) v_k = -\epsilon_{ihk}\epsilon_{ilj}\frac{\partial \Pi_j}{\partial x_l}v_k.$$

See (17.174).

16. Verify formula (17.61).

Hint: We must prove that (to simplify we are supposing $m = 1$)

$$F_h = -\epsilon_{hjl}\dot{\omega}_j x_l - \omega_h \omega_j x_j + \omega^2 x_h - 2\epsilon_{hjk}\omega_j v_k$$

$$= -\frac{\partial \hat{U}}{\partial x_h} + \left(\frac{\partial \Pi_h}{\partial x_k} - \frac{\partial \Pi_k}{\partial x_h}\right) v_k + \frac{\partial \Pi}{\partial t},$$

where

$$\hat{U} = \frac{1}{2}(\omega_i x_i)^2 - \frac{1}{2}\omega^2 x_i x_i,$$

$$\Pi_h = -\epsilon_{hjl}\omega_j x_l.$$

Now, we have

$$\frac{\partial \hat{U}}{\partial x_h} = (\omega_i x_i)\omega_h - \omega^2 x_h,$$

$$\frac{\partial \Pi}{\partial t} = -\epsilon_{hjl}\dot{\omega}_j x_l,$$

$$\left(\frac{\partial \Pi_h}{\partial x_k} - \frac{\partial \Pi_k}{\partial x_h}\right) v_k = -2\epsilon_{hjk}\omega_j v_k. \qquad (17.177)$$

Note that $\dot{x}_k = 0$ since (x_k) is a point of the frame $Ox_1x_2x_3$ and its velocity relative to this frame vanishes.

Chapter 18
Hamiltonian Dynamics

Abstract In this chapter we introduce the phase space and its symplectic structure. Then, canonical coordinates and generating functions are defined. The fundamental symplectic 2-form is introduced to define an isomorphism between vector fields and differential forms. Further, Hamiltonian vector fields and Poisson's brackets are presented. The relation between first integrals and symmetries is analyzed together with Poincaré's absolute and relative integral invariants. Liouville's theorem and Poincaré's theorem are proved and the volume form on invariant submanifolds is defined. Time-dependent Hamiltonian mechanics and contact manifolds are introduced together with contact coordinates and locally and globally Hamiltonian fields.

18.1 Symplectic Structure of the Phase Space

Let V_n be the configuration space of a dynamical system S subject to fixed and smooth constraints and to conservative forces. In the last section of the preceding chapter, we proved that the Legendre map transfers the dynamics of S from the tangent fiber bundle $M_{2n} = TV_n$ to the cotangent fiber bundle $M_{2n}^* = T^*V_n$. Henceforth we call M_{2n}^* the **phase space** of the dynamical system S. In this space, the dynamical trajectories of S are solutions of the Hamilton equations (17.157) and (17.158).[1]

We recall some results regarding the tangent and cotangent fiber bundles (Chap. 6). If $(U, \mathbf{q})$ is a chart on the configuration space V_n, then $(\partial/\partial q^h)_{\mathbf{q}}$ and $(dq^h)_{\mathbf{q}}$ denote a basis of the tangent vector space $T_{\mathbf{q}}(V_n)$ and a basis of the cotangent vector space $T_{\mathbf{q}}^*(V_n)$, respectively. An element $\mathbf{x} \in M_{2n}^*$ is a pair $(\mathbf{q}, \mathbf{p})$, with $\mathbf{q} \in V_n$ and $\mathbf{p} \in T_{\mathbf{q}}^*(V_n)$, that is determined by the $2n$ numbers $(\mathbf{q}, \mathbf{p}) \in U \times \Re^n$. The charts $(U \times \Re^n, (\mathbf{q}, \mathbf{p}))$ are called **natural charts**. It is evident that an atlas on V_n generates an atlas of natural charts on M_{2n}^*, called a **natural atlas** on M_{2n}^*.

---

[1] For the contents of Chapters 18–20 see also [1], [3], [19], [30].

© Springer International Publishing AG, part of Springer Nature 2018
A. Romano and A. Marasco, *Classical Mechanics with Mathematica®*,
Modeling and Simulation in Science, Engineering and Technology,
https://doi.org/10.1007/978-3-319-77595-1_18

A natural basis relative to a natural chart $(U \times \mathfrak{R}^n, (\mathbf{q}, \mathbf{p}))$ of M^*_{2n} is formed by the vectors $(\partial/\partial q^h, \partial/\partial p_h)$ tangent to the coordinate curves.

In a natural chart $(U \times \mathfrak{R}^n, (\mathbf{q}, \mathbf{p}))$ of M^*_{2n}, consider the differential form[2]

$$\omega = p_h \mathbf{d}q^h. \tag{18.1}$$

Similarly, in another natural chart $(\overline{U} \times \mathfrak{R}^n, (\overline{\mathbf{q}}, \overline{\mathbf{p}}))$, with $\overline{U} \cap U \neq \emptyset$, we define a new differential form

$$\overline{\omega} = \overline{p}_h \mathbf{d}\overline{q}^h. \tag{18.2}$$

Recalling that a transformation from natural coordinates to natural coordinates is expressed by the equations

$$\overline{q}^h = \overline{q}^h(\mathbf{q}), \tag{18.3}$$

$$\overline{p}_h = \frac{\partial q^k}{\partial \overline{q}^h}(\mathbf{q}) p_k, \tag{18.4}$$

in the intersection $(U \times \mathfrak{R}^n) \cap (\overline{U} \times \mathfrak{R}^n)$ between the two charts, we have that

$$\overline{\omega} = \omega, \tag{18.5}$$

and we can state that *the local differential forms* (18.1) *determine a unique differential form ω on the whole manifold* M^*_{2n}. Together with this *global* differential form, we consider on the *whole* M^*_{2n} the 2-form

$$\Omega = -\mathbf{d}\omega, \tag{18.6}$$

which in any natural chart has the following coordinate representation:

$$\mathbf{d}q^h \wedge \mathbf{d}p_h. \tag{18.7}$$

In view of (18.6), the 2-form Ω is *closed*, that is,

$$\mathbf{d}\Omega = \mathbf{0}. \tag{18.8}$$

Further, if Ω is intended as a skew-symmetric tensor, then in the natural basis $(\mathbf{d}q^h \otimes \mathbf{d}q^k, \mathbf{d}q^h \otimes \mathbf{d}p_k, \mathbf{d}p_h \otimes \mathbf{d}q^k, \mathbf{d}p_h \otimes \mathbf{d}p_k)$ of $T_{(\mathbf{q}, \mathbf{p})}M^*_{2n}$, its components are given by the $2n \times 2n$ skew-symmetric matrix

---

[2]The most general differential form on M^*_{2n} is

$$\omega = X_h(\mathbf{q}, \mathbf{p})\mathbf{d}q^h + X^h(\mathbf{q}, \mathbf{p})\mathbf{d}p_h.$$

Then (18.1) is a particular nonexact differential form.

$$\Omega = \begin{pmatrix} 0 & I \\ -I & 0 \end{pmatrix}, \tag{18.9}$$

where 0 and I denote, respectively, an $n \times n$ zero matrix and an $n \times n$ unit matrix.

Theorem 18.1. *The matrix Ω has the following properties:*

$$\det \Omega = 1, \quad \Omega^2 = -I, \quad \Omega^{-1} = -\Omega. \tag{18.10}$$

Proof. Equations $(18.10)_{1,2}$ follow at once from (18.9). Further, we also have that $\Omega(\Omega) = -I$, and $(18.10)_3$ is proved. $\qquad\square$

Definition 18.1. The pair (M^*_{2n}, Ω), where Ω is a global closed and nondegenerate 2-form, is called a *symplectic manifold*. It is the fundamental geometric structure of Hamiltonian mechanics independent of time.

18.2 Canonical Coordinates

Definition 18.2. A local chart $(U, (\mathbf{q}, \mathbf{p}))$ on M^*_{2n} is a *canonical* or *symplectic chart* if in the coordinates $(\mathbf{q}, \mathbf{p})$ the components of Ω are given by matrix (18.9).

In the preceding section, it was proved that natural coordinates are canonical. To determine other canonical coordinates, denote by $(\mathbf{q}, \mathbf{p})$ natural coordinates and consider the coordinate transformation

$$\overline{\mathbf{q}} = \overline{\mathbf{q}}(\mathbf{q}, \mathbf{p}), \tag{18.11}$$

$$\overline{\mathbf{p}} = \overline{\mathbf{p}}(\mathbf{q}, \mathbf{p}). \tag{18.12}$$

We prove the following theorem.

Theorem 18.2. *The coordinate transformation (18.11), (18.12) is canonical if and only if the Jacobian matrix*

$$\mathbf{J} = \begin{pmatrix} \dfrac{\partial \overline{\mathbf{q}}}{\partial \mathbf{q}} & \dfrac{\partial \overline{\mathbf{q}}}{\partial \mathbf{p}} \\[2ex] \dfrac{\partial \overline{\mathbf{p}}}{\partial \mathbf{q}} & \dfrac{\partial \overline{\mathbf{p}}}{\partial \mathbf{p}} \end{pmatrix} \tag{18.13}$$

satisfies one of the following conditions:

$$\Omega = (\mathbf{J}^{-1})^T \Omega \mathbf{J}^{-1}, \tag{18.14}$$

$$\mathbf{J}^T \Omega = \Omega \mathbf{J}^{-1}, \tag{18.15}$$

$$\Omega = \mathbf{J}^T \Omega \mathbf{J}, \quad \Omega = \mathbf{J} \Omega \mathbf{J}^T. \tag{18.16}$$

Proof. Introducing the notations $(\overline{x}^\alpha) = (\overline{\mathbf{q}}, \overline{\mathbf{p}})$, $(x^\alpha) = (\mathbf{q}, \mathbf{p})$, $\alpha = 1, \dots, 2n$, the transformation (18.11), (18.12) can be written as follows:

$$\overline{x}^\alpha = \overline{x}^\alpha(x^\beta). \tag{18.17}$$

Since $\boldsymbol{\Omega}$ is a $(0, 2)$-tensor, it is transformed according to the rule

$$\overline{\Omega}_{\alpha\beta} = \frac{\partial x^\lambda}{\partial \overline{x}^\alpha} \frac{\partial x^\mu}{\partial \overline{x}^\beta} \Omega_{\lambda\mu} = \frac{\partial x^\lambda}{\partial \overline{x}^\alpha} \Omega_{\lambda\mu} \frac{\partial x^\mu}{\partial \overline{x}^\beta}.$$

Then transformation (18.13) is canonical if and only if $\overline{\Omega}_{\alpha\beta} = \Omega_{\alpha\beta}$, where $\Omega_{\alpha\beta}$ is given by (18.9), and (18.14) is proved. Multiplying (18.14) on the left by $\mathbf{J}^T$, we obtain (18.15). Further, the first condition (18.16) follows from (18.15) multiplying on the right by $\mathbf{J}$. Finally, from condition (18.14) we have $\boldsymbol{\Omega}^{-1} = \mathbf{J} \boldsymbol{\Omega}^{-1} \mathbf{J}^T$ and recalling that $\boldsymbol{\Omega}^{-1} = -\boldsymbol{\Omega}$, the second condition (18.16) is proved.

Definition 18.3. Let $f_h(\mathbf{q}, \mathbf{p})$, $g_h(\mathbf{q}, \mathbf{p})$ be $h = 1, \dots, n$, $2n$ C^1 functions on an open region of M_{2n}^*. We call the function

$$[q^h, p_k] \equiv \sum_{j=1}^{n} \left(\frac{\partial f_j}{\partial q^h} \frac{\partial g_j}{\partial p_k} - \frac{\partial f_j}{\partial p_k} \frac{\partial g_j}{\partial q^h} \right) \tag{18.18}$$

a *Lagrange bracket* of $2n$ functions with respect to a pair of variables (q^h, p_k). In a similar way, we define Lagrange's bracket with respect to a pair (q^h, q^k) or (p_h, p_k).

Definition 18.4. Let $f(\mathbf{q}, \mathbf{p})$ and $g(\mathbf{q}, \mathbf{p})$ be two C^1 functions on an open region of M_{2n}^*. We call the function

$$\{f, g\} \equiv \sum_{j=1}^{n} \left(\frac{\partial f}{\partial q^j} \frac{\partial g}{\partial p_j} - \frac{\partial f}{\partial p_j} \frac{\partial g}{\partial q^j} \right). \tag{18.19}$$

a *Poisson bracket*.

It is a simple exercise to prove the following theorem.

Theorem 18.3. *Conditions (18.14)–(18.16) are equivalent, respectively, to*

$$[q^h, q^k] = 0, \quad [q^h, p_k] = \delta_k^h, \quad [p_h, p_k] = 0; \tag{18.20}$$

$$\frac{\partial \overline{p}_k}{\partial q^h} = -\frac{\partial p_h}{\partial \overline{q}^k}, \quad \frac{\partial \overline{q}^k}{\partial q^h} = \frac{\partial p_h}{\partial \overline{p}_k},$$

$$\frac{\partial \overline{p}_k}{\partial p_h} = \frac{\partial q^h}{\partial \overline{q}^k}, \quad \frac{\partial \overline{q}^k}{\partial p_h} = -\frac{\partial q^h}{\partial \overline{p}_k}; \tag{18.21}$$

$$\{\overline{q}^h, \overline{q}^k\} = 0, \ \{\overline{q}^h, \overline{p}_k\} = \delta_k^h, \ \{\overline{p}_h, \overline{p}_k\} = 0. \tag{18.22}$$

We conclude by noting that the composition of two symplectic transformations is still symplectic; the identity is symplectic, and the inverse of a symplectic transformation is also symplectic. In other words, the set of symplectic transformations is a group with respect to the composition. Such a group is called a *symplectic group of canonical transformations*.

18.3 Generating Functions of Canonical Transformations

For now, we do not know if there are canonical charts besides the natural ones. Further, although the preceding criteria are necessary and sufficient for the canonicity of a transformation, it is not an easy task to obtain by them a canonical chart. In this section, we show a different approach to overcoming this difficulty.

We have already stated that the charts $(U, (\mathbf{q}, \mathbf{p}))$ and $(\overline{U}, (\overline{\mathbf{q}}, \overline{\mathbf{p}}))$ of M_{2n}^* are symplectic if and only if [see (18.7) and (18.9)]

$$\mathbf{d}q^k \wedge \mathbf{d}p_h = \mathbf{d}\overline{q}^k \wedge \mathbf{d}\overline{p}_h \tag{18.23}$$

in $U \bigcap \overline{U}$. This condition is equivalent to

$$\mathbf{d}(p_h \mathbf{d}q^h) = \mathbf{d}(\overline{p}_h \mathbf{d}\overline{q}^h),$$

which, in turn, can also be written as

$$\overline{p}_h \mathbf{d}\overline{q}^h = p_h \mathbf{d}q^h - \mathbf{d}f, \tag{18.24}$$

where f is a C^1 function in $U \bigcap \overline{U}$. We show that this condition, which is equivalent to (18.23), allows us to generate a symplectic transformation for a convenient choice of the function f. For this reason, f is called a *generating function* of the canonical transformation.

First, we note that system (18.11), (18.12) defines a canonical transformation if it is invertible, i.e., if

$$\det \begin{pmatrix} \dfrac{\partial \overline{\mathbf{q}}}{\partial \mathbf{q}} & \dfrac{\partial \overline{\mathbf{q}}}{\partial \mathbf{p}} \\ \dfrac{\partial \overline{\mathbf{p}}}{\partial \mathbf{q}} & \dfrac{\partial \overline{\mathbf{p}}}{\partial \mathbf{p}} \end{pmatrix} \neq 0, \tag{18.25}$$

and it confirms condition (18.23) or, equivalently, (18.24). To understand how we can generate a symplectic transformation starting from a function f, we must consider system (18.11), (18.12) as a system of $2n$ equations in $4n$ unknowns. In particular, we already know that this system, in view of (18.25), establishes a one-to-one correspondence between the $2n$ variables $(\overline{\mathbf{q}}, \overline{\mathbf{p}})$ and the variables $(\mathbf{q}, \mathbf{p})$. Now, we suppose that *this system also defines a one-to-one correspondence between a suitable choice of $2n$ variables among the $4n$ variables* $(\mathbf{q}, \mathbf{p}, \overline{\mathbf{q}}, \overline{\mathbf{p}})$

For instance, suppose that (18.11) and (18.12) satisfy the further condition

$$\det \left(\frac{\partial \overline{q}^h}{\partial p_k} \right) \neq 0. \tag{18.26}$$

Then, locally, (18.11) and (18.12) can be written as

$$\mathbf{p} = \mathbf{p}(\mathbf{q}, \overline{\mathbf{q}}), \tag{18.27}$$
$$\overline{\mathbf{p}} = \overline{\mathbf{p}}(\mathbf{q}, \overline{\mathbf{q}}), \tag{18.28}$$

and the canonicity condition (18.24) becomes

$$\overline{p}_h d\overline{q}^h = p_h dq^h - df(\mathbf{q}, \mathbf{p}(\mathbf{q}, \overline{\mathbf{q}})) \equiv p_h dq^h - d\hat{f}(\mathbf{q}, \overline{\mathbf{q}}). \tag{18.29}$$

Expanding the differential $d\hat{f}$, we obtain the following equality:

$$\left(\overline{p}_h + \frac{\partial \hat{f}}{\partial \overline{q}^h} \right) d\overline{q}^h - \left(p_h - \frac{\partial \hat{f}}{\partial q^h} \right) dq^h = 0,$$

in which dq^h and $d\overline{q}^h$ are arbitrary. Therefore, (18.29) is equivalent to the following conditions:

$$\overline{p}_h = -\frac{\partial \hat{f}}{\partial \overline{q}^h}(\mathbf{q}, \overline{\mathbf{q}}), \tag{18.30}$$

$$p_h = \frac{\partial \hat{f}}{\partial q^h}(\mathbf{q}, \overline{\mathbf{q}}). \tag{18.31}$$

To put the canonical transformation (18.30), (18.31) into the form (18.11), (18.12), we must suppose that $\hat{f}(\mathbf{q}, \overline{\mathbf{q}})$ satisfies the following condition:

$$\det \left(\frac{\partial^2 \hat{f}}{\partial \overline{q}^h \partial q^k} \right) \neq 0. \tag{18.32}$$

As a further example, we assume $(\mathbf{q}, \overline{\mathbf{p}})$ as fundamental variables and suppose that (18.11), (18.12) can be written as

$$\mathbf{p} = \mathbf{p}(\mathbf{q}, \overline{\mathbf{p}}),\qquad(18.33)$$

$$\overline{\mathbf{q}} = \overline{\mathbf{q}}(\mathbf{q}, \overline{\mathbf{p}}).\qquad(18.34)$$

Then, (18.24) gives the condition

$$\overline{p}_h \mathbf{d}\overline{q}^h = p_h \mathbf{d}q^h - \mathbf{d}f(\mathbf{q}, \mathbf{p}(\mathbf{q}, \overline{\mathbf{p}})) \equiv p_h \mathbf{d}q^h - \mathbf{d}\hat{f}(\mathbf{q}, \overline{\mathbf{p}}),$$

which is equivalent to

$$p_h \mathbf{d}q^h + \overline{q}^h \mathbf{d}\overline{p}_h = \mathbf{d}\tilde{f}(\mathbf{q}, \overline{\mathbf{p}}),\qquad(18.35)$$

where

$$\tilde{f}(\mathbf{q}, \overline{\mathbf{p}}) = \hat{f}(\mathbf{q}, \overline{\mathbf{p}}) + \overline{q}^h(\mathbf{q}, \overline{\mathbf{p}})\overline{p}_h.$$

Expanding the differential of $\tilde{f}(\mathbf{q}, \overline{\mathbf{p}})$ and recalling the arbitrariness of $\mathbf{d}q^h$ and $\mathbf{d}\overline{p}_h$, we obtain that (18.29) is equivalent to the system

$$p_h = \frac{\partial \tilde{f}}{\partial q^h}(\mathbf{q}, \overline{\mathbf{p}}),\qquad(18.36)$$

$$\overline{q}^h = \frac{\partial \tilde{f}}{\partial \overline{p}_h}(\mathbf{q}, \overline{\mathbf{p}}).\qquad(18.37)$$

To put the canonical transformation (18.36), (18.37) into the form (18.11), (18.12), we now suppose that $\tilde{f}(\mathbf{q}, \overline{\mathbf{p}})$ satisfies the condition

$$\det\left(\frac{\partial^2 \tilde{f}}{\partial q^h \partial \overline{p}_k}\right) \neq 0.\qquad(18.38)$$

In conclusion, there are infinite canonical charts on M^* besides the natural ones.

The notebook **TransfCan**, containing the programs **TransfCan** and **FunGen**, allows to verify if an independent on time coordinate transformation is canonical or not.

18.4 Isomorphism Between Vector Fields and Differential Forms

Up to now, the skew-symmetric 2-tensor or the 2-form $\mathbf{\Omega}$ has played a fundamental role in selecting the canonical charts on M_{2n}^*. Further, at any point $x \in M_{2n}^*$, it defines an antiscalar product $\mathbf{\Omega}_x(\mathbf{X}, \mathbf{Y})$, $\forall \mathbf{X}, \mathbf{Y} \in T_x M_{2n}^*$, in any vector tangent space $T_x M_{2n}^*$

that becomes a symplectic space. In this section, we present another fundamental role played by $\boldsymbol{\Omega}$.

We already know (Chap. 6) that on a Riemannian manifold V_n the metric tensor $\mathbf{g}$ defines at any point $x \in V_n$ an isomorphism between the tangent space $T_x(V_n)$ and its dual space $T_x^*(V_n)$. Similarly, on the symplectic manifold $(M_{2n}^*, \boldsymbol{\Omega})$ we consider the linear map

$$\mathbf{X} \in T_x M_{2n}^* \to \omega \in T_x^* M_{2n}^*$$

such that

$$\omega(\mathbf{Y}) = \Omega_x(\mathbf{X}, \mathbf{Y}), \quad \forall \, \mathbf{Y} \in T_x M_{2n}^*. \tag{18.39}$$

Let (U, x^α), $\alpha = 1, \ldots, 2n$, be any local chart on M_{2n}^* in a neighborhood of x. If we denote by $(\partial/\partial x^h)$ the corresponding natural basis at x and by $\mathbf{d}x^\alpha$ the dual basis, then (18.39) has the following coordinate representation:

$$\omega_\alpha = (\Omega_x)_{\beta\alpha} X^\beta. \tag{18.40}$$

Since, in view of (18.10), $\det((\Omega_x)_{\alpha\beta}) = 1$, the linear mapping (18.40) is an isomorphism.

In a symplectic chart $(U, x^\alpha) = (U, (\mathbf{q}, \mathbf{p}))$, we have that

$$\left(\frac{\partial}{\partial x^\alpha} \right) = \left(\frac{\partial}{\partial q^h}, \frac{\partial}{\partial p_h} \right),$$
$$(\mathbf{d}x^\alpha) = (\mathbf{d}q^h, \mathbf{d}p_h),$$
$$\mathbf{X} = X^h \frac{\partial}{\partial q^h} + X_h \frac{\partial}{\partial p_h},$$
$$\omega = \omega_h \mathbf{d}q^h + \omega^h \mathbf{d}p_h,$$

and (18.40) becomes

$$(\omega_1, \ldots, \omega_n, \omega^1, \ldots, \omega^n) = (X^1, \ldots, X^n, X_1, \ldots, X_n) \begin{pmatrix} \mathbf{0} & \mathbf{I} \\ -\mathbf{I} & \mathbf{0} \end{pmatrix},$$

so that

$$\omega_h = -X_h, \tag{18.41}$$
$$\omega^h = X^h. \tag{18.42}$$

Henceforth we denote the 1-form corresponding to $\mathbf{X}$ by isomorphism (18.39) and the vector $\mathbf{X}$ corresponding to ω by the inverse isomorphism, respectively, with the notations ($\mathbf{X}$ flat, ω sharp)

$$\omega = \mathbf{X}^\flat, \quad \mathbf{X} = \omega^\sharp. \tag{18.43}$$

We conclude this section by noting that, since $\boldsymbol{\Omega}$ is a closed 2-form, from (8.28) we obtain

$$L_{\mathbf{X}}\boldsymbol{\Omega} = \mathbf{d}(i_{\mathbf{X}}\boldsymbol{\Omega}). \tag{18.44}$$

But in any system of coordinates, we have that

$$(i_{\mathbf{X}}\boldsymbol{\Omega})_\alpha = X^\beta \Omega_{\beta\alpha},$$

so that

$$i_{\mathbf{X}}\boldsymbol{\Omega} = \mathbf{X}^\flat. \tag{18.45}$$

In conclusion, from (18.44) and (18.45) we obtain the useful result

$$L_{\mathbf{X}}\boldsymbol{\Omega} = \mathbf{d}(\mathbf{X}^\flat), \quad \forall \mathbf{X} \in \chi M_{2n}^*. \tag{18.46}$$

18.5 Global and Local Hamiltonian Vector Fields

Definition 18.5. Let M_{2n}^* be the phase state. A vector field $\mathbf{X} \in \chi M_{2n}^*$ is **globally Hamiltonian** if there exists a differentiable function $H(x) \in \mathcal{F}M_{2n}^*$ such that

$$\mathbf{X}_H = (\mathbf{d}H)^\sharp. \tag{18.47}$$

The function $H(x)$ is called a **Hamiltonian function** of $\mathbf{X}$.

The following theorem is evident.

Theorem 18.4. *The set $\chi_H M_{2n}^*$ of Hamiltonian fields on M_{2n}^* coincides with the image of the exact 1-forms on M_{2n}^* by the linear isomorphism $\sharp$. In particular, $\chi_H M_{2n}^*$ is a subspace of χM_{2n}^*.*

In symplectic coordinates (q^h, p_h), from (18.41) and (18.42) we obtain

$$\mathbf{d}H = \begin{pmatrix} \dfrac{\partial H}{\partial \mathbf{q}} \\ \dfrac{\partial H}{\partial \mathbf{p}} \end{pmatrix} \xrightarrow{\sharp} \mathbf{X}_H = \begin{pmatrix} \dfrac{\partial H}{\partial \mathbf{p}} \\ -\dfrac{\partial H}{\partial \mathbf{q}} \end{pmatrix}. \tag{18.48}$$

The integral curves $\gamma(t)$ of a globally Hamiltonian vector field satisfy the equation

$$\dot{\gamma}(t) = \mathbf{X}_H(\gamma(t)) = (\mathbf{d}H)^\sharp(\gamma(t)), \tag{18.49}$$

which in symplectic coordinates assumes the canonical form

$$\dot{q}^h = \frac{\partial H}{\partial p_h},$$

$$\dot{p}^h = -\frac{\partial H}{\partial q^h}.$$

Definition 18.6. A *dynamic Hamiltonian system* is a triad (M, Ω, H).

Definition 18.7. A vector field $\mathbf{X} \in \chi M_{2n}^*$ is said to be *locally Hamiltonian* if

$$L_{\mathbf{X}} \Omega = 0, \tag{18.50}$$

i.e., if Ω is invariant along the integral curves of the vector field $\mathbf{X}$ or, equivalently, if it is invariant with respect to the one-parameter group of diffeomorphisms generated by $\mathbf{X}$.

Theorem 18.5. *If* $\mathbf{X}$ *is globally Hamiltonian, then it is locally Hamiltonian. If* $\mathbf{X}$ *is locally Hamiltonian, then* $\forall x \in M_{2n}^*$ *there exist a neighborhood* U *of* x *and a function* H *of class* $C^1(U)$ *such that*

$$\mathbf{X} = (\mathbf{d}H)^\sharp, \quad \forall \, x \in U. \tag{18.51}$$

Proof. The hypothesis that $\mathbf{X}$ is globally Hamiltonian implies $\mathbf{X} = (\mathbf{d}H)^\sharp$, and then $\mathbf{X}^\flat = \mathbf{d}H$. From (18.46) it follows that $L_{\mathbf{X}} \Omega = 0$. Conversely, in view of (18.46), the condition $L_{\mathbf{X}} \Omega = 0$ is equivalent to $\mathbf{d}(\mathbf{X}^\flat) = 0$, and this condition, in turn, implies the existence of a function H for which $\mathbf{X}^\flat = \mathbf{d}H$ or $\mathbf{X} = (\mathbf{d}H)^\sharp$ only locally. □

Definition 18.8. A one-parameter group of diffeomorphisms φ_t of M_{2n}^* is said to be *symplectic* if Ω is invariant with respect to any diffeomorphism φ_t, that is, if (18.50) holds.

18.6   Poisson Brackets

Definition 18.9. $\forall f, g \in \mathcal{F} M_{2n}^*$, the function

$$\{f, g\} = \Omega(\mathbf{d}f^\sharp, \mathbf{d}g^\sharp) = \Omega(\mathbf{X}_f, \mathbf{X}_g) \tag{18.52}$$

is the *Poisson brackets* of f and g.

Theorem 18.6. *In a symplectic system of coordinates, (18.52) gives*

$$\{f, g\} = \sum_{h=1}^{n} \left(\frac{\partial f}{\partial q^h} \frac{\partial g}{\partial p_h} - \frac{\partial f}{\partial p_h} \frac{\partial g}{\partial q^h} \right). \tag{18.53}$$

Proof. In fact,

$$\mathbf{d}f = \frac{\partial f}{\partial q^h}\mathbf{d}q^h + \frac{\partial f}{\partial p_h}\mathbf{d}p_h,$$

$$\mathbf{d}g = \frac{\partial g}{\partial q^h}\mathbf{d}q^h + \frac{\partial g}{\partial p_h}\mathbf{d}p_h$$

Owing to (18.41) and (18.42), in symplectic coordinates we have that

$$\mathbf{d}f^\sharp = \frac{\partial f}{\partial p_h}\frac{\partial}{\partial q^h} - \frac{\partial f}{\partial q^h}\frac{\partial}{\partial p_h},$$

$$\mathbf{d}g^\sharp = \frac{\partial g}{\partial p_h}\frac{\partial}{\partial q^h} - \frac{\partial g}{\partial q^h}\frac{\partial}{\partial p_h},$$

and in matrix form (18.52) gives

$$\{f, g\} = \left(\begin{array}{cc} \dfrac{\partial f}{\partial p_h} & -\dfrac{\partial f}{\partial q^h} \end{array}\right)\left(\begin{array}{cc} \mathbf{0} & \mathbf{I} \\ -\mathbf{I} & \mathbf{0} \end{array}\right)\left(\begin{array}{c} \dfrac{\partial g}{\partial p_h} \\ -\dfrac{\partial g}{\partial q^h} \end{array}\right)$$

$$= \left(\begin{array}{cc} \dfrac{\partial f}{\partial p_h} & -\dfrac{\partial f}{\partial q^h} \end{array}\right)\left(\begin{array}{c} -\dfrac{\partial g}{\partial q^h} \\ -\dfrac{\partial g}{\partial p_h} \end{array}\right), \tag{18.54}$$

and (18.53) is proved.   □

Theorem 18.7. $\forall f, g \in \mathcal{F}M_{2n}^*$, *we have that*

$$\{f, g\} = L_{\mathbf{X}_g}f = -L_{\mathbf{X}_f}g, \tag{18.55}$$

$$\left[\mathbf{X}_f, \mathbf{X}_g\right] = -\mathbf{X}_{\{f,g\}}. \tag{18.56}$$

Proof. In view of (18.39) and (18.43), we have that

$$\{f, g\} = \mathbf{\Omega}(\mathbf{d}f^\sharp, \mathbf{d}g^\sharp) = \mathbf{d}f(\mathbf{X}_g) = \mathbf{X}_g f = L_{\mathbf{X}_g}f.$$

From the skew symmetry of $\mathbf{\Omega}$ follows (18.55). We omit the proof of (18.56).   □

Theorem 18.8. $\forall f, g, h \in \mathcal{F}M_{2n}^*$ *and* $\forall a, b \in \mathfrak{R}$, *the following identities hold:*

$$\{f, f\} = 0, \tag{18.57}$$

$$\{f, g\} = -\{g, f\}, \tag{18.58}$$

$$\{f, ag + bh\} = a\{f, g\} + b\{f, h\}, \tag{18.59}$$

$$\{f, \{g, h\}\} + \{g, \{h, f\}\} + \{h, \{f, g\}\} = 0. \tag{18.60}$$

Proof. Equations (18.57)–(18.59) follow at once from (18.52). Further, in view of (18.55) and (18.56), we have

$$\{f, \{g, h\}\} = -L_{\mathbf{X}_f}\{g, h\} = L_{\mathbf{X}_f}L_{\mathbf{X}_g}h,$$
$$\{g, \{h, f\}\} = -\{g, \{f, h\}\} = -L_{\mathbf{X}_g}L_{\mathbf{X}_f}h,$$

$$\{h, \{f, g\}\} = L_{\mathbf{X}_{\{f,g\}}}h = \mathbf{X}_{\{f,g\}}h$$
$$= -[\mathbf{X}_f, \mathbf{X}_g]h = -(L_{\mathbf{X}_f}L_{\mathbf{X}_g} - L_{\mathbf{X}_g}L_{\mathbf{X}_f})h,$$

and (18.60) is proved.   □

Remark 18.1. We know (Theorem 6.4) that the binary operation

$$(\mathbf{X}, \mathbf{Y}) \in \chi M^*_{2n} \times \chi M^*_{2n} \to [\mathbf{X}, \mathbf{Y}] \in \chi M^*_{2n} \tag{18.61}$$

gives the $\Re$ vector space χM^*_{2n} a Lie algebra structure. Further, in view of (18.56), this operation associates a globally Hamiltonian field to a pair of globally Hamiltonian fields. In other words, these fields form a Lie subalgebra with respect to the bracket operation. On the other hand, Theorem 18.8 states that the $\Re$ vector space $\mathcal{F}M^*_{2n}$ is also a Lie algebra with respect to the Poisson bracket

$$(f, g) \in \mathcal{F}M^*_{2n} \times \mathcal{F}M^*_{2n} \to \{f, g\} \in \mathcal{F}M^*_{2n}.$$

Finally, from (18.56) it follows that the mapping

$$\phi : f \in \mathcal{F}M^*_{2n} \to \mathbf{X} = (\mathbf{d}f)^\sharp \in \chi M^*_{2n} \tag{18.62}$$

is a morphism between the preceding two Lie algebras. It is not an isomorphism because the kernel of ϕ is given by locally constant functions since

$$\phi(f) = (\mathbf{d}f)^\sharp = 0$$

implies $\mathbf{d}f = 0$.

18.7 First Integrals and Symmetries

Definition 18.10. Let $f \in \mathcal{F}M^*_{2n}$ be a differentiable function such that $\mathbf{d}f \neq 0$, $\forall x \in M^*_{2n}$. Then, f is said to be a *first integral* of the dynamical system $(M^*_{2n}, \mathbf{\Omega}, H)$ if

$$f(\gamma(t)) = const \tag{18.63}$$

along the integral curves of the Hamiltonian field $\mathbf{X}_H$.

In other words, f is a first integral if any integral curve $\gamma(t)$ of the Hamiltonian vector field $\mathbf{X}_H$ that crosses the level surface $f(x) = const$ is fully contained in this surface S. Equivalently, the restriction to S of the vector field $\mathbf{X}_H$ is tangent to S. From a physical point of view, a first integral gives a conservation law.

Condition (18.63) can be equivalently expressed by the conditions

$$\dot{f} = \mathbf{X}_H f = L_{\mathbf{X}_H} f = 0. \tag{18.64}$$

Recalling (18.55), we can state the following theorem.

Theorem 18.9. *The function $f \in \mathcal{F}M_{2n}^*$, such that $\mathbf{d}f \neq 0$, is a first integral if and only if*

$$\{f, H\} = 0. \tag{18.65}$$

Theorem 18.10. *We have that*

1. *H is a first integral of the dynamical system (M_{2n}^*, Ω, H);*
2. *If f is a first integral of (M_{2n}^*, Ω, H), then H is a first integral of (M_{2n}^*, Ω, f);*
3. *If f and g are first integrals of (M_{2n}^*, Ω, H), then $\{f, g\}$ is also a first integral of (M_{2n}^*, Ω, H).*

Proof. Since $\{H, H\} = -\{H, H\}$, we have that $\{H, H\} = 0$, and the first statement is proved. Again, property 2 follows from the skew symmetry of the Poisson brackets. Finally, the third statement follows at once from (18.60) and the conditions $\{f, H\} = \{g, H\} = 0$. $\square$

Remark 18.2. We underline that the first integral $\{f, g\}$ could be dependent on the first integrals f and g.

Remark 18.3. The first integral $H(x) = const$ states the conservation of the mechanical energy of the *mechanical* system (M_{2n}^*, Ω, H) with forces and constraints independent of time. In fact, for such a system we have that [see (17.156)]

$$H(q^h, p_h) = \left[p_h v^h - L(\mathbf{q}, \mathbf{v})\right]_{\mathbf{v}=\mathbf{v}(\mathbf{q},\mathbf{p})}$$

in any system of natural coordinates. But in our hypotheses,

$$L = T - U = \frac{1}{2}a_{hk}v^h v^k - U(\mathbf{q}),$$

where $U(\mathbf{q})$ is the potential energy of the acting forces. Further, we have that $p_h = \partial L/\partial v^h = a_{hk}v^k$. If we denote by (a^{hk}) the inverse matrix of (a_{hk}), then $v^h = a^{hk}p_k$, and the Hamiltonian function assumes the form

$$H = T + U = \frac{1}{2} a^{hk} p_h p_k + U(\mathbf{q}).$$

Remark 18.4. If in a system of symplectic coordinates $H(\mathbf{q}, \mathbf{p})$ does not depend on one of the coordinates q^h, for instance on q^1, then the coordinate q^h is said to be *cyclic*. In this case, the Hamiltonian equations show that the conjugate momentum is constant.

Definition 18.11. Let $f \in \mathcal{F} M_{2n}^*$ be a function such that $\mathbf{d} f \neq \mathbf{0}$. Suppose that the globally Hamiltonian field $\mathbf{X}_f = (\mathbf{d} f)^{\sharp}$ is complete on M_{2n}^* and denote by $\varphi_s \colon \mathfrak{R} \times M_{2n}^* \to M_{2n}^*$ the symplectic group generated by $\mathbf{X}_f$ (Definition 18.8). We say that φ_t is a *symmetry group* for the Hamiltonian function H of the dynamical system $(M_{2n}^*, \mathbf{\Omega}, H)$ if

$$H(x) = H(\varphi_s(x)), \quad \forall s \in \mathfrak{R}, \ \forall x \in M_{2n}^* \tag{18.66}$$

or, equivalently, if

$$L_{\mathbf{X}_f} H = 0. \tag{18.67}$$

The following theorem extends Noether's theorem to the Hamiltonian formalism.

Theorem 18.11. *The function f is a first integral of the dynamical system $(M_{2n}^*, \mathbf{\Omega}, H)$ if and only if H admits a symmetry group generated by the complete globally Hamiltonian field $\mathbf{X}_f$.*

Proof. It is sufficient to note that (18.64), (18.65), and (18.67) imply the following chain of identities:

$$\{f, H\} = 0 \Leftrightarrow L_{\mathbf{X}_f} H = 0 \Leftrightarrow H(x) = H(\varphi_s(x)). \tag{18.68}$$

$\square$

Remark 18.5. The last identity holds if $\mathbf{X}_f$ is complete. If M_{2n}^* is compact, then any Hamiltonian vector field is complete. In general, all Hamiltonian fields that are not complete can be only locally associated with a symmetry of H.

Remark 18.6. In Theorem 18.11, it is fundamental to suppose that the infinitesimal generator $\mathbf{X}$ of a symplectic group is Hamiltonian. In fact, if $\mathbf{X}$ generates a symplectic group of symmetry of H but it is not Hamiltonian, then we have that

$$L_{\mathbf{X}} H = 0, \quad L_{\mathbf{X}} \mathbf{\Omega} = \mathbf{0}.$$

The second condition, in view of (18.50), allows us to state that $\mathbf{X}$ is locally Hamiltonian, so that the existence of the first integral is only locally proved.

Remark 18.7. Noether's theorem states the *equivalence* between first integrals and symplectic symmetries generated by complete and globally Hamiltonian vector fields. This theorem is different from Theorem 17.5, in which it is proved that the symmetry of a Lagrangian function generated by a vector field $\mathbf{X}$ on the configuration space V_n only *implies* the existence of a first integral given by the formula

$$f = \frac{\partial L}{\partial \dot{q}^h} X^h.$$

This is due to a different definition of symmetry, which in Theorem 17.5 involves only transformations of the Lagrangian coordinates on V_n instead of transformations of the coordinates $(\mathbf{q}, \mathbf{v})$ on the fiber bundle TV_n. However, when this reduced class of transformations is taken into account, we have the advantage that the preceding formula explicitly gives the first integral. In contrast, in Theorem 18.11, the first integral is obtained by integrating the system of partial differential equations

$$(\mathbf{d}f)^\sharp = \mathbf{X}_f,$$

which in symplectic coordinates is written

$$X^h = \frac{\partial f}{\partial p_h}, \quad X_h = -\frac{\partial f}{\partial q^h}.$$

18.8 Poincaré's Absolute and Relative Integral Invariants

Definition 18.12. A k-form $\alpha \in \Lambda_k M_{2n}^*$ is an *absolute integral invariant* of the complete vector field $\mathbf{X}$ on M_{2n}^* if

$$L_{\mathbf{X}}\alpha = 0. \tag{18.69}$$

In other words, denoting by $\varphi_s(x)$ the one-parameter group of diffeomorphisms generated by $\mathbf{X}$, we can say that α is an absolute integral invariant if

$$\varphi_s^*(\alpha(\varphi_s(x))) = \alpha(x), \quad \forall x \in M_{2n}^*, \forall s \in \mathfrak{R}, \tag{18.70}$$

where φ_s^* is the codifferential of φ_s (Fig. 18.1).

Theorem 18.12. *Let Σ be an arbitrary differentiable k-dimensional submanifold of M_{2n}^*. Then α is an absolute integral invariant of the complete vector field $\mathbf{X}$ if and only if*

$$\int_\Sigma \alpha = \int_{\varphi_s(\Sigma)} \alpha, \quad \forall s \in \mathfrak{R}. \tag{18.71}$$

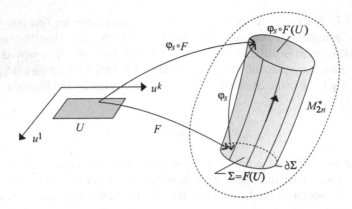

Fig. 18.1 Absolute integral invariant α

Proof. To simplify the proof, we suppose that Σ is a k-cube (U, σ, F) (Sect. 8.4), so that $\Sigma = F(U)$. From the definition of the integral of a k-form α we obtain

$$\int_{\varphi_s \circ \Sigma} \alpha = \int_U (\varphi_s \circ F)^* \alpha = \int_U F^* \varphi_s^* \alpha.$$

Consequently, if α is an integral invariant, then (18.70) implies that

$$\int_{\varphi_s \circ \Sigma} \alpha = \int_U F^* \alpha = \int_\Sigma \alpha,$$

and the first part of the theorem is proved. Conversely, if (18.71) holds for any U, then

$$\int_U F^* \alpha = \int_U F^* \varphi_s^* \alpha.$$

From the arbitrariness of U and the continuity of the functions under the integral we obtain $F^* \circ \varphi_s^* \alpha = F^* \alpha$, that is, $\varphi_s^* \alpha = \alpha$.   □

Definition 18.13. The $(k-1)$-form α is a *relative integral invariant* of the complete vector field $\mathbf{X}$ on M_{2n}^* if

$$\mathbf{d}(L_\mathbf{X} \alpha) = 0. \tag{18.72}$$

Recalling (8.30), the preceding condition becomes

$$L_\mathbf{X}(\mathbf{d}\alpha) = 0, \tag{18.73}$$

and, in view of (18.69), we can say that α is a relative integral invariant if and only if $\mathbf{d}\alpha$ is an absolute integral invariant.

Theorem 18.13. *The* $(k - 1)$*-form* α *on* M_{2n}^* *is a relative integral invariant of the complete vector field* $\mathbf{X}$ *if and only if*

$$\int_{\partial\Sigma} \alpha = \int_{\varphi_s(\partial\Sigma)} \alpha, \tag{18.74}$$

where Σ *is an arbitrary* k*-dimensional submanifold of* M_{2n}^* *and* $\partial\Sigma$ *is its boundary.*

Proof. Again, we suppose that Σ is a k-cube (U, σ, F). Then, from (18.73) and Theorem 18.12, we have the condition

$$\int_{\Sigma} \mathbf{d}\alpha = \int_{\varphi_s(\Sigma)} \mathbf{d}\alpha,$$

which, applying Stokes' theorem (Chap. 8), becomes

$$\int_{\partial\Sigma} \alpha = \int_{\varphi_s(\partial\Sigma)} \alpha,$$

and the theorem is proved.   □

Theorem 18.14. *Let* $\mathbf{X}$ *be a complete vector field on* M_{2n}^*. *Then the following properties are equivalent:*

1. $\mathbf{X}_H$ *is locally Hamiltonian.*
2. *The* 2*-form* Ω *is an absolute integral invariant of* $\mathbf{X}$.
3. *The* 1*-form* ω, *such that* $\Omega = -\mathbf{d}\omega$, *is a relative integral invariant.*

Proof. If $\mathbf{X}$ is locally Hamiltonian, then (18.50) holds, so that Ω is an absolute integral invariant [see (18.69)] and ω is a relative integral invariant [see (18.73)]. The other implications are evident.   □

The preceding theorem allows us to define the complete locally Hamiltonian vector fields $\mathbf{X}$ on M_{2n}^* as vector fields characterized by one of the following conditions:

$$\int_{\Sigma} \Omega = \int_{\varphi_s(\partial\Sigma)} \Omega, \tag{18.75}$$

$$\int_{\Sigma} \omega = \int_{\varphi_s(\partial\Sigma)} \omega, \tag{18.76}$$

where Σ is an arbitrary two-dimensional submanifold of M_{2n}^* and φ_s is the (symplectic) one-parameter group of diffeomorphisms generated by $\mathbf{X}$.

In many books, the Hamiltonian fields are introduced starting with one of the foregoing properties. This is possible only if we refer to a coordinate neighborhood, not to the whole manifold M_{2n}^*.

18.9   Two Fundamental Theorems

Theorem 18.15 (Liouville's Theorem). *Let (M_{2n}^*, Ω, H) be a Hamiltonian dynamical system, and suppose that the Hamiltonian vector field $\mathbf{X}_H$ is complete. Then the $2n$-form*

$$\hat{\Omega} = \frac{(-1)^{\frac{n(n-1)}{2}}}{n!} \Omega^n \tag{18.77}$$

is an absolute integral invariant of $\mathbf{X}_H$, that is, it confirms one of the following conditions:

$$L_{\mathbf{X}_H} \hat{\Omega} = 0, \quad \int_W \hat{\Omega} = \int_{\varphi_s(W)} \hat{\Omega}, \tag{18.78}$$

where W is an arbitrary $2n$-dimensional submanifold of M_{2n}^* and φ_s a one-parameter symplectic group of diffeomorphisms generated by $\mathbf{X}_H$. Further, in symplectic coordinates,

$$\hat{\Omega} = \mathbf{d}q^1 \wedge \cdots \wedge \mathbf{d}q^n \wedge \mathbf{d}p_1 \wedge \cdots \wedge \mathbf{d}p_n. \tag{18.79}$$

Remark 18.8. Let (q^h, p_h) be a symplectic system of coordinates of M_{2n}^*, and consider the parallelepiped having edges $(\partial/\partial q^1, \ldots, \partial/\partial q^n, \partial/\partial p_1, \ldots, \partial/\partial p_n)$. We can state that $\hat{\Omega}$ is a volume n-form for which the volume of the preceding parallelepiped is equal to one. Consequently, if W is a $2n$-cube (U, σ, F), where $U \subset \mathfrak{R}^{2n}$, then the volume of the $2n$-dimensional submanifold W is given by

$$\int_{F(U)} \mathbf{d}q^1 \cdots \mathbf{d}q^n \mathbf{d}p_1 \cdots \mathbf{d}p_n. \tag{18.80}$$

Finally, regard W as a set of initial data (q_0^h, p_{0h}) of the Hamilton equations of the system (M_{2n}^*, Ω, H). Since φ_s is generated by $\mathbf{X}_H$, its orbits coincide with the dynamical trajectories of the Hamilton equations. Therefore, $\varphi_s \circ W$ is the submanifold formed by the configurations at the instant s of all the motions starting from the points of W. In conclusion, the preceding theorem states that the volume of this region does not vary over time.

Proof. Since $L_{\mathbf{X}_H}$ is a derivation (Chap. 8) and (18.50) holds, we have that

$$
\begin{aligned}
L_{\mathbf{X}_H} \hat{\Omega} &= \frac{(-1)^{\frac{n(n-1)}{2}}}{n!} L_{\mathbf{X}_H} \Omega^n \\
&= \frac{(-1)^{\frac{n(n-1)}{2}}}{n!} (L_{\mathbf{X}_H} \Omega \wedge \Omega \cdots \wedge \Omega + \cdots + \Omega \wedge \cdots \wedge L_{\mathbf{X}_H} \Omega) = 0,
\end{aligned}
$$

and $\hat{\Omega}$ is an absolute integral invariant. Further, in symplectic coordinates,

$$\boldsymbol{\Omega}^n = (\mathbf{d}q^1 \wedge \mathbf{d}p_1 + \cdots + \mathbf{d}q^n \wedge \mathbf{d}p_n) \wedge \\ \cdots \wedge (\mathbf{d}q^1 \wedge \mathbf{d}p_1 + \cdots + \mathbf{d}q^n \wedge \mathbf{d}p_n). \tag{18.81}$$

The only nonvanishing terms of the preceding product have the form $\mathbf{d}q^{i_1} \wedge \mathbf{d}p_{i_1} \wedge \cdots \wedge \mathbf{d}q^{i_n} \wedge \mathbf{d}p_{i_n}$, where each of the indices $i_1, \ldots, i_n$ is different from the others. Consequently, the right-hand side of (18.81) contains $n!$ terms. Consider the term $\mathbf{d}q^1 \wedge \mathbf{d}p_1 \wedge \cdots \wedge \mathbf{d}q^n \wedge \mathbf{d}p_n$. It assumes the form $\mathbf{d}q^1 \wedge \mathbf{d}q^n \wedge \cdots \wedge \mathbf{d}p_1 \wedge \mathbf{d}p_n$ by $1 + 2 + \cdots + (n-1) = n(n-1)/2$ inversions in each of which the term changes its sign. On the other hand, any other term can be reduced to the form $\mathbf{d}q^1 \wedge \mathbf{d}p_1 \wedge \cdots \wedge \mathbf{d}q^n \wedge \mathbf{d}p_n$ by suitable exchanges of the factors $\mathbf{d}q^i \wedge \mathbf{d}p_i$ that do not modify the sign of the considered term. In conclusion, (18.81) can be put in the form

$$\boldsymbol{\Omega}^n = n!(-1)^{\frac{n(n-1)}{2}} \mathbf{d}q^1 \wedge \cdots \wedge \mathbf{d}q^n \wedge \mathbf{d}p_1 \wedge \cdots \wedge \mathbf{d}p_n,$$

and (18.80) is proved. □

Theorem 18.16. *The equilibrium positions of a Hamiltonian system cannot be asymptotically stable.*

Proof. In fact, if $x^* \in M_{2n}^*$ is an asymptotically stable position of a Hamiltonian system, then there exists a neighborhood I of x^* such that the dynamical trajectories starting from the points $x_0 \in I$ tend to x^* when $t \to \infty$. In other words, the volume of $\varphi_s(I)$ goes to zero when $t \to \infty$, against Liouville's theorem. □

Example 18.1. The phase portrait of a simple pendulum is represented in Fig. 18.2 (Sect. 14.9). Since the lower the pendulum period the higher the amplitude of its oscillations, the dynamical trajectories starting from initial data contained in region W and closer to the origin describe a greater fraction of the oscillation than trajectories starting from points of W farther from the origin. Therefore, the evolution of the set of initial data W is that sketched in Fig. 18.2 for the time instants $s_1 < s_2$.

Definition 18.14. Let $(M_{2n}^*, \boldsymbol{\Omega}, H)$ be a Hamiltonian dynamical system, and denote by φ_t a symplectic group generated by the complete Hamiltonian vector field $\mathbf{X}_H$. We say that the submanifold W of M_{2n}^* is ***invariant*** with respect to φ_t if

$$\varphi_t(W) = W, \quad \forall t \in \mathfrak{R}. \tag{18.82}$$

For instance, the level manifolds $f = const$, where f is a first integral of $(M_{2n}^*, \boldsymbol{\Omega}, H)$, are examples of invariant submanifolds of M_{2n}^*.

Theorem 18.17 (Poincaré). *Let $W \subset M_{2n}^*$ be a compact and invariant submanifold with respect to φ_t. Then, for any submanifold $S_0 \subset W$ and $\forall \tau > 0$, there exists an instant $t_1 > \tau$ such that $S_0 \cap \varphi_{t_1}(S_0) \neq \emptyset$.*

Proof. Owing to Liouville's theorem, for an arbitrary value $\tau > 0$ of t, the regions $S_n = \varphi_{n\tau}(S_0)$, $n = 0, 1, \ldots, n, \ldots$, have the same volume (Fig. 18.3) and are contained

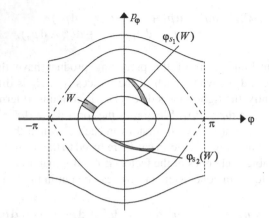

Fig. 18.2 Phase portrait of a simple pendulum

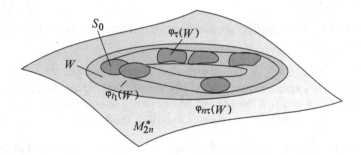

Fig. 18.3 Poincaré's theorem

in W. First, we show that there are two positive integers n_0 and n_{0+k} such that $S_{n_0} \cap S_{n_0+k} \neq \emptyset$. In fact, in the contrary case, all the domains S_n should be distinct. Since they have the same volume $\mathrm{vol}(S_0)$, the total volume of all the sets $S_1, \ldots, S_n, \ldots$ should be equal to infinity, against the hypothesis that W is invariant and compact. If in the condition $S_{n_0} \cap S_{n_0+k} \neq \emptyset$ we have $n_0 = 0$, then the theorem is proved. If $n_0 \geq 1$, since φ_τ is a diffeomorphism and $\varphi_{\tau+s} = \varphi_\tau \circ \varphi_s$, we obtain the results

$$
\begin{aligned}
S_{n_0} \cap S_{n_0+k} \neq \emptyset &\Rightarrow \varphi_\tau \circ \varphi_{(n_0-1)\tau}(S_0) \cap \varphi_\tau \circ \varphi_{(n_0+k-1)\tau}(S_0) \neq \emptyset \\
&\Rightarrow \varphi_{(n_0-1)\tau}(S_0) \cap \varphi_{(n_0+k-1)\tau}(S_0) \neq \emptyset \Rightarrow \\
&\cdots \Rightarrow S_0 \cup \varphi_{n_0\tau}(S_0) \neq \emptyset,
\end{aligned}
$$

and the theorem is proved.   □

Theorem 18.18. *All trajectories starting from the points of the region $S_0 \subset W$ (except for the points of a subset with measure equal to zero), where W is an invariant region of the Hamiltonian vector field, go back to S_0 an infinite number of times.*

Proof. In fact, if a set $U_0 \subset S_0$ exists for which this property is not verified, we obtain an absurd result in view of the previous theorem.   □

The preceding theorems show that, when orbits are not periodic, the phase portrait is very complex, especially if we take into account that, applying the uniqueness theorem to Hamiltonian equations, trajectories corresponding to different initial data cannot intersect each other.

18.10   Volume Form on Invariant Submanifolds

Let (M_{2n}^*, Ω, H) be a Hamiltonian system, and denote by Σ_E the level $(2n-1)$ submanifold of M_{2n}^* defined by the equation

$$H(x) = E. \tag{18.83}$$

We have already said that all dynamical trajectories with initial data belonging to Σ_E lie on Σ_E. Equivalently, the restriction $\mathbf{X}_{\Sigma_E}$ of $\mathbf{X}_H$ to the points of Σ_E is tangent to Σ_E.

Theorem 18.19. *If Σ_E is a regular submanifold of M_{2n}^*, then there exists a $(2n-1)$-form on Σ_E that is an invariant volume form of the vector field $\mathbf{X}_{\Sigma_E}$.*

Proof. Introduce into a neighborhood U of a point $x \in \Sigma_E$ the system of coordinates $(x^1, ..., x^{2n-1}, H)$ such that $(x^1, ..., x^{2n-1})$ are coordinates in $U \cap \Sigma_E$ and the value $H = E$ corresponds to $U \cap \Sigma_E$. Denoting by σ_0 a region contained in $U \cap \Sigma_E$, we consider the domain $\Delta_0 = \{x | (x^1 ..., x^{2n-1}) \in \sigma_0, E \leq H \leq E + \Delta E\} \subset M_{2n}^*$ (Fig. 18.4). Liouville's theorem states that the volume V_0 of Δ_0 is equal to the volume V_t of $\Delta_t = \varphi_t(\Delta_0)$, where φ_t is the one-parameter group generated by the Hamiltonian field $\mathbf{X}_H$. If we denote by $(\mathbf{q}, \mathbf{p})$ canonical coordinates in U, we can write

$$V_t = \int_{\Delta_t} \mathbf{d}q^1 \cdots \mathbf{d}q^n \mathbf{d}p_1 \cdots \mathbf{d}p_n = \int_E^{E+\Delta E} \mathbf{d}H \int_{\varphi_t(\sigma_0)} J \mathbf{d}x^1 \cdots \mathbf{d}x^{2n-1},$$

where J is the determinant of the Jacobian matrix of the coordinate transformation $(\mathbf{q}, \mathbf{p}) \rightarrow (x^1, ..., x^{2n-1}, H)$. Applying the mean value theorem to the integral, we obtain

$$V_t = \Delta E \int_{\varphi_t(\sigma_0)} \overline{J} \mathbf{d}x^1 \cdots \mathbf{d}x^{2n-1},$$

where $\overline{J}$ denotes the value of J at a suitable point of the interval $(E, E + \Delta E)$. Finally, the invariant volume of $\varphi_t(\sigma_0)$ is

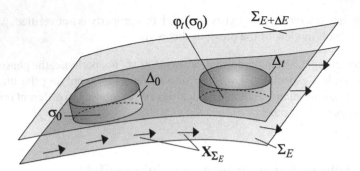

Fig. 18.4 Invariant volume form on submanifold

$$V(\varphi_t(\sigma_0)) = \lim_{\Delta E \to 0} \frac{V_t}{\Delta E} = \int_{\varphi_t(\sigma_0)} J_{H=E} \mathbf{d}x^1 \cdots \mathbf{d}x^{2n-1}. \tag{18.84}$$

In conclusion, the invariant volume $(2n-1)$-form of the vector field $\mathbf{X}_{\Sigma_E}$ is

$$\Omega_E = J_{H=E} \mathbf{d}x^1 \wedge \cdots \wedge \mathbf{d}x^{2n-1}. \tag{18.85}$$

$$\square$$

Now, we determine a more useful form of (18.85). Let $(\mathbf{q}, \mathbf{p})$ be a symplectic system of coordinates in M_{2n}^* and suppose that the equation of the level submanifold $H(\mathbf{q}, \mathbf{p}) = E$ can be written in the following parametric form:

$$\begin{aligned}
q^h &= q^h, \quad h = 1, \ldots, n, \\
p_\alpha &= p_\alpha, \quad \alpha = 1, \ldots, n-1, \\
p_n &= p_n(q^h, p_\alpha, E).
\end{aligned} \tag{18.86}$$

Consider the coordinate change $(\mathbf{q}, \mathbf{p}) \to (q^h, p_\alpha, H)$ such that

$$\begin{aligned}
q^h &= q^h, \quad h = 1, \ldots, n, \\
p_\alpha &= p_\alpha, \quad \alpha = 1, \ldots, n-1, \\
p_n &= p_n(q^h, p_\alpha, H).
\end{aligned} \tag{18.87}$$

It is evident that the determinant J^* of the Jacobian matrix of (18.87) is

$$J^* = \frac{\partial p_n}{\partial H}.$$

In conclusion, in these coordinates Ω_E assumes the form

$$\Omega_E = \left(\frac{\partial p_n}{\partial H} \right)_{H=E} \mathbf{d}q^1 \wedge \cdots \wedge \mathbf{d}q^n \wedge \mathbf{d}p_1 \wedge \cdots \mathbf{d}p_{n-1}. \tag{18.88}$$

Example 18.2. The Hamiltonian function of the simple pendulum with mass and length equal to 1 is

$$H = \frac{1}{2}p^2 - g \cos q.$$

Then, (18.86) assume the form

$$q = q, \quad p = \mp\sqrt{2(H + g \cos q)}$$

and $\boldsymbol{\Omega}_E$ becomes

$$\boldsymbol{\Omega}_E = \frac{1}{\sqrt{2(E + g \cos q)}}\mathrm{d}q.$$

Remark 18.9. In many books (e.g., [31]), instead of (18.84), the following alternative definition of invariant volume $V(\sigma)$ on the submanifold $H(\mathbf{q},\mathbf{p}) = E$ is proposed:

$$V(\sigma) = \int_\sigma \frac{\mathrm{d}\sigma}{|\nabla H|}, \tag{18.89}$$

where σ is any measurable region on the submanifold Σ_E and

$$|\nabla H| = \sqrt{\sum_{h=1}^{n}\left[\left(\frac{\partial H}{\partial q^h}\right)^2 + \left(\frac{\partial H}{\partial p_h}\right)^2\right]}.$$

Formula (18.89) is proved under the following assumptions.

- The symplectic coordinates $(\mathbf{q}, \mathbf{p})$ are Cartesian coordinates in M_{2n}^* that is identified with $\mathfrak{R}^{2n}$.
- The level submanifold $H(\mathbf{q},\mathbf{p}) = E$ is a Riemannian manifold equipped with the Riemannian metric (g_{ij}) induced by M_{2n}^*.

Regarding the first assumption, we note that if the symplectic coordinates $(\mathbf{q}, \mathbf{p})$ are supposed to be Cartesian, the components of the metric tensor in the Euclidean space M_{2n}^* are $g_{ij} = \delta_{ij}$. It is very easy to prove that if we introduce new symplectic coordinates $(\mathbf{q}', \mathbf{p}')$, then the components of the metric tensor cannot be equal to δ_{ij}.

The second assumption implies that, if $p_n = p_n(q^h, p_1, ..., p_{n-1}, E)$ is the local equation of Σ_E, then the induced Riemannian metric (g_{ij}) on Σ_E leads us to the following results:[3]

$$g = \det(g_{ij}) = 1 + \sum_{h=1}^{n}\left(\frac{\partial p_n}{\partial q^h}\right)^2 + \sum_{\alpha=1}^{n-1}\left(\frac{\partial p_n}{\partial p_\alpha}\right)^2, \tag{18.90}$$

---

[3] See Exercise 7.

$$d\sigma = \sqrt{g}dq^1 \cdots dq^n dp_1 \cdots dp_{n-1}. \tag{18.91}$$

On the other hand, the relations between the partial derivatives appearing in (18.90) and the implicit form $H(q^h, p_h) = E$ of Σ_E are

$$\frac{\partial p_n}{\partial q^h} = -\frac{\partial H/\partial q^h}{\partial H/\partial p_n}, \quad \frac{\partial p_n}{\partial p_\alpha} = -\frac{\partial H/\partial p_\alpha}{\partial H/\partial p_n}. \tag{18.92}$$

Introducing (18.92) into (18.90) and (18.91), we can write (18.89) as

$$V(\sigma) = \int_\sigma \frac{\partial p_n}{\partial H} dq^1 \cdots dq^n dp_1 \cdots dp_{n-1},$$

and we find again the invariant volume form (18.88). In conclusion, although (18.89) is based on wrong and useless hypotheses, it leads to the right definition of volume form.

18.11 Time-Dependent Hamiltonian Mechanics and Contact Manifolds

Let $V_n \times \Re$ be the configuration space–time of a mechanical system S with time-dependent constraints and acted upon by conservative forces depending on time. We define **phase space–time** the manifold

$$M_{2n}^* \times \Re = (T^* V_n) \times \Re. \tag{18.93}$$

If $(U, \mathbf{q})$ is a chart of V_n then $(U \times \Re^n, (\mathbf{q}, \mathbf{p}))$ is a *natural* chart of M_{2n}^*, where the components of the covector $\mathbf{p}$ are evaluated in the natural basis $(d\mathbf{q})$. Owing to definition (18.93), a chart of $M_{2n}^* \times \Re$ is given by $(U \times \Re^{n+1}, (\mathbf{q}, \mathbf{p}, t))$. Consequently, $M_{2n}^* \times \Re$ is a $2n+1$-dimensional manifold. Further, the local vector fields $(\partial/\partial\mathbf{q}, \partial/\partial\mathbf{p}, \partial/\partial t)$ are tangent to the coordinate curves of the chart $(U \times \Re^{n+1}, (\mathbf{q}, \mathbf{p}, t))$ and they form a basis of $T_{(x,t)}(M_{2n}^* \times \Re)$, where $x = (\mathbf{q}, \mathbf{p}) \in M_{2n}^*$. Finally, the 1-forms $(d\mathbf{q}, d\mathbf{p}, dt)$ are a basis of $T_{(x,t)}^*(M_{2n}^* \times \Re)$.

In the natural chart $(U \times \Re^{n+1}, (\mathbf{q}, \mathbf{p}, t))$ we define the 1-form

$$\omega_H = p_h dq^h - H dt, \tag{18.94}$$

where $H(\mathbf{q}, \mathbf{p}, t)$ is the Hamiltonian function of the mechanical system S.

In the intersection of the domains of two natural charts, the coordinate transformation is

$$\overline{q}^h = \overline{q}^h(\mathbf{q}, t), \quad \overline{p}_h = \frac{\partial q^k}{\partial \overline{q}^h} p_k, \quad \overline{t} = t. \tag{18.95}$$

Further, under a coordinate change (18.95), the 1-form (18.94) transforms as follows:

$$\omega_H = p_h dq^h - H(\mathbf{q}, \mathbf{p}, t)dt$$
$$= \overline{p}_h d\overline{q}^h - \overline{H}(\overline{\mathbf{q}}, \overline{\mathbf{p}}, \overline{t})d\overline{t} = \overline{\omega}_{\overline{H}}, \tag{18.96}$$

where $\overline{H}(\overline{\mathbf{q}}, \overline{\mathbf{p}}, t)$ is the Hamiltonian H expressed in terms of the variables $\overline{\mathbf{q}}, \overline{\mathbf{p}}$, and $\overline{t}$.

In conclusion, defining the 1-form ω_H in the domain of any chart of a *natural* atlas of $M_{2n}^* \times \mathfrak{R}$ we obtain a *global* 1-form, that is a 1-form defined over all the manifold. Also the 2-form

$$\Omega_H = -d\omega_H \tag{18.97}$$

is globally defined. Further, it is closed

$$d\Omega_H = -d^2\omega_H = 0 \tag{18.98}$$

and in natural coordinates assumes the following form:

$$\Omega_H = dq^h \wedge dp_h + dH \wedge dt. \tag{18.99}$$

We denote by the same symbol Ω_H both the 2-form (18.97) and the skew-symmetric $(0, 2)$-tensor whose strict components are the components of the 2-form. The components of the $(0, 2)$-tensor Ω_H relative to the basis $(dq^h \otimes dq^k, dq^h \otimes dp_k, dq^h \otimes dt, dp_h \otimes dq^k, dp_h \otimes dp_k, dp_h \otimes dt, dt \otimes dq^k, dt \otimes dp_k, dt \otimes dt)$ of $T_2(M_{2n}^* \times \mathfrak{R})$ are given by the elements of the following matrix:

$$\mathbb{E}_H = \begin{pmatrix} \mathbb{O} & \mathbb{I} & H_{\mathbf{q}} \\ -\mathbb{I} & \mathbb{O} & H_{\mathbf{p}} \\ -H_{\mathbf{q}} & -H_{\mathbf{p}} & 0 \end{pmatrix}, \tag{18.100}$$

where $\mathbb{O}$ and $\mathbb{I}$ are the $n \times n$ zero matrix and the $n \times n$ unit matrix, respectively, and $f_a = \partial f / \partial a$.

The determinant of $\mathbb{E}_H$ vanishes

$$\det \mathbb{E}_H = 0 \tag{18.101}$$

since from the skew symmetry of $\mathbb{E}_H$ there follows that

$$\det \mathbb{E}_H = \det(\mathbb{E}_H)^T = \det(-\mathbb{E}_H) = (-1)^{(2n+1)} \det \mathbb{E}_H = -\det \mathbb{E}_H.$$

It is evident that this result does not depend on the particular form of $\mathbb{E}_H$ since it holds for any skew-symmetric matrix with an odd number of rows and columns. We conclude noticing that

$$\text{rank}\,\boldsymbol{\Omega}_H = \text{rank}\,\mathbb{E}_H = 2n. \tag{18.102}$$

Let us adopt a *natural* atlas on the $(2n + 1)$-manifold $M_{2n}^* \times \mathfrak{R}$. We have shown that it is possible to define on this manifold a closed 2-form $\boldsymbol{\Omega}_H$ of rank $2n$. The pair $(M_{2n}^* \times \mathfrak{R}, \boldsymbol{\Omega}_H)$ is called a *contact manifold*.

18.12   Contact or Canonical Coordinates

A chart $(V, (\mathbf{q}, \mathbf{p}, t))$ of $M_{2n}^* \times \mathfrak{R}$ is said to be a *canonical or contact chart* if $\boldsymbol{\Omega}_H$ assumes the following coordinate form:

$$\boldsymbol{\Omega}_K = p_h dq^h - dK \wedge dt, \tag{18.103}$$

where K is a function of $\mathbf{q}, \mathbf{p}$, and t. In particular, if $\mathbf{q}, \mathbf{p}$, and t are natural coordinates, then K is equal to the Hamiltonian H (see (18.96)). A transformation from canonical coordinates to canonical ones is said to be a *canonical transformation*.

We prove that there are other canonical coordinates on $M_{2n}^* \times \mathfrak{R}$ besides the natural ones. To verify this statement, consider the coordinate transformation

$$\begin{aligned}
\overline{\mathbf{q}} &= \overline{\mathbf{q}}(\mathbf{q}, \mathbf{p}, t), \\
\overline{\mathbf{p}} &= \overline{\mathbf{p}}(\mathbf{q}, \mathbf{p}, t), \\
\overline{t} &= t,
\end{aligned} \tag{18.104}$$

where $(\mathbf{q}, \mathbf{p}, t)$ are natural coordinates, and introduce the notations

$$\mathbb{S} = \begin{pmatrix} \dfrac{\partial \overline{\mathbf{q}}}{\partial \mathbf{q}} & \dfrac{\partial \overline{\mathbf{q}}}{\partial \mathbf{p}} & \dfrac{\partial \overline{\mathbf{q}}}{\partial t} \\[2mm] \dfrac{\partial \overline{\mathbf{p}}}{\partial \mathbf{q}} & \dfrac{\partial \overline{\mathbf{p}}}{\partial \mathbf{p}} & \dfrac{\partial \overline{\mathbf{p}}}{\partial t} \\[2mm] 0 & 0 & 1 \end{pmatrix}, \quad \mathbb{S}^{-1} = \begin{pmatrix} \dfrac{\partial \mathbf{q}}{\partial \overline{\mathbf{q}}} & \dfrac{\partial \mathbf{q}}{\partial \overline{\mathbf{p}}} & \dfrac{\partial \mathbf{q}}{\partial \overline{t}} \\[2mm] \dfrac{\partial \mathbf{p}}{\partial \overline{\mathbf{q}}} & \dfrac{\partial \mathbf{p}}{\partial \overline{\mathbf{p}}} & \dfrac{\partial \mathbf{p}}{\partial \overline{t}} \\[2mm] 0 & 0 & 1 \end{pmatrix}. \tag{18.105}$$

Transformation (18.104) is canonical if and only if in the coordinates $(\overline{\mathbf{q}}, \overline{\mathbf{p}}, \overline{t})$ the matrix $\mathbb{E}_H$ assumes the form (18.100) where $H(\mathbf{q}, \mathbf{p}, t)$ is substituted by a new function $K(\overline{\mathbf{q}}, \overline{\mathbf{p}}, \overline{t})$.

Consequently, the coordinates $(\overline{\mathbf{q}}, \overline{\mathbf{p}}, \overline{t})$ are canonical if

$$\mathbb{E}_K = (\mathbb{S}^{-1})^T \mathbb{E}_H \mathbb{S}^{-1}, \tag{18.106}$$

or, equivalently, if

$$
\begin{pmatrix} \mathbb{O} & \mathbb{I} & K_{\overline{q}} \\ -\mathbb{I} & \mathbb{O} & K_{\overline{p}} \\ -K_{\overline{q}} & -K_{\overline{p}} & 0 \end{pmatrix} = \begin{pmatrix} \dfrac{\partial \mathbf{q}}{\partial \overline{\mathbf{q}}} & \dfrac{\partial \mathbf{p}}{\partial \overline{\mathbf{q}}} & 0 \\[2mm] \dfrac{\partial \mathbf{q}}{\partial \overline{\mathbf{p}}} & \dfrac{\partial \mathbf{p}}{\partial \overline{\mathbf{p}}} & 0 \\[2mm] \dfrac{\partial \mathbf{q}}{\partial \overline{t}} & \dfrac{\partial \mathbf{p}}{\partial \overline{t}} & 1 \end{pmatrix} \times
$$

$$
\times \begin{pmatrix} \mathbb{O} & \mathbb{I} & H_{\mathbf{q}} \\ -\mathbb{I} & \mathbb{O} & H_{\mathbf{p}} \\ -H_{\mathbf{q}} & -H_{\mathbf{p}} & 0 \end{pmatrix} \begin{pmatrix} \dfrac{\partial \mathbf{q}}{\partial \overline{\mathbf{q}}} & \dfrac{\partial \mathbf{q}}{\partial \overline{\mathbf{p}}} & \dfrac{\partial \mathbf{q}}{\partial \overline{t}} \\[2mm] \dfrac{\partial \mathbf{p}}{\partial \overline{\mathbf{q}}} & \dfrac{\partial \mathbf{p}}{\partial \overline{\mathbf{p}}} & \dfrac{\partial \mathbf{p}}{\partial \overline{t}} \\[2mm] 0 & 0 & 1 \end{pmatrix}. \tag{18.107}
$$

It is easy to verify that the above product of matrices leads to the result

$$
\begin{pmatrix} \mathbb{O} & \mathbb{I} & K_{\overline{q}} \\ -\mathbb{I} & \mathbb{O} & K_{\overline{p}} \\ -K_{\overline{q}} & -K_{\overline{p}} & 0 \end{pmatrix} =
$$

$$
\begin{pmatrix} [\overline{q}^h, \overline{q}^k] & [\overline{q}^h, \overline{p}_k] & [\overline{q}^h, \overline{t}] + \overline{H}_{\overline{q}^h} \\ -[\overline{q}^h, \overline{p}_k] & [\overline{p}_h, \overline{p}_k] & [\overline{p}_h, \overline{t}] + \overline{H}_{\overline{p}_h} \\ -[\overline{q}^h, \overline{t}] - \overline{H}_{\overline{q}^h} & -[\overline{p}_h, \overline{p}_k] - \overline{H}_{\overline{p}_h} & 0 \end{pmatrix}, \tag{18.108}
$$

where $[f, g]$ denotes the Lagrange parenthesis of the $2n+1$ functions $(\overline{q}^h, p_k, t)$ with respect to the variables f and g of (18.104) and $\overline{H} = H(\mathbf{q}(\overline{\mathbf{q}}, \overline{\mathbf{p}}, \overline{t}), \mathbf{p}(\overline{\mathbf{q}}, \overline{\mathbf{p}}, \overline{t}), \overline{t})$.

We now prove the following theorem.

Theorem 18.20. *Transformation (18.104) in $M_{2n}^* \times \Re$ is canonical if and only if it is symplectic in M_{2n}^* for any value of time t.*

Proof. First, (18.108) is equivalent to the following relations:

$$
[\overline{q}^h, \overline{q}^k] = 0, \quad [\overline{q}^h, \overline{p}_k] = \delta_k^h, \quad [\overline{p}_h, \overline{p}_k] = 0, \tag{18.109}
$$

and

$$
K_{\overline{q}^h} = [\overline{q}^h, \overline{t}] + \overline{H}_{\overline{q}^h}, \quad K_{\overline{p}_h} = [\overline{p}_h, \overline{t}] + \overline{H}_{\overline{p}_h}. \tag{18.110}
$$

Conditions (18.109) are equivalent to state that the coordinate transformation (18.104) is symplectic for any t. It remains to prove that (18.110) imply the existence of a function W such that

$$
K = W + \overline{H}. \tag{18.111}
$$

We begin noticing that (18.110) can also be written as follows:

$$d(K - \overline{H}) = [\overline{q}^h, \overline{t}]d\overline{q}^h + [\overline{p}_h, \overline{t}]d\overline{p}_h \equiv A_\alpha d\overline{x}_\alpha, \qquad (18.112)$$

where $(A_\alpha) = ([\overline{q}, \overline{t}], [\overline{p}, \overline{t}])$ and $(x_\alpha) = (\overline{q}, \overline{p})$. Therefore, the theorem is proved if the differential form on the right-hand side of (18.112) is integrable since, in this case, $A_\alpha d\overline{x}_\alpha = dW$ and (18.111) follows. On the other hand, in terms of 2-form

$$\Omega = \begin{pmatrix} \mathbb{O} & \mathbb{I} \\ -\mathbb{I} & \mathbb{O} \end{pmatrix},$$

where $\mathbb{O}$ is the $n \times n$ zero matrix and $\mathbb{I}$ is the $n \times n$ unit matrix, we can write the symplectic conditions (18.109) at any instant in the usual form

$$\Omega_{\alpha\beta} = J_{\alpha\lambda}\Omega_{\lambda\mu}J_{\mu\beta}, \qquad (18.113)$$

where $(J_{\alpha\beta})$ is the Jacobian matrix

$$\mathbb{J} = \left(\frac{\partial x_\alpha}{\partial \overline{x}_\beta} \right).$$

It is a simple exercise to verify that

$$A_\alpha = \frac{\partial x_\lambda}{\partial \overline{x}_\alpha}\Omega_{\lambda\mu}\frac{\partial x_\mu}{\partial \overline{t}} \qquad (18.114)$$

so that the integrability of the 1-form (18.112) is ensured if

$$A_{\alpha,\beta} = A_{\beta,\alpha}.$$

On the other hand we have

$$A_{\alpha,\beta} = \frac{\partial x_\lambda}{\partial \overline{x}_\beta \partial^2 \overline{x}_\alpha}\Omega_{\lambda\mu}\frac{\partial x_\mu}{\partial \overline{t}} + \frac{\partial x_\lambda}{\partial \overline{x}_\alpha}\Omega_{\lambda\mu}\frac{\partial^2 x_\mu}{\partial \overline{x}_\beta \partial \overline{t}}$$

and the first term on the right-hand side is symmetric with respect to the indices α and β. Therefore, it remains to verify that also the second term is symmetric with respect to these indices. In matrix form this term can be written as follows:

$$\mathbb{B} \equiv \mathbb{J}^T \Omega \frac{\partial \mathbb{J}}{\partial \overline{t}}.$$

Differentiating (18.113) with respect to $\overline{t}$ we obtain

$$\mathbb{O} = \frac{\partial \mathbb{J}^T}{\partial \overline{t}}\Omega\mathbb{J} + \mathbb{J}^T \Omega \frac{\partial \mathbb{J}}{\partial \overline{t}}$$

so that

$$\mathbb{B} = \mathbb{J}^T \Omega \frac{\partial \mathbb{J}}{\partial t} = -\frac{\partial \mathbb{J}^T}{\partial t} \Omega \mathbb{J} = \mathbb{B}^T$$

since Ω is skew-symmetric.   □

18.13 Locally Hamiltonian Fields

Let $(U, (\mathbf{q}, \mathbf{p}, t) \equiv (U, (x_\alpha, t))$ be a *natural* chart of $M_{2n}^* \times \Re$. We say that a vector field $\mathbf{X}$ of $M_{2n}^* \times \Re$ is a *local Hamiltonian field* if a function $H(x, t)$ exists in the open set U such that

$$X_\alpha = \Omega_{\alpha\beta} \frac{\partial H}{\partial x_\beta}, \quad X_{2n+1} = 1. \tag{18.115}$$

The function $H(x_\alpha, t)$ is called the Hamiltonian of $\mathbf{X}$.

The integral curves $(x_\alpha(\tau), t(\tau))$ of a locally Hamiltonian field are solutions of the following differential system:

$$\frac{dx_\alpha}{d\tau} = \Omega_{\alpha\beta} \frac{\partial H}{\partial x_\beta}, \quad \frac{dt}{d\tau} = 1. \tag{18.116}$$

Assuming t as a parameter, the integral curves $(x_\alpha(t), t)$ of a locally Hamiltonian field are solutions of the Hamiltonian system

$$\frac{dx_\alpha}{dt} = \Omega_{\alpha\beta} \frac{\partial H}{\partial x_\beta}. \tag{18.117}$$

Relevant are the following theorems

Theorem 18.21. *A vector field $\mathbf{X}$ is a locally Hamiltonian field if and only if*

$$\frac{\partial X_\alpha}{\partial x_\beta} = \Omega_{\alpha\lambda} B_{\lambda\beta}, \tag{18.118}$$

where $(B_{\lambda\beta})$ is a symmetric $2n \times 2n$ matrix.

Proof. The condition is necessary since, if $\mathbf{X}$ is a local Hamiltonian field, then condition (18.115) holds and we conclude that

$$\frac{\partial X_\alpha}{\partial x_\beta} = \Omega_{\alpha\lambda} \frac{\partial^2 H}{\partial x_\lambda \partial x_\beta}.$$

Inversely, if (18.118) holds and recall that $\boldsymbol{\Omega}^2 = -\mathbf{I}$, then the 1-form $Y_\alpha dx_\alpha = \Omega_{\alpha\beta} X_\beta dx_\alpha$ is exact since

$$\frac{\partial Y_\alpha}{\partial x_\lambda} = \Omega_{\alpha\beta}\frac{\partial X_\beta}{\partial x_\lambda} = \Omega_{\alpha\beta}\Omega_{\beta\mu}B_{\mu\lambda} = -\delta_{\alpha\mu}B_{\mu\lambda} = -B_{\alpha\lambda},$$

$$\frac{\partial Y_\lambda}{\partial x_\alpha} = \Omega_{\lambda\beta}\frac{\partial X_\beta}{\partial x_\alpha} = \Omega_{\lambda\beta}\Omega_{\beta\mu}B_{\mu\alpha} = -\delta_{\lambda\mu}B_{\mu\alpha} = -B_{\lambda\alpha}.$$

Then, a function $f(x)$ exists such that $Y_\alpha = \partial f/\partial x_\alpha$ and we obtain the result

$$Y_\alpha = \Omega_{\alpha\beta}X_\beta = \frac{\partial f}{\partial x_\alpha}$$

Multiplying this relation by $(\Omega^{-1})_{\lambda\alpha}$ and recalling that $(\Omega^{-1})_{\lambda\alpha} = -\Omega_{\lambda\alpha}$, we obtain $X_\alpha = -\Omega_{\alpha\beta}\partial f/\partial x_\beta$ and the theorem is proved. □

Theorem 18.22. *Under a canonical transformation (18.104) Hamilton's equations preserve their form, possibly with a different Hamiltonian that is uniquely determined by the Hamiltonian H relative to coordinates (x_α) and the transformation (18.104).*

Proof. Adopting the notations $(x_\alpha, t) = (\mathbf{q}, \mathbf{p}, t)$ and $(\overline{x}_\alpha, t) = (\overline{\mathbf{q}}, \overline{\mathbf{p}}, \overline{t})$ canonical transformation (18.104) can be written as follows:

$$\overline{x}_\alpha = \overline{x}_\alpha(x_\beta, t) \quad \overline{t} = t, \tag{18.119}$$

where (x_α, t) are supposed to be natural coordinates so that Hamilton's equation in these coordinates has the form

$$\dot{x}_\alpha = \Omega_{\alpha\beta}\frac{\partial H}{\partial x_\beta}. \tag{18.120}$$

In the coordinates $(\overline{x}_\alpha, t)$ it results

$$\dot{\overline{x}}_\alpha = \frac{\partial \overline{x}_\alpha}{\partial x_\beta}\dot{x}_\beta + \frac{\partial \overline{x}_\alpha}{\partial t} = \frac{\partial \overline{x}_\alpha}{\partial x_\beta}\Omega_{\beta\lambda}\frac{\partial H}{\partial x_\lambda} + \frac{\partial \overline{x}_\alpha}{\partial t}$$

so that it is

$$\dot{\overline{x}}_\alpha = \frac{\partial \overline{x}_\alpha}{\partial x_\beta}\Omega_{\beta\lambda}\frac{\partial \overline{H}}{\partial \overline{x}_\mu}\frac{\partial \overline{x}_\mu}{\partial x_\lambda} + \frac{\partial \overline{x}_\alpha}{\partial t}, \tag{18.121}$$

where $\overline{H}(\overline{x}, t) = H(x(\overline{x}, t), t)$. Since transformation (18.119) is canonical, it is simplectic for any t (see Theorem 18.20 and (18.16)) so that

$$\Omega_{\alpha\mu} = \frac{\partial \overline{x}_\alpha}{\partial x_\beta}\Omega_{\beta\lambda}\frac{\partial \overline{x}_\mu}{\partial x_\lambda}, \tag{18.122}$$

and (18.121) assumes the form

$$\dot{\overline{x}}_\alpha = \Omega_{\alpha\mu}\frac{\partial \overline{H}}{\partial \overline{x}_\mu} + \frac{\partial \overline{x}_\alpha}{\partial t}(\overline{x}, t), \tag{18.123}$$

The theorem is proved after showing that $\partial \overline{x}_\alpha/\partial t$ is a local Hamiltonian field. In view of Theorem 18.21, this field is Hamiltonian if it is

$$\frac{\partial}{\partial \overline{x}_\beta}\left(\frac{\partial \overline{x}_\alpha(\overline{x},t)}{\partial t}\right) = \Omega_{\alpha\lambda}B_{\lambda\beta}, \tag{18.124}$$

where $B_{\lambda\beta}$ is symmetric. Multiplying on the left by $\Omega_{\mu\alpha}$ and remembering that $-\Omega_{\mu\alpha} = (\Omega^{-1})_{\mu\alpha}$, from (18.124) we obtain

$$-B_{\mu\beta} = \Omega_{\mu\alpha}\frac{\partial}{\partial \overline{x}_\beta}\left(\frac{\partial \overline{x}_\alpha(\overline{x},t)}{\partial t}\right) = \Omega_{\mu\alpha}\frac{\partial}{\partial t}\frac{\partial \overline{x}_\alpha}{\partial x_\nu}\frac{\partial x_\nu}{\partial \overline{x}_\beta}. \tag{18.125}$$

Introducing the notation $\mathbb{J} = (\partial \overline{x}_\alpha/\partial x_\nu)$ and adopting the matrix form, (18.125) becomes

$$-\mathbb{B} = \Omega\frac{\partial \mathbb{J}}{\partial t}\mathbb{J}^{-1}. \tag{18.126}$$

To prove the symmetry of $\mathbb{B}$ we notice that, in view of (18.122), it is

$$\Omega = \mathbb{J}\Omega\mathbb{J}^T \Leftrightarrow \mathbb{J}^T\Omega = \Omega\mathbb{J}^{-1} \Leftrightarrow \Omega = \mathbb{J}^T\Omega\mathbb{J}.$$

Differentiating the last relation with respect to t gives

$$\frac{\partial \mathbb{J}^T}{\partial t}\Omega\mathbb{J} + \mathbb{J}^T\Omega\frac{\partial \mathbb{J}}{\partial t} = 0,$$

and multiplying this condition on the left by $(\mathbb{J}^{-1})^T$, on the right by $\mathbb{J}^{-1}$, and remembering that $\Omega^T = -\Omega$, we finally obtain

$$-\mathbb{B} = \Omega\frac{\partial \mathbb{J}}{\partial t}\mathbb{J}^{-1} = (\mathbb{J}^{-1})^T\Omega^T\frac{\partial \mathbb{J}^T}{\partial t} = -\mathbb{B}^T$$

and the vector field $\partial \overline{x}_\alpha/\partial t$ is locally Hamiltonian

$$\frac{\partial \overline{x}_\alpha}{\partial t} = \Omega_{\alpha\beta}\frac{\partial W}{\partial \overline{x}_\beta}$$

where the function $W(\overline{x},t)$ is obtained integrating the differential form

$$dW = -\Omega_{\lambda\alpha}\frac{\partial \overline{x}_\alpha}{\partial t}d\overline{x}^\lambda.$$

$\square$

We conclude this section presenting a procedure to generate canonical transformations that extends the procedure developed in Section 18.3 to generate simplectic transformations.

We defined a canonical transformation as a coordinate transformation (18.104) such that

$$d\bar{q}^h \wedge d\bar{p}_h + dK \wedge d\bar{t} = dq^h \wedge dp_h + dH \wedge dt. \tag{18.127}$$

This condition is equivalent to the following one:

$$\bar{p}_h d\bar{q}^h - K dt = p_h dq^h - H dt - df. \tag{18.128}$$

As in Section 18.3, if we suppose that f is a function of the variables $(\bar{q}, q, t)$ such that

$$\det \left(\frac{\partial f}{\partial \bar{q}^h \partial q^k} \right) \neq 0, \tag{18.129}$$

then (18.128) is equivalent to the following conditions:

$$\bar{p}_h = -\frac{\partial f}{\partial \bar{q}^h}(\bar{q}, q, t), \tag{18.130}$$

$$p_h = \frac{\partial f}{\partial q^h}(\bar{q}, q, t), \tag{18.131}$$

$$K = H + \frac{\partial f}{\partial t}(\bar{q}, q, t). \tag{18.132}$$

Further, if f is an arbitrary function of the variables $(q, \bar{p}, t)$ satisfying to the condition

$$\det \left(\frac{\partial f}{\partial \bar{p}^h \partial q^k} \right) \neq 0, \tag{18.133}$$

then (18.127) is equivalent to the conditions

$$\bar{q}_h = \frac{\partial f}{\partial \bar{p}^h}(\bar{p}, q, t), \tag{18.134}$$

$$p_h = \frac{\partial f}{\partial q^h}(\bar{p}, q, t), \tag{18.135}$$

$$K = H + \frac{\partial f}{\partial t}(\bar{p}, q, t). \tag{18.136}$$

18.14   Globally Hamiltonian Fields

In Section 18.4 an isomorphism

$$\flat : \mathbf{X} \in TM^*_{2n} \rightarrow \omega \in T^*M^*_{2n}$$

was defined by the following correspondence law:

$$\omega(\mathbf{Y}) = \boldsymbol{\Omega}(\mathbf{X}, \mathbf{Y}), \quad \forall \mathbf{Y} \in TM^*_{2n}. \tag{18.137}$$

In particular, the globally Hamiltonian fields $\mathbf{X}_f$ were introduced by the condition

$$\mathbf{d}f(\mathbf{Y}) = \boldsymbol{\Omega}(\mathbf{X}_f, \mathbf{Y}), \quad \forall \mathbf{Y} \in TM^*_{2n}. \tag{18.138}$$

In the preceding section we defined the locally Hamiltonian fields in terms of canonical coordinates. Here we propose an intrinsic definition of globally Hamiltonian fields on $TM^*_{2n} \times \Re$, that is a definition that does not resort to canonical charts. It is evident that the linear map

$$\mathbf{X} \in (TM^*_{2n}) \times \Re \rightarrow \omega \in (T^*M^*_{2n}) \times \Re$$

defined as follows

$$\omega(\mathbf{Y}) = \boldsymbol{\Omega}_H(\mathbf{X}, \mathbf{Y}), \quad \forall \mathbf{Y} \in TM^*_{2n} \times \Re \tag{18.139}$$

is not an isomorphism since $\det \boldsymbol{\Omega}_H = 0$. However, $\operatorname{rank} \boldsymbol{\Omega}_H = 2n$ so that the dimension of $\ker \flat = 1$. In arbitrary coordinates (x_α) on the manifold $M^*_{2n} \times \Re$ the map (18.139) has the following coordinate form

$$\omega_\alpha = \Omega_{\beta\alpha} X^\beta. \tag{18.140}$$

We want to show that the vector fields belonging to $\ker \flat$ are Hamiltonian fields. To prove this statement, we introduce a canonical chart $(U, (\mathbf{q}, \mathbf{p}, t))$ in $M^*_{2n} \times \Re$ and recall that any vector field can be written as follows

$$\mathbf{X} = X^h \frac{\partial}{\partial q^h} + X_h \frac{\partial}{\partial p_h} + X \frac{\partial}{\partial t}. \tag{18.141}$$

Then, in view of (18.100), (18.140) becomes

$$(w_h, w^h, w) = (X^h, X_h, X) \begin{pmatrix} \mathbb{O} & \mathbb{I} & H_{\mathbf{q}} \\ -\mathbb{I} & \mathbb{O} & H_{\mathbf{p}} \\ -H_{\mathbf{q}} & -H_{\mathbf{p}} & 0 \end{pmatrix}. \tag{18.142}$$

Consequently, the vector fields belonging to $\ker \flat$ are determined by the condition

$$(X^h, X_h, X) \begin{pmatrix} \mathbb{O} & \mathbb{I} & H_{\mathbf{q}} \\ -\mathbb{I} & \mathbb{O} & H_{\mathbf{p}} \\ -H_{\mathbf{q}} & -H_{\mathbf{p}} & 0 \end{pmatrix} = (\mathbb{O}, \mathbb{O}, 0) \tag{18.143}$$

that gives

$$- X_h - X H_{q^h} = 0, \tag{18.144}$$
$$X^h - X H_{p_h} = 0, \tag{18.145}$$
$$X^h H_{q^h} + X_h H_{p_h} = 0. \tag{18.146}$$

Equation (18.146) is a linear combination of (18.144), (18.145) as a consequence of the fact that $\mathrm{rank}\mathbb{E}_H = 2n$. In conclusion, the vector fields belonging to the one-dimensional vector space $\ker \flat$ have the following components:

$$\hat{\mathbf{X}}_H = (H_{p_h}, -H_{q^h}, X), \tag{18.147}$$

in any canonical chart of $M^*_{2n} \times \mathfrak{R}$. All these fields, when they are evaluated at the same point x of the canonical chart, belong to the same one-dimensional space. Therefore, when H is given in the chart $(U, (\mathbf{q}, \mathbf{p}, t))$, all these vectors define the same direction at the point $x \in U$. In the sequel, we choose at x the value $X = 1$ for the $(n + 1)$th component. The direction field defined by the vector field (18.147) is called the *characteristic directions* of Ω_E. The integral curves of this field are solutions of the Hamilton equations

$$\dot{q}^h = H_{p_h}, \quad \dot{p}_h = -H_{q^h}, \quad \dot{t} = 1. \tag{18.148}$$

A *first integral* of the dynamical system $M^*_{2n} \times \mathfrak{R}$ is a function $f : M^*_{2n} \times \mathfrak{R} \to \mathfrak{R}$ that is constant along the solutions of Hamilton equations (18.148), i.e.,

$$\dot{f} = L_{\hat{\mathbf{X}}_H} = \hat{\mathbf{X}}_H f = 0. \tag{18.149}$$

In canonical coordinates the above condition becomes

$$\dot{f} = \frac{\partial f}{\partial x_\alpha} \Omega_{\alpha\beta} \frac{\partial H}{\partial x_\beta} + \frac{\partial f}{\partial t} = \{f, H\} + \frac{\partial f}{\partial t} = 0. \tag{18.150}$$

18.15 Exercises

1. Verify that in any system of coordinates (x^α), $\alpha = 1, \ldots, 2n$ on M^*_{2n}, Hamilton's equations assume the form

$$\dot{x}^\alpha = \Omega_{\alpha\beta} \frac{\partial H}{\partial x^\beta}. \tag{18.151}$$

   Hint: See (18.49) and (18.47).
2. Verify that in any system of coordinates (x^α), $\alpha = 1, \ldots, 2n$ on M^*_{2n}, Poisson's brackets assume the form

$$\{f, g\} = \frac{\partial f}{\partial x^\alpha} \Omega_{\alpha\beta} \frac{\partial g}{\partial x^\beta}. \tag{18.152}$$

3. Verify that if the forces acting on a mechanical system $H(q^h, p_h)$ derive from a generalized potential, then $H = const$ reduces to $T + \hat{U}(q^h, p_h) = const$ [see (17.77)].

4. Show Noether's theorem starting from (18.66) and adopting arbitrary coordinates (x^α) on M^*_{2n}.

   Hint: Differentiating (18.66) with respect to s we obtain (see preceding exercises)

$$\dot{f} = \frac{\partial f}{\partial x^\alpha} \Omega_{\alpha\beta} \frac{\partial H}{\partial x^\beta} = \{f, H\}.$$

5. Suppose that the Hamiltonian function $H(q^h, p_h)$ does not depend on the symplectic coordinate q^1. Verify, by Noether's theorem, that the corresponding conjugate momentum is constant.

   Hint: The one-parameter group of diffeomorphisms

$$q'^h = q^h + s\delta^{1h}, \quad p'_h = p_h, \quad h = 1, \dots, n,$$

   is a symmetry group of H whose infinitesimal generator is the vector field

$$\mathbf{X} = (1, 0, \dots, 0).$$

   This field is globally Hamiltonian since it is defined by the system

$$\delta^{1h} = \frac{\partial f}{\partial p_h}, \quad 0 = \frac{\partial f}{\partial q^h}$$

   whose solution $f = p_1$ is a first integral.

6. Prove that for the Hamiltonian system for which

$$H = \frac{1}{2}\left(p_1^2 + p_2^2\right) + \frac{1}{2}\left((q^1)^2 + (q^2)^2\right)$$

   the volume 3-form on the invariant submanifold $H = E$ is

$$\Omega_E = \frac{1}{\sqrt{2E - (p_1^2 + (q^1)^2 + (q^2)^2)}} dq^1 \wedge dq^2 \wedge dp_1.$$

7. Given in $\Re^3$ the surface S with parametric equations

$$x^1 = x^1,$$
$$x^2 = x^2,$$
$$x^3 = f(x^1, x^2),$$

prove that the induced metric on S is

$$ds^2 = \left(1 + (f_{,x^1})^2\right)(dx^1)^2 + \left(1 + (f_{,x^2})^2\right)(dx^2)^2 + 2f_{,x^1}\,f_{,x^2}\,dx^1 dx^2,$$

whereas the determinant g of the matrix of metric coefficients is

$$g = 1 + (f_{,x^1})^2 + (f_{,x^2})^2$$

Hint:

$$ds^2 = (dx^1)^2 + (dx^2)^2 + (dx^3)^2,$$

where

$$dx^3 = f_{,x^1}\,dx^1 + f_{,x^2}\,dx^2.$$

Chapter 19
The Hamilton–Jacobi Theory

Abstract We are able to solve Hamilton's equations only in a few cases. Consequently, we must search for information about their solutions following other paths. In this chapter, we show that the Hamilton–Jacobi theory allows, at least in principle, to determine a set of canonical coordinates in which the Hamiltonian equations can be integrated at once. However, these coordinates can be determined solving the Hamilton–Jacobi nonlinear partial differential equation. Usually, to solve this equation is more difficult than to integrate Hamiltonian equations, but Jacobi proposed the method of separated variables that in some cases allows to solve the Hamilton–Jacobi equation.

19.1 The Hamilton–Jacobi Equation

We have repeatedly emphasized that determining the explicit solutions of Hamiltonian equations is a task beyond the capabilities of mathematics. This situation justifies all attempts to obtain information about these solutions. For instance, a knowledge of the first integrals allows us to localize the dynamical trajectories in the phase space M_{2n}^*. In turn, Nöther's theorem proves that the presence of first integrals is strictly related to the existence of symmetries. In this chapter, we analyze the Hamilton–Jacobi theory whose purpose it is to determine a set of canonical coordinates in which the Hamiltonian equations have such a simple form that we can obtain solutions without effort. However, determining this set of coordinates requires solving a nonlinear partial differential equation. Although, in general, this problem could be more difficult than the direct integration of the Hamiltonian equations, there are some interesting cases in which we are able to exhibit its solution by adopting the method, proposed by Jacobi, of separated variables.

© Springer International Publishing AG, part of Springer Nature 2018   381
A. Romano and A. Marasco, *Classical Mechanics with Mathematica®*,
Modeling and Simulation in Science, Engineering and Technology,
https://doi.org/10.1007/978-3-319-77595-1_19

Let us consider the Hamiltonian equations

$$\dot{q}^h = \frac{\partial H}{\partial p_h}, \quad \dot{p}_h = -\frac{\partial H}{\partial q^h}, \quad h = 1, \ldots, n, \tag{19.1}$$

in the symplectic coordinates $(\mathbf{q}, \mathbf{p})$ of M_{2n}^*.

We wish to determine a symplectic transformation of coordinates

$$\overline{\mathbf{q}} = \overline{\mathbf{q}}(\mathbf{q}, \mathbf{p}), \quad \overline{\mathbf{p}} = \overline{\mathbf{p}}(\mathbf{q}, \mathbf{p}) \tag{19.2}$$

such that, in the new coordinates, the Hamiltonian function depends only on the coordinates $\overline{\mathbf{p}}$. Recalling that, under a transformation of symplectic coordinates (19.2), the new Hamiltonian function is obtained by applying transformation (19.2) to the variables $\mathbf{q}, \mathbf{p}$ in the old Hamiltonian function $H(\mathbf{q}, \mathbf{p})$, we are searching for a transformation of symplectic coordinates for which

$$\overline{H}(\overline{\mathbf{q}}, \overline{\mathbf{p}}) = H(\mathbf{q}(\overline{\mathbf{q}}, \overline{\mathbf{p}}), \mathbf{p}(\overline{\mathbf{q}}, \overline{\mathbf{p}})) = \overline{H}(\overline{\mathbf{p}}). \tag{19.3}$$

In these new coordinates, the Hamiltonian equations become

$$\dot{\overline{\mathbf{q}}} = \frac{\partial \overline{H}}{\partial \overline{\mathbf{p}}} \equiv \nu(\overline{\mathbf{p}}), \quad \dot{\overline{\mathbf{p}}} = 0, \tag{19.4}$$

and their solutions are

$$\overline{\mathbf{q}} = \nu t + \alpha, \quad \overline{\mathbf{p}} = \beta, \tag{19.5}$$

where α and β are constant vectors depending on the initial data in the coordinates $(\overline{\mathbf{q}}, \overline{\mathbf{p}})$. We obtain the solutions of the Hamiltonian equations in the coordinates $(\mathbf{q}, \mathbf{p})$ by introducing (19.5) into (19.2)

$$\mathbf{q} = \mathbf{q}(\nu t + \alpha, \beta), \quad \mathbf{p} = \mathbf{p}(\nu t + \alpha, \beta). \tag{19.6}$$

Recognizing the advantage of such a transformation of coordinates, we must suggest a way to determine it. To this end, we recall that we can obtain a symplectic transformation of coordinates by the relations (Sect. 18.3)

$$\mathbf{p} = \frac{\partial W(\mathbf{q}, \overline{\mathbf{p}})}{\partial \mathbf{q}}, \quad \overline{\mathbf{q}} = \frac{\partial W(\mathbf{q}, \overline{\mathbf{p}})}{\partial \overline{\mathbf{p}}}, \tag{19.7}$$

where $W(\mathbf{q}, \overline{\mathbf{p}})$ is the generating function of (19.2) satisfying the condition

$$\det \left(\frac{\partial^2 W}{\partial \mathbf{q} \partial \overline{\mathbf{p}}} \right) \neq 0. \tag{19.8}$$

Introducing $(19.7)_1$ into (19.3), we can state that the generating function $W(\mathbf{q}, \overline{\mathbf{p}})$ must be a solution of the **Hamilton–Jacobi partial differential equation**

$$H\left(\mathbf{q}, \frac{\partial W(\mathbf{q}, \overline{\mathbf{p}})}{\partial \mathbf{q}}\right) = \overline{H}(\overline{\mathbf{p}}). \tag{19.9}$$

Remark 19.1. The first-order partial differential equation (19.9) is nonlinear since it depends quadratically on the first derivatives $\partial W / \partial q^h$. Consequently, determining its solution is more difficult than solving Hamiltonian equations. Further, before Jacobi, the method proposed by Cauchy, Monge, and Ampére to determine the solution of a first-order partial differential equation required the integration of a system of ordinary differential equations, the *characteristic system*, which can be associated with the starting partial differential equation (Appendix A). Now, in the case of (19.9), this characteristic system coincides with the Hamiltonian equations we wish to solve. In other words, we are faced with a vicious circle. However, in the next section, we present an alternative method proposed by Jacobi that allows us to overcome this gap.

Remark 19.2. On the right-hand side of (19.9) there is a Hamiltonian function in the new symplectic coordinates we are searching for. It is unknown since its form depends on the transformation itself. In other words, there are infinite symplectic transformations of coordinates satisfying (19.9), so we can arbitrarily choose the form of $\overline{H}(\overline{\mathbf{p}})$, and our choice will influence the transformation of coordinates we are searching for. In particular, Jacobi adopted the following choice of this function:

$$\overline{H}(\overline{\mathbf{p}}) = \overline{p}_n \equiv E, \tag{19.10}$$

where E is the total mechanical energy.

Remark 19.3. The solution of a partial differential equation depends on an *arbitrary function*, not on n *arbitrary constants*, as is required by (19.9). In fact, the simple first-order partial differential equation

$$\frac{\partial u}{\partial x} = xy \tag{19.11}$$

in the unknown $u(x, y)$ has the general solution

$$u(x, y) = \frac{1}{2}x^2 y + F(y),$$

where $F(y)$ is an arbitrary function of the variable y. A particular solution of (19.11) can be obtained, for instance, using the boundary datum $u_0(y) = u(0, y)$. In fact, taking into account this condition, we obtain

$$u(x, y) = \frac{1}{2}x^2 y + u_0(y).$$

More generally, we consider the equation

$$\frac{\partial u}{\partial x} + \frac{\partial u}{\partial y} = 0,$$

which, introducing the constant vector field $\mathbf{X} = (1,1)$, can also be written in the form

$$\mathbf{X} \cdot \nabla u = 0. \qquad (19.12)$$

Consider the system of ordinary differential equations (see Appendix A)

$$\dot{x} = 1, \quad \dot{y} = 1$$

whose solution is given by the family Γ of straight lines $x = x_0 + t$, $y = y_0 + t$, where x_0 and y_0 are arbitrary constants. In view of (19.12), we can say that the function u is constant along the family Γ of straight lines:

$$\frac{\mathrm{d}u}{\mathrm{d}t} = 0.$$

Let $(x_0(s), y_0(s))$ be a curve that intersects at any point one and only one straight line of Γ. Then, if the values $u_0(s)$ of $u(x,y)$ are given along the curve $(x_0(s), y_0(s))$, the function $u(x,y)$ is completely determined. Once again, the solution of (19.12) depends on an arbitrary function $u_0(s)$.

The preceding considerations show that we are interested not in finding the general integral of the Hamilton–Jacobi equation but in determining a family of its solutions, depending on n arbitrary constants. In the next section, we show that such a family of solutions, called a **complete integral**, exists for a wide class of Hamiltonian functions. We do not prove that, in general, the complete integral of a partial differential equation exists and that the general integral is the envelope of the family of solutions contained in the complete integral.

19.2   Separation Method

In the preceding section, we showed that knowledge of a complete integral of the Hamilton–Jacobi equation (19.9) leads us to a solution of the Hamiltonian equations. In this section, we present the **separation method** or the **method of separation of variables** to obtain a complete integral for a wide class of Hamiltonian functions. The mechanical systems whose Hamiltonian functions belong to this class are said to be **separable systems**.

We say that the separation method is applicable to Eq. (19.9) if its complete integral has the form

$$W(\mathbf{q}, \overline{\mathbf{p}}) = W_1(q^1, \overline{\mathbf{p}}) + \cdots + W_n(q^n, \overline{\mathbf{p}}). \tag{19.13}$$

It is evident that, in general, the solution of (19.9) cannot be given in the form (19.13) for any Hamiltonian function. However, we can prove that for a wide class of Hamiltonian functions it is possible to obtain the complete integral of (19.9) starting from (19.13).

Theorem 19.1. *If the Hamiltonian function has the form*

$$H(\mathbf{q}, \mathbf{p}) = g_n(\ldots g_3(g_2(g_1(q^1, p_1), q^2, p_2), q^3, p_3)\ldots, q^n, p_n), \tag{19.14}$$

then the complete integral of (19.9) can be obtained by (19.13).

Proof. If (19.14) holds, then the Hamilton–Jacobi equation (19.9) can be written as follows:

$$g_n\left(\ldots g_3\left(g_2\left(g_1\left(q^1, \frac{\partial W}{\partial q^1}\right), q^2, \frac{\partial W}{\partial q^2}\right), q^3, \frac{\partial W}{\partial q^3}\right)\ldots, q^n, \frac{\partial W}{\partial q^n}\right) = \overline{p}_n. \tag{19.15}$$

This equation is satisfied if

$$g_1\left(q^1, \frac{\partial W}{\partial q^1}\right) = \overline{p}_1,$$

$$g_2\left(q^2, \frac{\partial W}{\partial q^2}, \overline{p}_1\right) = \overline{p}_2,$$

$$\cdots\cdots\cdots\cdots\cdots\cdots\cdots\cdots\cdots\cdots\cdots\cdots\cdots\cdots$$

$$g_n\left(q^n, \frac{\partial W}{\partial q^n}, \overline{p}_1, \ldots, \overline{p}_{n-1}\right) = \overline{p}_n$$

Supposing that the conditions

$$\frac{\partial g_i}{\partial p_i} \neq 0, \quad i = 1, \ldots, n, \tag{19.16}$$

hold, the preceding system assumes the equivalent form

$$\frac{\partial W}{\partial q^1} = G_1(q^1, \overline{p}_1),$$

$$\frac{\partial W}{\partial q^2} = G_2(q^2, \overline{p}_1, \overline{p}_2),$$

$$\cdots\cdots\cdots\cdots\cdots\cdots\cdots\cdots\cdots\cdots\cdots\cdots\cdots\cdots$$

$$\frac{\partial W}{\partial q^n} = G_n(q^n, \overline{p}_1, \ldots, \overline{p}_n). \tag{19.17}$$

From the first equation we obtain

$$W = W_1(q^1, \overline{p}_1) + \hat{W}_1(q^2, \ldots, q^n, \overline{p}_1, \ldots, \overline{p}_n), \tag{19.18}$$

where

$$W_1(q^1, \overline{p}_1) = \int G_1(q^1, \overline{p}_1)dq^1,$$

and $\hat{W}_1(q^2, \ldots, q^n, \overline{p}_1, \ldots, \overline{p}_n)$ is an arbitrary function. Substituting (19.18) into
(19.17)$_2$ we have that

$$W = W_1(q^1, \overline{p}_1) + W_2(q^2, \overline{p}_1, \overline{p}_2) + \hat{W}_2(q^3, \ldots, q^n, \overline{p}_1, \ldots, \overline{p}_n), \tag{19.19}$$

where

$$W_2(q^2, \overline{p}_1, \overline{p}_2) = \int G_2(q^2, \overline{p}_1, \overline{p}_2)dq^2. \tag{19.20}$$

Proceeding in the same way, we finally obtain

$$W = W_1(q^1, \overline{p}_1) + W_2(q^2, \overline{p}_1, \overline{p}_2) + \cdots + W_n(q^n, \overline{p}_1, \ldots, \overline{p}_h), \tag{19.21}$$

where

$$W_h(q^h, \overline{p}_1, \ldots, \overline{p}_h) = \int G_h(q^h, \overline{p}_1, \ldots, \overline{p}_h)dq^h, \quad h = 1, \ldots, n, \tag{19.22}$$

and the theorem is proved.   □

The implicit form of the equations of motion is obtained by substituting (19.5)
into the symplectic transformation of coordinates

$$p_h = \frac{\partial W_h}{\partial q^h}(q^h, \overline{p}_1, \ldots, \overline{p}_h), \quad \overline{q}^h = \frac{\partial W_h}{\partial \overline{p}_h}(q^h, \overline{p}_1, \ldots, \overline{p}_h), \ h = 1, \ldots, n.$$
$$\tag{19.23}$$

19.3 Examples of Separable Systems

1. The Hamiltonian function of a harmonic oscillator S is

$$H(q, p) = \frac{p^2}{2m} + \frac{kq^2}{2}, \tag{19.24}$$

where m is the mass of S and k is the elastic constant of the force acting on S. Then, the Hamilton–Jacobi equation is

$$\frac{1}{2m}\left(\frac{\partial W}{\partial q}\right)^2 + \frac{kq^2}{2} = \overline{p} \equiv E. \tag{19.25}$$

The complete integral of (19.25)

$$W(q, \overline{p}) = \int \sqrt{2mE - mkq^2} dq$$

generates the canonical transformation $(\overline{q}, \overline{p}) \rightarrow (q, p)$ given by

$$p = \frac{\partial W}{\partial q} = \sqrt{2mE - mkq^2},$$

$$\overline{q} = \frac{\partial W}{\partial \overline{p}} = \int \frac{mdq}{\sqrt{2mE - mkq^2}}. \tag{19.26}$$

On the other hand, in the coordinates $(\overline{q}, \overline{p})$ the equations of motion are [see (19.5)]

$$\overline{q} = t + \alpha, \quad \overline{p} = E. \tag{19.27}$$

Introducing these functions into (19.26), we obtain

$$t + \alpha = \int \frac{m}{\sqrt{2mE - mkq^2}} dq = \sqrt{\frac{m}{k}} \int \frac{dq}{\sqrt{\frac{2E}{k} - q^2}},$$

so that

$$q = \sqrt{\frac{2E}{k}} \sin \sqrt{\frac{k}{m}}(t + \alpha).$$

2. Let P be a material point of mass m moving in a plane π, in which we adopt polar coordinates (r, ϕ), under the action of a Newtonian force with potential energy $U(r)$. Then, the Lagrangian function and the Hamiltonian function of P are given by

$$L = \frac{1}{2}m(\dot{r}^2 + r^2\dot{\varphi}^2) - U(r),$$

$$H = \frac{1}{2m}\left(p_r^2 + \frac{p_\varphi^2}{r^2}\right) + U(r). \tag{19.28}$$

Consequently, the Hamilton–Jacobi equation becomes

$$\frac{1}{2m}\left[\left(\frac{\partial W}{\partial r}\right)^2 + \frac{1}{r^2}\left(\frac{\partial W}{\partial \varphi}\right)^2\right] + U(r) = \overline{p}_2 \equiv E. \tag{19.29}$$

Comparing (19.15) and (19.29), we obtain that

$$g_1 = p_\varphi = \overline{p}_1, \quad g_2 = \frac{1}{2m}\left(p_r^2 + \frac{\overline{p}_1^2}{r^2}\right) + U(r),$$

and (19.13) becomes

$$W(r, \varphi, \overline{p}_1, E) = \overline{p}_1\varphi + \int \sqrt{2m(E - U(r)) - \frac{\overline{p}_1^2}{r^2}}\,dr.$$

The canonical transformation generated by this function is

$$p_r = \frac{\partial W}{\partial r} = \sqrt{2m(E - U(r)) - \frac{\overline{p}_1^2}{r^2}}, \tag{19.30}$$

$$p_\varphi = \frac{\partial W}{\partial \varphi} = \overline{p}_1, \tag{19.31}$$

$$\overline{q}^1 = \frac{\partial W}{\partial \overline{p}_1} = -\int \frac{\overline{p}_1}{r^2\sqrt{2m(E - U(r)) - \frac{\overline{p}_1^2}{r^2}}}\,dr + \varphi, \tag{19.32}$$

$$\overline{q}^2 = \frac{\partial W}{\partial \overline{p}_2} = \int \frac{m}{\sqrt{2m(E - U(r)) - \frac{\overline{p}_1^2}{r^2}}}\,dr. \tag{19.33}$$

On the other hand, in the coordinates $(\overline{\mathbf{q}}, \overline{\mathbf{p}})$ the equations of motion have the form

$$\overline{q}^1 = \alpha^1,\ \overline{q}^2 = t + \alpha,$$
$$\overline{p}_1 = \beta_1,\ \overline{p}_2 = E. \tag{19.34}$$

Introducing (19.34) into (19.32) and (19.33), we obtain the implicit form of the orbit $r = r(\varphi)$ and of the function $r = r(t)$.

3. Consider a material point P with mass m that is constrained to move on a surface of revolution S:

$$x = r \cos \varphi,$$
$$y = r \sin \varphi,$$
$$z = f(r),$$

where x, y, and z are the Cartesian coordinates of P with respect to a frame of reference $Oxyz$, φ is the angle between the plane π, containing P and the axis Oz, and the plane Oxz, and, finally r is the distance of P from Oz. If no active force is acting on P, since the square of velocity is $v^2 = \dot{x}^2 + \dot{y}^2 + \dot{z}^2$, then the Lagrangian function of P becomes

$$L = \frac{1}{2}m\left[(1 + f'^2(r))\dot{r}^2 + r^2\dot{\varphi}^2\right]. \tag{19.35}$$

Since

$$p_r = \frac{\partial L}{\partial \dot{r}} = m(1 + f'^2(r))\dot{r},$$
$$p_\varphi = \frac{\partial L}{\partial \dot{\varphi}} = mr^2\dot{\varphi},$$

the Hamiltonian function can be written as

$$H(r, \varphi, p_r, p_\varphi) = \frac{1}{2m}\left(\frac{p_r^2}{1 + f'^2} + \frac{p_\varphi^2}{r^2}\right).$$

Verify that the solution of the corresponding Hamilton–Jacobi equation is

$$W = \sqrt{p_1}\varphi + \int \sqrt{\left(2mE - \frac{p_1}{r^2}\right)(1 + f'^2)}\,dr.$$

19.4 Hamilton's Principal Function

In Chap. 17, we considered the derivative at $s = 0$ of the integral (17.119). In this integral $\mathbf{f}(s, t)$, $(s, t) \in [-\epsilon, \epsilon] \times [t_1, t_2] \to V_n$, denotes a one-parameter family of curves, starting from a point $\mathbf{q}_1$ at the instant t_1 and arriving at the point $\mathbf{q}_2$ at the instant t_2. In other words, the initial and final points, as well as the time interval $[t_1, t_2]$, are the same for all the curves $\mathbf{f}(s, t)$. In this section, we consider the wider family of curves such that, $\forall s \in [-\epsilon, \epsilon]$, the initial and final points and the time interval in which each curve is defined

$$\mathbf{f}(s, t) : t \in [t_1(s), t_2(s)] \to V_n \tag{19.36}$$

depend on s. Then, instead of (17.119), we take into account the other integral:

$$I_f(s) = \int_{t_1(s)}^{t_2(s)} L(\mathbf{f}(s,t), \dot{\mathbf{f}}(s,t), t) \, dt. \tag{19.37}$$

Repeating the calculations from Sect. 17.14 and taking into account the Leibniz formula, we obtain

$$\left(\frac{dI_f}{ds}\right)_0 = \left[L(\mathbf{q}(\bar{t}_i), \dot{\mathbf{q}}(\bar{t}_i), \bar{t}_i) \left(\frac{dt_i}{ds}\right)_0 \right]_1^2 + \left[\frac{\partial L}{\partial \dot{\mathbf{q}}}(\mathbf{q}(\bar{t}_i), \dot{\mathbf{q}}(\bar{t}_i), \bar{t}_i) \frac{\partial \mathbf{f}}{\partial s}(0, \bar{t}_i) \right]_1^2$$

$$+ \int_{\bar{t}_1}^{\bar{t}_2} \left[\frac{\partial L}{\partial \mathbf{q}} - \frac{d}{dt} \frac{\partial L}{\partial \dot{\mathbf{q}}} \right] \frac{\partial \mathbf{f}}{\partial s}(0, t) \, dt, \tag{19.38}$$

where

$$\mathbf{q}(t) = \mathbf{f}(0, t), \quad \bar{t}_i = t_i(0), \quad i = 1, 2. \tag{19.39}$$

This expression, introducing the Hamiltonian function [see (17.156)]

$$H(\mathbf{f}(s,t), \dot{\mathbf{f}}(s,t), t) = \mathbf{p} \cdot \dot{\mathbf{f}} - L, \tag{19.40}$$

assumes the equivalent form

$$\left(\frac{dI_f}{ds}\right)_0 = \left[\mathbf{p}(\mathbf{q}(\bar{t}_i), \dot{\mathbf{q}}(\bar{t}_i), \bar{t}_i) \xi \right]_1^2 - \left[H(\mathbf{q}(\bar{t}_i), \dot{\mathbf{q}}(\bar{t}_i), \bar{t}_i) \frac{dt_i}{ds}(0) \right]_1^2$$

$$+ \int_{\bar{t}_1}^{\bar{t}_2} \left[\frac{\partial L}{\partial \mathbf{q}} - \frac{d}{dt} \frac{\partial L}{\partial \dot{\mathbf{q}}} \right] \frac{\partial \mathbf{f}}{\partial s}(0, t) \, dt, \tag{19.41}$$

where we used the notation

$$\xi = \frac{\partial \mathbf{f}}{\partial s}(0, \bar{t}_i) + \frac{\partial \mathbf{f}}{\partial t}(0, \bar{t}_i) \frac{dt_i}{ds}(0). \tag{19.42}$$

In Sect. 17.4, we remarked that the boundary value problem of finding a solution $\mathbf{q}(t)$ of the Lagrange equations, satisfying the boundary conditions $\mathbf{q}(t_1) = \mathbf{q}_1$ and $\mathbf{q}(t_2) = \mathbf{q}_2$, is a problem profoundly different from the initial value problem relative to the same equations. In fact, for the former a general theorem insures the existence and uniqueness of the solution, whereas for the latter no general theorem holds. Consequently, the boundary value problem could have no solution, one solution, or infinite solutions (focal points). Regarding this problem, we assume that *for any choice of* $\mathbf{q}_1 \in D_1 \subseteq V_n, \mathbf{q}_2 \in D_2 \subseteq V_n$, *there is one and only one solution of Lagrange's equations starting from* $\mathbf{q}_1$ *at the instant* $\bar{t}_1$ *and arriving at* $\mathbf{q}_2$ *at the instant* $\bar{t}_2$. If this hypothesis is verified, then for any choice of $\mathbf{q}_1 \in D_1, \mathbf{q}_2 \in D_2$, we can evaluate the integral (19.37) along the unique solution $\mathbf{q}(t)$ of Lagrange's equations such that $\mathbf{q}(\bar{t}_1) = \mathbf{q}_1$ and $\mathbf{q}(\bar{t}_2) = \mathbf{q}_2$. In this way, the integral (19.37) defines a *function*

$$S : (\bar{t}_1, \bar{t}_2, \mathbf{q}_1, \mathbf{q}_2) \in \mathfrak{R}^2 \times D_1 \times D_2 \rightarrow \int_{\bar{t}_1}^{\bar{t}_2} L(\mathbf{q}(t), \dot{\mathbf{q}}(t), t) \, dt, \qquad (19.43)$$

which is called **Hamilton's principal function**. It is evident that the differential dS of S at the point $(\bar{t}_1, \bar{t}_2, \mathbf{q}_1, \mathbf{q}_2) \in \mathfrak{R}^2 \times D_1 \times D_2$ coincides with the differential of the function $I_f(s)$ evaluated at $s = 0$, *provided that all the curves* $\mathbf{f}(s,t)$ *are solutions of Lagrange's equations (effective motions)*. Then, in view of (19.41), we obtain

$$dS = [\mathbf{pdq}]_1^2 - [Hdt]_1^2, \qquad (19.44)$$

where [see (19.42)]

$$d\mathbf{q} = \xi ds = \frac{\partial \mathbf{f}}{\partial s}(0, \bar{t}_i) ds + \frac{\partial \mathbf{f}}{\partial t}(0, \bar{t}_i) d\bar{t}_i, \qquad (19.45)$$

$$d\bar{t}_i = \frac{dt_i}{ds}(0) ds. \qquad (19.46)$$

The initial and final points of the motions $\mathbf{f}(s,t)$ are given by $\mathbf{q}_i(s) = \mathbf{f}(s, t_i(s))$. Consequently, the differential (19.45) gives the variations of these points in going from a given motion to a close one. Finally, changing our notations as follows

$$\bar{t}_1 = t_0, \; \bar{t}_2 = t, \; \mathbf{q}_1 = \mathbf{q}_0, \; \mathbf{q}_2 = \mathbf{q}, \; \mathbf{p}_1 = \mathbf{p}_0, \; \mathbf{p}_2 = \mathbf{p}, \qquad (19.47)$$

from (19.44) we obtain

$$\frac{\partial S}{\partial \mathbf{q}}(t_0, t, \mathbf{q}_0, \mathbf{q}) = \mathbf{p}, \qquad (19.48)$$

$$\frac{\partial S}{\partial \mathbf{q}_0}(t_0, t, \mathbf{q}_0, \mathbf{q}) = -\mathbf{p}_0, \qquad (19.49)$$

$$\frac{\partial S}{\partial t}(t_0, t, \mathbf{q}_0, \mathbf{q}) = -H(\mathbf{q}, \mathbf{p}, t), \qquad (19.50)$$

$$\frac{\partial S}{\partial t_0}(t_0, t, \mathbf{q}_0, \mathbf{q}) = H(\mathbf{q}_0, \mathbf{p}_0, t_0). \qquad (19.51)$$

If Hamilton's principal function satisfies the condition

$$\det\left(\frac{\partial S}{\partial q^h \partial q_0^k}\right) \neq 0, \qquad (19.52)$$

then we can give (19.49) the form

$$\mathbf{q} = \mathbf{q}(t_0, t, \mathbf{q}_0, \mathbf{p}_0). \qquad (19.53)$$

In conclusion, if Hamilton's principal function $S(t_0, t, \mathbf{q}_0, \mathbf{q})$ is known and satisfies (19.52), then

- The equation of motion (19.53) is given in implicit form by (19.48) and (19.49);
- The correspondence between the initial data $(\mathbf{q}_0, \mathbf{p}_0) \in D_0$ and $(\mathbf{q}, \mathbf{p}) \in D$ is symplectic, that is, the motion in the phase state defines a sequence of symplectic transformations between the initial data and the corresponding positions at the instant t;
- In view of (19.48) and (19.50), the function $S(t_0, t, \mathbf{q}_0, \mathbf{q})$ is a complete integral of the equation

$$
\frac{\partial S}{\partial t} + H\left(\mathbf{q}, \frac{\partial S}{\partial \mathbf{q}}, t\right) = 0. \tag{19.54}
$$

Now we suppose that the Hamiltonian function H does not depend on time. If we denote by $E = H(\mathbf{q}_0, \mathbf{p}_0)$ the value of the total energy corresponding to the initial data and we recall that the total energy is constant during motion, then the function

$$
S(t_0, t, \mathbf{q}_0, \mathbf{q}) = -E(t - t_0) + W(\mathbf{q}_0, \mathbf{q}) \tag{19.55}
$$

is a solution of (19.54) provided that the function W is a complete integral of the Hamilton–Jacobi equation

$$
H\left(\mathbf{q}, \frac{\partial W}{\partial \mathbf{q}}(\mathbf{q}_0, \mathbf{q})\right) = E. \tag{19.56}
$$

We conclude this section with a geometric description of the Hamilton–Jacobi equation. Let V_n be the configuration space of the dynamical system S described by the Lagrangian function

$$
L(\mathbf{q}, \dot{\mathbf{q}}) = \frac{1}{2} a_{hk} \dot{q}^h \dot{q}^k - U(\mathbf{q}), \tag{19.57}
$$

where $U(\mathbf{q})$ is the potential energy of the forces acting on S, and let V_n be equipped with the Riemannian metric (Sect. 17.15)

$$
\mathrm{d}s^2 = a_{hk}(\mathbf{q}) \mathrm{d}q^h \mathrm{d}q^k. \tag{19.58}
$$

Before proceeding, we need to make clear the meaning of the **normal speed** c_n of a moving surface $\Sigma(t)$ in the Riemannian space V_n. Denoting by $f(\mathbf{q}, t) = 0$ the implicit equation of $\Sigma(t)$, the unit normal vector $\mathbf{N}$ to $\Sigma(t)$ has covariant components given by

$$
N_i = \frac{f_{,i}}{\sqrt{a^{hk} f_{,h} f_{,k}}} = \frac{f_{,i}}{|\nabla f|}, \tag{19.59}
$$

Fig. 19.1 Moving surface Σ

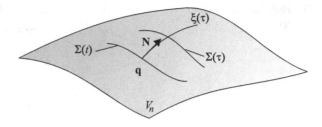

where $f_{,h} = \partial f / \partial q^h$ and (a^{hk}) is the inverse matrix of (a_{hk}). Now, at the arbitrary point $\mathbf{q} \in \Sigma(t)$ we consider a curve $\boldsymbol{\xi}(\tau)$ such that (Fig. 19.1)

$$\boldsymbol{\xi}(t) = \mathbf{q}. \tag{19.60}$$

The intersection points of the curve $\boldsymbol{\xi}(\tau)$ with $\Sigma(\tau)$, $\tau > t$, satisfy the equation $f(\boldsymbol{\xi}(\tau), \tau) = 0$. Differentiating this equation with respect to τ, setting $\tau = t$, and recalling (19.59), we obtain the condition

$$f_{,h}\dot{\xi}^h + f_{,t} = 0,$$

which can also be written as

$$c_n \equiv \mathbf{N} \cdot \dot{\boldsymbol{\xi}} = -\frac{f_{,t}}{|\nabla f|}. \tag{19.61}$$

It is evident that c_n is the speed of $\Sigma(t)$ along the normal $\mathbf{N}$ at $\mathbf{q}$, i.e., it is the normal velocity of $\Sigma(t)$. Since the surface is moving, we necessarily have $\partial f / \partial t \neq 0$. Consequently, we can locally put the equation of $\Sigma(t)$ in the form

$$t = \psi(\mathbf{q}), \tag{19.62}$$

where, in view of (19.61), the function $\psi(\mathbf{q})$ is a solution of the **eikonal equation**

$$|\nabla \psi|^2 = a^{hk}\psi_{,h}\psi_{,k} = \frac{1}{c_n^2(\mathbf{q})}. \tag{19.63}$$

We set $t_0 = 0$ in (19.55) and suppose that $\mathbf{q}_0$ is a fixed point of V_n (Fig. 19.2). Then we consider the moving surface $\Sigma^*(t)$ of the equation

$$t = \frac{1}{E}W(\mathbf{q}_0, \mathbf{q}) \equiv \psi^*(\mathbf{q}_0, \mathbf{q}). \tag{19.64}$$

It is evident that $\Sigma^*(t)$ reduces to the point $\mathbf{q}_0$ for $t = 0$. Further, omitting the dependency on $\mathbf{q}_0$, from (19.63) it follows that its normal speed c_n is given by

Fig. 19.2 Moving surface $t = \psi(\mathbf{q})$

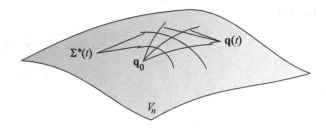

$$c_n = \frac{1}{|\nabla \psi^*(\mathbf{q})|} = \frac{E}{\sqrt{a^{hk} W_{,h} W_{,k}}}.$$

In view of (19.48) and (19.55),

$$\mathbf{p} = \frac{\partial W}{\partial \mathbf{q}}, \tag{19.65}$$

and we have that

$$a^{hk} W_{,h} W_{,k} = a^{hk} p_h p_k = 2T = 2(E - U(\mathbf{q})). \tag{19.66}$$

Finally, the normal speed can also be written as

$$c_n = \frac{E}{\sqrt{2(E - U(\mathbf{q}))}}. \tag{19.67}$$

In conclusion, Eq. (19.64) defines a moving surface in V_n advancing with normal speed (19.67).

Let $\mathbf{q}(t)$ be a dynamical trajectory starting from the point $\mathbf{q}_0$. The contravariant components of the vector tangent to this trajectory are $(\dot{q}^h)$, whereas in metric (19.58), the covariant components of this vector are $p_h = a_{hk}\dot{q}^h$. Comparing this result with (19.65), we can state that *the surfaces $\Sigma^*(t)$ are orthogonal to the trajectories sorting from $\mathbf{q}_0$.*

Remark 19.4. The normal speed c_n of a point $\mathbf{q} \in \Sigma^*(t)$ is evaluated along the normal $\mathbf{N}$ to $\Sigma^*(t)$ at $\mathbf{q}$. On the other hand, in view of (19.55), also the trajectory containing $\mathbf{q}$ is orthogonal to $\Sigma^*(t)$ at that point $\mathbf{q}$. However, c_n is not equal to the velocity of the mechanical system along the dynamical trajectory. In fact, the velocity $\mathbf{v}$ of a point moving along a dynamical trajectory has contravariant components $(\dot{q}^h)$, so that

$$v^2 = a_{hk}\dot{q}^h \dot{q}^k = 2T = 2(E - U(\mathbf{q})).$$

Consequently, in view of (19.67), we obtain the relation

Fig. 19.3 Huygens'
principle

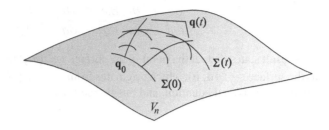

$$c_n = \frac{E}{v}.$$

We now prove **Huygens' principle**, which makes the following statement (Fig. 19.3).

Let Σ_0 be an $(n-1)$-dimensional surface of the initial data $\mathbf{q}_0(\mathbf{u})$, $\mathbf{u} = (u^1, \ldots u^{n-1})$, of the configuration space V_n. Then the moving surface $\Sigma^*(t)$ of equation $t = \psi^*(\mathbf{q}_0, \mathbf{q})$, $\mathbf{q}_0 \in \Sigma_0$, is the envelope of all the surfaces $\Sigma^*(t)$ originating from the different points of Σ_0.

First, we denote the collection of all the surfaces originating from an arbitrary but fixed point of Σ_0 by the equation

$$\hat{\psi}(\mathbf{u}, \mathbf{q}) = \psi^*(\mathbf{q}_0(\mathbf{u}), \mathbf{q}).$$

To prove the preceding statement, it is sufficient to recall that the envelope of the surfaces $t - \hat{\psi}(\mathbf{u}, \mathbf{q})$ upon varying the parameters $(u^1, \ldots, u^{n-1})$ can be obtained by eliminating these parameters in the system

$$t = \hat{\psi}(\mathbf{u}, \mathbf{q}), \tag{19.68}$$

$$\frac{\partial \hat{\psi}}{\partial \mathbf{u}}(\mathbf{u}, \mathbf{q}) = 0. \tag{19.69}$$

Supposing that (19.69) can be solved with respect to the parameters $u^1, \ldots, u^{n-1}$

$$u^\alpha = u^\alpha(\mathbf{q}), \quad \alpha = 1, \ldots, n-1,$$

substitution of these equations into (19.68) yields

$$t - \hat{\psi}(\mathbf{u}, \mathbf{q}) = t - \psi(\mathbf{q}) = 0. \tag{19.70}$$

To prove that the function $\psi(\mathbf{q})$ is a solution of the eikonal equation, we note that from (19.69) we obtain

$$\frac{\partial \psi}{\partial q^h} = \frac{\partial \hat{\psi}}{\partial u^\alpha} \frac{\partial u^\alpha}{\partial q^h} + \frac{\partial \hat{\psi}}{\partial q^h} = \frac{\partial \hat{\psi}}{\partial q^h}.$$

This result states that, at the point $\mathbf{q}$, the surface $t = \psi(\mathbf{q})$ has the same tangent plane as the surface $t - \hat{\psi}(\mathbf{u}, \mathbf{q})$, which originates from a point $\mathbf{q}_0 \in \Sigma_0$. Further, this surface satisfies the eikonal equation, and this is true for $\psi(\mathbf{q})$.

19.5   Time-dependent Hamilton–Jacobi Equation

In this section we formulate the Hamilton–Jacobi equation for time-dependent Hamiltonian dynamics. In canonical coordinates $(U, (\mathbf{q}, \mathbf{p}, t))$ on $M_{2n}^\star \times \Re$, Hamilton's equations have the canonical form

$$\dot{q}^h = \frac{\partial H}{\partial p_h}(\mathbf{q}, \mathbf{p}, t), \quad \dot{p}_h = -\frac{\partial H}{\partial q^h}(\mathbf{q}, \mathbf{p}, t). \tag{19.71}$$

We search for a canonical transformation

$$\begin{aligned}
\overline{\mathbf{q}} &= \overline{\mathbf{q}}(\mathbf{q}, \mathbf{p}, t), \\
\overline{\mathbf{p}} &= \overline{\mathbf{p}}(\mathbf{q}, \mathbf{p}, t), \\
\overline{t} &= t,
\end{aligned} \tag{19.72}$$

to new coordinates $(U, (\overline{\mathbf{q}}, \overline{\mathbf{p}}, t))$ in which the Hamiltonian $K(\overline{\mathbf{q}}, \overline{\mathbf{p}}, t)$ reduces to

$$K(\overline{\mathbf{q}}, \overline{\mathbf{p}}, t) = 0. \tag{19.73}$$

If we reach this result, then in the new coordinates Hamilton's equations become

$$\dot{\overline{q}}^h = 0, \quad \dot{\overline{p}}_h = 0, \tag{19.74}$$

and their solutions are

$$\overline{q}^h = \alpha^h, \quad \overline{p}_h = \beta_h, \tag{19.75}$$

where α^h and β_h, $h = 1, \ldots, n$, are constants representing the initial data in the coordinates $(\overline{\mathbf{q}}, \overline{\mathbf{p}}, t)$. From (19.75) and the inverse transformation of (19.72) we obtain the equations of motion

$$q^h = q^h(\alpha, \beta, t), \quad p_h = p_h(\alpha, \beta, t). \tag{19.76}$$

It remains to determine a function $S(\mathbf{q}, \overline{\mathbf{p}}, t)$ generating a canonical transformation by the relations

$$\overline{q}^h = \frac{\partial S}{\partial \overline{p}_h}(\mathbf{q}, \overline{\mathbf{p}}, t), \quad p_h = \frac{\partial S}{\partial q_h}(\mathbf{q}, \overline{\mathbf{p}}, t), \tag{19.77}$$

$$\det\left(\frac{\partial^2 S}{\partial q^h \partial \overline{p}_k}\right) \neq 0, \tag{19.78}$$

and satisfying (18.136) with $K = 0$, that is the **time-dependent Hamilton–Jacobi equation**

$$\frac{\partial S}{\partial t}(\mathbf{q}, \overline{\mathbf{p}}, t) + H\left(\mathbf{q}, \frac{\partial S}{\partial \mathbf{q}}(\mathbf{q}, \overline{\mathbf{p}}, t), t\right) = 0. \tag{19.79}$$

In particular, if H does not depend on time, a solution of (19.79) is given by

$$S(\mathbf{q}, \overline{\mathbf{p}}, t) = W(\mathbf{q}, \overline{\mathbf{p}}) - Et, \tag{19.80}$$

where E is a constant and W satisfies the stationary Hamilton–Jacobi equation

$$H\left(\mathbf{q}, \frac{\partial W}{\partial \mathbf{q}}\right) = E. \tag{19.81}$$

Chapter 20
Completely Integrable Systems

Abstract This chapter contains advanced topics of Hamiltonian mechanics. Arnold–Liouville's theorem shows that the motion of completely integrable systems can be determined. These systems describe the behavior of many physical systems and represent the starting point for analyzing more complex mechanical systems. The study of completely integrable systems leads to the introduction of very convenient canonical coordinates: the angle–action variables that are fundamental in celestial mechanics. Further, they were used by Bohr–Sommerfeld to formulate the quantization rules that allowed a first interpretation of hydrogenoid atomic spectra. A sketch of the Hamiltonian perturbation theory and an overview of KAM theorem conclude the chapter.

20.1 Arnold–Liouville's Theorem

This chapter contains some of the most advanced topics in analytical mechanics. *Due to the introductory character of the book, theorems requiring a deep knowledge of geometry and algebra are partially proved or only stated.*

Definition 20.1. Let (M_{2n}^*, Ω, H) be a Hamiltonian system that admits $p \leq n$ first integrals $f_1(x), \ldots, f_p(x)$ of class C^1. We say that these first integrals are **independent** if the differentials $\mathbf{d}f_1, \ldots, \mathbf{d}f_p$ are linearly independent at any point $x \in M_{2n}^*$.

Theorem 20.1. *Let M_{2n}^* be of class C^k. If the independent first integrals $f_1, \ldots, f_p$, $p \leq n$, are of class C^k, $k \geq 1$, then the subset of M_{2n}^**

$$M_{f(\mathbf{c})} = \left\{ x \in M_{2n}^*, \, f_h(x) = c_h, \, h = 1, \ldots, p \right\}, \tag{20.1}$$

where $c_1, \ldots, c_p$ are given constants, is a $(2n\text{-}p)$-dimensional C^k manifold.

Proof. Let x_0 be a point of $M_{f(\mathbf{c})}$, and denote by (x^α), $\alpha = 1, \ldots, 2n$, a system of coordinates of M_{2n}^* in a neighborhood of x_0. Then the coordinate representation

© Springer International Publishing AG, part of Springer Nature 2018

A. Romano and A. Marasco, *Classical Mechanics with Mathematica®*, Modeling and Simulation in Science, Engineering and Technology, https://doi.org/10.1007/978-3-319-77595-1_20

$\hat{f}_1(x^\alpha), \ldots, \hat{f}_p(x^\alpha)$ of the functions $f_1(x), \ldots, f_p(x)$ is of class C^k and the set $M_{f(c)}$ is locally defined by the system $\hat{f}_h = c_h$, $h = 1, \ldots, p$. Further, owing to the linear independence of the differentials $\mathbf{d}f_1, \ldots, \mathbf{d}f_p$, the Jacobian matrix $(\partial \hat{f}_h / \partial x^\alpha)$ has a rank equal to p. From what we proved in Chap. 6, it follows that M_{2n}^* is a *(2n-p)*-dimensional C^k manifold.   □

Definition 20.2. The manifold $M_{f(c)}$, determined by the p independent first integrals $f_1(x), \ldots, f_p(x)$, is called a *level manifold*.

Definition 20.3. The C^k first integrals $f_1(x), \ldots, f_p(x)$ are said to be in *involution* if

$$\{f_h, f_k\} = 0, \quad h, k = 1, \ldots, p, \ h \neq k. \tag{20.2}$$

Definition 20.4. *A Hamiltonian system (M_{2n}^*, Ω, H) is a **completely integrable system** if it admits n independent first integrals in involution.*

We have already discussed (Sect. 18.7) the importance of first integrals in localizing the dynamical trajectories of a Hamiltonian system (M_{2n}^*, Ω, H). In fact, if $f(x)$ is a first integral of (M_{2n}^*, Ω, H), then a dynamical trajectory starting from a point of the manifold $f(x) = c$ is fully contained in this manifold. This result, which still holds when the Hamiltonian system possesses p first integrals, pushes us to determine the structure of $M_{f(c)}$ in order to localize the region of M_{2n}^* in which the dynamical trajectories of (M_{2n}^*, Ω, H) are contained. In this regard, the following theorem is fundamental.

Theorem 20.2 (Arnold–Liouville). *Let (M_{2n}^*, Ω, H) be a completely integrable Hamiltonian system, and denote by $\mathbf{X}_{f_h}$ the Hamiltonian vector fields $(\mathbf{d}f)_h^\sharp$, $h = 1, \ldots, n$. If the level manifold*

$$M_{f(c^0)} = \{x \in M_{2n}^*, \ f_h(x) = c_h^0, \ h = 1, \ldots, n\}, \tag{20.3}$$

corresponding to the constants $\mathbf{c}^0 = (c_1^0, \ldots, c_n^0)$, is connected and compact, then the following statements hold.

1. *$M_{f(c^0)}$ is invariant with respect to the one-parameter group of diffeomorphisms $\phi_h^t(x)$ generated by the vector fields $\mathbf{X}_{f_h}$.*
2. *$M_{f(c^0)}$ is an n-dimensional torus*

$$T^n(\mathbf{c}^0) = \{(\varphi^1, \ldots, \varphi^n) \mod 2\pi\}, \tag{20.4}$$

where $\varphi^1, \ldots, \varphi^n$ are angular coordinates on $T^n(\mathbf{c}^0)$.
3. *There exist an open 2n-dimensional neighborhood U of $M_{f(c^0)}$ and a diffeomorphism $F : U \to T^n(\mathbf{c}^0) \times I(\mathbf{c}^0)$, where $I(\mathbf{c}^0) = (c_1^0 - \delta_1, c_1^0 + \delta_1) \times \cdots \times (c_n^0 - \delta_n, c_n^0 + \delta_n)$, $\delta_h > 0$; in other words, the variables $(\varphi^1, \ldots \varphi^n, c_1, \ldots, c_n)$, for*

$(c_1, \ldots, c_n) \in I(\mathbf{c}^0)$, are local coordinates on M_{2n}^* in a neighborhood of $M_{f(\mathbf{c}^0)}$ and the level manifolds $M_{f(\mathbf{c})}, \mathbf{c} \in I(\mathbf{c}^0)$ are tori.

4. The dynamical trajectories of (M_{2n}^*, Ω, H) that lie on $M_f(\mathbf{c})$ in the local coordinates $(\varphi^1, \ldots \varphi^n, c_1, \ldots, c_n)$ have the following parametric equations:

$$\varphi^h = w^h(\mathbf{c})t + \varphi_0^h, \quad c_h = \text{const}, \quad h = 1, \ldots, n, \tag{20.5}$$

where the quantities φ_0^h are constant.

Proof. We limit ourselves to sketching the proof of this very difficult theorem.

First, the Hamiltonian fields $\mathbf{X}_{f_h}, h = 1, \ldots, n$, are independent since the map $\sharp$ is an isomorphism (Sect. 18.4), and the differentials $\mathbf{d}f_h, h = 1, \ldots, n$, are linearly independent. Further, the function ϕ defined by (18.62) maps the Lie algebra of differentiable functions on M_{2n}^*, equipped with Poisson's bracket, into the Lie algebra of vector fields equipped with bracket. Therefore, from $\{f_h, f_k\} = 0$ there follows $\left[\mathbf{X}_{f_h}, \mathbf{X}_{f_k}\right] = \mathbf{0}$. Let $\phi_h^t(x)$ be a one-parameter group of diffeomorphisms generated by $\mathbf{X}_{f_h}$. Since $M_{f(\mathbf{c}_0)}$ is compact, the vector fields $\mathbf{X}_{f_h}$ are complete and the corresponding groups $\phi_h^t(x)$ are global. Along any orbit of $\phi_k^t(x)$ we have that

$$\frac{\mathbf{d}f_h}{\mathbf{d}t}(\phi_k^t(x)) = L_{\mathbf{X}_{f_k}} f_h = \{f_h, f_k\} = 0,$$

so that $f_h(\phi_h^t(x)) = \text{const}$. Consequently, if for $t = 0$ we have $f_h(\phi_k^0(x)) = f_h(x) = c_h$, then, $\forall t > 0$, we also have $f_h(\phi_k^t(x)) = c_h$ and the orbit starting from $x \in M_{f(\mathbf{c}^0)}$ belongs to $M_{f(\mathbf{c}^0)}$. In other words, $M_{f(\mathbf{c}^0)}$ is invariant with respect to all the global groups $\phi_h^t(x), h = 1, \ldots, n$, and the infinitesimal generators $\mathbf{X}_{f_h}, h = 1, \ldots, n$, of these groups are tangent to $M_{f(\mathbf{c}^0)}$.

At this point, we can state that on $M_{f(\mathbf{c}^0)}$ there exist n independent tangent vector fields $\mathbf{X}_{f_h}$ satisfying the conditions $\left[\mathbf{X}_{f_h}, \mathbf{X}_{f_k}\right] = \mathbf{0}$. It is possible to prove that the only connected and compact n-dimensional manifold on which there are n vector fields with the foregoing properties is an n-dimensional torus $T^n(\mathbf{c}^0) = S^1 \times \cdots \times S^1$. In the angular coordinates $\varphi^1, \ldots, \varphi^n$ on $T^n(\mathbf{c}^0)$ the vector fields $\mathbf{X}_{f_h}$ have the coordinate representation

$$\mathbf{X}_{f_h} = X_{f_h}^k(\mathbf{c}^0)\frac{\partial}{\partial \varphi^k}, \tag{20.6}$$

where the components $X_{f_h}^k$ are constant. Further, the linear independence of the differentials $\mathbf{d}f_1 \ldots, \mathbf{d}f_n$, after a suitable renumbering of the symplectic coordinates, leads to put the system

$$f_h(\mathbf{q}, \mathbf{p}) = c_h, \quad h = 1, \ldots, n, \tag{20.7}$$

in the form

$$q = q,$$
$$p = p(q, c),$$

at least for c belonging to a suitable interval $I(c^0)$. Then it is possible to prove that, in this interval, the preceding system defines compact manifolds $T^n(c)$ that inherit all the properties of $T^n(c^0)$. In particular, on each torus $T^n(c)$, the vector fields X_{f_h} have the coordinate representation

$$X_{f_h} = X^k_{f_h}(c) \frac{\partial}{\partial \varphi^k}, \tag{20.8}$$

and the dynamical trajectories have the form (20.5).   □

Remark 20.1. Any Hamiltonian system with one degree of freedom ($n = 1$) is completely integrable since the energy integral $H(q,p) = E$ holds. If $n = 2$, then a Hamiltonian system is completely integrable provided that it admits a first integral $f(q, p) = c$ independent of the energy integral $H(q, p) = E$. In fact, since f is a first integral, $\{f, H\} = 0$. Finally, it is evident that a Hamiltonian system with $n = 3$ is completely integrable, provided that it admits two first integrals f_1, f_2 such that $\{f_1, f_2\} = 0$.

Example 20.1. The Hamiltonian function of a simple pendulum of unit mass and unit length is

$$H = \frac{1}{2}p^2 - g\cos\theta,$$

where the symbols have the usual meaning and the phase space M_2^* is a cylinder $S \times \Re$.

Its phase portrait is shown in Fig. 20.1, where the level curve γ, given by $H = E^*$, separates the oscillatory motions from the progressive ones. The curve γ is formed by three connected components, that is, the point $\theta = \pi$, which is identical to $\theta = -\pi$, and the two curves c and c', which are both diffeomorphic to $\Re$. The curves corresponding to values of energy less than E^* are one-dimensional tori (circumferences).

Example 20.2. Let P be a material point with unit mass moving on the straight line r under the action of a conservative force. The phase portrait of P, corresponding to a potential energy $U(q)$ represented on the left-hand side of Fig. 20.2, is shown on the right-hand side of the same figure. For values $E < 0$ of the total energy E we obtain one-dimensional tori.

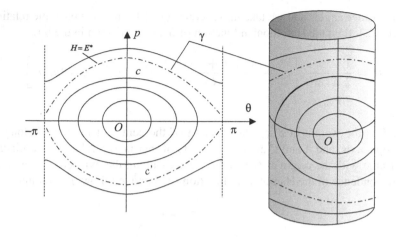

Fig. 20.1 Phase portrait of a simple pendulum

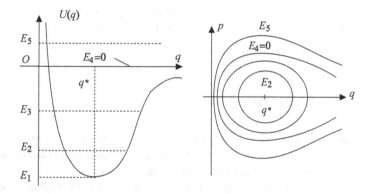

Fig. 20.2 Phase portrait of a one-dimensional motion

Example 20.3. Let S be a mechanical system with Hamiltonian function given by

$$H(\mathbf{q}, \mathbf{p}) = \frac{1}{2} \sum_{h=1}^{n} \left[p_h^2 + \lambda_h^2 (q^h)^2 \right].$$

It is easy to verify that the functions $f_h(\mathbf{q}, \mathbf{p}) = p_h^2 + \lambda_h^2 (q^h)^2$, $h = 1, \ldots, n$, are n independent first integrals such that $\{f_h, f_k\} = 0$, $h \neq k$. In other words, the system S is completely integrable. We notice that the Hamiltonian function is itself a first integral, but it depends on the preceding first integrals since it coincides with their sum. Each equation $f_h(\mathbf{q}, \mathbf{p}) = c_h$ defines an ellipsis in the plane (q^h, p_h). Consequently, the level manifold is the product of these ellipses, that is, it is an n-dimensional torus.

Example 20.4. Let S be a heavy gyroscope and denote by G its center of mass. Suppose that S is fixed at a point $O \neq G$ of the gyroscopic axis. Then, if φ, ψ, and

θ are the Euler angles and we take into account (15.21) and the kinematic relations of Sect. 12.4, then the Hamiltonian function of S can be written in the form

$$H = \frac{1}{2}\frac{1}{A\sin^2\theta}\left[p_\psi^2 + \frac{A\sin^2\theta + C\cos^2\theta}{C}p_\varphi^2 + p_\theta^2 - 2\cos\theta\, p_\psi p_\varphi\right]$$
$$+ Pz'_G\cos\theta,$$

where P is the weight of S, z'_G the abscissa of the center of mass of S along the gyroscopic axis, and A and C are the moments of inertia of S relative to a principal body frame with the origin at O and the Oz axis along the gyroscopic axis. From the independence of the Hamiltonian function of ψ and φ follow the first integrals

$$p_\psi = c_\psi, \quad p_\varphi = c_\varphi,$$

where c_ψ and c_φ are constant. Further, since in the Hamiltonian formalism the variables $\mathbf{q}$ and $\mathbf{p}$ are independent, we have also $\{p_\psi, p_\varphi\} = 0$. In conclusion, the first three integrals H, p_ψ, and p_φ are independent and in involution. By some calculations, it is possible to verify that the level manifolds are connected and compact so that they are tori.

20.2 Angle–Action Variables

Generally, the coordinates $(\varphi, \mathbf{c})$, defined by the Arnold–Liouville theorem, are not symplectic; in other words, functions (20.5) are not solutions of a Hamiltonian system. It must also be remarked that the aforementioned theorem proves the existence of these variables, but it does not specify their relation with the coordinates $(\mathbf{q}, \mathbf{p})$ in which the dynamical problem is initially formulated. In this section, we prove the existence in a neighborhood $V \subset M_{2n}^*$ of the level manifold $M_{f(\mathbf{c}^0)}$ of a new set of coordinates $(\boldsymbol{\theta}, \mathbf{I})$, called *angle–action variables*, such that the following statements hold.

- $(\boldsymbol{\theta}, \mathbf{I})$ are symplectic coordinates.
- The action variables $\mathbf{I}$ define the tori.
- The variables $\boldsymbol{\theta}$ are angular variables on the tori.
- The relation between the initial coordinates $(\mathbf{q}, \mathbf{p})$ and the new angle–action coordinates $(\boldsymbol{\theta}, \mathbf{I})$ is known.

To prove the existence of these coordinates, we consider the immersion map

$$i : T^n(\mathbf{c}) \to M_{2n}^*. \tag{20.9}$$

If $(\mathbf{q}, \mathbf{p})$ are local symplectic coordinates in M_{2n}^*, then the parametric equations of $T^n(\mathbf{c})$, i.e., the coordinate formulation of the mapping (20.9), have the form

$$\mathbf{q} = \mathbf{q}(\boldsymbol{\varphi}, \mathbf{c}), \tag{20.10}$$

$$\mathbf{p} = \mathbf{p}(\boldsymbol{\varphi}, \mathbf{c}). \tag{20.11}$$

Theorem 20.3. *Let ω be the Liouville 1-form. Then, the restriction $i^*\omega$ to $T^n(\mathbf{c})$ of the differential 1-form ω is closed on $T^n(\mathbf{c})$, i.e.,*

$$\mathbf{d}i^*\omega(\mathbf{u}, \mathbf{v}) = 0, \tag{20.12}$$

for any pair of vectors $(\mathbf{u}, \mathbf{v})$ tangent to $T^n(\mathbf{c})$.

Proof. In fact, in view of Theorem 8.2, we have that $\mathbf{d}(i^*\omega) = i^*(\mathbf{d}\omega)$, so that, denoting by σ any 2-chain on $T^n(\mathbf{c})$ (Chap. 8), Stokes' theorem gives

$$\int_{\partial\sigma} i^*\omega = \int_{\sigma} \mathbf{d}(i^*\omega) = \int_{\sigma} i^*(\mathbf{d}\omega).$$

On the other hand, $\mathbf{d}\omega = \Omega$, and the preceding equation yields

$$\int_{\partial\sigma} i^*\omega = \int_{\sigma} i^*\Omega. \tag{20.13}$$

Since the 2-form $i^*\Omega$ on $T^n(\mathbf{c})$ is the restriction to $T^n(\mathbf{c})$ of Ω, we have

$$i^*\Omega(\mathbf{u}, \mathbf{v}) = \Omega(\mathbf{u}, \mathbf{v}) \tag{20.14}$$

for any pair of vectors $\mathbf{u}, \mathbf{v}$, tangent to $T^n(\mathbf{c})$. In proving the Arnold–Liouville theorem, we showed that the vector fields $\mathbf{X}_{f_h}$ are independent and tangent to the tori. Consequently, at any point of $T^n(\mathbf{c})$, they determine a basis of the corresponding space that is tangent to $T^n(\mathbf{c})$, and we can write that

$$\mathbf{u} = u^h \mathbf{X}_{f_h}, \quad \mathbf{v} = v^h \mathbf{X}_{f_h}.$$

Introducing these representations of $\mathbf{u}$ and $\mathbf{v}$ into (20.14), we obtain

$$i^*\Omega(\mathbf{u}, \mathbf{v}) = u^h v^k \Omega\left(\mathbf{X}_{f_h}, \mathbf{X}_{f_k}\right) = u^h v^k \Omega\left((\mathbf{d}f_h)^\sharp, (\mathbf{d}f_k)^\sharp\right).$$

In view of Definition (18.52) of Poisson's bracket, we can state that

$$i^*\Omega(\mathbf{u}, \mathbf{v}) = u^h v^k \{f_h, f_k\} = 0$$

since the first integrals are in involution, and the theorem is proved. $\square$

The results (20.12) and (20.13) imply that

$$\int_{\partial\sigma} i^*\omega = 0 \tag{20.15}$$

along any closed curve $\gamma \equiv \partial\sigma$, which is the boundary of an oriented 2-chain on a torus.

In the set of curves on $T^n(\mathbf{c})$, we introduce the following equivalence relation: two curves γ and γ_1 are equivalent if they are homotopic.[1] We denote by $[\gamma]$ the equivalence class of all the curves homotopic to γ.

Let γ_h be a *closed* curve with parametric equations $\varphi^h = s$, $\varphi^k = a^k$, for all $k \neq h$, where $0 \leq s \leq 2\pi$ and $0 \leq a^k \leq 2\pi$ are constant, and denote by $[\gamma_h]$ the equivalence class of all the curves homotopic to γ_h. It is evident that none of the curves of $[\gamma_h]$ can be reduced to a point by a continuous transformation.

Theorem 20.4. *Let γ and γ_1 be two arbitrary curves of $[\gamma_h]$. Then*

$$\int_\gamma i^*\omega = \int_{\gamma_1} i^*\omega \equiv \int_{[\gamma_h]} i^*\omega. \tag{20.16}$$

Proof. Let σ be the oriented 2-chain on $T^n(\mathbf{c})$ whose boundary $\partial\sigma$ is $\gamma \cup \gamma_1$. Then, by Stokes' theorem and (20.12), we have that (Fig. 20.3)

$$\int_\sigma i^*\Omega = \int_\gamma i^*\omega - \int_{\gamma_1} i^*\omega = 0,$$

and the theorem is proved.   □

In view of this result, the following definition is consistent.

Definition 20.5. We define as *action variables* the n quantities

$$I_h = \frac{1}{2\pi} \int_{[\gamma_h]} i^*\omega, \quad h = 1, \dots, n. \tag{20.17}$$

Taking into account the meaning of the variables $(\varphi, \mathbf{c})$, the parametric equations of a torus $T^n(\mathbf{c})$ can be written as

$$\mathbf{q} = \mathbf{q}(\varphi, \mathbf{c}), \tag{20.18}$$
$$\mathbf{p} = \mathbf{p}(\varphi, \mathbf{c}), \tag{20.19}$$

where the variables $\mathbf{c}$ determine the torus $T^n(\mathbf{c})$ and the angles φ locate a point of it. Consequently, a curve $\gamma \in [\gamma_h]$ will have parametric equations

$$\mathbf{q}(s) = \mathbf{q}(\varphi(s), \mathbf{c}),$$
$$\mathbf{p}(s) = \mathbf{p}(\varphi(s), \mathbf{c}),$$

---

[1] A curve $\gamma_1 = \mathbf{x}_1(s)$, $s \in [a,b]$, on a manifold V_n is homotopic to a curve $\gamma = \mathbf{x}(s)$, $s \in [a,b]$, of V_n if there exists a continuous mapping $F : [a,b] \times [0,1] \to V_n$ such that $F(0,s) = \mathbf{x}(s)$ and $F(s,1) = \mathbf{x}_1(s)$, in other words, if γ_1 reduces to γ by a continuous transformation.

Fig. 20.3 Homotopic curves

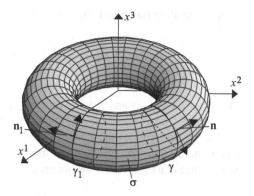

$\omega = p_h \mathbf{d}q^h$, and the action variables will only depend on the quantities $\mathbf{c}$, that is, on the torus. Formally, we can write that

$$\mathbf{I} = \mathbf{J(c)}. \qquad (20.20)$$

Since it is possible to prove that the Jacobian determinant of (20.20) is different from zero, there exists the inverse mapping

$$\mathbf{c} = \mathbf{c(I)}, \qquad (20.21)$$

which allows us to use the action variables to define the torus, instead of the quantities $\mathbf{c}$, as well as to write the parametric equations of the torus in the equivalent form

$$\mathbf{q} = \hat{\mathbf{q}}(\boldsymbol{\varphi}, \mathbf{I}), \qquad (20.22)$$
$$\mathbf{p} = \hat{\mathbf{p}}(\boldsymbol{\varphi}, \mathbf{I}). \qquad (20.23)$$

Now we search for a function $W(\mathbf{q}, \mathbf{I})$ generating a new set of symplectic coordinates $(\boldsymbol{\theta}\mathbf{I})$ by the usual formulae

$$\mathbf{p} = \frac{\partial W}{\partial \mathbf{q}}(\mathbf{q}, \mathbf{I}), \qquad (20.24)$$

$$\boldsymbol{\theta} = \frac{\partial W}{\partial \mathbf{I}}(\mathbf{q}, \mathbf{I}), \qquad (20.25)$$

where

$$\det\left(\frac{\partial W}{\partial q^h \partial I_k}\right) \neq 0. \qquad (20.26)$$

The coordinates defined by (20.25) are called ***angle variables***.

As generating function $W(\mathbf{q}, \mathbf{I})$ we take

$$W(\mathbf{q}, \mathbf{I}) = \int_\gamma i^*\omega = \int_{x_0}^{x} p_h(\mathbf{q}, \mathbf{I})dq^h, \tag{20.27}$$

where $x_0, x \in T^n(\mathbf{c})$ and γ is an arbitrary curve on the torus starting from x_0 and ending at x. Due to Theorem 20.3, integral (20.27), that is, the value of the function W, does not depend on the curve γ, provided that γ can be reduced to a point by a continuous transformation. In contrast, if γ contains a curve belonging to the equivalence class $[\gamma_h]$, then we obtain a variation ΔW of the value of W given by

$$\Delta W = 2\pi I_h. \tag{20.28}$$

In other words, the generating function W assumes infinite values at any point of the torus. As a consequence of this property of W, the values of the variables θ corresponding to a point of the torus $T^n(\mathbf{c})$ are infinite. The difference $\Delta_k \theta^h$ between two of these values is

$$\Delta_k \theta^h = \Delta_k \frac{\partial W}{\partial I_h} = \frac{\partial}{\partial I_h} \Delta_k W = 2\pi \frac{\partial I_k}{\partial I_h} = 2\pi \delta_k^h \tag{20.29}$$

when the curve γ contains a curve of the equivalence class $[\gamma_k]$.

We conclude by remarking that, since the coordinates $(\theta, \mathbf{I})$ are symplectic, the Hamiltonian function in these new variables is given by

$$\overline{H}(\theta, \mathbf{I}) = H(\mathbf{q}(\theta, \mathbf{I}), \mathbf{p}(\theta, \mathbf{I})). \tag{20.30}$$

On the other hand, a torus is determined by giving the action variables $\mathbf{I}$ and the total energy is constant on a torus. Consequently,

$$H = \overline{H}(\mathbf{I}), \tag{20.31}$$

and the Hamilton equations in the symplectic coordinates $(\theta, \mathbf{I})$ assume the form

$$\dot{\theta} = \frac{\partial \overline{H}}{\partial \mathbf{I}}, \tag{20.32}$$

$$\dot{\mathbf{I}} = \mathbf{0}. \tag{20.33}$$

Finally, the parametric equations of the dynamical trajectories are

$$\theta = \nu(\mathbf{I}_0)t + \theta_0, \tag{20.34}$$

$$\mathbf{I} = \mathbf{I}_0, \tag{20.35}$$

where θ_0 and $\mathbf{I}_0$ are constant.

The quantities

$$\nu(\mathbf{I}) = \frac{\partial H}{\partial \mathbf{I}}(\mathbf{I})$$

are the **fundamental frequencies** of a completely integrable Hamiltonian system.

Remark 20.2. From (20.30), (20.31), and (20.24) it follows that the generating function W is a complete integral of the Hamilton–Jacobi equation

$$H\left(\mathbf{q}, \frac{\partial W}{\partial \mathbf{q}}(\mathbf{q}, \mathbf{I})\right) = \overline{H}(\mathbf{I}). \qquad (20.36)$$

The following theorem shows that the angle–action variables are not uniquely defined.

Theorem 1. *Let $A = (A_k^h)$, $h, k = 1, \ldots, n$, be a numerical matrix such that $\det A = 1$ and denote by $(\boldsymbol{\theta}, \mathbf{J})$ the angle–action variables defined by (20.17) and (20.25). Then, the variables $(\overline{\boldsymbol{\theta}}, \overline{\mathbf{J}})$ defined by*

$$\overline{J}_h = A_h^k J_k, \quad \overline{\theta}^h = (A^{-1})_k^h \theta^k \qquad (20.37)$$

are new angle–action variables.

Proof. Since $\det A = 1$, the linear transformations (20.37) are invertible. To prove that (20.37) is a canonical transformation we evaluate their Poisson's brackets:

$$\{\overline{J}_h, \overline{J}_k\} = \sum_m \left[\frac{\partial \overline{J}_h}{\partial \theta^m} \frac{\partial \overline{J}_k}{\partial J_m} - \frac{\partial \overline{J}_h}{\partial J_m} \frac{\partial \overline{J}_k}{\partial \theta^m} \right] = 0,$$

$$\{\overline{\theta}^h, \overline{J}_k\} = \sum_m \left[\frac{\partial \overline{\theta}^h}{\partial \theta^m} \frac{\partial \overline{J}_k}{\partial J_m} - \frac{\partial \overline{\theta}^h}{\partial J_m} \frac{\partial \overline{J}_k}{\partial \theta^m} \right] = (A^{-1})_m^h A_k^m = \delta_k^h,$$

$$\{\overline{\theta}^h, \overline{\theta}^k\} = \sum_m \left[\frac{\partial \overline{\theta}^h}{\partial \theta^m} \frac{\partial \overline{\theta}^k}{\partial J_m} - \frac{\partial \overline{\theta}^h}{\partial J_m} \frac{\partial \overline{\theta}^k}{\partial \theta^m} \right] = 0.$$

Finally, the transformed Hamiltonian function depends only on the variables $\overline{\mathbf{J}}$.

20.3 Fundamental Frequencies and Orbits

Definition 20.6. Let $S = (M_{2n}^*, \Omega, H)$ be a completely integrable Hamiltonian system, and denote by $T^n(\mathbf{I})$ a torus corresponding to given values of the action variables. The **resonance module** of the vector $\nu(\mathbf{I}) \in \mathfrak{R}^n$ of the fundamental frequencies of S is the subset $\mathcal{M}$ of $\mathbb{Z}^n$ such that

$$\mathcal{M} = \{\mathbf{m} \in \mathbb{Z}^n, \mathbf{m} \cdot \boldsymbol{\nu}(\mathbf{I}) = 0\}. \tag{20.38}$$

The dimension of $\mathcal{M}$, called **resonance multiplicity**, gives the number of independent relations $\mathbf{m} \cdot \boldsymbol{\nu}(\mathbf{I}) = 0$, with integer coefficients, existing among the fundamental frequencies.

We omit the proof of the following theorem.

Theorem 20.5. *Let $(M_{2n}^*, \boldsymbol{\Omega}, H)$ be a completely integrable Hamiltonian system, and denote by $T^n(\mathbf{I})$ the torus corresponding to given values of the action variables. Then all the orbits belonging to $T^n(\mathbf{I})$ are*

- *Periodic if and only if* $\dim \mathcal{M} = n - 1$,
- *Everywhere dense on* $T^n(\mathbf{I})$ *if and only if* $\dim \mathcal{M} = 0$,
- *Everywhere dense on a torus with dimension* $n - \dim \mathcal{M}$ *contained in* $T^n(\mathbf{I})$ *if* $0 < \dim \mathcal{M} < n - 1$.

Definition 20.7. Let $\gamma(t) = \boldsymbol{\theta}(t)$ be an orbit on $T^n(\mathbf{I})$ corresponding to the initial datum $\boldsymbol{\theta}(0) = \boldsymbol{\theta}_0$. The **time average** of the function $F(\boldsymbol{\theta}, \mathbf{I})$ along $\gamma(t)$ is defined by the limit

$$\hat{F}(t_0, \boldsymbol{\theta}_0) = \lim_{T \to \infty} \int_{t_0}^{T} F(\boldsymbol{\theta}(t), \mathbf{I}) \, dt. \tag{20.39}$$

Theorem 20.6. *If the function $F(\boldsymbol{\theta}, \mathbf{I})$ is measurable with respect to the Lebesgue measure, then, for almost all the initial data $\boldsymbol{\theta}_0 \in T^n(\mathbf{I})$,[2] the time average exists and does not depend on t_0.*

Definition 20.8. Let $F(\boldsymbol{\theta}, \mathbf{I})$ be a function on $T^n(\mathbf{I})$ that is measurable with respect to the Lebesgue measure. Then the **space average** of $F(\boldsymbol{\theta}, \mathbf{I})$ is defined as follows:

$$\overline{F} = \frac{1}{\text{mis}(T^n(\mathbf{I}))} \int_{T^n(\mathbf{I})} F(\boldsymbol{\theta}, \mathbf{I}) \, d\sigma, \tag{20.40}$$

where $\text{mis}(T^n(\mathbf{I}))$ is the measure of $T^n(\mathbf{I})$.

Theorem 20.7. *If $\dim \mathcal{M} = 0$, that is, if the orbits are everywhere dense on the torus $T^n(\mathbf{I})$, then the time average does not depend on the orbits and is equal to the space average.*

A mechanical system for which the time and space averages are equal is said to be **ergodic**. We can say that a completely integrable Hamiltonian system is ergodic when the level manifolds defined by its n first integrals are tori and $\dim \mathcal{M} = 0$.

---

[2]That is, except for the points of $T^n(\mathbf{I})$ belonging to a subset with a Lebesgue measure equal to zero.

Theorem 20.8. *Let S be a completely integrable Hamiltonian system, and denote by* $(\boldsymbol{\theta}, \mathbf{I})$ *its angle–action variables and by* $\boldsymbol{\nu}$ *the vector of its fundamental frequencies. Suppose that* $\dim \mathcal{M} = l < n$, *i.e., that the fundamental frequencies confirm* l *independent linear relations*

$$m_h^\alpha \nu^h = 0, \quad \alpha = 1, \ldots, l, \quad h = 1, \ldots, n, \tag{20.41}$$

where the coefficients m_h^α *are integer numbers. Then there exists a canonical transformation* $(\boldsymbol{\theta}, \mathbf{I}) \to (\boldsymbol{\theta}', \mathbf{I}')$, *to new angle–action variables, in which the Hamiltonian function is*

$$H' = H' \left(I'_{l+1}, \ldots, I'_n \right). \tag{20.42}$$

Proof. Introduce the symplectic transformation $(\boldsymbol{\theta}, \mathbf{I}) \to (\boldsymbol{\theta}', \mathbf{I}')$ defined as follows:

$$\theta'^\alpha = \frac{\partial W'}{\partial I'_\alpha}, \tag{20.43}$$

$$\theta'^{l+\beta} = \frac{\partial W'}{\partial I'_{l+\beta}}, \tag{20.44}$$

$$I_h = \frac{\partial W'}{\partial \theta^h}, \tag{20.45}$$

where $\alpha = 1, \ldots, l, \beta = 1, \ldots, n - l, h = 1, \ldots, n$, and the generating function W' is

$$W' = m_h^\alpha \theta^h I'_\alpha + I'_{l+\beta} \theta^{l+\beta}. \tag{20.46}$$

In view of (20.46), system (20.43)–(20.45) becomes

$$\theta'^\alpha = m_h^\alpha \theta^h, \quad \theta'^{l+\beta} = \theta^{l+\beta}, \quad I_h = m_h^\alpha I'_\alpha + I'_{l+\beta} \delta_h^{l+\beta}. \tag{20.47}$$

Taking into account (20.34), (20.41), and (20.47), we obtain that

$$\dot{\theta}'^\alpha = m_h^\alpha \dot{\theta}^\alpha = m_h^\alpha \nu^h = 0, \tag{20.48}$$

$$\dot{\theta}'^{l+\beta} = \dot{\theta}^{l+\beta} = \nu^{l+\beta}. \tag{20.49}$$

On the other hand, recalling that the new coordinates are symplectic and taking into account (20.47), we have in these new coordinates the Hamiltonian function

$$H'(\boldsymbol{\theta}', \mathbf{I}') = H(\mathbf{I}(\boldsymbol{\theta}', \mathbf{I}')) = H'(\mathbf{I}'). \tag{20.50}$$

In view of (20.48) and (20.49), the Hamilton equations relative to this Hamiltonian function become

$$\dot{\theta}'^\alpha = \frac{\partial H'}{\partial I_{\alpha'}} = 0,$$

$$\dot{\theta}'^{l+\beta} = \frac{\partial H'}{\partial I'_{l+\beta}} = \nu'^{l+\beta},$$

$$\dot{I}'_h = -\frac{\partial H'}{\partial \theta'^h} = 0,$$

and the theorem is proved.   □

20.4   Angle–Action Variables of Harmonic Oscillator

The harmonic oscillator is a completely integrable Hamiltonian system
(Remark 20.1). Its Hamiltonian function is

$$H(q, p) = \frac{1}{2}(p^2 + q^2), \tag{20.51}$$

and the first integral $H = E$ defines a one-dimensional torus γ on the cylinder
$S \times \Re$ with the parametric equation

$$p = \mp\sqrt{2E - q^2}, \quad -\sqrt{2E} \le q \le \sqrt{2E}. \tag{20.52}$$

In view of (20.17), the action variable is

$$2\pi I(E) = \int_\gamma p dq = \frac{1}{\pi} \int_{-\sqrt{2E}}^{\sqrt{2E}} \sqrt{2E - q^2} \, dq = 2\pi E. \tag{20.53}$$

Since $E = I$, from (20.27) we obtain the generating function

$$W(q, I) = \int_0^q \sqrt{2E - u^2} du = \int_0^q \sqrt{2I - u^2} du. \tag{20.54}$$

The corresponding symplectic transformation $(q, p) \to (\theta, I)$ is given by the equations

$$p = \frac{\partial W}{\partial q} = \sqrt{2I - q^2}, \; \theta = \frac{\partial W}{\partial I} = \arcsin\left(\frac{q}{\sqrt{2I}}\right). \tag{20.55}$$

Finally, $H(I) = I$, and the Hamilton equations are

$$\dot{I} = 0, \; \dot{\theta} = 1. \tag{20.56}$$

More generally, the Hamiltonian function of an n-dimensional harmonic oscillator
(Example 20.3) is

$$H(\mathbf{q}, \mathbf{p}) = \frac{1}{2} \sum_{h=1}^{n} \left(p_h^2 + (\lambda^h q^h)^2 \right).$$ (20.57)

We have already shown that this Hamiltonian system is completely integrable and that the first integrals $f_h = p_h^2 + (\lambda^h q^h)^2 = c_h$ define a torus with parametric equations given by

$$p_h = \mp\sqrt{c_h - (\lambda^h q^h)^2}, \quad -\frac{\sqrt{c}}{\lambda^h} \le q^h \le \frac{\sqrt{c}}{\lambda^h}, \quad h = 1, \ldots, n.$$ (20.58)

Consequently, the action variables are defined by the relations

$$2\pi I_h(c_h) = 2 \int_{-\sqrt{c}/\lambda^h}^{\sqrt{c}/\lambda^h} \sqrt{c_h - (\lambda^h q^h)^2} dq^h,$$ (20.59)

which yield

$$I_h(c_h) = \frac{c_h}{2\lambda^h}, \quad h = 1, \ldots, n.$$ (20.60)

Finally, we have that

$$W(\mathbf{q}, \mathbf{I}) = \int_{q_0}^{q} \sum_{h=1}^{n} \sqrt{2\lambda^h I_h - (\lambda^h u^h)^2} du^h.$$ (20.61)

It is an easy exercise to determine the angle variables θ and the Hamiltonian equations in the variables $(\theta, \mathbf{I})$.

20.5 Angle–Action Variables for a Material Point in Newtonian Fields

Let P be a material point with mass m moving in the space under the action of a Newtonian force. If we adopt spherical coordinates (r, φ, θ) with the origin at the center of force, then the Lagrange function of P is

$$L = \frac{1}{2} \left(\dot{r}^2 + r^2 \dot{\theta}^2 + r^2 \sin^2\theta \dot{\varphi}^2 \right) + \frac{k}{r}.$$

Consequently, the Hamiltonian function can be written as

$$H = \frac{1}{2m}\left(p_r^2 + \frac{p_\theta^2}{r^2} + \frac{p_\varphi^2}{r^2\sin^2\theta}\right) - \frac{k}{r}, \tag{20.62}$$

where the momenta $p_h = \partial L/\partial \dot{q}^h$ are

$$p_r = m\dot{r}, \quad p_\varphi = mr^2\sin^2\theta\dot{\varphi}, \quad p_\theta = mr^2\dot{\theta}. \tag{20.63}$$

Since the coordinate φ is cyclic, the conjugate momentum p_φ is constant. To prove that P is a completely integrable Hamiltonian system, we must verify that there is a third integral that is in involution with p_φ (Remark 20.1). We can easily obtain this first integral recalling that in a central force field the total angular momentum $\mathbf{K}_O$ with respect to the center of force O vanishes. But $\mathbf{K}_O = \mathbf{r} \times m\dot{\mathbf{r}}$, where $\mathbf{r}$ is the position vector; further, in spherical coordinates (r, φ, θ), we have $\mathbf{r} = r\mathbf{e}_r$, $\dot{\mathbf{r}} = \dot{r}\mathbf{e}_r + r\sin\theta\dot{\varphi}\mathbf{e}_\varphi + r\dot{\theta}\mathbf{e}_\theta$, where $\mathbf{e}_r$, $\mathbf{e}_\varphi$, and $\mathbf{e}_\theta$ are unit vectors along the coordinate curves, and we can write the angular momentum as

$$\mathbf{K}_O = mr^2(\dot{\theta}\mathbf{e}_\varphi - \sin\theta\mathbf{e}_\theta). \tag{20.64}$$

If $\mathbf{K}_O$ is constant, in particular, its length is constant. Then, in view of (20.63), we at once verify that

$$|\mathbf{K}_O|^2 = p_\theta^2 + \frac{p_\varphi^2}{\sin^2\theta}. \tag{20.65}$$

In conclusion, we have the following three first integrals of our Hamiltonian system:

$$f_1 \equiv p_\varphi = \alpha_\varphi, \tag{20.66}$$

$$f_2 \equiv p_\theta^2 + \frac{p_\varphi^2}{\sin^2\theta} = \alpha_\theta^2, \tag{20.67}$$

$$f_3 \equiv H = \frac{1}{2m}\left(p_r^2 + \frac{p_\theta^2}{r^2} + \frac{p_\varphi^2}{r^2\sin^2\theta}\right) - \frac{k}{r} = \alpha_r. \tag{20.68}$$

It is not difficult to verify that the preceding first integrals are independent; further, $\{f_1, f_3\} = \{f_2, f_3\} = 0$ since f_1 and f_2 are first integrals and

$$\{f_1, f_2\} = \left\{p_\varphi, p_\theta^2 + \frac{p_\varphi^2}{\sin^2\theta}\right\}$$

$$= \sum_{h=1}^{3}\left[\frac{\partial p_\varphi}{\partial q^h}\frac{\partial}{\partial p_h}\left(p_\theta^2 + \frac{p_\varphi^2}{\sin^2\theta}\right)\right.$$

$$\left.- \frac{\partial}{\partial q^h}\left(p_\theta^2 + \frac{p_\varphi^2}{\sin^2\theta}\right)\frac{\partial p_\varphi}{\partial p_h}\right]$$

$$= -\frac{\partial}{\partial\theta}\left(p_\theta^2 + \frac{p_\varphi^2}{\sin^2\theta}\right)\frac{\partial p_\varphi}{\partial p_\theta} = 0,$$

since p_φ and p_θ are independent variables. In conclusion, f_1, f_2, and f_3 are in involution.

System (20.66)–(20.68) can also be written in the form

$$p_\varphi = \alpha_\varphi, \tag{20.69}$$

$$p_\theta = \mp\frac{1}{|\sin\theta|}\sqrt{\alpha_\theta^2\sin^2\theta - \alpha_\varphi^2}, \tag{20.70}$$

$$p_r = \mp\sqrt{2m\alpha_r r^2 - 2mkr - \alpha_\theta^2}, \tag{20.71}$$

provided that this yields

$$\alpha_\theta^2\sin^2\theta - \alpha_\varphi^2 \geq 0, \quad 2m\alpha_r r^2 - 2mkr - \alpha_\theta^2 \geq 0. \tag{20.72}$$

The first inequality in (20.72) is verified when

$$-\left|\frac{\alpha_\varphi}{\alpha_\theta}\right| \leq \sin\theta \leq \left|\frac{\alpha_\varphi}{\alpha_\theta}\right|,$$

i.e., when

$$\alpha_\varphi \leq \alpha_\theta, \quad -\theta_0 \leq \theta \leq \theta_0, \tag{20.73}$$

where $\theta_0 = \arcsin|\alpha_\varphi/\alpha_\theta|$. Further, if

$$\alpha_r \geq -\frac{mk^2}{2\alpha_\theta^2}, \tag{20.74}$$

then the roots of the trinomial $2m\alpha_r r^2 - 2mkr - \alpha_\theta^2$ are real and are equal to

$$r_{1,2} = -\frac{k}{2\alpha_r}\left[1 \mp \sqrt{1 + \frac{4\alpha_r\alpha_\theta^2}{mk^2}}\right]. \tag{20.75}$$

Applying the Cartesian rule to the trinomial, we recognize that, if $\alpha_r \geq 0$, then one of the roots (20.75) is positive and the other one is negative. Since r is a positive quantity, only the root r_2 is acceptable. Moreover, the trinomial is positive for $r > r_2$ and the orbit is unbounded. In contrast, if $\alpha_r < 0$, then both roots are positive, and the trinomial is not negative when

$$0 < r_1 \leq r \leq r_2. \tag{20.76}$$

Henceforth we refer to this case in which [see (20.73)]

$$-\frac{mk^2}{2\alpha_\theta^2} \leq \alpha_r < 0. \tag{20.77}$$

All the preceding considerations allow us to put the level manifold in the parametric form

$$\begin{aligned}
\varphi &= \varphi, \\
\theta &= \theta, \\
r &= r, \\
p_\varphi &= \alpha_\varphi, \\
p_\theta &= \mp\frac{1}{|\sin\theta|}\sqrt{\alpha_\theta^2\sin\theta - \alpha_\varphi^2}, \\
p_r &= \mp\frac{1}{r}\sqrt{2m\alpha_r r^2 - 2mkr - \alpha_\theta^2},
\end{aligned} \tag{20.78}$$

where

$$0 \leq \varphi \leq 2\pi, \quad -\theta_0 \leq \theta \leq \theta_0, \quad 0 < r_1 \leq r \leq r_2. \tag{20.79}$$

The preceding formulae show that the level manifold $T^3(\alpha_r, \alpha_\varphi, \alpha_\theta)$ is compact so that it is a torus. The coordinate curves obtained by varying one of the variables r, φ, and θ are closed curves defining three fundamental cycles $[\gamma_r]$, $[\gamma_\varphi]$, and $[\gamma_\theta]$.

The action variables are

$$\begin{aligned}
I_\varphi &= \frac{1}{2\pi}\int_{[\gamma_\varphi]} p_\varphi d\varphi = \alpha_\varphi\int_0^{2\pi} d\varphi = \alpha_\varphi, \\
I_\theta &= \frac{1}{2\pi}\int_{[\gamma_\theta]} p_\theta d\theta = \frac{1}{2\pi}\int_{-\theta_0}^{\theta_0}\sqrt{\alpha_\theta^2 - \frac{\alpha_\varphi^2}{\sin^2\theta}}\,d\theta, \\
I_r &= \frac{1}{2\pi}\int_{[\gamma_r]} p_r dr = \frac{1}{2\pi}\int_{r_1}^{r_2}\sqrt{2m\left(\alpha_r + \frac{k}{r}\right) - \frac{(\alpha_\theta + \alpha_\varphi)^2}{r^2}}.
\end{aligned} \tag{20.80}$$

It is possible to verify that the preceding relations yield the following expressions of the action variables:

$$I_\varphi = \alpha_\varphi, \; I_\theta = \alpha_\theta - \alpha_\varphi, \; I_r = -(I_\theta + I_\varphi) + k\sqrt{\frac{m}{-2\alpha_r}}, \qquad (20.81)$$

from which we obtain

$$\alpha_\varphi = I_\varphi, \; \alpha_\theta = I_\theta + I_\varphi, \; \alpha_r = -\frac{mk^2}{2(I_\theta + I_\varphi + I_r)^2}. \qquad (20.82)$$

Comparing (20.82) and (20.68), we deduce the Hamiltonian function in the angle–action variables

$$\overline{H}(I_\varphi, I_\theta, I_r) = -\frac{mk^2}{2(I_\theta + I_\varphi + I_r)^2}. \qquad (20.83)$$

If we denote by θ_φ, θ_θ, and θ_r the angle variables corresponding to I_φ, I_θ, and I_r, the Hamilton equations are

$$\dot{\theta}_\varphi = \dot{\theta}_\theta = \dot{\theta}_r = \frac{2mk^2}{(I_\theta + I_\varphi + I_r)^3}, \qquad (20.84)$$
$$\dot{I}_\varphi = \dot{I}_\theta = \dot{I}_r = 0.$$

Since all the fundamental frequencies are equal, from Theorem 20.5 we obtain that the orbits are closed on the torus and the motions are periodic.

20.6 Bohr–Sommerfeld Quantization Rules

It can be verified experimentally by a spectroscope that the radiation emitted by hot hydrogen gas is compounded by waves having discrete frequencies. Rydberg and Ritz empirically determined the following formula, which includes all the frequencies of the light emitted by hydrogen gas:

$$\tilde{\lambda} \equiv \frac{\nu}{c} = R_H \left[\frac{1}{p^2} - \frac{1}{q^2} \right], \qquad (20.85)$$

where $\tilde{\lambda}$ is the wave number of the radiation, ν its frequency, c the light velocity in a vacuum, and

$$R_H = 109{,}677.576 \, \text{cm}^{-1}$$

is the Rydberg constant. If in (20.85) we set $p = 1$, we obtain the frequencies of the Lyman series

$$\frac{\nu}{c} = R_H \left[1 - \frac{1}{q^2} \right], \quad q = 2, 3, \ldots.$$

For $p = 2$ we obtain the Balmer series

$$\frac{\nu}{c} = R_H \left[\frac{1}{2} - \frac{1}{q^2} \right], \quad q = 3, 4, \ldots.$$

Analogously, to $p = 3, 4, 5$ correspond the emission spectra of Paschen, Bracket, and Pfund, respectively.

This discrete character of the frequencies of the emitted light was not compatible with the physics of the beginning of 1900. In the attempt to give a theoretical justification of the experimental results, Bohr and Sommerfeld made the following assumptions:

- An electron rotating about the nucleus of a hydrogen atom cannot move along an arbitrary orbit compatible with classical mechanics. Further, the electron does not radiate electromagnetic waves during its accelerated motion on an allowed orbit, in disagreement with classical electrodynamics.
- Electromagnetic radiation is emitted or adsorbed by an electron only when it goes from an allowed orbit to another one. More precisely, let ΔE be the energy difference between the values of energy corresponding to two allowed orbits. Then, in going from one to the other, the electron radiates electromagnetic waves with frequency

$$\nu = \frac{|\Delta E|}{h}, \tag{20.86}$$

where h is Planck's constant.

At this point it is fundamental to assign a **selection rule** or a **quantization rule** to determine the allowed orbits and then to evaluate the possible values of ΔE. The rule proposed by Bohr and Sommerfeld refers to completely integrable systems. Therefore, it can be applied to a hydrogen atom.

Denote by $I_1, \ldots, I_n$ the action variables of a completely integrable Hamiltonian system S. The Bohr–Sommerfeld rule states that the allowed orbits satisfy the following conditions:

$$I_h = n_h \hbar, \quad n_h = 1, 2, \ldots, \quad \hbar = h/2\pi, \tag{20.87}$$

which, for a hydrogen atom (see preceding section), become

$$I_\varphi = n_\varphi \hbar, \quad I_\theta = n_\theta \hbar, \quad I_r = n_r \hbar. \tag{20.88}$$

Introducing these relations into (20.83), we obtain the possible values of the energy

$$E_n = -\frac{mk^2}{2n^2\hbar^2}, \quad n = n_\varphi + n_\theta + n_r. \tag{20.89}$$

On the other hand, if we denote by e the charge of the electron, we have

$$k = \frac{e^2}{4\pi\epsilon_0},$$

where ϵ_0 is the dielectric constant of the vacuum, and (20.89) becomes

$$E_n = -\frac{me^4}{8\epsilon_0^2 h^2} \frac{1}{n^2}. \tag{20.90}$$

In conclusion, we can evaluate the energy variation $|\Delta E| = |E_q - E_p|$ in going from one allowed orbit to another, and introducing this value into (20.86) we finally obtain the formula

$$\frac{\nu}{c} = \frac{me^4}{8c\epsilon_0 h^3} \left[\frac{1}{n^2} - \frac{1}{m^2}\right], \tag{20.91}$$

where n, m are integer and $n < m$. This formula agrees with (20.85) and allows one to evaluate the Rydberg constant.

In spite of the excellent results derived from this theory, it appears to be unsatisfactory for the following reasons:

- It introduces selection rules that are extraneous to the spirit of classical mechanics.
- It is applicable solely to completely integrable Hamiltonian systems.

20.7 A Sketch of the Hamiltonian Perturbation Theory

In this section we present a brief introduction to Hamiltonian perturbation theory. Our approach does not analyze the correctness of the mathematical procedures that are used. In this regard, we recall, without proof, the fundamental KAM theorem (Kolmogoroff–Arnold–Moser), which makes clear the hypotheses under which the results presented here hold.

The Hamiltonian theory of perturbation has as its aim the analysis of a Hamiltonian system S that is *close to a completely integrable Hamiltonian system* S_0. It is based on the assumption that the motions of S are close to the motions of S_0. In other words, it is based on the preconception that *small causes produce small effects*.

Let S_0 be a completely integrable Hamiltonian system, and denote by $H_0(\mathbf{I}_0)$ its Hamiltonian function depending only on the action variables $\mathbf{I}_0$ of S_0. Henceforth S_0 will be called an **unperturbed system**. Then we consider another Hamiltonian system S whose Hamiltonian function has the form

$$H(\theta_0, \mathbf{I}_0) = H_0(\mathbf{I}_0) + \epsilon H_1(\theta_0, \mathbf{I}_0), \qquad (20.92)$$

where θ_0 are the angle variables of S_0 and ϵ is a small nondimensional parameter. The system S is called a **perturbed system** or **quasi-integrable system**.

Example 20.5 (Fermi–Pasta–Ulam, 1955). Let S be a chain of $n+2$ material points $P_i, i = 0, \ldots, n + 1$, constrained to move on the Ox-axis. Suppose that any point P_i interacts with P_{i-1} and P_{i+1} by nonlinear elastic springs. The points P_0 and P_{n+1} are supposed to be fixed. Finally, all the particles have the same mass m. The Hamiltonian function of this system is

$$H(\mathbf{q}, \mathbf{p}) = \sum_{h=1}^{n} \frac{p_h^2}{2m} + \sum_{h=0}^{n} \left[\frac{k}{2}(q_{h+1} - q_h)^2 + \epsilon(q_{h+1} - q_h)^\alpha \right], \qquad (20.93)$$

where α is an integer greater than 2.

Example 20.6. Let P be a planet (e.g., the Earth) moving under the action both of the Sun S and a massive planet P_m (e.g., Jupiter). If we suppose that the motion of P_m is not influenced by P, then the position vector $\mathbf{r}_J$ of P_m is a known function of time. Let $\mathbf{q}, \mathbf{p}$ be the position vector and the momentum of P, and denote by $m = 1$, m_J, and m_S the masses of P, P_m, and the Sun, respectively. Then the Hamiltonian function of P is

$$H(\mathbf{q}, \mathbf{p}, t) = \frac{1}{2}|\mathbf{p}|^2 - \frac{K}{|\mathbf{q}|} - \epsilon K \frac{1}{|\mathbf{q} - \mathbf{r}_J|}, \qquad (20.94)$$

where K is a constant and $\epsilon = m_J/m_S$. This is an example of the restricted three-body problem (see Chapter 24)

It is quite natural to wonder if it is possible to determine a coordinate transformation $(\theta_0, \mathbf{I}_0) \to (\theta, \mathbf{I})$, where the variables θ are still periodic with a period 2π such that in the new variables the Hamiltonian function of the perturbed system S becomes a function of the only variables $\mathbf{I}$, *up to second-order terms in ϵ*, i.e.,

$$H(\theta, \mathbf{I}) = H_0^*(\mathbf{I}) + \epsilon H_1^*(\mathbf{I}) + \epsilon^2 H_2^*(\theta, \mathbf{I}). \qquad (20.95)$$

We can consider applying this procedure again searching for a new coordinate transformation $(\theta, \mathbf{I}) \to (\theta', \mathbf{I}')$ in which the Hamiltonian function becomes

$$H(\theta', \mathbf{I}') = H_0^*(\mathbf{I}') + \epsilon H_1^*(\mathbf{I}') + \epsilon^2 H_2^*(\mathbf{I}') + \epsilon^3 H_3^*(\theta', \mathbf{I}'). \qquad (20.96)$$

In other words, the idea underlying the Hamiltonian perturbation theory consists in *presupposing the existence of a sequence of completely integrable systems whose motions increasingly approximate the motions of the perturbed nonintegrable system S.* After making clear the aim of the theory, we show how to determine the first transformation $(\boldsymbol{\theta}_0, \mathbf{I}_0) \to (\boldsymbol{\theta}, \mathbf{I})$ without analyzing all the mathematical difficulties encountered. First, we search for a generating function $S(\boldsymbol{\theta}_0, \mathbf{I})$ of such a coordinate transformation. Since this function must generate an identity transformation when $\epsilon = 0$, we set

$$S(\boldsymbol{\theta}_0, \mathbf{I}) = \sum_{h=0}^{n} \theta_{0h} I_h + \epsilon S_1(\boldsymbol{\theta}_0, \mathbf{I}) + \cdots . \tag{20.97}$$

In fact, this function generates the symplectic transformation

$$I_{0h} = I_h + \epsilon \frac{\partial S_1}{\partial \theta_{0h}} + \cdots , \tag{20.98}$$

$$\theta_h = \theta_{0h} + \frac{\partial S_1}{\partial I_h} + \cdots , \tag{20.99}$$

which reduces to an identity where $\epsilon = 0$. Taking into account the preceding relations, we have the following Taylor expansions:

$$H_0(\mathbf{I}_0) = H_0(\mathbf{I}) + \epsilon \sum_{h=1}^{n} \left(\frac{\partial H_0}{\partial I_{0h}} \right)_{I_0 = I} \frac{\partial S_1}{\partial \theta_{0h}} + O(\epsilon^2), \tag{20.100}$$

$$\epsilon H_1(\boldsymbol{\theta}_0, \mathbf{I}_0) = \epsilon H_1(\boldsymbol{\theta}_0, \mathbf{I}) + O(\epsilon^2). \tag{20.101}$$

Introducing these relations into (20.95) we obtain the Hamiltonian function in terms of the new variables

$$H_0(\mathbf{I}) + \epsilon \left(\sum_{h=1}^{n} \nu_h(\mathbf{I}) \frac{\partial S_1}{\partial \theta_{0h}} + H_1(\boldsymbol{\theta}_0, \mathbf{I}) \right) + O(\epsilon^2), \tag{20.102}$$

where we have introduced the fundamental frequencies [see (20.35)] of the unperturbed system S_0

$$\nu_h(\mathbf{I}) = \left(\frac{\partial H_0}{\partial I_{0h}} \right)_{I_0 = I}. \tag{20.103}$$

In view of this condition, we see that (20.95) is confirmed if $S_1(\boldsymbol{\theta}_0, \mathbf{I})$ is a solution of the *fundamental equation of the first-order perturbation theory*:

$$\sum_{h=1}^{n} \nu_h(\mathbf{I}) \frac{\partial S_1}{\partial \theta_{0h}} (\boldsymbol{\theta}_0, \mathbf{I}) = H_1^*(\mathbf{I}) - H_1(\boldsymbol{\theta}_0, \mathbf{I}). \tag{20.104}$$

This equation contains two unknowns: the generating function $S_1(\theta_0, \mathbf{I})$ and the function $H_1^*(\mathbf{I})$. To determine the former function, we start by noting that to the values $(\theta_0, \mathbf{I})$ and $(\theta_0 + 2\pi, \mathbf{I})$ corresponds the same point of the phase state M_{2n}^*. Consequently, all the real functions on M_{2n}^* are periodic functions of the variables θ_0 with the same period 2π. In turn, this property implies that any function on M_{2n}^* admits a Fourier expansion in the variables θ_0 (Appendix B). In particular, the Fourier expansion of S_1 starts with the constant mean value. Since the derivatives of this value with respect to the variables θ_0 are zero, the mean value of the left-hand side of (20.104) vanishes and we have that

$$H_1^*(\mathbf{I}) = < H_1(\theta_0, \mathbf{I}) >, \qquad (20.105)$$

where $< H_1(\theta_0, \mathbf{I}) >$ is the mean value of the perturbation. Introducing the new function

$$\hat{H}(\theta_0, \mathbf{I}) = H_1(\theta_0, \mathbf{I}) - < H_1(\theta_0, \mathbf{I}) >, \qquad (20.106)$$

Eq. (20.104) becomes

$$\sum_{h=1}^{n} \nu_h(\mathbf{I}) \frac{\partial S_1}{\partial \theta_{0h}}(\theta_0, \mathbf{I}) = -\hat{H}_1(\theta_0, \mathbf{I}). \qquad (20.107)$$

To determine the solution $S_1(\theta_0, \mathbf{I})$ of this equation, we introduce the following Fourier expansions (Appendix B):

$$\hat{H}_1(\theta_0, \mathbf{I}) = \sum_{\mathbf{k}=-\infty}^{\infty} h_{\mathbf{k}}(\mathbf{I}) e^{i \mathbf{k} \cdot \theta_0}, \qquad (20.108)$$

$$S_1(\theta_0, \mathbf{I}) = \sum_{\mathbf{k}=-\infty}^{\infty} s_{\mathbf{k}}(\mathbf{I}) e^{i \mathbf{k} \cdot \theta_0}, \qquad (20.109)$$

where the components of the n-dimensional vector $\mathbf{k} = (k_1, \ldots, k_n)$ are integers. Introducing these expansions into (20.107) we have that

$$\sum_{h=1}^{n} \nu_h(\mathbf{I}) \frac{\partial S_1}{\partial \theta_{0h}}(\theta_0, \mathbf{I}) = i \sum_{h=1}^{n} \nu_h(\mathbf{I}) \sum_{\mathbf{k}=-\infty}^{\infty} s_{\mathbf{k}}(\mathbf{I}) e^{i \mathbf{k} \cdot \theta_0}$$

$$= i \sum_{\mathbf{k}=-\infty}^{\infty} \sum_{h=1}^{n} \nu_h k_h s_{\mathbf{k}}(\mathbf{I}) e^{i \mathbf{k} \cdot \theta_0}$$

$$= i \sum_{\mathbf{k}=-\infty}^{\infty} (\boldsymbol{\nu} \cdot \mathbf{k}) s_{\mathbf{k}}(\mathbf{I}) e^{i \mathbf{k} \cdot \theta_0}.$$

From these relations we derive the following expressions of the coefficients $s_{\mathbf{k}}(\mathbf{I})$ of expansion (20.109):

$$s_{\mathbf{k}}(\mathbf{I}) = \frac{i h_{\mathbf{k}}(\mathbf{I})}{\nu(\mathbf{I}) \cdot \mathbf{k}}. \tag{20.110}$$

This expression generates some perplexities regarding the convergence of series (20.109). In fact, if the fundamental frequencies are linearly dependent, then there exists at least one vector $\mathbf{k}$ such that

$$\nu \cdot \mathbf{k} = \nu_1 k_1 + \cdots + \nu_n k_n = 0,$$

and some denominators of (20.110) may vanish. Even if the fundamental frequencies are linearly independent, some denominators of (20.110) may assume values arbitrarily close to zero. For instance, suppose that we are concerned with a Hamiltonian system with two degrees of freedom such that the fundamental frequencies assume the values $\nu_1 = 1$ and $\nu_2 = \sqrt{2}$. Setting

$$a k_1 = \sqrt{2} k_1 + k_2$$

we obtain

$$\sqrt{2} = -\frac{k_2}{k_1} + a.$$

Consequently, with a convenient choice of the integers k_1 and k_2, we can make a as small as we wish. This means that some denominators of (20.110) could become very small.

Example 20.7. Consider a nonlinear oscillator with a Hamiltonian function given by

$$H(q, p) = \frac{1}{2}(p^2 + q^2) + \epsilon \frac{1}{4} q^4, \tag{20.111}$$

where q is the abscissa of the oscillator. In view of (20.56), $\nu = 1/2\pi$, $\theta = 2\pi\theta_0$, and (20.111) becomes

$$H(\theta_0, I_0) = \nu I_0 + \epsilon \nu^2 I_0^2 \sin^4 \theta_0 \equiv H_0(I_0) + \epsilon H_1(\theta_0, I_0), \tag{20.112}$$

where $-\pi \leq \theta_0 \leq \pi$.

The mean value of the perturbation term is [see (20.105)]

$$< H(\theta_0, I) > = \frac{1}{2\pi} \nu^2 I^2 \int_{-\pi}^{\pi} \sin^4 \theta_0 \, d\theta_0 = \frac{3}{8} \nu^2 I^2, \tag{20.113}$$

and then (see the notebook **Fourier.nb**)

$$\hat{H} = -\frac{1}{4} \nu^2 I^2 \left(e^{-i2\theta_0} + e^{i2\theta_0} - \frac{1}{4} e^{i4\theta_0} - \frac{1}{4} e^{-i4\theta_0} \right). \tag{20.114}$$

20.8   Overview of KAM Theorem

Let us return to action-angle coordinates. Theorem 20.5 shows that the flow on each torus crucially depends on the arithmetical properties of the fundamental frequencies ν. There are essentially two cases: (1) If the frequencies ν are nonresonant, or rationally independent on a torus, then each orbit is dense on this torus and the flow is ergodic. (2) If the frequencies ν are resonant, or rationally dependent, then the torus decomposes into an l-parameter family of invariant tori. Each orbit is dense on such a lower dimensional torus, but not in $T^n(\mathbf{I})$. A special case arises when there exist $l = n - 1$ independent resonant relations. Then each frequency $\nu_1, \ldots, \nu_n$ is an integer multiple of a fixed nonzero frequency, and the whole torus is filled by periodic orbits with the same period. Can we state that a small perturbation of an integrable system does not substantially modify the preceding description of an integrable system? The remarks at the end of the preceding section engender suspicions about the possibility of approximating the behavior of a perturbed system by the behavior of suitable completely integrable systems. These suspicions are strengthened by the following considerations.

Definition 20.9. Consider a perturbed Hamiltonian system with Hamiltonian function (20.92). We recall that the Hamiltonian function $H_0(\mathbf{I})$ of the unperturbed system is said to be *nondegenerate* if

$$\det \left(\frac{\partial \nu_h}{\partial I_k} \right) = \det \left(\frac{\partial^2 H_0(\mathbf{I})}{\partial I_h \partial I_k} \right) \neq 0 \qquad (20.115)$$

in a domain $D \subset \Re^n$ of the action variables.

In view of condition (20.115), the *frequency map* (see (20.103))

$$h : \mathbf{I} \in D \to \nu(\mathbf{I}) \in \Omega \qquad (20.116)$$

is a local diffeomorphism between D and some open *frequency domain* $\Omega \subset \Re^n$. In other words, the tori of the unperturbed system corresponding to $\mathbf{I} \in D$ are in a one-to-one correspondence with the frequencies $\nu \in \Omega$. From the preceding hypothesis it follows that nonresonant tori and resonant tori of all types form dense subsets in phase space. Indeed, the resonant ones sit among the nonresonant ones like rational numbers among irrational numbers.

Theorem 20.9 (Poincarè). *Let $H(\mathbf{I}, \theta, \epsilon)$ be periodic with respect to the variables θ and analytic with respect to ϵ. Then, if $H_0(\mathbf{I})$ is nondegenerate, there is no analytic first integral independent of H.*

The preceding theorem states that a perturbed Hamiltonian system cannot be integrable, at least if we consider *analytic* first integrals. Consequently, our request to approximate the behavior of a perturbed Hamiltonian system with the behavior of a sequence of completely integrable systems seems more and more vague.

Other negative results due to Poincarè prove that, also for small values of the parameter ϵ, invariant tori do not persist. More precisely, Poincarè proved that *in the presence of resonance*, invariant tori are in general destroyed by an arbitrarily small perturbation in the sense that, except for a torus with an $n-1$ family of periodic orbits, usually only a finite number of periodic orbits survive a perturbation and the others disintegrate, originating a *chaotic behavior*. In other words, in a nondegenerate system, a *dense* set of tori is usually destroyed.

Since dense sets of tori are destroyed, there seems to be no hope for other tori to survive. However, in 1954 Kolmogorov proved that the majority of tori survives. In fact, those tori survive whose frequencies ν are not only *nonresonant* but *strongly nonresonant* in the sense of the following definition.

Definition 20.10. Let $n > 1$. A vector $\nu \in \Re^n$ satisfies a **diophantine condition** (with constant $\gamma > 0$ and exponent $\mu \leq n - 1$) or a **small divisor condition** if, $\forall \mathbf{m} \in Z^n$, $\mathbf{m} \neq \mathbf{0}$,

$$|\mathbf{m} \cdot \nu| \geq \frac{\gamma}{|\mathbf{m}|^{\mu}}, \qquad (20.117)$$

where $|\mathbf{m}| = |m_1| + \cdots + |m_n|$.

Denote by $C_{\gamma, \mu} \subset \Omega$ a set, which can be proved to be nonempty, of all frequencies satisfying condition (20.117), for fixed γ and μ. Since the Hamiltonian function $H_0(\mathbf{I})$ is nondegenerate, there exists a map

$$h^{-1} : C_{\gamma, \mu} \to A_{\gamma, \mu}, \qquad (20.118)$$

where

$$A_{\gamma, \mu} = h^{-1}(C_{\gamma, \mu}) = \left\{ \mathbf{I}_0 : \nu(\mathbf{I}_0) \in C_{\gamma, \mu} \right\}. \qquad (20.119)$$

It can be proved that the set $A_{\gamma, \mu}$ has a very complex structure: it is closed, perfect, nowhere dense, and a Cantor set with positive measure.[3]

Now we can state the KAM theorem.

Theorem 20.10 (KAM). *Let S be a perturbed Hamiltonian system with Hamiltonian function $H(\mathbf{I}, \theta, \epsilon)$ analytic and nondegenerate. Denote by $\mu > n - 1 e \gamma$ given*

---

[3] An example of Cantor's set can be obtained as follows. Start with the unit interval $[0, 1]$ and choose $0 < k < 1$. Remove the middle interval of length $k/2$. The length of the two remaining intervals is $1 - k/2$. From each of them remove the middle interval of length $k/8$. The four remaining intervals will have a total length of $1 - k/2 - k/4$. From each of them remove the middle interval of length $k/32$. The remaining length will be $1 - k/2 - k/4 - k/8$, and so on. After infinitely many steps, the remaining set will have the length

$$1 - k/2 - k/4 - k/8 + \cdots = 1 - k > 0.$$

constants. Then a positive constant $\epsilon_c(\gamma)$ *exists such that,* $\forall\, \mathbf{I} \in A_{\gamma,\mu}$, *all the tori* $T^n(\mathbf{I})$ *are invariant, though slightly deformed. Moreover, they fill the region* $D \times T^n(\mathbf{I})$ *up to a set of measure* $O(\sqrt{\epsilon})$.

Remark 20.3. Since $A_{\gamma,\mu}$ has no interior points, it is impossible to state whether an initial position falls onto an invariant torus or into a gap between such tori. In other words, theorem KAM states that the probability that a randomly chosen orbit lies on an invariant torus is $1 - O(\sqrt{\epsilon})$.

Chapter 21
Elements of Statistical Mechanics of Equilibrium

Abstract Statistical mechanics has the purpose to show that, if the idea is accepted that bodies are formed by small particles with reasonable properties, then it is possible to derive the macroscopic behavior of the matter around us. This point of view was severely criticized by the academic world since at that time there were many doubts about the existence of elementary constituents of matter. This chapter is devoted to the study of statistical mechanics of equilibrium. We start with Maxwell's kinetic theory that is based on velocity distribution. Then, we formulate Gibb's distribution and analyze its consequences. Finally, the last sections of the chapter are devoted to an introduction to ergodic theory.

21.1 Introduction

Statistical mechanics is an important part of mechanics. Its origin can be found in the old Greek dream of describing the macroscopic behavior of real bodies starting from the properties of the elementary constituents of matter: the atoms (Democritus). If this approach were practicable, it would be possible to reduce the variety of macroscopic properties of real bodies to the few properties of the elementary components of matter. However, this point of view exhibits some apparently insuperable difficulties related to the huge number of atoms composing macroscopic bodies. These difficulties justify the opposite approach undertaken by continuum mechanics, which will be partially described in Chap. 23, devoted to fluid mechanics.[1]

---

[1] For the contents of this chapter see [10, 31, 50, 53].

© Springer International Publishing AG, part of Springer Nature 2018
A. Romano and A. Marasco, *Classical Mechanics with Mathematica®*,
Modeling and Simulation in Science, Engineering and Technology,
https://doi.org/10.1007/978-3-319-77595-1_21

Let S be a perfect gas *at equilibrium* in a container with adiabatic walls. Denote, respectively, by V, p, and θ the volume, pressure, and temperature of S. Experimentally, we verify that, at equilibrium, these quantities are *uniform* and *constant* and satisfy a *state equation*

$$p = p(\theta, V). \tag{21.1}$$

In a more complex system, for instance a mixture of two different gases S_1 and S_2, (21.1) is substituted by the more general state equation

$$p = p(\theta, V, c), \tag{21.2}$$

where c is the concentration of one of the gases S_1 constituting the mixture, i.e., the ratio between the mass m_1 of S_1 and the total mass m of the mixture. We recall that the concentrations c_1 and c_2 of the two gases confirm the condition $c_1 + c_2 = 1$. In any case, we can state that, from a macroscopic point of view, the uniform equilibrium of a gas S is described by a finite number of variables satisfying one or more state equations.

On the other hand, from a microscopic point of view, S is formed by a very large collection of (monoatomic, biatomic, ...) interacting molecules. Their number is so large that any mechanical description of their behavior in time appears rather utopian. The **statistical mechanics of equilibrium** sets itself the aim of deriving the macroscopic behavior of gases from their molecular structure simply by taking advantage of the large number of molecules whose evolution we are interested in describing.

Whatever the procedure we adopt to describe the microscopic behavior of S at equilibrium, we must

- Define the mechanical system $\mathcal{S}$ describing the body S microscopically,
- Obtain the macroscopic quantities as *mean values* of microscopic variables,
- Prove that these mean values satisfy the state equations.

21.2   Kinetic Theory of Gases

In the **kinetic theory** of the equilibrium of gases, the microscopic system $\mathcal{S}$ is formed by N molecules incessantly moving inside a container having a volume V. Each molecule, with f degrees of freedom, freely moves through the volume V until it encounters another molecule or hits the walls of the container. Any collision is supposed to be *elastic*, and the mean path between two subsequent collisions is called a *free mean path*. Further, a collision among more than two molecules is considered to be highly improbable, and the number of collisions per unit of time is supposed to be very large. The preceding considerations lead us to formulate the hypothesis of **molecular chaos**:

- *At macroscopic equilibrium, all molecules of S have the same probability of occupying an elementary volume $dV \subset V$.*
- *A possible velocity is equiprobable for all the molecules.*

For the sake of simplicity we consider the case $f = 3$, i.e., we suppose that the molecules are material points. From the hypothesis that, at equilibrium, the density of the gas S is uniform there follows that the number $dN(x)$ of molecules contained in an elementary volume having a vertex at $x \in V$ and edges dx, dy, and dz is given by the formula

$$dN(x) = \frac{N}{V} dx\, dy\, dz. \tag{21.3}$$

Besides the physical space V, we consider a three-dimensional Euclidean space $\Re^3$ in which we introduce a Cartesian frame of reference $Ov_x\, v_y\, v_z$. A possible velocity of a molecule is given by the position vector $\mathbf{v}$ of a point P of $\Re^3$ relative to O. It is evident that the state of a single particle is given by a point of $V \times \Re^3$. Let $dN(\mathbf{v})$ be the number of molecules with velocity $\overrightarrow{OQ} = \mathbf{v}$, where Q belongs to the parallelepiped having a vertex at P and edges dv_x, dv_y, and dv_z. Then, we call the function $g(\mathbf{v})$ a **velocity distribution function** if

$$dN(\mathbf{v}) = Ng(\mathbf{v})dv_x dv_y dv_z. \tag{21.4}$$

In other words, $dN(\mathbf{v})$ is the number of molecules having velocity $\mathbf{v}$, and $g(\mathbf{v})$ is a probability density in the velocity space $\Re^3$.

Since at equilibrium there is no preferred direction of velocity, the function $g(\mathbf{v})$ depends on the length of velocity, i.e., on v^2. To determine the function $g(v^2)$, we consider a collision between two molecules and denote by $\mathbf{u}$, $\mathbf{v}$ and $\mathbf{u}'$, $\mathbf{v}'$ their speeds before and after the collision, respectively. Applying the conservation of momentum and energy, and recalling that the two molecules have the same mass, we obtain

$$\mathbf{u} + \mathbf{v} = \mathbf{u}' + \mathbf{v}', \tag{21.5}$$
$$u^2 + v^2 = u'^2 + v'^2. \tag{21.6}$$

The total number of collisions per unit time is proportional to the product $g(u^2)g(v^2)$. Similarly, the inverse collision process, in which two molecules, after colliding with speeds $\mathbf{u}'$ and $\mathbf{v}'$, have speeds $\mathbf{u}$ and $\mathbf{v}$, has a probability proportional to the product $g(u'^2)g(v'^2)$. But at equilibrium the number of the first kind of collision must be equal to the number of the second type of collision to obtain a complete mixing as required by the hypothesis of molecular chaos. Therefore, we have that

$$g(u^2)g(v^2) = g(u'^2)g(v'^2). \tag{21.7}$$

In the preceding equation we set $\mathbf{v}' = \mathbf{0}$. Then we determine u'^2 in terms of u^2 and v^2 by (21.6), and we write (21.7) as

$$g(u^2)g(v^2) = g(u^2 + v^2)g(0).$$

After differentiating this relation with respect to u^2, we put $u^2 = 0$ into the result and obtain the condition

$$\frac{g'(v^2)}{g(v^2)} = \frac{g'(0)}{g(0)} \equiv -\alpha. \tag{21.8}$$

An elementary integration of this equation leads to the formula

$$g(v^2) = Ae^{-\alpha v^2}, \tag{21.9}$$

where A and α are integration constants. To determine these constants, we note that (21.4), in view of (21.9), assumes the form

$$dN(\mathbf{v}) = NAe^{-\alpha v^2}dv_x dv_y dv_z.$$

Integrating on the whole velocity space, we obtain the formula

$$1 = A \int_{\Re^3} e^{-\alpha v^2} dv_x dv_y dv_z,$$

which, when we adopt spherical coordinates in $\Re^3$, becomes

$$1 = 4\pi A \int_0^\infty v^2 e^{-\alpha v^2} dv. \tag{21.10}$$

We note that the integral on the right-hand side of (21.10) is convergent when $\alpha > 0$. Further, from the well-known formula

$$\int_0^\infty x^2 e^{-\alpha x^2} dx = \frac{1}{4\alpha}\left(\frac{\pi}{\alpha}\right)^{1/2} \tag{21.11}$$

we obtain

$$A = \left(\frac{\alpha}{\pi}\right)^{3/2},$$

so that we can write

$$dN(\mathbf{v}) = \left(\frac{\alpha}{\pi}\right)^{3/2} Ne^{-\alpha v^2} dv_x dv_y dv_z, \tag{21.12}$$

where the positive constant α is still to be determined.

Owing to (21.12), we can state that the mean value $\overline{f(\mathbf{v})}$ of any function $f(\mathbf{v})$ in the velocity space is given by the formula

$$\overline{f(\mathbf{v})} = \frac{1}{N}\int_{\Re^3} f(\mathbf{v})dN(\mathbf{v}) = \left(\frac{\alpha}{\pi}\right)^{3/2}\int_{\Re^3} f(\mathbf{v})e^{-\alpha v^2}dv_x dv_y dv_z. \tag{21.13}$$

In particular, the mean value $\bar{\epsilon}$ of the kinetic energy $\epsilon = mv^2/2$ of a molecule of mass m is given by the relation

$$\bar{\epsilon} = \frac{1}{2}\left(\frac{\alpha}{\pi}\right)^{3/2}\int_{\Re^3} mv^2 e^{-\alpha v^2}dv_x dv_y dv_z,$$

which in spherical coordinates becomes

$$\bar{\epsilon} = \frac{m}{2}\left(\frac{\alpha}{\pi}\right)^{3/2}4\pi\int_0^\infty v^4 e^{-\alpha v^2}dv.$$

Since it can be proved that

$$\int_0^\infty v^4 e^{-\alpha v^2}dv = \frac{3}{8\alpha^2}\left(\frac{\pi}{\alpha}\right)^{1/2},$$

we finally obtain the result

$$\bar{\epsilon} = \frac{3m}{4\alpha}. \tag{21.14}$$

On the other hand, from the macroscopic theory of perfect gases, we know that the total energy E is given by the formula

$$E = \frac{3}{2}Nk\theta, \tag{21.15}$$

where k is the Boltzmann constant and θ the absolute temperature of the gas. Since $\bar{\epsilon} = E/N$, by comparing (21.15) and (21.14), we determine the value of the constant α:

$$\alpha = \frac{m}{2k\theta}, \tag{21.16}$$

and (21.12) assumes the final form

$$dN(\mathbf{v}) = \left(\frac{m}{2\pi k\theta}\right)^{3/2}Ne^{-\frac{m}{2k\theta}v^2}dv_x dv_y dv_z. \tag{21.17}$$

Now we prove that (21.17) leads to the state equation of perfect gases. To prove this statement, it is sufficient to evaluate the pressure of the gas on the walls of the container. Microscopically, this pressure is due to the momentum absorbed by the walls during the many collisions of molecules with the walls themselves. We suppose that these collisions are *elastic* and consider a plane wall s orthogonal to the Ox-axis.

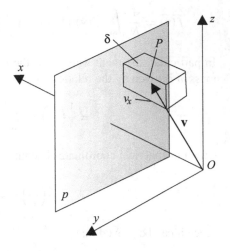

Fig. 21.1 Collisions on the wall p

Let $dZ(\mathbf{v})$ be the number of collisions per unit time and unit surface of s of the molecules with speed $\mathbf{v} = \overrightarrow{OP}$, with $v_x > 0$ and P belonging to the parallelepiped δ, which has a unit face on s and height v_x (Fig. 21.1). It is evident that all the molecules contained in δ at instant t reach the wall s in unit time so that $dZ(\mathbf{v})$ is equal to this number. Since the volume of δ is v_x, we obtain

$$dZ(\mathbf{v}) = \frac{v_x}{V} dN(\mathbf{v}). \tag{21.18}$$

On the other hand, in each collision a particle gives the wall s the momentum $2mv_x$, and then the total momentum transferred to the wall by all the molecules contained in δ with $v_x > 0$ is

$$2mv_x dZ(\mathbf{v}) = 2mv_x^2 \frac{dN(\mathbf{v})}{V}. \tag{21.19}$$

Consequently, the total momentum transferred to the wall s by all the particles contained in δ with any possible velocity $\mathbf{v}$ for which $v_x > 0$ is obtained by integrating (21.19):

$$p = \left(\frac{\alpha}{\pi}\right)^{3/2} \frac{2mN}{V} \int_{-\infty}^{\infty} e^{-\alpha v_y^2} dv_y \int_{-\infty}^{\infty} e^{-\alpha v_z^2} dv_z \int_{0}^{\infty} v_x^2 e^{-\alpha v_x^2} dv_x. \tag{21.20}$$

In view of (21.11) and recalling (21.16) and the formula

$$\int_{-\infty}^{\infty} e^{-\alpha x^2} dx = \left(\frac{\pi}{\alpha}\right)^{1/2},$$

from (21.20) we obtain that

$$p = \frac{kN}{V}\theta. \tag{21.21}$$

But Boltzmann's constant k is related to Avogadro's number N_0 and to the constant R of perfect gases by the relation $k = R/N_0$, and (21.21) assumes the form of the state equation of a perfect gas

$$p = \frac{N}{N_0}R\theta \equiv nR\theta. \tag{21.22}$$

21.3 Boltzmann–Gibbs Distribution

In this section we present an approach to statistical mechanics due to Gibbs. This approach is still based on statistical hypotheses but it resorts to the phase space of Hamiltonian mechanics. Let S be a gas at macroscopic equilibrium in a given volume V with adiabatic walls. As usual, we suppose that S contains a very large number N of identical molecules with n degrees of freedom. Finally, denote by $\mathcal{S}$ the microscopic system of these N molecules, M^*_{2nN} the phase space of $\mathcal{S}$, and by μ_{2n} the phase space of any molecule. In particular, if the particles of $\mathcal{S}$ are monoatomic, then $M^*_{2nN} = \mathfrak{R}^{6N}$ and $\mu_{2n} = \mathfrak{R}^6$; if the particles of $\mathcal{S}$ are biatomic, then the configuration space is $V_6 = \mathfrak{R}^4 \times P_2$, where P_2 is the two-dimensional projective space. Further, $M^*_{2nN} = (V_6 \times \mathfrak{R}^6)^N$ and $\mu_{2n} = V_6 \times \mathfrak{R}^6$. We can always order the symplectic coordinates $(\mathbf{q}, \mathbf{p})$ of M^*_{2nN} in such a way that the first $2n$ of them refer to the first particle, the coordinates from $2n$ to $4n$ refer to the second particle, and so on. We also suppose that the potential energy describing the interaction between the particles goes to infinity when the distance between any pair of molecules goes to zero and differs from zero only in a very small region about any particle. In this hypothesis there is no collision among the particles, the Hamiltonian formalism can be applied, and the dynamical orbit of $\mathcal{S}$ is a regular curve of the phase space M^*_{2nN}.

Since $\mathcal{S}$ is contained in a reservoir with adiabatic walls, its energy $H(\mathbf{q}, \mathbf{p})$ has a constant value E. Then, the admissible states of $\mathcal{S}$ are represented by the points of the level manifold Σ_E, defined by the equation $H(\mathbf{q}, \mathbf{p}) = E$, which is supposed to be *compact*. Owing to the large value of N, we have no possibility either of determining the initial data of the particles of $\mathcal{S}$ or of integrating the Hamiltonian equations of $\mathcal{S}$. Once again, we resort to statistical assumptions to have information about the evolution of $\mathcal{S}$.

In view of our choice of the symplectic coordinates of M^*_{2nN}, to any point $(\mathbf{q}, \mathbf{p}) \in M^*_{2nN}$ we can associate N points $(q_{(h-1)n+1}, \ldots, q_{hn}, p_{(h-1)n+1}, \ldots, p_{hn})$, $h = 1, \ldots, N$, belonging to the phase space μ_{2n}. These points represent the states of each particle when the whole system is in the state $(\mathbf{q}, \mathbf{p})$. When this state varies on Σ_E, the corresponding points of μ_{2n} vary in a set $\Gamma \subset \mu_{2n}$, which is bounded since Σ_E is compact. Let B be a cube containing the set Γ and consider a partition P of

B into elementary cells $C_i, i = 1, \ldots, L$, having edges $\Delta\omega$ and volume $\Delta\omega^{2n}$. We suppose that

1. Any molecule of S can be singled out with respect to the others;
2. The length $\Delta\omega$ of the edge of any cell C_i is small with respect to the edge of the cube B, so that we can suppose that all the particles inside C_i have the same energy, but it so large that any cell contains many points of Γ.

The first assumption states that it is possible to attribute a target to any molecule so that we can count them, distinguishing each particle from the others. The following considerations show the relevance of the second assumption.

Suppose that N_i particles of S are in the states represented by N_i points belonging to $C_i \cap \Gamma, i = 1, \ldots, L$. Remembering the meaning of the points of $C_i \cap \Gamma$, the numbers N_i must satisfy the following relations:

$$N_1 + \cdots + N_L = N, \tag{21.23}$$

$$N_1 \epsilon_1 + \cdots + N_L \epsilon_L = E. \tag{21.24}$$

It is evident that a single point on the surface Σ_E corresponds to such a distribution of particles in the sets $C_i \cap \Gamma, i = 1, \ldots, L$. Since we can distinguish one particle from all others, we can change the particles belonging to the set $C_i \cap \Gamma$ without changing their number. After this operation, we have the *same distribution* $N_1, \ldots, N_L$ of *different* particles in $C_i \cap \Gamma, i = 1, \ldots, L$. It is evident that a different point of Σ_E corresponds to this new distribution. Denoting by $D(N_1, \ldots, N_L)$ the number of possible ways to obtain a given distribution $N_1, \ldots, N_L$, we say that the distribution $N_1^*, \ldots, N_L^*$ is the *most probable* if it corresponds to the maximum of $D(N_1, \ldots, N_L)$ on varying $N_1, \ldots, N_L$.

The total number $D(N_1, \ldots, N_L)$ of possible ways to obtain a given distribution $N_1, \ldots, N_L$, is equal to the number of combinations obtained setting N_1 particle into $C_1 \cap \Gamma$, N_2 particles into $C_2 \cap \Gamma$, and so on. This number is given by the relation

$$D(N_1, \ldots, N_L) = \binom{N}{N_1} \binom{N - N_1}{N_2} \cdots \binom{N - N_1 - \cdots N_{L-1}}{N_L}$$

which yields the following result:

$$D(N_1, \ldots, N_L) = \frac{N!}{N_1! N_2! \cdots N_L!}. \tag{21.25}$$

Before going on, we recall Stirling's formula which states that, for the positive integer $n \to \infty$, it is

$$n! \approx \sqrt{2\pi} n^{n+\frac{1}{2}} e^{-n} \approx \sqrt{2\pi} n^n e^{-n}. \tag{21.26}$$

Consequently, we can state that

$$\ln n! = n \ln n - n + \ln \sqrt{2\pi} \approx n \ln n - n. \tag{21.27}$$

Taking into account (21.27) and (21.25), we obtain

$$\ln D(N_1, \ldots, N_L) = N \ln N - \sum_{i=1}^{L} N_i \ln N_i. \tag{21.28}$$

We are interested in evaluating the distribution $N_1, \ldots, N_L$ corresponding to the maximum value of $D(N_1, \ldots, N_L)$, under the constraints (21.23) and (21.24). It is evident that this distribution can also be obtained by evaluating the maximum value of $\ln D(N_1, \ldots, N_L)$, that is, of (21.28), under the same constraints. Introducing the Lagrangian multipliers α and β, we have to find the maximum of the function

$$F(N_1, \ldots, N_L) = N \ln N - \sum_{i=1}^{L} (N_i \ln N_i + \alpha N_i + \beta N_i \epsilon_i) + \alpha N + \beta E. \tag{21.29}$$

Equating to zero the partial derivatives of (21.29) with respect to the variables N_i, we obtain the system

$$\ln N_i + 1 + \alpha + \beta \epsilon_i = 0, \quad i = 1, \ldots, L, \tag{21.30}$$

so that the values of $N_1, \ldots, N_L$ at which $F(N_1, \ldots, N_L)$ reaches an extremum are

$$N_i^* = e^{-(1+\alpha)} e^{-\beta \epsilon_i}, \tag{21.31}$$

where the constants α and β can be obtained from (21.23) and (21.24). We omit the proof that the extremum (21.31) is a maximum.

Following Boltzmann and Gibbs, we state that S evolves in such a way that its states are always very close to the state of S determined by the most probable distribution $N_1^*, \ldots, N_L^*$. This assumption is supported by the following considerations.

Suppose that a distribution $N_1, \ldots, N_L$ of particles into the cells $C_1, \ldots, C_L$ is given, satisfying conditions (21.23) and (21.24). Then, since the coordinates of all the particles are assigned, a point of Σ_E is determined that belongs to a cell $\mathbb{C}$ of M_{2nN}^* intersecting Σ_E and having a volume $\Delta \omega^{2nN}$. It is important to remark that by changing the particles into the cells $C_1, \ldots, C_L$ without changing their number, we change their coordinates obtaining a different cell $\mathbb{C}_i$ of volume $\Delta \omega^{2nN}$. In conclusion, to all distributions corresponding to N_1 particles into C_1, N_2 particles in C_2, and so on, we can associate a family of cells of M_{2nN}^* with a volume

$$\Omega(N_1, \ldots, N_L) = D(N_1, \ldots, N_L) \Delta \omega^{2nN}. \tag{21.32}$$

In particular, the volume of the cells of M_{2nN}^* corresponding to distribution (21.31) is given by

$$\Omega(N_1^*, \ldots, N_L^*) = D(N_1^*, \ldots, N_L^*)\Delta\omega^{2nN}. \tag{21.33}$$

In order to evaluate the ratio

$$\frac{\Omega(N_1, \ldots, N_L)}{\Omega(N_1^*, \ldots, N_L^*)} = \frac{D(N_1, \ldots, N_L)}{D(N_1^*, \ldots, N_L^*)}, \tag{21.34}$$

we start considering the Taylor expansion of the logarithm of D

$$\ln D = \ln D^* + \sum_{i=1}^{L} \left(\frac{\partial}{\partial N_i} \ln D\right)_* (N_i - N_i^*)$$

$$+ \frac{1}{2} \sum_{i,h=1}^{L} \left(\frac{\partial^2}{\partial N_i \partial N_h} \ln D\right)_* (N_i - N_i^*)(N_h - N_h^*), \tag{21.35}$$

where, in view of (21.28), it is

$$\frac{\partial}{\partial N_i} \ln D = -\ln N_i - 1,$$

$$\frac{\partial^2}{\partial N_i \partial N_h} \ln D = -\frac{1}{N_h}\delta_{hi}.$$

Finally, by (21.30), (21.35) becomes

$$\ln D = \ln D^* + \sum_{i=1}^{L} (\alpha + \beta\epsilon_i)_* (N_i - N_i^*)$$

$$- \frac{1}{2} \sum_{i=1}^{L} \frac{1}{N_i^*}(N_i - N_i^*)^2.$$

When we take into account conditions (21.23) and (21.24), the above formula reduces to

$$\ln \frac{D}{D^*} = -\frac{1}{2} \sum_{i=1}^{L} \frac{1}{N_i^*}(N_i - N_i^*)^2$$

and (21.34) gives

$$\frac{\Omega}{\Omega^*} = e^{-\frac{1}{2}\sum_{i=1}^{L} \frac{1}{N_i^*}(N_i - N_i^*)^2} = e^{-\frac{1}{2}\sum_{i=1}^{L} N_i^*\left(\frac{N_i}{N_i^*} - 1\right)^2}, \tag{21.36}$$

which shows that the ratio Ω/Ω^* is very close to zero for any distribution $N_1, \ldots, N_L$ different from $N_1^*, \ldots, N_L^*$. The following remark shows the interest of the preceding result. We know that the orbit of the system S lies on the manifold $\Sigma_E \subset M_{2nN}^*$. In view of (21.36), we can state that this orbit is almost contained in those cells $\mathbb{C}_i^*$ corresponding to distribution (21.33). Further, the total volume of these cells is close to the volume of Σ_E.

Multiplying (21.31) by $\Delta\omega^{2n}$, in the limit $N \to \infty$, we obtain

$$dN(\mathbf{q}, \mathbf{p}) = a e^{-\beta\epsilon(\mathbf{q}, \mathbf{p})} d\mathbf{q} d\mathbf{p}, \tag{21.37}$$

where $a = e^{-(1+\alpha)}$. Integrating on the space μ_{2n}, we determine the value of a:

$$a = \frac{N}{\displaystyle\int_{\mu_{2n}} e^{-\beta\epsilon(\mathbf{q}, \mathbf{p})} d\mathbf{q} d\mathbf{p}}, \tag{21.38}$$

and (21.37) gives the **Boltzmann–Gibbs distribution**:

$$\frac{dN}{N} = \frac{1}{\displaystyle\int_{\mu_{2n}} e^{-\beta\epsilon(\mathbf{q}, \mathbf{p})} d\mathbf{q} d\mathbf{p}} e^{-\beta\epsilon(\mathbf{q}, \mathbf{p})} d\mathbf{q} d\mathbf{p}, \tag{21.39}$$

which allows us to evaluate the mean value $\overline{f}$ of any function $f(\mathbf{q}, \mathbf{p})$:

$$\overline{f} = \frac{1}{\displaystyle\int_{\mu_{2n}} e^{-\beta\epsilon(\mathbf{q}, \mathbf{p})} d\mathbf{q} d\mathbf{p}} \int_{\mu_{2n}} f(\mathbf{q}, \mathbf{p}) e^{-\beta\epsilon(\mathbf{q}, \mathbf{p})} d\mathbf{q} d\mathbf{p}. \tag{21.40}$$

Proceeding as in Sect. 21.2, we can prove that $\beta = 1/k\theta$.

Remark 21.1. Suppose that the degrees of freedom of any particle are equal to 3, the momentum $\mathbf{p} = m\dot{\mathbf{r}}$, and the energy ϵ reduces to the kinetic energy. Then, integrating (21.39) on the space V we obtain again the Maxwell distribution (21.17).

Remark 21.2. We notice that the Boltzmann–Gibbs distribution does not depend on the choice of canonical coordinates since the determinant of the Jacobian matrix of a canonical transformation of coordinates $(\mathbf{q}, \mathbf{p}) \to (\overline{\mathbf{q}}, \overline{\mathbf{p}})$ is equal to one.

21.4 Equipartition of Energy

Let S be a system of identical molecules $P_i, i = 1, \ldots, N$, at macroscopic equilibrium at the absolute temperature θ. Assume that any molecule has n degrees of freedom, and denote by μ_{2n} the phase state of each of them and by $(\mathbf{q}, \mathbf{p})$ canonical coordinates on μ_{2n}. For instance, if P_i is monoatomic, then it has three translational degrees of freedom; if P_i is biatomic, then it has three translational degrees of freedom, two rotational degrees of freedom, and one vibrational degree of freedom.

Further, we assume that the interaction among the different parts of P_i is described by a potential energy $U(\mathbf{q})$ that has a minimum at a position $\mathbf{0}$. Then we can suppose that the small oscillations of the molecule P_i are described by a Lagrangian function

$$L(\mathbf{q}, \dot{\mathbf{q}}) = \frac{1}{2} \sum_{i=1}^{n} \left(a_i (\dot{q}^i)^2 - \lambda_i (q^i)^2 \right) \tag{21.41}$$

[see (17.111)], where the quantities a_i are the coefficients of the kinetic energy and $\lambda_i > 0$ denotes the principal frequencies. Consequently, the Hamiltonian function is

$$H(\mathbf{q}, \dot{\mathbf{p}}) = \frac{1}{2} \sum_{i=1}^{n} \left(\frac{1}{a_i} (\dot{p}^i)^2 + \lambda_i (q^i)^2 \right). \tag{21.42}$$

Now we prove the fundamental theorem of the ***equipartition of energy***.

Theorem 21.1. *Let S be a system of N identical particles whose Hamiltonian function has the form (21.42) in a suitable system of symplectic coordinates. Then, if S is at macroscopic equilibrium at the absolute temperature θ, the mean energy $\bar{\epsilon}$ of any particle is obtained by supposing that to each quadratic term of (21.42) corresponds the mean energy $k\theta/2$, that is, $\bar{\epsilon}$ is given by*

$$\bar{\epsilon} = nk\theta. \tag{21.43}$$

Proof. The mean energy $\bar{\epsilon}$ is obtained by (21.40) with $f = \epsilon$, where f is given by (21.42). To evaluate the right-hand side of (21.40), we first note that

$$B = \int_{\mu_{2n}} e^{-\frac{1}{2k\theta} \sum_{i=1}^{n} \left(\frac{p_i^2}{a_i} + \lambda_i (q^i)^2 \right)} d\mathbf{q} d\mathbf{p}$$

$$= \prod_{i=1}^{n} \int_{\Re} e^{-\frac{p_i^2}{2a_i k\theta}} dp_i \prod_{i=1}^{n} \int_{\Re} e^{-\frac{\lambda_i (q^i)^2}{2k\theta}} dq^i. \tag{21.44}$$

Further, we have that

$$\int_{\mu_{2n}} \frac{1}{2} \sum_{h=1}^{n} \left(\frac{p_h^2}{a_h} + \lambda_h (q^h)^2 \right) e^{-\frac{1}{2k\theta} \sum_{i=1}^{n} \left(\frac{p_i^2}{a_i} + \lambda_i (q^i)^2 \right)} d\mathbf{q} d\mathbf{p}$$

$$= \int_{\mu_{2n}} \frac{1}{2} \sum_{h=1}^{n} \frac{p_h^2}{a_h} e^{-\frac{1}{2k\theta} \sum_{i=1}^{n} \left(\frac{p_i^2}{a_i} + \lambda_i (q^i)^2 \right)} d\mathbf{q} d\mathbf{p}$$

$$+ \int_{\mu_{2n}} \frac{1}{2} \sum_{h=1}^{n} \lambda_h (q^h)^2 e^{-\frac{1}{2k\theta} \sum_{i=1}^{n} \left(\frac{p_i^2}{a_i} + \lambda_i (q^i)^2 \right)} d\mathbf{q} d\mathbf{p}. \tag{21.45}$$

The first integral on the right-hand side of the preceding equation can be transformed as follows:

$$A \equiv \int_{\mu_{2n}} \frac{1}{2} \sum_{h=1}^{n} \frac{p_h^2}{a_h} e^{-\frac{1}{2k\theta} \sum_{i=1}^{n} \left(\frac{p_i^2}{a_i} + \lambda_i (q^i)^2\right)} d\mathbf{q} d\mathbf{p}$$

$$= \frac{1}{2} \sum_{h=1}^{n} \int_{\mathfrak{R}} \frac{p_h^2}{a_h} e^{-\frac{p_h^2}{2k\theta a_h}} dp_h \prod_{i \neq h} \int_{\mathfrak{R}} e^{-\frac{p_i^2}{2k\theta a_i}} dp_i \prod_{i=1}^{n} \int_{\mathfrak{R}} e^{-\frac{\lambda_i (q_i)^2}{2k\theta}} dq_i. \tag{21.46}$$

Taking into account (21.44) and (21.46) we obtain that

$$\frac{A}{B} = \frac{1}{2} \sum_{h=1}^{n} \frac{1}{\int_{\mathfrak{R}} e^{-\frac{p_h^2}{a_h}} dp_h} \int_{\mathfrak{R}} \frac{p_h^2}{a_h} e^{-\frac{p_h^2}{2k\theta a_h}} dp_h.$$

Finally, recalling that

$$\frac{1}{2} \int_{-\infty}^{\infty} \frac{p_h^2}{a_h} e^{-\frac{p_h^2}{2k\theta a_h}} dp_h = k\theta \frac{\sqrt{\pi a_h k\theta}}{2\sqrt{2}},$$

$$\int_{-\infty}^{\infty} e^{-\frac{p_h^2}{a_h}} dp_h = \sqrt{2\pi a_h k\theta}, \tag{21.47}$$

we can state that

$$n\frac{A}{B} = \frac{n}{2} k\theta.$$

The theorem is proved repeating the same calculations for the second integral on the right-hand side of (21.45). $\square$

The preceding theorem allows us to calculate the specific heat c of gases. In fact, in view of (21.43), the total energy E of a volume of gas containing N molecules with n degrees of freedom is given by

$$E = Nnk\theta, \tag{21.48}$$

and the specific heat is

$$c = \frac{dE}{d\theta} = Nnk. \tag{21.49}$$

In this chapter we have merely sketched the ideas underlying the statistical mechanics of equilibrium. More precisely, we have shown how it is possible to describe the equilibrium of macroscopic bodies adopting a simple mechanical microscopic model, satisfying classical mechanics, and some convenient statistical axioms that are necessary to manage the huge number of elementary particles (molecules) associated with the bodies. In doing this, we have omitted considerations of many important topics forming the wide subject of statistical mechanics because of the introductory character of this book. The reader interested in the many aspects of equilibrium and nonequilibrium statistical mechanics may refer to the extensive bibliography on the topic (see, for instance, [10, 31, 50, 53]).

21.5   Statistical Mechanics and Ergodic Theory

In Gibbs' approach to statistical mechanics, probabilistic hypotheses are added to the Hamiltonian formalism. In the ergodic theory, that we sketch in the remaining part of this chapter, it is proved that the ergodic hypothesis and Hamiltonian formalism lead to the same results of Gibbs' theory.

As we have already noticed, in statistical mechanics the microscopic system S is supposed to be formed by N_1 molecules, with f_1 degrees of freedom, N_2 molecules, with f_2 degrees of freedom, and so on. For instance, for a mixture of a monoatomic gas and biatomic gas it is $f_1 = 3$ and $f_2 = 6$. Consequently, the configuration space of S is a differentiable manifold V_n, where $n = f_1 N_1 + f_2 N_2 + \cdots$. We further suppose that the interaction forces among the particles of S are conservative with a potential energy $U(\mathbf{q})$, $\mathbf{q} \in V_n$. Under these hypotheses system S can be described by Hamiltonian formalism, and its evolution is represented by a curve $\gamma(t, \mathbf{x_0})$ in the phase space M_{2n}^* that is a solution of Hamilton's equations corresponding to the initial datum $\mathbf{x_0} \in M_{2n}^*$. In other words, we are stating that the evolution of S is described by the ordinary laws of mechanics, that is, the motion of molecules is represented by a regular solution of Hamilton's equations. Consequently, we need to prevent collisions between molecules and of molecules with the walls of the container. In conclusion, we must to prevent that the Lagrangian coordinates $(q^1, \cdots, q^n)$ assume values corresponding to the above collisions. That can be obtained requiring that the potential energy assumes very high values when the Lagrangian coordinates correspond to collisions. Further, since we are interested in the macroscopic equilibrium of S, the total energy E of S is supposed to be constant, that is the energy integral holds

$$H(\mathbf{x}) = E. \tag{21.50}$$

We also suppose that the level manifold $\Sigma_E \subset M_{2n}^*$, defined by the equation (21.50), is a $(2n - 1)$-dimensional manifold that is **regular** and **compact** for any value of E belonging to a convenient interval. The first integral (21.50) implies that the dynamical trajectories of S belong to Σ_E. Finally, we denote by $\mathbf{X}_H | \Sigma_E$ the restriction to Σ_E of the Hamiltonian field $\mathbf{X}_H$ and by $\gamma(t, \mathbf{x_0})$, $\mathbf{x_0} \in \Sigma_E$, the dynamical trajectories of $\mathbf{X}_H | \Sigma_E$. Applying Theorem 18.19 we can state that the set of dynamical trajectories $\gamma(t, \mathbf{x_0})$, describing the evolution of S on varying the initial datum $\mathbf{x_0} \in \Sigma_E$, determines regions that have an invariant volume provided that the volume of these regions is evaluated by the $(2n - 1)$-form Ω_E.

The value of any macroscopic quantity $\mathcal{F}$ should be deduced from a suitable function $F(\mathbf{x})$ on the phase space M_{2n}^*. The simplest hypothesis is that the value of $\mathcal{F}$ is equal to the time average $\hat{F}$ of F along a trajectory $\gamma(t, \mathbf{x}_0)$, i.e.,

$$\hat{F} = \lim_{T \to \infty} \frac{1}{T} \int_0^T F(\gamma(t, \mathbf{x}_0))\, dt. \tag{21.51}$$

The following theorem holds that we recall without the difficult proof.

Theorem 21.2 (Von Neumann, Birkhoff). *The time average $\hat{F}$ exists for any Lebesgue measurable $F(\mathbf{x})$ and any initial datum $\mathbf{x}_0 \in \Sigma_E$ except for a subset with zero Lebesgue measure.*

The first difficulty related to (21.51) is that the time for measuring the macroscopic quantity $\mathcal{F}$ is finite whereas in (21.51) T tends to infinity. This contradiction can be overcome by admitting that the microscopic evolution of S is very fast so that the time τ of a macroscopic measure is much larger than the characteristic times of the evolution of S. Consequently, τ is a good approximation to evaluate the right-hand side (21.51). The second difficulty exhibited by (21.51) is related to the fact that (21.51) requires the knowledge of trajectory $\gamma(t, \mathbf{x}_0)$ that is of the solution of Hamilton's equations corresponding to the initial datum $\mathbf{x}_0 \in \Sigma_E$. But this is impossible for the following two reasons:

1. The number of Hamilton's equations is very high;
2. The value of $\mathcal{F}$ seems to depend on the initial datum $\mathbf{x}_0 \in \Sigma_E$.

It is evident that (21.51) can be accepted as a right definition of time average provided that the right-hand side is the same for all the trajectories on Σ_E. In other words, *the limit in* (21.51) *must be independent of real microscopic motion or, equivalently, $\hat{F}$ must be almost everywhere constant on Σ_E.* This certainly happens if we accept the **ergodic hypothesis** according to which *the time average is equal to the space average*

$$\hat{F} = \bar{F} = \frac{1}{\omega(E)} \int_{\Sigma_E} F(\mathbf{x})\, \Omega_E, \tag{21.52}$$

where

$$\omega(E) = \int_{\Sigma_E} \Omega_E.$$

When the above hypothesis is confirmed, manifold Σ_E is said to be **ergodic**. The pair $(\Sigma_E, \omega(E))$ is called the **microcanonical statistical set** of S with **structure function** $\omega(E)$. It is evident that if

$$\hat{F} = \overline{F}$$

then the time average $\hat{F}$ is almost everywhere constant on Σ_E.

Formula (21.52), although it makes possible to evaluate $\hat{F}$ without knowing the microscopic motions of $\mathcal{S}$, requires the integration of $F(\mathbf{x})$ with respect to a very large number of variables. For this reason it is essential to determine asymptotic formulae to calculate $\overline{F}$ that become more accurate when the number of particles of $\mathcal{S}$ increases. Finding these formulae is just one of the main problems of equilibrium statistical mechanics.

When we accept the ergodic hypothesis, we can, at list in principle, evaluate the macroscopic value of any physical quantity $\mathcal{F}$ from the corresponding phase function $F(\mathbf{x})$, provided that the energy E is given. This implies that, given the volume V occupied by the system $\mathcal{S}$, the energy is *the only independent quantity*, and any other macroscopic quantity can be formulated in terms of E. As we have already proved in the kinetic theory, the temperature θ is associated with the energy E. Consequently, such a model can describe at most the macroscopic behavior of a perfect gas but it is not able to describe more complex systems, for instance a mixture of perfect gases. Therefore, in general we must suppose that there are others $k-1$ first integrals $f_1(\mathbf{x}), \ldots, f_{k-1}(\mathbf{x})$ that, together with the energy integral, determine a $(2n-k)$-dimensional manifold, contained in M_{2n}^* and obtained fixing the values $(C_1, \ldots, C_{k-1}, E)$ of the above first integrals. Then, the ergodic hypothesis will be supposed to hold on Σ. In this way any physical quantity will be a function of k independent parameters. Further, the already cited Theorem 18.19 applied to Σ allows to define on Σ a measure. We wish to conclude this section with some definitions. We define *microcanonical probability* $P(\sigma)$ that the state of $\mathcal{S}$ is contained in the region $\sigma(\mathbf{x}) \subset \Sigma_E$ the number

$$P(\sigma) = \frac{1}{\omega(E)} \int_\sigma \boldsymbol{\Omega}_E \equiv \frac{\omega(\sigma)}{\omega(E)}, \tag{21.53}$$

where $\omega(\sigma)$ is the measure of σ evaluated by $\boldsymbol{\Omega}_E$. This number belongs to the interval $[0, 1]$ and it is invariant during the evolution of $\mathcal{S}$. In particular, denoting by $d\sigma$ an elementary region of Σ_E about the point $\mathbf{x}$, the probability $dP(\mathbf{x})$ that the state of $\mathcal{S}$ is contained in $d\sigma$ is

$$dP(\mathbf{x}) = \frac{\omega(d\sigma)}{\omega(E)},$$

and (21.52) becomes

$$\hat{F} = \overline{F} = \int_{\Sigma_E} F(\mathbf{x}) \, dP(\mathbf{x}). \tag{21.54}$$

We conclude this section recalling some theorems without supplying a proof.[2]

---

[2]For a proof see, for instance, [64].

Definition 21.1. *The invariant compact manifold Σ_E is **metrically indecomposable** (or metrically transitive) if it cannot be decomposed into the union of two disjoint measurable subsets M_1 and M_2, each invariant and of positive Lebesgue measure, i.e., if do not exist two submanifolds Σ_1 and Σ_2 of Σ_E, invariant, disjoint, and with positive Lebesgue measure, such that*

$$\Sigma_E = \Sigma_1 \cup \Sigma_2.$$

Theorem 21.3. *Σ_E is ergodic if and only if it is metrically indecomposable.*

Theorem 21.4 (Birkhoff). *Let S be a Hamiltonian system and denote by Σ_E the compact level manifold $H = E$. Then Σ_E is ergodic if and only if every first integral of S different from H is almost everywhere constant.*

21.6 Fundamental Formulae Related to the Structure Function ω_E

Let S be a mechanical system with Hamiltonian

$$H = \frac{1}{2} a^{hk} p_h p_k + U(\mathbf{q}).$$

Here (a^{hk}) is the inverse matrix of (a_{hk}) that is the matrix formed with the coefficients of the Lagrangian kinetic energy. Denote by E a value of the total energy. From the equation $H = E$ of the level surface Σ_E we have

$$a^{hk} p_h p_k = 2 (E - U(\mathbf{q})) \geq 0. \tag{21.55}$$

Consider the submanifold $C_E = \{\mathbf{q} \in V_n, E - U(\mathbf{q}) \geq 0\}$ of the configuration space V_n. For any $\mathbf{q} \in C_E$, (21.55) determines a subset of $T_{\mathbf{q}}^* V_n$ that, provided that $T_{\mathbf{q}}^* V_n$ is equipped with the scalar product

$$\mathbf{p} \cdot \mathbf{p} = a^{hk} p_h p_k,$$

is a $(n-1)$-dimensional sphere $S(\mathbf{q}, E) \subset T_{\mathbf{q}}^* V_n$ with radius

$$r = \sqrt{2(E - U(\mathbf{q}))}. \tag{21.56}$$

The equations of a sphere referred to an orthonormal basis $(\mathbf{E}^i)$ of $T_{\mathbf{q}}^* V_n$ can be written in terms of $(n-1)$ angular parameters (φ_α), $\alpha = 1, \ldots, n-1$. We denote by P_h the components of the arbitrary covector $\mathbf{p} \in T_{\mathbf{q}}^* V_n$ with respect to the basis $(\mathbf{E}^i)$, in which $a^{hk} = \delta^{hk}$, and by

$$e^i = A^i_j E^j,$$

the basis in which (21.55) holds. Then, we have

$$a^{hk} = A^h_l A^k_m \delta^{lm} = \sum_{l=1}^n A^h_l A^k_l, \quad p_h = (A^{-1})^k_h P_k,$$

where P_h are functions of angular parameters φ_α and r. The above relations respectively imply

$$\det (a^{hk}) = (\det (A^h_k))^2, \tag{21.57}$$

$$\begin{aligned}
J^*\big|_{H=E} &= \det \left(\frac{\partial \mathbf{p}}{\partial \varphi_\alpha}, \frac{\partial \mathbf{p}}{\partial E} \right)\Big|_{H=E} \\
&= \det ((A^{-1})^k_h) \det \left(\frac{\partial \mathbf{P}}{\partial \varphi_\alpha}, \frac{\partial \mathbf{P}}{\partial r} \right) \frac{dr}{dH}\Big|_{H=E}.
\end{aligned} \tag{21.58}$$

On the other hand $\det (A^h_k) = 1/\det ((A^{-1})^k_h)$ and (21.57) gives

$$\sqrt{a} = \det ((A^{-1})^k_h),$$

where $a = \det (a_{hk}) = 1/\det (a^{hk})$. Since $(dr/dH)_{H=E} = 1/r$, formula (21.58) becomes

$$J^*\big|_{H=E} = \frac{1}{r} \sqrt{a} \mathcal{I},$$

with $\mathcal{I} = \det \left(\frac{\partial \mathbf{P}}{\partial \varphi_\alpha}, \frac{\partial \mathbf{P}}{\partial r} \right)\Big|_{H=E}$. In conclusion, (18.79) and (18.88) assume the form

$$\hat{\mathbf{\Omega}} = \frac{\sqrt{a}}{\sqrt{2(H - U(\mathbf{q}))}} dq^1 \wedge \cdots \wedge dq^n \wedge d\Sigma \wedge dH, \tag{21.59}$$

$$\mathbf{\Omega}_E = \frac{\sqrt{a}}{\sqrt{2(E - U(\mathbf{q}))}} dq^1 \wedge \cdots \wedge dq^n \wedge d\Sigma, \tag{21.60}$$

where $d\Sigma = \mathcal{I} d\varphi_1 \wedge \cdots \wedge d\varphi_{n-1}$ represents the $(n-1)$ volume form on the sphere $S(\mathbf{q}, E)$.

Let us suppose that the level manifold $\Sigma_E = \{\mathbf{x} \in V_n, H(\mathbf{x}) = E\}$ is compact when $E \in [0, \bar{E}]$. Then, by (21.60) the measure $\omega(E)$ is given by

$$w\left(E\right) = \int_{\Sigma_E} \boldsymbol{\Omega}_E = \int_{C_E} \int_{S(\mathbf{q},E)} \frac{\sqrt{a}}{\sqrt{2\left(E - U(\mathbf{q})\right)}} dq^1 \cdots dq^n \wedge d\Sigma. \qquad (21.61)$$

On the other hand the surface $\mathbb{S}$ of the $(n-1)$-dimensional sphere $S\left(\mathbf{q}, E\right)$ is

$$\mathbb{S} = 2\frac{\pi^{n/2}}{\Gamma\left(\frac{n}{2}\right)} r^{n-1} \qquad (21.62)$$

whereas its volume $\mathbb{V}$ is

$$\mathbb{V} = \frac{\pi^{n/2}}{\Gamma\left(\frac{n}{2}+1\right)} r^n = 2\frac{\pi^{n/2}}{n\Gamma\left(\frac{n}{2}\right)} r^n. \qquad (21.63)$$

Here the function Γ is defined as follows:

$$\Gamma\left(p\right) = \int_0^\infty x^{p-1} e^{-x} dx, \quad p > 0,$$

and verifies the properties

$$\Gamma\left(1\right) = 1, \quad \Gamma\left(1/2\right) = \sqrt{\pi}, \quad \Gamma\left(p+1\right) = p\Gamma\left(p\right).$$

Then (21.61) becomes

$$w\left(E\right) = \chi_n \int_{C_E} \left(E - U\right)^{n/2-1} \sqrt{a}\, dq^1 \cdots dq^n \qquad (21.64)$$

where

$$\chi_n = \frac{\left(2\pi\right)^{n/2}}{\Gamma\left(\frac{n}{2}\right)}. \qquad (21.65)$$

Let $\mathbb{V}\left(E\right)$ be the volume of the n-dimensional region

$$V_E = \left\{\mathbf{x} \in M_{2n}^*, 0 \le H\left(\mathbf{x}\right) \le E\right\},$$

belonging to the level manifold Σ_E and suppose that

$$\mathbb{V}\left(0\right) = 0,$$
$$\mathbb{V}\left(E'\right) < \mathbb{V}\left(E''\right), \quad \text{if } E' < E'', \text{ and } E', E'' \in \left[0, \bar{E}\right].$$

Under these hypothesis, when we take again into account (21.64) and integrate (21.59) on V_E, we obtain

$$\mathbb{V}(E) = \int\limits_{C_E} \int\limits_0^E \int\limits_{S(\mathbf{q},H)} \frac{\sqrt{a}}{\sqrt{2(H-U(\mathbf{q}))}} dq^1 \cdots dq^n d\Sigma dH \qquad (21.66)$$

$$= \chi_n \int\limits_{C_E} \int\limits_0^E (H-U(\mathbf{q}))^{n/2-1} \sqrt{a}\, dq^1 \cdots dq^n dH,$$

so that we have proved the formula

$$\mathbb{V}(E) = \frac{2}{n}\chi_n \int\limits_{C_E} (E-U(\mathbf{q}))^{n/2} \sqrt{a}\, dq^1 \cdots dq^n. \qquad (21.67)$$

If $F(\mathbf{x})$ is any summable function on V_E the following result holds:

$$\int\limits_{\Sigma_E} F\Omega_E = \frac{d}{dE} \int\limits_{V_E} F\hat{\Omega}. \qquad (21.68)$$

In fact, we have

$$\frac{d}{dE} \int\limits_{V_E} F\hat{\Omega} = \frac{d}{dE} \int\limits_{C_E} \int\limits_0^E \int\limits_{S(\mathbf{q},H)} F\frac{\sqrt{a}}{\sqrt{2(H-U(\mathbf{q}))}} dq^1 \cdots dq^n d\Sigma dH$$

$$= \int\limits_{C_E} \int\limits_{S(\mathbf{q},H)} F\frac{\sqrt{a}}{\sqrt{2(H-U(\mathbf{q}))}} dq^1 \cdots dq^n d\Sigma,$$

so that from (21.60) we obtain (21.68).

Formula (21.68), when $F=1$, gives

$$\omega(E) = \frac{d\mathbb{V}(E)}{dE} \iff \mathbb{V}(E) = \int\limits_0^E \omega(a)\, da. \qquad (21.69)$$

Analogously, (21.52) leads to the following result for the average of a function $F(\mathbf{q}, E)$ (that is independent of φ)

$$\bar{F} = \frac{1}{\omega(E)}\chi_n \int\limits_{C_E} F(E-U(\mathbf{q}))^{n/2-1} \sqrt{a}\, dq^1 \cdots dq^n, \qquad (21.70)$$

where $\omega(E)$ is defined by (21.64).

Let $\mathcal{S}_1$ and $\mathcal{S}_2$ be two mechanical systems, $M^*_{2n_1}$ and $M^*_{2n_2}$ their phase spaces of dimensions $2n_1$ and $2n_2$, respectively. Further, let $H_1(\mathbf{x}_1)$ and $H_2(\mathbf{x}_2)$ be the Hamiltonian of $\mathcal{S}_1$ and $\mathcal{S}_2$ in the form

$$H(\mathbf{x}) = T + U = \frac{1}{2}a^{hk}p_h p_k + U(\mathbf{q}).$$

Let $\mathcal{S}$ be the mechanical system whose phase space is $M^*_{2n} = M^*_{2n_1} \times M^*_{2n_2}$, where $2n = 2n_1 + 2n_2$, and the Hamiltonian function is

$$H(\mathbf{x}) = H_1(\mathbf{x}_1) + H_2(\mathbf{x}_2),$$

where $\mathbf{x} = (\mathbf{x}_1, \mathbf{x}_2)$. In this case we say that $\mathcal{S}$ is *decomposable in the two components* $\mathcal{S}_1$ and $\mathcal{S}_2$. We now introduce the notations

$$V_{n_1,E} = \left\{\mathbf{x}_1 \in M^*_{2n_1}, 0 \leq H_1(\mathbf{x}_1) \leq E\right\},$$

$$V_{n_2,E-E_1} = \left\{\mathbf{x}_2 \in M^*_{2n_2}, 0 \leq H_2(\mathbf{x}_2) \leq E - E_1\right\},$$

and $dV = dq^1_{(1)} \cdots dq^{n_1}_{(1)} dp_{(1)1} \cdots dp_{(1)n_1} dq^1_{(2)} \cdots dq^{n_2}_{(2)} dp_{(2)1} \cdots dp_{(2)n_2}$. Taking into account Liouville's theorem in the space M^*_{2n} the volume $\mathbb{V}(E)$ of the region V_E can be written as

$$\mathbb{V}(E) = \int_{V_E} dV = \int_{V_{n_1,E}} dV_1 \int_{V_{n_2,E-E_1}} dV_2 = \int_{V_{n_1,E}} \mathbb{V}_{n_2}(E - E_1)\, dV_1,$$

where $dV_1 = dq^1_{(1)} \cdots dq^{n_1}_{(1)} dp_{(1)1} \cdots dp_{(1)n_1}$, and $dV_2 = dq^1_{(2)} \cdots dq^{n_2}_{(2)} dp_{(2)1} \cdots dp_{(2)n_2}$. From the above identities and (21.69) it follows

$$\mathbb{V}(E) = \int_0^E \mathbb{V}_{n_2}(E - E_1)\,\omega_1(E_1)\,dE_1$$

$$= \int_0^\infty \mathbb{V}_{n_2}(E - E_1)\,\omega_1(E_1)\,dE_1,$$

since $\mathbb{V}_{n_2}(E - E_1) = 0$ for $E_1 > E$. Finally, differentiating $\mathbb{V}(E)$ with respect to E and taking into account (21.69) we obtain

$$\omega(E_1) = \int_0^\infty \omega_1(E_1)\,\omega_2(E - E_1)\,dE_1.$$

Extending this formula to the case in which S is decomposable into N components $S_1, \ldots, S_N$ we obtain

$$\omega(E) = \int_0^\infty \left[\prod_{i=1}^{N-1} \omega_i(E_i) \, dE_i \right] \omega_N \left(E - \sum_{i=1}^{N-1} E_i \right).$$

It is evident that any Hamiltonian system of N free particles is decomposable into N components. Later we see that this conclusion holds also for a system of N *weakly interacting* particles, at least if we are interested in the mean quantities. In fact, the above formula can be substituted by an approximate formula that is more accurate when the number of particles increases.

21.7 Some Simple Thermodynamic Applications

The *entropy* of system S is defined as follows:

$$S(V, E) = k \ln \frac{\omega(E)}{S_0}, \qquad (21.71)$$

where k is the Boltzmann constant and S_0 is an arbitrary suitable constant. If we introduce the *absolute temperature* θ and the *pressure* p by the relations

$$\theta = 1 \bigg/ \frac{\partial S}{\partial E}, \quad p = \theta \frac{\partial S}{\partial V} = \frac{\partial S}{\partial V} \bigg/ \frac{\partial S}{\partial E}, \qquad (21.72)$$

then, differentiating $S(V, E)$

$$dS = \frac{\partial S}{\partial E} dE + \frac{\partial S}{\partial V} dV,$$

we obtain the first principle of thermodynamics

$$\theta dS = dE + p dV. \qquad (21.73)$$

We note that these considerations do not depend on the particular form of the function $S(V, E)$.

More explicit results can be obtained by considering a system S of N *identical noninteracting particles* with mass m, that move freely in a container with volume V. One could think that this mechanical model describes the microscopic behavior of an ideal gas. It is easy to verify that this assumption is wrong. In fact, the Hamiltonian function of S is

$$H = \sum_{i=1}^N \frac{\mathbf{p}_i^2}{2m}, \qquad (21.74)$$

where $\mathbf{p}_i = m v_i$ is the momentum of ith particle, and the phase space is $V^{3N} \times \mathfrak{R}^{3N}$. Moreover, the corresponding Hamilton's equations admit the following phase trajectories:

$$\mathbf{q}_i = \frac{\mathbf{p}_i}{m} t + \mathbf{q}_i^0, \quad \mathbf{p}_i = \mathbf{p}_i^0, \quad i = 1, \ldots, N.$$

In other words, the particles move uniformly on straight lines. But this is impossible since particles collide each with other and with the walls of the container.

In order to have a *regular* behavior of trajectories we substitute the Hamiltonian (21.74) with

$$H = \sum_{i=1}^{N} \frac{\mathbf{p}_i^2}{2m} + \sum_{i \neq j} U\left(|\mathbf{r}_i - \mathbf{r}_j|\right) + \sum_{i=1}^{N} W, \tag{21.75}$$

where U is the interaction potential between two particles, and W is the interaction potential between a particle and the walls of the container. We suppose that these potentials assume very high values when the distance between two particles or between a particle and a wall became very small. Further, we assume that they go quickly to zero when the above distances increase. These assumptions allow to substitute $E - U$ with E in formulae (21.64), (21.67), (21.70) without producing a variation of the integrals.

If we take into account (21.64) in which $\sqrt{a} = m^{\frac{3N}{2}}$, the measure $\omega(E)$ of the compact manifold Σ_E becomes

$$\omega(E) = m V^N \chi_{3N} (m E)^{\frac{3N}{2} - 1}. \tag{21.76}$$

Now we can obtain explicit formulae for $\theta(V, E)$ and $p(V, E)$ from (21.76) and (21.71). In fact, it can be proved that the expansion for $x \to \infty$ of Γ-function is

$$\Gamma(x + 1) \approx \sqrt{2\pi} x^{x + 1/2} e^{-x},$$

so that, in view of (21.65), $S(V, E)$ assumes the form

$$S(V, E) = N k \ln\left[\lambda V \left(m\pi \frac{4E}{3N}\right)^{\frac{3}{2}}\right] + \frac{3}{2} k N, \tag{21.77}$$

where λ is an arbitrary constant.

Equation (21.77) can also be written as follows:

$$S(V, E) = k N \ln \frac{V}{N V_0} + \frac{3}{2} k N \ln \frac{E}{N E_0} + C, \tag{21.78}$$

where C is a constant and V_0, E_0 are suitable constants.

Now it is evident that

$$\frac{1}{\theta} = \frac{\partial S}{\partial E} = \frac{3}{2E}kN, \tag{21.79}$$

$$\frac{E}{N} = \frac{3}{2}k\theta, \quad P = k\frac{N}{V}\theta \equiv kn\theta,$$

and we obtain again the results of ideal gases that we obtained from the kinetic theory.

21.8 The Theorem of Equipartition of Kinetic Energy

In the previous section we have considered a mechanical system of N particles that interact by forces with a short action radius compared with the mean free path. This model, in which any particle has 3 degrees of freedom, allows to derive some fundamental thermodynamic relations of perfect gases. In this section we want to prove a result that holds for any mechanical system S.

Theorem 21.5 (Equipartition of kinetic energy). *Let S be a mechanical system with n degrees of freedom. Then, its mean kinetic energy is*

$$\bar{T} = \frac{n}{2}\frac{\mathbb{V}(E)}{\omega(E)} \equiv \frac{n}{2}\beta, \tag{21.80}$$

where β is a constant.

Proof. In view of (21.70) we have

$$2\bar{T} = \frac{1}{\omega(E)}\chi_n \int_{C_E} 2T\,(E - U(\mathbf{q}))^{n/2-1}\sqrt{a}\,dq^1\cdots dq^n,$$

so that, recalling $T = (E - U)$ and (21.67), we obtain

$$2\bar{T}\omega(E) = 2\chi_n \int_{C_E} (E - U(\mathbf{q}))^{n/2}\sqrt{a}\,dq^1\cdots dq^n = n\mathbb{V}(E),$$

and the theorem is proved.   □

For system S of N weakly interacting particles it is $E = T = \bar{T}$, where E is constant along the motions, and $n = 3N$, then (21.80) gives

$$E = \frac{3}{2}N\beta.$$

Comparing this formula with (21.15) we obtain

$$\beta = k\theta. \tag{21.81}$$

It is possible to prove that (21.81) still holds for a mechanical system S of N polyatomic molecules with m degrees of freedom. Then, $n = mN$ and (21.80) supplies

$$\bar{T} = \frac{m}{2} Nk\theta. \tag{21.82}$$

If we can neglect the interaction energy among the molecules, we have $E = T = \bar{T}$ and (21.82) becomes

$$E = \frac{m}{2} Nk\theta, \tag{21.83}$$

so that the specific heat c_V at constant volume of a perfect gas of polyatomic molecules is

$$c_V = \frac{1}{n} \frac{dE}{d\theta} = \frac{m}{2} R. \tag{21.84}$$

21.9 The Boltzmann Distribution

Definitions (21.71) and (21.72) in Section 21.7 allowed to obtain the constitutive equations of ideal gases since the Hamiltonian (21.74) has a very simple form. Then, it is quite natural to search for formulae that generalize (21.79) when the Hamiltonian contains interaction terms as in (21.75).

Let us consider an *isolated* system S formed by two systems S_1 and S_2 with N_1 and N_2 particles, respectively. We suppose $N_1 \ll N_2$ and that the interaction between S_1 and S_2 is very weak. In other words, S_1 is a physical system that is at equilibrium with a thermostat S_2 and the system $S = \{S_1, S_2\}$ is isolated. Hamilton's function of S is

$$H_\epsilon (\mathbf{q}, \mathbf{p}) = H_1 (\mathbf{q}_1, \mathbf{p}_1) + H_2 (\mathbf{q}_2, \mathbf{p}_2) + \epsilon U (\mathbf{q}). \tag{21.85}$$

where $\epsilon U (\mathbf{q})$ describes the weak interaction between S_1 and S_2, and $H_1 (\mathbf{q}_1, \mathbf{p}_1) = T_1 (\mathbf{q}_1, \mathbf{p}_1) + U_1 (\mathbf{q}_1)$ and $H_2 (\mathbf{q}_2, \mathbf{p}_2) = T_2 (\mathbf{q}_2, \mathbf{p}_2) + U_2 (\mathbf{q}_2)$.

We note that this weak interaction can produce relevant energy exchanges between S_1 and S_2 but these exchanges occur very slowly. We suppose that when $\epsilon \neq 0$, the system S is ergodic so that the time average of a function $F (\mathbf{x})$ defined on the phase spaces M_{2n}^* can be identified with its space average

$$\bar{F}_\epsilon = \frac{1}{\omega_\epsilon (E)} \int_{\Sigma_{\epsilon,E}} F (\mathbf{x}) \, \boldsymbol{\Omega}_{\epsilon,E}, \tag{21.86}$$

where $\boldsymbol{\Omega}_{\epsilon,E}$ is defined by (21.60), $U = U_1 + U_2 + \epsilon U$ and

$$\omega_\epsilon (E) = \int_{\Sigma_{\epsilon,E}} \mathbf{\Omega}_{\epsilon,E}, \tag{21.87}$$

is the volume of the level region $\Sigma_{\epsilon,E} = \{\mathbf{x} \in M^*_{2n}, H_\epsilon(\mathbf{x}) = E\}$.

It is possible to prove, at least for some systems S and functions $F(\mathbf{x})$, that

$$\lim_{\epsilon \to 0} \bar{F}_\epsilon = \frac{1}{\omega(E)} \int_{\Sigma_E} F(\mathbf{x}) \mathbf{\Omega}_E, \tag{21.88}$$

where $\omega(E)$, Σ_E, and $\mathbf{\Omega}_E$ are the value of $\omega_\epsilon(E)$, $\Sigma_{\epsilon,E}$, and $\mathbf{\Omega}_{\epsilon,E}$ when $\epsilon \to 0$.

In other words, when the interaction between S_1 and S_2 is very weak, we can neglect in (21.85) the interaction term $\epsilon U(\mathbf{q})$ if we are interested in evaluating the averages of functions defined on the phase space. Consequently, we suppose that S is described by the Hamiltonian

$$H(\mathbf{q}, \mathbf{p}) = H_1(\mathbf{q}_1, \mathbf{p}_1) + H_2(\mathbf{q}_2, \mathbf{p}_2) \tag{21.89}$$

so that $M^*_{2n} = M^*_{2n_1} \times M^*_{2n_2}$.

Let W_1 be a measurable set of $M^*_{2n_1}$ and introduce the other set

$$W = \{\mathbf{x} \equiv (\mathbf{x}_1, \mathbf{x}_2) \in M^*_{2n}, \mathbf{x}_1 \in W_1\}.$$

Then, the probability $P_1(W_1)$ that $\mathbf{x}_1 \in W_1$ is equal to the probability $P(W)$ that $\mathbf{x} \in W$. In view of (21.53) we have

$$P_1(W_1) = \frac{1}{\omega(E)} \int_{W \cap \Sigma_E} \mathbf{\Omega}_E = \frac{1}{\omega(E)} \int_{\Sigma_E} \varphi \mathbf{\Omega}_E, \tag{21.90}$$

where φ is the characteristic function of W, and E is the energy of the isolated system S.

When we suppose that hypotheses leading to (21.67) and (21.68) are verified, we obtain

$$P_1(W_1) = \frac{1}{\omega(E)} \frac{d}{dE} \int_{V_E} \varphi \hat{\mathbf{\Omega}} = \frac{1}{\omega(E)} \int_{V_E} \varphi dq^1 \ldots dp_n. \tag{21.91}$$

On the other hand it is

$$V_E = \{(\mathbf{x}_1, \mathbf{x}_2) \in M^*_{2n}, 0 \le H(\mathbf{x}) \le E\},$$

so that, introducing the notation

$$V_{1,E} = \left\{ \mathbf{x}_1 \in M_{2n_1}^*, 0 \le H_1 (\mathbf{x}_1) \le E \right\} \supseteq W_1,$$

we have that, $\forall \mathbf{x}_1 \in V_{1,E}$, and the value $E_1 = H_1 (\mathbf{x}_1)$ is determined. Therefore $(\mathbf{x}_1, \mathbf{x}_2) \in V_E$ if and only if $\mathbf{x}_2$ belongs to the set

$$V_{2,E-E_1} (\mathbf{x}_1) = \left\{ \mathbf{x}_2 \in M_{2n_2}^*, 0 \le H_2 (\mathbf{x}_2) \le E - E_1 \right\}.$$

The above consideration allows us to write (21.91) as follows:

$$
\begin{aligned}
P_1 (W_1) &= \frac{1}{w (E)} \frac{d}{dE} \int_{V_{1,E}} \varphi dq_{(1)}^1 \dots dp_{(1)n_1} \int_{V_{2,E-E_1}(\mathbf{x}_1)} dq_{(2)}^1 \dots dp_{(2)n_2} \\
&= \frac{1}{w (E)} \frac{d}{dE} \int_{V_{1,E}} \varphi V_2 (E - E_1) dq_{(1)}^1 \dots dp_{(1)n_1} \\
&= \frac{1}{w (E)} \frac{d}{dE} \int_{W_1} \varphi V_2 (E - E_1) dq_{(1)}^1 \dots dp_{(1)n_1},
\end{aligned}
$$

since φ is the characteristic function of W that vanishes when $(\mathbf{x}_1, \mathbf{x}_2) \notin W$, that is, if $\mathbf{x}_1 \notin W_1$.

Noticing that W_1 is a set independent of the value E, the differentiation and integration commute and then, when we recall (21.69), we obtain

$$P_1 (W_1) = \frac{1}{w (E)} \int_{W_1} w_2 (E - E_1) dV_1, \tag{21.92}$$

where $dV_1 = dq_{(1)}^1 \dots dp_{(1)n_1}$ is the elementary volume of $M_{2n_1}^*$.

Further, the probability density

$$p_1 (E_1) = \frac{w_2 (E - E_1)}{w (E)}$$

appearing into (21.92) can also be written in the form

$$p_1 (E_1) = AG (E_1), \tag{21.93}$$

where

$$A = \frac{w_2 (E)}{w (E)}, \quad G (E_1) = \frac{w_2 (E - E_1)}{w_2 (E)}. \tag{21.94}$$

We conclude that *the probability density of S_1 is a function of the total energy E_1 of S_1 whose form depends on the nature of S_2. Resorting to* (21.64), *formula* (21.94)$_2$ becomes

$$G(E_1) = \frac{1}{\omega_2(E)} \chi_n \int_{C_{2,E-E_1}} (E - E_1 - U_2)^{\frac{n_2}{2}-1} \sqrt{a_2} dq_{(2)}^1 \dots dq_{(2)}^{n_2},$$

where

$$C_{2,E-E_1} = \left\{ \mathbf{x}_2 \in M_{2n_2}^*, 0 \leq U_2(\mathbf{x}_2) \leq E - E_1 \right\}.$$

Therefore, $G(0) = 1$ and $G(E) = 0$.

It can also be proved that all the derivatives of order less or equal to $(n_2/2 - 1)$ vanish when $E_1 = E$ and that the function $G(E_1)$ is decreasing and differentiable.

Now we show that if $n_2 \gg n_1$, $G(E_1)$ can be approximated by an exponential function. First, we have

$$G(E_1 + \Delta E_1) = \frac{1}{\omega_2(E)} \left[\int_{C_{2,E-E_1}} (E - E_1 + \Delta E_1 - U_2)^{\frac{n_2}{2}-1} \sqrt{a_2} dq_{(2)}^1 \dots dq_{(2)}^{n_2} \right.$$

$$\left. + \Delta E_1 \int_{\partial C_{2,E-E_1}} (E - E_1 - U_2)^{\frac{n_2}{2}-1} \sqrt{a_2} d\Sigma_{(2)} \right],$$

where $d\Sigma_{(2)}$ is the surface element of $\partial C_{2,E-E_1}$. Since at any point of $\partial C_{2,E-E_1}$ it is $U_2 = E - E_2$ and the last term of the above equation vanishes identically, we can write

$$-\frac{G(E_1)}{G'(E_1)} = \frac{1}{\frac{n_2}{2} - 1} \frac{\displaystyle\int_{C_{2,E-E_1}} (E - E_1 - U_2)^{\frac{n_2}{2}-1} \sqrt{a_2} dq_{(2)}^1 \dots dq_{(2)}^{n_2}}{\displaystyle\int_{C_{2,E-E_1}} (E - E_1 - U_2)^{\frac{n_2}{2}-2} \sqrt{a_2} dq_{(2)}^1 \dots dq_{(2)}^{n_2}}.$$

This relation shows that the ratio on the left-hand side is a function of the two variables $E - E_1$ and n_2, that is $(E - E_1)/n_2$ and $1/n_2$

$$-\frac{G(E_1)}{G'(E_1)} = \psi\left(e - \frac{E_1}{n_2}, \frac{1}{n_2}\right), \tag{21.95}$$

where $e = E/n_2$.

When $n_2 \to \infty$ in such a way that e remains constant, (21.95) becomes

$$-\frac{G(E_1)}{G'(E_1)} = \beta(E),$$

and (21.93) writes

$$p_1(E_1) = Ae^{-\frac{E_1}{\beta}}, \tag{21.96}$$

where A and β do not depend on E_1.

Equation (21.96) gives the Boltzmann canonical probability distribution. The constant A is determined imposing the condition

$$\int_{V_{1,E}} p_1(E_1)\,dV_1 = 1,$$

that, in turn implies

$$A = \frac{1}{\displaystyle\int_{V_{1,E}} e^{-\frac{E_1}{\beta}}\,dV_1}. \tag{21.97}$$

If we proceed as we did to prove the Maxwell distribution, we again obtain that $\beta = k\theta$, with θ absolute temperature of S. Finally, the average of any function $F(\mathbf{x}_1)$ on $V_{1,E}$ can be evaluated by the formula

$$\bar{F} = A \int_{V_{1,E}} F e^{-\frac{H_1(\mathbf{x}_1)}{k\theta}}\,dV_1. \tag{21.98}$$

Chapter 22
Impulsive Dynamics

Abstract In this chapter, we formulate the equations of impulsive dynamics starting from the balance equations of momentum and angular momentum of rigid bodies. Then, the Lagrangian formulation of the impulsive dynamics is developed. Exercises conclude the chapter.

22.1 Balance Equations of Impulsive Dynamics

If $f(t)$ is a function of time, then we denote by

$$\Delta f = f(t_2) - f(t_1) \tag{22.1}$$

the variation of $f(t)$ in the time interval $[t_1, t_2]$.

Let $\mathcal{B}$ be a rigid body acted upon by the forces $(P_i, \mathbf{F}_i)$, $i = 1, \ldots, s$, where the force $\mathbf{F}_i$ is applied at the point P_i belonging to the region C occupied by $\mathcal{B}$. Integrating the balance equations (15.48) and (15.49) in the time interval $[t_1, t_2]$, we obtain

$$m\Delta \dot{\mathbf{r}}_G = \sum_{i=1}^{s} \int_{t_1}^{t_2} \mathbf{F}_i \, dt, \tag{22.2}$$

$$\Delta \mathbf{K}_O = \sum_{i=1}^{s} \int_{t_1}^{t_2} (P_i - O) \times \mathbf{F}_i \, dt, \tag{22.3}$$

where O is a fixed point or the center of mass G of $\mathcal{B}$. The integral

$$\mathbf{I}_i = \int_{t_1}^{t_2} \mathbf{F}_i \, dt \tag{22.4}$$

is called the **impulse** of the force $\mathbf{F}_i$ in the time interval $[t_1, t_2]$.

© Springer International Publishing AG, part of Springer Nature 2018
A. Romano and A. Marasco, *Classical Mechanics with Mathematica®*,
Modeling and Simulation in Science, Engineering and Technology,
https://doi.org/10.1007/978-3-319-77595-1_22

Definition 22.1. We say that the motion of the rigid body $\mathcal{B}$ is *impulsive* in the *short time interval* $[t_1, t_2]$ if in this interval there is a finite variation in the velocity field of $\mathcal{B}$ that is not accompanied by an appreciable variation in its position. When this happens, the short time interval $[t_1, t_2]$ is called an *exceptional interval*.

A typical example of an impulsive motion is given by the collision between two billiard balls. In this case, the exceptional interval coincides with the time interval during which the two balls are in contact with each other (a fraction of a second).

Definition 22.2. The velocity fields of the body $\mathcal{B}$ at the instants t_1 and t_2 are called the *anterior velocity field* and the *posterior velocity field*, respectively.

Impulsive dynamics is based on the following assumptions:

1. An exceptional time interval $[t_1, t_2]$ can be identified with the initial instant t_1. Then, such an instant is called an *exceptional instant*.
2. During an exceptional interval, the body $\mathcal{B}$ remains at the position it occupies at the initial instant t_1.

In the preceding assumptions, any quantity depending on the velocity field of $\mathcal{B}$ generally exhibits a finite discontinuity at an exceptional instant. This implies that in analyzing an impulsive motion we are forced to drop the hypothesis of regularity of the velocity field. On the other hand, assumption 1 requires particular behavior of the forces acting during the exceptional time interval $[t_1, t_2]$. In fact, in view of (22.2) and (22.4), a variation in the velocity field is compatible with a finite impulse of the acting forces only if these forces assume very large values in the time interval $[t_1, t_2]$. Forces having this characteristic are called *impulsive forces*.

Before proceeding, we make clear the preceding considerations using a simple example. A material point P of mass m that is constrained to move along the axis Ox is at rest at the initial time $t = 0$. In the time interval $[0, T]$ point P is acted upon by the sinusoidal force

$$F = A \sin \frac{\pi}{T} t, \tag{22.5}$$

so that in $[0, T]$ the velocity and position of P are given by the functions

$$\dot{x}(t) = \frac{AT}{\pi m} \left(1 - \cos \frac{\pi t}{T} \right), \tag{22.6}$$

$$x(t) = \frac{AT}{\pi m} t - \frac{AT^2}{\pi^2 m} \sin \frac{\pi t}{T}. \tag{22.7}$$

Further, since the impulse of the force F is

$$I = \int_0^T A \sin \frac{\pi}{T} t = \frac{2AT}{\pi}, \tag{22.8}$$

we obtain a constant value $2\alpha/\pi$ of I for any T, provided that we choose $A = \alpha/T$, that is, if we increase the applied force when the time interval $(0, T)$ reduces to zero. In the limit $T \to 0$, formulae (22.6) and (22.7) give

$$\Delta \dot{x} = \frac{2\alpha}{\pi m} = \frac{I}{m}, \quad \Delta x = 0. \tag{22.9}$$

Impulsive forces, as ordinary forces, can be active and reactive. In the first case, they are real impulsive forces applied to $\mathcal{B}$ at an exceptional instant. In the second case, they are produced by the introduction of new constraints at an exceptional instant. An example of a reactive impulsive force is a ball bouncing on the floor.

From (22.2) and (22.3) and assumptions 1 and 2 we can write the **balance equations of impulsive dynamics** of a rigid body $\mathcal{B}$ in the following form:

$$m\Delta \dot{\mathbf{r}}_G = \sum_{i=1}^{s} \mathbf{I}_i, \tag{22.10}$$

$$\Delta \mathbf{K}_O = \sum_{i=1}^{s} (P_i - O) \times \mathbf{I}_i, \tag{22.11}$$

where $\mathbf{I}_i$, $i = 1, \ldots, s$, are the *active* and *reactive* impulses acting on the body $\mathcal{B}$ at an exceptional instant.

For a system $\mathcal{S}$ of N rigid bodies $\mathcal{B}_i$, $i = 1, \ldots, N$, we must apply the preceding equations to any solid $\mathcal{B}_i$, recalling that the active and reactive impulses acting on $\mathcal{B}_i$ are external to $\mathcal{B}_i$. The reactive impulses can be external to $\mathcal{S}$ or determined by other solids of $\mathcal{S}$. In this last case, they satisfy the action and reaction principle.

In conclusion, if $\mathcal{S}$ is a constrained system of N rigid bodies $\mathcal{B}_1, \ldots, \mathcal{B}_N$, then for each body we can write

$$m\Delta \dot{\mathbf{r}}_{G_i} = \sum_{j=1}^{s_i} \mathbf{I}_{ij} \equiv \mathbf{I}_i, \tag{22.12}$$

$$\Delta \mathbf{K}_{O_i} = \sum_{j=1}^{s_i} (P_j - O_i) \times \mathbf{I}_{ij} \equiv \mathbf{M}_{O_i}^{(\mathrm{I})}, \tag{22.13}$$

where s_i denotes the number of impulses applied to the body $\mathcal{B}_i$, and $\mathbf{I}_i$ and $\mathbf{M}_{O_i}^{(\mathrm{I})}$ are, respectively, the total impulse and the total impulse torque of the active and reactive impulses acting on $\mathcal{B}_i$.

22.2 Fundamental Problem of Impulsive Dynamics

Let $S = \{B_1, \ldots, B_N\}$ be a system of constrained rigid bodies and suppose that, at the exceptional instant t, S is subject to active and reactive impulses. We make the following assumptions:

- Active and reactive impulses act on points of the bodies B_i of S.
- The constraints are smooth.
- The reactive impulses are due to *bilateral or unilateral* constraints introduced at the exceptional instant.
- In the collision between two solids of S or between a solid and an exterior obstacle, the reactive impulses verify the action and reaction principle.

The first assumption simplifies the analysis of the problem we are faced with. The second assumption implies that the reactive impulse acting at the point $P_j \in B_i$ is orthogonal to the boundary of the region C_i occupied by B_i at the exceptional instant. Regarding the third assumption we note that an example of a unilateral constraint is given by the collision of a body of S with an external obstacle or with another body of S. A bilateral constraint can be introduced by fixing one or two points of a body of S.

With the preceding assumptions we can formulate the fundamental problem of impulsive dynamics of constrained rigid bodies as follows:
Given the position of S, its anterior velocity, the active impulses acting on S, and the smooth constraints at an exceptional instant, determine the posterior velocity field of S.

To solve this problem we have at our disposal:

1. Equations (22.12) and (22.13) applied to each body B_i of S;
2. The restrictions on the reactive impulses deriving from the hypothesis that the constraints are smooth.

Simple examples (see Exercises 1–3 at the end of the chapter) show that the preceding data are not sufficient to determine the posterior velocity field in the presence of unilateral constraints, i.e., when at an exceptional instant there are collisions between bodies of S or between a body of S and an external obstacle. To solve the fundamental problem of the impulsive dynamics of rigid bodies, we need to add *Newton's laws on collisions between solids*. To formulate these laws, consider two solids B_i and B_j of S and denote by P_i and P_j the contact points of the bodies B_i and B_j colliding at an exceptional instant. Further, we denote by $\mathbf{n}_i$ the internal normal vector to the boundary of B_i at the point P_i and by $\mathbf{n}_j$ the internal normal vector to the boundary of B_j at the point P_j (Fig. 22.1). It is evident that $\mathbf{n}_i = -\mathbf{n}_j$. Finally, we denote by $\dot{\mathbf{r}}_i^-$ and $\dot{\mathbf{r}}_i^+$ the anterior and posterior speeds of P_i, respectively, that is, the velocities of P_i before and after the exceptional instant. Similarly, we denote by $\dot{\mathbf{r}}_j^-$ and $\dot{\mathbf{r}}_j^+$ the speeds of P_j before and after the collision. The two quantities

$$w_{ij}^- = \mathbf{n}_i \cdot (\dot{\mathbf{r}}_j^- - \dot{\mathbf{r}}_i^-), \quad w_{ij}^+ = \mathbf{n}_i \cdot (\dot{\mathbf{r}}_i^+ - \dot{\mathbf{r}}_j^+) \tag{22.14}$$

Fig. 22.1 Collision between
two solids

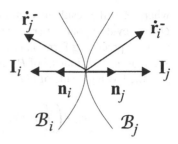

are called the ***speed of approach*** and the ***speed of separation*** of the two solids $\mathcal{B}_i$
and $\mathcal{B}_j$, respectively. Then, Newton's law states that:

*There exists a positive number $e_{ij} \in [0, 1]$, called the **coefficient of restitution**
and depending on the physical constitution of $\mathcal{B}_i$ and $\mathcal{B}_j$, such that*

$$w_{ij}^+ = e_{ij} w_{ij}^-. \tag{22.15}$$

The collision is said to be *perfectly elastic* if $e_{ij} = 1$ and *perfectly inelastic* if $e_{ij} = 0$.
It is evident that if the body $\mathcal{B}_i$ collides with an external obstacle, then Newton's laws
can still be applied. However, in this case, the speed $\dot{\mathbf{r}}_j$ in (22.14) coincides with the
velocity of the obstacle at the collision point with $\mathcal{B}_i$.

When we consider condition (22.15), for *any* contact between solids of S or of a
solid of S with an external obstacle, (22.12) and (22.13), and the restrictions on the
reactive impulses deriving from the hypothesis that the constraints are smooth, then
we obtain a system of equations that allow us to solve the fundamental problem of
impulsive dynamics. We verify this statement by solving some exercises at the end of
the chapter. The theoretical proof of the preceding statement requires the Lagrangian
formalism and it will be given in the next sections.

22.3 Lagrangian Formulation of Impulsive Dynamics

Let $\mathbf{u}(\mathbf{r})$ be an arbitrary vector field depending on the position vector $\mathbf{r}$ of a point
belonging to region C. Denote by

$$\delta \mathbf{r} = \delta \mathbf{r}_O + \delta \boldsymbol{\varphi} \times (\mathbf{r} - \mathbf{r}_O) \tag{22.16}$$

an infinitesimal rigid displacement of region C, where $\delta \mathbf{r}_O$ is the infinitesimal displacement of an arbitrary point $O \in C$ and $\delta \boldsymbol{\varphi}$ the infinitesimal rotation of C. The
following formula holds

$$\int_C \mathbf{u}(\mathbf{r}) \cdot \delta \mathbf{r} \, dC = \mathbf{R} \cdot \delta \mathbf{r}_O + \mathbf{M}_O \cdot \delta \boldsymbol{\varphi}, \tag{22.17}$$

where $\mathbf{R}$ is the resultant vector of the field $\mathbf{u}(\mathbf{r})$ and $\mathbf{M}_O$ its torque with respect to O. It is evident that if $\mathbf{u}(\mathbf{r})$ is only defined at a finite set of points $P_1, \ldots, P_s$ of C, then the preceding formula is valid provided that the integral is substituted with the summation over index i of P_i. To prove (22.17), it is sufficient to introduce (22.16) into the left-hand side of (22.17) to obtain

$$\delta\mathbf{r}_O \cdot \left(\int_C \mathbf{u}(\mathbf{r})dC \right) + \delta\varphi \cdot \left(\int_C (\mathbf{r} - \mathbf{r}_O) \times \mathbf{u}(\mathbf{r})dC \right),$$

and (22.17) is proved.

Now let $\mathcal{S} = \{\mathcal{B}_1, \ldots, \mathcal{B}_N\}$ be a system of N rigid bodies acted upon by forces satisfying the hypotheses of impulsive dynamics. Denote by C_i the region occupied by the body $\mathcal{B}_i$ at the exceptional instant $\overline{t}$ and ρ_i its mass density. The total force and torque of the vector field $\rho_i \Delta\dot{\mathbf{r}}dC$, that denotes the variation of momentum of the region $dC \subset C_1 \cup \cdots \cup C_N$ at the instant $\overline{t}$, are

$$\sum_{i=1}^{N} m_i \Delta\dot{\mathbf{r}}_{G_i}, \quad \sum_{i=1}^{N} \Delta\mathbf{K}_{G_i}, \tag{22.18}$$

where m_i is the mass of $\mathcal{B}_i$. By (22.17), for any infinitesimal rigid displacement of all the bodies of $\mathcal{S}$,[1] we obtain the result

$$\sum_{i=1}^{N} \int_{C_i} \rho_i \Delta\dot{\mathbf{r}} \cdot \delta\mathbf{r} \, dC = \sum_{i=1}^{N} \left(\delta\mathbf{r}_{G_i} \cdot m_i \Delta\dot{\mathbf{r}}_{G_i} + \delta\varphi_i \cdot \Delta\mathbf{K}_{G_i} \right)$$

$$= \sum_{i=1}^{N} \left(\delta\mathbf{r}_{G_i} \cdot \mathbf{I}_i + \delta\varphi_i \cdot \mathbf{M}_{G_i}^{(\mathrm{I})} \right). \tag{22.19}$$

This result allows us to conclude that (22.12) and (22.13) are equivalent to the condition

$$\sum_{i=1}^{N} \int_{C_i} \rho_i \Delta\dot{\mathbf{r}} \cdot \delta\mathbf{r} \, dC - \sum_{i=1}^{N} \sum_{j=1}^{s_i} \mathbf{I}_{ij} \cdot \delta\mathbf{r}_{ij} = 0. \tag{22.20}$$

In (22.20) $\delta\mathbf{r}$ is an arbitrary infinitesimal displacement that is rigid for any body $\mathcal{B}_i$ of $\mathcal{S}$ and $\delta\mathbf{r}_{ij}$ is the corresponding displacement of the point $\mathbf{r}_{ij}$ at which the impulse $\mathbf{I}_{ij}$ is applied at the exceptional instant.

Definition 22.3. An infinitesimal *rigid* displacement of $\mathcal{B}_i$ is **virtual** if it is compatible with the constraints of $\mathcal{S}$ at an exceptional instant.

---

[1] We explicitly note that these displacements could not satisfy the constraints of $\mathcal{S}$.

Although (22.20), for this class of displacements, is no longer equivalent to (22.12) and (22.13), it leads to a condition that supplies the Lagrangian formulation of the impulsive dynamics.

In fact, let V_n be the configuration space of the constrained system S at the exceptional instant $\bar{t}$, and denote by $q^1, \ldots, q^n$ arbitrary Lagrangian coordinates of V_n. Then the position vector $\mathbf{r}$ of any point of S in its configuration at the instant $\bar{t}$ is a function of the Lagrangian coordinates (Sect. 17.3), and any infinitesimal displacement

$$\delta \mathbf{r} = \sum_{h=1}^{n} \frac{\partial \mathbf{r}}{\partial q^h} \delta q^h \qquad (22.21)$$

is compatible with all the constraints of S, i.e., it is a virtual displacement. Introducing (22.21) into (22.20), we obtain the condition

$$\sum_{h=1}^{n} \left(\sum_{i=1}^{N} \int_{C_i} \rho_i \Delta \dot{\mathbf{r}} \cdot \frac{\partial \mathbf{r}}{\partial q^h} dC \right) \delta q^h = \sum_{h=1}^{n} I_h \delta q^h, \qquad (22.22)$$

where

$$I_h = \sum_{i=1}^{N} \sum_{j=1}^{s_i} \mathbf{I}_{ij} \cdot \left(\frac{\partial \mathbf{r}}{\partial q^h} \right)_{r_{ij}} \qquad (22.23)$$

are the **Lagrangian components of the impulses**.

Since with (17.45) we proved that

$$\frac{\partial \mathbf{r}}{\partial q^h} = \frac{\partial \dot{\mathbf{r}}}{\partial \dot{q}^h},$$

we can write

$$\sum_{i=1}^{N} \int_{C_i} \rho_i \Delta \dot{\mathbf{r}} \cdot \frac{\partial \dot{\mathbf{r}}}{\partial \dot{q}^h} dC = \Delta \frac{\partial T}{\partial \dot{q}^h} = \Delta p_h, \qquad (22.24)$$

where T is the kinetic energy of S and p_h, $h = 1, \ldots, n$, are the momenta of S (Chap. 17). Therefore, (22.22) can be written as follows:

$$\sum_{h=1}^{n} \Delta p_h \delta q^h = \sum_{h=1}^{n} I_h \delta q^h. \qquad (22.25)$$

22.4   Analysis of the Lagrange Equations

The impulses acting on the system $\mathcal{S}$ at the exceptional instant $\bar{t}$ are active and reactive. Further, the reactive impulses can be determined by bilateral or unilateral constraints. In other words, we can decompose I_h into the summation

$$I_h = I_h^a + I_h^{(rb)} + I_h^{(ru)}, \qquad (22.26)$$

where $I_h^{(a)}$, $h = 1, \ldots, n$, are the Lagrangian components of the active impulses, $I_h^{(rb)}$ are the Lagrangian components of the reactive impulses produced by bilateral constraints, and $I_h^{(ru)}$ are the Lagrangian components of the reactive impulses generated by unilateral constraints. On the other hand, if the constraints are smooth, $\mathbf{I}_i^{(rb)}$ is orthogonal to $\delta\mathbf{r}_i$, $i = 1, \ldots, N$, and all the corresponding Lagrangian components vanish.

Consequently, in the absence of unilateral constraints the variations δq_h are arbitrary and (22.25) gives the system

$$\Delta p_h = I_h^{(a)}, \quad h = 1, \ldots, n, \qquad (22.27)$$

that allows us to determine the quantities p_h^+ when the active impulses and the values p_h^- are known.

We now consider the case of unilateral constraints. First we note that these constraints do not modify the degree of freedom of the system $\mathcal{S}$. Therefore, we can eliminate unilateral constraints at the exceptional instant provided that we substitute them with the reactive impulses they generate. In this way, we can apply the above consideration relating to the case of bilateral constraints and the quantities δq_h can again be taken arbitrarily. Then, we have

$$\Delta p_h = I_h^{(a)} + I_h^{(ru)}, \quad h = 1, \ldots, n. \qquad (22.28)$$

It is important to remark that (22.28) do not allow to determine the quantities p_h^+ since the reactive impulses $I_h^{(ru)}$ due to the unilateral constraints are unknown.

Equations (22.27) and (22.28) are the ***Lagrangian equations of impulsive dynamics for bilateral constraints and for bilateral and unilateral constraints***, respectively.

Recalling that the Lagrangian velocities $\dot{q}^h$ and the kinetic momenta p_h are related by the relations (Chap. 17)

$$\dot{q}^h = a^{hk} p_k, \qquad (22.29)$$

we can conclude that (22.27) solve the fundamental problem of impulsive dynamics, in the absence of unilateral constraints, i.e., in the absence of collisions. We want to show that *we can solve the fundamental problem of impulsive dynamics also in the presence of unilateral smooth constraints, provided that we add Newton's laws of collision to the Eqs.* (22.28). For the sake of simplicity, we suppose that only the bodies $\mathcal{B}_i$ and $\mathcal{B}_j$ of the system $\mathcal{S}$ collide at the instant $\bar{t}$, when they occupy regions C_i and C_j, respectively. Denote by $P_i \in C_i$ and $P_j \in C_j$ the collision points. Then, since the constraints are smooth, the reactive impulses at the contact points are parallel to the internal unit normal vectors $\mathbf{n}_i$ and $\mathbf{n}_j$ to the surfaces of C_i and C_j, respectively. Further, these impulses verify the action and reaction principle and $\mathbf{n}_i = -\mathbf{n}_j$. In view of these remarks, we can write that

$$I_h^{(ru)} = I_i^{(ru)} \mathbf{n}_i \cdot \frac{\partial \mathbf{r}_i}{\partial q^h} + I_j^{(ru)} \mathbf{n}_j \cdot \frac{\partial \mathbf{r}_j}{\partial q^h} = I_i^{(ru)} \mathbf{n}_i \cdot \left(\frac{\partial \mathbf{r}_i}{\partial q^h} - \frac{\partial \mathbf{r}_j}{\partial q^h} \right). \tag{22.30}$$

Introducing (22.30) into (22.27), we obtain the system

$$p_h^+ = p_h^- + I_h^{(a)} + I_i^{(ru)} \mathbf{n}_i \cdot \left(\frac{\partial \mathbf{r}_i}{\partial q^h} - \frac{\partial \mathbf{r}_j}{\partial q^h} \right), \quad h = 1, \ldots, n. \tag{22.31}$$

System (22.31) shows that the collision between $\mathcal{B}_i$ and $\mathcal{B}_j$ introduces the only unknown $I_i^{(ru)}$. Therefore, we need a further equation to equate the number of equations and unknowns. In view of the Lagrangian formula of the Lagrangian velocity,

$$\dot{\mathbf{r}} = \frac{\partial \mathbf{r}}{\partial q^h} \dot{q}^h, \tag{22.32}$$

(22.14) and (22.29), Newton's law of collision (22.15) leads to the following equation:

$$\mathbf{n}_i \cdot \left(\frac{\partial \mathbf{r}_i}{\partial q^h} - \frac{\partial \mathbf{r}_j}{\partial q^h} \right) a^{hk} p_k^+ = e_{ij} \mathbf{n}_i \cdot \left(\frac{\partial \mathbf{r}_j}{\partial q^h} - \frac{\partial \mathbf{r}_i}{\partial q^h} \right) a^{hk} p_k^-, \tag{22.33}$$

which, together with (22.31), gives a system to determine the $n+1$ unknowns p_h^+, $h = 1, \ldots, n$, and $I_i^{(ru)}$. It is evident that the preceding considerations can be extended to the case of more collisions.

22.5 Smooth Collision of Two Solids

In this section we consider the smooth collision of two solids as an example of impulsive motion due to unilateral constraints. In this case we need Newton's law to determine the posterior speeds of the solids. Let $\mathcal{S}_1$ and $\mathcal{S}_2$ be the two solids colliding at the exceptional instant t_1 and denote by P_1 and P_2 the points of $\mathcal{S}_1$ and $\mathcal{S}_2$, respectively. Finally, let π be the tangent plane to the boundaries of $\mathcal{S}_1$ and $\mathcal{S}_2$ and let $\mathbf{n}_1$ and $\mathbf{n}_2 \equiv \mathbf{n}$ be the unit normals to π at the contact point $P_1 \equiv P_2$ (see Fig. 22.1).

Because of the absence of friction, the reactive impulse $\mathbf{I}_1$ due to the action of $\mathcal{S}_2$ on $\mathcal{S}_1$ and the reactive impulse $\mathbf{I}_2$ due to the action of $\mathcal{S}_1$ on $\mathcal{S}_2$ can be written as follows:

$$\mathbf{I}_1 = I\mathbf{n}_i, \tag{22.34}$$

where I is the unknown intensity that is the same for the two impulses since for the action and reaction principle it is $\mathbf{I}_1 = -\mathbf{I}_2$. Equation (22.12), when it is applied to each solid, gives

$$m_i \Delta \dot{\mathbf{r}}_{G_i} = I\mathbf{n}_i, \tag{22.35}$$

where m_i is the mass of S_i. Projecting $\dot{\mathbf{r}}_{G_i}$ along π and $\mathbf{n}$ we have

$$\dot{\mathbf{r}}_{G_i} = \tau_i + v_i \mathbf{n}, \tag{22.36}$$

so that from 22.35 we obtain

$$\Delta \tau_i = \mathbf{0}, \quad m_2 \Delta v_2 = -m_1 \Delta v_1 = I. \tag{22.37}$$

The first of the above equation shows that in the collision the tangential velocities of the center of mass of two solids do not change, whereas the second equation gives

$$m_2 \Delta v_2 + m_1 \Delta v_1 = 0. \tag{22.38}$$

From (22.13) applied to each solid we obtain

$$\Delta \mathbf{K}_i = I(P_i - G_i) \times \mathbf{n}_i, \tag{22.39}$$

where $\mathbf{K}_i$ is the angular momentum of S_i with respect to G_i. Let $G_i x_i y_i z_i$ be a central frame of inertia of S_i. Then, the components of $\mathbf{K}_i$ along the axes of $G_i x_i y_i z_i$ are

$$K_{x_i}^{(i)} = A_i p_i, \quad K_{x_i}^{(i)} = B_i q_i, \quad K_{x_i}^{(i)} = C_i r_i, \quad i = 1, 2, \tag{22.40}$$

where A_i, B_i, C_i are the central moments of inertia of S_i in the frame $G_i x_i y_i z_i$, and p_i, q_i, r_i are the components in the same frame of angular velocity ω_i of S_i. Inserting (22.40) in (22.39) we can express the variations of angular velocities ω_1 and ω_2 in the collision in terms of the unknown impulse I.

It is evident that to solve the problem we need to resort to Newton's law. To simplify the calculation we limit our attention to the case of *central collision*, i.e., when the centers of mass G_1 and G_2 of the solids belong to the normal to π at the contact point. This happens, for instance, in the collision between two homogeneous spheres. In this condition (22.39) reduces to

$$\Delta \mathbf{K}_i = \mathbf{0}, \tag{22.41}$$

and, in view of (22.40), we can state that the angular velocities of S_1 and S_2 do not vary in a central collision.

It remains to apply Newton's law to determine the variations Δv_1 and Δv_2 of the central velocities along $\mathbf{n}$. This law gives

$$\mathbf{n} \cdot (\dot{\mathbf{r}}_2^+ - \dot{\mathbf{r}}_1^+) = e\mathbf{n} \cdot (\dot{\mathbf{r}}_1^- - \dot{\mathbf{r}}_2^-), \tag{22.42}$$

where $\dot{\mathbf{r}}_2^+, \dot{\mathbf{r}}_1^+$ and $\dot{\mathbf{r}}_1^-, \dot{\mathbf{r}}_2^-$ denote the posterior and anterior speeds of P_1 and P_2, respectively. In the central collision we have

$$\dot{\mathbf{r}}_i = \dot{\mathbf{r}}_{G_i} + \boldsymbol{\omega}_i \times (P_i - G_i), \quad i = 1, 2, \tag{22.43}$$

so that, projecting along $\mathbf{n}$ we obtain

$$v_i = \dot{\mathbf{r}}_i \cdot \mathbf{n} = \dot{\mathbf{r}}_{G_i} \cdot \mathbf{n}, \quad i = 1, 2. \tag{22.44}$$

In conclusion, (22.44), (22.42), (22.38) lead to the system

$$\begin{cases} m_1 v_1^+ + m_2 v_2^+ = m_1 v_1^- + m_2 v_2^-, \\ (v_2^+ - v_1^+) = e(v_1^- - v_2^-), \end{cases} \tag{22.45}$$

from which we easily obtain

$$\begin{cases} m v_1^+ = (m_1 - e m_2) v_1^- + (1 + e) m_2 v_2^-, \\ m v_2^+ = (1 + e) m_1 v_1^- + (m_2 - e m_1) v_2^-, \end{cases} \tag{22.46}$$

where $m = m_1 + m_2$.

In particular, if the collision is anelastic, i.e. $e = 0$, (22.46) give

$$v_1^+ = v_2^+ = \frac{m_1 v_1^- + m_2 v_2^-}{m}, \tag{22.47}$$

and we can state that *in the central collision of two anelastic solids the normal components of central speeds are both equal to the normal component of the central speed of the system of two solids.*

Differently, when the collision is elastic, that is $e = 1$, and $m_1 = m_2$ we obtain

$$v_1^+ = v_2^-, \quad v_2^+ = v_1^-, \tag{22.48}$$

and we conclude that *in the central collision of two elastic solids with the same mass the normal components of central speeds exchange their values.*

In particular, if we consider a direct collision, that is a collision in which the tangential components of $\dot{\mathbf{r}}_{G_i}$ vanishes, the centers of mass G_1 and G_2 exchange their values so that if G_2 is at rest before the collision, then G_1 will be at rest after the collision.

Fig. 22.2 Homogeneous
disk on a plane

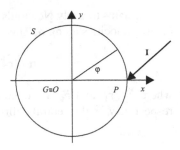

22.6 Exercises

1. Let S be a homogeneous circular disk at rest on a horizontal plane π. Suppose
   that at the exceptional instant t an active impulse $\mathbf{I}$, parallel to π, is applied at a
   point P on the boundary of S. Determine the posterior speed field assuming that
   π is smooth.
   We choose the Cartesian axes as in Fig. 22.2. First, we solve the problem by
   cardinal equations that, with our notations, give

$$
\begin{aligned}
m\Delta\dot{x}_G &= I_x, \\
m\Delta\dot{y}_G &= I_y, \\
C\Delta\dot{\varphi} &= rI_y,
\end{aligned}
$$

   where m is the mass of S, C the moment of inertia relative to the Oz-axis, and
   r the radius of S. Since $\dot{x}_G^- = \dot{y}_G^- = \dot{\varphi}^- = 0$, from the preceding equations we
   obtain

$$
\dot{x}_G^+ = \frac{I_x}{m},
$$

$$
\dot{y}_G^+ = \frac{I_y}{m},
$$

$$
\dot{\varphi}^+ = \frac{rI_y}{C}.
$$

2. Solve the preceding problem using the Lagrangian formalism.
   We choose as Lagrangian coordinates x_G, y_G, and φ, where x_G and y_G are Cartesian
   coordinates with the origin at a point $O \neq G$. Then the position vector of any point
   of S is

$$
\mathbf{r} = x_G\mathbf{i} + y_G\mathbf{j} + \rho(\cos\varphi\,\mathbf{i} + \sin\varphi\,\mathbf{j}), \tag{22.49}
$$

   where $\mathbf{i}$ and $\mathbf{j}$ are the unit vectors along the Cartesian axes and $0 \leq \rho \leq r$. Further,
   the kinetic energy of S can be written as

Fig. 22.3 Collision of a
homogeneous disk

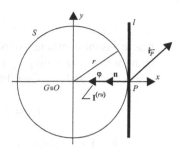

$$T = \frac{1}{2}m\left(\dot{x}_G^2 + \dot{y}_G^2\right) + \frac{1}{2}C\dot{\varphi}^2. \tag{22.50}$$

First, verify that

$$p_{x_G} = m\dot{x}_G, \quad p_{y_G} = m\dot{y}_G, \quad p_\varphi = C\dot{\varphi},$$

and the Lagrangian components of the active impulse are

$$I_{x_G} = \mathbf{I} \cdot \left(\frac{\partial \mathbf{r}}{\partial x_G}\right)_P = \mathbf{I} \cdot \mathbf{i} = I_x,$$

$$I_{y_G} = \mathbf{I} \cdot \left(\frac{\partial \mathbf{r}}{\partial y_G}\right)_P = \mathbf{I} \cdot \mathbf{j} = I_y,$$

$$I_\varphi = \mathbf{I} \cdot \left(\frac{\partial \mathbf{r}}{\partial \varphi}\right)_P = \mathbf{I} \cdot (-\rho \sin \varphi \mathbf{i} + \rho \cos \varphi \mathbf{j})_P = r I_y,$$

since at the point P it is $\rho = r$ and $\varphi = 0$.
Applying (22.27), we obtain the same equations we derived from the balance equations.

3. Let S be a homogeneous circular disk of radius r and mass m moving on a horizontal smooth plane π. Suppose that at instant t, disk S collides with a vertical plane whose intersection with π is the straight line l. Determine the posterior velocity field knowing the velocity field before the collision.

We refer to Fig. 22.3. The balance equations of impulsive dynamics give

$$m\Delta \dot{x}_G = -I^{(ru)}, \quad m\Delta \dot{y}_G = 0, \quad C\Delta \dot{\varphi} = 0,$$

where C is the moment of inertia relative to the axis orthogonal to π at point G. Equivalently, we have that

$$\dot{x}_G^+ = \dot{x}_G^- - \frac{I^{(ru)}}{m}, \quad \dot{y}_G^+ = \dot{y}_G^-, \quad \dot{\varphi}^+ = \dot{\varphi}^-. \tag{22.51}$$

On the other hand, since the velocity of the point $P \in l$ is zero, Newton's collision law (22.15) gives

$$\mathbf{n} \cdot \dot{\mathbf{r}}_P^+ = -e\mathbf{n} \cdot \dot{\mathbf{r}}_P^-, \tag{22.52}$$

where e is the coefficient of restitution relative to the collision between S and the straight line l. Denoting by $\mathbf{i}$ and $\mathbf{j}$ the unit vectors along the Ox- and Oy-axes, respectively, and recalling that

$$\dot{\mathbf{r}}_P = \dot{x}_G\mathbf{i} + \dot{y}_G\mathbf{j} + \dot{\varphi}\mathbf{k} \times (x_G\mathbf{i} + y_G\mathbf{j}),$$

where $k = \mathbf{i} \times \mathbf{j}$, relation (22.52), in view of (22.51), can also be written as

$$\dot{x}_G^+ = -e\dot{x}_G^-. \tag{22.53}$$

By this equation and (22.51), we solve the problem of determining the posterior speed of S.

4. Solve Problem 3 adopting the Lagrangian formalism.
   Introduce a new Cartesian frame Oxy, $O \neq G$, with axes parallel to the axes adopted in Fig. 22.3. Then, x_G, y_G, and φ are Lagrangian coordinates for S and the position vector $\mathbf{r}$ of the points of S and the kinetic energy T are given by (22.49) and (22.50), respectively. Consequently, we have that

$$p_{x_G} = m\dot{x}_G, \ p_{y_G} = m\dot{y}_G, \ p_\varphi = C\dot{\varphi}.$$

On the other hand, since $\mathbf{n} = -\mathbf{i}$ and the straight line l is at rest, the Lagrangian components of the reactive impulse $\mathbf{I}^{(ru)} = I^{(ru)}\mathbf{n}$, due to the unilateral constraint, are

$$I_{x_G}^{(ru)} = - I^{(ru)}\mathbf{i} \cdot \left(\frac{\partial \mathbf{r}}{\partial x_G}\right)_P = -I_x,$$

$$I_{y_G}^{(ru)} = I^{(ru)}\mathbf{j} \cdot \left(\frac{\partial \mathbf{r}}{\partial y_G}\right)_P = 0,$$

$$I_\varphi^{(ru)} = I^{(ru)}\mathbf{i} \cdot \left(\frac{\partial \mathbf{r}}{\partial \varphi}\right)_P$$

$$= I^{(ru)}\mathbf{i} \cdot (-\rho \sin \varphi\mathbf{i} + \rho \cos \varphi\mathbf{j})_P = 0.$$

In conclusion, (22.27) reduce to

$$m\Delta\dot{x}_G = -I^{(ru)}, \ \Delta\dot{y}_G = 0, \ \Delta\dot{\varphi} = 0.$$

It remains to apply the Lagrangian formulation of Newton's law (22.33), which is now written as

$$-\mathbf{i} \cdot \left(\frac{\partial \mathbf{r}}{\partial q^h}\right)_P (\dot{q}^h)^+ = e\mathbf{i} \cdot \left(\frac{\partial \mathbf{r}}{\partial q^h}\right)_P (\dot{q}^h)^-. \tag{22.54}$$

Fig. 22.4 Collision between
two disks

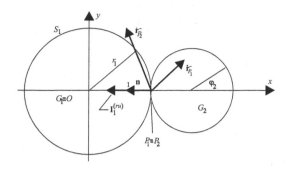

Recalling that $(\dot{q}^h) = (\dot{x}_G, \dot{y}_G, \dot{\varphi})$ and that at the point P it is $\varphi = 0$, it is a simple exercise to verify that this equation coincides with (22.53).

5. Determine the posterior speed field after the collision of two homogeneous circular disks S_1 and S_2 moving on a smooth horizontal plane π.
   Denote by m_1 and m_2 the masses of S_1 and S_2 and by r_1 and r_2 their radii. Finally, denote by C_1 and C_2 their moments of inertia relative to the centers of mass G_1 and G_2 (Fig. 22.4). The balance equations of impulsive dynamics applied to each disk give

$$m_i \Delta \dot{x}_{G_i} = -I_i^{(ru)},$$
$$m_i \Delta \dot{y}_{G_i} = 0,$$
$$C_i \Delta \varphi_i = 0,$$

$i = 1, 2$, where $I_1^{(ru)} = -I_2^{(ru)}$, owing to the action and reaction principle. Show that Newton's law of collision allows one to obtain the values of $\dot{x}_{G_i}^+$ after the collision.

6. Solve the preceding problem by the Lagrange equations.

Chapter 23
Introduction to Fluid Mechanics

Abstract In this chapter we consider the simplest model of a deformable body: a compressible or incompressible perfect fluid. We start analyzing the fundamental principles underlying mechanical models in which the deformability of bodies is taken into account. After an introduction to kinematics of deformable bodies, we determine Euler's equations of hydrodynamics and study their consequences at equilibrium and in dynamical conditions. In particular, we prove Archimedes's principle, Thomson's theorem, Lagrange's theorem, and the fundamental Bernoulli theorem. Then, we formulate the boundary value problems for Euler's equations and analyze some plane motions by using complex potentials, among which is the Joukovsky potential. After showing D'Alembert paradox, we prove Blausius formula. Finally, we analyze the propagation of small amplitude waves in a compressible fluid.

23.1 Kinematics of a Continuous System

In previous chapters, we only considered systems of point particles or rigid bodies. In this chapter, we analyze the fundamental principles underlying mechanical models in which, together with the extension of bodies, their deformability is taken into account.[1]

Let S be a deformable continuous system moving relative to a Cartesian frame of reference $Ox_1x_2x_2$. Denote by $C(0)$ and $C(t)$ the regions occupied by S at the initial time and at the arbitrary instant t, respectively. If $\mathbf{x} \in C(t)$ is the position vector at instant t of the point $\mathbf{X} \in C(0)$, then the equation of motion of S has the form

$$\mathbf{x} = \mathbf{x}(\mathbf{X}, t). \tag{23.1}$$

---

[1] For the contents of this chapter see [43, 44, 51].

© Springer International Publishing AG, part of Springer Nature 2018
A. Romano and A. Marasco, *Classical Mechanics with Mathematica®*,
Modeling and Simulation in Science, Engineering and Technology,
https://doi.org/10.1007/978-3-319-77595-1_23

We suppose that function (23.1) is of class C^2 with respect to its variables and confirms the condition

$$J = \det\left(\frac{\partial x_i}{\partial X_j}\right) > 0, \tag{23.2}$$

so that (23.1) is invertible for any instant t.

Any quantity of the continuous system S can be represented in **Lagrangian** or **Eulerian** form, depending on whether it is expressed as a function of $\mathbf{X}$, t, or $(\mathbf{x}, t)$, i.e., if it is a field assigned on the initial configuration $C(0)$ or on the current configuration $C(t)$:

$$\psi = \psi(\mathbf{x}, t) = \psi(\mathbf{x}(\mathbf{X}, t), t) = \widetilde{\psi}(\mathbf{X}, t). \tag{23.3}$$

For instance, the **velocity** and the **acceleration** of the particle $\mathbf{X} \in C(0)$ are given by the following partial derivatives:

$$\mathbf{v} = \tilde{\mathbf{v}}(\mathbf{X}, t) = \frac{\partial \mathbf{x}}{\partial t}, \quad \mathbf{a} = \tilde{\mathbf{a}}(\mathbf{X}, t) = \frac{\partial^2 \mathbf{x}}{\partial t^2}. \tag{23.4}$$

Because of the invertibility assumption of (23.1), the preceding fields can be represented in the Eulerian form as follows:

$$\mathbf{v} = \mathbf{v}(\mathbf{x}, t) = \tilde{\mathbf{v}}(\mathbf{x}(\mathbf{X}, t), t), \quad \mathbf{a} = \mathbf{a}(\mathbf{x}, t) = \tilde{\mathbf{a}}(\mathbf{x}(\mathbf{X}, t), t). \tag{23.5}$$

Consequently, the Eulerian form of acceleration is given by

$$\mathbf{a} = \frac{\partial \mathbf{v}}{\partial t} + \mathbf{v} \cdot \nabla \mathbf{v}, \tag{23.6}$$

where

$$\mathbf{v} \cdot \nabla \mathbf{v} = v_j \frac{\partial \mathbf{v}}{\partial x_j}. \tag{23.7}$$

More generally, the time derivative of the Eulerian field $\psi(\mathbf{x}, t)$ is

$$\dot{\psi} = \frac{\partial \psi}{\partial t} + \mathbf{v} \cdot \nabla \psi, \tag{23.8}$$

and we can say that the material derivative in the spatial representation contains two contributions: the first one is a local change expressed by the partial time derivative and the second one is the **convective derivative**.

Relevant features of motion are often highlighted by referring to **particle paths** or to **streamlines**. For this reason we introduce the following definitions.

The vector field $\mathbf{v}(\mathbf{x}, t)$, at a fixed time instant t, is called a **kinetic field**.

A *particle path* is the trajectory of an individual particle of S. In Lagrangian terms, particle paths can be obtained by integration of (23.4):

$$\mathbf{x} = \mathbf{x}_0 + \int_0^t \tilde{\mathbf{v}}(\mathbf{X}, t) dt. \tag{23.9}$$

If the velocity field is expressed in the *Eulerian* form $\mathbf{v} = \mathbf{v}(\mathbf{x}, t)$, the determination of particle paths requires the integration of a *nonautonomous* system of first-order differential equations:

$$\frac{d\mathbf{x}}{dt} = \mathbf{v}(\mathbf{x}, t). \tag{23.10}$$

A *streamline* is defined as the continuous line, *at a fixed instant t*, whose tangent at any point is in the direction of velocity at that point.

Based on this definition, streamlines represent the integral curves of the kinetic field, i.e., the solution of the *autonomous* system

$$\frac{d\mathbf{x}}{ds} = \mathbf{v}(\mathbf{x}, t), \qquad t = \text{const}, \tag{23.11}$$

where s is a parameter along the curve. We note that two streamlines cannot intersect; otherwise, one would face the paradoxical situation of a velocity with two directions at the intersection point.

The motion is defined as *stationary* if

$$\frac{\partial}{\partial t} \mathbf{v}(\mathbf{x}, t) = \mathbf{0}, \tag{23.12}$$

which is equivalent to saying that all particles of S that during the time evolution cross the position $\mathbf{x} \in C$ have the same velocity.

Since in the *stationary motion* the right-hand sides of both (23.11) and (23.12) are independent of t, the two systems of differential equations are equivalent to each other, so that the *particle paths and streamlines coincide*.

23.2 Velocity Gradient

Two relevant kinematic tensors will be used extensively. The first one is a symmetric tensor, defined as the *rate of deformation* or *stretching*

$$D_{ij} = \frac{1}{2} \left(\frac{\partial v_i}{\partial x_j} + \frac{\partial v_j}{\partial x_i} \right) = D_{ji}, \tag{23.13}$$

and the second one is a skew-symmetric tensor, defined as a *spin* or *vorticity tensor*

$$W_{ij} = \frac{1}{2}\left(\frac{\partial v_i}{\partial x_j} - \frac{\partial v_j}{\partial x_i}\right) = -W_{ji}. \tag{23.14}$$

By means of these two tensors, the **spatial gradient of velocity** can be conveniently additively decomposed into

$$\nabla \mathbf{v} = \mathbf{D} + \mathbf{W}. \tag{23.15}$$

By recalling the definition of differential

$$\mathbf{v}(\mathbf{x} + d\mathbf{x}, t) = \mathbf{v}(\mathbf{x}, t) + d\mathbf{x} \cdot \nabla \mathbf{v} \tag{23.16}$$

and using the decomposition (23.15), we obtain

$$\mathbf{v}(\mathbf{x} + d\mathbf{x}, t) = \mathbf{v}(\mathbf{x}, t) + \boldsymbol{\omega} \times d\mathbf{x} + \mathbf{D}d\mathbf{x}, \tag{23.17}$$

where $\boldsymbol{\omega}$ is a vector such that $\boldsymbol{\omega} \times d\mathbf{x} = \mathbf{W}d\mathbf{x}$. It can be verified that

$$\boldsymbol{\omega} = \frac{1}{2}\nabla \times \mathbf{v}, \tag{23.18}$$

which shows that, *in a neighborhood of* $\mathbf{x} \in C$, *the local kinetic field is composed by a rigid motion, with angular velocity given by* (23.17), *so that it is a function of time and* $\mathbf{x}$, *as well as of the contribution* $\mathbf{D}d\mathbf{x}$.

In the sequel, it will be useful to refer to the following expression for Eulerian acceleration, derived from (23.6):

$$\begin{aligned} a_i &= \frac{\partial v_i}{\partial t} + v_j\frac{\partial v_i}{\partial x_j} = \frac{\partial v_i}{\partial t} + v_j\frac{\partial v_i}{\partial x_j} + v_j\frac{\partial v_j}{\partial x_i} - v_j\frac{\partial v_j}{\partial x_i} \\ &= \frac{\partial v_i}{\partial t} + 2W_{ij}v_j + \frac{1}{2}\frac{\partial v^2}{\partial x_i}. \end{aligned}$$

Since $W_{ij}v_j = \epsilon_{ihj}\omega_h v_j$, we obtain

$$a_i = \frac{\partial v_i}{\partial t} + 2\epsilon_{ihj}\omega_h v_j + \frac{1}{2}\frac{\partial v^2}{\partial x_i},$$

and, referring to (23.17), we obtain the relation

$$\mathbf{a} = \frac{\partial \mathbf{v}}{\partial t} + (\nabla \times \mathbf{v}) \times \mathbf{v} + \frac{1}{2}\nabla v^2. \tag{23.19}$$

23.3 Rigid, Irrotational, and Isochoric Motions

For the analysis we are going to present, it is of interest to introduce Liouville's formula

$$\dot{J} = J \nabla \cdot \mathbf{v}. \tag{23.20}$$

To prove (23.20), first observe that, by definition,

$$
\dot{J} = \frac{d}{dt} \det \left(\frac{\partial x_i}{\partial X_L} \right) = \begin{vmatrix} \frac{\partial v_1}{\partial X_1} & \frac{\partial v_1}{\partial X_2} & \frac{\partial v_1}{\partial X_3} \\ \frac{\partial x_2}{\partial X_1} & \frac{\partial x_2}{\partial X_2} & \frac{\partial x_2}{\partial X_3} \\ \frac{\partial x_3}{\partial X_1} & \frac{\partial x_3}{\partial X_2} & \frac{\partial x_3}{\partial X_3} \end{vmatrix} + \cdots + \begin{vmatrix} \frac{\partial x_1}{\partial X_1} & \frac{\partial x_1}{\partial X_2} & \frac{\partial x_1}{\partial X_3} \\ \frac{\partial x_2}{\partial X_1} & \frac{\partial x_2}{\partial X_2} & \frac{\partial x_2}{\partial X_3} \\ \frac{\partial v_3}{\partial X_1} & \frac{\partial v_3}{\partial X_2} & \frac{\partial v_3}{\partial X_3} \end{vmatrix}.
$$

Moreover,

$$\frac{\partial v_h}{\partial X_L} = \frac{\partial v_h}{\partial x_j} \frac{\partial x_j}{\partial X_L}$$

and, since each determinant can be written as $J \partial v_i / \partial x_i$ (no summation on i), (23.20) is proved.

Theorem 23.1. *The motion of S is (globally) rigid if and only if*

$$\mathbf{D} = \mathbf{0}. \tag{23.21}$$

Proof. To prove that (23.21) is a necessary condition, we start from the velocity field of a rigid motion $v_i(\mathbf{x}, t) = v_i(\mathbf{x}_0, t) + \epsilon_{ijl}\omega_j(t)(x_l - x_{0l})$. Then the velocity gradient

$$\frac{\partial v_i}{\partial x_k} = \epsilon_{ijk}\omega_j$$

is skew-symmetric, and we obtain that

$$2D_{ik} = \frac{\partial v_i}{\partial x_k} + \frac{\partial v_k}{\partial x_i} = \left(\epsilon_{ijk} + \epsilon_{kji} \right) \omega_j = 0.$$

To prove that $\mathbf{D} = \mathbf{0}$ is a sufficient condition, by (23.17), we have that

$$\frac{\partial v_i}{\partial x_j} = W_{ij}, \qquad W_{ij} = -W_{ji}. \tag{23.22}$$

System (23.22) of nine differential equations with the three unknown functions $v_i(\mathbf{x})$ can be written in the equivalent form as follows:

$$dv_i = W_{ij}dx_j, \tag{23.23}$$

so that (23.21) has a solution if and only if we can integrate the differential forms (23.23). If region C of the kinetic field is simply linearly connected, a necessary and sufficient condition for (23.23) to be integrable is

$$\frac{\partial W_{ij}}{\partial x_h} = \frac{\partial W_{ih}}{\partial x_j}.$$

By cyclic permutation of indices, two additional conditions follow:

$$\frac{\partial W_{hi}}{\partial x_j} = \frac{\partial W_{hj}}{\partial x_i}, \qquad \frac{\partial W_{jh}}{\partial x_i} = \frac{\partial W_{ji}}{\partial x_h}.$$

Summing up the first two, subtracting the third one, and taking into account (23.22)$_2$, we obtain the condition

$$\frac{\partial W_{ij}}{\partial x_h} = 0,$$

which shows that the skew-symmetric tensor W_{ij} does not depend on spatial variables and eventually depends on time. Then, integration of (23.23) gives

$$v_i(\mathbf{x}, t) = v_{0i}(t) + W_{ij}(t)(x_j - x_{0j}),$$

and the motion is rigid.   □

The motion of S is **irrotational** if

$$\boldsymbol{\omega} = \frac{1}{2} \nabla \times \mathbf{v} = \mathbf{0}. \tag{23.24}$$

Again supposing that region C is simply linearly connected, it follows that the motion is irrotational if and only if

$$\mathbf{v} = \nabla \varphi, \tag{23.25}$$

where φ is a velocity **potential**, also defined as the **kinetic potential**.

Consider any region $c(t) \subset C(t)$, which is the mapping of $c(0) \subset C(0)$, throughout the equations of motion. This region is said to be a **material volume** because it is always occupied by the same particles. If during the motion of S the volume of any arbitrary material region does not change, then the motion is said to be **isochoric** or **isovolumic**, i.e.,

$$\frac{\mathrm{d}}{\mathrm{d}t} \int_{c(t)} \mathrm{d}c = 0.$$

By a change of variables $(x_i) \rightarrow (X_i)$, the previous requirement is also written as

$$\frac{\mathrm{d}}{\mathrm{d}t} \int_{c(0)} J \mathrm{d}c = 0,$$

and, since the volume $c(0)$ is fixed, differentiation and integration can be exchanged and, because of (23.20), it holds that

$$\int_{c(0)} J \nabla \cdot \mathbf{v} \, \mathrm{d}c = \int_{c(t)} \nabla \cdot \mathbf{v} \, \mathrm{d}c = 0, \qquad \forall c(t) \subset C(t).$$

Thus, the conclusion is reached that a motion is isochoric if and only if

$$\nabla \cdot \mathbf{v} = 0. \tag{23.26}$$

Finally, an irrotational motion is isochoric if and only if the velocity potential verifies *Laplace*'s *equation*

$$\Delta \varphi = \nabla \cdot \nabla \varphi = 0, \tag{23.27}$$

whose solutions are known as **harmonic functions**.

23.4 Mass Conservation

To derive the mass conservation law for a continuous system S, the basic assumption is introduced that the mass of S is continuously distributed over the region $C(t)$ it occupies at instant t. In other words, we assume the existence of a function $\rho(\mathbf{x}, t)$, called the **mass density**, which is supposed to be of class C^1 on $C(t)$.

According to this assumption, if any arbitrary region $c(0) \in C(0)$ of S is mapped onto $c(t)$ during motion $\mathbf{x}(\mathbf{X}, t)$, then the mass of $c(0)$ at instant t is given by

$$m(c(0)) = \int_{c(t)} \rho(\mathbf{x}, t) \, \mathrm{d}c, \tag{23.28}$$

and the **mass conservation principle** postulates that, during motion, the mass of any *material region* does not change over time

$$\frac{\mathrm{d}}{\mathrm{d}t} \int_{c(t)} \rho(\mathbf{x}, t) \, \mathrm{d}c = 0. \tag{23.29}$$

The mass conservation law for an *arbitrary fixed volume* v assumes the form as follows:

$$\frac{\mathrm{d}}{\mathrm{d}t} \int_v \rho(\mathbf{x}, t) \, \mathrm{d}c = - \int_{\partial v} \rho(\mathbf{x}, t) \, \mathbf{v} \cdot \mathbf{N} \mathrm{d}c. \tag{23.30}$$

Starting both from (23.29) and (23.30), when the density function is continuous in the region $C(t)$, we obtain that mass conservation is expressed by the following local equation:

$$\frac{\partial \rho}{\partial t} + \nabla \cdot (\rho \mathbf{v}) = \dot{\rho} + \rho \nabla \cdot \mathbf{v} = 0. \tag{23.31}$$

Before formulating the dynamical laws of a perfect fluid according to Euler, we prove the fundamental *transport theorem*:

Theorem 23.2. *If $\psi(\mathbf{x}, t)$ is a C^1 Eulerian field defined on the material region $C(t)$ occupied by the continuous system S, then the following result holds:*

$$\frac{\mathrm{d}}{\mathrm{d}t} \int_{c(t)} \rho \psi \, \mathrm{d}c = \int_{c(t)} \rho \dot{\psi} \, \mathrm{d}c, \tag{23.32}$$

for any material region $c(t) \subset C(t)$.

Proof. In fact, we have that

$$\frac{\mathrm{d}}{\mathrm{d}t} \int_{c(t)} \rho \psi \, \mathrm{d}c = \frac{\mathrm{d}}{\mathrm{d}t} \int_{c(0)} J \rho \psi \, \mathrm{d}c.$$

Taking into account (23.20) and (23.31), the preceding equation gives

$$\frac{\mathrm{d}}{\mathrm{d}t} \int_{c(t)} \rho \psi \, \mathrm{d}c = \int_{c(0)} (J\nabla \cdot \mathbf{v} \rho \psi + J \dot{\rho} \psi + J \rho \dot{\psi}) \mathrm{d}c = \int_{c(0)} J \rho \dot{\psi} \mathrm{d}c = \int_{c(t)} \rho \dot{\psi} \mathrm{d}c,$$

and the theorem is proved. $\qquad\qquad\qquad\qquad\qquad\qquad\qquad\qquad\qquad\qquad\qquad\qquad \Box$

23.5   Equations of Balance of a Perfect Fluid

In this section we introduce a model, due to Euler, that gives a good description of the behavior of gases and liquids, at least under suitable kinematic conditions clarified in the sequel. We do not present this model starting from the modern general principles of continuum mechanics since we prefer to follow a historical approach.

   We start with the first fundamental assumption on which Euler's model of a perfect fluid is based. Let $\partial c(t)$ be the boundary of an arbitrary region $c(t) \subset C(t)$, where $C(t)$ is the region occupied by the fluid S at instant t, and let $\mathbf{n}$ be the *external* unit normal to $\partial c(t)$. Then the total force $\mathbf{F}$ exerted by the fluid external to $c(t)$ on the fluid contained in $c(t)$ is given by

$$\mathbf{F}_{\partial c(t)} = -\int_{\partial c(t)} p(\mathbf{x}) \mathbf{n} \mathrm{d}\sigma, \tag{23.33}$$

where the *pressure* p depends only on the point $\mathbf{x} \in \partial c(t)$. In other words, (23.33) states that Pascal's law also holds in dynamical conditions. Besides the surface force (23.33), it is assumed that there exist forces called ***body forces*** that are distributed over the volume $C(t)$. Denoting by $\rho\mathbf{b}$ the force per unit mass, the total volume force acting on $c(t)$ is expressed by the integral

$$\mathbf{F}_{c(t)} = \int_{c(t)} \rho\mathbf{b} \, dc. \tag{23.34}$$

In particular, if the only volume force is the weight, then $\rho\mathbf{b} = \rho\mathbf{g}$, where $\mathbf{g}$ is the gravity acceleration.

The second fundamental assumption of fluid dynamics states that *the balance equations of mechanics hold for any material volume c of the fluid.*

Taking into account (23.33), we can write the balance equations of a fluid in the following form:

$$\frac{d}{dt} \int_{c(t)} \rho\mathbf{v} \, dc = -\int_{\partial c(t)} p\mathbf{n} \, d\sigma + \int_{c(t)} \rho\mathbf{b} \, dc, \tag{23.35}$$

$$\frac{d}{dt} \int_{c(t)} (\mathbf{x} - \mathbf{x}_0) \times \rho\mathbf{v} \, dc = -\int_{\partial c(t)} (\mathbf{x} - \mathbf{x}_0) \times p\mathbf{n} \, d\sigma$$

$$+ \int_{c(t)} (\mathbf{x} - \mathbf{x}_0) \times \rho\mathbf{b} \, dc, \tag{23.36}$$

where $\mathbf{x}_0$ is the position vector of an arbitrary pole.

From the law of balance (23.35), valid for any material volume $c(t) \subset C(t)$, when we assume that the functions under the integrals are continuous, take into account (20.30), and apply Gauss's theorem to the surface integral, we obtain the following local equation at any point $\mathbf{x} \in C(t)$:

$$\rho\mathbf{a} = -\nabla p + \rho\mathbf{b}. \tag{23.37}$$

Similarly, assuming that the pole $\mathbf{x}_0$ is fixed in the frame of reference R relative to which we evaluate the motion of the fluid S, the ith component of (23.36) gives the local condition

$$\epsilon_{ijh}(x_j - x_{0j})\rho a_h = -\epsilon_{ijh}\delta_{jh}p - \epsilon_{ijh}(x_j - x_{0j})\frac{\partial p}{\partial x_h}$$

$$+ \epsilon_{ijh}(x_j - x_{0j})\rho b_h, \tag{23.38}$$

which is identically satisfied in view of (23.37).

In conclusion, collecting the results of this and the previous section, we have the following equations to evaluate the dynamical evolution of a perfect fluid:

$$\dot{\rho} = -\rho \nabla \cdot \mathbf{v}, \tag{23.39}$$

$$\rho \dot{\mathbf{v}} = -\nabla p + \rho \mathbf{b}. \tag{23.40}$$

This is a system of nonlinear partial differential equations [see (23.6)] in the unknowns $\rho(\mathbf{x}, t)$, $\mathbf{v}(\mathbf{x}, t)$, and $p(\mathbf{x}, t)$. In other words, this system is not closed. This circumstance is quite understandable since the preceding equations hold for any perfect fluid S. We need to introduce in some way the nature of S to obtain the balancing between equations and unknowns.

If S is *incompressible*, i.e., if S is a **liquid**, then the mass density ρ is constant and system (23.39), (23.40) becomes

$$\nabla \cdot \mathbf{v} = 0, \tag{23.41}$$

$$\dot{\mathbf{v}} = -\frac{1}{\rho}\nabla p + \mathbf{b}. \tag{23.42}$$

In this case, we obtain a system of two nonlinear partial differential equations in the unknowns $\mathbf{v}(\mathbf{x}, t)$, and $p(\mathbf{x}, t)$. In other words, one liquid differs from another only in its density.

When the fluid S is *compressible*, i.e., if S is a **gas**, then the balancing between the equations and unknowns is obtained giving a **constitutive equation** relating the pressure to the density of the gas as follows:

$$p = p(\rho). \tag{23.43}$$

This equation can only be determined by experimentation. For instance, for an *ideal gas*

$$p = R\rho\theta, \tag{23.44}$$

where R is the universal gas constant and θ the absolute temperature. For a van der Waals gas S

$$p = \frac{R}{M}\frac{\theta}{v - b} - \frac{a}{v^2}, \tag{23.45}$$

where M is the molar mass of S, $v = 1/\rho$ is the specific volume, and a and b are two constants depending on the gas.

23.6   Statics of Fluids

In this section, we consider *conservative* body forces, i.e., forces for which a potential energy $U(\mathbf{x})$ exists such that

$$\mathbf{b} = -\nabla U(\mathbf{x}). \tag{23.46}$$

The statics of a perfect fluid subject to conservative body forces is governed by the following equation [see (23.39) and (23.40)]

$$\nabla p = -\rho \nabla U(\mathbf{x}), \tag{23.47}$$

where ρ is constant for a liquid and a given function $p = p(\rho)$ for a gas. Since experimental evidence leads us to assume that $dp/d\rho > 0$, the function $p = p(\rho)$ can be inverted.

If we introduce the notation

$$h(\rho) = \int \frac{dp}{\rho}, \tag{23.48}$$

then

$$\nabla h(p) = \frac{dh}{d\rho} \nabla p = \frac{1}{\rho} \nabla p,$$

and (23.47) becomes

$$\nabla(h(p) + U(\mathbf{x})) = 0. \tag{23.49}$$

This equation shows that in any connected region occupied by fluids, the following condition holds:

$$h(p) + U(\mathbf{x}) = \text{const.} \tag{23.50}$$

In particular, for a liquid, (23.50) becomes [see (23.48)]

$$\frac{p}{\rho} + U(\mathbf{x}) = \text{const.} \tag{23.51}$$

We now propose to analyze some consequences of (23.50) and (23.51).

- In a fluid S, the level surfaces of the potential energy coincide with the isobars.
- Let S be a liquid, subject to its weight, on whose boundary acts a uniform pressure p_0 (for instance, atmospheric pressure). If the vertical axis Oz is downward oriented, then the relation $U(\mathbf{x}) = -gz + \text{const}$ implies that the free surface of S is a horizontal plane.
- Let S be a liquid in a cylinder subject to its weight and to a uniform pressure on its boundary. If the cylinder containing S is rotating about a vertical axis a with uniform angular velocity ω, then its free surface is a paraboloid of rotation about a. In fact, the force acting on the arbitrary particle $P \in S$ is given by $\mathbf{g} + \omega^2 \overrightarrow{QP}$, where $\mathbf{g}$ is the gravity acceleration and Q the projection of P on a. It follows that

$$U = -gz - \frac{\omega^2}{2}(x^2 + y^2),$$

and (23.51) proves that the free surface is a paraboloid.

• Let S be a liquid in a container subject to its weight and to atmospheric pressure p_0 on its free surface. Assume that the vertical axis Oz is downward oriented, and choose the origin in such a way that the arbitrary constant of the potential energy is equal to zero when $z=0$. Then from (23.51) it follows that

$$\frac{p(z)}{\rho} - gz = \frac{p_0}{\rho},$$

i.e.,

$$p(z) = p_0 + \rho gz, \quad z > 0, \tag{23.52}$$

and **Stevin's law** is obtained: *Pressure increases linearly with depth by an amount equal to the weight of the liquid column acting on the unit surface.*

• Let S be an ideal gas at equilibrium at constant and uniform temperature when subject to its weight. Assuming that the vertical axis Oz is upward oriented, we get $U(z) = gz + \text{const}$, and, taking into account (23.44), (23.50) becomes

$$R\theta \int_{p_0}^{P} \frac{dp}{p} = -gz,$$

so that

$$p(z) = p_0 \exp\left(-\frac{gz}{R\theta}\right). \tag{23.53}$$

Finally, we prove an important consequence of (23.51).

Theorem 23.3. (Archimedes' principle). *The buoyant force on a body submerged in a liquid is equal to the weight of the liquid displaced by the body.*

Proof. To prove this statement, consider a body S submerged in a liquid, as shown in Fig. 23.1. If p_0 is the atmospheric pressure, then the force acting on S is given by

$$\mathbf{F} = -\int_{\sigma_e} p_0 \mathbf{N}\, d\sigma - \int_{\sigma_i} (p_0 + \rho gz)\mathbf{N}\, d\sigma, \tag{23.54}$$

where $\mathbf{N}$ is the outward unit vector normal to the body surface σ, σ_i is the submerged portion of σ, and σ_e is the portion above the waterline. By adding and subtracting on the right-hand side of (23.54) the integral of $-p_0\mathbf{N}$ over Ω (Fig. 23.1), (23.54) becomes

Fig. 23.1 Archimedes'
principle

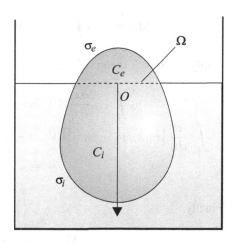

$$\mathbf{F} = -\int_{\partial C_e} p_0 \mathbf{N} \, d\sigma - \int_{\partial C_i} (p_0 + \rho gz)\mathbf{N} \, d\sigma, \qquad (23.55)$$

so that knowledge of $\mathbf{F}$ requires the computation of the integrals in (23.55). To do this, we define a virtual pressure field (continuous on Ω) in the interior of the body as follows:

$$p = p_0 \qquad \text{on } C_e,$$
$$p = p_0 + \rho gz \qquad \text{on } C_i.$$

Applying Gauss's theorem to the integrals in (23.55), we obtain

$$\int_{\partial C_e} p_0 \mathbf{N} \, d\sigma = \int_{C_e} \nabla p_0 \, dV = 0, \qquad \int_{\partial C_i} p_0 \mathbf{N} \, d\sigma = \int_{C_i} \nabla p_0 \, dV = 0,$$
$$\int_{\partial C_i} \rho gz \mathbf{N} \, d\sigma = \rho g \int_{C_i} \nabla z \, dV.$$

Finally,

$$\mathbf{F} = -\rho g V_i \mathbf{k}, \qquad (23.56)$$

where $\mathbf{k}$ is the unit vector associated with Oz and V_i is the volume of region C_i.

Equation (23.56) gives the resultant force acting on the body S. To complete the equilibrium analysis, the momentum $\mathbf{M}_O$ of pressure forces with respect to an arbitrary pole O must be explored. If $\mathbf{r} = \overrightarrow{OP}$, then

$$\mathbf{M}_O = -p_0 \int_{\sigma_e} \mathbf{r} \times \mathbf{N} \, d\sigma - \int_{\sigma_i} \mathbf{r} \times (p_0 + \rho gz)\mathbf{N} \, d\sigma. \qquad (23.57)$$

Again, by adding and subtracting on the right-hand side of (23.57) the integral of $-p_0 \mathbf{r} \times \mathbf{N}$ over Ω, (23.57) becomes

$$\mathbf{M}_O = -p_0 \int_{\partial C_e} \mathbf{r} \times \mathbf{N}\, d\sigma - \int_{\partial C_i} \mathbf{r} \times (p_0 + \rho g z)\mathbf{N}\, d\sigma.$$

Applying Gauss's theorem, we obtain

$$-p_0 \int_{C_e} \epsilon_{ijl} \frac{\partial x_j}{\partial x_l}\, dV = -p_0 \int_{C_e} \epsilon_{ijl}\delta_{jl}\, dV = 0,$$

$$\int_{C_i} \frac{\partial}{\partial x_l}\left[\epsilon_{ijl}x_j(p_0 + \rho g z)\right] dV = -\int_{C_i} \left[\epsilon_{ijl}\delta_{jl}(p_0 + \rho g z) + \epsilon_{ijl}x_j\rho g\delta_{3l}\right] dV$$

$$= -\rho g \int_{C_i} \epsilon_{ij3}x_j\, dV.$$

Finally,

$$\mathbf{M}_O = -\rho g\, [x_{2C}\mathbf{i} - x_{1C}\mathbf{j}]\, V_i, \tag{23.58}$$

where $\mathbf{i}$ and $\mathbf{j}$ are orthonormal base vectors on the horizontal plane and

$$x_{1C} V_i = \int_{C_i} x_1\, dV, \qquad x_{2C} V_i = \int_{C_i} x_2\, dV. \tag{23.59}$$

Expression (23.58) shows that the momentum of the pressure forces vanishes if the line of action of the buoyant force passes through the centroid of the body. The centroid of the displaced liquid volume is called the *center of buoyancy*.

In summary: *A body floating in a liquid is at equilibrium if the buoyant force is equal to its weight and the line of action of the buoyant force passes through the centroid of the body.* It can be proved that the equilibrium is stable if the center of buoyancy is above the centroid and is unstable if the center of buoyancy is below the centroid.   □

23.7 Fundamental Theorems of Fluid Dynamics

The momentum balance equation (23.40), when applied to a perfect fluid subjected to conservative body forces, is written as

$$\rho\dot{\mathbf{v}} = -\nabla p - \rho\nabla U; \tag{23.60}$$

with the additional introduction of (23.48), it holds that

$$\dot{\mathbf{v}} = -\nabla(h(p) + U). \tag{23.61}$$

Recalling the preceding definitions, we can prove the following theorems.

Theorem 23.4. (W. Thomson, Lord Kelvin). *In a barotropic flow under conservative body forces, the circulation around any closed material curve γ is preserved, i.e., it is independent of time:*

$$\frac{\mathrm{d}}{\mathrm{d}t} \int_\gamma \mathbf{v} \cdot \mathrm{d}\mathbf{s} = 0. \tag{23.62}$$

Proof. If γ is a material closed curve, then there exists a closed curve γ_0 in C_0 such that γ is the image of γ_0 under the motion equation $\gamma = \mathbf{x}(\gamma_0, t)$. It follows that

$$\frac{\mathrm{d}}{\mathrm{d}t} \int_\gamma v_i\, \mathrm{d}x_i = \frac{\mathrm{d}}{\mathrm{d}t} \int_{\gamma_0} v_i \frac{\partial x_i}{\partial X_j}\, \mathrm{d}X_j = \int_{\gamma_0} \frac{\mathrm{d}}{\mathrm{d}t}\left(v_i \frac{\partial x_i}{\partial X_j}\, \mathrm{d}X_j \right)$$

$$= \int_{\gamma_0} \left(\dot{v}_i \frac{\partial x_i}{\partial X_j} + v_i \frac{\partial \dot{x}_i}{\partial X_j} \right) \mathrm{d}X_j = \int_\gamma \left(\dot{v}_i + v_i \frac{\partial v_i}{\partial x_j} \right) \mathrm{d}x_j,$$

and, taking into account (23.61), it is proved that

$$\frac{\mathrm{d}}{\mathrm{d}t} \int_\gamma \mathbf{v} \cdot \mathrm{d}\mathbf{s} = \int_\gamma \left(\dot{\mathbf{v}} + \frac{1}{2}\nabla v^2 \right) \cdot \mathrm{d}\mathbf{s} = - \int_\gamma \nabla \left(h(p) + U - \frac{1}{2}v^2 \right) \cdot \mathrm{d}\mathbf{s} = 0.$$

$$\tag{23.63}$$

$\square$

Theorem 23.5. (Lagrange). *If at a given instant t_0 a motion is irrotational, then it continues to be irrotational at any $t > t_0$, or, equivalently, vortices cannot form.*

Proof. This can be regarded as a special case of Thomson's theorem. Suppose that in the region C_0 occupied by fluid at instant t_0 the condition $\omega = 0$ holds. Stokes's theorem requires that

$$\Gamma_0 = \int_{\gamma_0} \mathbf{v} \cdot \mathrm{d}\mathbf{s} = 0$$

for any material closed curve γ_0. But Thomson's theorem states that $\Gamma = 0$ for all $t > t_0$, so that also $\omega(t) = 0$ holds at any instant. $\square$

Theorem 23.6. (Bernoulli). *In a steady flow, along any particle path, i.e., along the trajectory of an individual element of fluid, the quantity*

$$H = \frac{1}{2}v^2 + h(p) + U \tag{23.64}$$

is constant. In general, the constant H changes from one streamline to another, but if the motion is irrotational, then H is constant in time and over the whole space of the flow field.

Proof. Recalling (23.19) and the time independence of the flow, we can write (23.61) as

$$\dot{\mathbf{v}} = (\nabla \times \mathbf{v}) \times \mathbf{v} + \frac{1}{2}\nabla v^2 = -\nabla(h(p) + U). \tag{23.65}$$

A scalar multiplication by $\mathbf{v}$ gives the relation

$$\mathbf{v} \cdot \nabla \left(\frac{1}{2}v^2 + h(p) + U \right) = 0,$$

which proves that (23.64) is constant along any particle path. If the steady flow is irrotational, then (23.65) implies $H = $ const through the flow field at any time.   □

In particular, if the fluid is incompressible, then Bernoulli's theorem states that in a steady flow along any particle path (or through the flow field if the flow is irrotational) the quantity H is preserved, i.e.,

$$H = \frac{1}{2}v^2 + \frac{p}{\rho} + U = \text{const.} \tag{23.66}$$

The Bernoulli equation is often used in another form, obtained by dividing (23.66) by the gravitational acceleration

$$h_z + h_p + h_v = \text{const},$$

where $h_z = U/g$ is the **gravity head** or **potential head**, $h_p = p/\rho g$ is the **pressure head**, and $h_v = v^2/2g$ is the **velocity head**.

In a steady flow, a **stream tube** is a tubular region Σ within a fluid bounded by streamlines. We note that streamlines cannot intersect each other. Because $\partial \rho/\partial t = 0$, the mass conservation (23.31) gives $\nabla \cdot (\rho \mathbf{v}) = 0$, and by integrating over a volume V defined by the sections σ_1 and σ_2 of a stream tube (Fig. 23.2), we obtain

$$Q = \int_{\sigma_1} \rho \mathbf{v} \cdot \mathbf{n} \, d\sigma = \int_{\sigma_2} \rho \mathbf{v} \cdot \mathbf{n} \, d\sigma. \tag{23.67}$$

This relation proves that the *flux* is constant across any section of the stream tube. If the fluid is incompressible, then (23.67) reduces to

$$\int_{\sigma_1} \mathbf{v} \cdot \mathbf{n} \, d\sigma = \int_{\sigma_2} \mathbf{v} \cdot \mathbf{n} \, d\sigma. \tag{23.68}$$

The local angular speed ω is also called the **vortex vector**, and the related integral curves are **vortex lines**; furthermore, a **vortex tube** is a surface represented by all vortex lines passing through the points of a (nonvortex) closed curve. Recalling that a vector field $\mathbf{w}$ satisfying the condition $\nabla \cdot \mathbf{w} = 0$ is termed *solenoidal* and that

Fig. 23.2 Stream tube

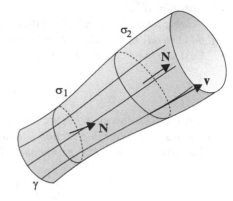

Fig. 23.3 Uniform motion

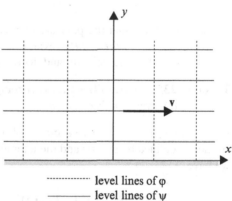

----------- level lines of φ
——— level lines of ψ

$2\nabla \cdot \boldsymbol{\omega} = \nabla \cdot \nabla \times \mathbf{v} = 0$, we conclude that the field $\boldsymbol{\omega}$ is solenoidal. Therefore, vortex lines are closed if they are limited, and they are open if unconfined. We observe that Fig. 23.3 can also be used to represent a vortex tube if the vector $\mathbf{v}$ is replaced by $\boldsymbol{\omega}$.

The following examples illustrate some relevant applications of Bernoulli's equation.

1. Consider an open vessel with an orifice at depth h from the free surface of the fluid. Suppose that fluid is added on the top to keep the height h constant. Under these circumstances, it can be proved that the *velocity of the fluid leaving the vessel through the orifice is equal to that of a body falling from the elevation h with initial velocity equal to zero* (this result is known as **Torricelli's theorem** because it was found long before Bernoulli's work). Assuming that at the free surface we have $\mathbf{v} = 0$ and $z = 0$, it follows that $H = p_0/\rho$, and applying (23.66) we derive the following relation:

$$H = \frac{p_0}{\rho} = \frac{v^2}{2} - gh + \frac{p_0}{\rho},$$

. so that $v = \sqrt{2gh}$.

2. *In a horizontal pipe of variable cross section, the pressure of an incompressible fluid in steady motion decreases in a converging section.*

First, the mass conservation (23.68) requires that

$$v_1\sigma_1 = v_2\sigma_2, \tag{23.69}$$

where the velocities v_1 and v_2 are supposed to be uniform, so that *the fluid velocity increases in a converging section and decreases in the diverging section.*

Furthermore, since $U = \rho z = const$ along the stream tube, (23.66) implies that

$$\frac{v_1^2}{2} + \frac{p_1}{\rho} = \frac{v_2^2}{2} + \frac{p_2}{\rho},$$

and this proves that the pressure decreases in a converging section. This result is applied in **Venturi's tube**, where a converging section acts as a nozzle by increasing the fluid velocity and decreasing its pressure.

Theorem 23.7. (Helmoltz's first theorem). *The flux of a vortex vector across any section of a vortex tube is constant.*

Proof. Let σ_1 and σ_2 be two sections of a vortex tube T, and consider the closed surface Σ defined by σ_1, σ_2, and the lateral surface of T. Applying Gauss's theorem we have

$$\int_\Sigma \omega \cdot \mathbf{N}\,d\sigma = \int_V \nabla \cdot \omega\,dV = \frac{1}{2}\int_V \nabla \cdot (\nabla \times \mathbf{v})\,dV = 0,$$

where $\mathbf{N}$ is the unit outward vector normal to Σ. The definition of a vortex tube implies that ω is tangent to Σ at any point, so that the theorem is proved since

$$\int_\Sigma \omega \cdot \mathbf{N}\,d\sigma = \int_{\sigma_1} \omega \cdot \mathbf{N}_1\,d\sigma = \int_{\sigma_2} \omega \cdot \mathbf{N}_2\,d\sigma = 0,$$

where $\mathbf{N}_1$ is the unit vector normal to σ_1, pointing toward the interior of the tube, and $\mathbf{N}_2$ is the outward unit vector normal to σ_2. □

From this theorem it also follows that the particle vorticity increases if the vortex curves are converging.

Theorem 23.8. (Helmoltz's second theorem). *Vortex lines are material lines.*

Proof. At the instant $t_0 = 0$, the vector ω is supposed to be tangent to the surface σ_0. Denote by $\sigma(t)$ the material surface defined by the particles lying upon σ_0 at the instant t_0. We must prove that $\sigma(t)$ is a vortex surface at any arbitrary instant. First, we verify that the circulation Γ along any closed line γ_0 on σ_0 vanishes. In fact, if A is the portion of σ_0 contained in γ_0, it holds that

$$\Gamma = \int_{\gamma_0} \mathbf{v} \cdot d\mathbf{s} = \int_A \nabla \times \mathbf{v} \cdot \mathbf{N} \, d\sigma = 2 \int_A \boldsymbol{\omega} \cdot \mathbf{N} \, d\sigma = 0$$

since $\boldsymbol{\omega}$ is tangent to A. According to Thomson's theorem, the circulation is preserved along any material curve, so that, if $\gamma(t)$ is the image of γ_0, it follows that

$$\int_{\gamma(t)} \mathbf{v} \cdot d\mathbf{s} = \int_{A(t)} \nabla \times \mathbf{v} \cdot \mathbf{N} \, d\sigma = 2 \int_{A(t)} \boldsymbol{\omega} \cdot \mathbf{N} \, d\sigma = 0.$$

Since $A(t)$ is arbitrary, $\boldsymbol{\omega} \cdot \mathbf{N} = 0$, and the theorem is proved. ☐

The theorem can also be stated by saying that vortex lines are constituted by the same fluid particles and are transported during motion. Examples include smoke rings, whirlwinds, and so on.

23.8 Boundary Value Problems for a Perfect Fluid

The motion of a perfect compressible fluid S subjected to body forces $\mathbf{b}$ is governed by the momentum equation [see (23.40)]

$$\dot{\mathbf{v}} = -\frac{1}{\rho} \nabla p(\rho) + \mathbf{b} \tag{23.70}$$

and the mass conservation

$$\dot{\rho} + \rho \nabla \cdot \mathbf{v} = 0. \tag{23.71}$$

Equations (23.70) and (23.71) are a first-order system for the unknowns $\mathbf{v}(\mathbf{x}, t)$ and $\rho(\mathbf{x}, t)$, and to find a unique solution, both initial and boundary conditions must be specified.

If we consider the motion in a fixed and compact region C of a space (e.g., a gas in a container with rigid walls), then the initial conditions are

$$\mathbf{v}(\mathbf{x}, 0) = \mathbf{v}_0(\mathbf{x}), \qquad \rho(\mathbf{x}, 0) = \rho_0(\mathbf{x}) \qquad \forall \mathbf{x} \in C, \tag{23.72}$$

and the boundary condition is

$$\mathbf{v} \cdot \mathbf{N} = 0 \qquad \forall \mathbf{x} \in \partial C, \quad t > 0. \tag{23.73}$$

This boundary condition states that the fluid can perform any tangential motion on a fixed surface whose unit normal is $\mathbf{N}$.

The problem is then to find in $C \times [0, t]$ the fields $\mathbf{v}(\mathbf{x}, t)$ and $\rho(\mathbf{x}, t)$ that satisfy the balance equations (23.70) and (23.71), the initial conditions (23.72), and the boundary condition (23.73).

If the fluid is incompressible (ρ = const), then Eq. (23.70) becomes

$$\dot{\mathbf{v}} = -\frac{1}{\rho} \nabla p + \mathbf{b}, \tag{23.74}$$

while the mass conservation (23.71) leads us to the condition

$$\nabla \cdot \mathbf{v} = 0. \tag{23.75}$$

The unknowns of system (23.74) and (23.75) are given by the fields $\mathbf{v}(\mathbf{x}, t)$ and $p(\mathbf{x}, t)$, and the appropriate initial and boundary conditions are

$$\begin{aligned}
\mathbf{v}(\mathbf{x}, 0) &= \mathbf{v}_0(\mathbf{x}) \quad \forall \mathbf{x} \in C, \\
\mathbf{v} \cdot \mathbf{N} &= 0 \quad \forall \mathbf{x} \in \partial C, \, t > 0.
\end{aligned} \tag{23.76}$$

A more complex problem arises when a part of the boundary is represented by a *moving* or *free surface* $f(\mathbf{x}, t) = 0$. In this case, finding the function f is a part of the boundary value problem. The moving boundary $\partial C'$, represented by $f(\mathbf{x}, t) = 0$, is a *material surface* since a material particle located on it must remain on this surface during the motion. This means that its velocity c_N along the unit normal $\mathbf{N}$ to the free surface is equal to $\mathbf{v} \cdot N$; that is, $f(\mathbf{x}, t)$ must satisfy the condition [see (4.32)]

$$\frac{\partial}{\partial t} f(\mathbf{x}, t) + \mathbf{v}(\mathbf{x}, t) \cdot \nabla f(\mathbf{x}, t) = 0.$$

In addition, on the free surface it is possible to prescribe the value of the pressure, so that the *dynamic boundary conditions* are

$$\begin{aligned}
\frac{\partial}{\partial t} f(\mathbf{x}, t) + \mathbf{v}(\mathbf{x}, t) \cdot \nabla f(\mathbf{x}, t) &= 0, \\
p &= p_e \quad \forall \mathbf{x} \in \partial C', \quad t > 0,
\end{aligned} \tag{23.77}$$

where p_e is the prescribed external pressure.

On the fixed boundary part, the no friction condition (23.73) applies, and if boundary C extends to infinity, then conditions related to the asymptotic behavior of the solution at infinity must be added.

23.9   Two-Dimensional Steady Flow of a Perfect Fluid

The following two conditions define an irrotational steady motion of an incompressible fluid S:

$$\nabla \times \mathbf{v} = \mathbf{0}, \qquad \nabla \cdot \mathbf{v} = 0, \tag{23.78}$$

where $\mathbf{v} = \mathbf{v}(\mathbf{x})$. The first condition allows us to deduce the existence of a **velocity** or **kinetic potential** $\varphi(\mathbf{x})$ such that

$$\mathbf{v} = \nabla\varphi, \tag{23.79}$$

where φ is a single- or a multiple-valued function, depending on whether or not the motion region C is connected.[2]

In addition, taking into account $(23.78)_2$, it holds that

$$\Delta\varphi = \nabla \cdot \nabla\varphi = 0. \tag{23.80}$$

Equation (23.80) is known as **Laplace's equation**, and its solution is a **harmonic function**.

Finally, in dealing with a *two-dimensional* (2D) flow, the velocity vector $\mathbf{v}$ at any point is parallel to a plane π and is independent of the coordinate normal to this plane. In this case, if a system $Oxyz$ is introduced, where the x- and y-axes are parallel to π and the z-axis is normal to this plane, then we have

$$\mathbf{v} = u(x, y)\mathbf{i} + v(x, y)\mathbf{j},$$

where u and v are the components of $\mathbf{v}$ on x and y, and $\mathbf{i}$ and $\mathbf{j}$ are the unit vectors of these axes.

If now C is a simply connected region of the plane Oxy, then conditions (23.78) become

$$-\frac{\partial u}{\partial y} + \frac{\partial v}{\partial x} = 0, \qquad \frac{\partial u}{\partial x} + \frac{\partial v}{\partial y} = 0. \tag{23.81}$$

These conditions allow us to state that the two differential forms $\omega_1 = u\,dx + v\,dy$ and $\omega_2 = -v\,dx + u\,dy$ are integrable, i.e., there is a function φ, called the **velocity potential** or the **kinetic potential**, and a function ψ, called the **stream potential** or the **Stokes potential**, such that

$$d\varphi = u\,dx + v\,dy, \qquad d\psi = -v\,dx + u\,dy. \tag{23.82}$$

From $(23.78)_2$ it follows that the curves $\varphi = \text{const}$ are at any point normal to the velocity field. Furthermore, since $\nabla\varphi \cdot \nabla\psi = 0$, the curves $\psi = \text{const}$ are flow lines.

It is relevant to observe that (23.82) suggest that the functions φ and ψ satisfy the Cauchy–Riemann conditions

---

[2]If C is not a simply connected region, then the condition $\nabla \times \mathbf{v} = \mathbf{0}$ does not imply that $\int_\gamma \mathbf{v} \cdot d\mathbf{s} = 0$ on any closed curve γ since this curve could not be the boundary of a surface contained in C. In this case, Stokes's theorem cannot be applied.

$$\frac{\partial \varphi}{\partial x} = \frac{\partial \psi}{\partial y}, \qquad \frac{\partial \varphi}{\partial y} = -\frac{\partial \psi}{\partial x}, \tag{23.83}$$

so that the complex function

$$F(z) = \varphi(x, y) + i\psi(x, y) \tag{23.84}$$

is **holomorphic** and represents a **complex potential**. Then the complex potential can be defined as a holomorphic function whose real and imaginary parts are the velocity potential φ and the stream potential ψ, respectively. The two functions φ and ψ are harmonic and the derivative of $F(z)$

$$V \equiv F'(z) = \frac{\partial \varphi}{\partial x} + i\frac{\partial \psi}{\partial x} = u - iv = |V| e^{-i\theta}, \tag{23.85}$$

represents the **complex velocity**, with V being the modulus of the velocity vector and θ the angle that this vector makes with the x-axis.

Within the context of considerations developed in the following discussion, it is relevant to remember that the line integral of a holomorphic function vanishes around any arbitrary closed path in a simply connected region since the Cauchy–Riemann equations are necessary and sufficient conditions for the integral to be independent of the path (and therefore it vanishes for a closed path).

The preceding remarks lead to the conclusion that a 2D irrotational flow of an incompressible fluid is completely defined if a harmonic function $\varphi(x, y)$ or a complex potential $F(z)$ is prescribed, as is shown in the following examples.

Example 23.1. (Uniform motion). Given the complex potential

$$F(z) = U_0(x + iy) = U_0z, \tag{23.86}$$

it follows that $V = U_0$, and the 2D motion

$$\mathbf{v} = U_0\mathbf{i} \tag{23.87}$$

is defined, where $\mathbf{i}$ is the unit vector of the Ox-axis. The kinetic and Stokes potentials are $\varphi = U_0x$ and $\psi = U_0y$, and the curves $\varphi = $ const and $\psi = $ const are parallel to Oy and Ox (Fig. 23.3), respectively. This example shows that the complex potential (23.86) can be introduced to describe a 2D uniform flow, parallel to the wall $y = 0$.

Example 23.2. (Vortex potential). Let a 2D flow be defined by the complex potential

$$F(z) = -i\frac{\Gamma}{2\pi} \ln z = -i\frac{\Gamma}{2\pi} \ln re^{i\theta} = \frac{\Gamma}{2\pi}\theta - i\frac{\Gamma}{2\pi} \ln r, \tag{23.88}$$

where r and θ are polar coordinates. It follows that

Fig. 23.4 Vortex

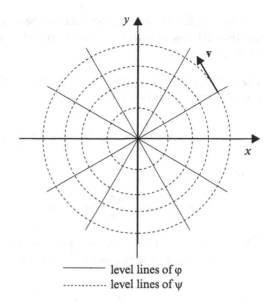

level lines of φ

level lines of ψ

$$\varphi = \frac{\Gamma}{2\pi}\theta = \frac{\Gamma}{2\pi}\arctan\frac{y}{x},$$
$$\psi = -i\frac{\Gamma}{2\pi}\ln r = -\frac{\Gamma}{2\pi}\ln(x^2 + y^2),$$

and the curves φ=const are straight lines through the origin, while the curves ψ=const are circles whose center is the origin (Fig. 23.4).

Accordingly, the velocity components become

$$u = \frac{\partial\varphi}{\partial x} = -\frac{\Gamma}{2\pi}\frac{y}{x^2 + y^2} = -\frac{\Gamma}{2\pi}\frac{\sin\theta}{r},$$
$$v = \frac{\partial\varphi}{\partial y} = \frac{\Gamma}{2\pi}\frac{x}{x^2 + y^2} = \frac{\Gamma}{2\pi}\frac{\cos\theta}{r}.$$

It is relevant to observe that the circulation around a path γ bordering the origin is given by

$$\int_\gamma \mathbf{v}\cdot d\mathbf{s} = \Gamma,$$

so that it does not vanish if $\Gamma \neq 0$. This does not contradict the condition $\nabla \times \mathbf{v} = \mathbf{0}$ since a plane without an origin is no longer a simply connected region. In Sect. 23.10, further arguments will be addressed to explain why the circulation around an obstacle does not vanish in a 2D flow.

The complex potential (23.88) can then be used with the advantage to describe the uniform 2D flow of particles rotating around an axis through the origin and normal to the plane Oxy.

Example 23.3. (Sources and sinks). For pure radial flow in the horizontal plane, the complex potential is taken to be

$$F(z) = \frac{Q}{2\pi}\ln z = \frac{Q}{2\pi}\ln re^{i\theta} = \frac{Q}{2\pi}(\ln r + i\theta), \qquad Q > 0,$$

so that

$$\varphi = \frac{Q}{2\pi}\ln r = \frac{Q}{2\pi}\log(x^2 + y^2), \qquad \psi = \frac{Q}{2\pi}\theta = \frac{Q}{2\pi}\arctan\frac{y}{x}.$$

The curves $\varphi=$const are circles, whereas the curves $\psi=$const are straight lines through the origin. The velocity field is given by

$$\mathbf{v} = \frac{Q}{2\pi r}\mathbf{e}_r,$$

where $\mathbf{e}_r$ is the unit radial vector (Fig. 23.5). Given an arbitrary closed path γ around the origin, the radial flow pattern associated with the preceding field is said to be either a *source*, if

$$\int_\gamma \mathbf{v}\cdot\mathbf{n}\,d\sigma = Q > 0, \tag{23.89}$$

or a *sink*, if $Q < 0$. The quantity $Q/2\pi$ is called the *strength* of a source or sink.

Example 23.4. (Doublet). A *doublet* is the singularity obtained by taking to zero the distance between a source and a sink having the same strength. More precisely, consider a source and a sink of equal strength, the source placed at a point A, with $z_1 = -ae^{i\alpha}$, and the sink placed at B, with $z_2 = ae^{i\alpha}$. The complex potential for the combined flow is then

$$F(z) = \frac{Q}{2\pi}\ln(z + ae^{i\alpha}) - \frac{Q}{2\pi}\ln(z - ae^{i\alpha}) \equiv f(z, a). \tag{23.90}$$

The reader is invited to find the kinetic and Stokes potential (Fig. 23.6).

If points A and B are very near to each other, i.e., $a \simeq 0$, it follows that

$$F(z) = f(z, 0) + f'(z, 0)a = \frac{Q}{2\pi}\frac{1}{z}2ae^{i\alpha}.$$

In particular, if $a \to 0$, so that $Qa/\pi \to m$, then we obtain the potential of a doublet (source–sink):

Fig. 23.5 Source

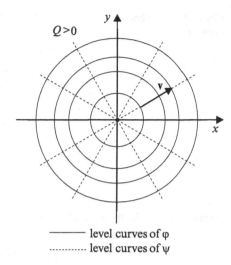

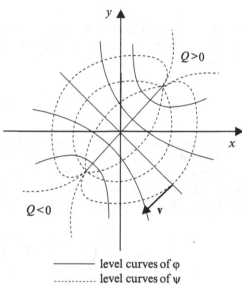

Fig. 23.6 Doublet

$$F(z) = \frac{m}{z} e^{i\alpha}. \tag{23.91}$$

Furthermore, if $\alpha = 0$, e.g., if the source and the sink are on the x-axis, then

$$F(z) = \frac{m}{z} = \frac{my}{x^2 + y^2} - i \frac{mx}{x^2 + y^2} \equiv \varphi + i\psi, \tag{23.92}$$

and the flow lines $\psi = $ const are circles through the origin.

Fig. 23.7 Flow about a
cylinder

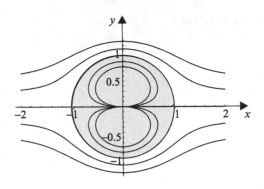

Example 23.5. Consider a 2D flow with complex potential

$$F(z) = U_0 z + \frac{U_0 a^2}{z}, \tag{23.93}$$

i.e., the flow is obtained as the superposition of a uniform flow parallel to the x-axis (in the positive direction) and the flow due to a dipole (source–sink) system of strength $m = U_0 a^2$. Assuming $z = x + iy$, then from (23.93) it follows that

$$\varphi = U_0 x \left(1 + \frac{a^2}{x^2 + y^2}\right), \qquad \psi = U_0 y \left(1 - \frac{a^2}{x^2 + y^2}\right).$$

The flow lines $\psi = $ const are represented in Fig. 23.7 (for $U_0 = a = 1$).

Note that the condition $x^2 + y^2 = a^2$ corresponds to the line $\psi = 0$. Furthermore, if the region internal to this circle is substituted by the cross section of a cylinder, then (23.93) can be assumed to be the complex potential of the velocity field around a cylinder. The components of the velocity field are given by

$$u = \frac{\partial \varphi}{\partial x} = U_0 \left(1 - a^2 \frac{x^2 - y^2}{r^4}\right) = U_0 \left(1 - \frac{a^2}{r^2} \cos 2\theta\right),$$
$$v = \frac{\partial \varphi}{\partial y} = U_0 \frac{a^2}{r^2} \sin 2\theta.$$

It should be noted that, although singular points associated with a doublet do not actually occur in real fluids, they are interesting because the flow pattern associated with a doublet is a useful approximation far from singular points, and it can be combined to advantage with other nonsingular complex potentials.

Example 23.6. All the previous examples, with the exception of Example 23.2, refer to 2D irrotational flows whose circulation along an arbitrary closed path is vanishing. We now consider the complex potential

Fig. 23.8 Flow about a cylinder with vorticity

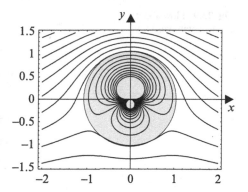

$$F(z) = U_0 z + \frac{U_0 a^2}{z} + i\frac{\Gamma}{2\pi}\ln\frac{z}{a}. \tag{23.94}$$

This represents a flow around a cylinder of radius a obtained by superposing a uniform flow, a flow generated by a dipole, and the flow due to a vortex. The dipole and the vortex are supposed to be located on the center of the section. In this case, according to (23.89), the vorticity is Γ, and it can be supposed that the vortex is produced by a rotation of the cylinder around its axis. Flow lines are given by the complex velocity

$$V = \frac{dF}{dz} = U_0(1 - \frac{a^2}{z^2}) + i\frac{\Gamma}{2\pi z}.$$

To find **stagnation points** ($V = 0$), we must solve the equation $dF/dz = 0$. Its roots are given by

$$z_{v=0} = \frac{1}{4U_0\pi}\left(-i\Gamma \pm \sqrt{\left(16U_0^2\pi^2 a^2 - \Gamma^2\right)}\right). \tag{23.95}$$

Three cases can be distinguished: $\Gamma^2 < 16\pi^2 a^2 U^2$, $\Gamma^2 = 16\pi^2 a^2 U^2$, and $\Gamma^2 > 16\pi^2 a^2 U^2$, whose corresponding patterns are shown in Figs. 23.8–23.10 (assuming $U_0 = a = 1$).

At this stage the reader should be aware that these examples, although of practical relevance, are based on equations of steady-state hydrodynamics, which in some cases give rise to inaccurate or paradoxical results. As an example, in the next section we will analyze D'Alembert's paradox, according to which the drag of a fluid on an obstacle is zero. To remove this paradox, it will be necessary to improve the model of a fluid taking into account the effects of viscosity.

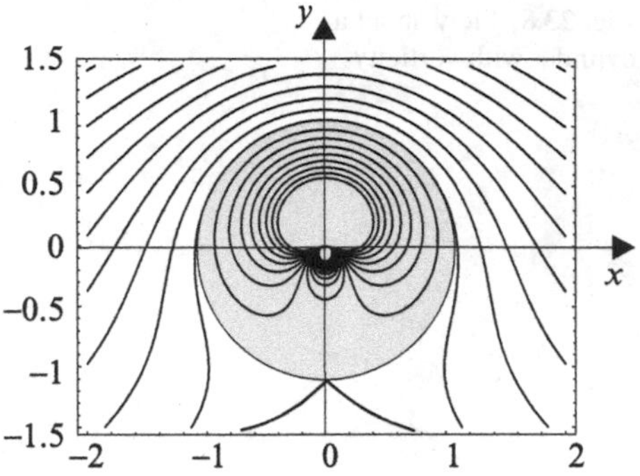

Fig. 23.9 Flow about a cylinder with vorticity

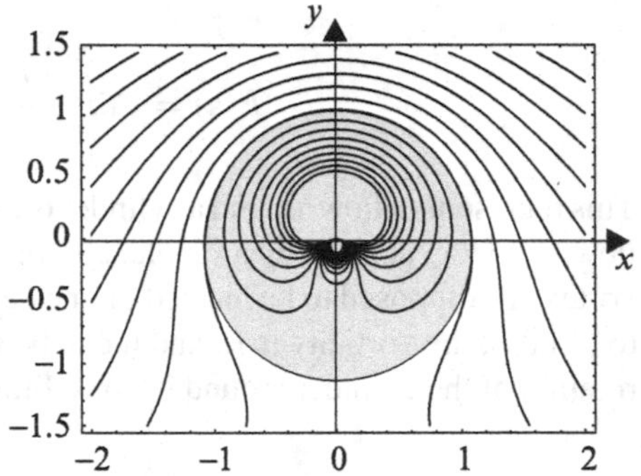

Fig. 23.10 Flow about a cylinder with vorticity

23.10   D'Alembert's Paradox and the Kutta–Joukowky Theorem

In this section, we prove that the model of a perfect liquid leads to the paradox that a body experiences no resistance while moving through a liquid (D'Alembert's paradox). This paradox shows that, although the model of a perfect liquid leads to results that are in agreement with many experimental results, in some cases it fails. This is due to the fact that this model does not take into account the viscosity of liquids. In other words, there are some cases where it is possible to neglect the effects of viscosity on motion and the model of a perfect liquid can be used. In other cases, it is fundamental to take into account these effects to obtain again agreement between theory and experimental results.

Let S be a solid of volume C placed in an *irrotational and steady* flow of a liquid in the absence of body forces (Fig. 23.11). Then (23.74) becomes

$$\rho v_j \frac{\partial v_i}{\partial v_j} = -\frac{\partial p}{\partial x_i}.$$

Further, taking into account the incompressibility condition (23.75), the preceding equation gives

Fig. 23.11 Flow about a
cylinder with vorticity

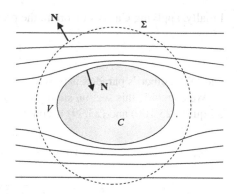

$$\frac{\partial}{\partial x_j}(p + v_i v_j) = 0. \tag{23.96}$$

Let V be an arbitrary *fixed* volume V containing the body S. Then, integrating (23.96) on the region $V - C$, applying Gauss's theorem, and recalling that $\mathbf{v}\cdot\mathbf{N} = 0$ on the whole boundary of C, we obtain the integral condition

$$\int_{\partial C} p\mathbf{N}\,d\sigma + \int_{\partial V}(p\mathbf{I} + \rho\mathbf{v} \otimes \mathbf{v})\cdot\mathbf{N}\,d\sigma = \mathbf{0}. \tag{23.97}$$

But the first integral is the opposite of the force $\mathbf{F}$ acting on S, so that the preceding condition gives

$$\mathbf{F} = -\int_{\partial V}(p\mathbf{I} + \rho\mathbf{v} \otimes \mathbf{v})\cdot\mathbf{N}\,d\sigma. \tag{23.98}$$

On the other hand, Bernoulli's theorem (23.66), in the absence of body forces, states that

$$p = p_0 + \frac{1}{2}\rho(V^2 - v^2), \tag{23.99}$$

where p_0 and V denote, respectively, the pressure and the *uniform* velocity in a liquid at infinity. We do not prove that in our hypothesis, liquid flow has the following asymptotic behavior:

$$\mathbf{v} = \mathbf{V} + O(r^{-3}), \tag{23.100}$$

where r is the distance of a point in the flow relative to an arbitrary origin. Introducing (23.100) into (23.98), taking into account the asymptotic behavior (23.99), and assuming that V is a sphere, we obtain that

$$\mathbf{F} = -(p_0\mathbf{I} - \rho\mathbf{V} \otimes \mathbf{V})\int_{\partial V}\mathbf{N}\,d\sigma + O(r^{-1}). \tag{23.101}$$

Finally, applying Gauss's theorem, the preceding condition gives

$$\mathbf{F} = \mathbf{0},\tag{23.102}$$

and D'Alembert's paradox is proved.

We conclude this section stating without proof (see [43]) that in a planar flow of a liquid, (23.100) and (23.99) become

$$\mathbf{v} = V\mathbf{i} + \frac{\Gamma}{2\pi r^2}(-y\mathbf{i} + x\mathbf{j}) + O(r^{-2}),\tag{23.103}$$

$$p = p_0 + \frac{1}{2}\rho\frac{\Gamma}{2\pi r^2}yV + O(r^{-2}),\tag{23.104}$$

where $\mathbf{i}$ and $\mathbf{j}$ are unit vectors along the Ox- and Oy-axes in the plane of the flow, the speed $\mathbf{V}$ at infinity of the flow is parallel to $\mathbf{i}$, and

$$\Gamma = \oint_\gamma \mathbf{v} \cdot \mathrm{d}\mathbf{s}$$

is the circulation along any curve γ surrounding body C. Then it is possible to prove that, instead of (23.103), we obtain the **Blausius formula**

$$\mathbf{F} = -\rho\Gamma V\mathbf{j}.\tag{23.105}$$

This formula shows that *although a steady flow of an inviscid fluid predicts no drag on an obstacle in the direction of relative velocity in an unperturbed region, it can predict a force normal to this direction.* This is a result obtained independently by W.M. Kutta in 1902 and N.E. Joukowski (sometimes referred as Zoukowskii) in 1906, known as the **Kutta–Joukowski theorem**. Such a force is called **lift**, and it is important for understanding why an airplane can fly.

Before closing this section, we observe that our inability to predict the drag of an inviscid fluid in the direction of relative velocity does not mean we should abandon the perfect fluid model. Viscosity plays an important role around an obstacle, but, far from the obstacle, the motion can still be conveniently described according to the assumption of an inviscid fluid.

23.11   Waves in Perfect Gases

This section is devoted to wave propagation in perfect fluids. Waves are considered to be small perturbations of an undisturbed state, corresponding to a homogeneous fluid at rest. This assumption allows us to linearize the motion equations and to apply elementary methods.[3]

---

[3]For an analysis of wave propagation in nonlinear media, see [44].

Consider a *compressible* perfect fluid at rest, with uniform mass density ρ_0, in the absence of body forces. Assume that motion is produced by a small perturbation, so that it is characterized by a velocity $\mathbf{v}$ and a density ρ that are only slightly different from $\mathbf{0}$ and ρ_0, respectively. More precisely, such an assumption states that $\mathbf{v}$ and $\sigma = \rho - \rho_0$ are first-order quantities together with their first-order derivatives.

In this case, the motion equations

$$\rho \dot{\mathbf{v}} = -\nabla p(\rho) = -p'(\rho)\nabla\rho, \tag{23.106}$$
$$\dot{\rho} + \rho\nabla \cdot \mathbf{v} = 0$$

can be linearized. In fact, neglecting the first-order terms of $\mathbf{v}$ and σ, we obtain

$$p'(\rho) = p'(\rho_0) + p''(\rho_0)\sigma + \cdots ,$$
$$\dot{\rho} = \dot{\sigma} = \frac{\partial\sigma}{\partial t} + \mathbf{v}\cdot\nabla\sigma = \frac{\partial\sigma}{\partial t} + \cdots ,$$
$$\dot{\mathbf{v}} = \frac{\partial\mathbf{v}}{\partial t} + \mathbf{v}\cdot\nabla\mathbf{v} = \frac{\partial\mathbf{v}}{\partial t} + \cdots ,$$

so that (23.106) become

$$\rho_0\frac{\partial\mathbf{v}}{\partial t} = -p'(\rho_0)\nabla\sigma, \tag{23.107}$$
$$\frac{\partial\sigma}{\partial t} + \rho_0\nabla \cdot \mathbf{v} = 0.$$

Consider now a sinusoidal 2D wave propagating in the direction of the vector $\mathbf{n}$ with velocity U and wavelength λ:

$$\mathbf{v} = \mathbf{a}\sin\frac{2\pi}{\lambda}(\mathbf{n}\cdot\mathbf{x} - Ut), \ \sigma = b\sin\frac{2\pi}{\lambda}(\mathbf{n}\cdot\mathbf{x} - Ut).$$

We want a solution to the previous system with this form. To find this solution, we note that

$$\frac{\partial\mathbf{v}}{\partial t} = -\frac{2\pi}{\lambda}U\mathbf{a}\cos\frac{2\pi}{\lambda}(\mathbf{n}\cdot\mathbf{x} - Ut), \ \frac{\partial\sigma}{\partial t} = -\frac{2\pi}{\lambda}Ub\cos\frac{2\pi}{\lambda}(\mathbf{n}\cdot\mathbf{x} - Ut),$$
$$\nabla\cdot\mathbf{v} = \frac{2\pi}{\lambda}\mathbf{n}\cdot\mathbf{a}\cos\frac{2\pi}{\lambda}(\mathbf{n}\cdot\mathbf{x} - Ut), \ \nabla\sigma = \frac{2\pi}{\lambda}\mathbf{n}b\cos\frac{2\pi}{\lambda}(\mathbf{n}\cdot\mathbf{x} - Ut),$$

so that system (23.107) becomes

$$\rho_0 U\mathbf{a} = p'(\rho_0)b\mathbf{n},$$
$$-Ub + \rho_0\mathbf{a}\cdot\mathbf{n} = 0,$$

i.e.,

$$(p'(\rho_0)\mathbf{n} \otimes \mathbf{n} - U^2)\mathbf{I}\mathbf{a} = 0,$$
$$(p'(\rho_0) - U^2\mathbf{b} = 0.$$

Assuming a reference frame whose axis Ox is oriented along $\mathbf{n}$, we obtain

$$\begin{aligned} U &= 0, \quad \mathbf{a} \perp \mathbf{n}, \\ U &= \pm\sqrt{p'(\rho_0)}, \quad \mathbf{a} = a\mathbf{n}, \end{aligned} \tag{23.108}$$

where the eigenvalue 0 has a multiplicity of 2. Thus we prove the existence of *dilational waves* propagating at a speed

$$U = \sqrt{p'(\rho_0)}.$$

The velocity U is called the **sound velocity**, and the ratio $m = v/U$ is known as the **Mach number**. In particular, the motion is called **subsonic** or **supersonic**, depending on whether $m < 1$ or $m > 1$.

Chapter 24
An Introduction to Celestial Mechanics

Abstract Celestial mechanics is one of the most interesting applications of classical mechanics. This topic answers one of the oldest questions of facing humankind: what forces govern the motion of celestial bodies? How do celestial bodies move under the action of these forces? In this chapter we discuss the foundations of this subject. We first recall the two-body problem and introduce the orbital elements. Then, we analyze the restricted three-body problem in which the third body has a very small mass compared with the masses of the other two bodies. In particular, we describe the Lagrange equilibrium positions and their stability. Then, we consider the N-body problem showing that the Hamiltonian of this system is obtained by adding the Hamiltonian of the two-body problem, that describes a completely integrable system, to another term that can be regarded as a perturbation term. In the remaining part of the chapter, we show how the gravitation law for mass points can be extended to continuous mass distributions, and we evaluate the asymptotic behavior of the gravitational field produced by extended bodies. Then, we define the local inertial frames and tidal forces. Finally, we pose the problem of determining the form of self-gravitating bodies.

24.1 A Brief Historical Overview of Celestial Mechanics

Celestial mechanics, as a scientific theory, began when Kepler formulated the famous three laws about the planetary motion with respect to the Sun. Newton realized the second fundamental step in the development of celestial mechanics. In fact, he stated the fundamental laws of dynamics and the gravitation law showing that Kepler's laws were a consequence of these theoretical statements. However, more accurate astronomical observations showed that the planetary motion has a behavior much more complex than the behavior deduced by Kepler's laws. In other words, Kepler's laws represented a first approximation of this motion and, consequently, Newton's laws seemed to not describe the motion of the planets with sufficient accuracy. At

© Springer International Publishing AG, part of Springer Nature 2018   505
A. Romano and A. Marasco, *Classical Mechanics with Mathematica®*,
Modeling and Simulation in Science, Engineering and Technology,
https://doi.org/10.1007/978-3-319-77595-1_24

that time nobody questioned Newton's laws and therefore everybody was sure that the discrepancies between experimental and theoretical results could be explained supposing that the motion of a planet depends not only on the solar action but also on the interaction of the planet with the other ones.

The observed discrepancies were of two kinds: first, there were disturbances which have no cumulative effect since they righted themselves after a fixed time. For this reason they were called *periodic* discrepancies. Much more serious were those anomalies that proceeded in the same sense, always increasing the departure from the Keplerian motion. They were called *secular* anomalies. Before Laplace, the best known of them was the *great anomaly of Jupiter and Saturn.*

Lagrange formulated the Newtonian equations of motion of a planet interacting with the Sun and the other planets in terms of orbital characteristics (Sections 24.3, 24.4). The right-hand sides of these equations contain the first derivatives of the *perturbation function* that is formulated in terms of the orbital parameters and takes into account the interactions among planets. This function can be developed as Fourier series and approximate solutions of Lagrange's equations can be evaluated. Laplace, starting from these equations, computed solutions to a higher order than his predecessors, and he found that in the final expression for the effect of Jupiter's disturbing action upon the mean motion of Saturn, the terms canceled each other out. The same result held for the effect of any planet upon the mean motion of any other: thus the mean motions of the planets cannot have any secular modification as a result of their mutual attractions. Laplace showed that the Jupiter and Saturn *great anomaly* is not a secular inequality, since it describes periodic anomalies of motion of both the planets with the very long period of 929 years. On the contrary, a typical secular effect is represented by the anomalous motion of the perihelion of the planets.

The above-mentioned approach allowed to explain also this secular effect, and the theoretical results were in agreement with the experimental ones except for Mercury for which there was a difference of 42" for a century between the theoretical and experimental data.

In analyzing the restricted three-body problem (Section 24.5) Poincarè discovered that, in spite of determinism of mechanical laws, chaos could be present in the motion of the body with the smallest mass. Systems of N mass points interacting by mutual gravitational forces can be studied only if $N = 2$ and partially if $N = 3$. For more complex systems, we must resort to the perturbation theory that supplies approximate motion solutions. In the first half of the 20th century, the perturbation theory became increasingly popular, and one of the main results of this period was KAM theorem that made the relation between regular motion and chaotic motions in the presence of a gravitational perturbation more precise.

24.2   The Orbital Elements

In this section we analyze again (see Section 14.3) the motion of an isolated system S of two material points P_1 and P_2 subject to the action of the mutual Newtonian gravitational force since this is a prototype problem of celestial mechanics. This is an important model since it gives a simplified description of the system Sun and any planet, a planet and its major satellite, and a system of isolated interacting double stars. Since P_1 and P_2 are arbitrary mass points, not necessarily Earth and Sun, we use **apocenter** instead of aphelion and **pericenter** instead of perihelion. From the hypothesis that the system S is isolated, it follows that the center of mass G of S moves uniformly on a straight line relative to an inertial frame I. Consequently, the frame of reference $I_G \equiv Gx'y'z'$, with the origin at the center of mass G and the axes parallel to the axes of I, is inertial. Instead of studying the motion of P_2 relative to P_1 (see Section 14.3), we now adopt $Gx'y'z'$ as a frame of reference and denote by $\mathbf{r}_1$ and $\mathbf{r}_2$ the position vectors relative to the origin G of I_G. Then we have

$$m_1\mathbf{r}_1 + m_2\mathbf{r}_2 = (m_1 + m_2)\mathbf{r}_G = \mathbf{0}, \tag{24.1}$$

where m_1, m_2 are the masses of P_1 and P_2, respectively. Further, Newton's equations of P_1 and P_2 in the inertial frame I_G are

$$m_1\ddot{\mathbf{r}}_1 = \mathfrak{h}\, m_1 m_2 \frac{\mathbf{r}_2 - \mathbf{r}_1}{|\mathbf{r}_2 - \mathbf{r}_1|^3}, \tag{24.2}$$

$$m_2\ddot{\mathbf{r}}_2 = -\mathfrak{h}\, m_1 m_2 \frac{\mathbf{r}_2 - \mathbf{r}_1}{|\mathbf{r}_2 - \mathbf{r}_1|^3}, \tag{24.3}$$

where $\mathfrak{h}$ is the gravitational constant.

Since from (24.1) we obtain

$$\mathbf{r}_2 = -\frac{m_1}{m_2}\mathbf{r}_1, \quad |\mathbf{r}_2 - \mathbf{r}_1| = \frac{m_1 + m_2}{m_2}|\mathbf{r}_1|,$$

we can give (24.2), (24.3) the following form:

$$(m_1 + m_2)^2\ddot{\mathbf{r}}_1 = -\mathfrak{h}\, m_2^3 \frac{\mathbf{r}_1}{|\mathbf{r}_1|^3}, \tag{24.4}$$

$$(m_1 + m_2)^2\ddot{\mathbf{r}}_2 = -\mathfrak{h}\, m_1^3 \frac{\mathbf{r}_2}{|\mathbf{r}_2|^3}. \tag{24.5}$$

These equations show that P_1 and P_2 move, respectively, along two conics Γ_1 and Γ_2 with a focus at the center of mass G. In view of (24.1), both the conics are fixed assigning the initial conditions only for one of the points P_1 and P_2. We suppose that the initial conditions are assigned in such a way that both the conics are ellipses. From Chapter 14 we know that each ellipse Γ_i, $i = 1, 2$, is contained in a *fixed* plane π_i. On the other hand (24.1) states that, at any instant, the points P_1, P_2, and G are collinear. Consequently, the planes π_1 and π_2 coincide and the points P_1 and P_2 move along Γ_1 and Γ_2 with the same angular velocity. Introduce polar coordinates (r, φ) with center at G and polar axis along the major axis of Γ_1 and suppose that the apocenter and the pericenter of Γ_1 correspond to $\varphi = 0$ and $\varphi = \pi$, respectively (see Fig. 24.1).

Fig. 24.1 Polar coordinates

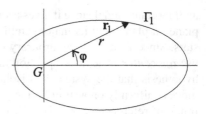

Fig. 24.2 A possible
configuration

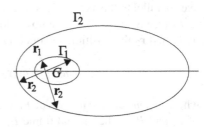

Fig. 24.3 Another possible
configuration

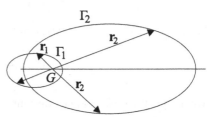

During the motion of P_1 along an ellipse (that does not reduce to a circle), the velocity verifies the condition $\mathbf{r}_1 \cdot \dot{\mathbf{r}}_1 = 0$ only at the apocenter and pericenter. From (24.1) we obtain that, if this condition is satisfied, then also the other condition $\mathbf{r}_2 \cdot \dot{\mathbf{r}}_2 = 0$ holds and we can state that P_2 is at its apocenter or pericenter. Further, still from (24.1), we obtain that $|\mathbf{r}_1|$ and $|\mathbf{r}_2|$ together assume the maximum or minimum value. Consequently, when P_1 is at its apocenter also P_2 is at its apocenter. The same situation holds at the pericenter. In other words, the major axes of Γ_1 and Γ_2 coincide. In Figs. 24.2 and 24.3 two possible configurations of Γ_1 and Γ_2 are shown.

In particular, if P_1 moves along a circumference with radius r_1, then also P_2 describes a circumference with radius $r_2 = (m_1/m_2)r_1$. It is possible to prove that if a_1 and a_2 are the semimajor axes of Γ_1 and Γ_2, then the following relation holds:

$$a_2 = \frac{m_1}{m_2} a_1. \tag{24.6}$$

We omit to prove that the masses m_1 and m_2 are related to a_1, a_2 and the common value T of the period by the formulae

$$m_1 = 4\pi^2 a_2 \frac{(a_1 + a_2)^2}{T^2}, \tag{24.7}$$

$$m_2 = 4\pi^2 a_1 \frac{(a_1 + a_2)^2}{T^2}. \tag{24.8}$$

These formulae are important since they allow to determine the masses m_1 and m_2 of P_1 and P_2 by measurements of the period and their distances from G. Finally, it is possible to prove that the ellipses intersect when their eccentricities e_i, $i = 1, 2$, verify the condition

$$e_i > \frac{|m_1 - m_2|}{m_1 + m_2}.$$

24.3 Orbital Elements a, e, and E

Let S be an isolated system of two mass points P_1 and P_2 with mass m_1 and m_2, respectively. In Section 14.3 we proved that the motion of P_2 relative to P_1, that is, the motion of P_2 in the *noninertial* frame of reference R with the origin at P_1 and the axes parallel to the axes of an inertial frame I, is described by the equation

$$\ddot{\mathbf{r}} = -\mathfrak{h} \frac{(m_1 + m_2)}{r^3} \mathbf{r}, \tag{24.9}$$

where

$$\mathbf{r} = \mathbf{r}_2 - \mathbf{r}_1 \tag{24.10}$$

is the position vector of P_2 with respect to P_1. We showed that (24.9) can be solved exactly and admits one and only one solution when the initial conditions $\mathbf{r}(0)$ and $\dot{\mathbf{r}}(0)$ are given. If the total energy is negative, then the orbit is elliptic. Now it is fundamental to introduce the **orbital elements**, that, is quantities that give the geometry of the ellipse, the position of P_2 on the ellipse, and the orientation of the ellipse relative to the noninertial frame to which (24.9) refers.

Let us first introduce quantities that describe the shape of the ellipse $\mathfrak{E}$. The shape of $\mathfrak{E}$ is completely described by the semimajor axis a and the semiminor axis b or, equivalently, by a and the eccentricity e of $\mathfrak{E}$. In Section 14.2 a conic Γ was defined as the locus of points for which the ratio between the distance of $P \in \Gamma$ from a straight line (the directrix) and a point (focus) is constant. This constant ratio was called the eccentricity of the conic. For an ellipse $\mathfrak{E}$ the eccentricity can also be defined in other ways. For instance, we prove that

$$e = \frac{CF}{a}, \tag{24.11}$$

where CF is the distance between the focus F and center C of the ellipse (see Fig. 24.4).

Fig. 24.4 Definition of a, e, and E

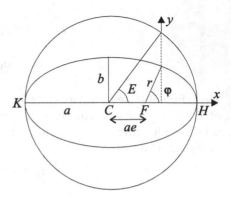

In fact, from the equation of the ellipse in polar coordinates with center at the focus F

$$r = \frac{p}{1 + e\cos\varphi},$$ (24.12)

we have that

$$FH = \frac{p}{1 + e}, \quad FK = \frac{p}{1 - e}.$$ (24.13)

Consequently, it is

$$2a = FH + FK = \frac{2p}{1 - e^2}$$

and we can write that

$$p = a(1 - e^2).$$ (24.14)

On the other hand, by (24.13) and (24.14) we obtain that

$$CF = a - FH = a - a(1 - e) = e\,a$$ (24.15)

and (24.11) is proved.

Let us denote by $\overline{\varphi}$ the angle for which the semiminor axis is given by $b = r\sin\overline{\varphi}$. This value is obtained by the condition $dy/d\varphi = d(r\sin\varphi)/d\varphi = 0$ that, in view of (24.12), gives

$$\cos\overline{\varphi} = -e, \quad \sin\overline{\varphi} = \sqrt{1 - e^2}.$$

Therefore, it is

$$b = r\sin\overline{\varphi} = \frac{p\sin\overline{\varphi}}{1 + e\cos\overline{\varphi}} = \frac{p\sqrt{1 - e^2}}{1 - e^2}$$

In view of (24.14) we obtain the relation between the semiminor axis b, the semimajor axis a and the eccentricity

$$b = a\sqrt{1 - e^2}.$$ (24.16)

Recall that the parametric equations of the ellipse with the center at C, when we take into account (24.15) and (24.16), are

$$x + ea = a \cos E, \qquad (24.17)$$

$$y = a\sqrt{1 - e^2} \sin E, \qquad (24.18)$$

where $E \in [0, 2\pi]$ is called **eccentric anomaly** and its meaning is shown in Fig. 24.4.

We can now find the position of P_2 on the ellipse in terms of E. To this end, it is sufficient to express r and φ in terms of a, e, and E. First, from (24.16), (24.17) we have

$$r^2 = x^2 + y^2 = a^2 - a^2 e^2 \sin^2 E - 2ea^2 \cos E$$
$$= a^2 + a^2 e^2 (1 - \sin^2 E) - 2ae \cos E = a^2 (1 - e \cos E)^2.$$

Since $\cos \varphi = x/r$, we finally we obtain

$$r = a(1 - e \cos E), \qquad (24.19)$$

$$\cos \varphi = \frac{\cos E - e}{1 - e \cos E}. \qquad (24.20)$$

In conclusion, the shape of the ellipse is determined by a and e and the position of P_2 on the ellipse is determined by E. Therefore, we need an equation determining the eccentric anomaly E as a function of time.

Before formulating this equation we introduce the orbital frequency or **mean motion** n and the **mean anomaly** M of the moving point P_2 by the positions

$$n = \frac{2\pi}{T}, \qquad (24.21)$$

$$M = n(t - t_0), \qquad (24.22)$$

where T is the period of P_2 and t_0 is the time of passage to pericenter. From Kepler's third law $T^2/a^3 = 4\pi^2/\hbar(m_1 + m_2)$ and (24.21), we derive that

$$n = \sqrt{\hbar(m_1 + m_2)} a^{-3/2}. \qquad (24.23)$$

Now, it is possible to prove the following **Kepler's equation**

$$n(t - t_0) = E - e \sin E, \qquad (24.24)$$

that, at least in principle, allows to evaluate the eccentric anomaly as a function of t, when the shape of the ellipse (a and e) is known starting from the initial data.

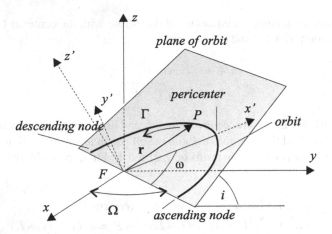

Fig. 24.5 Definition of the orbital elements i, Ω, and ω

24.4   Orbital Elements i, Ω, and ω

We now characterize the orientation of the ellipse Γ in space with respect to the (non-inertial) frame R with the origin at a focus F at which there is the central point P_1. We reach this objective by introducing the **orbital elements** that are three additional angles. Denote by $Fxyz$ Cartesian axes at rest in R. The first angle i gives the inclination of the orbital plane with respect to the reference plane Fxy (see Fig. 24.5). If $i \neq 0$, then the orbit intersects Fxy in two points called the **nodes** of the orbit. Astronomers distinguish between an *ascending* node, where the point P_2 passes from negative to positive values of z, and a *descending* node when P_2 moves toward negative values of z. The angle Ω between the axis Fx and the ascending node is called the *longitude of nodes*. The last angle ω gives the orientation of the ellipse Γ in the plane of orbit. This angle is called *argument of pericenter* and defines the angular position of pericenter with respect to the line connecting the focus F with the ascending node.

Consider a Cartesian frame of reference $Fx'y'z'$ with the plane $Fx'y'$ coinciding with the plane containing Γ and the axis Fx' containing the pericenter. Finally, the axis Fz' is orthogonal to the plane of orbit. It is important to evaluate how the components of the position vector $\mathbf{r}$ and velocity $\dot{\mathbf{r}}$ of P_2 with respect to $Fx'y'z'$ are related to the components of these vectors relative to the frame $Fxyz$.

First, in view of (24.17) and (24.18), the components of $\mathbf{r}$ relative to $Fx'y'z'$ are

$$x' = a(\cos E - e), \tag{24.25}$$

$$y' = a\sqrt{1 - e^2} \sin E. \tag{24.26}$$

$$z' = 0. \tag{24.27}$$

Further, from Kepler's equation (24.24)

$$f(t, E) = E - e \sin E - n(t - t_0) = 0,$$

that implicitly defines function $E(t)$, and Dini's theorem we obtain

$$\dot{E} = -\frac{\partial f / \partial t}{\partial f / \partial E} = \frac{n}{1 - e \cos E}. \tag{24.28}$$

Taking into account this result, we easily derive the velocity components in the frame $Fx'y'z'$

$$\dot{x}' = -n \frac{a \sin E}{1 - e \cos E}, \tag{24.29}$$

$$\dot{y}' = n \frac{a\sqrt{1 - e^2} \cos E}{1 - e \cos E}. \tag{24.30}$$

$$\dot{z}' = 0. \tag{24.31}$$

We obtain the components of $\mathbf{r}$ and $\dot{\mathbf{r}}$ relative to the Cartesian frame $Fxyz$ if we determine the orthogonal matrix Q such that

$$\begin{pmatrix} x \\ y \\ z \end{pmatrix} = Q \begin{pmatrix} x' \\ y' \\ 0 \end{pmatrix}. \tag{24.32}$$

The matrix Q can easily be obtained by noticing that the angles Ω, i, and ω have, respectively, the same meaning of Euler's angles ψ, θ, and φ that we met in the kinematics of a rigid body with a fixed point (see Section 12.4, formulae 12.32).

We conclude with some formulae that will be useful later. We start proving that the energy conservation relative to equation (24.9)

$$\mathfrak{H} = \frac{1}{2}|\dot{\mathbf{r}}|^2 - \mathfrak{h}\frac{m_1 + m_2}{|\mathbf{r}|}, \tag{24.33}$$

can be written in the form

$$\mathfrak{H} = -\mathfrak{h}\frac{m_1 + m_2}{2a}. \tag{24.34}$$

In fact, in view of (24.25)–(24.27), (24.29)–(24.31), we have

$$|\dot{\mathbf{r}}|^2 = \dot{x}'^2 + \dot{y}'^2 = n^2 \frac{a^2 \sin^2 E}{(1 - \cos e)^2} + n^2 \frac{a^2(1 - e^2)\cos^2 E}{(1 - \cos e)^2}$$

$$= n^2 a^2 \frac{1 - e^2 \cos^2 E}{(1 - E \cos E)^2} = n^2 a^2 \frac{1 + e \cos E}{1 - e \cos E}.$$

Recalling (24.23) we obtain

$$\frac{1}{2}|\dot{\mathbf{r}}|^2 = \mathfrak{h}\frac{m_1 + m_2}{2a}\frac{1 + e\cos E}{1 - e\cos E}. \tag{24.35}$$

On the other hand, in view of (24.19), the second term on the right-hand side of (24.33) becomes

$$-\mathfrak{h}\frac{m_1 + m_2}{|\mathbf{r}|} = -\mathfrak{h}\frac{m_1 + m_2}{a}\frac{1}{1 - e\cos E}. \tag{24.36}$$

Finally, introducing (24.35) into (24.33) and taking into account (24.36), we easily prove (24.34).

We now evaluate the modulus of angular momentum $\mathbf{K}_F$ in the frame $Fxyz$ and its component along the axis Oz. We first notice that the angular momentum $\mathbf{K}_F$ is orthogonal to the orbit plane so that it has only a component along the axis Fz'. On the other hand, the modulus of $\mathbf{K}_F$ coincides with the absolute value of its component $K_{z'}$ along Fz'. Further, $|\mathbf{K}_F|$ is invariant in the frame change $Fx'y'z' \rightarrow Fxyz$. In conclusion, in view of (24.25)–(24.27), (24.29)–(24.31), we have that

$$K_{z'} = x'\dot{y}' - y'\dot{x}'$$
$$= \frac{a^2 n\sqrt{1 - e^2}}{1 - e\cos E}[(\cos E - e)\cos E + \sin^2 E] = a^2 n\sqrt{1 - e^2} > 0.$$

Recalling (24.23) we finally obtain

$$|\mathbf{K}_F| = \sqrt{\mathfrak{h}(m_1 + m_2)a}, \tag{24.37}$$

and

$$K_z = \cos i\sqrt{\mathfrak{h}(m_1 + m_2)a}. \tag{24.38}$$

24.5 The Restricted Three-Body Problem

In Chapter 14 and the preceding sections, we have shown that the motion of two mass points acted upon by mutual gravitational forces can completely be solved. If we consider a system of more than two mass points, then we are able to exhibit a closed solution only in some particular cases. In general, for systems of more than two mass points we only succeed in obtaining approximate solutions.

We now consider a system S of three mass points P_1, P_2, and P_3 with masses m_1, m_2, and m_3, respectively, moving under the action of the mutual gravitational forces. The **restricted three-body problem** consists in analyzing the motion of S when the following conditions are verified:

- $m_2 > m_1$, $m_1 + m_2 \gg m_3$;
- The orbits of P_1 and P_2 are circular;
- The trajectory of P_3 is contained in the plane of the orbits of the binary system P_1, P_2.

This mathematical model supplies a good description of the system Earth–Moon-spacecraft or Sun-Jupiter-asteroid.

Owing to the first hypothesis, we can state that the motion of P_1 and P_2 is not modified by the presence of P_3 that becomes a test particle in the gravitational field produced by P_1 and P_2. Then, we can neglect the action of P_3 on P_1, P_2 and the motion of these two mass points is completely determined. We know that their trajectories are ellipses with a focus at their center of mass G. On the other hand, the hypothesis that these orbits are circular leads to the conclusion that P_1 and P_2 describe two circumferences Γ_1 and Γ_2 with center at G and radii $|\mathbf{r}_1|$ and $|\mathbf{r}_2|$ given by (see (24.1))

$$|\mathbf{r}_1| = \frac{m_2}{m_1 + m_2} d, \quad |\mathbf{r}_2| = \frac{m_1}{m_1 + m_2} d, \tag{24.39}$$

where $d = |\mathbf{r}_1| + |\mathbf{r}_2|$ is the distance between P_1 and P_2. Moreover, these points rotate about G with the same constant angular velocity (see Fig. 24.6).

To simplify calculations, we introduce a convenient system of units. First, we adopt the mass unit so that

$$M \equiv m_1 + m_2 = 1, \tag{24.40}$$

and introduce the notations

$$m_1 = \mu, \quad m_2 = 1 - \mu, \tag{24.41}$$

where $\mu < \frac{1}{2}$, since $m_2 > m_1$. Then, we adopt as unit length the distance d, that is we put

$$d = 1. \tag{24.42}$$

Fig. 24.6 Orbits in the restricted three-body system

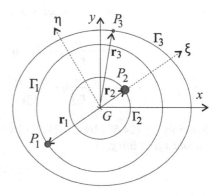

Finally, we adopt a time unit for which

$$\mathfrak{h} = 1. \tag{24.43}$$

To explain this choice, we notice that from Kepler's third law in the form (24.23) we obtain

$$n^2 d^3 = \mathfrak{h}(m_1 + m_2) \Rightarrow n = 1 \tag{24.44}$$

and (24.39) becomes

$$|\mathbf{r}_1| = 1 - \mu, \quad |\mathbf{r}_2| = \mu. \tag{24.45}$$

Since P_1 and P_2 rotate uniformly with angular velocity n about the center of mass G and, at any instant, belong to a straight line containing G, we can introduce a new system of Cartesian axes $G\xi\eta$ such that $G\xi$ contains P_1 and P_2 (*sinodic system*). In this frame of reference P_1 and P_2 occupy fixed positions on $G\xi$. In writing the equation of motion relative to $G\xi\eta$ of P_3 we must take into account the fictitious forces. Therefore, we have

$$\ddot{\mathbf{r}}_3 = -\mathfrak{h}\frac{m_1}{|\mathbf{r}_3 - \mathbf{r}_1|^3}(\mathbf{r}_3 - \mathbf{r}_1) - \mathfrak{h}\frac{m_2}{|\mathbf{r}_3 - \mathbf{r}_2|^3}(\mathbf{r}_3 - \mathbf{r}_2) + n^2\mathbf{r}_3 - 2n\mathbf{k} \times \dot{\mathbf{r}}_3, \tag{24.46}$$

where $\mathbf{k}$ is a unit vector orthogonal to the plane $G\xi\eta$. Denote by $\mathbf{i}$ and $\mathbf{j}$ the unit vectors along the axes $G\xi$ and $G\eta$, respectively. In the frame $G\xi\eta$, when we recall the measure units we have adopted, we have

$$\mathbf{r}_1 = -(1 - \mu)\mathbf{i}, \quad \mathbf{r}_2 = \mu\mathbf{i}, \quad \mathbf{r}_3 = \xi\mathbf{i} + \eta\mathbf{j}, \tag{24.47}$$

$$r_{13} \equiv |\mathbf{r}_3 - \mathbf{r}_1| = \sqrt{(\xi + 1 - \mu)^2 + \eta^2}, \tag{24.48}$$

$$r_{23} \equiv |\mathbf{r}_3 - \mathbf{r}_2| = \sqrt{(\xi - \mu)^2 + \eta^2}. \tag{24.49}$$

Since $\mathfrak{h} = n = 1$, the components of (24.46) along the axes of $G\xi\eta$ are

$$\ddot{\xi} = -\frac{\mu}{r_{13}^3}(\xi + 1 - \mu) - \frac{1 - \mu}{r_{23}^3}(\xi - \mu) + \xi + 2\dot{\eta}, \tag{24.50}$$

$$\ddot{\eta} = -\frac{\mu}{r_{13}^3}\eta - \frac{1 - \mu}{r_{23}^3}\eta + \eta - 2\dot{\xi}. \tag{24.51}$$

Multiplying (24.50) by $\dot{\xi}$, (24.51) by $\dot{\eta}$ and adding the obtained results, we find *Jacobi's first integral*

$$\mathfrak{J} \equiv \frac{1}{2}(\dot{\xi}^2 + \dot{\eta}^2) + W, \tag{24.52}$$

where $\mathfrak{J}$ is constant and

$$W = -\frac{1}{2}(\xi^2 + \eta^2) - \frac{\mu}{r_{13}} - \frac{1-\mu}{r_{23}}. \tag{24.53}$$

The first integral (24.52) implies that during the motion the following condition must be verified

$$\mathfrak{J} - W \geq 0 \Rightarrow \mathfrak{J} \geq W. \tag{24.54}$$

It is evident that the value of $\mathfrak{J}$ is determined by giving the initial conditions. Then, inequality (24.54) defines the regions of the plane $G\xi\eta$ that are accessible to P_3. The curves

$$W(\xi, \eta) = c, \tag{24.55}$$

obtained on varying the constant c, have the property that along them the velocity of P_3 relative to the frame $G\xi\eta$ vanishes. They are called **Hill's curves** and Fig. 24.7 shows these curves for $\mu = 0.2$. Figure 24.8 shows the 3D-plot of the surface $\zeta = W(\xi, \eta)$ and its intersection with the plane $\zeta = -2$. It is evident that if the initial data, that determine the value of c, are such that the point P_3 occupies a position inside a closed Hill's curve Γ, then P_3 cannot leave the region inside Γ.

Fig. 24.7 Hill's curve for $\mu = 0.2$

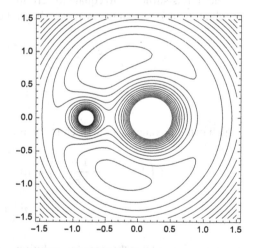

Fig. 24.8 Hill's surface for $\mu = 0.2$ and $c = -2$

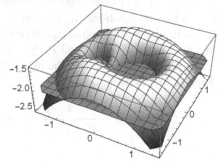

Fig. 24.9 Lagrange's
equilibrium positions

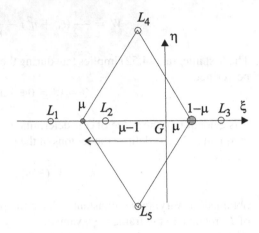

24.6 Equilibria of the Restricted Three-Body Problem

In order to find if there are *relative* equilibrium positions in the restricted three-body
problem, it is sufficient to equate to zero the right-hand sides of (24.50) and (24.51)
in which $\dot{\xi} = \dot{\eta} = 0$. In this way we obtain the system

$$-\frac{\mu}{r_{13}^3}(\xi + 1 - \mu) - \frac{1 - \mu}{r_{23}^3}(\xi - \mu) + \xi = 0, \qquad (24.56)$$

$$-\frac{\mu}{r_{13}^3}\eta - \frac{1 - \mu}{r_{23}^3}\eta + \eta = 0, \qquad (24.57)$$

which can equivalently be written in the form

$$\xi\left(1 - \frac{\mu}{r_{13}^3} - \frac{1 - \mu}{r_{23}^3}\right) + \mu(1 - \mu)\left(\frac{1}{r_{23}^3} - \frac{1}{r_{13}^3}\right) = 0, \qquad (24.58)$$

$$\eta\left(1 - \frac{\mu}{r_{13}^3} - \frac{1 - \mu}{r_{23}^3}\right) = 0. \qquad (24.59)$$

Lagrange proved that there are five solutions of this system: three of them correspond
to equilibrium positions L_1, L_2, and L_3 located on the axis $G\xi$ whereas the other
two solutions correspond to equilibrium positions L_4 and L_5 located in such a way
that the triangles $P_1 P_2 L_4$ and $P_1 P_2 L_5$ are equilateral.

To find the equilibria on the axis $G\xi$ it is sufficient to put $\eta = 0$ into (24.59).
Then, from (24.58) we obtain

$$\xi = \frac{1 - \mu}{|\xi - \mu|(\xi - \mu)} + \frac{\mu}{|\xi + 1 - \mu|(\xi + 1 - \mu)}. \qquad (24.60)$$

Fig. 24.10 Plots of the
right-hand side of (24.60)
and the function $\eta = \xi$

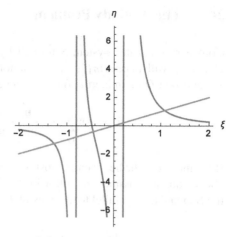

Figure (24.10) shows the plots of the function on the right-hand side of (24.60) and $\eta = \xi$ for $\mu = 0.2$. It is evident that there are three real roots of (24.60) located at the points qualitatively shown in Fig. 24.10. We omit the explicit evaluation of these roots. Let us now consider the equilibrium positions that do not belong to the axis $G\xi$. Since $\eta \neq 0$, from (24.59) we have that

$$1 - \frac{\mu}{r_{13}^3} - \frac{1-\mu}{r_{23}^3} = 0, \tag{24.61}$$

whereas (24.58) gives

$$r_{13} = r_{23}. \tag{24.62}$$

Introducing (24.62) into (24.61), we finally obtain $r_{13} = r_{23} = 1$. This result states that the equilibrium points L_4 and L_5 belong to the intersection of two circumferences with unit radius and centers at points P_1 and P_2. It is easy to verify that the coordinates of these points are

$$L_4 = \left(-\frac{1}{2} + \mu, \frac{\sqrt{3}}{2}\right), \quad L_5 = \left(-\frac{1}{2} + \mu, -\frac{\sqrt{3}}{2}\right). \tag{24.63}$$

We conclude without proving that the collinear equilibrium positions are unstable. Further, the two triangular equilibrium positions are unstable when $\mu/(1-\mu)$ exceeds the value 0.03852.

24.7   The N–Body Problem

Consider an isolated system S formed by the Sun, with mass m_0, and N planets $P_1 \ldots P_N$, with masses $m_1, \ldots, m_N$. Denoting by $\mathbf{u}_i$ the position vector of P_i relative to the origin O of an inertial frame I, the equations of motion of S in I are

$$\ddot{\mathbf{u}}_i = -\hbar \sum_{i \neq j} m_j \frac{\mathbf{u}_i - \mathbf{u}_j}{|\mathbf{u}_i - \mathbf{u}_j|^3}, \quad i, j = 0, \ldots N. \tag{24.64}$$

If we introduce the heliocentric position vectors $\mathbf{r}_i = \mathbf{u}_i - \mathbf{u}_0$ of the planets, then the above equations assume the following form in the frame of reference with origin at the Sun and axes parallel to the axes of I

$$\ddot{\mathbf{r}}_i = -\hbar \frac{m_0 + m_i}{|\mathbf{r}_i|^3} + \sum_{j=1, j \neq i}^{N} \hbar m_j \left(\frac{\mathbf{r}_j - \mathbf{r}_i}{|\mathbf{r}_j - \mathbf{r}_i|^3} - \frac{\mathbf{r}_j}{|\mathbf{r}_j|^3} \right), \tag{24.65}$$

$i = 1 \ldots, N$. After determining the vector functions $\mathbf{r}_i(t)$, we obtain the motion of Sun in the inertial frame I by the relation

$$\mathbf{u}_0 = -\frac{\sum_{i=1}^{N} m_i \mathbf{r}_i}{\sum_{i=0}^{N} m_i}. \tag{24.66}$$

Equations (24.65) constitute what is called the *(N + 1)–body problem*.

Suppose the mass m_1 of the mass point $P_1 \in S$ is so small with respect to the masses of $m_0, m_2, \ldots, m_N$ of the other points of S that P_1 can be considered as a test particle in the gravitational field of the other points. In other words, P_1 is acted upon by the gravitational action of the other mass points but it does not influence the motion of $P_0, P_2, \ldots, P_N$. Further, we suppose that the position vectors $\mathbf{r}_i, i = 0, 2, \ldots, N$, are known functions of time; in other words, we suppose that the motion of $P_0, P_2, \ldots, P_N$ is known. Then, from (24.65) we have the following equation in the only unknown $\mathbf{r}_1(t)$:

$$\ddot{\mathbf{r}}_1 = -\hbar \frac{m_0}{|\mathbf{r}_1|^3} \mathbf{r}_1 + \sum_{j=0, j \neq 1}^{N} \hbar m_j \left(\frac{\mathbf{r}_j - \mathbf{r}_1}{|\mathbf{r}_j - \mathbf{r}_1|^3} - \frac{\mathbf{r}_j}{|\mathbf{r}_j|^3} \right). \tag{24.67}$$

It is important to notice that this equation is nonautonomous since the right-hand side is a known function of time. Equation (24.67) describes the *(N + 1)–body restricted problem*.

Searching for solutions of (24.65) and (24.67) in a closed form is hopeless. We must search for approximate solutions of (24.65) and (24.67) resorting to a convenient mathematical approach. A nondimensional analysis of (24.65) and (24.67) shows us that we can attempt with a perturbation method since the second terms on the right-hand side of (24.65) and (24.67) are much smaller than the first term on the same side. To carry out this nondimensional analysis, we must introduce convenient reference quantities. To this end we introduce as length of reference L the semimajor axis of the terrestrial orbit and as time of reference the terrestrial revolution period T. In view of Kepler's third law, we have

$$\frac{T^2}{L^3} = \frac{4\pi^2}{\mathfrak{h}\, m_0},$$ (24.68)

where m_0 is the solar mass. Then, if we denote with the same symbols the nondimensional quantities and the dimensional ones and suppose that $m_i \ll m_0, i = 1, \ldots, N$, then (24.65) assumes the form

$$\frac{L}{T^2}\ddot{\mathbf{r}}_i = -\frac{1}{L^2}\mathfrak{h}\,\frac{m_0 + m_i}{|\mathbf{r}_i|^3}\mathbf{r}_i + \frac{1}{L^2}\sum_{j=1, j\neq i}^{N}\mathfrak{h}\, m_j\left(\frac{\mathbf{r}_j - \mathbf{r}_i}{|\mathbf{r}_j - \mathbf{r}_i|^3} - \frac{\mathbf{r}_j}{|\mathbf{r}_j|^3}\right).$$ (24.69)

Taking into account (24.68), we finally have

$$\ddot{\mathbf{r}}_i = -4\pi^2\frac{\mathbf{r}_i}{|\mathbf{r}_i|^3} + \frac{4\pi^2}{m_0}\sum_{j=1, j\neq i}^{N} m_j\left(\frac{\mathbf{r}_j - \mathbf{r}_i}{|\mathbf{r}_j - \mathbf{r}_i|^3} - \frac{\mathbf{r}_j}{|\mathbf{r}_j|^3}\right).$$ (24.70)

A similar result can be obtained for (24.67). If we suppose that *all the quantities* $|\mathbf{r}_i|$ *and* $|\mathbf{r}_i - \mathbf{r}_j|$ *are of the same order at any instant*, then the second term on the right-hand side of (24.70) is much smaller than the first one because of the ratio m_j/m_0. This result suggests to regard the second terms on the right-hand side of (24.65) and (24.67) as a small perturbation of the first term.

24.8 Hamiltonian Form of N–Body Problem

We want to apply Hamiltonian perturbation method to N–body problem. To this end we have first to find the right Hamiltonian functions corresponding to equations (24.65), (24.67), respectively. Then, we must determine the angle–action variables for the unperturbed problems relative to the above equations and show that both the corresponding Hamiltonian functions can be written as the sum of a function only depending on the action variables and a small function depending on all the angle–action variables.

We begin analyzing the restricted N–body problem (24.67). First, we can easily verify that the following function

$$U(\mathbf{r}, t) = -\mathfrak{h}\frac{m_0}{|\mathbf{r}|} - \mathfrak{h}\sum_{j=1}^{N} m_j\left(\frac{1}{|\mathbf{r}_j - \mathbf{r}|} - \frac{\mathbf{r}\cdot\mathbf{r}_j}{|\mathbf{r}_j|}\right)$$ (24.71)

is the potential energy of the point P since the force on the right-hand side of (24.67)
is given by $-\nabla_{\mathbf{r}} U$. Then, the Hamiltonian function of the restricted problem is

$$\mathfrak{H} = \left(\frac{1}{2}|\mathbf{v}|^2 - \mathfrak{h}\frac{m_0}{|\mathbf{r}|}\right) - \mathfrak{h}\sum_{j=1}^{N} m_j\left(\frac{1}{|\mathbf{r}_j - \mathbf{r}|} - \frac{\mathbf{r}\cdot\mathbf{r}_j}{|\mathbf{r}_j|}\right). \tag{24.72}$$

Adopting Cartesian coordinates in the frame R, the Lagrangian coordinates are the
components r_1, r_2, r_3 of $\mathbf{r}$ and the corresponding momenta are $p_1 = \dot{r}_1, \ldots, p_3 = \dot{r}_3$.

We now consider the planetary system S, that is, the $(N + 1)$–body problem
described by (24.64). After multiplying both sides of this equation by m_i, we can
easily see that the force on the right-hand side is given by $-\nabla U(\mathbf{u}_0, \ldots, \mathbf{u}_N)$, where
U is the following potential energy:

$$U(\mathbf{u}_0, \ldots, \mathbf{u}_N) = -\mathfrak{h}\sum_{j=1}^{N}\sum_{i=0}^{j-1}\frac{m_i m_j}{|\mathbf{u}_i - \mathbf{u}_j|}. \tag{24.73}$$

Therefore, the Hamiltonian function of (24.64) is

$$\mathfrak{H} = T + U = \frac{1}{2}\sum_{j=0}^{N} m_j|\dot{\mathbf{u}}_j|^2 - \mathfrak{h}\sum_{j=1}^{N}\sum_{i=0}^{j-1}\frac{m_i m_j}{|\mathbf{u}_i - \mathbf{u}_j|}, \tag{24.74}$$

so that the Lagrangian coordinates are the components in the inertial frame I
of the position vectors $\mathbf{u}_0, \ldots, \mathbf{u}_N$ and the corresponding momenta are $(p_\alpha) =
(m_0\dot{\mathbf{u}}_0, \ldots, m_N\dot{\mathbf{u}}_N)$, $\alpha = 0, \ldots, 3N$. We can also adopt as Lagrangian coordi-
nates the variables $(q^\alpha) = (\mathbf{r}_0, \ldots, \mathbf{r}_N)$, $\alpha = 0, \ldots, 3N$. However, if we want
that in the new variables Hamilton's equations preserve the canonical form, we have
to find the momenta $(p'_\alpha) = (\mathbf{p}'_0, \ldots, \mathbf{p}'_N)$ in such a way that the transformation
$(q^\alpha, p_\alpha) = (\mathbf{u}_i, m_i\dot{\mathbf{u}}_i) \rightarrow (q'^\alpha, p'_\alpha) = (\mathbf{r}_i, \dot{\mathbf{p}}_i)$ is canonical. To solve this problem,
it is sufficient to notice that $(\mathbf{u}_i, m_i\dot{\mathbf{u}}_i)$ are natural canonical coordinates. Further,
changing the Lagrangian coordinates, we obtain new natural canonical coordinates
by the transformation

$$q'^\alpha = q'^\alpha(\mathbf{q}), \quad p'_\alpha = \frac{\partial q^\beta}{\partial q'^\alpha}p_\beta. \tag{24.75}$$

In our case the Lagrangian coordinate transformation is

$$\mathbf{r}_0 = \mathbf{u}_0, \ \mathbf{r}_1 = \mathbf{u}_1 - \mathbf{u}_0, \ldots, \mathbf{r}_N = \mathbf{u}_N - \mathbf{u}_0. \tag{24.76}$$

whose inverse is

$$\mathbf{u}_0 = \mathbf{r}_0, \ \mathbf{u}_1 = \mathbf{r}_1 + \mathbf{r}_0, \ldots, \mathbf{u}_N = \mathbf{r}_N + \mathbf{r}_0. \tag{24.77}$$

Consequently, in view of (24.77), (24.75), we obtain the new momenta

$$\mathbf{v}_0 = m_0\dot{\mathbf{u}}_0 + \cdots + m_N\dot{\mathbf{u}}_N, \quad \mathbf{v}_i = m_i\dot{\mathbf{u}}_i, \quad i = 1, \ldots, N. \tag{24.78}$$

Introducing the new canonical coordinates $(\mathbf{r}_0, \ldots, \mathbf{r}_N, \mathbf{v}_0, \ldots, \mathbf{v}_N)$ into (24.74), we obtain

$$\mathfrak{H} = \frac{|\mathbf{v}_0|^2}{2m_0} - \sum_{j=1}^{N} \frac{\mathbf{v}_0 \cdot \mathbf{v}_j}{m_0} + \frac{1}{2}\sum_{j=1}^{N} |\mathbf{v}_j|^2\left(\frac{1}{m_j} + \frac{1}{m_0}\right)$$

$$+ \sum_{j=1}^{N}\sum_{i=1}^{j-1} \frac{\mathbf{v}_i \cdot \mathbf{v}_j}{m_0} - \mathfrak{h}\sum_{j=1}^{N}\sum_{i=1}^{j-1} \frac{m_i m_j}{|\mathbf{r}_i - \mathbf{r}_j|} - \mathfrak{h}\sum_{j=1}^{N} \frac{m_0 m_j}{|\mathbf{r}_j|}. \tag{24.79}$$

Since the Hamiltonian does not depend on $\mathbf{r}_0$, Hamilton's equations give $\dot{\mathbf{v}}_0 = 0$. Since $\mathbf{v}_0$ is the opposite of barycenter velocity, this means that the center of mass keeps its inertial motion. Without loss of generality, we put $\mathbf{v}_0 = 0$ and drop all the terms containing $\mathbf{v}_0$ from (24.79). In this way we obtain

$$\mathfrak{H} = \sum_{j=1}^{N}\left(\frac{m_0 + m_j}{2m_0 m_j}|\mathbf{v}_j|^2 - \mathfrak{h}\frac{m_0 m_j}{|\mathbf{r}_j|}\right)$$

$$+ \sum_{j=1}^{N}\sum_{i=1}^{j-1}\left(\frac{\mathbf{v}_i \cdot \mathbf{v}_j}{|\mathbf{r}_i - \mathbf{r}_j|} - \mathfrak{h}\frac{m_0 m_j}{|\mathbf{r}_j|}\right). \tag{24.80}$$

We conclude this section noticing that the Hamiltonian of the restricted N–body problem (24.72) is the sum of a term that is the Hamiltonian of a mass point acted upon by a Newtonian gravitational field

$$\mathfrak{H}_0^{(r)} = \left(\frac{1}{2}|\mathbf{v}|^2 - \mathfrak{h}\frac{m_0}{|\mathbf{r}|}\right) \tag{24.81}$$

perturbed by the remaining terms on the right-hand side of (24.72). Similarly, the Hamiltonian (24.80) of the $(N + 1)$–body problem is the sum of the Hamiltonian of a two-body problem

$$\mathfrak{H}_0 = \frac{|\mathbf{v}_j|^2}{2\mu_j} - \mathfrak{h}\frac{(m_0 + m_j)\mu_j}{|\mathbf{r}_j|}, \tag{24.82}$$

with reduced mass $\mu_j = m_0 m_j/(m_0 + m_j)$, perturbed by terms whose size is proportional to the mass of planets relative to that of the Sun.

24.9 Angle–Action Variables for N–Body Problem

To apply the Hamiltonian perturbation method that we discussed in Section 20.7, we must determine the angle–action variables (θ, I) for the mechanical systems with Hamiltonian functions (24.81) and (24.82). These variables have already been determined for the first Hamiltonian in Section 20.5. However, in the same section we proved that these variables are not uniquely determined since any transformation

$$I_i' = A_i^j I_j, \quad \theta'^i = (A^{-1})_j^i \theta^j, \ i, j = 1, 2, 3, \tag{24.83}$$

where $\mathbf{A}$ is a numerical nonsingular matrix, leads to new angle–action variables. In particular, for the Hamiltonian (24.81) it is convenient to introduce the **Delaunay variables** defined as follows:

$$
\begin{aligned}
L &= I_1 + I_2 + I_3, \ l = \theta^3, \\
G &= I_1 + I_2, \qquad g = \theta^2 - \theta^3, \\
H &= I_1, \qquad\quad h = \theta^1 - \theta^2.
\end{aligned}
\tag{24.84}
$$

With this choice, the Hamiltonian (24.81) becomes

$$\mathfrak{H}_0^{(r)} = -\mathfrak{h}^2 \frac{m_0^2}{2L^2}. \tag{24.85}$$

Further, taking into account (24.34), (24.37), (24.38), it is possible to prove that the Delaunay variables are related to the orbital elements by the following relations:

$$
\begin{aligned}
L &= \sqrt{\mathfrak{h} m_0 a}, \quad l = M, \\
G &= L\sqrt{1 - e^2}, \ g = \omega, \\
H &= G \cos i, \quad h = \Omega.
\end{aligned}
\tag{24.86}
$$

Finally, the Hamiltonian of the restricted problem can be rewritten in Delaunay's variables as

$$\mathfrak{H} = -\mathfrak{h}^2 \frac{m_0^2}{2L^2} + \mathfrak{H}_1(L, G, H, l, g, h, t), \tag{24.87}$$

where $\mathfrak{H}_1(L, G, H, l, g, h, t)$ is obtained by rewriting the perturbation

$$-\mathfrak{h} \sum_{j=1}^{N} m_j \left(\frac{1}{|\mathbf{r}_j - \mathbf{r}|} - \frac{\mathbf{r} \cdot \mathbf{r}_j}{|\mathbf{r}_j|} \right)$$

(see (24.72)) in terms of Delaunay variables.

For the Hamiltonian (24.82) of the planetary problem, the introduction of Delaunay variables is a little more complicated.

First, the Hamiltonian (24.80) contains unperturbed Hamiltonian functions (24.82) that differ from (24.81) since they are two-body problem Hamiltonians. However, multiplying (24.82) by μ_j, it becomes formally equal to (24.81), with $\mu_j^2 \mathfrak{h}$ replacing $\mathfrak{h}$, so that we can apply to this Hamiltonian the procedure we followed for the restricted problem. Further, the momenta in (24.82) are not given by $\mu_j \dot{\mathbf{r}}_j$. However, this circumstance does not change the relationships between $\mathbf{v}_j$, $\mathbf{r}_j$, and the Delaunay variables l_j, g_j, h_j, L_j, G_j, H_j with respect to those that hold for a point acted upon by a Newtonian force where $\mathbf{v}_j = m_j \dot{\mathbf{r}}_j$. Consequently, by using (24.32) with $\mathbf{v}_j / \mu_j$ instead of $\dot{\mathbf{r}}_j$, we can define formal orbital elements by the formulae

$$
\begin{aligned}
L_j &= \mu_j \sqrt{\mathfrak{h}(m_0 + m_j) a_j}, & l_j &= M_j, \\
G_j &= L_j \sqrt{1 - e_j^2}, & g_j &= \omega_j, \\
H_j &= G_j \cos i_j, & h_j &= \Omega_j.
\end{aligned}
\tag{24.88}
$$

In terms of these variables the Hamiltonian becomes

$$
\mathfrak{H} = -\sum_j \mathfrak{h}^2 \frac{(m_0 + m_j)^2 \mu_j^3}{2 L_j^2} + \mathfrak{H}_1,
\tag{24.89}
$$

where $\mathfrak{H}_1$ can explicitly be written as a function of Delaunay's variables (24.88).

The reader interested in the analysis of the Hamiltonians (24.87) and (24.89) by the Hamiltonian perturbation theory is invited to consult textbooks in celestial mechanics (for instance, [57, 58]).

24.10 Newton's Gravitational Law

In the above sections, we have studied the motion of N material points acted upon by Newtonian forces. From now on we evaluate the gravitational field produced by extended bodies.

Let $S = (P_1, \ldots, P_n)$ be a system of n mass points gravitationally interacting and let $\mathbf{x}_i$ denote the position vector of P_i with respect to an inertial frame I. Assuming that each P_i has a *passive gravitational mass* μ_i, besides its inertial mass m_i, the Newton gravitational law states that any point P_i of S is acted upon by a force, due to the gravitational attraction of the other mass points of S, given by

$$
\mathbf{F}_i = -\mathfrak{h}\, \mu_i \sum_{j=1, j \neq i}^{n} \frac{\mu_j}{r_{ij}^3} \mathbf{r}_{ij},
\tag{24.90}
$$

where $\mathbf{r}_{ij} = \mathbf{x}_i - \mathbf{x}_j$, $r_{ij} = |\mathbf{r}_{ij}|$, and $\mathfrak{h}$ is the universal gravitational constant. The minus sign is due to the attractive character of the gravitational force.

Suppose that a new mass point P, with position vector $\mathbf{x}$ and mass μ, is placed at an arbitrary point $P \neq P_j$, $j = 1, \dots, n$, of the space around S and the notation $\mathbf{r}_j = \mathbf{x} - \mathbf{x}_j$ is introduced.. Then, P is acted upon by a gravitational force $\mathbf{F}$ given by

$$\mathbf{F} = -\hbar \mu \sum_{j=1}^{n} \frac{\mu_j}{r_j^3} \mathbf{r}_j. \tag{24.91}$$

We define *gravitational field* $\mathbf{g}$ produced by S at the point P the following limit:

$$\mathbf{g} = \lim_{\mu \to 0} \frac{\mathbf{F}}{\mu}. \tag{24.92}$$

In other words, the gravitational field at P is the force acting on the unit mass at P. From (24.91) and (24.92) we deduce that

$$\mathbf{g}(\mathbf{x}) = -\hbar \sum_{j=1}^{n} \frac{\mu_j}{r_j^3} \mathbf{r}_j. \tag{24.93}$$

This field is defined at any point P which is not occupied by one of the particles P_i of S. Finally, introducing the *gravitational potential energy*

$$U(\mathbf{x}) = -\hbar \sum_{j=1}^{n} \frac{\mu_j}{r_j}, \tag{24.94}$$

we have that

$$\mathbf{g} = -\nabla_{\mathbf{x}} U. \tag{24.95}$$

24.11   Newton's Gravitation Theory for Extended Body

The above considerations can be extended to a continuous distribution of matter. Let C be a bounded region of the space $\mathfrak{R}^3$ in which matter is distributed with a continuous (gravitational) mass density $\varrho(\mathbf{x})$ (see Figure 24.11) and introduce the notations $\mathbf{x} \in \mathfrak{R}^3$, $\mathbf{x}' \in C$, and $r = |\mathbf{x} - \mathbf{x}'|$. Then, in view of (24.94), it is natural to call the function

$$U(\mathbf{x}) = -\hbar \int_C \frac{\varrho(\mathbf{x}')}{r(\mathbf{x}, \mathbf{x}')} dv' \tag{24.96}$$

the *gravitational potential energy* of the continuous mass distribution.

Fig. 24.11 Continuous mass distribution

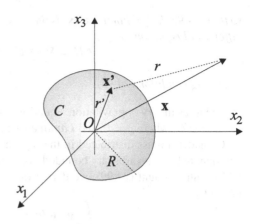

Note that (24.96) has been obtained by a formal extension of (24.94) to a continuous distribution of matter. At first sight, (24.96) generates some suspicions about its validity. First, the function under the integral becomes infinity at any point $\mathbf{x} \in C$. In fact, for any $\mathbf{x} \in C$, the difference $r = |\mathbf{x} - \mathbf{x}'|$ vanishes when $\mathbf{x}' = \mathbf{x}$ so that it seems that the gravitational potential is not defined at any point of the region occupied by the matter. Further, can we state that

$$\mathbf{g}(\mathbf{x}) = -\nabla_{\mathbf{x}} U = -\mathfrak{h} \int_C \varrho(\mathbf{x}') \frac{1}{r^3} \mathbf{r} \, dv' \quad ? \tag{24.97}$$

It is evident that (24.97) holds if we can differentiate under the integral sign. But the standard theorem ensuring this operation requires the differentiability of the function $1/r(\mathbf{x}, \mathbf{x}')$ with respect to $\mathbf{x}$ at any point of C, and this is not the case. However, well-known integrability criteria allow us to conclude that the functions under the integrals on the right-hand side of (24.96) and (24.97) are integrable so that the integrals are finite functions of $\mathbf{x}, \forall \mathbf{x} \in \mathfrak{R}^3$. Further, it is possible to prove the following theorem[1]

Theorem 24.1. *Let C be a bounded and measurable region of $\mathfrak{R}^3$. Then, the function $U(\mathbf{x})$ is of class $C^1(\mathfrak{R}^3)$, if the function $\varrho(\mathbf{x})$ is bounded in C, and of class $C^2(\mathfrak{R}^3 - \partial C)$, if $\varrho(\mathbf{x})$ is of class C^1 in C. For large values of r, this function verifies the condition*

$$|U(\mathbf{x})| < \frac{K}{r}, \tag{24.98}$$

where K is a positive constant (i.e., $|U(\mathbf{x})|$ is of the order of $1/r$ when $r \to \infty$). Further, at any point $\mathbf{x} \in \mathfrak{R}^3$, it is

$$\mathbf{g} \equiv -\nabla_{\mathbf{x}} U(\mathbf{x}) = \mathfrak{h} \int_C \varrho(\mathbf{x}') \nabla_{\mathbf{x}} \frac{1}{r(\mathbf{x}, \mathbf{x}')} dv', \tag{24.99}$$

---

[1]For the difficult proof see, for instance, [61], [62].

so that (24.97) holds. Finally, $U(\mathbf{x})$ is the only solution of the following elliptic partial differential equation

$$\Delta U = \nabla \cdot \nabla U(\mathbf{x}) = 4\mathfrak{h}\pi\varrho \qquad (24.100)$$

satisfying condition (24.98).

In Newton's theory of gravitation, equation (24.100) is named **Poisson's equation** in the region C, where $\varrho \neq 0$, and **Laplace's equation** in all the points at which $\varrho = 0$.

Consider a mass distributed in the region C with continuous mass density $\varrho(\mathbf{x})$ and denote by V an arbitrary bounded domain of $\mathfrak{R}^3$. Applying Gauss' theorem and taking into account (24.99), we derive the following fundamental consequence of (24.100):

$$\int_{\partial V} \mathbf{g} \cdot \mathbf{n} d\sigma = -4\mathfrak{h}\pi M(V), \qquad (24.101)$$

where $\mathbf{n}$ is the exterior unit normal to the boundary ∂V of V, $d\sigma$ is the element of area of ∂V and

$$M(V) = \int_V \varrho dV, \qquad (24.102)$$

is the total mass contained in V. Suppose that the mass distribution inside a sphere C_R is spherically symmetric about the center O of C_R and introduce spherical coordinates (ρ, φ, θ) having their origin at O. Then, we have that

$$\varrho = \varrho(r), \quad \mathbf{g}(\mathbf{x}) = -g(r)\mathbf{n}. \qquad (24.103)$$

Denote by C_r any sphere centered at O of radius r and let V_r be its volume. Introducing (24.103) into (24.101), we obtain that

$$g(r) = \mathfrak{h}M(V_r)/r^2. \qquad (24.104)$$

This relation shows that the gravitational field produced by the spherically symmetric mass distribution at any point of ∂V_r is equal to the gravitational field produced by the total mass contained in V_r when it is concentrated at the center O of V_r. In particular, if the mass density ϱ is supposed to be constant in V_r, then it is

$$M(V_r) = \frac{4}{3}\pi\varrho r^3, \quad r \in [0, R], \qquad M(V_r) = M(V_R), \quad r \in [R, \infty). \quad (24.105)$$

Introducing these results into (24.101), we obtain that

$$g(r) = \frac{4}{3}\mathfrak{h}\pi\varrho r, \quad r \in [0, R], \qquad g(r) = h\frac{M(V_R)}{r^2}, \quad r \in [R, \infty). \quad (24.106)$$

Fig. 24.12 Spherically symmetric mass distribution

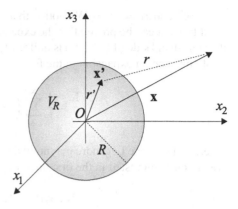

Fig. 24.13 Arbitrary continuous mass distribution

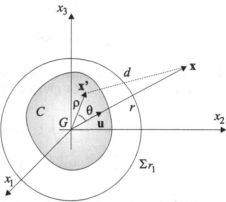

24.12 Asymptotic Behavior of Gravitational Potential

Consider now an *assigned* mass distribution S occupying the region C with a continuous mass density $\varrho(\mathbf{x})$, $\mathbf{x} \in C$. Let G be the center of mass of S and let Σ_{r_1} be a sphere with center at G and radius r_1 containing C (see Fig. 24.13). If d denotes the distance between G and an arbitrary point P external to Σ_{r_1}, $\mathbf{u}$ the unit vector along GP, and θ the angle between $\mathbf{u}$ and $\mathbf{x}'$, then the gravitational potential energy $U(\mathbf{x})$ of the mass contained into C is

$$U(\mathbf{x}) = -\hbar \int_C \frac{\varrho(\mathbf{x}')}{d(\mathbf{x}, \mathbf{x}')} \, dv', \tag{24.107}$$

where

$$\frac{1}{d} = \frac{1}{\sqrt{r^2 + r'^2 - 2rr' \cos \theta}} = \frac{1}{r} \frac{1}{\sqrt{1 + \alpha^2 - 2\alpha \cos \theta}}, \tag{24.108}$$

$r' = |\mathbf{x}'|$ and $\alpha = r'/r$. We notice that, if $\mathbf{x} \in \mathfrak{R}^3 - \Sigma_{r_1}$, then $\alpha < 1$. In these conditions, it can be proved that the expansion as a power series in the variable α of the right-hand side of (24.108) is uniformly convergent inside $\mathfrak{R}^3 - \Sigma_{r_1}$. It is simple to obtain such an expansion in the form

$$\frac{1}{d} = \frac{1}{r} \sum_{n=0}^{\infty} \left(\frac{r'}{r}\right)^n P_n(\cos \theta), \tag{24.109}$$

where $P_n(x)$ is the Legendre polynomial of order n. In the following considerations, we are only interested in the first three Legendre polynomials that are

$$
\begin{aligned}
P_0(\cos \theta) &= 1, \\
P_1(\cos \theta) &= \cos \theta, \\
P_2(\cos \theta) &= \frac{1}{2}(3 \cos^2 \theta - 1).
\end{aligned}
\tag{24.110}
$$

Since series (24.109) is uniformly convergent, it can be integrated term by term. Consequently, if we insert (24.109) into (24.107) we obtain

$$
\begin{aligned}
U(\mathbf{x}) = &-\frac{\hbar}{r} \int_C \varrho \, dv' - \frac{\hbar}{r^2} \int_C \varrho r' \cos \theta \, dv' + \\
&- \frac{\hbar}{2r^3} \int_C \varrho r'^2 (3 \cos^2 \theta - 1) \, dv' - \cdots \equiv U_1 + U_2 + U_3 \cdots
\end{aligned}
\tag{24.111}
$$

The first term, that can also be written as follows:

$$U_1 = -\hbar \frac{M}{r}, \quad \text{with} \quad M = \int_C \varrho \, dv', \tag{24.112}$$

can be regarded as the gravitational potential produced by the total mass of S supposed to be concentrated at the center of mass of S. In other words, at very large distances from G, the gravitational potential reduces to that of a mass point located at G with mass M.

Taking into account that

$$r' \cos \theta = \frac{1}{r} r \, r' \cos \theta = \frac{1}{r}(x_1 x_1' + x_2 x_2' + x_3 x_3'), \tag{24.113}$$

the second term U_2 assumes the form

$$U_2 = -\frac{\hbar}{r^3} \sum_{i=1}^{3} x_i \int_C \varrho x_i' \, dv'. \tag{24.114}$$

But the integrals on the right-hand side of (24.114) give the coordinates of the center of mass of S, and these coordinates vanish owing to our choice of the origin of the axes. Then, it is

$$U_2 = 0. \tag{24.115}$$

In order to evaluate the third term U_3, we recall (24.111) to obtain

$$\frac{\hbar}{r^3} \varrho r'^2 \frac{3\cos^2\theta - 1}{2} = \frac{\hbar}{2r^3} \varrho \frac{3\sum(x_i x_i')^2 - r'^2 r^2}{r^2}$$
$$= \frac{\hbar}{2r^5} \varrho \left[3\sum(x_i x_i')^2 - r'^2 r^2 \right]. \tag{24.116}$$

Elementary but tedious calculations lead us to the following expression of U_3:

$$U_3 = -\frac{\hbar}{2r^5} \left[(x_1^2 - x_2^2)(I_{22} - I_{11}) + (x_2^2 - x_3^2)(I_{33} - I_{22}) \right.$$
$$\left. + (x_3^2 - x_1^2)(I_{11} - I_{33}) + 6(x_1 x_2 I_{12} + x_2 x_3 I_{23} + x_1 x_3 I_{13}) \right], \tag{24.117}$$

where I_{11}, I_{22}, I_{33} are the momenta of inertia of S with respect to the axes and I_{12}, I_{23}, I_{13} are the products of inertia. In particular, if the axes of the frame of reference are principal of inertia, then $I_{12} = I_{23} = I_{13} = 0$. The contribution (24.117) to the gravitational potential energy U is called **quadrupole potential energy**.

Finally, if S has a symmetry of revolution about the axis Oz, and the axes Ox_1, Ox_2 are principal of inertia, then $I_{11} = I_{22}$ and (24.117) becomes

$$U_3 = -\frac{\hbar}{2r^5}(I_{33} - I_{11}) \left[(x_2^2 - x_3^2) - (x_3^2 - x_1^2) \right]. \tag{24.118}$$

When S is spherically symmetric about G, $U_3 = 0$ and we find again the result of the above section.

24.13 Local Inertial Frames and Tidal forces

The second law of dynamics for a test particle P moving under the action of a gravitational field **g** is

$$m_i \ddot{\mathbf{x}} = -m_g \nabla U, \tag{24.119}$$

where m_i and m_g are the *inertial* mass and the *passive gravitational* mass of P, respectively. Extremely accurate experiments carried out by Eötvösz, Dicke, Roll, Krotkov, and Braginski show the *universal equality of inertial and passive gravitational mass*,

$$m_i = m_g, \tag{24.120}$$

for any test particle. Then, equation (24.119) becomes

$$\ddot{\mathbf{x}} = -\nabla U. \qquad (24.121)$$

In conclusion, the experimental result (24.120) and the consequent equation (24.121) lead us to formulate the following.

Weak principle of equivalence – *The orbits of all freely falling test particles in a given gravitational field are independent of the mass and composition of the test particles being uniquely determined by initial conditions.*

Consider a mass point acted upon by a constant and uniform gravitational field $\mathbf{g} = -\nabla U$. Then, its motion relative to an inertial frame I is described by

$$\ddot{\mathbf{x}} = \mathbf{g}. \qquad (24.122)$$

Let R be a frame of reference whose motion relative to I is translational but accelerated with acceleration $\mathbf{g}$. Then, the equation of motion of P with respect to R is

$$\ddot{\mathbf{x}} = \mathbf{g} - \mathbf{g} = \mathbf{0}. \qquad (24.123)$$

This equation implies that, from a gravitational point of view, this frame seems to be inertial to the observer at rest in R, since the external gravitational field has been balanced by the inertial force per unit mass $-\mathbf{g}$.

When the gravitational field $\mathbf{g}(\mathbf{x})$ is not uniform, we can consider the position $\mathbf{x}_0$ of the mass point P at the instant t_0 in the inertial frame I. Then, in the accelerated frame R, whose motion relative to I is translational with acceleration $\mathbf{g}(\mathbf{x}_0)$, we have, at the point $\mathbf{x}_0$ and at the instant t_0, that the motion equation of P in the frame R is

$$\ddot{\mathbf{x}}(t_0) = \mathbf{g}(\mathbf{x}_0) - \mathbf{g}(\mathbf{x}_0) = \mathbf{0}, \qquad (24.124)$$

and in R there is no gravitational field, but only at the point $\mathbf{x}_0$ and at the instant t_0. In other words, R is a *local inertial frame* both in space and in time.

Consider now two mass points P and Q moving relative to the inertial frame I in a nonuniform gravitational field $\mathbf{g}$. If $\mathbf{x}$ and $\mathbf{y}$ denote the position vectors of P and Q, respectively, then the equations of motion of P and Q relative to I are

$$\ddot{\mathbf{x}} = \mathbf{g}(\mathbf{x}), \quad \ddot{\mathbf{y}} = \mathbf{g}(\mathbf{y}). \qquad (24.125)$$

Denote by $\boldsymbol{\zeta} = \mathbf{y} - \mathbf{x}$ the position vector of Q relative to P. From (24.125) we have that

$$\ddot{\boldsymbol{\zeta}} = \mathbf{g}(\mathbf{y}) - \mathbf{g}(\mathbf{x}).$$

Considering the series expansion of $\mathbf{g}(\mathbf{y})$ at the point $\mathbf{x}$ and supposing that $|\boldsymbol{\zeta}|$ is a first-order quantity, we can write the above equation as follows:

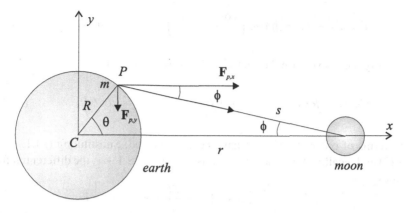

Fig. 24.14 Tidal forces on the Earth due to the Moon

$$\ddot{\zeta}_i = \frac{\partial g_i}{\partial y_j}(\mathbf{x})\zeta_j = -\frac{\partial^2 U}{\partial y_i \partial y_j}(\mathbf{x})\zeta_j, \qquad (24.126)$$

so the nonuniformity of the gravitational field generates a *differential force*.

A familiar gravitational phenomenon due to differential force is the **gravitational tidal force**. This force arises because any body has a finite size and gravitational fields are never uniform. Hence, different points of extended bodies undergo a different gravitational action. Under ordinary conditions tidal forces are, of course, negligible and unnoticeable. However, tidal forces are proportional to body size, hence, they become appreciable when we consider, for instance, the Earth in the gravitational field of the Moon. In fact, these forces produce tides in the oceans, and this phenomenon justifies the name of these forces.

The shape of tidal bulges on the Earth can be understood by analyzing the difference in the gravitational force acting at the center of the planet and at a point P on the surface (see Figure 24.14).

For simplicity, we will consider only forces in the x-y plane. At the center C of the Earth, the x- and y-components of the gravitational force on a test mass m due to the Moon are given by

$$F_{C,x} = \frac{\hbar M m}{r^2}, \qquad F_{C,y} = 0,$$

where M is the mass of the Earth. Further, at point P the components of the attractive force are

$$F_{P,x} = \frac{\hbar M m}{s^2} \cos \phi, \qquad F_{P,y} = -\frac{\hbar M m}{s^2} \sin \phi.$$

Therefore, denoting by $\hat{\mathbf{i}}$ and $\hat{\mathbf{j}}$ the unit vectors along the axes, the differential force between Earth's center and its surface is:

$$\Delta \mathbf{F} \equiv \mathbf{F}_P - \mathbf{F}_C = \hbar Mm \left(\frac{\cos \phi}{s^2} - \frac{1}{r^2} \right) \hat{\mathbf{i}} - \frac{\hbar Mn}{s^2} \sin \phi \hat{\mathbf{j}}. \tag{24.127}$$

To simplify the above formula, we write s in terms of r, R, and θ,

$$s^2 = (r - R \cos \theta)^2 + (R \sin \theta)^2 \simeq r^2 \left(1 - \frac{2R}{r} \cos \theta \right), \tag{24.128}$$

where terms of order $R^2/r^2 \ll 1$ have been neglected. Substituting (24.128) into (24.127) and recalling that when $x \ll 1$ it is $(1 - x)^{-1} \simeq 1 + x$, the differential force becomes

$$\begin{aligned}\Delta \mathbf{F} \simeq &\frac{\hbar Mm}{r^2} \left[\cos \phi \left(1 + \frac{2R}{r} \cos \theta \right) - 1 \right] \hat{\mathbf{i}} \\ &- \frac{\hbar Mm}{r^2} \left[1 + \frac{2R}{r} \cos \theta \right] \sin \phi \hat{\mathbf{j}}\end{aligned} \tag{24.129}$$

Finally, using the first-order approximations $\cos \phi \simeq 1$, $\sin \phi \simeq (R \sin \theta)/r$, we have

$$\Delta \mathbf{F} \simeq \frac{\hbar MmR}{r^3} \left(2 \cos \theta \hat{\mathbf{i}} - \sin \theta \hat{\mathbf{j}} \right). \tag{24.130}$$

The gravitational forces due to the Moon are directed toward the center of mass of the Moon, but the *differential* forces act to compress the Earth along the y-direction and elongate it along the line between their centers of mass, producing the tidal bulges. It is the symmetry of the bulges that produces two high tides in a 25-hour period during which Earth rotates about an orbiting Moon.

We conclude noting that Earth's bulges are not directly aligned with the Moon because the rotation period of the Earth is shorter than the Moon's orbital period and the frictional forces on the surface of the Earth drag the bulge axis ahead of the Earth–Moon line. Because friction is a dissipative force, rotational kinetic energy is constantly being lost and Earth's spin rate is continuously decreasing. At the present time, Earth's rotation period is lengthening at the rate of 0.0016 sec. century^{-1}, which although slow, is measurable.

24.14   Self-Gravitating Bodies

All the considerations developed in the previous sections of this chapter refer to a *rigid* mass distribution S and to the gravitational field it produces. In other words, the Newtonian gravitational force among parts of S does not modify the form and the volume occupied by S. In these conditions, we are only interested in the effects produced by S on the motion of mass points acted upon by the gravitational field of S.

In some cases this hypothesis cannot be accepted. For instance, the Earth is not a sphere because of the combined effect of the gravitational action of the mass distributed over its volume and the inertial forces due to its rigid rotation about the terrestrial axis. Another example is given by a star modeled as a rotating fluid under the combined action of the expansive forces due to pressure and temperature inside the star and the contractive gravitational forces. Continuous systems in which the effects of the auto-gravitation must be taken into account are said to be *self-gravitating bodies*.

A well-known self-gravitating system is the solar system whose mathematical description originated the N-body problem we sketched in the above sections. We here wish to introduce the equations governing the evolution of a *continuous* self-gravitating body S. With the aim to simplify the final equations, we suppose that S is a *perfect fluid*. Then, the mass conservation and the local momentum equation are (see 23.39–23.40)

$$\dot{\rho} + \rho \nabla \cdot \mathbf{v} = 0, \tag{24.131}$$

$$\rho \dot{\mathbf{v}} = -\nabla p - \rho \nabla U, \tag{24.132}$$

where ρ is the gas density, $\mathbf{v}$ the velocity field, p the pressure, and U the gravitational potential energy that is a solution of Poisson's equation

$$\Delta U = 4h\pi\rho. \tag{24.133}$$

Since the temperature θ inside S must be determined, we consider the local balance of energy in the form (see [43])

$$\rho \dot{\epsilon} = -p \nabla \cdot \mathbf{v} - \nabla \cdot \mathbf{h} + \rho r, \tag{24.134}$$

where ϵ is the specific internal energy, $\mathbf{h}$ the heat flux vector, and r the specific energy supply.

Equations (24.131)–(24.134) form a system of six scalar equations in the unknowns ρ, $\mathbf{v}$, θ, p, U, ϵ, $\mathbf{h}$, and r. To obtain a closed system, we must assign the constitutive equations for p, ϵ, $\mathbf{h}$, and r. For a perfect gas we have

$$p = p(\rho, \theta), \; \epsilon = \epsilon(\rho, \theta), \; r = r(\rho, \theta), \tag{24.135}$$

and

$$\mathbf{h} = -h(\rho, \theta)\nabla\theta. \tag{24.136}$$

Introducing these constitutive equations into (24.131)–(24.134) we obtain a closed system in the unknowns ρ, $\mathbf{v}$, U, and θ. This system is very difficult to solve and therefore we consider two simple interesting cases. First, we search for the equilibrium configurations of the fluid S. At equilibrium the above equations become

Fig. 24.15 Rotating fluid

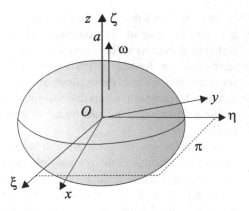

$$\nabla p + \rho \nabla U = 0, \tag{24.137}$$

$$\Delta U = 4h\pi\rho. \tag{24.138}$$

$$\nabla \cdot \mathbf{h} = \rho r, \tag{24.139}$$

and must be solved with respect to the unknowns ρ, U, and θ. If we search for solutions with spherical symmetry, the problem can be easily solved.

Now, we want to verify if it is possible that a fluid S rotates uniformly with constant angular velocity ω about an axis a. We suppose that in such a motion S has a symmetry plane π orthogonal to a, and denote by O the point at which a intersects π. Consider an inertial frame of reference I and let $O\xi\eta\zeta$ be Cartesian axes at rest relative to I with the axes $O\xi$ and $O\eta$ in the plane π. Then, we introduce a frame of reference $Oxyz$ co-moving with S and having the axis Oz along the direction of ω (see Fig. 24.15). In this new frame of reference the fluid S is at rest, but to evaluate the possible equilibrium configuration we must add to the equation of momentum the drag force due to the uniform rotation. Then, we have (see 23.60 and 17.63)

$$\frac{1}{\rho}\nabla p + \nabla U + \omega \times (\omega \times \mathbf{r}) = \mathbf{0}, \tag{24.140}$$

where $\mathbf{r} = \overrightarrow{CP}$ is the position vector of any particle $P \in S$. On the other hand it is

$$\omega \times (\omega \times \mathbf{r}) = (\omega \cdot \mathbf{r})\omega - \omega^2\mathbf{r}, \tag{24.141}$$

so that the drag force is a conservative force admitting a generalized potential energy given by

$$V = -\frac{1}{2}((\omega \cdot \mathbf{r})^2 + |\omega|^2|\mathbf{r}|^2). \tag{24.142}$$

Suppose that the constitutive equation $p = p(\rho, \theta)$ can be inverted with respect to ρ so that the function $\rho = \rho(p, \theta)$ exists. Then, equation (24.140) assumes the following form

$$\nabla(\mathfrak{H} + U + V) = \mathbf{0}, \tag{24.143}$$

where

$$\mathfrak{H}(p, \theta) = \int \frac{p}{\rho(p, \theta)} dp, \tag{24.144}$$

is the specific enthalpy of S. In conclusion, the equations we have to solve are

$$\mathfrak{H} + U + V = const, \tag{24.145}$$

$$\Delta U = 4h\pi\rho, \tag{24.146}$$

$$-\nabla \cdot \mathbf{h} + \rho r = 0, \tag{24.147}$$

in the unknowns ρ, U, and θ. Since we are searching for equilibrium configurations with a cylindrical symmetry, it is convenient to write the above system in cylindrical coordinates. This problem received great attention and many possible configurations have been found.

Chapter 25
One-Dimensional Continuous Systems

Abstract In this chapter, a brief introduction to one-dimensional continuous systems is presented. First, the balance equations relative to these systems are formulated: the mass conservation and the balance of momentum and angular momentum. Then, these equations are applied to describe the behavior of Euler's beam. Further, the simplified model of wires is introduced to describe the suspension bridge and catenary. Finally, the equation of vibrating strings is derived and the corresponding Sturm–Liouville problem is analyzed. In particular, the homogeneous string is studied by D'Alembert solution.

25.1 The One-Dimensional Model and the Linear Mass

Bodies in which one dimension prevails over the others are often used for applications. Examples of these bodies are wires, rods, and beams. Continua S with this property have been widely studied in *linear* elasticity supposing that they are subjected to loads acting only on the bases and resorting to the Saint Venant principle.[1] This analysis cannot be extended to the case in which forces act on the whole lateral boundary of S and the deformation is finite. In this more general situation, we are compelled to follow a different approach. The conformation of a body S in which a dimension prevails over the others suggests the possibility to describe its deformations under applied forces by resorting to simplified models that reduce the analysis of deformation to the behavior of the line intercepting all the centers of mass of the transversal sections of S. This line l is called the *directrix*. In other words, a one-dimensional mathematical model is proposed with mechanical properties that, in a certain way, are derived by the three-dimensional properties of the body S. Whether the substitution of the real body S with this model is acceptable or not can only be verified by comparing the theoretical results with the experimental data. It has to be noticed that many authors tried to obtain this model or a generalization of it

---

[1] This principle postulates that if forces act only on the bases of S, then the deformation of S depends only on the total force and torque acting on the bases of S, not on their distribution, except for the regions very close to the bases.

© Springer International Publishing AG, part of Springer Nature 2018   539
A. Romano and A. Marasco, *Classical Mechanics with Mathematica®*,
Modeling and Simulation in Science, Engineering and Technology,
https://doi.org/10.1007/978-3-319-77595-1_25

(one-dimensional system with directors) starting from the three-dimensional body and using suitable mean processes (see [59, 60]).

Let S be a one-dimensional continuous system and denote by Γ_* its reference configuration relative to the frame of reference $Oxyz$. For the sake of simplicity, we suppose that Γ_* is an interval $[0, L_*]$ of the axis Ox. Let $s_* \in [0, L_*]$ be the abscissa of a point $P_* \in \Gamma_*$. After the deformation due to applied loads, the point P_* will occupy a position P relative to $Oxyz$ so that the actual configuration $\Gamma(t)$ of S, i.e., its configuration at the instant t, will be defined by the vector function $\mathbf{r} = \hat{\mathbf{r}}(s_*, t)$, where $\mathbf{r}$ is the position vector of P relative to $Oxyz$. We denote by $L(t)$ the length of S in the actual configuration and by $s \in [0, L(t)]$ the curvilinear abscissa of a point $P \in \Gamma(t)$. If the curve occupied by S in the actual configuration is a simple regular curve, then, for any $t \in [0, T]$, there is an invertible C^1-function $s = s(s_*, t)$ and the actual configuration $\Gamma(t)$ is given by the regular curve

$$\mathbf{r} = \hat{\mathbf{r}}(s_*, t) = \mathbf{r}(s, t). \tag{25.1}$$

We suppose that the orientation of $\Gamma(t)$ is such that

$$\frac{\partial s}{\partial s_*} > 0. \tag{25.2}$$

We call *material part* of S any curve $\gamma(t) \subseteq \Gamma(t)$ that is the image by (25.1) of an arbitrary interval $[s_{*1}, s_{*2}] \subseteq [0, L_*]$. In other words a material part of S is a curve $\gamma(t)$ that represents the evolution of an interval of Γ_*. We define the linear mass densities $\mu_*(s_*)$ and $\mu(s, t)$, in Γ_* and $\Gamma(t) \times [0, T]$, two C^1-functions that define the mass per unit length of Γ_* and $\Gamma(t)$, respectively. The mass $m(s_{*1}, s_{*2})$ of the part $[s_{*1}, s_{*2}]$ is given by

$$m(s_{*1}, s_{*2}) = \int_{s_{*1}}^{s_{*2}} \mu_* ds_* = \int_{s_1(t)}^{s_2(t)} \mu(s, t) ds, \tag{25.3}$$

where $s_1(t) = s(s_{*1}, t)$ and $s_2(t) = s(s_{*2}, t)$. From the arbitrariness of $[s_{*1}, s_{*2}]$ and the continuity of μ_* and μ, we obtain the *Lagrangian form* of the linear mass conservation

$$\mu \frac{\partial s}{\partial s_*} = \mu_*. \tag{25.4}$$

We now prove the *Eulerian form* of linear mass conservation

$$\dot{\mu} + \mu \frac{\partial \dot{s}}{\partial s} = 0, \tag{25.5}$$

where

$$\dot{\mu} = \frac{\partial \mu}{\partial t} + \frac{\partial s}{\partial t} \frac{\partial \mu}{\partial s}.$$

In fact, if γ_* is an arbitrary material curve and $\gamma(t)$ its image by (25.1), we have

$$\frac{d}{dt} \int_{\gamma(t)} \mu(s,t) ds = \frac{d}{dt} \int_{\gamma_*} \mu \frac{\partial s}{\partial s_*} ds_* = \int_{\gamma_*} \left(\dot{\mu} \frac{\partial s}{\partial s_*} + \mu \frac{\partial \dot{s}}{\partial s_*} \right) ds_*$$

$$= \int_{\gamma(t)} \left(\dot{\mu} + \mu \frac{\partial \dot{s}}{\partial s} \right) ds.$$

From the arbitrariness of $\gamma(t)$ and the continuity of the functions under the integral we obtain (25.5).

Taking into account (25.5), it can easily be proved that if $\psi(s,t)$ is an arbitrary C^1−function, then the following result holds:

$$\frac{d}{dt} \int_{\gamma(t)} \mu \psi \, ds = \int_{\gamma(t)} \mu \dot{\psi} \, ds. \tag{25.6}$$

25.2 Integral and Local Laws of Balance

Let $\mathbf{r}(\bar{s}, t)$ be a point of $\Gamma(t)$.[2] We suppose that the mechanical actions of the part $\Gamma^+(t)$ formed by the point of $\Gamma(t)$ with curvilinear abscissa $s \in [\bar{s}, L(t)]$ are summarized by a force $\mathbf{T}(\bar{s}, t)$ and a torque $\mathbf{M}(\bar{s}, t)$ applied at $\mathbf{r}(\bar{s}, t)$. $\mathbf{T}$ and $\mathbf{M}$ are called the *stress* and the *stress torque* at $\mathbf{r}(\bar{s}, t)$. In particular, the component of $\mathbf{T}$ along the unit tangent $\mathbf{t}$ to $\gamma(t)$ is the *normal stress* and the component orthogonal to $\mathbf{t}$ is the *shear stress*. Similarly, the components of $\mathbf{M}$ along $\mathbf{t}$ and orthogonal to $\mathbf{t}$ are called *intrinsic torque* and *bending torque*. Then, we suppose that the following integral momentum balance holds for any arbitrary material part $\gamma(t) \subseteq \Gamma(t)$:

$$\frac{d}{dt} \int_{\gamma(t)} \mu \dot{\mathbf{r}} \, ds = \mathbf{T}(s'', t) - \mathbf{T}(s', t) + \int_{\gamma(t)} \mu \mathbf{F} \, ds + \sum_{i=1}^{N} \mathbf{F}_i, \tag{25.7}$$

where s' and s'' are the arc lengths of the initial and final points of $\gamma(t)$ respectively, $\mathbf{F}$ is the specific body force acting along $\gamma(t)$, and $\mathbf{F}_i$ are concentrated forces applied at the points $\mathbf{r}_i = \mathbf{r}(s_i) \in \gamma(t)$, $i = 1, \ldots, N$.

Since it is

$$\mathbf{T}(s'', t) - \mathbf{T}(s', t) = \mathbf{T}(s'', t) - \sum_{i=1}^{N}[[\mathbf{T}]]_{s_i} - \mathbf{T}(s', t) + \sum_{i=1}^{N}[[\mathbf{T}]]_{s_i}$$

$$= \int_{s'}^{s''} \frac{\partial \mathbf{T}}{\partial s} \, ds + \sum_{i=1}^{N}[[\mathbf{T}]]_{s_i}, \tag{25.8}$$

---

[2]For the sake of simplicity we omit the dependence of $\bar{s}$ on s_* and t.

where $[[\mathbf{T}]]_{s_i}$ is the jump of $\mathbf{T}$ at the point $\mathbf{r}_i$ due to the presence of the concentrated forces $\mathbf{F}_i$, (25.7) assumes the form

$$\frac{d}{dt} \int_{\gamma(t)} \mu \dot{\mathbf{r}} \, ds = \int_{\gamma(t)} \frac{\partial \mathbf{T}}{\partial s} \, ds + \int_{\gamma(t)} \mu \mathbf{F} \, ds + \sum_{i=1}^{N} (\mathbf{F}_i + [[\mathbf{T}]]_{s_i}). \qquad (25.9)$$

In view of (25.6), the arbitrariness of $\gamma(t)$, and the continuity of the functions under the integrals, we obtain the following local conditions

$$\mu \ddot{\mathbf{r}} = \frac{\partial \mathbf{T}}{\partial s} + \mu \mathbf{F}, \quad \forall s \neq s_i, \qquad (25.10)$$

$$[[\mathbf{T}]]_{s_i} = -\mathbf{F}_i, \quad i = 1, \dots, N, \qquad (25.11)$$

that are equivalent to the integral balance law (25.7).

We now postulate for any material curve $\gamma(t)$ the balance of angular momentum in the form

$$\frac{d}{dt} \int_{\gamma(t)} (\mathbf{r} - \mathbf{r}') \times \mu \dot{\mathbf{r}} \, ds = (\mathbf{r}'' - \mathbf{r}') \times \mathbf{T}(s'') + \int_{\gamma(t)} (\mathbf{r} - \mathbf{r}') \times \mu \mathbf{F} \, ds$$

$$+ \mathbf{M}(s'') - \mathbf{M}(s') + \sum_{i=1}^{N} (\mathbf{r}_i - \mathbf{r}') \times \mathbf{F}_i, \qquad (25.12)$$

where $\mathbf{r}$ is the position vector of the running point of $\gamma(t)$, $\mathbf{r}' = \mathbf{r}(s', t)$, and $\mathbf{r}'' = \mathbf{r}(s'', t)$.

By simple calculations we derive the following local conditions

$$\frac{\partial \mathbf{M}}{\partial s} + \mathbf{t} \times \mathbf{T} = \mathbf{0}, \qu\forall s \neq s_i, \qquad (25.13)$$

$$[[\mathbf{M}]]_{s_i} = \mathbf{0}, \quad i = 1, \dots, N, \qquad (25.14)$$

that are equivalent to the integral balance law (25.12).

Finally, we must consider the following *continuity conditions*

$$[[\mathbf{r}_i]] = \mathbf{0}, \quad i = 1, \dots, N, \qquad (25.15)$$

and the *boundary conditions*

$$\mathbf{T}_{s=0} = -\mathbf{F}_A, \quad \mathbf{M}_{s=0} = -\mathbf{M}_A, \qquad (25.16)$$

$$\mathbf{T}_{s=L} = \mathbf{F}_B, \quad \mathbf{M}_{s=L} = \mathbf{M}_B, \qquad (25.17)$$

where $\mathbf{F}_A$ and $\mathbf{M}_A$ are the force and torque applied at the initial point A ($s = 0$) of $\Gamma(t)$ and $\mathbf{F}_B$ and $\mathbf{M}_B$ are the force and torque applied at the final point B ($s = L$) of $\Gamma(t)$.

The unknowns appearing in equations (25.10) and (25.13) are $\mathbf{r}(s, t)$, $\mathbf{T}(s, t)$, $\mathbf{M}(s, t)$, and $s = s(s_*, t)$. In other words, we have 10 scalar unknowns and 6 scalar equations. Then, equations (25.10) and (25.13) are not sufficient to determine the evolution of $\Gamma(t)$. This is quite understandable since these equations, that hold for any one-dimensional system, do not contain its physical characteristics. These characteristics are specified by constitutive equations that relate $\mathbf{T}(s, t)$ and $\mathbf{M}(s, t)$ to the deformation $\hat{\mathbf{r}}(s_*, t)$.

25.3 An Example: Euler's Beam

In this section we discuss a simple application of the equations of the preceding section. We want to determine the *equilibrium* configuration Γ of **Euler's beam** that is a one-dimensional system S confirming the following conditions:

1. The reference configuration is rectilinear;
2. S is inextensible, that is $s(s_*, t) = s_*$;
3. Γ is contained in a plane π;
4. The applied forces are contained in π whereas the stress torque $\mathbf{M}$ is orthogonal to π;
5. The intensity M of $\mathbf{M}$ is proportional to the curvature c of Γ.

Let L be the invariant length of Γ and introduce a frame of reference $Oxyz$ such that the axes Ox and Oy lie in the plane π and Ox contains the reference configuration $\Gamma_* = [0, L]$. The axis Oz is orthogonal to π and we denote by $\mathbf{i}$, $\mathbf{j}$, and $\mathbf{k}$ the unit vectors along the axes Ox, Oy, and Oz, respectively. Then, from item 4 we have

$$\mathbf{M} = M\mathbf{k}, \tag{25.18}$$

where, in view of item 5 it is

$$M = Kc, \tag{25.19}$$

and K is a constant that defines the physical nature of the beam. On the other hand, the curvature c at a point of Γ is given by

$$c = \left| \frac{d\mathbf{t}}{ds} \right|, \tag{25.20}$$

with $\mathbf{t}$ the unit vector tangent to Γ. If we denote by θ the angle between $\mathbf{t}$ and the unit vector $\mathbf{i}$, then $\mathbf{t} = \cos\theta\mathbf{i} + \sin\theta\mathbf{j}$ and we obtain that

$$c = \frac{d\theta}{ds} \equiv \theta'(s). \tag{25.21}$$

Finally, in view of (25.18), (25.19), and (25.21), it is

$$\mathbf{M} = K\theta'\mathbf{k} \tag{25.22}$$

and the equilibrium equations become (see (25.10), (25.13))

$$\frac{d\mathbf{T}}{ds} + \mu\mathbf{F} = \mathbf{0}, \quad K\theta''\mathbf{k} + \mathbf{t} \times \mathbf{T} = \mathbf{0}. \tag{25.23}$$

Now we solve a particular equilibrium problem corresponding to assigned specific forces $\mathbf{F}$ and boundary conditions. First, we neglect the weight of the beam ($\mathbf{F} = \mathbf{0}$). Then, remembering that $\mathbf{t} = \cos\theta\mathbf{i} + \sin\theta\mathbf{j}$, $\mathbf{T} = T_x\mathbf{i} + T_y\mathbf{j}$ and projecting (25.23) along the axes Ox and Oy, we have

$$\frac{dT_x}{ds} = 0, \quad \frac{dT_y}{ds} = 0, \tag{25.24}$$

$$K\theta''(s) = T_x \sin\theta - T_y \cos\theta. \tag{25.25}$$

To give the boundary conditions, we consider the case of a beam with the initial point $s = 0$ wedged in a wall and a force $\mathbf{F}_B = -\hat{F}\mathbf{j}$ applied at the final point $s = L$. Since there is no torque applied at $s = L$, the corresponding boundary conditions are (see (25.22))

$$T_x(L) = 0, \quad T_y = -\hat{F}, \quad \theta(0) = \theta'(L) = 0. \tag{25.26}$$

Equation (25.24) and the first two boundary conditions in (25.26) imply that

$$T_x = 0, \quad T_y = -\hat{F}. \tag{25.27}$$

Introducing this result into (25.25) we have

$$K\theta'' = \hat{F}\cos\theta. \tag{25.28}$$

This equation does not admit a closed solution. Therefore we suppose that the deformation is small so that $\cos\theta \simeq 1$. In this hypothesis (25.28) becomes

$$\theta''(s) = \frac{\hat{F}}{K}, \tag{25.29}$$

and its solution verifying the boundary conditions (25.26) is

$$\theta(s) = -\frac{\hat{F}(2Ls - s^2)}{2K}. \tag{25.30}$$

Now, we are in condition to determine the profile of the beam. In fact from $dx/ds = \cos\theta$ and $dy/ds = \sin\theta$, we obtain

$$x(s) = \int_0^s \cos\theta(u)\, du \simeq \int_0^s du, \tag{25.31}$$

$$y(s) = \int_0^s \sin\theta(u)\, du \simeq \int_0^s \theta(u)\, du \tag{25.32}$$

and the profile for small deformations has the following parametric equations

$$x = s, \quad y = -\frac{FLs^2}{2K} + \frac{Fs^3}{6K}. \tag{25.33}$$

25.4 Wires

A one-dimensional continuous system is said to be a **wire** if

$$\mathbf{M} = 0, \quad \mathbf{T} = T\mathbf{t}, \quad T \geq 0. \tag{25.34}$$

The function T is called *tension*. In view of these conditions, equation (25.13) is identically satisfied whereas (25.10) assumes the form

$$\mu\ddot{\mathbf{r}} = \frac{\partial}{\partial s}(T\mathbf{t}) + \mu\mathbf{F}. \tag{25.35}$$

The unknowns of (25.35) are the parametric equations $x(s, t)$, $y(s, t)$, $z(s, t)$ of the actual configuration $\Gamma(t)$ of the wire, the tension $T(s, t)$, and the function $s = s(s_*, t)$ (see Section 25.1). Since the functions $x(s, t)$, $y(s, t)$, $z(s, t)$ are not independent in view of the following condition:

$$|\mathbf{t}|^2 = \left(\frac{\partial x}{\partial s}\right)^2 + \left(\frac{\partial y}{\partial s}\right)^2 + \left(\frac{\partial z}{\partial s}\right)^2 = 1, \tag{25.36}$$

the number of equations is equal to the number of unknowns only if the wire is inextensible. In general, we need a further constitutive equation relating the tension to the elongation of the wire when it passes from the reference configuration Γ_* to the actual configuration $\Gamma(t)$.

It will be useful to write (25.35) relative to the Frenèt triad $(\mathbf{t}, \mathbf{n}, \mathbf{b})$ at $\mathbf{r} \in \Gamma(t)$, where $\mathbf{t}$ is the unit tangent vector, $\mathbf{n}$ is the unit principal normal to $\Gamma(t)$, respectively, and $\mathbf{b}$ is the unit binormal, that is the unit vector orthogonal to the plane generated by $\mathbf{t}$ and $\mathbf{n}$. First, we recall Frenèt's formula

Fig. 25.1 Wire acted upon
by parallel forces

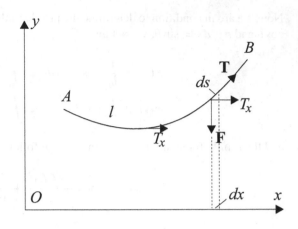

$$\frac{\partial \mathbf{t}}{\partial s} = c\mathbf{n}, \tag{25.37}$$

where c is the curvature at the considered point of $\Gamma(t)$. Second, in view of (11.10) we have

$$\ddot{\mathbf{r}} = \ddot{s}\mathbf{t} + c|\dot{s}|^2\mathbf{n} \tag{25.38}$$

and we can give (25.35) the following form

$$\mu\left(\ddot{s}\,\mathbf{t} + c|\dot{s}|^2\mathbf{n}\right) = \frac{\partial T}{\partial s}\mathbf{t} + c\,T\mathbf{n} + \mu\mathbf{F}, \tag{25.39}$$

whose projections on the triad $(\mathbf{t}, \mathbf{n}, \mathbf{b})$ are

$$\mu\ddot{s} = \frac{\partial T}{\partial s} + \mu F_t, \quad \mu c|\dot{s}|^2 = cT + \mu F_n, \quad F_b = 0. \tag{25.40}$$

In the foregoing sections we consider some applications of equations (25.35) and (25.40).

25.5　Suspension Bridge

Let AB be an inextensible wire with fixed end points A and B (see Fig. 25.1) and denote its length by l. We suppose that AB is at equilibrium under the action of a distributed force $\mathbf{F}$ per unit length having a constant direction, for instance along the vertical straight line. We choose a frame of reference $Oxyz$ with the axis Oy parallel to $\mathbf{F}$ and upward directed. Finally, Ox is chosen to be horizontal and such that Oxy contains the tangent to AB at A.

Owing to our choice of $Oxyz$ we have

$$\mu F_x = 0, \quad \mu F_y = -\mu F, \quad \mu F_z = 0, \tag{25.41}$$

and

$$z_A = \left(\frac{dz}{ds}\right)_A = 0. \tag{25.42}$$

In view of (25.41) and (25.42), equation (25.35) implies that

$$T\frac{dz}{ds} = 0. \tag{25.43}$$

Since T cannot be identically equal to zero, we can state that $dz/ds = 0$, that is z is constant along the wire AB. Taking into account (25.42) we conclude that $z = 0$ and the equilibrium configuration of AB is contained in the plane Oxy.

Further, (25.41) and (25.37) lead to the result

$$T_x \equiv T\frac{dx}{ds} = const, \tag{25.44}$$

so that *the component T_x of the tension is constant along all the wire AB*. It is evident that $x = const$ if and only if the configuration of the wire is a vertical segment. If we exclude this trivial case, we can always suppose that $T > 0$.

Finally, from (25.38) and (25.44) we obtain the equation

$$T_x\frac{d}{ds}\left(\frac{dy}{dx}\right) = \mu F \tag{25.45}$$

that can also be written as follows:

$$\frac{d^2y}{dx^2} = \frac{1}{T_x}\frac{ds}{dx}\mu F. \tag{25.46}$$

Denote by $y = y(x)$ the Cartesian equation of the equilibrium configuration of the wire AB. Then,

$$\frac{ds}{dx} = \sqrt{1 + \left(\frac{dy}{dx}\right)^2}$$

and (25.46) becomes

$$\frac{d^2y}{dx^2} = \frac{\mu F}{T_x}\sqrt{1 + \left(\frac{dy}{dx}\right)^2}. \tag{25.47}$$

If F depends on x, y, and dy/dx, then (25.47) is a second-order differential equation in normal form. For this equation we have to determine a solution satisfying the following boundary conditions:

Fig. 25.2 Suspension bridge

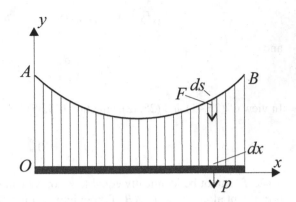

$$y(x_A) = y_A, \quad y(x_B) = y_B. \tag{25.48}$$

We conclude this section by solving the boundary value problem (25.47), (25.48) relative to the suspension bridge. Suppose that the distributed force F depends only on x and verifies the condition

$$\mu F(x)ds = p(x)dx. \tag{25.49}$$

An example of such a condition is shown in Fig. 25.2 where a cable AB, schematized as a wire, holds a suspension bridge by a large number of tie rods. In this case, $p(x)$ is the weight of that part of the bridge contained between two tie rods. Introducing (25.49) into (25.46) we obtain

$$\frac{d^2y}{dx^2} = \frac{p(x)}{T_x}. \tag{25.50}$$

In particular, if $p(x)$ is constant, then a double integration of (25.50) gives

$$y - y_* = \frac{p}{2T_x}(x - x_*)^2, \tag{25.51}$$

where x_* and y_* are two integration constants that can be determined imposing that (25.51) satisfies the boundary conditions (25.48).

25.6 Homogeneous Catenary

In this section we consider a second application of (25.47). Let AB be a wire acted upon by its weight. Then, the distributed force F per unit length is given by

$$\mu F = \mu g, \tag{25.52}$$

where μ is the linear mass density, here supposed to be constant, and g the gravity acceleration. When (25.52) holds, equation (25.47) becomes

$$\frac{d^2y}{dx^2} = \frac{1}{\lambda}\sqrt{1 + \left(\frac{dy}{dx}\right)^2},$$

(25.53)

with (see (25.44))

$$\lambda = \frac{T_x}{\mu g} = const.$$

(25.54)

All the equilibrium configurations of a wire acted upon by its weight are solutions $y(x)$ of (25.53) and are called *homogeneous catenaries*. To integrate (25.53) we introduce the notation

$$\eta = \frac{dy}{dx}$$

(25.55)

by which (25.53) assumes the form

$$\frac{d\eta}{dx} = \frac{1}{\lambda}\sqrt{1 + \eta^2}.$$

(25.56)

Integrating (25.56) we have

$$\int \frac{d\eta}{\sqrt{1 + \eta^2}} = \frac{1}{\lambda}\int dx$$

(25.57)

and putting

$$\eta = \sinh \xi,$$

(25.58)

we obtain

$$\int \frac{d(\sinh \xi)}{\sqrt{1 + \sinh^2 \xi}} = \int \frac{\cosh \xi}{\cosh \xi}d\xi = \int d\xi = \frac{x - x_*}{\lambda},$$

where x_* is an integration constant. Therefore, its is

$$\xi = \frac{x - x_*}{\lambda}$$

and taking into account (25.55) and (25.58) we can write the equation

$$\frac{dy}{dx} = \sinh \frac{x - x_*}{\lambda}$$

(25.59)

whose integration gives

$$y - y_* = \lambda \cosh \frac{x - x_*}{\lambda},$$

(25.60)

Fig. 25.3 Homogeneous
catenary

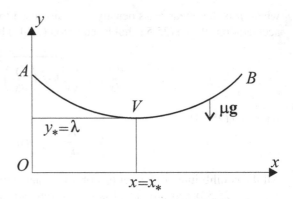

where x_* and y_* are integration constants.

(25.60) is the equation of a homogeneous catenary that is a curve with the upward concavity, symmetric with the straight line $x = x_*$, and having the vertex V at a distance $y_* = \lambda$ from the axis Ox (see Fig. 25.3). It is evident that we know the equilibrium configuration (25.60) after determining the integration constants x_*, y_* and the quantity λ. We omit to show that these quantities can be determined by the boundary conditions.

25.7 The Vibrating String Equation

We want to analyze the *small oscillations of an inextensible and uniform string AB about a rectilinear equilibrium configuration AB*. Introduce a frame of reference $Oxyz$ and suppose that the reference configuration of AB coincides with the interval $0 \leq x \leq L$. We say that the oscillations about AB are small if the angle α between the tangent vector at any point of the actual configuration of the string and Ox is small, that is if

$$\frac{\partial x}{\partial s} = \cos \alpha \simeq 1. \tag{25.61}$$

Integrating this relation we obtain

$$x = s. \tag{25.62}$$

From the inextensibility it follows that $\mu(s) = \mu(x) = const$. Moreover, we suppose that the distributed force acting on the string depends on x and is parallel to AB. Consequently, we have

$$F_y = F_z = 0. \tag{25.63}$$

In view of the above hypotheses equations (25.35) gives

$$\frac{\partial T}{\partial x} + \mu F_x = 0, \tag{25.64}$$

$$\mu \frac{\partial^2 y}{\partial t^2} = \frac{\partial}{\partial x}\left(T\frac{\partial y}{\partial x}\right), \tag{25.65}$$

$$\mu \frac{\partial^2 z}{\partial t^2} = \frac{\partial}{\partial x}\left(T\frac{\partial z}{\partial x}\right). \tag{25.66}$$

Integrating the first equation we obtain

$$T(x) \equiv f(x) = T_A - \mu \int_0^x F_x dx = T_B + \mu \int_x^L F_x dx, \tag{25.67}$$

where T_A and T_B are the values of the tension T at A and B, respectively. If $\mathbf{F}_A$ and $\mathbf{F}_B$ are the forces parallel to Ox applied at A and B, then we have

$$T_A = -F_A, \quad T_B = F_B. \tag{25.68}$$

Introducing the function (25.67) inside (25.65), (25.66) we obtain two partial differential equations in the unknowns $y(x, t)$ and $z(x, t)$. Since these equations are identical, we suppose that the oscillations are contained in the plane Oxz and we limit our analysis to the equation

$$\mu \frac{\partial^2 z}{\partial t^2} = \frac{\partial}{\partial x}\left(f(x)\frac{\partial z}{\partial x}\right) \tag{25.69}$$

that is called the **equation of vibrating string** or the **wave equation**.

In particular, if $F_x = 0$, then in view of (25.67) it is $f = const$ and the above equation reduces to

$$\mu \frac{\partial^2 z}{\partial t^2} = f \frac{\partial^2 z}{\partial x^2}. \tag{25.70}$$

25.8 Boundary Value Problems and Uniqueness Theorems

We search for a solution of (25.69) verifying the following boundary conditions and initial data

$$z(0, t) = 0, \quad z(L, t) = 0, \tag{25.71}$$

$$z(x, 0) = z_0(x), \quad \left(\frac{\partial z}{\partial t}\right)_{t=0} = \dot{z}_0(x). \tag{25.72}$$

The first two conditions state that the the string has fixed end points whereas (25.72) assign the initial configuration and initial velocity field.

In other words, we are searching for a solution in the strip of Cartesian plane Oxt defined by

$$0 \le x \le L, \quad t \ge 0, \tag{25.73}$$

such that

• On the sides $x = 0$ and $x = L$ of the strip confirms the boundary data (25.71);
• On the side $t = 0$ confirms the initial data (25.72).

Before proving that a solution of this problem exists, we show that, if it exists, then it is unique.

In fact, suppose that two solutions $z_1(x, t)$ and $z_2(x, t)$ exist that satisfy the above boundary and initial conditions. Then, since equation (25.69) is linear, the difference $\Delta z = z_1 - z_2$ is a solution of the equation

$$\mu \frac{\partial^2 \Delta z}{\partial t^2} = \frac{\partial}{\partial x}\left(f(x)\frac{\partial \Delta z}{\partial x}\right). \tag{25.74}$$

that satisfies the following homogeneous data

$$\Delta z(0, t) = 0, \quad \Delta z(L, t) = 0, \tag{25.75}$$

$$\Delta z(x, 0) = 0, \quad \left(\frac{\partial \Delta z}{\partial t}\right)_{t=0} = 0. \tag{25.76}$$

Multiplying (25.74) by $\partial \Delta z/\partial t$, we obtain the equation

$$\frac{\partial \Delta z}{\partial t}\frac{\partial}{\partial x}\left(f(x)\frac{\partial \Delta z}{\partial x}\right) = \frac{\partial}{\partial t}\left[\frac{1}{2}\mu\left(\frac{\partial \Delta z}{\partial t}\right)^2\right]$$

whose integration gives

$$\int_0^L \frac{\partial \Delta z}{\partial t}\frac{\partial}{\partial x}\left(f(x)\frac{\partial \Delta z}{\partial x}\right) dx = \frac{1}{2}\frac{d}{dt}\int_0^L \mu\left(\frac{\partial \Delta z}{\partial t}\right)^2 dx. \tag{25.77}$$

Further, integrating by parts the integral on the left-hand side, when we take into account (25.75), we obtain

$$\int_0^L \frac{\partial \Delta z}{\partial t}\frac{\partial}{\partial x}\left(f\frac{\partial \Delta z}{\partial x}\right) dx = \left[f\frac{\partial \Delta z}{\partial t}\frac{\partial \Delta z}{\partial x}\right]_{x=0}^{x=L} - \int_0^L f\frac{\partial \Delta z}{\partial x}\frac{\partial^2 \Delta z}{\partial x \partial t} dx$$

$$= -\frac{1}{2}\frac{d}{dt}\int_0^L f\left(\frac{\partial \Delta z}{\partial x}\right)^2 dx.$$

Introducing this result into (25.77) we have

$$\frac{d}{dt} \int_0^L \left[f \left(\frac{\partial \Delta z}{\partial x} \right)^2 + \mu \left(\frac{\partial \Delta z}{\partial t} \right)^2 \right] dx = 0$$

so that, in view of (25.76), we obtain the result

$$\int_0^L \left[f \left(\frac{\partial \Delta z}{\partial x} \right)^2 + \mu \left(\frac{\partial \Delta z}{\partial t} \right)^2 \right] dx = 0. \tag{25.78}$$

Since the functions f and μ are positive, we can state that

$$\frac{\partial \Delta z}{\partial x} = \frac{\partial \Delta z}{\partial t} = 0$$

and the function Δz is constant in all over the strip. Finally, (25.76) implies that this constant is equal to zero so that $z_1 = z_2$.

There are other interesting boundary problems that can be formulated for the string. For instance, suppose that the initial point A of the string is fixed, the string is acted upon by a force $\mathbf{F}_A$ parallel to Ox whereas the final point $x = L$ is free. In this case the component of the tension T along the axis Oz vanishes and we have

$$T \left(\frac{\partial z}{\partial x} \right)_{x=L} = 0 \Rightarrow \left(\frac{\partial z}{\partial x} \right)_{x=L} = 0. \tag{25.79}$$

If the point $x = L$ is constrained to move in an assigned way, then at the point $x = L$ the following boundary condition holds

$$z(L, t) = z_B(t). \tag{25.80}$$

25.9 The Method of Separated Variables

In this section we search for a solution of equation (25.69) supposing that it can be written in the form

$$z(x, t) = y(x)\rho(t), \tag{25.81}$$

that is, as the product of a function depending on x and a function depending on t (separated variables). Then, introducing (25.81) into (25.69), we have the equation

$$\rho(t) \frac{d}{dx} \left[f(x) \frac{dy}{dx} \right] = \mu(x)y(x) \frac{d^2\rho}{dt^2}$$

that can also be written as follows:

$$\frac{1}{\mu(x)y(x)} \frac{d}{dx} \left[f(x)\frac{dy}{dx} \right] = \frac{1}{\rho(t)} \frac{d^2\rho}{dt^2}. \tag{25.82}$$

The left-hand side of this equation depends on x whereas the right-hand side depends on t. Then equation (25.82) holds if and only if both the sides are equal to a same constant λ. Consequently, we have

$$\frac{d}{dx} \left[f(x)\frac{dy}{dx} \right] + \lambda\mu y = 0, \tag{25.83}$$

$$\frac{d^2\rho}{dt^2} + \lambda\rho = 0. \tag{25.84}$$

(25.83) is called the **Sturm–Liouville equation**, and (25.84) is the equation of harmonic motions provided that $\lambda > 0$.[3] We recall that we are searching for a solution of one of the boundary problems listed in the preceding section. Therefore, the solution $y(x)$ of (25.83) must satisfy the following boundary conditions (see (25.71), (25.79)):

- for the string with fixed end points

$$y(0) = y(L) = 0, \tag{25.85}$$

- for the string with the point A fixed and the point B free

$$y(0) = y'(L) = 0. \tag{25.86}$$

It is important to remark that the method of separated variables works only if the boundary data are homogeneous. Consequently, the problem of a string with A fixed and B moving with an assigned law cannot be discussed with this method.

25.10   Eigenvalues and Eigenfunctions of Sturm–Liouville Boundary Value Problems

We now consider the Sturm–Liouville boundary value problems (25.83), (25.85) and (25.83), (25.86). Any real value of λ for which one of these problems has a solution $y(x)$ is called an **eigenvalue** of the linear operator

$$\frac{d}{dx} \left[f(x)\frac{d}{dx} \right] + \lambda\mu \tag{25.87}$$

---

[3]The reader is invited to analyze the case $\lambda < 0$.

acting on the functions of $C^2[0, L]$. The corresponding solutions $y(x) \neq 0$ are called
eigenfunctions belonging to the eigenvalue λ. The number of linearly independent
functions belonging to the same eigenvalue λ is the *multiplicity* of λ. In particular,
λ is said to be *simple* if the eigenfunctions belonging to it differ for a multiplicative
constant.

It is fundamental the following result:

The eigenvalues of the **Sturm–Liouville boundary problems** (25.83), (25.85) *and*
(25.83), (25.86) *are positive, simple, and can be ordered in an increasing sequence*

$$\lambda_1, \ldots, \lambda_n, \ldots \qquad (25.88)$$

We just show that the eigenvalues are positive and simple. Multiplying (25.83) by y
and integrating on the interval $[0, L]$ we obtain the equation

$$\int_0^L y \frac{d}{dx}(fy')dx + \lambda \int_0^L \mu y^2 \, dx = 0$$

that, integrating by parts, gives

$$\left[fyy'\right]_{x=0}^{x=L} - \int_0^L fy'^2 \, dx + \lambda \int_0^L \mu y^2 \, dx = 0.$$

Taking into account (25.85), (25.86), we have

$$\lambda = \frac{\displaystyle\int_0^L fy'^2 dx}{\displaystyle\int_0^L \mu y^2 dx} \qquad (25.89)$$

and the eigenvalues are positive.

To prove that any eigenvalue is simple, denote by y and y_* two eigenfunctions
belonging to the same eigenvalue λ. From (25.83) we have

$$y_* \frac{d}{dx}(fy') + \lambda \mu y y_* = 0,$$
$$y \frac{d}{dx}(fy'_*) + \lambda \mu y y_* = 0,$$

and subtracting these equations we obtain

$$\frac{d}{dx}\left[f(y_*y' - yy'_*)\right] = 0,$$

so that $f(y_*y' - yy'_*) = const.$ When (25.85), (25.86) are taken into account we
conclude that this constant vanishes. Therefore, it is

$$y_* y' - y y'_* = 0 \Rightarrow \frac{d}{dx}\left(\frac{y}{y_*}\right) = 0.$$

This result shows that the ratio y/y_* is constant so that λ is simple.

After proving that all the eigenvalues are positive and simple, we can denote the sequence of the corresponding eigenfunctions as follows:

$$y_1(x), y_2(x), \ldots, y_n(x), \ldots, \tag{25.90}$$

where the arbitrary multiplicative constant of any eigenfunction is fixed by imposing the condition

$$\int_0^L \mu y_n^2 dx = 1. \tag{25.91}$$

We now prove the following important property of eigenfunctions (25.90)

$$\int_0^L \mu y_m y_n \, dx = 0, \quad m \neq n. \tag{25.92}$$

In fact, multiplying (25.83) written for y_n by the eigenfunction y_m, we obtain

$$\int_0^L y_m \frac{d}{dx}\left(f y_n'\right) dx + \lambda_n \int_0^L \mu y_m y_n \, dx = 0.$$

Integrating this equation by parts we can write

$$\left[f y_m y_n'\right]_{x=0}^{x=L} - \int_0^L f y_m' y_n' \, dx + \lambda_n \int_0^L \mu y_m y_n \, dx = 0.$$

so that, recalling (25.85), (25.86), we obtain the identity

$$- \int_0^L f y_m' y_n' \, dx + \lambda_n \int_0^L \mu y_m y_n \, dx = 0. \tag{25.93}$$

Exchanging y_m with y_n in (25.93) gives

$$- \int_0^L f y_m' y_n' \, dx + \lambda_m \int_0^L \mu y_m y_n \, dx = 0, \tag{25.94}$$

and subtracting (25.94) from (25.93) we finally have that

$$(\lambda_m - \lambda_n) \int_0^L \mu y_m y_n \, dx = 0. \tag{25.95}$$

Condition (25.92) is proved when $\lambda_m \neq \lambda_n$.

We don't prove the following fundamental theorem.

Theorem 25.1 *Any function $g(x)$, that is continuous in $[0, L]$, piecewise of class C^2 in this interval, and confirming the boundary conditions (25.85), (25.86), can be written as an absolutely and uniformly convergent series*

$$g(x) = \sum_{n=1}^{\infty} c_n y_n(x), \qquad (25.96)$$

where

$$c_n = \int_0^L \mu g y_n \, dx. \qquad (25.97)$$

25.11 Initial Value Problems

In the preceding section we have proved that $z(x, t) = y(x)\rho(t)$ is a solution of the equation of vibrating string (25.69) and of the relative boundary conditions if $y(x)$ is eigenfunction of one of the Sturm–Liouville boundary problem (25.83), (25.85) or (25.83), (25.86). If the function $y(x)$ in (25.81) is identified with the eigenfunction $y_n(x)$ belonging to the eigenvalue λ_n, then the function $\rho(t)$, that we now denote by $\rho_n(t)$, must be a solution of the equation

$$\frac{d^2 \rho_n}{dt^2} + \lambda_n \rho_n = 0. \qquad (25.98)$$

This is the equation of harmonic motions since $\lambda_n > 0$. Therefore, its general integral is

$$\rho_n(t) = a_n \cos \sqrt{\lambda_n} t + b_n \sin \sqrt{\lambda_n} t, \qquad (25.99)$$

where a_n and b_n are integration constants. A solution of (25.69), satisfying the boundary conditions (25.71) for the string with fixed end points, or (25.79) for a string with the initial fixed point A and free final point B is given by

$$z_n(x, t) = (a_n \cos \sqrt{\lambda_n} t + b_n \sin \sqrt{\lambda_n} t) y_n(x), \qquad (25.100)$$

and it represents a vibrating state of the string in which all the point oscillate harmonically with frequency $\nu_n = \sqrt{\lambda_n}/2\pi$. All the vibrations (25.100) are called *principal vibrations*. The solution for $n = 1$ is the *first harmonic*, the solution for $n = 2$ is the *second harmonic*, and so on.

All these solutions confirm the boundary data but they do not satisfy the initial data. On the other hand, since (25.69) is linear, any linear combination of solutions is still a solution of (25.69). Suppose that the series

$$z(x, t) = \sum_{n=1}^{\infty} z_n(x, t) = \sum_{n=1}^{\infty} (a_n \cos \sqrt{\lambda_n} t + b_n \sin \sqrt{\lambda_n} t) y_n(x) \qquad (25.101)$$

is absolutely and uniformly convergent (so that it can be derived term by term). Then, series (25.101) is a solution of (25.69) and we prove that this more general solution can satisfy the initial data (25.72). In fact, for $t = 0$, (25.101) gives

$$z_0(x) = \sum_{n=1}^{\infty} a_n y_n(x), \quad \dot{z}_0(x) = \sum_{n=1}^{\infty} \sqrt{\lambda_n} b_n y_n(x) \qquad (25.102)$$

and Theorem 25.1 allows to express the functions $z_0(x)$ and $\dot{z}_0(x)$ by absolutely and uniformly convergent series (25.96).

In conclusion, the method of separated variables allows us to obtain a solution of the boundary and initial data in the form (25.101), where the functions $y_n(x)$ are eigenfunctions of the Sturm–Liouville homogeneous problems and the coefficients a_n and b_n are given by the following formulae

$$a_n = \int_0^L \mu z_0 y_n \, dx, \quad b_n = \frac{1}{\sqrt{\lambda_n}} \int_0^L \mu \dot{z}_0 y_n \, dx. \qquad (25.103)$$

25.12 Homogeneous String at Constant Tension and Fixed End Points

In this section, the Sturm–Liouville method is applied to the simple case of a homogeneous string with fixed end points along which there is no distributed force. In this case the tension T is constant and the equation to solve is (25.70).

The method of separated variables, instead of giving (25.83) and (25.84), leads to the following equations:

$$f \frac{d^2 y}{dx^2} + \lambda \mu y = 0, \qquad (25.104)$$

$$\frac{d^2 \rho}{dt^2} + \lambda \rho = 0. \qquad (25.105)$$

Introducing the notation

$$v = \sqrt{\frac{f}{\mu}}, \qquad (25.106)$$

we can write (25.104) as follows:

$$\frac{d^2 y}{dx^2} = -\frac{\lambda}{v^2} y. \qquad (25.107)$$

Since $\lambda > 0$, the solutions of (25.107) are

$$y(x) = c_1 \cos \left(\frac{\sqrt{\lambda}}{v} x \right) + c_2 \sin \left(\frac{\sqrt{\lambda}}{v} x \right), \tag{25.108}$$

where c_1 and c_2 are arbitrary integration constants. Imposing the boundary conditions (25.85) to (25.108), we obtain

$$c_1 = 0, \quad c_2 \sin \left(\frac{\sqrt{\lambda}}{v} L \right) = 0. \tag{25.109}$$

Therefore, a solution different from zero is obtained provided that

$$\frac{\sqrt{\lambda}}{v} L = n\pi, \quad n = 1, 2, \ldots \tag{25.110}$$

so that the eigenvalues of the linear Sturm–Liouville operator

$$\frac{d^2 y}{dx^2} + \frac{\lambda}{v^2}$$

are given by

$$\lambda_n = \frac{n^2 \pi^2 v^2}{L^2}, \quad n = 1, 2, \ldots \tag{25.111}$$

whereas the corresponding eigenfunctions are

$$y_n(x) = c_2 \sin \frac{n\pi}{L} x, \quad n = 1, 2, \ldots \tag{25.112}$$

To assign a fixed value to the arbitrary constant c_2, we impose the condition

$$\int_0^L \mu y_n^2 \, dx = 1. \tag{25.113}$$

Introducing (25.112) into (25.113), we obtain $c_2 = \sqrt{2/\mu L}$. In conclusion, the normalized eigenfunctions are

$$y_n(x) = \sqrt{\frac{2}{\mu L}} \sin \frac{n\pi}{L} x, \quad n = 1, 2, \ldots \tag{25.114}$$

The orthogonality conditions (25.92) are confirmed in view of the elementary result

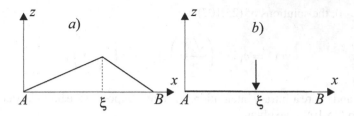

Fig. 25.4 a Pinched string **b** Struck string

$$\int_0^L \sin\frac{m\pi x}{L} \sin\frac{n\pi x}{L}\, dx = 0, \quad m \neq n, \tag{25.115}$$

and Theorem 25.1 reduces to Fourier's expansion. In view of (25.111), the general solution (25.101) satisfying the boundary and the initial data (25.102) assumes the following form

$$z(x, t) = \sum_{n=1}^{\infty} \left(a_n' \cos\frac{n\pi v}{L}t + b_n' \sin\frac{n\pi v}{L}t \right) \sin\frac{n\pi x}{L}, \tag{25.116}$$

where (see (25.103))

$$a_n' = \sqrt{\frac{2}{\mu L}}a_n = \frac{2}{L}\int_0^L z_0(x) \sin\frac{n\pi v}{L}x\, dx, \tag{25.117}$$

$$b_n' = \sqrt{\frac{2}{\mu L}}b_n = \frac{2}{n\pi v}\int_0^L \dot z_0(x) \sin\frac{n\pi v}{L}x\, dx, \tag{25.118}$$

25.13   Pinched or Struck String

In this section we apply solution (25.116), (25.117), (25.118) to the following cases:

1. The string AB is *pinched* at the point $x = \xi$, that is, it is released with zero speed from the configuration shown in Fig. 25.4a. This happens, for instance, in the harp, cithara, and guitar.
2. the string AB is *struck* at the point $x = \xi$ starting from a rectilinear configuration. This is the case of a piano (see Fig. 25.4b).

In the case 1 the functions $z_0(x)$ and $\dot z_0(x)$, that give the initial configuration and velocity field of the string, are

$$
z_0(x) = \begin{cases} h\dfrac{x}{\xi}, & x \le \xi \\[2ex] h\dfrac{L-x}{L-\xi}, & x \ge \xi \end{cases} \tag{25.119}
$$

$$
\dot z_0(x) = 0. \tag{25.120}
$$

(25.118) and (25.120) imply that $b'_n = 0$, for any integer n. Further, we have

$$
\begin{aligned}
a'_n &= \frac{2h}{L\xi} \int_0^\xi x \sin \frac{n\pi x}{L}\, dx + \frac{2h}{L(L-\xi)} \int_\xi^L (L-x) \sin \frac{n\pi x}{L}\, dx \\
&= -\frac{2h}{n\pi\xi} \left(\left[x \cos \frac{n\pi x}{L} \right]_{x=0}^{x=\xi} - \int_0^\xi \cos \frac{n\pi x}{L}\, dx \right) \\
&\quad - \frac{2h}{n\pi(L-\xi)} \left(\left[(L-x) \cos \frac{n\pi x}{L} \right]_{x=\xi}^{x=L} + \int_\xi^L \cos \frac{n\pi x}{L}\, dx \right).
\end{aligned}
$$

Finally, we get the following expression of the coefficients a_n

$$
\begin{aligned}
a'_n &= \frac{2hL}{n^2\pi^2} \left(\frac{1}{\xi} + \frac{1}{L-\xi} \right) \sin \frac{n\pi\xi}{L} \\
&= \frac{2hL^2}{n^2\pi^2\xi(L-\xi)} \sin \frac{n\pi\xi}{L}. \tag{25.121}
\end{aligned}
$$

The final solution

$$
z(x,t) = \frac{2hL^2}{\pi^2\xi(L-\xi)} \sum_{n=1}^{n} \frac{1}{n^2} \sin \frac{n\pi\xi}{L} \sin \frac{n\pi x}{L} \cos \frac{n\pi vt}{L} \tag{25.122}
$$

that is obtained taking into account (25.121) and (25.116), shows that the terms of the series decrease as $1/n^2$ so that the first oscillation is slightly modified by the presence of the other terms. In particular, if $\xi = L/2$, then $\sin(n\pi\xi/L) = 0$ for $n = 2, 4, \ldots$ and all the corresponding terms vanish. This result explains why the sound of the harp, for instance, is so pure. In Fig. 25.5 the string configurations are shown for different values of time.

In case 2 of the struck string, $z_0(x) = 0$ whereas $\dot z_0(x)$ is different from zero only in a very small interval $(\xi - \epsilon, \xi + \epsilon)$, where ξ is the abscissa of the point at which the string is struck. Denoting by I the intensity of the impulse supposed to act along the Oz axis, we have

$$
\mu \int_{\xi-\epsilon}^{\xi+\epsilon} \dot z_0(x)\, dx = I. \tag{25.123}
$$

Fig. 25.5 Configurations of a pinched string for $L = 1$, $\xi = 0.5$, $v = 1$ on varying time

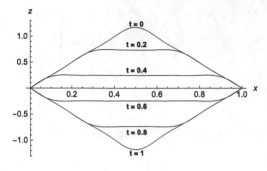

Fig. 25.6 Configurations of a struck string for $L = 1$, $\xi = 0.5$, $v = 1$ on varying time

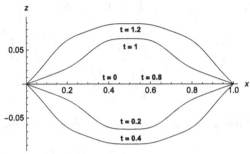

Since in the interval $(\xi - \epsilon, \xi + \epsilon)$ it is

$$\sin \frac{n\pi x}{L} \simeq \sin \frac{n\pi \xi}{L}$$

(25.117) and (25.118) give

$$a'_n = 0,$$

$$b'_n = \frac{2}{n\pi v} \int_{\xi-\epsilon}^{\xi+\epsilon} \dot{z}_0(x) \sin \frac{n\pi x}{L} \, dx = \frac{2I}{n\pi v \mu} \sin \frac{n\pi \xi}{L}$$

and (25.116) becomes

$$z(x, t) = \frac{2I}{\pi v \mu} \sum_{n=1}^{\infty} \frac{1}{n} \sin \frac{n\pi \xi}{L} \sin \frac{n\pi x}{L} \sin \frac{n\pi v t}{L}. \tag{25.124}$$

The terms of this series decrease as $1/n$ so that the sound of a piano is less pure than the sound of a harp. In Fig. 25.6 the configuration of a struck string on varying time is shown.

25.14 D'Alembert's Integral

The method we consider in this section is due to D'Alembert, and it can be applied only to a homogeneous string along which the tension is constant. Under these conditions the evolution equation is (25.70).

The method consists in applying the variable change

$$\xi = x - vt, \quad \eta = x + vt, \tag{25.125}$$

to equation (25.70). Since it is

$$\frac{\partial}{\partial x} = \frac{\partial}{\partial \xi} + \frac{\partial}{\partial \eta}, \quad \frac{1}{v}\frac{\partial}{\partial t} = \frac{\partial}{\partial \eta} - \frac{\partial}{\partial \xi}$$

we have that

$$\frac{\partial^2}{\partial x^2} - \frac{1}{v^2}\frac{\partial^2}{\partial t^2} = \left(\frac{\partial}{\partial x} + \frac{1}{v}\frac{\partial}{\partial t}\right)\left(\frac{\partial}{\partial x} - \frac{1}{v}\frac{\partial}{\partial t}\right) = 4\frac{\partial^2}{\partial \xi \partial \eta},$$

and equation (25.70) assumes the form

$$\frac{\partial^2 z}{\partial \xi \partial \eta} = 0, \tag{25.126}$$

that can easily be integrated. In fact, after a first integration we obtain that $\partial z/\partial \eta$ is the sum of a function of ξ and a function of η. Therefore, the general integral of (25.126) is

$$z(\xi, \eta) = f_1(\xi) + f_2(\eta) \tag{25.127}$$

and the general integral of (25.70) is given by

$$z(x, t) = f_1(x - vt) + f_2(x + vt). \tag{25.128}$$

This solution is called ***D'Alembert's integral.***

To understand the meaning of the two functions f_1 and f_2 in (25.128), suppose that the string has an infinite length. Then, the function f_1 and f_2 have the following properties:

$$f_1(x + \Delta x - v(t + \Delta t)) = f_1(x - vt) \Leftrightarrow \Delta x = v\Delta t, \tag{25.129}$$
$$f_2(x + \Delta x + v(t + \Delta t)) = f_2(x + vt) \Leftrightarrow \Delta x = -v\Delta t. \tag{25.130}$$

Property (25.129) shows that the form of the string at the instant $t + \Delta t$ is the same as the form at the instant t but it is translated of the quantity $\Delta x = v\Delta t$. In other words, f_1 represents a wave moving along the positive direction of the axis

Fig. 25.7 Plot of $\sin(x - t)$

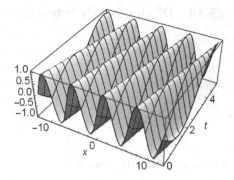

Fig. 25.8 Plot of $\exp(-(x - t)^2) + \exp(-(x + t)^2)$

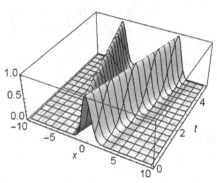

Ox without modifying its form. For this reason f_1 is called a *progressive wave*. Similarly, property (25.130) represents a wave moving along the negative axis Ox without modifying its form. Therefore, f_2 is said to be a *regressive wave*. Fig. 25.7 shows an example of f_1 and Fig. 25.8 shows an example of f_1 and f_2.

In conclusion, D'Alembert integral supplies the general solution of equation (25.70) as the sum of a progressive wave and a regressive one. The forms f_1 and f_2 of these two waves, that for the present are quite arbitrary, must be determined by the initial data (notice that in an infinite string there are no boundary data).

It is interesting to notice that the function f_1 is constant along the straight lines $x - vt = const$ whereas f_2 is constant along the straight lines $x + vt = const$. These straight lines are called *characteristic curves* of wave equation.

Imposing the initial data to (25.128) we obtain

$$f_1(x) + f_2(x) = z_0(x), \quad -f_1'(x) + f_2'(x) = \frac{1}{v}\dot{z}_0(x), \qquad (25.131)$$

where $x \in \mathfrak{R}$. Differentiating the first of (25.131) we have

$$f_1'(x) + f_2'(x) = z_0'(x), \qquad (25.132)$$

so that, in view of (25.131), (25.132), we can write

$$f_1'(x) = \frac{1}{2}\left[z_0'(x) - \frac{1}{v}\dot{z}_0(x)\right], \quad f_2'(x) = \frac{1}{2}\left[z_0'(x) + \frac{1}{v}\dot{z}_0(x)\right],$$

and f_1, f_2 are given by

$$f_1(x) = \frac{1}{2}z_0(x) - \frac{1}{2v}\int_0^x \dot{z}_0(\xi)\,d\xi + c, \tag{25.133}$$

$$f_2(x) = \frac{1}{2}z_0(x) + \frac{1}{2v}\int_0^x \dot{z}_0(\xi)\,d\xi - c, \tag{25.134}$$

where c is an arbitrary constant that disappears when we notice that the first equation (25.131) implies

$$f_1(0) + f_2(0) = z_0(0). \tag{25.135}$$

Finally, for an infinite string, D'Alembert's solution satisfying the initial data in $(-\infty, \infty)$ can be written as follows:

$$z(x, t) = \frac{1}{2}\left[z_0(x - vt) + z_0(x + vt)\right] + \frac{1}{2v}\int_{x-vt}^{x+vt} \dot{z}_0(\xi)\,d\xi. \tag{25.136}$$

In order to reach an interesting interpretation of (25.136), we choose a point (x_0, t_0) of the plane Oxt and draw the two characteristic curves through this point. The equations of these curves are

$$x - vt = x_0 - vt_0, \quad x + vt = x_0 + vt_0$$

so that they intersect the Ox axis ($t = 0$) at the points $x_0 - vt_0$ and $x_0 + vt_0$, respectively. Further, from (25.136) we obtain that the solution at (x_0, t_0) is given by

$$z(x_0, t_0) = \frac{1}{2}\left[z_0(x_0 - vt_0) + z_0(x_0 + vt_0)\right] + \frac{1}{2v}\int_{x_0-vt_0}^{x_0+vt_0} \dot{z}_0(\xi)\,d\xi.$$

This equation tells us that the solution at (x_0, t_0) is obtained by averaging the values of z_0 at the points $x_0 - vt_0$ and $x_0 + vt_0$ and integrating $\dot{z}_0(x)$ in the interval $[x_0 - vt_0, x_0 + vt_0]$. In other words, the value of the solution at (x_0, t_0) depends only on the values of the initial data in the interval $[x_0 - vt_0, x_0 + vt_0]$. For this reason, this interval is called the *domain of dependence* of (x_0, t_0). Another important remark is this: the values in $[x_0 - vt_0, x_0 + vt_0]$ of the initial data can influence the solution only in the triangular region formed by the two characteristics through (x_0, t_0) and the interval $[x_0 - vt_0, x_0 + vt_0]$.

25.15   D'Alembert's Solution for a String with Fixed End Points

Let S be a string of length L and fixed end points. In this section, we prove that it is possible to derive from (25.136) a solution of (25.70) that describes the evolution of S and confirms arbitrary initial data.

It will be sufficient to verify that there are suitable initial data for the infinite string such that $z(0, t) = z(L, t) = 0$ for any $t \in [0, \infty)$ and $z_0(x)$, $\dot{z}_0(x)$ are arbitrary in $[0, L]$. In fact, the corresponding solution of the infinite string verifies equation (25.70); its restriction to $[0, L]$ verifies the boundary conditions $z(0, t) = z(L, t) = 0$ and the initial data. Further, in view of the uniqueness theorem, it is the only solution of the boundary problem of the string with fixed end points.

In order to find the aforesaid initial data for the infinite string, we start requiring that (25.136) satisfies the condition $z(0, t) = 0$ for any $t \in [0, \infty)$:

$$z_0(-vt) + z_0(vt) + \frac{1}{v} \int_{-vt}^{vt} \dot{z}_0(\xi)\, d\xi = 0. \tag{25.137}$$

This condition is certainly satisfied if separately we have

$$z_0(-vt) + z_0(vt) = 0, \qquad \int_{-vt}^{vt} \dot{z}_0(\xi)\, d\xi = 0. \tag{25.138}$$

The first condition (25.138) imposes that $z_0(x)$ is an odd function of x. Further, differentiating the second condition (25.138) with respect to time we obtain

$$\dot{z}_0(-vt) + \dot{z}_0(vt) = 0, \tag{25.139}$$

so that $\dot{z}_0(x)$ also is an odd function.

Now, imposing that $x = L$ is a fixed point, we have that

$$z_0(L - vt) + z_0(L + vt) + \frac{1}{v} \int_{L-vt}^{L+vt} \dot{z}_0(\xi)\, d\xi = 0. \tag{25.140}$$

Again this condition is satisfied if it separately results

$$z_0(L - vt) + z_0(L + vt) = 0, \qquad \int_{L-vt}^{L+vt} \dot{z}_0(\xi)\, d\xi = 0, \tag{25.141}$$

so that, differentiating with respect to t the second condition we have

$$z_0(L - vt) + z_0(L + vt) = 0, \quad \dot{z}_0(L - vt) + \dot{z}_0(L + vt) = 0. \tag{25.142}$$

On the other hand, we have already proved that $z_0(x)$ and $\dot{z}_0(x)$ are odd functions of x. Consequently, it is $-z_0(L - vt) = z_0(-L + vt)$, $-\dot{z}_0(L - vt) = \dot{z}_0(-L + vt)$ and conditions (25.142) assume the equivalent form

$$z_0(L + vt) = z_0(-L + vt), \quad \dot{z}_0(L + vt) = \dot{z}_0(-L + vt), \qquad (25.143)$$

so that the functions $z_0(x)$ and $\dot{z}_0(x)$, that we have already proved to be odd functions of x, are also periodic with period $2L$.

In conclusion, to obtain a solution of the boundary and initial problem of a string S with fixed end points, it is sufficient to take the solution of the infinite string in which the initial data are obtained by extending the initial data of the string S to the whole real straight line $(-\infty, \infty)$ in such a way that the extension is an odd function with a period $2L$.

Chapter 26
An Introduction to Special Relativity

Abstract It is well known that at the beginning of the 20th century, a deep fracture was detected between classical mechanics and electrodynamics. It was evident that if classical mechanics is accepted without modification, then electrodynamics is valid only in a frame of reference: the ether frame. All the attempts to localize this frame of reference failed. Einstein with the special theory of relativity overcame all the difficulties accepting electrodynamics and modifying mechanics. In this chapter, adopting Einstein's original approach, we analyze the foundations of this theory together with some applications. Then, we discuss the four-dimensional formulation of the special relativity proposed by Minkowski and write the physical laws in tensor form.

26.1 Galilean Relativity

A relativistic theory has as its aim to show that it is possible to give the general laws of physics a form independent of any observer, that is, of the adopted frame of reference. Up to now, this objective has not been completely realized, even from the general theory of relativity.[1] In this chapter, a relativistic theory is intended as a theory for which we have defined:

1. The class $\mathfrak{R}$ of the frames of reference in which the physical laws have the same form,
2. The transformations of space and time coordinates in going from a frame of reference belonging to $\mathfrak{R}$ to another one in the same class,
3. The physical laws relative to any frame of reference of $\mathfrak{R}$,
4. The transformation formulae of the physical quantities appearing in physical laws upon changing the frame of reference in the class $\mathfrak{R}$.

---

[1] See, for instance, [37].

© Springer International Publishing AG, part of Springer Nature 2018
A. Romano and A. Marasco, *Classical Mechanics with Mathematica®*,
Modeling and Simulation in Science, Engineering and Technology,
https://doi.org/10.1007/978-3-319-77595-1_26

For instance, in Chap. 13, we showed that Galilean relativity assumes that $\mathfrak{R}$ is the class of inertial frames of reference, the space–time coordinate transformations are the Galilean ones, i.e., formulae (12.5) with a constant matrix Q_{ij}, the physical laws are Newtonian laws, and the mass and forces are objective.

After the discovery of the Maxwell equations governing all optical and electromagnetic phenomena, it was quite natural to verify if these equations could be accepted within the framework of Galilean relativity. We now verify that the Maxwell equations do not satisfy the principle of Galilean relativity.[2]

Let S be a charged continuous medium moving relative to an inertial observer I. If we denote by ρ the charge density and by $\mathbf{J}$ the current density, then the charge conservation is expressed by the following *continuity equation*:

$$\frac{\partial \rho}{\partial t} + \nabla \cdot \mathbf{J} = 0. \tag{26.1}$$

Further, in a vacuum, the Maxwell equations have the form

$$\nabla \times \mathbf{H} = \epsilon_0 \frac{\partial \mathbf{E}}{\partial t}, \tag{26.2}$$

$$\nabla \times \mathbf{E} = -\mu_0 \frac{\partial \mathbf{H}}{\partial t}, \tag{26.3}$$

where $\mathbf{H}$ is the magnetic field, $\mathbf{E}$ the electric field, ϵ_0 the dielectric constant, and μ_0 the magnetic permeability of a vacuum. Further, applying the divergence operator $\nabla \cdot$ to both sides of (26.2) and (26.3), we obtain

$$\nabla \cdot \mathbf{E} = 0, \quad \nabla \cdot \mathbf{H} = 0, \tag{26.4}$$

provided that we suppose that the electromagnetic fields vanish before a given instant in the past.

It is a very simple exercise to verify that, under a Galilean transformation (12.5) (with a constant matrix Q_{ij}), leading from inertial frame I to another inertial frame I', we have that

$$\nabla = \nabla', \tag{26.5}$$

$$\frac{\partial}{\partial t} = \frac{\partial}{\partial t'} - \mathbf{u} \cdot \nabla', \tag{26.6}$$

where $\mathbf{u}$ is the uniform velocity of I' relative to I. Then, in view of (26.5) and (26.6), under a Galilean transformation $I \to I'$, (26.1) becomes

$$\frac{\partial \rho}{\partial t'} + \nabla'(\mathbf{J} - \rho \mathbf{u}) = 0.$$

---

[2]For the topics of this chapter see [15, 22, 27, 36, 38, 39, 41, 42, 47, 48, 56].

This equation again assumes the form (26.1) in the inertial frame I' if and only if ρ and $\mathbf{J}$ are transformed according to the formulae

$$\rho' = \rho, \quad \mathbf{J}' = \mathbf{J} - \rho\mathbf{u}. \tag{26.7}$$

Consequently, *if these transformation formulae were experimentally confirmed*, we could conclude that the continuity equation satisfies the Galilean principle of relativity.

On the other hand, if the foregoing procedure is applied to the Maxwell equations (26.2) and (26.3), we obtain

$$\nabla' \times \mathbf{H} = \epsilon_0 \frac{\partial \mathbf{E}}{\partial t'} - \epsilon_0 \mathbf{u} \cdot \nabla\mathbf{E},$$

$$\nabla' \times \mathbf{E} = -\mu_0 \frac{\partial \mathbf{H}}{\partial t'} + \mu_0 \mathbf{u} \cdot \nabla\mathbf{H}.$$

Taking into account the vector identity

$$\nabla \times (\mathbf{a} \times \mathbf{b}) = \mathbf{b} \cdot \nabla\mathbf{a} - \mathbf{a} \cdot \nabla\mathbf{b} + \mathbf{a}\nabla \cdot \mathbf{b} - \mathbf{b}\nabla \cdot \mathbf{a},$$

where $\mathbf{a}$ and $\mathbf{b}$ are arbitrary regular fields, recalling (26.4) and that the relative velocity $\mathbf{u}$ is constant, the preceding equations become

$$\nabla' \times (\mathbf{H} - \epsilon_0\mathbf{u} \times \mathbf{E}) = \epsilon_0 \frac{\partial \mathbf{E}}{\partial t'},$$

$$\nabla' \times (\mathbf{E} + \mu_0\mathbf{u} \times \mathbf{H}) = -\mu_0 \frac{\partial \mathbf{H}}{\partial t'}.$$

It is evident that there is no transformation formula of the fields $\mathbf{E}$ and $\mathbf{H}$ for which these equations assume the same form in the inertial frame I'. In conclusion, Maxwell's equations do not agree with Galilean relativity.

We now derive a further disconcerting consequence of Maxwell's equations. Applying the operator $\nabla \times$ to both sides of (26.2) and (26.3), recalling the vector identity

$$\nabla \times (\nabla \times \mathbf{a}) = \nabla(\nabla \cdot \mathbf{a}) - \Delta\mathbf{a},$$

where Δ is the Laplace operator, and taking into account (26.4), we obtain the following equations:

$$\Delta\mathbf{H} = \epsilon_0\mu_0 \frac{\partial^2 \mathbf{H}}{\partial t^2},$$

$$\Delta\mathbf{E} = \epsilon_0\mu_0 \frac{\partial^2 \mathbf{E}}{\partial t^2},$$

showing that the electric and magnetic fields propagate in a vacuum with a speed

$$c = \frac{1}{\sqrt{\epsilon_0 \mu_0}},$$

independently of the motion of electromagnetic field's sources. In particular, since we have proved that Maxwell's equations are not compatible with Galilean relativity, we must conclude that they are valid at most in one inertial frame. Finally, we note that all the preceding results apply to light waves since Maxwell proved that optical waves are special electromagnetic waves.

26.2 Optical Isotropy Principle

The wave character of light propagation was established during the 18th century, when scientists became convinced that all physical phenomena could be described by mechanical models. Consequently, it appeared natural to the researchers of that time to suppose that empty space was filled by an isotropic and transparent medium, the *ether*, which supported light waves. This hypothesis seemed to be confirmed by the circumstance that the forces acting on charges and currents could be evaluated by supposing that electromagnetic fields generated a deformation state of the ether, which was described by the Maxwell stress tensor. However, this description was unacceptable for the following reasons:

- The high value of light speed required a very high density of ether.
- Such a high density implied the existence of both longitudinal and transverse waves, whereas Maxwell's equations showed the transverse character of electromagnetic waves.

Also damaging to the hypothesis of the ether as a material medium was the fact that Maxwell's equations were not covariant, i.e., invariant in form, under Galilean transformations, as we saw in the previous section. Consequently, electromagnetic phenomena did not confirm a relativity principle, so they were assumed to hold only in *one* frame of reference, the *optically isotropic frame*. It is well known that any attempt to localize this frame of reference failed (e.g., the Michelson, Morley, Kennedy, Fitzgerald experiments), so that researchers were faced with a profound physical inconsistency.

Einstein gave a brilliant and revolutionary solution to the preceding problem by accepting the existence of an optically isotropic system and exploiting the consequences of this assumption. More precisely, he postulated that:

There exists at least an optically isotropic frame of reference.

Owing to the preceding postulate, it is possible to find a frame of reference I in which light propagates in empty space with constant speed along straight lines in any direction. In particular, in this frame I it is possible to define a global time t by choosing an arbitrary value c of the speed of light in empty space. In fact, the time measured by a clock located at a point $O \in I$ will be accepted if a light signal that is sent from O at $t = t_0$ toward an arbitrary point P, where it is again reflected by a mirror toward O, reaches O at the instant $t = t_0 + 2OP/c$, for any t_0. Let us suppose that a set of identical clocks is distributed among different points in space. A clock at any point P can be synchronized with a clock located at the fixed point O by sending a light signal at the instant t_0 from O and requiring that it arrive at P at the instant $t = t_0 + OP/c$. We do not prove that the time variable defined by the foregoing procedure is independent of the initial point O and the initial instant t_0.

In view of what was stated in Sect. 26.1, if we accept the Galilean transformations in passing from a frame of reference I to another frame of reference I', which moves with respect to I with constant velocity $\mathbf{u}$, then there exists at most one isotropic optical frame. Conversely, if we drop the assumptions on which the Galilean transformations are based, then it is possible to prove (see [22], p. 403) that there are infinite optically isotropic frames of reference. More precisely, if (x^1, x^2, x^3, t) and (x'^1, x'^2, x'^3, t') are the spatial and temporal coordinates associated with the same event by two observers I and I', then the *finite* relations between these coordinates are expressed by the **Lorentz transformations**:

$$x'^i = x'^i_O + Q^i_j \left(\delta^j_h + (\gamma - 1) \frac{u^j}{u^2} u_h \right) x^h - Q^i_j \gamma u^j t, \qquad (26.8)$$

$$t' = t_0 - \gamma \frac{u_h x^h}{c^2} + \gamma t, \qquad (26.9)$$

where x^i_O, t_0, and u^i are constant, Q^i_j are the coefficients of a constant orthogonal matrix, $u = |\mathbf{u}|$, and

$$\gamma = \frac{1}{\sqrt{1 - \left(\frac{u}{c}\right)^2}} \equiv \frac{1}{\sqrt{1 - \beta^2}} > 1. \qquad (26.10)$$

Remark 26.1 It is not an easy task to prove (26.8) and (26.9) starting from the preceding postulate. However, if we add the hypothesis that the required transformations are linear, then it becomes very simple to derive the Lorentz transformations, as is proved in most books on special relativity.[3] *This assumption is equivalent to requiring that optically isotropic frames also are inertial frames.*

Remark 26.2 Let us consider any event that has constant space–time coordinates in the frame I'. Then, differentiating (26.8), we obtain

$$\left(\delta^i_j + (\gamma - 1) \frac{u^i}{u^2} u_j \right) \frac{dx^j}{dt} = \gamma u^i. \qquad (26.11)$$

---

[3] A simple proof is shown at the end of this section.

It is an easy exercise to verify that the preceding system admits the following solution:

$$\frac{dx^i}{dt} = u^i,$$ (26.12)

so that we can say that any point of I' moves with respect to I with constant velocity (u^i). Therefore, the relative motion of I' with respect to I is a uniform translatory motion with velocity (u^i). Similarly, we can prove that any point of I moves with respect to I' with constant velocity $-\mathbf{u}$. In particular, $x_O'^i$ denotes the coordinates of the origin of I with respect to I', when $t = 0$.

Remark 26.3 In the limit $c \to \infty$, the Lorentz transformations reduce to the Galilean ones.

We now introduce some particular Lorentz transformations. We define as a *Lorentz transformation without rotation* any Lorentz transformation obtained from (26.8) by supposing that the 3×3 orthogonal matrix (Q_j^i) reduces to the identity matrix

$$x'^i = x_O'^i + \left(\delta_h^i + (\gamma - 1) \frac{u^i}{u^2} u_h \right) x^h - \gamma u^i t,$$ (26.13)

$$t' = t_0 - \gamma \frac{u_h x^h}{c^2} + \gamma t.$$ (26.14)

To understand the meaning of the preceding definition, we denote by (S_j^i) the constant 3×3 orthogonal matrix defining the rotation $\overline{x}^i = S_j^i x^j$ of the spatial axes of I, in which the uniform velocity $\mathbf{u}$ of I' with respect to I becomes parallel to the axis $O\overline{x}^1$:

$$\begin{pmatrix} \overline{u}^1 \\ 0 \\ 0 \end{pmatrix} = \left(S_j^i \right) \left(u^j \right).$$ (26.15)

If we note that $u^2 = u_h u^h$ and $u_h x^h$ are invariant with respect to a change in the spatial axes, then, applying the rotation (S_j^i) to both sides of (26.13), that is, by performing the *same* rotation on the spatial axes of I' and I, we obtain

$$\overline{x}'^i = \overline{x}_O'^i + \left(\delta_h^i + (\gamma - 1) \frac{\overline{u}^i}{u^2} \overline{u}_h \right) \overline{x}^h - \gamma \overline{u}^i t,$$ (26.16)

$$t' = t_0 - \gamma \frac{\overline{u}_h \overline{x}^h}{c^2} + \gamma t.$$ (26.17)

These formulae, when we recall (26.10), presuppose that $x_O'^i = t_0 = 0$, and, for the sake of simplicity, we omit the overline, lead us to the *special Lorentz transformations*

$$x'^1 = \gamma \left(x^1 - ut\right), \tag{26.18}$$

$$x'^2 = x^2, \tag{26.19}$$

$$x'^3 = x^3, \tag{26.20}$$

$$t' = \gamma \left(t - \frac{u}{c^2}x^1\right), \tag{26.21}$$

provided that the two inertial frames I and I' have the same origin when $t = t' = 0$, the axes $O'x'^1$ are laid upon one another during the relative motion with uniform velocity $\mathbf{u}$ parallel to Ox^1, $O'x'^2$ is parallel to Ox^2, and $O'x'^3$ is parallel to Ox^3. These transformations include all the significant relativistic aspects since the most general Lorentz transformations can be obtained by arbitrary rotations of the spatial axes of the frames I and I' and arbitrary translations of the origins O and O'.

We conclude this section with an elementary proof of (26.18)–(26.21) supposing that the transformation between the two inertial frames I and I' is *linear*. First of all, we note that, since the motion of I' relative to I is along the Ox^1-axis, the coordinates x^2 and x^3 are transformed according to (26.19) and (26.20). Owing to the supposed linearity of the transformations, for the remaining variables x^1 and t we have that

$$x'^1 = Ax^1 + Bt, \tag{26.22}$$

$$t' = Cx^1 + Dt, \tag{26.23}$$

where the constants A, B, C, and D must be determined. First, the origin O' of I' moves relative to I with uniform velocity u parallel to Ox^1. Consequently, when $x'^1 = 0$, (26.22) gives

$$\frac{x^1}{t} = -\frac{B}{A} = u. \tag{26.24}$$

On the other hand, the origin of I moves relative to I' with velocity $-u$. Therefore, from (26.22), in which $x^1 = 0$, we obtain that

$$\frac{x'^1}{t'} = \frac{B}{D} = -u. \tag{26.25}$$

Comparing (26.24) and (26.25), we have that

$$B = -uA, \quad D = A,$$

and (26.22) and (26.23) become

$$x'^1 = A(x^1 + ut), \tag{26.26}$$

$$t' = Cx^1 + At. \tag{26.27}$$

To find the remaining constants A and C, we consider a planar light wave propagating along the Ox^1-axis with speed c. According to the optical isotropy principle, the propagation of this wave is described by the equations

$$(x^1)^2 - c^2 t^2 = 0, \quad (x'^1)^2 - c^2 (t')^2 = 0 \tag{26.28}$$

in the frames I and I', respectively. Inserting (26.26) and (26.27) into (26.28)$_2$, we obtain

$$(A^2 - c^2 C^2)(x^1)^2 - 2A(Au + c^2 C)x^1 t + A^2(u^2 - c^2)t^2 = 0. \tag{26.29}$$

This equation reduces to (26.28)$_1$ if

$$A^2 - c^2 C^2 = 1, \quad Au + c^2 C = 0, \quad A^2(u^2 - c^2) = -c^2,$$

i.e., if

$$A = \frac{1}{\sqrt{1 - \dfrac{u^2}{c^2}}} \equiv \gamma, \quad C = -\gamma \frac{u}{c^2},$$

and (26.18) and (26.21) are proved.

26.3   Simple Consequences of Lorentz's Transformations

In this section, we show that the Lorentz transformations lead to the conclusion that the concepts of space and time intervals are relative to the observer. Let x'^1_A and $x'^1_B > x'^1_A$ be the abscissas of two points A' and B' belonging to the axis $O'x'^1$ of the inertial frame I', and denote by $\Delta x'^1 = x'^1_B - x'^1_A$ the distance between these points, measured by rulers at rest relative to I'. The length $\Delta x'^1$ will be called the *rest or proper length* of the segment $A'B'$. Now we define the length Δx^1 of the segment $A'B'$ with respect to an inertial frame I, relative to which this segment moves with uniform velocity u along the axis $Ox^1 \equiv O'x'^1$. The quantity Δx^1 is identified by the distance between the two points A and B of the Ox^1-axis that are occupied by A' and B', respectively, at the same instant t in the inertial frame I. For t constant, (26.18) gives

$$\Delta x'^1 = \gamma \Delta x^1 > \Delta x^1, \tag{26.30}$$

so that the length of $A'B'$ relative to I is $1/\gamma$ times smaller than the length relative to I. Since the dimensions of the lengths in the direction perpendicular to the velocity do not change, a volume V is connected with the proper volume V' by the relation

$$V' = \gamma V. \tag{26.31}$$

Consider now a *rest or proper* time interval $\Delta t' = t'_2 - t'_1$ evaluated by a clock C at rest at the point x'^1_A of the inertial frame I'. To obtain the corresponding instants t_1 and t_2 evaluated in the inertial frame I, it is sufficient to apply the inverse of (26.21) and remember that I moves relative to I' with uniform velocity $-u$. In this way we have

$$t_1 = \gamma \left(t'_1 + \frac{u}{c^2} x'^1_A \right),$$
$$t_2 = \gamma \left(t'_2 + \frac{u}{c^2} x'^1_A \right).$$

Consequently, we have

$$\Delta t' = \frac{1}{\gamma} \Delta t < \Delta t, \tag{26.32}$$

where $\Delta t = t_2 - t_1$. In other words, *a moving clock goes slower than a clock at rest.* Now we present two simple but interesting applications of (26.31) and (26.32).

- Let P be an unstable particle produced in the atmosphere at a distance l from the terrestrial surface. Denote by I' and I the rest inertial frame of P and a terrestrial frame, respectively. Suppose that the mean life $\Delta t'$ (evaluated in I') and the speed v of P (evaluated in I) are such that $l > v\Delta t'$. If we assume classical kinematics, then it is impossible for P to reach the surface of the Earth. Conversely, resorting to (26.31) and (26.32), we can justify the arrival of P at the terrestrial surface. In fact, for the observer at rest relative to P, the particle must cover a shorter distance $l' = l/\gamma < l$ in the time $\Delta t'$. In contrast, for an observer at rest on the Earth's surface, the particle P must cover the distance l but lives longer, i.e., $\gamma \Delta t'$.
- Suppose that at the points $x^1_A < x^1_B$ of the Ox^1-axis of the inertial frame I two events happen at the instants t_A and t_B, respectively. For an observer in the inertial frame I' moving relative to I with velocity u, the same events happen at the instants t'_A and t'_B, given by

$$t'_A = \gamma \left(t_A - \frac{u}{c^2} x_A \right), \quad t'_B = \gamma \left(t_B - \frac{u}{c^2} x_B \right),$$

so that

$$t'_B - t'_A = \gamma \left(t_B - t_A - u \frac{x_B - x_A}{c^2} \right). \tag{26.33}$$

This condition shows that two events that are simultaneous for I are not simultaneous for I' if $x_B \neq x_A$. The following question arises: supposing that $t_B > t_A$, is it possible to find an observer I' for which the order of the events is inverted? This circumstance is confirmed if and only if the following condition is satisfied:

$$t_B - t_A < \frac{u}{c^2}(x_B - x_A). \tag{26.34}$$

Because $u/c \leq 1$, then also $u/c^2 \leq 1/c$, and the preceding condition gives

$$t_B - t_A < \frac{x_B - x_A}{c}. \tag{26.35}$$

This necessary condition for an inversion of the time order of two events allows us to state that *if in an inertial frame I' we have an inversion of the time order, then the time interval between the two events is necessarily less than the time taken by light to cover the distance $x_B - x_A$.* Conversely, from (26.35) follows (26.34), at least for a certain value of u. We remark that (26.35) cannot be satisfied for two events that are related to each other by a cause-and-effect relationship. In other words, *in an inertial frame, no physical perturbation propagates with a speed greater than the light speed c.*

26.4   Relativistic Composition of Velocities and Accelerations

From (26.21) we obtain that

$$\frac{\mathrm{d}t}{\mathrm{d}t'} = \left[\gamma \left(1 - \frac{\mathbf{u} \cdot \dot{\mathbf{r}}}{c^2} \right) \right]^{-1} \tag{26.36}$$

where $\mathbf{r}$ is the position vector with respect to the origin of the inertial frame I.

Owing to the preceding relation, by simple but tedious calculations, it is possible to derive from (26.18)–(26.21) the transformation formulae for the velocity and the acceleration under a Lorentz transformation. For instance, starting from the special Lorentz transformations, we obtain the following special formulae of the velocities:

$$\dot{x}'^1 = \frac{\dot{x}^1 - u}{1 - \dfrac{u\dot{x}^1}{c^2}}, \tag{26.37}$$

$$\dot{x}'^2 = \frac{1}{\gamma} \frac{\dot{x}^2}{1 - \dfrac{u\dot{x}^1}{c^2}}, \tag{26.38}$$

$$\dot{x}'^3 = \frac{1}{\gamma} \frac{\dot{x}^3}{1 - \dfrac{u\dot{x}^1}{c^2}}. \tag{26.39}$$

By a further derivation with respect to time, we can derive the following formulae for acceleration in going from inertial frame I to inertial frame I':

$$\ddot{x}'^1 = \frac{\left(1 - \dfrac{u^2}{c^2}\right)^{3/2}}{\left(1 - \dfrac{\dot{x}^1 u}{c^2}\right)^3} \ddot{x}^1,$$
(26.40)

$$\ddot{x}'^2 = \frac{1 - \dfrac{u^2}{c^2}}{\left(1 - \dfrac{\dot{x}^1 u}{c^2}\right)^2} \left(\ddot{x}^2 + \frac{\dfrac{\dot{x}^2 u}{c^2}}{1 - \dfrac{\dot{x}^1 u}{c^2}} \ddot{x}^1 \right),$$
(26.41)

$$\ddot{x}'^3 = \frac{1 - \dfrac{u^2}{c^2}}{\left(1 - \dfrac{\dot{x}^1 u}{c^2}\right)^2} \left(\ddot{x}^3 + \frac{\dfrac{\dot{x}^3 u}{c^2}}{1 - \dfrac{\dot{x}^1 u}{c^2}} \ddot{x}^1 \right).$$
(26.42)

Remark 26.4 It is evident that the preceding formulae reduce to Galilean ones when $c \to \infty$.

Now we present some simple consequences of the preceding formulae.

- It is a simple exercise to verify that a spherical light wave in I,

$$(x^1)^2 + (x^2)^2 + (x^3)^2 - c^2 t^2 = 0,$$

is described in the frame I' by the equation

$$(x'^1)^2 + (x'^2)^2 + (x'^3)^2 - c^2 t'^2 = 0,$$

in agreement with the principle of optical isotropy.
- Let a be a light ray emitted by a star, and let us suppose that it reaches the terrestrial surface in a direction orthogonal to the translational velocity u of the Earth (Fig. 26.1). I' is the rest inertial frame of the Earth, and I is the rest inertial frame of the star. Then, in (26.37)–(26.39) we have that

$$\dot{x}^1 = \dot{x}^2 = 0, \quad \dot{x}^3 = -c.$$

Introducing into (26.37)–(26.39) the preceding components of speed relative to the star frame I, we obtain the components of the ray speed in the terrestrial frame

$$\dot{x}'^1 = -u, \quad \dot{x}'^2 = 0, \quad \dot{x}'^3 = -\frac{c}{\gamma}.$$

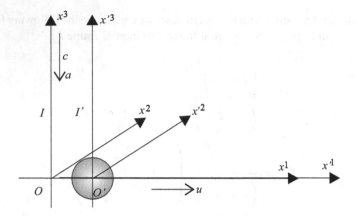

Fig. 26.1 Star aberration

Consequently, the angle θ between the velocity $\dot{\mathbf{r}}'$ in the frame I' and $O'x'^3$ is given by the relation

$$\tan \theta = \frac{\dot{x}'^1}{\dot{x}'^3} = \gamma \frac{u}{c} \simeq \frac{u}{c}\left(1 + \frac{u^2}{2c^2}\right) \simeq \frac{u}{c}.$$

This result explains the aberration of fixed stars.

- Let S be a homogeneous and isotropic optical medium, and let n be its refractive index. Suppose that S moves uniformly with a speed u along the Ox^1-axis of the inertial frame I and denote by I' the rest frame of S, with $O'x'^1$ parallel to Ox^1. Consider a light ray a propagating along $O'x'^1$ with speed c/n. Then, applying the inverse formulae of (26.37)–(26.39), the speed of a along Ox^1 is given by

$$\dot{x}^1 = \frac{\dfrac{c}{n} + u}{1 + \dfrac{u}{cn}} \simeq \left(\frac{c}{n} + u\right)\left(1 - \frac{u}{cn},\right)$$

that is,

$$\dot{x}^1 = \frac{c}{n} + u\left(1 - \frac{1}{n^2}\right).$$

This result represents a simple explanation of Fizeau's experiment, which, before special relativity, required very difficult theoretical interpretations that had not been experimentally confirmed. One of these attempts of interpretation required the partial drag of the ether by a moving medium (Sect. 26.2).

26.5   Principle of Relativity

In classical mechanics a relativity principle holds in the set $\Re$ of inertial frames. These frames are related to each other by the Galilean transformations. Einstein extends this principle to any field of physics. More precisely, he supposes that

The fundamental equations of physics have the same form in the whole class of the inertial frames. Analytically, the fundamental equations of physics must be covariant under Lorentz transformations.

To make clear the deep meaning of this principle, let us consider a physical law that in the inertial frame I is expressed by the *differential* relation

$$F\left(A, B, \ldots, \frac{\partial A}{\partial x^i}, \frac{\partial B}{\partial x^i}, \ldots, \frac{\partial A}{\partial t}, \frac{\partial B}{\partial t}, \ldots\right) = 0, \qquad (26.43)$$

where $A, B, \ldots$ are physical fields depending on the spatial coordinates and time. If we denote by $A', B', \ldots$ the corresponding fields evaluated by the inertial observer I', then this law satisfies the relativity principle if in the new frame I' it assumes the form

$$F\left(A', B', \ldots, \frac{\partial A'}{\partial x'^i}, \frac{\partial B'}{\partial x'^i}, \ldots, \frac{\partial A'}{\partial t'}, \frac{\partial B'}{\partial t'}, \ldots\right) = 0, \qquad (26.44)$$

where $A', B', \ldots$ are the quantities $A, B, \ldots$ evaluated by the inertial observer I'.

We underline that the principle of relativity does not state that the evolution of a physical phenomenon is the same in all inertial frames. It only states that the fundamental physical laws, which are expressed by *differential equations*, have the same form in any inertial frame. As a consequence, two inertial observers who repeat an experiment in the *same* initial and boundary conditions arrive at the same results.

We note that the possibility to verify the covariance of the physical law (26.43) is related to knowledge of the transformation law of the quantities $A, B, \ldots$ in going from inertial frame I to inertial frame I'. In other words, from a mathematical viewpoint, a physical law could be covariant with respect to more transformation groups, provided that the transformed quantities $A', B', \ldots$ are suitably defined. Therefore, we can establish the covariance of a physical law only by verifying that the assumed transformation laws for the quantities $A, B, \ldots$ are experimentally confirmed.

To make clear this aspect of the theory, we recall that in Sect. 26.1 we proved that the continuity equation is covariant with respect to Galilean transformations provided that the charge density ρ and the current density $\mathbf{J}$ are transformed according to (26.7). In contrast, if we start from the special Lorentz transformations (26.18)–(26.21), instead of (26.5) and (26.6), then we derive the formulae

$$\frac{\partial}{\partial x^1} = \gamma \left(\frac{\partial}{\partial x'^1} - \frac{u}{c^2} \frac{\partial}{\partial t'} \right),$$ (26.45)

$$\frac{\partial}{\partial x^2} = \frac{\partial}{\partial x'^2},$$ (26.46)

$$\frac{\partial}{\partial x^3} = \frac{\partial}{\partial x'^3},$$ (26.47)

$$\frac{\partial}{\partial t} = \gamma \left(\frac{\partial}{\partial t'} - u \frac{\partial}{\partial x'^1} \right).$$ (26.48)

Then, under a Lorentz transformation, (26.1) becomes

$$\frac{\partial}{\partial t'} \left[\gamma \left(\rho - \frac{u}{c^2} J^1 \right) \right] + \frac{\partial}{\partial x'^1} \left[\gamma \left(J^1 - \rho u \right) \right]$$
$$+ \frac{\partial}{\partial x'^2} J^2 + \frac{\partial}{\partial x'^3} J^3 = 0.$$ (26.49)

We see that the continuity equation is covariant under Lorentz transformations if and only if

$$\rho' = \gamma \left(\rho - \frac{u}{c^2} J^1 \right), \quad J'^1 = \gamma(J^1 - \rho u), \quad J'^2 = J^2, \quad J'^3 = J^3.$$ (26.50)

Denoting by $\mathbf{a}_\parallel$ the component of vector $\mathbf{a}$ parallel to the relative velocity $\mathbf{u}$ and by $\mathbf{a}_\perp$ the component orthogonal to $\mathbf{u}$, the preceding relations can also be written as follows:

$$\rho' = \gamma \left(\rho - \frac{\mathbf{J} \cdot \mathbf{u}}{c^2} \right), \quad \mathbf{J}'_\parallel = \gamma(\mathbf{J}'_\parallel - \rho \mathbf{u}), \quad \mathbf{J}'_\perp = \mathbf{J}_\perp.$$ (26.51)

We conclude that, a priori, i.e., from a mathematical point of view, the continuity equation of charge could be covariant under both Galilean and Lorentz transformations, provided that we assume different transformation properties for the charge density and the current vector. However, the experiment will compel us to decide which of (26.7) and (26.50) we must choose. We emphasize that it could happen that, still from a mathematical point of view, the considered equations would be covariant under only one of the two transformation groups. In fact, in Sect. 26.1, we showed that Maxwell's equations in a vacuum, (26.2) and (26.3), are not covariant under Galilean transformations. However, in the next section, we verify that they are covariant under Lorentz transformations, provided that the electromagnetic fields are transformed in a suitable way in going from an inertial frame to another inertial frame.

26.6 Covariance of Maxwell's Equations

In this section, we verify the covariance of Maxwell's equations in a vacuum under Lorentz transformations. In an inertial frame I, $(26.3)_2$ and $(26.4)_2$ have the following coordinate form:

$$\frac{\partial E^3}{\partial x^2} - \frac{\partial E^2}{\partial x^3} = -\mu_0 \frac{\partial H^1}{\partial t},$$
$$\frac{\partial E^1}{\partial x^3} - \frac{\partial E^3}{\partial x^1} = -\mu_0 \frac{\partial H^2}{\partial t},$$
$$\frac{\partial E^2}{\partial x^1} - \frac{\partial E^1}{\partial x^2} = -\mu_0 \frac{\partial H^3}{\partial t},$$
$$\frac{\partial H^1}{\partial x^1} + \frac{\partial H^2}{\partial x^2} + \frac{\partial H^3}{\partial x^3} = 0.$$

Inserting (26.45)–(26.48) into the preceding system, we obtain

$$\frac{\partial E^3}{\partial x'^2} - \frac{\partial E^2}{\partial x'^3} = -\mu_0 \gamma \frac{\partial H^1}{\partial t'} + \mu_0 \gamma u \frac{\partial H^1}{\partial x'^1},$$
$$\frac{\partial E^1}{\partial x'^3} - \frac{\partial}{\partial x'^1}\left[\gamma(E^3 + \mu_0 u H^2)\right] = -\mu_0 \frac{\partial}{\partial t'}\left[\gamma\left(H^2 + \frac{u E^3}{\mu_0 c^2}\right)\right],$$
$$\frac{\partial}{\partial x'^1}\left[\gamma\left(E^2 - \mu_0 u H^3\right)\right] - \frac{\partial E^1}{\partial x'^2} = -\mu_0 \frac{\partial}{\partial t'}\left[\gamma\left(H^3 - \frac{u E^2}{\mu_0 c^2}\right)\right],$$
$$\gamma\frac{\partial H^1}{\partial x'^1} - \gamma\frac{u}{c^2}\frac{\partial H^1}{\partial t'} + \frac{\partial H^2}{\partial x'^2} + \frac{\partial H^3}{\partial x'^3} = 0,$$

where $c = 1/\sqrt{\epsilon_0 \mu_0}$ is the light velocity in a vacuum. Solving the fourth of the preceding equations with respect to $\gamma\, \partial H^1/\partial x'^1$ and inserting the obtained expression into the first equation, we have the following system:

$$\frac{\partial}{\partial x'^2}\left[\gamma\left(E^3 + \mu_0 u H^2\right)\right] - \frac{\partial}{\partial x'^3}\left[\gamma\left(E^2 - \mu_0 u H^3\right)\right] = -\mu_0 \frac{\partial H^1}{\partial t'},$$
$$\frac{\partial E^1}{\partial x'^3} - \frac{\partial}{\partial x'^1}\left[\gamma(E^3 + \mu_0 u H^2)\right] = -\mu_0 \frac{\partial}{\partial t'}\left[\gamma\left(H^2 + \frac{u E^3}{\mu_0 c^2}\right)\right],$$
$$\frac{\partial}{\partial x'^1}\left[\gamma\left(E^2 - \mu_0 u H^3\right)\right] - \frac{\partial E^1}{\partial x'^2} = -\mu_0 \frac{\partial}{\partial t'}\left[\gamma\left(H^3 - \frac{u E^2}{\mu_0 c^2}\right)\right],$$
$$\gamma\frac{\partial H^1}{\partial x'^1} - \gamma\frac{u}{c^2}\frac{\partial H^1}{\partial t'} + \frac{\partial H^2}{\partial x'^2} + \frac{\partial H^3}{\partial x'^3} = 0.$$

By adopting the same notations of (26.51), we can state that Maxwell's equations in a vacuum are covariant under Lorentz transformations if the electromagnetic field is transformed according to the following formulae:

$$\mathbf{E}'_\parallel = \mathbf{E}_\parallel, \quad \mathbf{E}'_\perp = \gamma \left(\mathbf{E}_\perp + \mu_0 \mathbf{u} \times \mathbf{H}_\perp \right), \tag{26.52}$$

$$\mathbf{H}'_\parallel = \mathbf{H}_\parallel, \quad \mathbf{H}'_\perp = \gamma \left(\mathbf{H}_\perp - \mu_0 \mathbf{u} \times \mathbf{E}_\perp \right). \tag{26.53}$$

We conclude our considerations about relativistic electromagnetism in a vacuum[4] by proving **charge conservation**. Consider a volume V occupied by a charge q, distributed inside V with a charge density $\rho(\mathbf{r}, t)$. Let V move relative to an inertial frame I with velocity $\dot{\mathbf{r}}$. Denote by $I_0(\mathbf{x}, t)$ the proper inertial frame of the point $\mathbf{x} \in V$ at an arbitrary instant t. Because the current density $\mathbf{J}$ becomes $\mathbf{J} = \rho \dot{\mathbf{r}}$ and the velocity $\mathbf{u}$ of $I_0(\mathbf{x}, t)$ relative to I at the instant t is $\dot{\mathbf{r}}(\mathbf{x}, t)$, condition $(26.50)_1$ gives

$$\rho_0 = \frac{\rho}{\gamma}.$$

Taking into account (26.31), we have that

$$q = \int_V \rho \, dV = \int_{V_0} \rho_0 \, dV_0 = q_0, \tag{26.54}$$

and we can state that the total charge is invariant with respect to Lorentz transformations.

26.7   Relativistic Dynamics

In this section we derive a relativistic equation describing the dynamics of a material point in inertial frames following the approach proposed by Einstein.

Let us suppose the material point P with mass m and charge q moving relative to the inertial frame I with velocity $\dot{\mathbf{r}}$ in the presence of an electromagnetic field $(\mathbf{E}, \mathbf{H})$. Then P is acted upon by the **Lorentz force**

$$\mathbf{F} = q(\mathbf{E} + \mu_0 \dot{\mathbf{r}} \times \mathbf{H}). \tag{26.55}$$

Denote by I_0 the **proper frame of P at an arbitrary instant** t, that is, the inertial frame that moves relative to I with a *uniform* velocity equal to the velocity $\dot{\mathbf{r}}$ at instant t. In the proper frame I_0, the Lorentz force reduces to the electrostatic force

$$\mathbf{F}_0 = q \mathbf{E}_0 \tag{26.56}$$

---

[4]For the difficult topic of electromagnetism in matter, see, for instance, [27, 39, 44].

since the charge q carried by P is invariant [see (26.54)]. It is a simple exercise to verify that from (26.52) and (26.53) we obtain the following transformation formulae for electric force under a special Lorentz transformation $I \to I_0$

$$F_0^1 = F^1, \tag{26.57}$$

$$F_0^2 = \gamma F^2, \tag{26.58}$$

$$F_0^3 = \gamma F^3, \tag{26.59}$$

which, adopting the notation of (26.51), can be summarized as follows:

$$\mathbf{F}_{0\|} = \mathbf{F}_\|, \quad \mathbf{F}_{0\perp} = \gamma \mathbf{F}_\perp. \tag{26.60}$$

Einstein makes the following assumptions:

1. *All the forces, regardless of their nature, transform according to (26.60) under a Lorentz transformation.*

   Also, for the change of inertial frame $I \to I_0$, formulae (26.40)–(26.42) give

$$\ddot{x}_0^1 = \gamma^3 \ddot{x}^1, \tag{26.61}$$

$$\ddot{x}_0^2 = \gamma^2 \ddot{x}^2, \tag{26.62}$$

$$\ddot{x}_0^1 = \gamma^2 \ddot{x}^3, \tag{26.63}$$

which can equivalently be written as

$$\ddot{\mathbf{r}}_{0\|} = \gamma^3 \ddot{\mathbf{r}}_\|, \quad \ddot{\mathbf{r}}_{0\perp} = \gamma^2 \ddot{\mathbf{r}}_\perp. \tag{26.64}$$

The second assumption on which Einstein bases his relativistic dynamics is

2. *The Newtonian equation of motion is valid in the proper frame I_0, that is, we have that*

$$m_0 \ddot{\mathbf{r}}_0 = \mathbf{F}_0. \tag{26.65}$$

Applying (26.60) and (26.64)–(26.65), we obtain the following equations in any inertial frame I:

$$m_0 \gamma^3 \ddot{\mathbf{r}}_\| = \mathbf{F}_\|, \quad m_0 \gamma \ddot{\mathbf{r}}_\perp = \mathbf{F}_\perp. \tag{26.66}$$

Now we prove that the preceding equations are the projections parallel and orthogonal to $\dot{\mathbf{r}}$, respectively, of the equation

$$\frac{d}{dt}(m\dot{\mathbf{r}}) = \mathbf{F}, \tag{26.67}$$

where

$$m = m_0 \gamma = m_0 \frac{1}{\sqrt{1 - \dfrac{|\dot{\mathbf{r}}|^2}{c^2}}} \tag{26.68}$$

is the *relativistic mass* of the material point P.

To prove the preceding statement, we first note that

$$\frac{d\gamma}{dt} = \frac{d}{dt}\left(1 - \frac{|\dot{\mathbf{r}}|^2}{c^2}\right)^{-1/2} = \gamma^3 \frac{\dot{\mathbf{r}} \cdot \ddot{\mathbf{r}}_{\parallel}}{c^2}, \tag{26.69}$$

$$\gamma^2 = 1 + \gamma^2 \frac{|\dot{\mathbf{r}}|^2}{c^2}. \tag{26.70}$$

Then also

$$\frac{d}{dt}(\gamma \dot{\mathbf{r}}) = \gamma\left(\frac{\gamma^2}{c^2} \dot{\mathbf{r}} \cdot \ddot{\mathbf{r}}_{\parallel}\dot{\mathbf{r}} + \ddot{\mathbf{r}}_{\parallel} + \ddot{\mathbf{r}}_{\perp}\right), \tag{26.71}$$

so that we have

$$\left(\frac{d}{dt}(\gamma\dot{\mathbf{r}})\right)_{\parallel} = \gamma\left(\frac{\gamma^2}{c^2}\dot{\mathbf{r}} \cdot \ddot{\mathbf{r}}_{\parallel}\dot{\mathbf{r}}_{\parallel} + \ddot{\mathbf{r}}_{\parallel}\right), \tag{26.72}$$

$$\left(\frac{d}{dt}(\gamma\dot{\mathbf{r}})\right)_{\perp} = \gamma\ddot{\mathbf{r}}_{\perp}. \tag{26.73}$$

Introducing the unit vector $\mathbf{u}$ along $\dot{\mathbf{r}}$ we have that $\dot{\mathbf{r}} = |\dot{\mathbf{r}}|\mathbf{u}$, $\ddot{\mathbf{r}}_{\parallel} = |\ddot{\mathbf{r}}_{\parallel}|\mathbf{u}$, and recalling (26.70), we can write (26.72) as follows:

$$\left(\frac{d}{dt}(\gamma\dot{\mathbf{r}})\right)_{\parallel} = \gamma\left(\frac{\gamma^2}{c^2}|\dot{\mathbf{r}}|^2 + 1\right)\ddot{\mathbf{r}}_{\parallel} = \gamma^3\ddot{\mathbf{r}}_{\parallel}. \tag{26.74}$$

This equation, together with (26.73), proves the equivalence between (26.66) and (26.67).

Now we derive an important consequence of the relativistic equation (26.67). By a scalar product of (26.67) and $\dot{\mathbf{r}}$, we obtain the equation

$$\dot{\mathbf{r}} \cdot \frac{d}{dt}(m\dot{\mathbf{r}}) = \dot{\mathbf{r}} \cdot \mathbf{F}.$$

Taking into account (26.74), the preceding equation becomes

$$\dot{\mathbf{r}} \cdot \frac{d}{dt}(m\dot{\mathbf{r}}) = \dot{\mathbf{r}} \cdot \left(\frac{d}{dt}(m\dot{\mathbf{r}})\right)_{\parallel} = m_0\gamma^3\dot{\mathbf{r}} \cdot \ddot{\mathbf{r}}_{\parallel} = \dot{\mathbf{r}} \cdot \mathbf{F}. \tag{26.75}$$

In view of (26.69), we finally obtain the equation

$$\frac{dE}{dt} = \dot{\mathbf{r}} \cdot \mathbf{F},\tag{26.76}$$

where the scalar quantity

$$E = mc^2 = m_0 \gamma c^2\tag{26.77}$$

is called the **relativistic energy** of the material point P. Also, we call the scalar quantities

$$T = E - m_0 c^2, \quad E_0 = m_0 c^2\tag{26.78}$$

the **relativistic kinetic energy** and the **rest or proper relativistic energy**. From the approximate equality

$$\gamma - 1 \simeq \frac{1}{2} \frac{|\dot{\mathbf{r}}|^2}{c^2}$$

we derive the condition

$$T = \frac{1}{2} m_0 |\dot{\mathbf{r}}|^2 + O\left(\frac{|\dot{\mathbf{r}}|^2}{c^2}\right).$$

This relation shows that for a velocity much smaller than the light speed c, the relativistic kinetic energy reduces to the classical one. In other words, (26.75) is the relativistic formulation of the classical theorem on the variation of kinetic energy.

26.8 Minkowski's Space–Time

In the preceding section, we described Einstein's formulation of special relativity, which starts from both physical considerations and Maxwell's equations. Henceforth we discuss the elegant and useful geometric formulation of the special theory of relativity supplied by Minkowski. This formulation represented the bridge that Einstein crossed to arrive at a geometric formulation of gravitation.

We start by analyzing the geometric structure of the four-dimensional space underlying this model. We denote by V_4 a generalized Euclidean four-dimensional space in which an orthonormal frame of reference $(O, \mathbf{e}_\alpha)$, $\alpha = 1, \ldots, 4$, can be found such that the coefficients $\eta_{\alpha\beta}$ of the scalar product, which is defined by a symmetric covariant 2-tensor $\mathbf{g}$, are given by the following matrix:

$$(\eta_{\alpha\beta}) = \begin{pmatrix} 1 & 0 & 0 & 0 \\ 0 & 1 & 0 & 0 \\ 0 & 0 & 1 & 0 \\ 0 & 0 & 0 & -1 \end{pmatrix}. \tag{26.79}$$

The space V_4 is called Minkowski's *space–time* and any point $P \in V_4$ is an *event*. Moreover, any vector $\mathbf{v}$ of the vector space associated with V_4 is said to be a *four-vector* or 4-*vector*.

We call a *Lorentz frame* any frame of reference of V_4 in which the coefficients $\eta_{\alpha\beta}$ of the scalar product assume the form (26.79). Finally, a *Lorentz transformation* is any (linear) coordinate transformation

$$x'^{\alpha} = x_0'^{\alpha} + A_{\beta}^{\alpha} x^{\beta} \tag{26.80}$$

relating the coordinates (x^{α}) of any event in the Lorentz frame $(O, \mathbf{e}_{\alpha})$ to the coordinates $(O', \mathbf{e}_{\alpha}')$ of the same event in the Lorentz frame $(O', \mathbf{e}_{\alpha}')$. The definition of Lorentz frame requires that the matrix A_{β}^{α} satisfies the conditions

$$\eta_{\alpha\beta} = A_{\alpha}^{\lambda} A_{\beta}^{\mu} \eta_{\lambda\mu}. \tag{26.81}$$

Since in any Lorentz frame the square of the distance s between two events $x_{(1)}^{\alpha}$ and $x_{(2)}^{\alpha}$ assumes the form

$$\begin{aligned} s^2 &= \eta_{\alpha\beta} \left(x_{(2)}^{\alpha} - x_{(1)}^{\alpha} \right) \left(x_{(2)}^{\beta} - x_{(1)}^{\beta} \right) \\ &= \sum_{i=1}^{3} \left(x_{(2)}^{i} - x_{(1)}^{i} \right)^2 - \left(x_{(2)}^{4} - x_{(1)}^{4} \right)^2, \end{aligned} \tag{26.82}$$

we can say that the Lorentz transformations are orthogonal transformations of V_4, provided that the orthogonality is evaluated by the scalar product $\eta_{\alpha\beta}$. Owing to (26.81) or (26.82), we can conclude that the set of all these transformations is a *group*.

The square $\mathbf{v} \cdot \mathbf{v} = \mathbf{g}(\mathbf{v}, \mathbf{v})$ of the norm of a 4-vector $\mathbf{v}$ of V_4 can be positive, negative, or zero. In the first case, we say that $\mathbf{v}$ is a *spacelike* 4-vector. In the second case, $\mathbf{v}$ is said to be a *timelike* 4-vector. Finally, if $\mathbf{v} \cdot \mathbf{v} = 0$, then $\mathbf{v}$ is a *null* 4-vector. At any point $O \in V_4$, the set of the events $P \in V_4$ such that the 4-vectors $\overrightarrow{OP} = \mathbf{v}$ are null, i.e., the set of the 4-vectors for which

$$\overrightarrow{OP} \cdot \overrightarrow{OP} = 0, \tag{26.83}$$

is a cone C_O, with its vertex at O, which is called a *light cone* at O. Condition (26.83) defines a cone since, if $\overrightarrow{OP} \in C_O$, then also $\lambda \overrightarrow{OP} \in C_O$. In a Lorentz frame $(O, \mathbf{e}_{\alpha})$, the cone (26.83) is represented by the equation

$$\sum_{i=1}^{3} (x^i)^2 - (x^4)^2 = 0. \tag{26.84}$$

The 4-vectors $\mathbf{v} = \overrightarrow{OP}$, corresponding to events P, which are internal to C_O, are timelike 4-vectors, whereas those corresponding to events P, which are external to C_O, are spacelike 4-vectors.

Let us fix an arbitrary timelike 4-vector $\hat{\mathbf{e}}$, and let $(O, \hat{\mathbf{e}})$ be a uniform vector field of V_4. We say that, at any $O \in V_4$, the 4-vector $\hat{\mathbf{e}}$ defines the *direction of the future* at O. Moreover, the internal region C_O^+ of C_O, to which $\hat{\mathbf{e}}$ belongs, is said to be the *future* of O, whereas the remaining internal region C_O^- of C_O is the *past* of O. Finally, the set of events external to C_O is defined as the *present* of O.

The first three axes of a Lorentz frame $(O, \mathbf{e}_i)$ are spacelike 4-vectors, whereas the fourth axis $\mathbf{e}_4$ is a timelike 4-vector. Moreover, let $\sigma_{O,\mathbf{e}_4}$ be the three-dimensional space formed by all the 4-vectors generated by linear combinations of the three 4-vectors $\mathbf{e}_1, \mathbf{e}_2, \mathbf{e}_3$. It is easy to verify that $\sigma_{O,\mathbf{e}_4}$ is a properly Euclidean space, with respect to the scalar product $\mathbf{g}$, and that its elements are spacelike 4-vectors.

Before making clear the physical meaning of all the preceding definitions, we must prove the following propositions.

Proposition 26.1. *Let* $\mathbf{u}$ *be any timelike 4-vector, and let us denote by* $\sigma_{O,\mathbf{u}}$ *the three-dimensional space of all 4-vectors that are orthogonal to* $\mathbf{u}$. *Then, any* $\mathbf{v} \in \sigma_{O,\mathbf{u}}$ *is a spacelike 4-vector.*

Proof. Let $(O, \mathbf{e}_\alpha)$ be a Lorentz frame, and let $\mathbf{u} = u^i \mathbf{e}_i + u^4 \mathbf{e}_4 \equiv \mathbf{u}_\perp + u^4 \mathbf{e}_4$ be the decomposition of the 4-vector $\mathbf{u}$ in the basis $(\mathbf{e}_\alpha)$. Since $\mathbf{u}_\perp \cdot \mathbf{u}_\perp = \sum_{i=1}^{3} (u^i)^2$, the 4-vector $\mathbf{u}_\perp$ is spacelike. With the same notation, we set $\mathbf{v} = \mathbf{v}_\perp + v^4 \mathbf{e}_4$, so that the following relations hold:

$$\mathbf{u} \cdot \mathbf{u} = \mathbf{u}_\perp \cdot \mathbf{u}_\perp - (u^4)^2 < 0,$$
$$\mathbf{u} \cdot \mathbf{v} = \mathbf{u}_\perp \cdot \mathbf{v}_\perp - u^4 v^4 = 0,$$
$$\mathbf{v} \cdot \mathbf{v} = \mathbf{v}_\perp \cdot \mathbf{v}_\perp - (v^4)^2.$$

Consequently,

$$\mathbf{v} \cdot \mathbf{v} = \mathbf{v}_\perp \cdot \mathbf{v}_\perp - \frac{(\mathbf{u}_\perp \cdot \mathbf{v}_\perp)^2}{(u^4)^2},$$

so that, by the application of Schwartz's inequality to $\mathbf{u}_\perp \cdot \mathbf{v}_\perp$, we obtain

$$\mathbf{v} \cdot \mathbf{v} \geq \mathbf{v}_\perp \cdot \mathbf{v}_\perp - \frac{(\mathbf{u}_\perp \cdot \mathbf{u}_\perp)(\mathbf{v}_\perp \cdot \mathbf{v}_\perp)}{(u^4)^2}.$$

Finally, this inequality implies that

$$\mathbf{v} \cdot \mathbf{v} \geq -\frac{\mathbf{v}_\perp \cdot \mathbf{v}_\perp}{(u^4)^2}[\mathbf{u}_\perp \cdot \mathbf{u}_\perp - (u^4)^2] > 0,$$

and the proposition is proved.   □

Remark 26.5. Owing to the preceding proposition, we can say that any timelike 4-vector $\mathbf{u}$ defines infinite Lorentz frames at event O. In fact, it is sufficient to consider the frame $(O, \mathbf{e}_1, \mathbf{e}_2, \mathbf{e}_3, \mathbf{u}/|\mathbf{u}|)$, where the mutually orthogonal unit vectors $\mathbf{e}_i$, $i = 1, 2, 3$, belong to σ_O, $\mathbf{u}$.

Proposition 26.2. *If $\mathbf{v}$ is a spacelike 4-vector at $O \in V_4$, then it is possible to find at least a timelike 4-vector such that $\mathbf{u} \cdot \mathbf{v} = 0$.*

Proof. Let $(O, \mathbf{e}_\alpha)$ be a Lorentz frame at O. Then, by adopting the same notations used in the preceding proposition, we have that

$$\mathbf{v}_\perp \cdot \mathbf{v}_\perp - (v^4)^2 > 0. \tag{26.85}$$

We must now prove that there exists at least a 4-vector $\mathbf{u}$ such that

$$\mathbf{u} \cdot \mathbf{u} = \mathbf{u}_\perp \cdot \mathbf{u}_\perp - (u^4)^2 < 0, \tag{26.86}$$

$$\mathbf{u} \cdot \mathbf{v} = \mathbf{u}_\perp \cdot \mathbf{v}_\perp - u^4 v^4 = 0. \tag{26.87}$$

From (26.87) we derive that the 4-vector $\mathbf{u}$ is orthogonal to $\mathbf{v}$ if we choose arbitrarily the components u^1, u^2, u^3, i.e., the spacelike 4-vector $\mathbf{u}_\perp$, provided that the component u^4 is given by

$$u^4 = \frac{\mathbf{u}_\perp \cdot \mathbf{v}_\perp}{v^4}.$$

For this choice of u^4, condition (26.86) becomes

$$\mathbf{u} \cdot \mathbf{u} = \mathbf{u}_\perp \cdot \mathbf{u}_\perp - \frac{(\mathbf{u}_\perp \cdot \mathbf{v}_\perp)^2}{(v^4)^2}.$$

If we choose $\mathbf{u}_\perp = \mathbf{v}_\perp$, then we can write the preceding equation as

$$\mathbf{u} \cdot \mathbf{u} = \mathbf{v}_\perp \cdot \mathbf{v}_\perp \left(1 - \frac{\mathbf{v}_\perp \cdot \mathbf{v}_\perp}{(v^4)^2}\right) \tag{26.88}$$

so that, in view of (26.85), the vector

$$\mathbf{u} = \mathbf{v}_\perp + \frac{\mathbf{v}_\perp \cdot \mathbf{v}_\perp}{(v^4)^2}\mathbf{e}_4$$

confirms (26.86) and (26.87).   □

Remark 26.6. For any spacelike vector $\mathbf{v}$ at event O, we can find infinite Lorentz frames $(O, \mathbf{e}_\alpha)$ for which $\mathbf{e}_1 = \mathbf{v}/|\mathbf{v}|$. In fact, it is sufficient to take one of the existing timelike 4-vectors $\mathbf{e}_4$ orthogonal to $\mathbf{v}$ and choose in the three-dimensional space $\sum \sigma_{O,\mathbf{e}_4}$ to which $\mathbf{v}$ belongs two other unit vectors that are orthogonal to each other and orthogonal to $\mathbf{v}$.

Let us suppose that we have introduced a direction of the future by the uniform 4-vector $\hat{\mathbf{u}}$, so that it is possible to define the cones C_O^- and C_O^+ at any event O.

Proposition 26.3. *Let* $\mathbf{u} \in C_O^+$. *Then,* $\mathbf{v} \in C_O^+$ *if and only if*

$$\mathbf{u} \cdot \mathbf{v} < 0. \tag{26.89}$$

Proof. In fact, $\mathbf{u} \cdot \mathbf{u} < 0$ since $\mathbf{u} \in C_O^+$. But $\mathbf{u} \cdot \mathbf{w}$ is a continuous function of $\mathbf{w} \in C_O^+$, which is a connected set. Consequently, if, for a 4-vector $\mathbf{v} \in C_O^+$, we had $\mathbf{u} \cdot \mathbf{v} > 0$, then there would exist a timelike 4-vector $\mathbf{u}^* \in C_O^+$ such that $\mathbf{u} \cdot \mathbf{u}^* = 0$. But this is impossible for Proposition 26.1.

Conversely, owing to the remark about Proposition 26.1, we can find a Lorentz frame at O having $\mathbf{e}_4 = \mathbf{u}/|\mathbf{u}|$ as a timelike axis. In this frame, we have that $\mathbf{u} \cdot \mathbf{u} = -u^4 v^4$. On the other hand, since $\mathbf{u}, \mathbf{u} \in C_O^+$, we have $u^4 > 0$ and $v^4 > 0$, and the proposition is proved. $\square$

We denote by L_O^+ the set of Lorentz frames $(O, \mathbf{e}_\alpha)$ at O whose axis $\mathbf{e}_4$ belong to the positive cone C_O^+. The Lorentz transformation between two Lorentz frames in L_O^+ is said to be **orthochronous**. It is evident that the totality of these transformations is a group.

Proposition 26.4. *Let* $(O, \mathbf{e}_\alpha) \in L_O^+$ *be a Lorentz frame. Then the Lorentz frame* $(O, \mathbf{e}_\alpha')$ *belongs to* L_O^+ *if and only if*

$$A_4^4 > 0. \tag{26.90}$$

Proof. Since $\mathbf{e}_4' = A_4^\alpha \mathbf{e}_\alpha$, we have that $\mathbf{e}_4' \cdot \mathbf{e}_4 = A_4^4$. The proposition is proved when we take into account Proposition 26.3. $\square$

26.9 Physical Meaning of Minkowski's Space–Time

To attribute a physical meaning to *some* geometrical objects associated with the Minkowski space–time, we begin with the following remark. Let I and I' be two inertial frames of reference. We have already said that the relation between the coordinates (x_A^i, t_A) and $(x_A'^i, t_A')$, which observers I and I' associate with the same event A, is a Lorentz transformation. It is a very simple exercise to verify that any Lorentz transformation leaves invariant the following quadratic form:

$$s^2 \equiv \sum_{i=1}^{3} (x_A^i - x_B^i)^2 - c^2(t_A - t_B)^2$$

$$= \sum_{i=1}^{3} (x_A'^i - x_B'^i)^2 - c^2(t_A' - t_B')^2, \tag{26.91}$$

relating the space–time coordinates (x_A^i, t_A) and (x_B^i, t_B) of event A in the Lorentz frame I and the corresponding coordinates $(x_A'^i, t_A')$, $(x_B'^i, t_B')$ of the same events in I'

Moreover, from (26.21) we obtain that

$$\frac{dt'}{dt} = \gamma > 0. \tag{26.92}$$

Furthermore, if we introduce the notation $x^4 = ct$, then the quadratic form (26.91) becomes identical to (26.82).

Let us introduce a direction of the future at any point of V_4 by the uniform timelike field $\mathbf{e}_4$. Owing to Proposition 26.2 and the associated remark, we can define at any point of V_4 a Lorentz frame $(O, \mathbf{e}_\alpha) \in L_O^+$. Let us introduce a one-to-one correspondence among the inertial frames I in the physical space and the orthogonal or Lorentz frames in L_O^+ in the following way. First, we associate a fixed inertial frame I to the Lorentz frame $(O, \mathbf{e}_\alpha)$ in L_O^+. Let I' be any inertial frame whose relation with I is expressed by (26.8) and (26.9). Then to I' we associate the Lorentz frame $(O', \mathbf{e}_\alpha')$, whose transformation formulae (26.80) with respect to $(O, \mathbf{e}_\alpha)$ are given by

$$(x_{O'}^\alpha) = (x_{O'}^i, ct_0), \tag{26.93}$$

$$(A_\beta^\alpha) = \begin{pmatrix} Q_j^i \left(\delta_h^j + (\gamma - 1) \dfrac{u^j u_h}{u^2} \right) & -\gamma Q_j^i \dfrac{u^j}{c} \\ -\gamma \dfrac{u_i}{c} & \gamma \end{pmatrix}. \tag{26.94}$$

Since

$$A_4^4 = \gamma > 0, \tag{26.95}$$

the Lorentz frame $(O', \mathbf{e}_\alpha')$ belongs to L_O^+.

Conversely, let us assign the Lorentz frame $(O', \mathbf{e}_\alpha') \in L_O^+$ by (26.80), where $A_4^4 > 0$. To determine the corresponding inertial frame I', we must determine the right-hand sides of (26.93) and (26.94), i.e., the quantities t_0, u^i, Q_j^i, and $x_{O'}^i$. Owing to the orthogonality of the matrix (Q_j^i), we must evaluate ten quantities starting from the four coordinates $(x_{O'}^\alpha)$ and the six independent coefficients A_β^α [see (26.81)].

In particular, if I and I' are related by a special Lorentz transformation, then (26.93) and (26.94) reduce to the following formulae:

$$(x_{O'}^{\alpha}) = (0, 0), \tag{26.96}$$

$$(A_{\beta}^{\alpha}) = \begin{pmatrix} \gamma & 0 & 0 & -\gamma\dfrac{u}{c} \\ 0 & 1 & 0 & 0 \\ 0 & 0 & 1 & 0 \\ -\gamma\dfrac{u}{c} & 0 & 0 & \gamma \end{pmatrix}. \tag{26.97}$$

The preceding considerations allow us to make clear the physical meaning of the definitions we gave in the preceding section. In fact, to any point $P \in V_4$, with coordinates (x^{α}) in a Lorentz frame $(O, \mathbf{e}_{\alpha}) \in L_O^+$, there corresponds an event that has coordinates $(x^i, x^4/c)$ in the inertial frame I. In particular, all the events belonging to the light cone C_O have coordinates verifying (26.84). To these points of V_4 correspond all the events whose coordinates in the inertial frame I satisfy the equation

$$\sum_{i=1}^{3} (x^i)^2 - c^2 t^2 \equiv r^2 - c^2 t^2 = 0 \tag{26.98}$$

or, equivalently, the two equations $r - ct = 0$ and $r + ct = 0$, which represent, respectively, a spherical light wave expanding from O and a spherical light wave contracting toward O.

Moreover, owing to Proposition 26.1–26.3 and the related remark, for any point $P \in C_O^+$ it is possible to find a Lorentz frame in L_O^+ and then an inertial frame I such that event P has in I the coordinates $(0, 0, 0, t)$, $t > 0$, so that it appears to happen after the event at the origin. In contrast, if $P \in C_O^-$, then the corresponding event appears to happen before the event at O for observer I. In other words, any event belonging to C_O^+ happens after the event at O, for some inertial frames, and any event in C_O^- happens before the event at O, for some inertial frames.

In contrast, an event that belongs to the present of O, owing to Proposition 26.2 and the related remark, appears to happen at the same time to some inertial observer I.

We conclude this section with a very important remark regarding a consequence of the correspondence existing among the inertial frames and the Lorentz frames in V_4. *If we succeed in formulating the physical laws by tensor relations in V_4, then they will be covariant with respect to Lorentz's transformations; in other words, they will satisfy the principle of relativity.*

26.10 Four-Dimensional Equation of Motion

In the preceding sections, we presented Einstein's approach to the relativistic dynamics of a material point in any inertial frame I. We can summarize the results of this approach stating that the Newtonian equations must be substituted by the following ones:

$$\frac{d}{dt}(m\mathbf{v}) = \mathbf{F},\qquad(26.99)$$

$$\frac{d}{dt}(mc^2) = \mathbf{F}\cdot\dot{\mathbf{r}},\qquad(26.100)$$

where

$$m = \frac{m_0}{\sqrt{1 - \dfrac{|\dot{\mathbf{r}}|^2}{c^2}}}\qquad(26.101)$$

is the relativistic mass of P, m_0 its rest mass, and $\dot{\mathbf{r}}$ the velocity of P. In this section, we recall the four-dimensional formulation of (26.99), (26.100) in the space–time V_4.

Let

$$x^i = x^i(t)\qquad(26.102)$$

be the equation of motion of a particle with respect to an inertial frame I. To the trajectory (26.102) we can associate a curve σ of V_4, which in the Lorentz frame $\Im \equiv (O, \mathbf{e}_\alpha)$ corresponding to I has equations

$$x^i = x^i(t),\quad x^4 = ct.\qquad(26.103)$$

The curve σ is called a **world trajectory** or **world line** of P. It defines a curve of V_4 that does not depend on the adopted Lorentz frame. The square norm of the tangent vector $(\dot{x}^i, c)$ to σ is

$$\sum_{i=1}^{3}(\dot{x}^i)^2 - c^2 < 0,\qquad(26.104)$$

and it is negative since the velocity of P in any inertial frame is less than the velocity c of light in a vacuum. In other words, the curve σ is timelike and its tangent vector at any point P is contained in the map C_P^+ of the light cone at P.

Let $x^i(\bar{t})$ be the position of P at the instant $\bar{t}$ in the inertial frame I. We recall that the rest frame or proper frame $\bar{I}$ of P at the instant $\bar{t}$ is an inertial frame that has its origin at $x^i(\bar{t})$ and moves with velocity $\bar{\mathbf{v}}$ with respect to I. The corresponding Lorentz frame $((x^i(\bar{t}), c\bar{t}), \mathbf{e}_\alpha) \equiv \bar{\Im}$ has a time axis $\mathbf{e}_4$ tangent to σ at the point $x^i(\bar{t})$ since in $\bar{\Im}$ the vector $\mathbf{e}_4$ must have components $(0,0,0,1)$. In going from $\Im$ to $\bar{\Im}$, the infinitesimal distance between two events on σ is invariant, so that we have

$$ds^2 = \left(\sum_{i=1}^{3}(\dot{x}^i)^2 - c^2\right)dt^2 = -c^2 d\tau^2,\qquad(26.105)$$

where τ is the **proper time**, i.e., the time evaluated by observer $I(\bar{t})$. From (26.105) we derive the relation

$$\frac{dt}{d\tau} = \gamma. \tag{26.106}$$

If we adopt this time along σ, then the parametric equations (26.103) become

$$x^\alpha = x^\alpha(\tau). \tag{26.107}$$

We define as a **world velocity** or **4-velocity** the 4-vector

$$U^\alpha = \frac{dx^\alpha}{d\tau}, \tag{26.108}$$

which, in view of (26.106), has the following components in the Lorentz frame I:

$$U^\alpha = (\gamma\dot{\mathbf{r}}, \gamma c). \tag{26.109}$$

Moreover,

$$U^\alpha U_\alpha = \gamma^2(|\dot{\mathbf{r}}|^2 - c^2) = -c^2 < 0. \tag{26.110}$$

It is a simple exercise to verify that we can write (26.99) and (26.100) in the covariant form

$$m_0 \frac{dU^\alpha}{d\tau} = \phi^\alpha, \tag{26.111}$$

where the 4-**force** is given by

$$(\phi^\alpha) = \gamma\left(\mathbf{F}, \frac{\mathbf{F} \cdot \dot{\mathbf{r}}}{c}\right). \tag{26.112}$$

26.11 Tensor Formulation of Electromagnetism in Vacuum

Let S be a continuous system of moving charges, and denote by ρ and $\dot{\mathbf{r}}$ the charge density and the velocity field of S, respectively. If S moves in a vacuum, the continuity equation (26.1) becomes

$$\frac{\partial\rho}{\partial t} + \nabla \cdot (\rho\dot{\mathbf{r}}) = 0. \tag{26.113}$$

Consider a transformation (26.80) from a Lorentz frame I to another one I', and suppose that the quantities

$$J^\alpha = (\rho\dot{\mathbf{r}}, c\,\rho) \tag{26.114}$$

are the components of a 4-vector, called 4-*current*. In other words, we are assuming that, in going from I to I', we obtain the following results:

$$J'^\alpha = A_\beta^\alpha J^\beta. \tag{26.115}$$

Then (26.113) can be written as the divergence in V_4 of the 4-current

$$\frac{\partial J^\alpha}{\partial x^\alpha} = 0. \tag{26.116}$$

The form of this equation is independent of the Lorentz frame and, consequently, verifies a principle of relativity.

Exercise 26.1. Verify that from (26.115) and (26.97) we obtain again (26.50).

Similarly, the Maxwell equations (26.2) and (26.4)$_1$ can be written in the form

$$\frac{\partial F^{\alpha\beta}}{\partial x^\beta} = 0, \tag{26.117}$$

where

$$F^{\alpha\beta} = \begin{pmatrix} 0 & H^3 & -H^2 & -c\epsilon_0 E^1 \\ -H^3 & 0 & H^1 & -c\epsilon_0 E^2 \\ H^2 & -H^1 & 0 & -c\epsilon_0 E^3 \\ c\epsilon_0 E^1 & c\epsilon_0 E^2 & c\epsilon_0 E^3 & 0 \end{pmatrix}. \tag{26.118}$$

Consequently, if $F^{\alpha\beta}$ are supposed to be the components of a (skew-symmetric) 2-tensor, then (26.117) has the same form in any Lorentz frame and the relativity principle is satisfied by (26.2) and (26.4)$_1$. $F^{\alpha\beta}$ is called an *electromagnetic tensor*.

If we introduce the adjoint tensor of $F^{\alpha\beta}$

$$F_{\alpha\beta}^\star = -\frac{1}{2}\epsilon_{\alpha\beta\lambda\mu} F^{\lambda\mu}, \tag{26.119}$$

which is obtained from $F^{\alpha\beta}$ with the substitution $c\mathbf{E} \rightarrow -\mathbf{H}$, then Maxwell's equations (26.2) and (26.4)$_1$ assume the covariant form[5]

$$\frac{\partial F^{\star\alpha\beta}}{\partial x^\beta} = 0. \tag{26.121}$$

Exercise 26.2. Verify that from

$$F'^{\alpha\beta} = A_\lambda^\alpha A_\mu^\beta F^{\lambda\mu}$$

and (26.97) we obtain (26.52) and (26.53).

Exercise 26.3. Verify that the components of the Lorentz force (26.55) are equal to the first three components of the 4-vector

$$F^\alpha = \mu_0 q \, F^{\alpha\beta} U_\beta.$$

---

[5]If moving charge are present, then, instead of (26.117), we have

$$\frac{\partial F^{\star\alpha\beta}}{\partial x^\beta} = J^\alpha. \tag{26.120}$$

Chapter 27
Variational Calculus with Applications

Abstract In Chapter 17, Hamilton's principle was proved by using Euler's simple approach according to which a one-parameter family of curves is introduced that reduces the search for a minimum of a functional to that of a minimum of a function depending on a single variable. In this chapter, we present a deeper approach to variational calculus that shows how to extend to functionals many concepts of differential calculus relative to functions. Further, we formulate the problems of constrained stationary points of functionals and show how to obtain the momentum equation of continuum mechanics by variational principle.

27.1 Lagrange's Functional Space

Let $\mathfrak{F}$ be a suitable set of smooth functions $f(\mathbf{x})$, $\mathbf{x} \in D \subset \mathfrak{R}^n$, that satisfy specific conditions in the closed and compact set D. A functional F is a map

$$F : \mathfrak{F}^m \to \mathfrak{R} \tag{27.1}$$

that associates a real number to any m-tuple of functions $f = (f_1(\mathbf{x}), \ldots, f_m(\mathbf{x})) \in \mathfrak{F}^m$. We say that a physical law is formulated in a variational form when the fundamental fields describing the phenomenon minimize a suitable functional F defined in $\mathfrak{F}^m$. Usually physical laws are expressed by differential equations. In this case, we search for functions that satisfy relations between the functions themselves and their derivatives. In Chapter 17 we have already noticed that the variational formulation leads to a mathematical problem that is quite different in nature from the initial value problem that Cauchy stated for differential equations. We shall see that many physical phenomena are naturally described in a variational form. The idea that the physical laws could be formulated in such a way to minimize something, that is, the idea that Nature evolves in such a way to reduce the consumption of something was considered a proof of its divine origin, and so a metaphysical meaning was attributed

© Springer International Publishing AG, part of Springer Nature 2018
A. Romano and A. Marasco, *Classical Mechanics with Mathematica®*,
Modeling and Simulation in Science, Engineering and Technology,
https://doi.org/10.1007/978-3-319-77595-1_27

to variational principles. This attitude pushed many researchers to convert physical laws described by differential equations into variational principles. Afterward it was remarked that it is not possible to attribute a physical meaning to any variational principle. In conclusion, they only demonstrate the ability of the human mind to synthesize physical laws.

It is important to note that the first researches of Euler and Lagrange showed that searching for a minimum of a functional was equivalent to solving a suitable set of differential equations, the Euler–Lagrange equations. In other words, the global statement of searching for a minimum of a functional was again converted into a boundary value problem for differential equations. Today, direct methods to determine the minimum of a functional are known. Therefore, the inverse path is followed: given a system of differential equations, we search for its solutions by proving that these equations are the Euler–Lagrange equations of a suitable functional.

In this chapter we analyze the following functionals

-
$$J[\mathbf{y}(x)] = \int_a^b L(x, \mathbf{y}(x), \mathbf{y}'(x))dx, \tag{27.2}$$

where $\mathbf{y}(x) = (y_1(x), \ldots, y_m(x))$ is an m−tuple of functions of the variable $x \in [a, b]$.

-
$$J[\mathbf{u}(x)] = \int_\Omega L(\mathbf{x}, \mathbf{u}(\mathbf{x}), \nabla\mathbf{u})d\Omega, \tag{27.3}$$

where Ω is a closed and bounded region of $\Re^n$ and $\mathbf{u}(\mathbf{x})$ is an m−tuple of functions $(u_1(\mathbf{x}), \ldots, u_m(\mathbf{x}))$ of the point $\mathbf{x} \in \Omega$.

It is quite natural to answer the following questions about the functionals (27.2), (27.3):

- For which functions L, $\mathbf{y}(x)$, and $\mathbf{u}(\mathbf{x})$, functionals (27.2), (27.3) are defined?
- Is it possible to extend the ordinary definitions of continuity and differentiability of functions of one or more real variables to a functional?
- Is it possible to determine a minimum of a functional by this new definition of differentiability?

We answer the above questions starting from the simple case of a functional (27.2) in which $\mathbf{y}(x)$ reduces to a single function $y(x)$ of the real variable $x \in [a, b]$:

$$J[y(x)] = \int_a^b L(x, y(x), y'(x))dx. \tag{27.4}$$

The results we obtain in this case will easily be extended to more general functionals (27.2) and (27.3).

Denote by A the set of $C^1[a, b]$ functions, that is the set of functions that are continuous together with their first derivatives in the interval $[a, b]$, and verify the following boundary conditions:

$$y(a) = y_a, \quad y(b) = y_b. \tag{27.5}$$

It is evident that A is a vector space that becomes a metric space when it is equipped with the *first-order Lagrange metric*

$$d(y_1(x), y_2(x)) = \max_{x \in [a,b]} \left\{ |y_1(x) - y_2(x)| + |y_1'(x) - y_2'(x)| \right\}. \tag{27.6}$$

Two functions $y_1(x)$ and $y_2(x)$ are "nearby" for the metrics (27.6) if the values of the functions and their first derivatives are nearby in the whole interval $[a, b]$.

To understand the meaning of this metric, we consider a sequence $y_1(x), \ldots,$ $y_n(x), \ldots$ of functions belonging to A. We say that it converges to the function $y(x) \in A$ with respect to the metric (27.6) and we write

$$\lim_{n \to \infty} y_n(x) = y(x), \tag{27.7}$$

if

$$\lim_{n \to \infty} d(y_n, y) = 0. \tag{27.8}$$

Let $y_1(x), \ldots, y_n(x), \ldots$ be a Cauchy sequence, that is a sequence for which $\forall \epsilon > 0, \exists \nu(\epsilon) \in N$ such that $\forall n, m > \nu(\epsilon)$ it is

$$d(y_n(x), y_m(x)) < \epsilon. \tag{27.9}$$

In view of (27.6), this inequality implies that $\forall n, m > \nu(\epsilon)$,

$$\max_{x \in [a,b]} |y_n(x) - y_m(x)| < \epsilon, \quad \max_{x \in [a,b]} |y_n'(x) - y_m'(x)| < \epsilon, \tag{27.10}$$

so that it is also

$$|y_n(x) - y_m(x)| \leq \max_{x \in [a,b]} |y_n(x) - y_m(x)| < \epsilon, \tag{27.11}$$

$$|y_n'(x) - y'(x)| \leq \max_{x \in [a,b]} |y_n'(x) - y'(x)| < \epsilon. \tag{27.12}$$

For Cauchy's convergence criterion, we can state that the convergence of a Cauchy sequence according to the metric (27.6) is equivalent to the *uniform* convergence in the interval $[a, b]$ of the sequences $y_1(x), \ldots, y_n(x), \ldots$ and $y_1'(x), \ldots, y_n'(x), \ldots$. Therefore, the limit function $y(x)$ also belongs to A. In conclusion, the space A, with the metric (27.6), is *complete* since any Cauchy sequence converges to an element of A.

Finally, the vector space A becomes a normed space when the norm $||y(x)||$ of the function $y(x) \in A$ is defined as follows:

$$||y(x)|| = d(y(x), 0) = \max_{x \in [a,b]} \left\{ |y(x)| + |y'(x)| \right\}. \tag{27.13}$$

27.2 Fréchet's Differential and Gâteaux's Differential

Functional $J[y(x)] : A \to \Re$ is said to be *continuous* at $y_0(x) \in A$ if

$$\lim_{y(x) \to y_0(x)} J[y(x)] = J[y_0(x)], \tag{27.14}$$

where the limit $y(x) \to y_0(x)$ is evaluated by the metric (27.6).

Functional $J[y(x)] : A \to \Re$ is said to be **Fréchet differentiable** at $y_0(x) \in C^1([a, b])$ if a *linear* functional $D_{y_0(x)}J : A \to \Re$ exists such that, $\forall y_0(x) + h(x) \in A$, [1]it is

$$J[y_0(x) + h(x)] = J[y_0(x)] + D_{y_0(x)}J[h(x)] + r[y_0(x) + h(x)], \tag{27.15}$$

where the functional $r[y_0(x), h(x)]$ verifies the condition

$$\lim_{h(x) \to 0} \frac{r[y_0(x) + h(x)]}{||h(x)||} = 0. \tag{27.16}$$

The linear operator $D_{y_0(x)}J$ is called the **Fréchet derivative** of J at $y_0(x)$.

Functional $J[y(x)] : A \to \Re$ is said to be **Gâteaux differentiable** at $y_0(x) \in A$ if the limit

$$D_G J_{y_0(x)} = \lim_{t \to 0} \frac{J[y_0(x) + th(x)] - J[y_0(x)]}{t}, \tag{27.17}$$

where t is a real variable and is finite $\forall y_0(x)$, $y_0(x) + h(x) \in A$. The operator $D_G J_{y_0(x)}$ is called the **Gâteaux derivative** of J at $y_0(x)$.

A functional is Fréchet (Gâteaux) differentiable in A if it is Fréchet (Gâteaux) differentiable at any $y_0(x) \in A$.

It can be proved that if J is Fréchet differentiable it is also Gâteaux differentiable; generally the opposite implication is not true. The Gâteaux differentiability implies the Fréchet differentiability if the Gateau derivative is continuous in a neighborhood of $y_0(x)$. In this case, the Gâteaux differential depends linearly on $h(x)$ and coincides with the Fréchet derivative.

Functional J has a *local minimum* at $y_0(x) \in A$, if a neighborhood $I(y_0(x))$ of $y_0(x)$ (with respect to the metric (27.6)) exists such that

$$J[y_0(x)] \le J[y(x)], \quad \forall y(x) \in I(y_0(x)). \tag{27.18}$$

A local maximum can be defined in a similar way.

For real functions $f(\mathbf{x}) \in C^1(\Omega)$, $\Omega \subset \Re^n$, the existence of a maximum or a minimum at a point $\mathbf{x}_0$ necessarily implies that its differential vanishes at that point:

$$(df)_{\mathbf{x}_0} = (\nabla_{\mathbf{x}} f)_{\mathbf{x}_0} \cdot \mathbf{h} = 0, \quad \forall \mathbf{h} \in \Re^n.$$

---

[1]When the functions of A verify the boundary conditions (27.5), then $h(a) = h(b) = 0$.

Similarly, in view of (27.15), $y_0(x)$ is a minimum or a maximum for a Fréchet differentiable functional $J[y(x)]$, then necessarily

$$D_{y_0(x)} J[h(x)] = 0, \qquad (27.19)$$

$\forall h(x) \in C^1[a, b]$ and vanishing at a and b when the conditions (27.5) must be satisfied. Any function $y_0(x)$ confirming (27.19) is said to be an **extremal** or a **stationary point** of J.

We want to determine the Fréchet differential of functional (27.4) supposing that the function L depends continuously on its variables and has continuous first and second derivatives in the domain

$$a \le x \le b, \quad -\infty \le y \le \infty, \quad -\infty \le y' \le \infty.$$

First, we note that

$$J[y(x) + h(x)] - J[y(x)] = \int_a^b [L(x, y(x) + h(x), y'(x) + h'(x)) - L(x, y(x), y'(x))]dx.$$
$$(27.20)$$

Applying Taylor's formula to L,

$$L = \frac{\partial L}{\partial y} h(x) + \frac{\partial L}{\partial y'} h'(x) + \hat{r}(x + \xi h),$$

where $\xi \in [0, 1]$, the above relation becomes

$$J[y(x) + h(x)] - J[y(x)] = \int_a^b \left[\frac{\partial L}{\partial y} h(x) + \frac{\partial L}{\partial y'} h'(x) \right] dx + \int_a^b \hat{r} dx. \quad (27.21)$$

If the second derivatives of L with respect to the variables y e y' are bounded in $[a, b]$ by a real positive number K, then it is possible to prove that the second integral on the right-hand side of (27.21) is less than

$$K ||h(x)||^2.$$

Since the first integral is a linear functional of $h(x)$ and condition (27.16) is verified, we conclude that the Fréchet differential of J is

$$D_{y(x)} J[h(x)] = \int_a^b \left[\frac{\partial L}{\partial y} h(x) + \frac{\partial L}{\partial y'} h'(x) \right] dx. \qquad (27.22)$$

It is easy to verify that, after integrating by part the second term under the integral, we can write (27.22) in the form

$$D_{y(x)}J[h(x)] = \left[\frac{\partial L}{\partial y'}h(x)\right]_a^b + \int_a^b \left[\frac{\partial L}{\partial y} - \frac{d}{dx}\frac{\partial L}{\partial y'}\right]h(x)dx. \qquad (27.23)$$

In conclusion, the function $y(x)$ is an extremal of J if and only if (27.23) vanishes for any choice of $h(x)$.

In order to determine an equivalent form of (27.23), we resort to the following result that we state without proof:

Theorem 27.1. *Let $f(x)$ and $h(x)$ be continuous functions. Then, the condition*

$$\int_a^b f(x)h(x)dx = 0, \qquad (27.24)$$

is verified for any choice of the continuous function $h(x)$ if and only if $f(x) = 0$ in $[a, b]$.

We want to find necessary conditions for a stationary point of J on varying $y(x)$:

- In the space $C^1([a, b])$, without boundary conditions;
- In the set

$$A = \left\{y(x) \in C^1([a, b]) : y(a) = y_a, y(b) = y_b\right\}. \qquad (27.25)$$

In the first case, taking into account (27.23), (27.24), and the arbitrariness of $h(x)$, function $y(x) \in C^1([a, b])$ is an extremal of J if and only if it is a solution of the Euler–Lagrange equation

$$\frac{\delta L}{\delta y} \equiv \frac{\partial L}{\partial y} - \frac{d}{dx}\left(\frac{\partial L}{\partial y'}\right) = 0, \qquad (27.26)$$

satisfying the boundary conditions

$$\left(\frac{\partial L}{\partial y'}\right)_a = \left(\frac{\partial L}{\partial y'}\right)_b = 0. \qquad (27.27)$$

We call such a kind of extremal as "free".

In the second case, from (27.25) there follows that $h(x)$ must satisfy the following boundary conditions:

$$h(a) = h(b) = 0. \qquad (27.28)$$

Therefore, from (27.23) and (27.28) there follows that the function $y(x) \in A$ is an extremal if and only if it is a solution of the Lagrange equation (27.26) belonging to the set (27.25).

27.3 Particular Cases and Examples

- Suppose that the function L in (27.4) does not depend on y'. Then, Euler–Lagrange equation (27.26) assumes the form

$$\frac{\partial L(x, y(x))}{\partial y} \equiv F(x, y(x)) = 0,$$

and this condition implicitly defines the extremals. Therefore these extremals, in general, do not satisfy the boundary conditions $y(a) = y_a$ and $y(b) = y_b$.
- The function L depends only on y', that is $L = L(y'(x))$, and $\partial^2 L/\partial y'^2 \neq 0$. In this case the Euler–Lagrange equation becomes

$$\frac{d}{dx}\left(\frac{\partial L}{\partial y'}\right) = \frac{\partial^2 L}{\partial y'^2} y'' = 0 \Rightarrow y''(x) = 0,$$

and the extremals are the straight lines

$$y(x) = Ax + B,$$

with A and B constant. Imposing the boundary conditions $y(a) = y_a$ and $y(b) = y_b$, we finally obtain

$$y = \frac{y_b - y_a}{b - a}(x - a) + y_a.$$

- The function L does not depend on y, that is $L = L(x, y'(x))$. Then, the Euler–Lagrange equation assumes the form

$$\frac{\partial L}{\partial y'} = C,$$

where C is an arbitrary constant. This equation is a first-order differential equation so that it, in general, cannot satisfy the boundary conditions (27.5) or (27.28).

In order to illustrate the above concepts, we consider the following examples.

1. Determine the extremals of the functional

$$J[y(x)] = \int_1^2 (y'^2 - 2xy)dx$$

that satisfy the boundary conditions

$$y(1) = 0, \quad y(2) = -1.$$

The Euler–Lagrange equation becomes

$$y'' + x = 0$$

and its general integral is

$$y = -\frac{x^3}{6} + Ax + B. \qquad (27.29)$$

Imposing the boundary conditions we have

$$y = \frac{x}{6}(1 - x^2).$$

If we search for "free" extremals and notice that

$$\frac{\partial L}{\partial y'} = 2y',$$

then the general solution (27.29) cannot satisfy both the boundary conditions (see (27.27))

2. Determine the extremals of the functional

$$J[y(x)] = \int_0^{2\pi} (y'^2 - y^2)dx, \quad y(0) = 1, \quad y(2\pi) = 1.$$

The Euler–Lagrange equation becomes $y'' + y = 0$ and its general integral is

$$y(x) = A\cos x + B\sin x.$$

Using the boundary conditions we obtain

$$y(x) = \cos x + C\sin x,$$

where C is an arbitrary constant. This boundary problem has infinite solutions.

3. Prove that the extremals of the functional

$$J[y(x)] = \int_a^b \sqrt{1 + y'^2}dx, \quad y(a) = y_a, \quad y(b) = y_b$$

are the straight lines

$$y = \frac{y_b - y_a}{b - a}(x - a) + y_a.$$

27.4 Generalization of the Above Results

In Section 27.2 we searched for the stationary points of functional (27.4) and we showed that this is equivalent to searching for the solution of the relative Euler–Lagrange equation equipped with suitable boundary conditions. If we apply the same procedure to functional (27.2)

$$J[\mathbf{y}(x)] = \int_a^b L(x, \mathbf{y}(x), \mathbf{y}'(x))dx, \quad \mathbf{y}(a) = \mathbf{y}_a, \ \mathbf{y}(b) = \mathbf{y}_b, \tag{27.30}$$

where $\mathbf{y}(x) = (y_1(x), \dots, y_m(x))$,

$$\mathbf{y}(a) = \mathbf{y}_a \Leftrightarrow (y_1(a) = y_{1a}, \dots, y_m(a) = y_{ma})$$

$$\mathbf{y}(b) = \mathbf{y}_b \Leftrightarrow (y_1(b) = y_{1b}, \dots, y_m(b) = y_{mb}),$$

then, we have to determine a vector function $\mathbf{y}(x)$ to obtain the extremals of (27.30). The components of this function must satisfy the system of Euler–Lagrange equations

$$\frac{\partial L}{\partial y_1} - \frac{d}{dx}\frac{\partial L}{\partial y_1'} = 0,$$
$$\cdots\cdots\cdots\cdots$$
$$\frac{\partial L}{\partial y_m} - \frac{d}{dx}\frac{\partial L}{\partial y_m'} = 0, \tag{27.31}$$

as well as the boundary conditions (27.30). It is evident that the vector form of this system is

$$\frac{\delta L}{\delta \mathbf{y}} \equiv \frac{\partial L}{\partial \mathbf{y}} - \frac{d}{dx}\frac{\partial L}{\partial \mathbf{y}'} = 0.$$

Finally, we consider the functional (27.3)

$$J[\mathbf{u}(x)] = \int_\Omega L(\mathbf{x}, \mathbf{u}(\mathbf{x}), \nabla\mathbf{u})d\Omega, \tag{27.32}$$

where $\mathbf{u}(\mathbf{x}) = (u_1(\mathbf{x}), \dots, u_m(\mathbf{x})) \in C^2(\Omega)$ is a function that depends on the vector variable $\mathbf{x} \in \Omega \subset \Re^n$ and reduces to the function $\mathbf{u}_0$ on the part $\partial\Omega_1$ of the boundary $\partial\Omega$ of Ω. Proceeding as in Section 27.3, instead of (27.22), we find

$$D_\mathbf{u} J[h(\mathbf{x})] = \int_\Omega \left(\frac{\partial L}{\partial u_\alpha}h_\alpha + \frac{\partial L}{\partial u_{\alpha,i}}h_{\alpha,i} \right) d\Omega,$$

where the indices α and i are summed, the first from 1 to m and the second from 1 to n. Integrating by parts the second term under the integral and applying Gauss' theorem we have

$$D_{\mathbf{u}} J[h(\mathbf{x})] = \int_{\Omega} \left[\left(\frac{\partial L}{\partial u_{\alpha}} - \frac{\partial}{\partial x_i} \frac{\partial L}{\partial u_{\alpha,i}} \right) h_{\alpha} + \frac{\partial}{\partial x_i} \left(\frac{\partial L}{\partial u_{\alpha,i}} h_{\alpha} \right) \right] d\Omega$$

$$= \int_{\Omega} \left(\frac{\partial L}{\partial u_{\alpha}} - \frac{\partial}{\partial x_i} \frac{\partial L}{\partial u_{\alpha,i}} \right) h_{\alpha} \, d\mathbf{x} + \int_{\partial \Omega} \frac{\partial L}{\partial u_{\alpha,i}} n_i h_{\alpha} d\Omega,$$

with $\mathbf{n}$ the unit normal to the boundary $\partial\Omega$.

Then, $\mathbf{u}(\mathbf{x})$ is an extremal for (27.32) if it is a solution of the following vector equation

$$\frac{\delta L}{\delta \mathbf{u}} \equiv \frac{\partial L}{\partial \mathbf{u}} - \sum_{i=1}^{n} \frac{\partial}{\partial x_i} \frac{\partial L}{\partial \mathbf{u}_{,x_i}} = 0, \tag{27.33}$$

and verifies the boundary conditions $\mathbf{u}(\mathbf{x}) = \mathbf{u}_0$ on $\partial\Omega_1$ and

$$\frac{\partial L}{\partial u_{\alpha,i}} n_i = 0, \quad \alpha = 1, \ldots, m, \tag{27.34}$$

on $\partial\Omega - \partial\Omega_1$, where (n_i) is the unit normal to $\partial\Omega - \partial\Omega_1$.

We recall the following identities

$$\frac{\partial}{\partial x_j} \left(\frac{\partial L}{\partial u_{\alpha,j}} u_{\alpha,i} - L\delta_{ij} \right) = -\frac{\delta L}{\delta u_{\alpha}} u_{\alpha,i} - L_{,i}, \quad \alpha = 1, \ldots, m, \tag{27.35}$$

that can be checked by direct inspection.[2]

In particular, if the function L does not depend on $\mathbf{x}$, then *any solution* of (27.33) satisfies the equations

$$\frac{\partial}{\partial x_j} \left(\frac{\partial L}{\partial u_{\alpha,j}} u_{\alpha,i} - L\delta_{ij} \right) = 0, \quad \alpha = 1, \ldots, m. \tag{27.36}$$

In particular, if index i runs through four values, three marking the spatial coordinates and one the time, then the corresponding equations are the momentum balance and energy balance. The tensor

$$T_{ij} = \frac{\partial L}{\partial u_{\alpha,j}} u_{\alpha,i} - L\delta_{ij} \tag{27.37}$$

is called the ***momentum-energy tensor***.

---

[2] To verify (27.35) it is important to notice that $\partial L / \partial x_j$ denotes the total derivative of L with respect to x_j whereas $L_{,j}$ denotes the partial derivative of L, if L explicitly depends on x_j.

Remark 27.1. Many physical laws cannot be expressed by a variational principle. However, they can be derived from a **variational equation**, i.e., from an equation stating that the variation of a functional J is not equal to zero but it is equal to the variation of a *linear* functional. In other words, we have to consider the following functional equation

$$\int_\Omega L(\mathbf{x}, \mathbf{u}(\mathbf{x}), \nabla\mathbf{u})d\Omega = \sum_{\alpha=1}^m \int_\Omega F_\alpha(\mathbf{x})u_\alpha d\Omega + \sum_{\alpha=1}^m \int_{\partial\Omega-\partial\Omega_1} t_\alpha(\mathbf{x})u_\alpha d\sigma. \quad (27.38)$$

where $d\sigma$ is the elementary area of the boundary $\partial\Omega$ of Ω. Such a variational equation locally implies

$$\frac{\delta L}{\delta u_\alpha} = F_\alpha(\mathbf{x}), \quad \frac{\partial L}{\partial u_{\alpha,i}}n_i = t_\alpha(\mathbf{x}), \quad \alpha = 1, \ldots, m, \quad (27.39)$$

instead of (27.33) and (27.34). As we see in a next section, in continuum mechanics (27.33) and (27.34) do not hold when there are external forces acting on the continuous system.

27.5 Conditional Extrema

Up to now, we have determined necessary conditions that must be satisfied by an extremal of a functional assuming given values on the boundary of the region in which the extremals are defined. Now, we propose to determine the extremals of a functional satisfying further constraints, besides the boundary data.

We are now interested in searching for the stationary point of a functional

$$J[y(x)] = \int_a^b L(x, y, y')dx, \quad (27.40)$$

satisfying the boundary conditions

$$y(a) = y_a \quad y(b) = y_b, \quad (27.41)$$

and the integral condition

$$K[y(x)] = \int_a^b G(x, y, y')dx = 2l. \quad (27.42)$$

The variational problem (27.40), (27.41), (27.42) is called a problem of **conditional extremum** or an **isoperimetric problem**.

The following theorem holds that we state without proof.

Fig. 27.1 Dido's problem

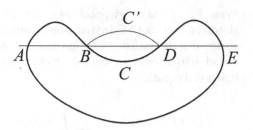

Theorem 27.2. *If the function $y(x)$ is an extremal of functional (27.40), satisfying conditions (27.41), (27.42), and if $y(x)$ is not an extremal of $K[y(x)]$, then a constant λ exists such that the function $y(x)$ is an extremal of*

$$\Phi = \int_a^b \left[L(x, y, y') + \lambda G(x, y, y') \right] dx. \tag{27.43}$$

Example 27.1. (Dido's problem) Among all the plane and closed curves γ with a length $2l$, determine the curve enclosing the greatest area.

First, the curve γ must be convex. In fact, if it is not convex, there is a straight line AE (see Fig. 27.1) such that if the part BCD of the boundary is reflected with respect to AE, then a region with area greater than the initial one is obtained whose boundary has still the length of γ.

Further, suppose that γ has a length $2l$ and encloses the greatest area. Then, any straight line r that divides γ into two parts with the same length divides the area A inside γ in two equal parts. In fact, in the opposite case, a part A_1 of A should have greater area than the remaining part A_2. Then, after a reflection of A_1 with respect to r, we would obtain a region with an area greater than the area of A inside a curve with the same length of γ.

These considerations imply that, if we adopt one of the above-mentioned straight lines as axis Ox, the posed problem reduces to searching for a function $y(x)$, $x \in [-a, a]$, that minimizes the functional

$$J[y(x)] = \int_{-a}^a y(x)dx, \tag{27.44}$$

verifies the boundary conditions

$$y(-a) = 0, \quad y(a) = 0, \tag{27.45}$$

and the integral condition

$$K[y(x)] = \int_{-a}^a \sqrt{1 + y'^2(x)}dx = l. \tag{27.46}$$

In view of Theorem 27.2, we must search for a function $y(x)$, $x \in [-a, a]$ that satisfies conditions (27.45) and is an extremal of the functional

$$\Phi[y(x)] = \int_{-a}^{a} \left[y(x) + \lambda\sqrt{1 + y'^2(x)} \right] dx, \tag{27.47}$$

where λ is an unknown constant. In other words, we must search for a solution of the Euler–Lagrange equation relative to the functional (27.47)

$$\frac{d}{dx} \left(\frac{\lambda y'}{\sqrt{1 + y'^2(x)}} \right) = 1, \tag{27.48}$$

and satisfying conditions (27.45). Integrating we have that

$$\frac{\lambda y'}{\sqrt{1 + y'^2(x)}} = x - C_1, \tag{27.49}$$

where C_1 is a constant. We omit to prove that by integrating (27.49) we obtain the following result

$$(x - C_1)^2 + (y - C_2)^2 = \lambda^2 \tag{27.50}$$

so that the extremals are circumferences with the center at (C_1, C_2) and radius λ. It is also possible to prove that the boundary conditions (27.45) and the isoperimetric condition (27.41) allow to determine the constants C_1, C_2 and λ.

The previous considerations about the extremal curve γ were intended to simplify the search of the curve itself. On the other side, it is spontaneous to give the closed curve γ in the plane Oxy by two parametric equations $x(t)$ e $y(t)$, $t \in [-a, a]$, such that

$$x(-a) = x(a) \quad y(-a) = y(a). \tag{27.51}$$

It remains to formulate the area inside this curve by the parametric equations. To this end, we consider the vector field that gives the position of any point of Oxy with respect to the origin O, that is the vector field $\mathbf{r} = (x, y)$. Applying the Gauss theorem to the vector field $\mathbf{r}/2$ in the region Ω bounded by the curve γ, we have

$$\frac{1}{2} \int_{\Omega} \nabla \cdot \mathbf{r} \, dx \, dy = \frac{1}{2} \int_{\gamma} \mathbf{r} \cdot \mathbf{n} \, ds, \tag{27.52}$$

where $\mathbf{n}$ is the unit vector orthogonal to the curve γ. Since $\nabla \cdot \mathbf{r} = 2$,

$$ds = \sqrt{\dot{x}^2 + \dot{y}^2} dt$$

and **n** has components

$$\mathbf{n} = \frac{1}{\sqrt{\dot{x}^2 + \dot{y}^2}} (\dot{y}, -\dot{x}),$$

(27.52) becomes

$$\int_\Omega dxdy = \frac{1}{2} \int_{-a}^{a} (x\dot{y} - y\dot{x}) \, dt,$$ (27.53)

where the integral on the left hand side gives the area inside γ. In conclusion, Dido's isoperimetric problem is equivalent to searching for the extremals of functional

$$\Phi = \frac{1}{2} \int_{-a}^{a} \left[(x\dot{y} - y\dot{x}) + \lambda\sqrt{\dot{x}^2 + \dot{y}^2} \right] dt,$$ (27.54)

satisfying the boundary conditions (27.51).

The isoperimetric problem is not the only example of a conditional extremum. In fact, the isoperimetric problem requires that the extremal of a functional $J[y(x)]$ verifies an integral condition. Differently, there are situations in which it is required that the extremals satisfy local conditions.

We denote by $\mathbf{y}(x)$ an m−tuple of functions $(y_1(x), \ldots, y_m(x))$ (see Section 27.4). Let

$$J[\mathbf{y}(x)] = \int_a^b L(x, \mathbf{y}(x), \mathbf{y}'(x)) \, dx,$$ (27.55)

be a functional for which we search extremals satisfying the boundary conditions

$$\mathbf{y}(a) = \mathbf{y}_a, \quad \mathbf{y}(b) = \mathbf{y}_b,$$ (27.56)

and the following independent equations

$$\varphi_i(x, \mathbf{y}) = 0, \ i = 1, \ldots, p, \ p < m.$$ (27.57)

Theorem 27.3. *If the m−tuple of functions $\mathbf{y}(x)$ is an extremal of the variational problem (27.55), (27.56), (27.57), it is necessarily an extremal of the functional*

$$\Phi[\mathbf{y}(x)] = \int_a^b \left[L(x, \mathbf{y}) + \sum_{i=1}^{p} \lambda_i(x)\varphi_i(x, \mathbf{y}) \right] dx,$$ (27.58)

for a suitable choice of functions $\lambda_i(x)$.

27.6 Euler's Method and Rietz Method

We search for the stationary points of functional

$$J[y(x)] = \int_a^b L(x, y, y') \, dx, \qquad (27.59)$$

satisfying the boundary conditions

$$y(a) = y_a, \quad y(b) = y_b. \qquad (27.60)$$

Divide the interval $[a, b]$ into n partial intervals $[x_i, x_{i+1}]$, $i = 0, \ldots, (n-1)$,

$$x_{i+1} - x_i = h, \quad h = \frac{b-a}{n}, \qquad (27.61)$$

where

$$x_0 = a, \quad x_n = b. \qquad (27.62)$$

Then (27.59) can be written as follows

$$J[y(x)] = \sum_{i=0}^{n-1} \int_{x_i}^{x_i+h} L(x, y, y') dx. \qquad (27.63)$$

If h is sufficiently small, any term of summation (27.63) can be written in the approximate form

$$\int_{x_i}^{x_i+h} L(x, y, y') dx = L(x_i, y_i, y_i') h, \qquad (27.64)$$

where

$$y_i = y(x_i), \quad y_i' = \frac{y_{i+1} - y_i}{h}. \qquad (27.65)$$

In view of (27.64) and (27.65), formula (27.63) becomes

$$J[y(x)] \simeq h \sum_{i=0}^{n-1} L\left(x_i, y_i, \frac{y_{i+1} - y_i}{h}\right) dx \equiv \Phi(y_1, \ldots, y_{n-1}). \qquad (27.66)$$

(27.66) approximates the value of $J[y(x)$ by a *function* of the $n-1$ values that $y(x)$ assumes at the points $x_1, \ldots, x_{n-1}$. Therefore, an approximation of the extremal $y(x)$ is obtained considering the system

$$\frac{\partial \Phi}{\partial y_i} = 0, \quad i = 1, \ldots, n-1. \qquad (27.67)$$

Solving this system of $n - 1$ equations in the $n - 1$ unknowns $y_1, \ldots, y_{n-1}$ and adding the boundary conditions, we obtain a broken line (x_i, y_i), $i = 0, \ldots, n$, that approximates the extremal $y(x)$ in the interval $[a, b]$. Note that **Euler's method** gives the values of the extremal $y(x)$ at the points x_i. Therefore, it does not allow to determine an analytic form of $y(x)$. The advantage of the method is represented by the circumstance that it only requires the numeric solution of system (27.67) and this solution can be obtained by many numeric procedures.

Differently, **Rietz's method** *allows to obtain an approximate analytic form of the stationary point provided that we succeed in determining the integral of a convenient family of functions.* Consider the variational problem (27.59) with *homogeneous* boundary data

$$y(a) = 0, \quad y(b) = 0. \tag{27.68}$$

Choose a set of n independent polynomials such that each of them satisfies conditions (27.68), for instance the polynomials

$$\varphi_i(x) = (x - a)(x - b)^i, \quad i = 1, \ldots, n, \tag{27.69}$$

and consider the function

$$y_n(x) = \sum_{i=1}^{n} \alpha_i \varphi_i(x), \tag{27.70}$$

where $\alpha_1, \ldots, \alpha_n$ are arbitrary constants. Introducing function (27.70) into (27.59), we obtain the function

$$\Phi(\alpha_1, \ldots, \alpha_n) = \int_a^b L(x, y_n, y_n') \, dx, \tag{27.71}$$

that will be known *when we are able to evaluate the integral on the right-hand side of* (27.71). Solving the system

$$\frac{\partial \Phi}{\partial \alpha_i} = 0, \quad i = 1, \ldots, n, \tag{27.72}$$

we determine the constants α_i that, in view of (27.70), give the approximate expression of the extremal.

Among the programs written using *Mathematica* that accompany this book, there are the programs **Euler** and **Ritz**, contained in the notebook **Euler**, that apply the methods explained above. In particular, the program **Euler**, adopting the interpolation methods of *Mathematica*, supplies a polynomial approximation of the stationary point that, when the boundary data are homogeneous, can be compared with the polynomial approximation of the stationary point obtained by Ritz's method.

Fig. 27.2 The set Ω and the unit vectors normal to $\partial\Omega$

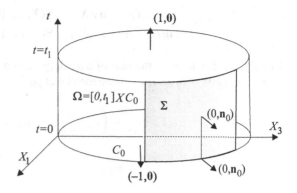

27.7 Applications to Continuum Mechanics

In mechanics of material points and rigid bodies, we have formulated the variational Hamilton's principle and Maupertuis' principle. In this section we show that also the momentum balance equation of continuous bodies can be derived by a variational principle. In this section we adopt the notations of Chapter 23.

Let S be a continuous system and denote by C_0 and $C(t)$ the initial and actual configurations of S, respectively, and by $\mathbf{u}(t, \mathbf{X})$ the equation of motion of S relative to a frame of reference $O X_1 X_2 X_3$ (see Fig. 27.2). Then, adopt the following notations

$$\Omega = [0, t_1] \times C_0, \quad (X_a) = (X_0, X_1, X_2, X_3) = (t, X_1, X_2, X_3). \tag{27.73}$$

We now consider the functional equation

$$\int_\Omega \rho_0 L(\mathbf{u}, _{X_a})\, d\Omega = \int_\Omega \rho_0 f_i(X_a) u_i d\Omega + \int_{\partial\Omega - \partial\Omega_1} t_{0i}(X_a) u_i d\sigma, \tag{27.74}$$

where ρ_0 is the mass density in the initial configuration,

$$L(\mathbf{u}, _{X_a}) = \frac{1}{2} u_{i, X_0} u_{i, X_0} + \Psi(\nabla_{X_i} \mathbf{u}) = \frac{1}{2} \dot{u}_i \dot{u}_i + \Psi(\mathbf{u}, _{X_i}), \tag{27.75}$$

$\dot{u}_i = \partial u_i(t, X_j)/\partial t$, $i = 1, 2, 3$, Ψ is the *elastic potential*, (f_i) the specific force, and (t_{0i}) the force acting on the unit area of a part of the boundary ∂C_0 at any instant $t \in [0, t_1]$. Further, we assign the following boundary data

$$u_i(\mathbf{X}) = 0, \quad \forall \mathbf{X} \in \partial\Omega - (\{t_1\} \times C_0) \cup (\{0\} \times C_0) \cup \Sigma. \tag{27.76}$$

$$\mathbf{u}(0, X_i) = \mathbf{u}_0(X_i), \quad \forall(X_i) \in \{0\} \times C_0, \tag{27.77}$$

$$\mathbf{u}(t_1, X_i) = \mathbf{u}_1(X_i), \quad \forall(X_i) \in \{t_1\} \times C_0. \tag{27.78}$$

Finally, we denote by Σ the part of boundary $\partial\Omega$ on which $u_i \neq 0$.

From (27.38), (27.39), when we recall that ρ_0 does not depend on time t, we have

$$\rho_0 \ddot{u}_i - \frac{\partial}{\partial X_j}\left(\rho_0 \frac{\partial \Psi}{\partial u_{i,j}}\right) = \rho_0 f_i, \quad \forall \mathbf{X} \in \Omega \tag{27.79}$$

$$\frac{\partial}{\partial u_{i,j}}(\rho_0 \Psi)n_{0j} = t_{0i}, \quad \forall \mathbf{X} \in \Sigma. \tag{27.80}$$

In theory of elasticity the tensor

$$T_{*ij} = \rho_0 \frac{\partial \Psi}{\partial u_{i,j}}, \tag{27.81}$$

is called *Piola–Kirkhoff stress tensor* and (27.79) is the *Lagrangian form* of momentum equation for elastic continua. It is possible to formulate (27.79) in terms of Eulerian quantities, that is, in terms of fields defined in the actual configuration. Recalling that the mass conservation is equivalent to state that (see Chapter 23)

$$J\rho = \rho_0, \tag{27.82}$$

where

$$J = \det(u_{i,j}) \equiv \det(F_{ij}), \tag{27.83}$$

we can write (27.79) in the form

$$\rho \ddot{u}_i = \frac{1}{J}\frac{\partial}{\partial X_j}\left(\rho_0 \frac{\partial \Psi}{\partial F_{ij}}\right) + \rho f_i = \frac{1}{J}F_{hj}\frac{\partial}{\partial u_h}\left(\rho_0 \frac{\partial \Psi}{\partial F_{ij}}\right) + \rho f_i$$

$$= \frac{\partial}{\partial u_h}\left(\frac{1}{J}F_{hj}T_{*ij}\right) - T_{*ij}\frac{\partial}{\partial u_h}\left(\frac{1}{J}F_{hj}\right) + \rho f_i.$$

On the other hand,[3] it is

$$\frac{\partial}{\partial u_h}\left(\frac{1}{J}F_{hj}\right) = \frac{\partial F_{km}}{\partial u_h}\frac{\partial}{\partial F_{km}}\left(\frac{1}{J}F_{hj}\right)$$

$$= \frac{\partial F_{km}}{\partial u_h}\left[-\frac{1}{J^2}J(F^{-1})_{mk}F_{hj} + \frac{1}{J}\delta_{hk}\delta_{mj}\right]$$

$$= \frac{1}{J}(F^{-1})_{nh}\frac{\partial F_{km}}{\partial X_n}\left[\delta_{hk}\delta_{mj} - (F^{-1})_{mk}F_{hj}\right]$$

---

[3] We use the formula $\dfrac{\partial J}{\partial F_{km}} = J(F^{-1})_{mk}.$

$$= \frac{1}{J} \frac{\partial F_{km}}{\partial X_n} \left[(F^{-1})_{nk} \delta_{mj} - (F^{-1})_{mk} \delta_{nj} \right]$$

and finally, we obtain

$$\frac{\partial}{\partial u_h} \left(\frac{1}{J} F_{hj} \right) = \frac{1}{J} \left[(F^{-1})_{nk} \frac{\partial^2 u_k}{\partial X_j \partial X_n} - (F^{-1})_{mk} \frac{\partial^2 u_k}{\partial X_j \partial X_m} \right] = 0.$$

If we introduce the *Cauchy stress tensor*

$$T_{hi} = \frac{1}{J} F_{hj} T_{*ij} = \rho F_{hj} \frac{\partial \Psi}{\partial u_{i,j}} \tag{27.84}$$

taking into account (27.81), (27.82), then (27.79) reduces to the Eulerian momentum equation of elastic continua [4]

$$\rho \ddot{u}_i = \frac{\partial T_{ij}}{\partial x_j} + \rho f_i. \tag{27.85}$$

If we suppose that $\Psi = \hat{\Psi}(J) = \Psi(\rho)$, then, from (27.84) we have

$$T_{hi} = \rho F_{hj} \frac{\partial \hat{\Psi}(J)}{\partial J} \frac{\partial J}{\partial F_{ij}} = \rho F_{hj} J (F^{-1})_{ji} \frac{\partial \Psi}{\partial \rho} \frac{\partial \rho}{\partial J}$$

that is

$$T_{hi} = -\rho^2 \frac{\partial \Psi}{\partial \rho} \delta_{hi} \equiv -p(\rho) \delta_{hi}. \tag{27.86}$$

In other words, S is a perfect fluid, $p(\rho)$ is the pressure inside S, and the momentum equation becomes the Euler equation of a perfect fluid

$$\rho \dot{v}_i = -p_{,i} + \rho f_i, \tag{27.87}$$

where (v_i) is the velocity field.

---

[4]The Eulerian formulation of the boundary condition (27.80) is more difficult. See, for instance, [44], Section 1.2.

Exercises

1. Prove that the extremal $u(x, y)$ of the functional

$$J[u(x, y)] = \frac{1}{2}\int_\Omega [(u,_x)^2 + (u,_y)^2]dxdy, \quad u_{\partial\Omega} = u_0,$$

is the solution $u(x, y)$ of Laplace's equation

$$\Delta u = 0,$$

   that verifies the boundary condition $u_{\partial\Omega} = u_0$.

2. Prove that the extremals of the functional

$$J[u(x, y)] = \frac{1}{2}\int_\Omega [(u,_x)^2 - (u,_y)^2]dxdy$$

   are the solutions of D'Alembert equation

$$\frac{\partial^2 u}{\partial x^2} - \frac{\partial^2 u}{\partial y^2} = 0.$$

3. Find the extremal of functional

$$J = \int_1^3 (3x - y)y\, dx$$

   that satisfies the boundary conditions $y(1) = 1$, $y(3) = 2$.

   Hint: Euler's equation gives $y(x) = 3x/2$ that verifies no boundary condition. Therefore, the variational problem has no solution.

4. Find the extremal of the functional

$$J = \int_0^{2\pi} (y'^2 - y^2)dx$$

   that satisfies the boundary conditions $y(0) = 1$, $y(2\pi) = 1$.

   Hint: Euler's equation is $y'' + y = 0$ and its solutions are $y(x) = C_1 \cos x + C_2 \sin x$. Imposing the boundary conditions we obtain $y(x) = \cos x + C \sin x$, where C is an arbitrary constant. Then, the given boundary problem has infinite solutions.

5. Solve the following variational problem

$$J = \int_1^e (xy' + yy')dx, \quad y(1) = 0, \ y(e) = 1.$$

6. Verify that Euler's equation of the functional

$$J = \int_\Omega \left[\sum_{i=1}^{2} \left(\frac{\partial u}{\partial x_i \partial x_j} \right)^2 - 2uf(x_1, x_2) \right] dx_1 \, dx_2$$

is

$$\frac{\partial^4 u}{\partial x_1^4} + 2 \frac{\partial^4 u}{\partial x_1^2 \partial x_2^2} + \frac{\partial^4 u}{\partial x_2^4} = f(x_1, x_2).$$

7. Find the extremal of the functional

$$J[u(x, y)] = \int_0^1 \int_0^1 e^{u, y} \sin u,_y \, dx \, dy,$$

satisfying the boundary conditions

$$u(x, 0) = 0, \quad u(x, 1) = 1.$$

8. Find the extremal of the functional

$$J = \int_0^1 y'^2 \, dx$$

verifying the boundary condition

$$y(0) = 1, \quad y(1) = 6,$$

and the integral condition

$$\int_0^1 y \, dx = 3.$$

9. Find the extremal of the functional

$$J = \int_0^1 (x^2 + y'^2) \, dx$$

verifying the boundary condition

$$y(0) = 0, \quad y(1) = 0,$$

and the integral condition

$$\int_0^1 y^2 \, dx = 2.$$

Appendix A
First-Order PDE

A.1 Monge's Cone

In this appendix, we give a sketch of the method proposed by Monge, Ampere, and Cauchy to reduce the integration of a first-order PDE to the integration of a system of ordinary equations.

Let $F(\mathbf{x}, u, \mathbf{p})$ be a function of class $C^2(\Re^{2n+1})$ verifying the following conditions:

1. The set $\mathfrak{F} = \{(\mathbf{x}, u, \mathbf{p}) \in \Re^{2n+1}, \ F(\mathbf{x}, u, \mathbf{p}) = 0\}$ is not empty.
2. $\sum_{i=1}^n F_{p_i}^2 \neq 0, \ \forall (\mathbf{x}, u, \mathbf{p}) \in \mathfrak{F}.$

Under these hypotheses, the following general first-order PDE:

$$F(\mathbf{x}, u, \nabla u) = 0 \tag{A.1}$$

in the unknown $u = u(\mathbf{x}) \in C^2(D)$, with $D \subset \Re^n$, will be considered.

First, any solution $u = u(\mathbf{x})$ of (A.1) defines a surface Σ that is called an *integral surface* of (A.1). Moreover, the vector $\mathbf{N} \equiv (\nabla u, -1) \equiv (\mathbf{p}, -1)$ of $\Re^{n+1}$ is normal to the integral surface Σ. Consequently, (A.1) expresses the relation existing between the normal vectors to all the integral surfaces at any point $(\mathbf{x}, u)$.

To understand the geometrical meaning of this relation, we start by noting that the quantities $(\mathbf{x}_0, u_0, \mathbf{p}_0)$ completely define a plane containing the point $(\mathbf{x}_0, u_0)$ and having a normal vector with components $(\mathbf{p}_0, -1)$. Then, for a fixed point $(\mathbf{x}_0, u_0)$, the equation

$$F(\mathbf{x}_0, u_0, \mathbf{p}_0) = 0 \tag{A.2}$$

defines a set Π_0 of planes containing $(\mathbf{x}_0, u_0)$ and tangent to the integral surfaces of (A.1), to which $(\mathbf{x}_0, u_0)$ belongs. Owing to conditions 1 and 2, Π_0 is not empty, and

© Springer International Publishing AG, part of Springer Nature 2018
A. Romano and A. Marasco, *Classical Mechanics with Mathematica®*,
Modeling and Simulation in Science, Engineering and Technology,
https://doi.org/10.1007/978-3-319-77595-1

we can suppose that $F_{p_n}(\mathbf{x}_0, u_0, \mathbf{p}_0) \neq 0$. In turn, this result implies that (A.2), at least locally, can be written in the form

$$p_n = p_n(\mathbf{x}_0, u_0, p_1, \ldots, p_{n-1}).$$

Consequently, the set Π_0 is formed by a family of planes depending on the parameters $(p_1, \ldots, p_{n-1})$. Any plane $\pi \in \Pi_0$ contains the point $(\mathbf{x}_0, u_0)$ and has $(p_1, \ldots, p_{n-1}, p_n(p_1, \ldots, p_{n-1}), -1)$ as a normal vector, i.e., it is represented by the equation

$$f(\mathbf{X}, U, p_1, \ldots, p_{n-1}) \equiv \sum_{\alpha=1}^{n-1} p_\alpha(X_\alpha - x_{0\alpha}) + p_n(p_\alpha)(U - u_0) = 0, \qquad (\text{A.3})$$

where $(\mathbf{X}, U)$ are the coordinates of any point of π.

The envelope of all these planes is defined by the system formed by (A.3) and the equation

$$(X_\alpha - x_{0\alpha}) + \frac{\partial p_n}{\partial p_\alpha}(X_n - x_{0n}) = 0, \qquad \alpha = 1, \ldots, n-1, \qquad (\text{A.4})$$

which is obtained by differentiating (A.3) with respect to p_α, $\alpha = 1, \ldots, n-1$. By Dini's theorem, $\partial p_n / \partial p_\alpha = -F_{p_\alpha}/F_{p_n}$, and (A.4) becomes

$$F_{p_n}(X_\alpha - x_{0\alpha}) - F_{p_\alpha}(X_n - x_{0n}) = 0, \qquad \alpha = 1, \ldots, n-1. \qquad (\text{A.5})$$

Equations (A.3) and (A.5) constitute a linear system of n equations in the unknowns $((X_\alpha - x_{0\alpha}), (X_n - x_{0n}))$. The determinant of the matrix of this system's coefficients is

$$\Delta = \begin{pmatrix} p_1 & p_2 & \cdots & p_{n-1} & p_n \\ F_{p_n} & 0 & \cdots & 0 & -F_{p_1} \\ 0 & F_{p_n} & \cdots & 0 & -F_{p_2} \\ \cdots & \cdots & \cdots & \cdots & \cdots \\ 0 & 0 & \cdots & F_{p_n} & -F_{p_{n-1}} \end{pmatrix} = (-1)^{n-1} F_{p_n}^{n-2}(\mathbf{p} \cdot \mathbf{F_p}) \neq 0.$$

Consequently, the parametric equations of the envelope of Π_0 become

$$X_i - x_{0i} = \frac{F_{p_i}}{\mathbf{p} \cdot \mathbf{F_p}}(U - u_0), \qquad i = 1, \ldots, n. \qquad (\text{A.6})$$

Fig. A.1 Monge's cone

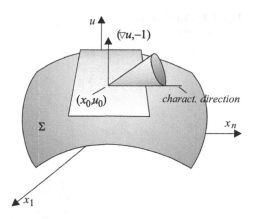

These equations define a cone C_0 since, if the vector $\mathbf{V} = (\mathbf{X} - \mathbf{x}_0, U - u_0)$ is a solution of (A.6), then also the vector $\lambda \mathbf{V}$, where $\lambda \in \Re$, is a solution. This cone C_0 is called **Monge's cone** at $(\mathbf{x}_0, u_0)$.

A.2 Characteristic Strips

The preceding considerations allow us to state that, if $u = u(\mathbf{x})$ is an integral surface Σ of (A.1), then a plane tangent to Σ at the point $(\mathbf{x}_0, u_0)$ belongs to Π_0, i.e., it is tangent to Monge's cone at that point (Fig. A.1). Moreover, the characteristic directions of Monge's cone, which are tangent to Σ, define a field of tangent vectors on Σ whose integral curves are called **characteristic curves**.

We can associate with any point of this curve $(\mathbf{x}(s), u(s))$ a plane tangent to Σ having director cosines $(\mathbf{p}, -1) \equiv (\nabla u(\mathbf{x}(s)), -1)$. The 1-parameter family of these planes along a characteristic curve is said to be a **characteristic strip** (Fig. A.2).

We now write the system of ordinary differential equations in the unknowns $(\mathbf{x}(s), u(s), \mathbf{p}(s))$ that define a characteristic strip on the integral surface Σ. Since the tangent to a characteristic curve at a point belongs to Monge's cone at that point, from (A.6) we have that

$$\frac{d\mathbf{x}}{ds} = F_{\mathbf{p}}, \tag{A.7}$$

$$\frac{du}{ds} = \mathbf{p} \cdot F_{\mathbf{p}}. \tag{A.8}$$

Moreover, along a characteristic curve $(\mathbf{x}(s), u(s))$ we have $\mathbf{p}(s) = \nabla u(\mathbf{x}(s))$, and then

Fig. A.2 Characteristic
curve and strip

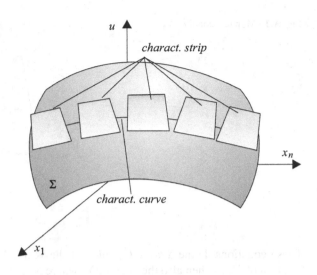

$$\frac{dp_i}{ds} = u_{x_i x_j} F_{p_j}.$$

On the other hand, the differentiation of (A.1) with respect to x_i gives

$$F_{x_i} + u_{x_i} F_u + u_{x_i x_j} F_{p_j} = 0.$$

Comparing the last two relations, we obtain the following system of $2n+1$ equations:

$$\frac{d\mathbf{x}}{ds} = F_{\mathbf{p}}, \tag{A.9}$$

$$\frac{d\mathbf{p}}{ds} = -(F_{\mathbf{x}} + F_u \mathbf{p}), \tag{A.10}$$

$$\frac{du}{ds} = \mathbf{p} \cdot F_{\mathbf{p}}, \tag{A.11}$$

in the $2n+1$ unknowns $(\mathbf{x}(s), u(s), \mathbf{p}(s))$, which is called a ***characteristic system*** of
(A.1).

We have proved that, if $u(\mathbf{x})$ is an integral surface of (A.1), then its characteristic strips satisfy system (A.9)–(A.11). Moreover, any solution of (A.9)–(A.11) is
a characteristic strip of an integral surface of (A.1), as is proved by the following
theorem.

Theorem A.1. *Let* $(\mathbf{x}_0, u_0, \mathbf{p}_0 = \nabla u(\mathbf{x}_0))$ *be a plane tangent to an integral surface*
Σ *at the point* $\mathbf{x}_0, u_0$*). Then the solution of (A.9)–(A.11) corresponding to the initial
datum* $(\mathbf{x}_0, u_0, \mathbf{p}_0)$ *is a characteristic strip of* Σ.

Proof. It is sufficient to recall what we have already proved and to remark that the characteristic strip determined by the initial datum $(\mathbf{x}_0, u_0, \mathbf{p}_0)$ is unique for the uniqueness theorem.   □

Taking into account the function $F(\mathbf{x}(s), u(s), \mathbf{p}(s))$ and recalling (A.9)–(A.11), we verify that

Theorem A.2.

$$F(\mathbf{x}(s), u(s), \mathbf{p}(s)) = \text{const.} \tag{A.12}$$

Remark A.1. If Eq. (A.1) is quasilinear

$$F(\mathbf{x}, u, \nabla u) = \mathbf{a}(\mathbf{x}, u) \cdot \nabla u - b(\mathbf{x}, u) = 0, \tag{A.13}$$

then system (A.9)–(A.11) becomes

$$\frac{d\mathbf{x}}{ds} = \mathbf{a}(\mathbf{x}, u), \tag{A.14}$$

$$\frac{du}{ds} = b(\mathbf{x}, u). \tag{A.15}$$

This is a system of $n + 1$ equations in the unknowns $(\mathbf{x}(s), u(s))$ that can be solved without the help of (A.10). In particular, if (A.13) is linear, then $\mathbf{a}$ and $\mathbf{b}$ depend only on $\mathbf{x}$. Therefore, (A.14) supplies the projection of the characteristic curves in $\Re^n$ and (A.15) gives the remaining unknown $u(s)$.

Remark A.2. Let us suppose that (A.1) has the form

$$F(\mathbf{x}, \nabla u) = 0. \tag{A.16}$$

Denoting by t the variable x_n, by p the derivative F_{u_t}, by $\mathbf{x}$ the vector $(x_1, \ldots, x_{n-1})$, and by $\mathbf{p}$ the vector $(p_1, \ldots, p_{n-1})$, the preceding equation becomes

$$F(\mathbf{x}, t, \mathbf{p}, p) = 0. \tag{A.17}$$

If $F_p \neq 0$, then (A.17) can be written as a Hamilton–Jacobi equation

$$p + H(\mathbf{x}, t, \mathbf{p}) = 0, \tag{A.18}$$

whose characteristic system is

$$\frac{d\mathbf{x}}{ds} = H_{\mathbf{p}}, \tag{A.19}$$

$$\frac{d\mathbf{p}}{ds} = -H_{\mathbf{x}}, \tag{A.20}$$

$$\frac{du}{ds} = \mathbf{p} \cdot H_{\mathbf{p}} + p, \tag{A.21}$$

$$\frac{dt}{ds} = 1, \tag{A.22}$$

$$\frac{dp}{ds} = -H_t. \tag{A.23}$$

Owing to (A.22), we can identify s with t and the preceding system becomes

$$\frac{d\mathbf{x}}{dt} = H_{\mathbf{p}}, \tag{A.24}$$

$$\frac{d\mathbf{p}}{dt} = -H_{\mathbf{x}}, \tag{A.25}$$

$$\frac{du}{dt} = \mathbf{p} \cdot H_{\mathbf{p}} - H, \tag{A.26}$$

$$\frac{dp}{ds} = -H_t. \tag{A.27}$$

We note that (A.24) and (A.25) are Hamiltonian equations that can be solved independently of the remaining ones.

A.3 Cauchy's Problem

The Cauchy problem relative to (A.1) can be formulated as follows.

Let Γ be an $(n-1)$-dimensional manifold, contained in a region D of $\Re^n$, and let $u_0(\mathbf{x}) \in C^2(\Gamma)$ be an assigned function on Γ. Determine a solution $u(\mathbf{x})$ of (A.1) whose restriction to Γ coincides with $u_0(\mathbf{x})$.

In other words, if

$$\mathbf{x}_0 = \mathbf{x}_0(v_\alpha), \quad (v_\alpha) \in V \subset \Re^{n-1}, \tag{A.28}$$

is a parametric representation of Γ, then Cauchy's problem consists in finding a solution $u(\mathbf{x})$ of (A.1) such that

Fig. A.3 Cauchy data for
(A.9)–(A.11)

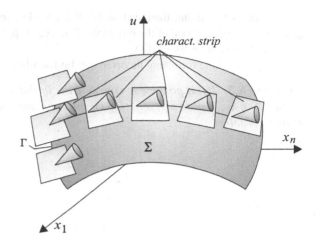

$$u(\mathbf{x}_0(v_\alpha)) = u_0(\mathbf{x}_0(v_\alpha)). \tag{A.29}$$

Geometrically, we can say that we want to determine an n-dimensional manifold $u(\mathbf{x})$ satisfying (A.1) and containing the initial $(n-1)$-dimensional manifold Γ.

We prove that the requested solution can be obtained as the envelope of a suitable family of characteristic strips. In this regard, we first note that the points of Γ, given by the initial datum (A.29), supply the initial data for the unknowns $\mathbf{x}(s), u(s)$ of (A.9), (A.11). However, up to now, we have had no initial data for the unknown $\mathbf{p}(s)$. We show that these data can also be obtained by (A.29). In fact, we recall that the variable $\mathbf{p}$, together with -1, defines the normal vector to the plane tangent to the integral surface Σ at the point $(\mathbf{x}, u(\mathbf{x})) \in \Sigma$. Moreover, Σ must be tangent to the Monge cone along a characteristic direction at any point and, in particular, at any point of Γ (Fig. A.3).

Consequently, we take $\mathbf{p}_0$ in such a way that, at any point of Γ, $(\mathbf{p}_0, -1)$ is orthogonal to the vector $(\partial \mathbf{x}/\partial v_\alpha, \partial u/\partial v_\alpha)$, which is tangent to Γ. Expressed as formulae, we have that

$$\frac{\partial \mathbf{x}_0}{\partial v_\alpha} \cdot \mathbf{p}_0 - \frac{\partial u_0}{\partial v_\alpha} = 0, \tag{A.30}$$

$$F(\mathbf{x}_0(v_\alpha), u_0(v_\alpha), \mathbf{p}_0(v_\alpha)) = 0, \tag{A.31}$$

$\alpha = 1, \ldots, n-1$.

If $(\overline{\mathbf{x}}_0(\overline{v}_\alpha), \overline{u}(\overline{v}_\alpha), \overline{\mathbf{p}}_0(\overline{v}_\alpha)) \in \Gamma$ verifies the preceding system and the condition

$$J = \det\left(\frac{\partial \mathbf{x}_0}{\partial v_\alpha} \; F_{\mathbf{p}}\right) \neq 0, \tag{A.32}$$

is satisfied at this point, then system (A.30), (A.31) can be solved with respect to $\mathbf{p}_0(v_\alpha)$ in a neighborhood of the point $(\overline{\mathbf{x}}_0(\overline{v}_\alpha), \overline{u}(\overline{v}_\alpha), \overline{\mathbf{p}}_0(\overline{v}_\alpha))$. In this way we obtain the initial data for (A.9)–(A.11).

The proof of the following theorem can be found in any book on PDEs:

Theorem A.3. *If it is possible to complete the initial datum (A.29) by solving, with respect to* $\mathbf{p}_0(v_\alpha)$, *system (A.30), (A.31); then one and only one solution exists of the Cauchy problem.*

Appendix B
Fourier Series

A function $f(t)$ is said to be **periodic** of **period** $2T$ if

$$f(t + 2T) = f(t), \quad \forall t \in \Re. \tag{B.1}$$

The **Fourier series** of the periodic function $f(t)$ is the series

$$\frac{a_0}{2} + \sum_{k=1}^{\infty} \left(a_k \cos \frac{k\pi t}{T} + b_k \sin \frac{k\pi t}{T} \right), \tag{B.2}$$

where

$$a_0 = \frac{1}{T} \int_{-T}^{T} f(f)\, dt, \tag{B.3}$$

$$a_k = \frac{1}{T} \int_{-T}^{T} f(t) \cos \frac{k\pi t}{T}\, dt, \tag{B.4}$$

$$b_k = \frac{1}{T} \int_{-T}^{T} f(t) \sin \frac{k\pi t}{T}\, dt \tag{B.5}$$

are the **Fourier coefficients** of the function $f(t)$. In particular, $a_0/2$ is equal to the mean value of the function $f(t)$ in the period $2T$. The following fundamental theorem can be proved.

Theorem B.1. *Let $f: \Re \to \Re$ be a periodic function of period $2T$, which is almost continuous together with its first derivative. If the discontinuities of $f(t)$ and $f'(t)$ are finite, then the series (B.2) converges to $f(t)$ at any point $t \in \Re$ at which $f(t)$ is continuous, whereas it converges to*

© Springer International Publishing AG, part of Springer Nature 2018
A. Romano and A. Marasco, *Classical Mechanics with Mathematica®*,
Modeling and Simulation in Science, Engineering and Technology,
https://doi.org/10.1007/978-3-319-77595-1

$$\frac{f(c^+) - f(c^-)}{2} \tag{B.6}$$

at any discontinuity point c.

Definition B.1. The function $f(t)$ is *even* if

$$f(t) = f(-t), \tag{B.7}$$

and it is *odd* if

$$f(t) = -f(-t), \tag{B.8}$$

$\forall t \in \mathfrak{R}$.

Remark B.1. If the function $f(t)$ is even, then the Fourier coefficients b_k vanish for any integer k. In fact, it is sufficient to decompose integral (B.5) into the sum of the two integrals over the intervals $(-T, 0)$ and $(0, T)$ and recall that $\sin t$ is an odd function. Similarly, if $f(t)$ is odd, then all the Fourier coefficients a_k vanish.

Remark B.2. A Fourier series can be extended to any function $f(t)$ that in a bounded interval $[0, T]$ is continuous with its first derivative almost everywhere in $[0, T]$, provided that the discontinuities of $f(t)$ and $f'(t)$ are finite. In fact, it is sufficient to apply series (B.2) to one of the following periodic functions that extend the function $f(t)$ in the interval $[-T, T]$ as follows (Figs. B.1 and B.2):

$$F_1(t) = \begin{cases} f(t), & t \in [0, T], \\ f(-t), & t \in [-T, 0], \end{cases}$$
$$F_2(t) = \begin{cases} f(t), & t \in [0, T], \\ -f(-t), & t \in [-T, 0]. \end{cases}$$

It is more convenient for applications to resort to the complex form of a Fourier series. This form is obtained as follows. From Euler's identity

$$e^{\mp it} = \cos t \mp i \sin t \tag{B.9}$$

there follows

$$\sum_{k=-\infty}^{\infty} A_k e^{i\frac{k\pi t}{T}} = A_0 + \sum_{k=1}^{\infty} \left(A_k e^{i\frac{k\pi t}{T}} + A_{-k} e^{-i\frac{k\pi t}{T}} \right)$$
$$= A_0 + \sum_{k=1}^{\infty} \left[A_k \left(\cos \frac{k\pi t}{T} + i \sin \frac{k\pi t}{T} \right) \right.$$

$$+ A_{-k} \left(\cos \tfrac{k\pi t}{T} - i \sin \tfrac{k\pi t}{T} \right) \Big]$$

$$= A_0 + \sum_{k=1}^{\infty} \Big[(A_k + A_{-k}) \cos \tfrac{k\pi t}{T}$$

$$+ i \, (A_k - A_{-k}) \sin \tfrac{k\pi t}{T} \Big].$$

This result is identical to (B.2) if we set

$$A_k + A_{-k} = a_k, \tag{B.10}$$

$$A_k - i A_{-k} = -i b_k, \tag{B.11}$$

that is, if and only if

$$A_0 = \frac{a_0}{2}, \tag{B.12}$$

$$A_k = \frac{1}{2}(a_k - i b_k), \tag{B.13}$$

$$A_{-k} = \frac{1}{2}(a_k + i b_k). \tag{B.14}$$

Equations (B.9)–(B.11) allow us to write (B.2) as

$$\sum_{-\infty}^{\infty} A_k e^{i \frac{k\pi t}{T}}, \tag{B.15}$$

where

$$A_k = \frac{1}{2T} \int_{-T}^{T} f(t) e^{-i \frac{k\pi t}{T}} \, dt, \tag{B.16}$$

$$A_{-k} = \frac{1}{2T} \int_{-T}^{T} f(t) e^{i \frac{k\pi t}{T}} \, dt = \overline{A}_k. \tag{B.17}$$

The complex form (B.15) can be at once extended to the case of a function $f(t_1, \ldots, t_n)$ of n real variables $(t_1, \ldots, t_n)$, which is periodic with the same period T with respect to each variable. In fact, since $f(t_1, \ldots, t_n)$ is periodic with respect to t_1, we have that

$$f(t_1, \ldots, t_n) = \sum_{k_1 = -\infty}^{\infty} A_{k_1}^{(1)}(t_2, \ldots, t_n) e^{i \frac{\pi k_1 t_1}{T}},$$

Fig. B.1 First periodic
extension of $f(t)$

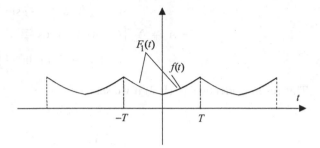

Fig. B.2 Second periodic
extension of $f(t)$

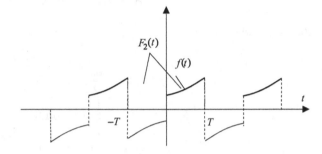

where, in view of (B.16) and (B.17), the functions $A_{k_1}^{(1)}(t_2, \ldots, t_n)$ are periodic in the
variables $t_2, \ldots, t_n$. Applying (B.15) to these functions, we obtain

$$f(t_1, \ldots, t_n) = \sum_{k_2=-\infty}^{\infty} \sum_{k_1=\infty}^{\infty} A_{k_1 k_2}^{(2)}(t_3, \ldots, t_n) e^{i\frac{\pi k_1 t_1}{T}} e^{i\frac{\pi k_2 t_2}{T}}.$$

Finally, iterating the foregoing procedure, we can write that

$$f(\mathbf{t}) = \sum_{k_1, \ldots, k_n=-\infty}^{\infty} A_{\mathbf{k}} e^{i\frac{\pi \mathbf{k} \cdot \mathbf{t}}{T}}, \tag{B.18}$$

where $\mathbf{t} = (t_1, \ldots, t_n)$ and $\mathbf{k} = (k_1, \ldots, k_n)$ is a vector whose components are integer
numbers.

The notebook **Fourier** gives both the real and complex Fourier series of a function
of one or more variables; in addition, it contains many worked-out examples. More
precisely, the notebook **Fourier** contains the programs **SymbSerFour** giving the
formal Fourier expansion of a function and **NSerFour** that supplies the numeric
Fourier expansion.

References

1. Abraham, R., Marsden, J.E.: Foundation of Mechanics. Benjamin Cummings, Reading, Mass. (1978)
2. Arnold, V.L.: Equazioni Differenziali Ordinarie. Mir, Moscow (1979)
3. Arnold, V.L.: Metodi Matematici della Meccanica Classica. Editori Riuniti, Rome (1979)
4. Barger, V., Olsson, M.: Classical Mechanics: A Modern Perspective. McGraw-Hill, New York (1995)
5. Bellomo, N., Preziosi, L., Romano, A.: Mechanics and Dynamical Systems with Mathematica. Birkhäuser, Basel (2000)
6. Birkhoff, G., Rota, G.C.: Ordinary Differential Equations. Wiley, New York (1989)
7. Bishop, R.L., Goldberg, S.I.: Tensor Analysis on Manifolds. MacMillan, New York (1968)
8. Boothby, W.M.: An Introduction to Differential Manifolds and Riemannian Geometry. Academic, San Diego (1975)
9. Borrelli, R.L., Coleman, C.S.: Differential Equations: A Modeling Approach. Prentice-Hall, Englewood Cliffs, NJ (1987)
10. Cercignani, C., Illner, R., Pulvirenti, M.: The Mathematical Theory of Diluite gases. Springer, Berlin Heidelberg New York (1994)
11. Choquet-Bruhat, Y.: Gèometrie Diffèrentielle et Systèmes Extèrieurs. Dunod, Paris (1968)
12. Chow, T.L.: Classical Mechanics. Wiley, New York (1995)
13. Crampin, M., Pirani, F.: Applicable Differential Geometry. Cambridge University Press, Cambridge (1968)
14. de Carmo, Manfredo P.: Differential Geometry of Curves and Surfaces. Prentice-Hall, Englewood Cliffs, NJ (1976)
15. Dixon, W.G.: Special Relativity: The Foundations of Macroscopic Physics. Cambridge University Press, Cambridge (1978)
16. Dubrovin, B.A., Novikov, S.P., Fomenko, A.T.: Geometria delle Superfici, dei Gruppi di Trasformazioni e dei Campi. Editori Riuniti, Rome (1987)
17. Dubrovin, B.A., Novikov, S.P., Fomenko, A.T.: Geometria e Topologia delle Varietà. Editori Riuniti, Rome (1988)
18. Dubrovin, B.A., Novikov, S.P., Fomenko, A.T.: Geometria Contemporanea. Editori Riuniti, Rome (1989)
19. Fasano, A., Marmi, S.: Meccanica Analitica. Bollati Boringhieri, Turin (1994)
20. Fawles, G., Cassidy, G.: Analytical Dynamics. Saunders, Ft. Worth, TX (1999)
21. Flanders, H.: Differential Forms with Applications to the Physical Sciences. Academic, New York (1963)

© Springer International Publishing AG, part of Springer Nature 2018
A. Romano and A. Marasco, *Classical Mechanics with Mathematica®*,
Modeling and Simulation in Science, Engineering and Technology,
https://doi.org/10.1007/978-3-319-77595-1

22. Fock, V.: The Theory of Space, Time and Gravitation, 2nd edn. Pergamon, Oxford (1964)
23. Goldstein, H.: Classical Mechanics, 2nd edn. Addison-Wesley, Reading, MA (1981)
24. Goldstein, H., Poole, C., Safko, J.: Classical Mechanics, 3rd edn. Addison Wesley, Reading, MA (2002)
25. Hale, J., Koçak, H.: Dynamics and Bifurcations. Springer, Berlin Heidelberg New York (1991)
26. Hill, R.: Principles of Dynamics. Pergamon, Oxford (1964)
27. Hutter, K., van der Ven, A.A., Ursescu, A.: Electromagnetic Field Matter Interactions in Thermoelastic Solids and Viscous Fluids. Springer, Berlin Heidelberg New York (2006)
28. Jackson, E.A.: Perspectives of Nonlinear Dynamic, vol. I. Cambrige University Press, Cambridge (1990)
29. Jackson, E.A:, Perspectives of Nonlinear Dynamic, vol. II. Cambrige University Press, Cambridge (1990)
30. Josè, J.V., Saletan, E.: Classical Dynamics: A Contemporary Approach. Cambridge University Press, Cambridge (1998)
31. Khinchin, A.I.: Mathematical Foundations of Statistical Mechanics. Dover, New York (1949)
32. Landau, L., Lifshitz, E.: Mechanics, 3rd edn. Pergamon, Oxford (1976)
33. MacMillan, W.D.: Dynamics of Rigid Bodies. Dover, New York (1936)
34. Marasco, A.: Lindstedt-Poincarè method and mathematica applied to the motion of a solid with a fixed point. Int. J. Comput. Math. Appl. **40**, 333 (2000)
35. Marasco, A., Romano, A.: Scientific Computing with Mathematica. Birkhauser, Basel (2001)
36. Möller, C.: The Theory of Relativity, 2nd edn. Clarendon, Gloucestershire, UK (1972)
37. Norton, J.D.: General covariance and the foundations of general relativity: eight decades of dispute. Rep. Prog. Phys. **56**, 791–858 (1993)
38. Panowsky, W., Phyllips, M.: Classical Electricity and Magnetism. Addison-Wesley, Reading, MA (1962)
39. Penfield, P., Haus, H.: Electrodynamics of Moving Media. MIT Press, Cambridge, MA (1967)
40. Petrovski, I.G.: Ordinary Differential Equations. Prentice-Hall, Englewood Cliffs, NJ (1966)
41. Resnick, R.: Introduction to Special Relativity. Wiley, New York (1971)
42. Rindler, W.: Introduction to Special Relativity. Clarendon Press, Gloucestershire, UK (1991)
43. Romano, A., Lancellotta, R., Marasco, A.: Continuum Mechanics using Mathematica, Fundamentals, Applications, and Scientific Computing. Birkhauser, Basel (2006)
44. Romano, A., Marasco, A.: Continuum Mechanics, Advanced Topics and Research Trends. Birkhauser, Basel (2010)
45. Saletan, E.J., Cromer, A.H.: Theoretical Mechanics. Wiley, New York (1971)
46. Scheck, F.: Mechanics: From Newton's Laws to Deterministic Chaos. Springer, Berlin Heidelberg New York (1990)
47. Stratton, J.A.: Electromagnetic Theory. McGraw-Hill, New York (1952)
48. Synge, J.: Relativity: The Special Theory, 2nd edn. North-Holland, Amsterdam (1964) (Reprint 1972)
49. Synge, J.L., Griffith, B.A.: Principles of Mechanics. McGraw-Hill, New York (1959)
50. Thomson, C.J.: Mathematical Statistical Mechanics. Princeton University Press, Princeton, NJ (1972)
51. Truesdell, C., Noll, W.: The Nonlinear Field Theories of Mechanics. Handbuch der Physik, vol. III/3. Springer, Berlin Heidelberg New York (1965)
52. von Westenholz, C.: Differential Forms in Mathematical Physics. North-Holland, New York (1981)
53. Wang, K.: Statistical Mechanics, 2nd edn. Wiley, New York (1987)
54. Whittaker, E.T.: A Treatise on the Analytical Dynamics of Particles and Rigid Bodies. Cambridge University Press, Cambridge (1989)
55. Zhang, W-B.: Differential Equations, Bifurcations, and Chaos in Economy. World Scientific, Singapore (2005)
56. Zhong Zhang, Y.: Special Relativity and Its Experimental Foundations. Advanced Series on Theoretical Physical Science, vol. 4. World Scientific, Singapore (1997)
57. Morbidelli A.: Modern Celestial Mechanics, Taylor & Francis, London and New York (2002)

58. Brumberg V.A.: Analytical Techniques of Celestial Mechanics, Springer (1995)
59. Kafadar C.B.: *On the Nonlinear Theory of Rods*, Int. J. Engng Sci., vol.10, (1972)
60. Rubin M.B.: Cosserat Theories: Shells, Rods and Points, Springer, (2000)
61. Kellog O. D., Foundations of Potential Theory, Springer Verlag, (1929)
62. Tichonov A., Samarskij A., The Equations of Mathematical Physics. Mir, (1981)
63. Berdichevsky V., Variational Principles of Continuum Mechanics, I. Fundamentals, Springer (2009)
64. Fasano A., Marmi S., Classical Mechanics. An introduction, Oxford University Press (2002)
65. Chester C.R., Techniques in partial differential equations, McGraw-Hill, 1971
66. Smirnov V. I., A course of higher mathematics, Pergamon Student Editions, 1964
67. Krasnov M. I., Makarenko G.I., Kiselev A. I., Problems and exercises in the calculus of variations, MIR publisher Moskov, 1975

Index

© Springer International Publishing AG, part of Springer Nature 2018
A. Romano and A. Marasco, *Classical Mechanics with Mathematica®*,
Modeling and Simulation in Science, Engineering and Technology,
https://doi.org/10.1007/978-3-319-77595-1

Printed in the United States
By Bookmasters